OXIDATIVE STRESS IN PLANTS
Causes, Consequences and Tolerance

OXIDATIVE STRESS IN PLANTS

Causes, Consequences and Tolerance

Editors

NASER A. ANJUM

Centre for Environmental and Marine Studies (CESAM)
& Department of Chemistry, University of Aveiro, Aveiro, Portugal

SHAHID UMAR

Department of Botany, Faculty of Science, Hamdard University,
Hamdard Nagar, New Delhi, India

and

ALTAF AHMAD

Department of Botany, Faculty of Science, Hamdard University,
Hamdard Nagar, New Delhi, India

I.K. International Publishing House Pvt. Ltd.

NEW DELHI • BANGALORE

Published by

I.K. International Publishing House Pvt. Ltd.
S-25, Green Park Extension
Uphaar Cinema Market
New Delhi–110 016 (India)
E-mail: info@ikinternational.com
Website: www.ikbooks.com

ISBN 978-93-81141-02-1

10 9 8 7 6 5 4 3 2 1

Published by Krishan Makhijani for I.K. International Publishing House Pvt. Ltd., S-25, Green Park Extension, Uphaar Cinema Market, New Delhi–110 016 and Printed by Rekha Printers Pvt. Ltd., Okhla Industrial Area, Phase II, New Delhi–110 020.

Dedicated to

ZOYA
and
Affectionate Parents & the Teachers

FOREWORD

The study of environmental stress factors (such as, biotic and abiotic) has become a subject of ecosystem health and sustainable development in the era of climate change. Oxidative stress is a central factor in a number of biotic and abiotic stress phenomena, when there is a serious imbalance in any plant cell compartment between the production of reactive oxygen species (ROS) and antioxidant defence, leading to physiological imbalance. The presence of high concentration of ROS in cells causes oxidative damage to photosynthetic pigments, biomolecules such as lipids, proteins and nucleic acids, leakage of electrolytes via lipid peroxidation resulting in the disruption of cellular metabolism. Recent investigation based on photo-oxidative damage to plants showed that O_2^- is the main ROS responsible for oxidative damage in plants tissues. To counteract this damage, cells are equipped with a large set of enzymatic and non-enzymatic defence mechanisms. Some of these are constitutive and others that are only activated when a stress-specific signal is perceived. Enzymatic scavenger of ROS includes superoxide dismutase, catalase, guaicol peroxidase, ascorbate peroxidase and glutathione reductase. Non-enzymatic antioxidants include glutathione, ascorbic acid, α-tocopherol, flavonoids and carotenoids. Some metabolites like polyphenols, polysugars, proline, glycinebetaine and trehalose (probably) have been found to be accumulated during stress conditions. Abiotic stresses may also induce the emission of multiple biogenic volatile organic compounds (BVOCs) such as terpenes, oxylipins, methanol, ethanol, formaldehyde, acetaldehyde, etc. Many of the emitted compounds are synthesized from the degradation of cellular structures. Excessive amounts of stresses trigger a wide range of cellular responses, synthesis of metal-detoxifying peptides and induce a cascade pathway against ROS generation. Recent investigations have confirmed major roles of MAPK (mitogen-activated protein kinase) cascades via the regulation of development, growth, programmed cell death in both biotic and abiotic stress signal transduction of plants. MAPK pathways transfer information from sensors to cellular responses and represent an evolutionary conserved signalling mechanism in eukaryotes.

Although there has been a rapid progress in recent years in the major theme of the present book, but there are still many uncertainties and gaps in our understanding of sources and mechanism of ROS generation and oxidative stress, the major consequences of oxidative stress, and the potential strategies to enhance oxidative stress-tolerance in plants. I understand that the current edited volume will elucidate recent advances in most of these facts in detail through significant contributions from over the globe to the book.

M.N.V. Prasad
Professor
Department of Plant Sciences, University of Hyderabad
Prof. C.R. Rao Road Gachhibowli
Central University P.O., Hyderabad-500046, India
e-mail: prasad_mnv@yahoo.com, mnvsl@uohyd.ernet.in

PREFACE

Plant growth, development and productivity are affected by different adverse environments. The majority of adverse environments cause multifarious undesirable effects in plants. Of them, production of reactive oxygen species (ROS) is a common phenomenon. These ROS are highly reactive because they can interact with a number of cellular molecules and metabolites thereby leading to a number of destructive processes causing cellular damage. Plants possess to a variable extent antioxidant metabolites, enzymes and non-enzymes that have the ability to detoxify ROS thus, to protect the photosynthetic and cellular functions against ROS-accrued oxidative stress and to maintain optimum growth, development and productivity under changing environment.

In the present volume, the emphasis of discussion has been on understanding the causes of oxidative stress, its consequences and the biological significance of various antioxidant molecules in plants. In particular, the role of different antioxidants as potential selection criteria for improving plant tolerance to oxidative stress has been critically discussed at length in different chapters contributed by eminent researchers and scientists from over the globe. However, occasional overlaps of information between chapters could not be avoided; moreover, they reflect the central and multiple aspects of the causes, consequences of oxidative stress and strategies for enhancing tolerance to oxidative stress in detail in the light of the advances in molecular biology.

Chapter 1 introduces the concept of oxidative stress in plants and the role of important antioxidants in plant tolerance. Chapters 2 and 3 discuss various environmental factors affecting different components of ascorbate-glutathione pathway and the regulatory role of components of the pathway in plant tolerance to oxidative stress in detail. Chapters 4 to 13 describe the insights of the significance of enzymatic and non-enzymatic antioxidants including ascorbate and glutathione in the protection of plant growth and metabolism against various stress factors including heavy metals, salinity and UV-B-accrued oxidative stress. Chapter 14 deals with the role of different secondary metabolites as non-enzymatic plant protectors against oxidative stress in plants. Chapter 15 is concerned with the role of ascorbate in the growth and development of cells during the formation of annual rings in coniferous trees. Chapter 16 deals mainly with the modulation of glutathione, its redox couple and related major enzymes in plants under abiotic stress conditions whereas the last Chapter 17 summarizes major heavy metals in plants and soil and their effects on important components of plant antioxidant defense system.

Like all creative works, completing a book in any discipline is really a hard work to accomplish. We are thankful to contributors for their significant manuscripts and for their full faith and cooperation that eventually made the present volume possible.

We offer our sincere thanks to Mr. Krishan Makhijani, Managing Director, I.K. International Publishing House Pvt. Ltd., New Delhi, India for their efficient and friendly cooperation which made our efforts successful in bringing out the present treatise.

Thanks are also due to all the well-wishers, teachers, seniors, research students and family. Without their unending support, motivation and encouragement the present grueling task would have never been completed.

We are also thankful to the Foundation for Science & Technology (FCT), PORTUGAL (SFRH/BPD/64690/2009), Council of Scientific & Industrial Research (CSIR), New Delhi, INDIA [(9/112(0401)2K8-EMR-I (312208/2K7/1)], Agricultural Biotechnology Research Center (ABRC), Academia Sinica, TAIWAN, the Department of Science and Technology (DST), New Delhi, INDIA, Potash Research Institute of India (PRII), Gurgaon, INDIA and International Potash Institute (IPI), SWITZERLAND for providing financial assistance to our research work.

Last but not least, we bow in reverence to Allah, who blessed us with an innumerable favour of the academic work.

Editors
Naser A. Anjum
Shahid Umar
Altaf Ahmad

CONTRIBUTORS

Agrawal, Madhoolika
Lab of Air Pollution and Global Climate Change, Ecology Research Circle, Department of Botany, Banaras Hindu University, Varanasi–221005, India

Agrawal, SB
Lab of Air Pollution and Global Climate Change, Ecology Research Circle, Department of Botany, Banaras Hindu University, Varanasi–221005, India
E-mail: sbagrawal56@gmail.com

Ahmad, Altaf
Department of Botany, Faculty of Science, Hamdard University, Hamdard Nagar, New Delhi–110062, India

Ahmad, Iqbal
Center for Environmental and Marine Studies (CESAM) & Department of Biology and Chemistry, University of Aveiro, 3810-193 Aveiro, Portugal

Anjum Naser A.
Academia Sinica, Agricultural Biotechnology Research Center, Tainan City, Taiwan; Division of Plant Physiology and Plant Biochemistry, Department of Botany, Aligarh Muslim University, Aligarh–202 002, India. ***Present address:*** Center for Environmental and Marine Studies (CESAM) & Department of Chemistry, University of Aveiro, 3810-193 Aveiro, Portugal
E-mail: anjum@ua.pt naser.a.anjum@gmail.com

Ansari, Mohd. Kafeel A.
Department of Botany, Faculty of Science, Hamdard University, Hamdard Nagar, New Delhi–110062, India. ***Present address:*** Centre for Environmental Risk Assessment and Remediation, University of South Australia, Mawson Lakes, SA 5095, Australia
E-mail: kafeelansari123@gmail.com

Antonova, GF
VN Sukachev Institute of Forestry, Siberian Branch of Russian Academy of Sciences, Akademgorodok, Krasnoyarsk–660036, Russia
E-mail: institute_forest@ksc.krasn.ru

Apostolova, Pavlina
Acad. M. Popov Institute of Plant Physiology, Bulgarian Academy of Sciences, Acad. G. Bonchev Street, Bldg. 21, 1113, Sofia, Bulgaria

Bashmakov, Dmitrij I
Department of Botany and Plant Physiology, Mordovian N.P. Ogariov State University, Bolhevistskaja Str., 68. Saransk–430005, Russia

Basu, Supratim
Department of Botany, Bose Institute, 93/1, A.P.C. Road, Kolkata–700009, West Bengal, India

Chen, Kun-Ming
College of Life Sciences, Northwest A&F University, Yangling–712100, Shaanxi, China
E-mail: kunmingchen@163.com

Chu, Li-Ye
Institute for Life Sciences, Qingdao University of Science & Technology (QUST), Qingdao–266042, China

Duarte, Armando C.
Center for Environmental and Marine Studies (CESAM) & Department of Chemistry, University of Aveiro, 3810-193 Aveiro, Portugal

Fujita, Masayuki
Laboratory of Plant Stress Responses, Faculty of Agriculture, Kagawa University, Miki-cho, Kita-gun, Kagawa 761-0795, Japan
E-mail: fujita@ag.kagawa-u.ac.jp

Gong, Hai-Jun
College of Horticulture, Northwest A&F University, Yangling–712100, Shaanxi, China

Hossain, Mohammad A
Laboratory of Plant Stress Responses, Faculty of Agriculture, Kagawa University, Miki-cho, Kita-gun, Kagawa 761-0795, Japan; Department of Genetics & Plant Breeding, Bangladesh Agricultural University, Mymensingh–2202, Bangladesh

Huerta-Heredia, Ariana A
Departamento de Biotecnología y Bioingeniería. Centro de Investigación y de Estudios Avanzados del Instituto Politécnico Nacional (CINVESTAV-IPN). Av. Instituto Politécnico Nacional 2508. Col. San Pedro Zacatenco, 07360 México, D.F., Mexico

Iqbal, Muhammad
Department of Botany, Faculty of Science, Hamdard University, Hamdard Nagar, New Delhi–110062, India

Janda, Tibor
Agricultural Research Institute of the Hungarian Academy of Sciences, H-2462, Martonvásár, POB. 19, Hungary

Kocsy, Gábor
Agricultural Research Institute of the Hungarian Academy of Sciences, H-2462, Martonvásár, POB. 19, Hungary

Krishnan, PN
Plant Biotechnology and Bioinformatics Division, Tropical Botanic Garden and Research Institute, Kerala-695562, India
E-mail: peringattulli@yahoo.co.uk

Liu, Zeng-Hui
Institute for Life Sciences, Qingdao University of Science & Technology (QUST), Qingdao-266042, China

Lukatkin, Alexander S
Department of Botany and Plant Physiology, Mordovian N.P. Ogariov State University, Bolhevistskaja Str., 68. Saransk-430005, Russia
E-mail: aslukatkin@yandex.ru

Oztetik, Elif
Anadolu University, Science Faculty, Department of Biology, 26470 Eskisehir/Turkey
E-mail: eoztetik@anadolu.edu.tr

Pacheco, Mário
Center for Environmental and Marine Studies (CESAM) & Department of Biology, University of Aveiro, 3810-193 Aveiro, Portugal

Pereira, Eduarda
Center for Environmental and Marine Studies (CESAM) & Department of Chemistry, University of Aveiro, 3810-193 Aveiro, Portugal

Popova, Losanka P
Acad. M. Popov Institute of Plant Physiology, Bulgarian Academy of Sciences, Acad. G. Bonchev Street, Bldg. 21, 1113, Sofia, Bulgaria
E-mail: lpopova@bio21.bas.bg

Prasad, Vishal
Department of Botany, Banaras Hindu University, Varanasi-221005 India

Ramos-Valdivia, Ana C
Departamento de Biotecnología y Bioingeniería. Centro de Investigación y de Estudios Avanzados del Instituto Politécnico Nacional (CINVESTAV-IPN). Av. Instituto Politécnico Nacional 2508. Col. San Pedro Zacatenco, 07360 México, D.F., Mexico
E-mail: aramos@cinvestav.mx

Rojas, Carlos M. Cerda-García
Departamento de Química. Centro de Investigación y de Estudios Avanzados del Instituto Politécnico Nacional (CINVESTAV-IPN). Av. Instituto Politécnico Nacional 2508. Col. San Pedro Zacatenco, 07360 México, D.F., Mexico

Roychoudhury, Aryadeep
Post Graduate Department of Biotechnology, St. Xavier's College, 30, Mother Teresa Sarani, Park Street, Kolkata–700016, West Bengal, India
E-mail: aryadeep19@rediffmail.com

Shao, Hong-Bo
Institute for Life Sciences, Qingdao University of Science & Technology (QUST), Qingdao–266042, China; Institute of Soil and Water Conservation, Chinese Academy of Sciences (CAS), Yangling–712100, China; Yantai Institute of Coastal Zone Research, Chinese Academy of Sciences (CAS), Yantai–264003, China
E-mail: shaohongbochu@126.com

Shi, Weiming
State Key Laboratory of Soil and Sustainable Agriculture, Institute of Soil Science, Chinese Academy of Sciences, Nanjing–210008, China
E-mail: wmshi@issas.ac.cn

Singh, Suruchi
Lab of Air Pollution and Global Climate Change, Ecology Research Circle, Department of Botany, Banaras Hindu University, Varanasi–221005, India

Sreedevi, S
Department of Botany, Sree Narayana College, Kollam–691001, Kerala, India

Stepanov, Mikhail E
Department of Botany and Plant Physiology, Mordovian N.P. Ogariov State University, Bolhevistskaja Str., 68. Saransk–430005, Russia

Szalai, Gabriella
Agricultural Research Institute of the Hungarian Academy of Sciences, H-2462, Martonvásár, POB. 19, Hungary

Trejo-Tapia, Gabriela
Departamento de Biotecnología, Centro de Desarrollo de Productos Bióticos, Instituto Politécnico Nacional (CEPROBI-IPN), P.O. Box 24 62730, Yautepec, Morelos, Mexico

Tripathi, Ruchika
Lab of Air Pollution and Global Climate Change, Ecology Research Circle, Department of Botany, Banaras Hindu University, Varanasi–221005, India

Umar, Shahid
Department of Botany, Faculty of Science, Hamdard University, Hamdard Nagar, New Delhi-110062, India
E-mail: s.umar9@hotmail.com

Upadhyay, RS
Department of Botany, Banaras Hindu University, Varanasi-221005, India
E-mail: upadhyay_bhu@yahoo.co.uk

Xu, Weifeng
State Key Laboratory of Soil and Sustainable Agriculture, Institute of Soil Science, Chinese Academy of Sciences, Nanjing-210008, China

Yadav, Sudesh K
Biotechnology Division, Institute of Himalayan Bioresource Technology, CSIR, Palampur-176061 (HP), India
E-mail: skyt@rediffmail.com, skyicgeb@yahoo.com

Yan, Kun
The CAS/Shandong Provincial Key Laboratory of Coastal Environmental Process, Yantai Institute of Coastal Zone Research, Chinese Academy of Sciences (CAS), Yantai-264 003, China

Zhang, Li-Hua
Institute for Life Sciences, Qingdao University of Science & Technology (QUST), Qingdao-266 042, China

Zhao, Zhongqiu
School of Land Science and Technology, China University of Geosciences, No. 29 Xueyuan Rd, Haidian District, Beijing-100083, China, Key Laboratory of Land Regulation, Ministry of Land and Resources, Beijing-100035, China
E-mail: zhongqiuzhao@163.com

CONTENTS

1

An Introduction to Oxidative Stress in Plants and the Role of Non-Enzymatic Antioxidants

Elif Oztetik*

Anadolu University, Science Faculty, Department of Biology, 26470 Eskisehir/Turkey
***Corresponding author:** eoztetik@anadolu.edu.tr*

ABSTRACT

Oxidative stress occurs when there is an imbalance between the production of reactive oxygen species (ROSs) and antioxidant defence in any cellular compartment. ROSs are produced in both unstressed and stressed cells and in different locations. As different ROSs have a common feature of having capacity to cause oxidative damage to proteins, lipids and DNA, plant cells equipped with several means of detoxification systems, like enzymatic or non-enzymatic antioxidants. Recently, it has become apparent that ROSs are actively produced by plants as signalling molecules, which coordinate a various range of plant processes such as growth, development, programmed cell death and some other physiological responses. With respect to most recent information, this review discusses the well known non-enzymatic antioxidants in plants. Special attention has been given to several environmental stress factors together with their relation to ROS production, different types of ROS damages to biologically important components, also first and second lines of defence against the maleficent effects of ROS, including antioxidant enzymes in brief. Finally, chapter is completed with the detailed functions of non-enzymatic antioxidants against oxidative stress in plants, including their structures and biosynthetic pathways.

Keywords: Antioxidants, oxidative stress, plants, reactive oxygen species.

1. OXIDATIVE STRESS AND ROS PRODUCTION IN PLANTS

Oxygen is necessary for plants. However, since its appearance in the Earth atmosphere, molecular oxygen became a two-edged sword due to unwelcome companions of reactive oxygen species (ROS) into the aerobic life (Halliwell and Gutteridge 1989). Oxidative stress can simply be defined as an imbalance between production and elimination of ROS, leading to several physiological challenges in any cellular compartment. The supposed fine-tuned balance could be disturbed because of an excess accumulation of ROS and diminishing of antioxidants which determine the fate of the plant (Fig. 1).

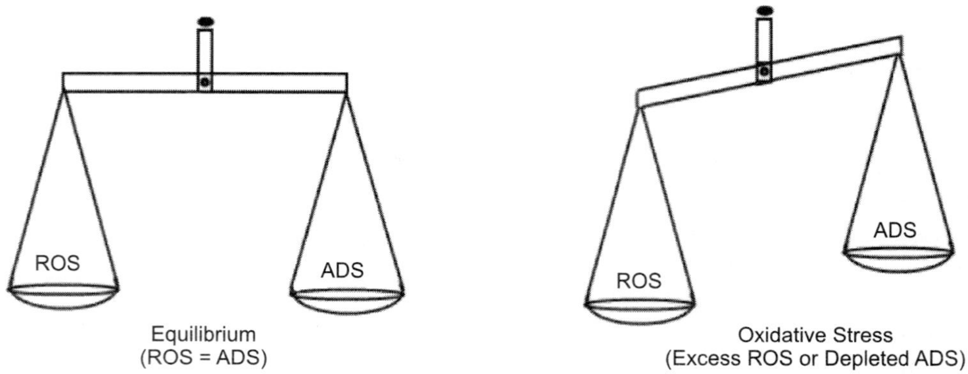

Fig. 1. Oxidative stress results from imbalance between ROS and antioxidant defence system (ADS). The redox equilibrium changes when ROS levels are increased or ADS are depleted.

Generally, molecular oxygen (O_2) is an inert molecule in the property of any catalysts. Higher organisms use it in the pathway of four-electron reduction, leading to production of water (H_2O). However, sometimes one-, two- and three-electron reduction of O_2 or excitation of triplet oxygen ($^3O_2$) occurs and cause the formation of superoxide radical ($O_2^{\bullet-}$) or hydroperoxyl radical ($HO^{\bullet}_2$), hydrogen peroxide (H_2O_2), hydroxyl radical ($OH^{\cdot}$) and singlet oxygen ($^1O_2$), respectively (Apel and Hirt 2004) (Fig. 2). By comparing to the molecular oxygen, all these species are more toxic and ready to take a part in the unlimited chemical oxidation reactions, therefore, they are collectively termed **R**eactive **O**xygen **S**pecies (**ROSs**) (Bartosz 1997).

1.1 Specific Biochemical Properties of ROS in Plants

ROS, being the derivatives of oxygen, are the most important groups of free radicals in living organisms. Each one has specific properties, for example, H_2O_2 can easily diffuse the membranes and react at distant sites by comparing to charged $O_2^{\bullet-}$. Therefore, it can be more important than $O_2^{\bullet-}$ for initiating the oxidative damage in distant cellular components, even less reactive than $O_2^{\bullet-}$. On the other hand, the protonated form of $O_2^{\bullet}$, hydroperoxyl radical ($HO_2^{\bullet-}$), can cross-biological membranes as a neutral molecule and propagate lipid oxidation by extracting protons from polyunsaturated fatty acids (PUFA) (Gechev et al., 2006).

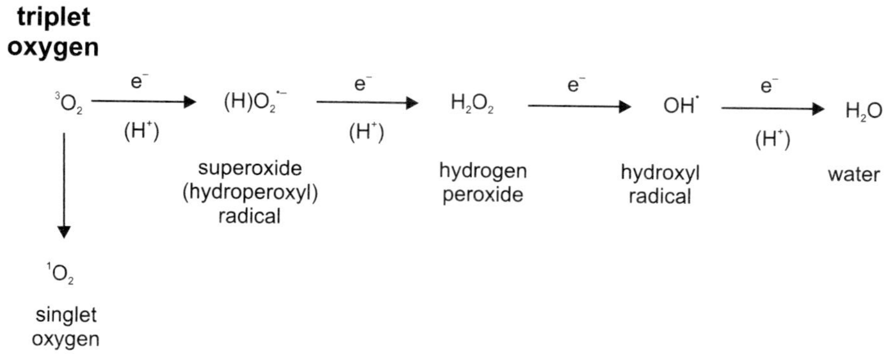

Fig. 2. Production of various ROS by excitation or sequential reduction of triplet oxygen.

$O_2^{\bullet-}$ is quickly converted into H_2O_2 by the enzyme called superoxide dismutase (SOD). This conversion can occur spontaneously by dismutation of $O_2^{\bullet-}$, but the velocity of the reaction is greatly increased with the presence of SOD in most of the living organisms. H_2O_2 is an oxidizing agent, but not very reactive. Concentrations of H_2O_2 are in the micro- to millimolar ranges in the plant tissues (Cheeseman 2006; Halliwell and Gutteridge 1999; Puntorula et al., 1988). $O_2^{\bullet-}$ and H_2O_2 by themselves are relatively harmless and removed from the biological systems in the absence of any metal ions. However, their damaging effects become more explicit when they interact to form highly reactive hydroxyl radicals (OH·) in the presence of metal ions (Kehrer 2000; Moller et al., 2007). For a historical view, towards to the end of nineteenth century, Fenton defined the oxidizing potential of H_2O_2 with ferrous salts (Fe^{2+}) (Fenton 1894, 1899).

$$Fe^{2+} + H_2O_2 \rightarrow Fe^{3+} + OH^{\bullet} + OH^{-} \text{ (Fenton reaction)} \quad [1]$$

The bioavailability of ferrous ion may limit the above reaction, but ferric ion (Fe^{3+}) can be recycled to reduced ferrous state by reducing agents such as $O_2^{\bullet-}$.

$$O_2^{\bullet-} + Fe^{3+} \rightarrow O_2 + Fe^{2+} \quad [2]$$

About 40 years later on Fenton's discovery, Haber and Weiss (1934) identified $OH^{\bullet}$ as the oxidizing species in these reactions. The sum of reactions [1] + [2] gives a formation of $OH^{\bullet}$ from $O_2^{\bullet-}$ and H_2O_2 catalysed by transition metal ions. Today, $OH^{\bullet}$ radicals are known to be generated by Haber-Weiss reaction [3] and other transition metal ions, like Cu^{+} and Cu^{2+} can replace Fe^{2+} and Fe^{3+} in these reactions (Bartosz 1997).

$$H_2O_{2\,+}\, O_2^{\bullet-} \rightarrow OH^{\bullet} + OH^{-} + O_2 \text{ (Haber-Weiss Reaction)} \quad [3]$$

$OH^{\bullet}$ is the most reactive product of partial oxygen reduction in biological systems and has a half-life of 1ns (Pryor 1986). As a result, it can react with and damage any type of molecule with which it comes into contact in a cell.

The important feature of ROS chemistry originates from their interconversion of one ROS into another. In this manner, the Fenton and Haber-Weiss reactions were considered to be the key events in ROS interconversions. Besides the above-mentioned reactions of ROS, ROS can interact with each other which leads to production of a third group of ROS species and those are more toxic than the primordial ones (Dat et al., 2000a). For example, $O_2^{\bullet-}$ can react with nitric oxide ($NO^{\bullet}$) to produce peroxynitrite ($ONOO^-$) (Bartosz 1996; Beligni and Lamattina 1999). The protonated form of $ONOO^-$, peroxynitrous acid (ONOOH), is a strong oxidizing agent and causes formation of other devastating radical and non-radical reactive species (Halliwell 2006). As discussed shortly, different ROS have very diverse properties and reactivities. The order of reactivity is being as $OH^{\bullet} > O_2^{\bullet-} > H_2O_2$ (Sweetlove and Moller 2009).

1.2 Sites and Sources of ROS in Plants

There are several sites and sources of ROS present in the plants. This multiplicate situation causes an increased complexity in ROS production. ROSs are produced in both unstressed and stressed cells (Halliwell and Gutteridge, 1989). During normal physiological conditions, ROS are mainly produced in the chloroplasts, mitochondria, and peroxisomes as unavoidable by-products of aerobic metabolic processes, such as photosynthesis, photorespiration and respiration. Primary ROSs are given according to their cellular locations in Table 1.

Table 1: Primary ROS and their cellular locations in plants.

Primary ROS	Cellular Locations[a]
$^1O_2$	Chl
$O_2^{\bullet-}$	Chl, Mit, PM, Per, CW
H_2O_2	Per, Apo, CW
$OH^{\bullet}$	Many locations[b]

[a] Mittler 2002 is used as reference for cellular locations of ROS. The abbreviations are: Apo, apoplast; Chl, chloroplast; CW, cell wall; Mit, mitochondria; Per, peroxisome; $^1O_2$, singlet oxygen; $O_2^{\bullet-}$, superoxide radical; H_2O_2, hydrogen peroxide; OH·, hydroxyl radical. [b] In all locations where other species are available.

In green parts of the plants, the chloroplast and peroxisomes are the main producers of ROS in the light (Foyer and Noctor, 2003). However, in the dark or in non-green plant parts mitochondria take over this duty (Moller 2001). While chloroplast produce $^1O_2$ at photosystem II (PSII) and $O_2^{\bullet-}$ at photosystem I (PSI) (Asada 2006) and PSII (Pospisil et al., 2004), mitochondria produce $O_2^{\bullet-}$ at complexes I and III (Foyer and Noctor 2003; Moller 2001). Even marked supply of ROS are produced in mitochondria, due to lack of light energy absorbing chlorophyll pigments, mitochondrial ROS production is found to be much lower by comparing to chloroplasts. At least in C3 plants, the approximate H_2O_2 production in mitochondria may be 20 times lower than in the chloroplasts (Foyer and Noctor 2003). However, this condition does not preclude to mitochondrial ROS as being the important regulators of a number of cellular processes, such as stress adaptation and programmed cell death (PCD) (Robson and Vanlerberghe 2002; Gechev et al., 2006).

Plant peroxisomes also generate $O_2^{\bullet-}$ and H_2O_2 in diverse metabolic reactions (del Rio et al., 2006).

In addition to the metabolic processes which produce ROS, new sources of ROS have been identified in plants recently. Some of those ROS-generating enzymes are NADPH oxidases, oxalate oxidases, peroxidases and few others (Mittler 2002).

1.3 Environmental Stress and ROS in Plants

Plants are exposed to many harsh and unavoidable environmental stress conditions, which affect their developmental, physiological, biochemical, morphological and molecular integrity. Under normal physiological conditions, ROS are generally pursued at stable basal levels in the cells. However, their levels persistently or transiently are perturbed by a number of different environmental factors (Apel and Hirt 2004; Gechev et al., 2006). It is generally observed that intracellular levels of ROS and ROS-induced damage increase due to the action of these disturbances (Noctor and Foyer 1998; Dat et al., 2000a; Desikan et al., 2001). For example, under stress condition $O_2^{\bullet-}$ production enhances from 240 to 720 μM s^{-1} and a steady-state level of H_2O_2 in chloroplast from 0.5 to 5-15 μM (Polle, 2001).

The above-mentioned unfavourable environmental stress conditions can be abiotic or biotic stresses (Fig. 3). Abiotic stress arises through redundancy or deficiency in the chemical or physical environment, like drought, salinity, high light, UV radiation, extreme temperatures (heat shock, chilling), heavy metals and air pollutants (ozone, SO_2), mechanical stress, xenobiotics (including herbicides), toxins, nutrient deprivation and many others (Chaitanya et al., 2002; Reddy et al., 2004; Huang and Guo 2005;

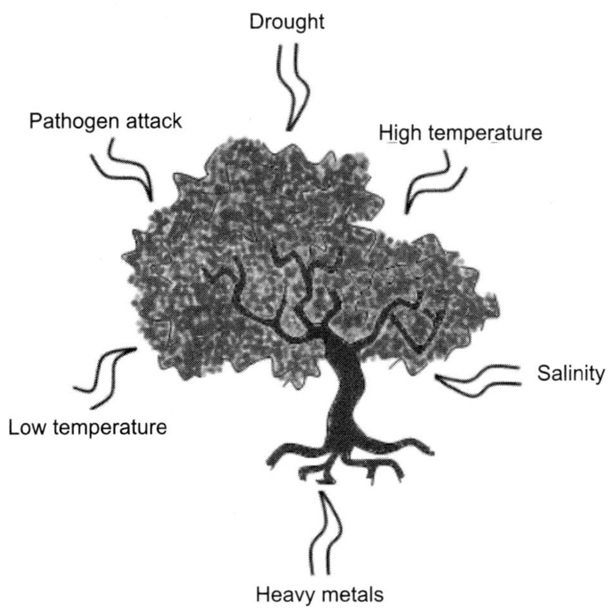

Fig. 3. Some common environmental factors that affect plants.

Sumithra et al., 2006; Agarwal 2007). On the other hand, biotic stress is imposed by other organisms, such as pathogens attack (Wojtaszek 1997; Hammond-Kosack and Jones 2000; Dangl and Jones 2001). These stress factors limit the productivity of plants by individually or collectively (Wang et al., 2001). Now we will examine some of them very briefly to justify why ROS scavenging enzymatic and non-enzymatic antioxidants have to be evolved.

1.3.1 Drought

A drought is an extended period of months or even years when a region shows shortage in its water supply. Drought stress is one of the most converse agents of plant growth and productivity among other environmental stresses. Water deficiency (shortage) and desiccation can be categorized under general term of drought stress. While water deficit can be defined as a moderate loss of water, desiccation is an excessive loss of water which can potentially lead to entirely disruption of metabolism and cell structure and eventually to the cessation of enzyme-catalysed reactions (Smirnoff, 1993). Drought stress is seen in virtually all plants, however, its extent varies between the species (Reddy et al., 2004).

The drought caused inhibition of photosynthesis leads to increased production of ROS (Smirnoff 1998; Peltzer et al., 2002) which results from the higher leakage of electrons to O_2, due to inhibition of CO_2 assimilation which is coupled with the changes in photosynthetic electron transport capacity (Robinson and Bunce 2000). There are many reports in the literature about increased oxidative stress in plants during drought conditions (Sgherri et al., 1993; Sgherri et al., 1996; Boo and Jung 1999; Robinson and Bunce 2000; Langenkaneper et al., 2001; Mano et al., 2002).

In plants, several ROS have been shown to increase as a result of drought stress. Increased production of $O_2^{\bullet-}$ and H_2O_2 in wheat chloroplast (Price et al., 1989; Sairam et al., 1998) and H_2O_2 generation in cotton (Burke et al., 1985) can be given as examples and there are many others. On the other hand, some ROS-dependent changes are interconnected in plants and often unified with drought stress, like lipid peroxidation and changes in antioxidant levels. Lipid peroxidation increases in pea (Gogorcena et al., 1995; Iturbe-Ormaetxe et al., 1998), wheat (Sairam et al., 1998) and *Brassica napus* (Aziz and Larher, 1998) tissues during drought. There are also some reports about the high activities of several ROS scavenging enzymes (see Section 2.2) or their enhanced rate of synthesis during drought stress in different plants, such as in pea (Mittler and Zilinskas 1992, 1994), maize (Zhu and Scandalios 1994; Pastori and Trippi 1992), wheat and cotton (Burke et al., 1985), mulberry (*Morus alba* L) Reddy et al. (2000). More recently, Lee et al. (2009) have determined the development of the antioxidant enzymes induced by drought stress in white clover (*Trifolium repens* L.) leaves.

All of the above discussions are in rapport with the report of Mano (2002). As in plants, the consecutive steps of events under drought stress are: (i) enhanced generation of ROS, (ii) higher expression of genes with antioxidant functions and enhancement in the concentrations of antioxidants, and (iii) enhanced scavenging capacity for ROS, resulting in tolerance against the drought stress. However, there are also other records in the literature, which report the down-regulation of activities (Smirnoff and Colombe

1988; Zhang and Kirkham 1994; Foyer and Noctor 2000; Jiang and Huang 2001). Therefore, it should be noted that, the responses of plants to drought stress are highly complex and depend on many factors, such as plant species, organ and cell type, cellular compartment, age and stage of development of plants and the length and intensity of the stress (Bray, 1997).

1.3.2 Salinity

Salinity in soil or water is one of the major environmental stress factors and can limit growth and productivity of plants (Allakhverdiev et al., 2000). It reduces water potential and causes disturbances in osmotic and ionic homeostasis and toxicity. Together with those effects, the interactions of salts with mineral nutrition may result in nutrient imbalances and deficiencies. These changes in water status lead to initial growth reduction, limitation of plant productivity and can ultimately to plant death. According to their capacity to grow on high salt medium, plants have been classified as glycophytes or halophytes. As indicated by Sairam and Tyagi (2004), most plants are glycophytes and cannot tolerate salt stress. On the other hand, halophytes, depending on their salt tolerating capacity, are either called obligate or facultative halophytes.

The production of ROS during salt stress is probably similar to that during drought and is mainly attributed to increased leakage of electrons to O_2, due to a decline in CO_2 fixation. H_2O_2 concentrations were found to increase upon salt stress in rice shoot tissues (Fadzillah et al., 1997). Salt stress was demonstrated to enhance $O_2^{\bullet-}$ production by submitochondrial particles and lipid peroxidation of mitochondrial membranes in pea (Hernandez et al., 1993). During the last 30 years, research on salinity in plants has produced a vast number of literature. In some of them salt stress, its physiological traits and plant responses to high salinity have been discussed (Rains 1972; Greenway and Munns 1980; Kirst 1989; Ehret and Plant 1999; Munns 2002), while in others molecular aspects and genetic information related to the salt stress have been examined (Hasegawa et al., 2000; Zhu 2002; Xiong and Zhu 2002; Tester and Davenport 2003).

There are many studies showing the participation of oxidative stress through the way of differences in antioxidant levels during salt stress conditions. For example, there are reports about the increased isozymes of ROS scavenging enzymes or reduced glutathione (GSH) levels in pea (Hernandez et al., 1993, 1995) and soybean root nodules (Comba et al., 1998). By Mittowa and associates, the response of the chloroplastic, mitochondrial and peroxisomal antioxidative systems of the cultivated tomato *Lycopersicon esculentum* (Lem) and the wild salt tolerant related species *Lycopersicon pennellii* (Lpa) to salt stress have been investigated (Mittowa et al., 2002, 2003). Results indicated that the salt induced increase in the antioxidant enzyme activities in the Lpa plants confer cross-tolerance toward enhanced mitochondrial and peroxisomal ROS production.

Plants develop many biochemical and molecular strategies to cope with salt stress (Parida and Das 2005). All these strategies are employed to fulfil the three categories of salinity tolerance mechanisms in plants, namely; (i) osmotic stress tolerance, (ii) Na^+ or Cl^- exclusion and (iii) tissue tolerance (Munns and Tester 2008). Moreover, the plant response is dependent on the extent and intensity of the salt stress, age and stage of plant development, organ and cell type and species of the plant (Parvaiz and Satyawati, 2008).

1.3.3 Extreme temperatures

The changes in ambient temperature affect plants more than animals. Temperature extremes experienced by plants can be grouped as high temperatures, chilling at temperatures above 0°C and freezing at temperatures below 0°C. However, adding 'heat shock' to this classification is not a wrong attempt. Because, both permanent and transient extreme temperatures cause subversive influences on cellular homeostasis and plant metabolism, which affect plant growth and development. The severity of imposed temperature stress is dependent on both duration and intensity of stress and also the type of plant, tissue, cell or cellular components.

There are many different responses of plants to temperature stress, such as morphological, anatomical, phenological, physiological and molecular responses (Wahid et al. 2007). Today, it is known that membrane systems and proteins are the primary sites which are strongly affected by temperature changes in plants. High temperature accelerates the kinetic energy and motion of molecules across membranes. This situation leads to biological membranes to become more fluid (leaky) due to denaturation of proteins or an increase in unsaturated fatty acids (Savchenko et al., 2002). Even the leakage of electrolytes is influenced by different variables (plant species, age of plant or tissue, sampling organ, developmental stage, degree of hardening and others), the decrease in cell membrane thermostability (CMT) has been reported in many plant species (soya bean, potato and tomato, wheat, cotton, sorghum, cowpea and barley) as an indirect measure of heat stress tolerance (see individual references in Wahid et al., 2007). On the other hand, both low temperatures (Gomes et al., 2000) and freezing injury also affect membranes (Steponkus et al., 1993; Uemura et al., 1995).

In the case of proteins, high temperature stress hinders cell recovery due to permanent denaturation of proteins (Martin and Hartl 1997). Therefore, when plants are subjected to excess heat, a characteristic set of cellular and metabolic response is triggered which leads to synthesis of stress related proteins (Iba 2002). These specific sets of stress proteins are called 'heat shock proteins' (HSPs), because they are made in plenty when plants are exposed to high temperature and are not present or in small quantities in unstressed organisms (Sridevi et al., 1999). However, HSPs are not only expressed under high temperature stress, but also in response to water stress, salinity and osmotic, cold and oxidative stress (Waters et al., 1996; Boston et al., 1996). HSPs, as molecular chaperons, are assigned in intracellular distribution, assembly and degradation of proteins, folding, stabilization of proteins/membranes and refolding of proteins under stress conditions (Baniwal et al., 2004; Ahn and Zimmerman, 2006).

There is significant literature about a direct link between the alleviation of ROS and plant tolerance to temperature stress. Similar to high temperature, chilling and freezing temperature exposures involve ROS accumulation (Dat et al., 2000a). ROS generated by these stresses have been shown to injure cell membranes and proteins (Larkindale and Knight 2002; O'Kane et al., 1996). Moreover, it has been shown that high light stress has an additional effect to enhance the production of ROS and ROS related injury during temperature stresses (Niyogi 1999; Larkindale and Knight 2002). The ROS generation mechanisms under temperature stress conditions are many, but show some similarities to the previously mentioned ROS production mechanisms of drought and salinity stresses. At high temperature exposure, besides NAD(P)H dependent oxidase, the prime site of ROS production is probably the chloroplast, through leakage of electron following a decline in CO_2 fixation (Dat et al., 1998a). However, during the low temperature exposure, not only the chloroplast, but also mitochondrial ROS accumulation has been reported (Gonzalez-Meler et al., 1999; De Santis et al., 1999).

Under heat stress conditions, an accumulation of H_2O_2 was measured in mustard seedlings (Dat et al., 1998b) and similarly exposure of whole tobacco seedlings to 40°C for 1 h in the light induced a significant increase in H_2O_2 (Dat et al., 2000b). In other studies, the enhancement of specific antioxidants was reported for different plants species under high temperature treatments, such as in wheat (Kraus and Fletcher, 1994), in alfalfa (Badiani et al., 1997) and in mustard seedlings (Dat et al., 1998a). Protection against oxidative stress is an important component in determining the survival of a plant under heat stress. Studies on heat acclimated versus non-acclimated cool season turfgrass species suggested that the former had lower production of ROS as a result of enhanced synthesis of ascorbate (AsA) and glutathione (GSH) (Xu et al., 2006). Heat stress response and defence mechanisms, involving HSPs and heat shock transcription factors (HSFs) are thought to be associated with ROS (Pnueli et al., 2003). Several studies (Panchuk et al., 2002; Davletova et al., 2005) have indicated that HSFs are involved in the ROS sensing.

Similar to heat stress, chilling stress was shown to enhance H_2O_2 accumulation in different plant species, such as winter wheat, cucumber, maize seedlings, *A. thaliana* callus tissue, rice and coffee (see individual references in Dat et al., 2000a). The increased activities of various antioxidant enzymes under chilling stress have been correlated with tolerance to the stress (O'Kane et al., 1996; Prasad et al., 1994; Saruyama and Tanida 1995; Sato et al., 2001). For example, in cucumber radicles, higher contents of defence enzymes were correlated with higher chilling tolerance (Kang and Saltveit 2002).

1.3.4 Heavy metals

Heavy metals are important environmental pollutants and are defined as the group of elements with a density higher than about 5 g/cm^3. According to Weast (1984), 53 out of 90 naturally occurring elements are heavy metals and 17 of them are of importance for organisms and/or ecosystems, based on their solubility under physiological conditions. These are namely; Fe, Mo, Mn, Zn, Ni, Cu, V, Co, W, Cr, As, Hg, Ag, Sb, Cd, Pb and U. The excessive concentration of all these heavy metals can be toxic to most plants. Physiological processes, like cell elongation, photosynthesis, respiration, nitrogen metabolism and mineral nutrition are inhibited by heavy metals in plants (Nies 1999; Zornoza et al., 2002).

Plants' capacity to heavy metal tolerance varies for different species. Heavy metal tolerant plants can be examined under three groups: (i) metal excluders, (ii) metal indicators and (iii) hyperaccumulators. The latter group has higher importance for its feasibility in phytoremediation processes. Phytoremediation is a process that can be defined as the usage of green plants for their abilities of take up, extract, sequester, accumulate and detoxify different heavy metals to clean up the soil or water environments (Salt et al., 1998). However, such natural hyperaccumulating plants are uncommon, as they have low biomass or are restricted to certain growth areas.

When they reach the cytosol of the cell, toxic heavy metals become a strong sinister to plants. Several strategies have been developed by plants to respond and control the uptake and accumulation of heavy metals. Plants' heavy metal tolerance mechanisms include, metal binding to wall, reduced transport through the cell membrane, active efflux and compartmentalization of metals, chelation and sequestration of heavy metals. In the case of loss in any of those processes, plants suffer from hypersensitivity to heavy metal ions (Cobbet and Goldsbrough, 2002). Some plants, such as *Anthyllis vulnenaria* and *Biscutella leavigata,* can transport the excess of metals to aging organs and apoplast of leaves and remove them seasonally (Siedlecka et al., 2001).

Another strategy which contributes to heavy metal tolerance is the chelation of the metals by various ligands/chelators. Chalators include phytochelatins (PCs), metallothioneins (MTs), organic acids and amino acids (Rauser 1999; Clemens 2001; Christopher 2000). PCs can firmly bind and sequester metals. They are sulfhydryl (SH) containing peptides with basic structure of $(\gamma\text{-Glu-Cys})_n$-Gly (n is generally in the range of 2 to 11) (Rauser 1990). PCs are rapidly induced by a wide range of heavy metal ions and enzymatically synthesized from GSH by the enzyme called PC synthase (EC 2.3.2.15) (Grill et al., 1989; Cobbett 2000). The report of Clemens et al. (1999) has declared the presence of PC synthase in different higher plants. PCs, as metal binding proteins, chelate and inactivate all toxic metal ions entering the cytosol before they can inactivate the enzymes of important metabolic pathways. The metal-PC complex is subsequently actively transported from the cytosol to the vacuole (Salt and Rauser, 1995). During oxidative stress, the reason of depletion in GSH concentration in some plants, such as *Silene cucubalus* (De Vos et al., 1992) and *Holcus lanatus* L. (Hartley-Whitaker et al., 2001), was attributed to heavy metal induced PCs synthesis. MTs are low molecular weight cystein rich metal binding proteins (Robinson et al., 1993). Besides their role in detoxification by sequestering metals (Cu in particular) in plant cells, they also participate in maintenance of essential transition metals' homeostasis and protection against intracellular oxidative damage (Garcia-Hernandez et al., 1998). The first plant MT was the wheat Ec (eraly cystein labelled) protein (Lane et al., 1987). Today, a number of genes encoding MT-like proteins have been found in plants (Rauser 1999). PCs and MTs are the most studied and best characterized groups among heavy metal binding ligands in plant cells. However, there are also other ligands as candidate of cellular chelators, these are some low molecular weight organic acids, such as citrate, malate and oxalate (Ma et al., 2000; Li et al., 2000; Yang et al., 2001) and amino acids, such as nicotinamine, histidine and proline (Rai 2002; Kerkeb and Kramer 2003).

Redox active heavy metals show their toxic effects by ROS generation through autooxidation and Fenton reaction. As reported by Sahw et al. (2004), oxidative stress has been introduced to plant cells in several ways by the actions of reactive heavy metals. Like, transition metals (such as Fe, Cu, etc.) accept or donate single electrons which is promoting the ROS interconversion (Elstner et al., 1988; Weckx and Clijsters 1996; Kurepa et al., 1997; Caro and Puntarulo 1996) and metabolic pathways were disturbed by metals that again enhanced production of ROS (Richards et al., 1998). For example, zinc treatment of *Phaseolus vulgaris* increases the levels of H_2O_2 in roots and lipid peroxidation in primary leaves (Weckx and Clijsters 1997). On the other hand, heavy metals generally inactivate the antioxidant enzyme activities (Somashekaraiah et al., 1992) and due to loss of enzyme activities, heavy metal accumulation results in the depletion of some well-known antioxidants, such as GSH, which is used for PCs production (Hartley-Whitaker et al., 2001). This severe depletion of GSH is accepted as a very common response of plant exposure to heavy metals (di Toppi et al., 1999; Rao and Sresty 2000; Schützendübel et al., 2001). However, when plants face heavy metal stress, the direction of the response was dependent on the metal and tissue type, the intensity of the stress and the plant species. Therefore, it is not surprising to find literature reporting the induced activities of that antioxidant enzyme, as well as synthesis of antioxidant molecules under heavy metal stress (Pereira et al., 2002; Fornazier et al., 2002; Skorzynska-Polit et al., 2003-2004; Li et al., 2006; Dixit et al., 2001).

1.3.5 Pathogen attack

In nature, biotic stress occurs as a result of damage to plants by other living organisms, including pathogens, herbivores (i.e., insects) and even other plants (i.e., weeds) and the consequences of these interactions can be very complex (Poschenrieder et al., 2006). For a pathogen, its capability to generate a disease in the host plant is the exception rather than the rule, because of elaborate defence mechanisms developed by plants against pathogen attacks. Therefore, a limited number of true/successful pathogens produce diseases, because they are able to evade detection or suppress host defence mechanisms **'compatible response'**, whereas most potential invaders are recognised and hindered in their penetration by plant defences **'incompatible response'** (De Gara et al., 2003). In compatible response, pathogens grow exhaustively within the plant, but without causing any necrosis to it. As a result, the plant exhibits susceptibility and the extensive growth of the pathogen can cause varying degrees of damage or disease. On the other hand, in the incompatible responses the recognition event causes a rapid tissue necrosis at the site of pathogen infection. This rapid and localized incident creates inconvenient conditions for pathogen growth and called **h**ypersensitive **r**esponse (HR). Finally, HR causes a localized programmed cell death (PCD) (De Gara et al., 2003).

The production of ROS is one of the earliest cellular responses following successful pathogen recognition within minutes of an attack (Adam et al., 1989). Avirulent pathogens are recognized by disease resistance (R) gene products in plants which cause a two-phased ROS accumulation. The first phase of ROS generation, which is weak and transient, after some hours is followed by a second, massive and sustained phase of ROS production called **'oxidative burst'**. This biphasic kinetics of ROS production is typical of incompatible plant-pathogen interactions which are characterized by HR (Lamb and Dixon 1997; De Gara et al., 2003) and therefore may be an important element contributing to disease resistance. On the other hand, virulent pathogens avoid host recognition and involve only in the first phase of this response (Torres et al., 2006).

The plant responses to pathogens by generation of ROS have been observed and reported in the literature with different plant species (monocotyledonous and dicotyledonous) and varying pathogens (Doke 1983 a and b; Auh and Murphy 1995;

Grant et al., 2000b). In the pioneer research of Doke (1983a), the author observed that inoculation of aged potato discs with a potato blight fungus (*Phytophthora infestans*) results in an HR increased the reduction of both Cyt*c* and the dye nitroblue tetrazolium in the bathing solution. However, the addition of SOD to the medium inhibited those reactions. On the contrary, it has been realized that another fungal race that infects tubers without eliciting the HR did not increase Cyt*c* reducing activity. Therefore, this report was the first observation of ROS generation during plant-pathogen interactions and his findings suggest that $O_2^{\bullet -}$ generation may be involved in host cell hypersensitivity. During the years, similar plant-pathogen interactions which triggered to ROS production came from not only from fungal pathogens (Vera-Estrella et al., 1992; Apostol et al., 1989), but also from bacterial pathogens (Adam et al., 1989; Keppler et al., 1989; Baker et al., 1993) and viruses (Doke and Ohashi 1988).

Today, several researches propose that ROS can orchestrate the establishment of plant defence responses and HR after successful pathogen recognition (Levine et al., 1994). They can act directly against phytopathogen attack by killing the microorganism or reducing its viability (Collins et al., 2003). However, some pathogens (such as necrotrophs) induce production of ROS to their own advantage and cause cell death in the infected tissue (Govrin and Levine 2000). ROS can directly lead strengthening of host cell walls by facilitating lignin polymerization (Lamb and Dixon 1997; Ros Barcelo 1997) and also initiate the stimulation of phytoalexin biosynthesis (Low and Merida 1996).

There are many other stressors also present in the environment for the plants as mentioned above, due to the space restrictions only some of the abiotic causes and pathogen attack as examples of biotic stress condition were briefly examined in this review. Abiotic and biotic stresses often affect the same metabolic pathways and show some commonalities. However, it should be noted that, increases in ROS concentration triggered by either biotic or abiotic stresses are generally attributed to different mechanisms. During biotic stresss, the activities of ROS detoxifying enzymes are inhibited by SA and NO (Klessig et al., 2000). Thus, plant simultaneously produces more ROS and decreases its own ROS scavenging capacities which result in the accumulation of ROS and the activation of PCD. However, ROS production at the apoplast alone without suppression of ROS detoxification, does not result in the induction of PCD. This situation indicates an absolute necessity for the concurrent production of ROS and inhibition of ROS scavenging mechanisms (Delledonne et al., 2001). On the other hand, in the case of abiotic stresses, ROS induces ROS scavenging mechanisms which decrease the steady-state level of ROS in cells. Therefore, the differences in the function of ROS between abiotic and biotic stresses might result from the action of hormones (SA and NO) and crosstalk between varying signalling pathways or from differences in the locations where ROS are produced or accumulated during different stresses (Mittler 2002).

1.4 Damages caused by ROS in Plants

ROS have many important roles in plants metabolism. For example, in recent years, a new role for ROS has been identified as signalling molecules, if they are produced within specific compartments under tightly regulated manner. The use of ROS for signalling purposes by plant cells means the control and regulation of biological processes, such as growth, cell cycle, hormone signalling, development and stress responses (Mullineaux and Karpinski 2002; Foreman et al., 2003; Overmyer et al., 2003; Kwak et al., 2003). This suggests that during the course of evolution plants were able to achieve a high degree of control over ROS toxicity, as they were originally considered to be detrimental to cells. On the other hand, according to the traditional concept, if ROS are produced in excess amounts, such as under the abiotic and biotic stress conditions, the emerging excess ROS cause damage to proteins, lipids, carbohydrates and DNA and eventually result in cell death (Noctor and Foyer 1998). Now, we will examine the harmful effects of ROS to biologically important components very shortly.

Protein oxidation is described as covalent modification of a protein induced by ROS or by-products of oxidative stress. Most types of protein oxidations are essentially irreversible, (with those side groups of His, Arg, Lys, Pro, Tyr and Trp), whereas a few involving sulphur containing amino acids (with side groups of Cys and Met) are reversible (Ghezzi and Bonetto 2003) due to enzyme catalyzed re-reduction (Bechtold et al., 2004; Rouhier et al., 2006). The oxidation by ROS is found to be only triggered by the highly reactive $OH^{\bullet}$ or $^{1}O_{2}$. This indicates that oxidation events only occur in proteins which are localized highly close to the site of production of these radicals (Sweetlove and Moller 2009). Protein oxidation is prevalent and results in fragmentation of the peptide chain, aggregation of cross-linked reaction products and often used as a diagnostic marker for oxidative stress (Moller et al., 2007). It is generally thought that oxidized proteins are degraded relatively rapidly by proteases. A number of proteases are found in all the major cellular compartments, such as in cytosol (Vierstra 2003), chloroplast (Adam et al., 2006; Sakamoto 2006) and in mitochondria (Janska 2005; Moller and Kristensen 2004). However, massive protein damage can lead to the formation of protein aggregates that apparently cannot be degraded (Davies and Shringarpure 2006) or are degraded by autophagy (Xiong et al., 2007).

Peroxidation of membrane lipids occurs when ROS reacts with polyunsaturated fatty acids (PUFAs). PUFAs are particularly susceptible to attack by $OH^{\bullet}$ and $^{1}O_{2}$ which leads to complex mixtures of lipid hydroperoxides (Halliwell and Gutteridge 1999; Mueller 2004). Extensive PUFA peroxidation decreases the fluidity of the membrane, increases leakiness, causes rapid desiccation and secondary damage to membrane proteins and to cell death (Halliwell 2006). The harmful effect of ROS is due primarily to their ability to initiate a variety of autoxidative chain reactions on PUFAs (Smirnoff 1993). By-products of some aldehydes, hydroxyl and keto fatty acids, which are formed through the PUFA peroxidation, can make conjugates with DNA and proteins and some of the other PUFA oxidation products function as secondary messengers in plant cells (Mueller 2004).

Free carbohydrates, like sugars and polyols react with $OH^{\bullet}$ (Smirnoff and Cumbes 1989). $OH^{\bullet}$ actually acts as a wall loosening agent *in vivo* and are diagnostic of an $OH^{\bullet}$ attack, that can be found in cell wall polysaccharides of ripening fruit (Fry 2004). These oxidative modifications are not damaging as they are not detrimental to the modified component or the cell. On the other hand, formic acid, which is released as a product of sugar oxidation with $OH^{\bullet}$ (Isbell et al., 1973), is the subtrate for the formate dehydrogenase enzyme (Igamberdiev et al., 1999; Juszczuk et al., 2007).

DNA can be modified by ROS in many different ways, such as deletions, mutations, strand breaks and by other lethal genetic effects (Srivalli et al., 2003). Withinside of ROS, while $OH^{\bullet}$ is the most reactive one, $^1O_2$ primarily attacks guanine and 8-hydroxyguanine is the most commonly observed modification. On the other hand, some oxidative modifications of DNA cause to changes in the methylation of cytosines. These are very important for the regulation of gene expressions (Halliwell 2006). Even though, there is no gene defined as sensitive to ROS damage, according to Halliwell and Gutteridge (1999), at hot spots, some mutation clusters have been identified which indicate that ROS damage to both mtDNA and nDNA is not completely random. ROS can modify DNA both by directly (through oxidation) or indirectly, through the conjugation of PUFA products with guanine (Jeong et al., 2005).

2. SCAVENGING OF ROS IN PLANTS

As it is explained previously (in Section 1.4), ROS can react with proteins, lipids and DNA which cause decreased enzyme activities, increased membrane permeability and mutations, respectively (Halliwell and Gutteridge 1999). Therefore, both the production and removal of ROS must be precisely controlled. Even it is outside the scope of this review to discuss their properties in detail, we will mention the avoidance mechanisms and enzymatic antioxidants for ROS very briefly.

2.1 Avoidance

To keep the ROS induced damages within a narrow ranges in the cell, it is clear that there is a need to limit or minimize the production of ROS, to detoxify ROS once formed and to repair damage caused by ROS efficiently. Therefore, plant cell compartments should have some mechanisms to protect themselves from the deleterious efffects caused by ROS. In this concept, avoidance can be viewed as the first line of defence against the potential maleficent effects of ROS. For example, plant mitochondria have systems for limiting ROS generation by recruiting the alternative oxidases (AOX). AOX is a single homodimeric protein (Umbach and Siedow 1993) and first suggested by Purvis and Shewfelt (1993) for their helps in maintenance of limited ROS generation by electron transport chain (ETC). In the following years, Maxwell et al. (1999) have declared that transgenic cultured tobacco cells in which the AOX had been overexpressed contained half as much ROS as control cells. On the contrary, cells in which the expression of AOX had been reduced by antisensing contained five times more ROS than control cells. These results are evidence that AOX plays a role in lowering mitochondrial ROS formation in plant cells as an alternative pathway during ETC. There are also several other mechanisms for removing ROS once formed, such as activation of the uncoupling proteins (Casolo et al., 2000; Korshunov et al., 1998) and internal rotenone-insensitive NADH dehydrogenases [NDin(NADH)] (Moller and Palmer 1982). The studies relating to these mechanisms are well documented in the literature (Laloi et al., 1997; Vercesi et al., 1995; Cook-Johnson et al., 2000; Moller 2001; Mittler 2002).

2.2 Scavenging by Antioxidant Enzymes

Detoxification constitutes the second line of defence against the maleficent effects of ROS. Therefore, once formed, the ROS must be detoxified as effectively as possible to minimize damage (Moller 2001). Efficient destruction of ROS requires the action of several antioxidant enzymes acting in synchronicity with each other (Noctor and Foyer 1998). Plants have also developed very efficient enzymatic antioxidant scavenging system to protect themselves against destructive ROS reactions (Mittler et al., 2004). In the cell, toxic effects of ROS are counteracted by several scavenging and detoxifying enzymes, such as superoxide dismutase (SOD), catalase (CAT), ascorbate peroxidase (APX), glutathione peroxidases (GPX), glutathione-*S*-transferases (GST), monodehydroascorbate reductases (MDHAR), dehydroascorbate reductases (DHAR) and glutathione reductases (GR) (see Table 2). While some of these enzymes are entirely dedicated to ROS homeostasis, others are involved also in other processes related to control of development, redox regulation of target proteins and detoxification reactions (Gechev et al., 2006). In cells, different scavenging enzymes encoded by the ROS network can be found in almost every subcellular compartment. Besides, each cellular compartment contains more than one enzymatic activity that detoxifies a particular ROS (Mittler et al., 2004). For example, the cytosol contains at least three different enzymatic activities that scavenge H_2O_2 (Nobuhiro and Mittler 2006). It is clear that the presence of different enzymes in various cellular compartments disclosures their significant role in ROS detoxification for the survival of the plant (Mittler et al., 2004). In Table 2, ROS scavenging enzymes, their abbreviations, EC numbers, catalyzed reactions and cellular localizations are given.

Table 2. Major ROS scavenging and detoxifying enzymes in plants.

Enzyme (Abbreviation)	EC Number	Reaction catalysed	Localization[a]
Superoxide dismutase (SOD)	1.15.1.1	$O_2^{\bullet-} + O_2^{\bullet-} + 2H^+ \rightarrow H_2O_2 + O_2$	Cyt, Chl, Mit, Per
Ascorbate peroxidase (APX)	1.11.1.11	2 L-AA $+ H_2O_2 \rightarrow 2$ MDA $+ 2\ H_2O$	Cyt, Chl, Mit, Per,
Catalase (CAT)	1.11.1.6	$2\ H_2O_2 \rightarrow 2H_2O + O_2$	Mit, Per, Gly
Glutathione peroxidase (GPX)	1.11.1.9	$H_2O_2 + 2$GSH $\rightarrow 2H_2O +$ GSSG	Cyt, Chl, Mit, ER
Glutathione reductase (GR)	1.6.4.2	GSSG + NADPH $\rightarrow$ $NADP^+$ + 2GSH	Cyt, Chl, Mit, Per
Monodehydroascorbate reductase (MDHAR)	1.6.5.4	2 MDHA + NADH $\rightarrow$ 2 L-AA + NAD^+	Cyt, Chl, Mit
Dehydroascorbate reductase (DHAR)	1.8.5.1	DHA + 2GSH $\rightarrow$ L-AA + GSSG	Cyt, Chl, Mit
Glutathione *S*-transferases (GST)	2.5.1.18	RX + GSH $\rightarrow$ HX + R-S-GSH*	Apo, Cyt, Chl, Mit, Nuc
Guaiacol type peroxidases (POX)	1.11.1.7	Donor + $H_2O_2 \rightarrow$ oxidized donor + $2H_2O$	CW, Cyt, Mit, Vac

[a] Gechev et al. 2006 is used as reference for localization of enzymes. The abbreviations are: Apo, apoplast; Cyt, cytosol; Chl, chloroplasts; CW, cell wall; ER, endoplasmatic reticulum; Gly, glyoxysomes; Mit, mitochondria; Nuc, nucleus; Per, peroxisomes; Vac, vacuole. *R may be an aliphatic, aromatic or heterocyclic group; X may be a sulphate, nitrite or halide group.

3. NON-ENZYMATIC ANTIOXIDANTS IN PLANTS

Under normal physiological conditions, cells are able to balance their oxidant and antioxidant productions. However, ROS generated by either abiotic or biotic stresses can be toxic to the cell and a strict control of ROS levels is essential to prevent this toxicity. Therefore, plants have evolved not only enzymatic (see above), but also non-enzymatic antioxidant systems to limit their production and/or scavenge them to prevent their excessive accumulation. As a term, **'antioxidants'** can be described as any compound that is capable of quenching ROS without themselves undergoing conversion to a destructive radical (Nishikimi and Yagi 1996; Rose and Bode 1993). So, together with antioxidant enzymes, non-enyzmatic antioxidants function to interrupt the cascades of uncontrolled oxidation (Noctor and Foyer 1998). In the cell, several compounds could serve as antioxidants and contribute an electron to neutralize free radicals to non-reactive species. However, as indicated by Rose and Bode (1993), the effective biological antioxidant is supposed to do more than simply react with free radicals. Therefore, it must be present in an adequate amount in the cell, it must react with a variety of free radicals and it must be suitable for regeneration. The compounds having this combination of properties are considered perfect antioxidants for the cells, such as L-ascorbic acid (L-AA, vitamin C), reduced glutathione (GSH), α-tocopherol (vitamin E), carotenoids, flavonoids and some compatible solutes (Fig. 4). These most abundant water- and lipid-soluble antioxidants have central and interrelated roles, acting both non-enzymatically and as substrates in enzyme catalysed detoxification reactions (Winkler et al., 1994; Chaudiere and Ferrari-Iliou 1999; Dellapenna and Pogson 2006). As we will see in the next section, in some scavenging steps (like in ascorbate-glutathione cycle), it is impossible to separate enzymatic and non-enyzmatic antioxidants/pathways from each other due to regeneration purposes.

3.1 L-Ascorbic Acid (vitamin C)

L-ascorbic acid (also known as L-AA, ascorbate, AsA, Asc, vitamin C) is the trivial name for the six-carbon sugar derivative L-threo-hex-2-enono-1,4-lactone and is one of the simplest vitamins. The de novo biosynthesis of L-AA in animals was established over four decades ago (Burns 1960). L-AA can be synthesized by the D-glucuronic acid (hexuronic acid) pathway intermediates utilization and biosynthesis take place either in the liver or the kidney of species having L-gulono-1,4-lactone oxidase activity (Braun et al., 1994; Banhegyi et al., 1997). However, humans, non-human primates and a few other mammals cannot synthesize L-AA due to the enzyme catalysing the last step in the biosynthesis (L-gulono-1,4-lactone oxidase) is highly mutated and non-functional (Stone 1972; Chatterjee 1973; Sato and Udenfriend 1978; Nishikimi and Yagi 1991). Therefore, they require L-AA as an essential component of their diets (Dabrowski 1994; Nishikimi and Yagi 1996).

L-AA is apparently a ubiquitous molecule in eukaryotes and is one of the best characterised components contributing to the nutritive value of plants (Horemans et al., 2000; Smirnoff and Wheeler 2000). In the literature, there are a number of reviews on ascorbate in plants have been published in previous years (Loewus and Loewus 1987;

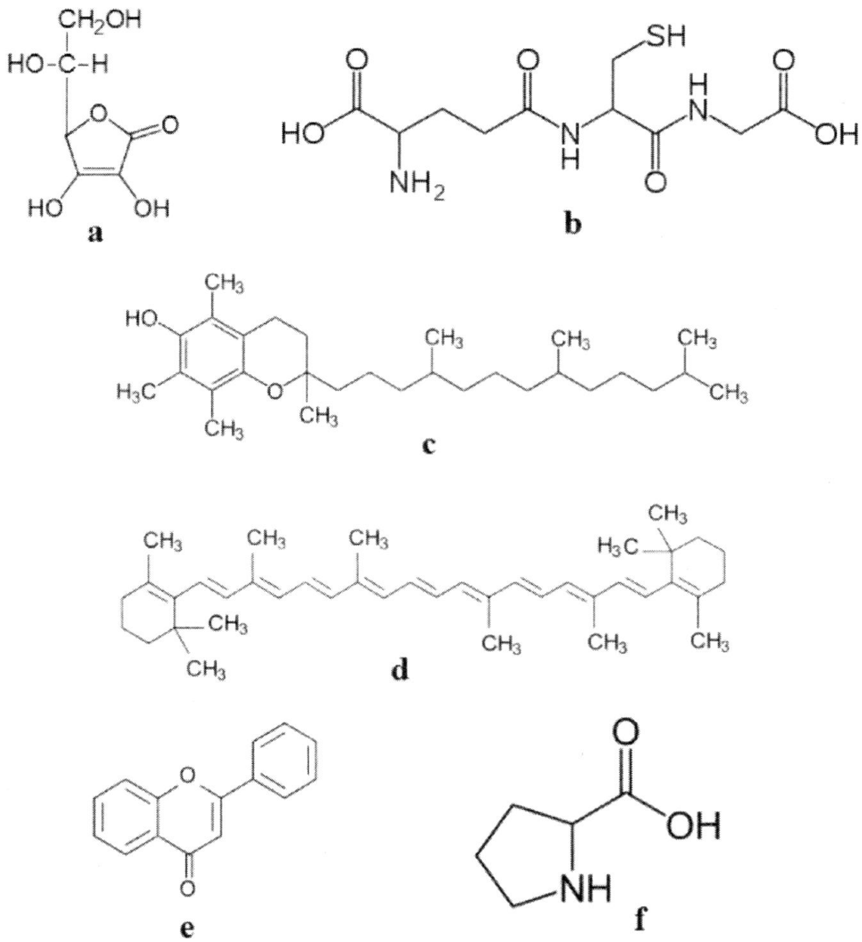

Fig. 4 Structures of the antioxidant compounds referred to in this review. a, L-AA; b, GSH; c, α-tocopherol; d, β-carotene; e, basic skeleton of the flavonoids; f, proline.

Loewus 1988; Arrigoni 1994; Smirnoff 1996; Noctor and Foyer 1998; Smirnoff and Wheeler 2000). However, much of the advances have been gained in the past few years about their biosynthesis and functions in plants.

A biosynthetic pathway for L-AA biosynthesis in plants was proposed by Wheeler and his colleagues (1998). This pathway (known as Smirnoff–Wheeler pathway) proposes that the conversion from GDP-D-mannose to L-galactono-1,4-lactone proceeds via L-galactose. L-galactono-1,4-lactone is converted to L-AA by action of a dehydrogenase. In plants, virtually all cells synthesise L-AA (Arrigoni and De Tullio 2002). However, L-AA contents vary considerably between tissues and depend on the physiological status of the plant, such as the age and the leaf surface light intensity (Foyer 1993; Davey et al., 2000), as well as on environmental factors (Loewus 1988; Smirnoff 1996; Horemans et al., 2000). For example, L-AA contents are generally higher in younger tissues than in older ones and L-AA accumulates in actively growing tissues (i.e., meristems) (Arrigoni 1994; Arrigoni and De Tullio 2002; De Gara and Tommasi 1999; Luwe et al., 1993). In plants, the first steps of L-AA synthesis are likely to be cytosolic, whereas L-galactono-1,4-lactone dehydrogenase, catalysing L-galactono-1,4-lactone conversion into L-AA, is located at the inner mitochondrial membrane (Bartoli et al., 2000). Since L-AA is a monovalent anion at physiological pHs, it will not freely diffuse through membranes and it must be transported across the membranes of all other subcellular compartments where it is found, including chloroplasts, cell wall, cytosol, peroxisomes and vacuole (Anderson et al., 1983; Smirnoff 1996; Rautenkranz et al., 1994; Foyer and Lelandais 1996; Jimenez et al., 1997).

Today, it is clear that L-AA has roles in plant stress physiology, as well as growth (cell expansion and division), development and synthesis of hydroxyproline-rich proteins (Horemans et al., 2000; Padh 1990; Conklin 2001). It has also importance in photoprotection and the regulation of photosynthesis (Forti and Elli 1995; Foyer and Harbinson 1994). As an enzyme cofactor, it modulates a number of important ezymatic reactions (Davey et al., 2000). L-AA is also a powerful secondary antioxidant, reducing the oxidized form of α-tocopherol (an antioxidant in non-aqueous phases) (Padh 1990; Thomas et al., 1992) (see Section 3.3). Finally, L-AA appears to be the precursor for oxalate and tartrate biosynthesis (for reviews, see Davey et al., 2000; Horemans et al., 2000). In addition to the above functions, most importantly, L-AA is the most effective and least toxic primary antioxidant (Nijs and Kelley 1991) protecting cells against oxidative stress and plays a crucial role in ascorbate-glutathione cycle (or so called Halliwell-Asada pathway) together with glutathione (GSH) (see Section 3.2).

Ascorbate reacts rapidly with superoxide, singlet oxygen, hydroxyl radicals and hydrogen peroxide (Padh 1990). It thus participates in removal of these reactive forms of oxygen which are generated during aerobic metabolism (Smirnoff 1996; Buettner and Jurkiewicz 1996). The ascorbate redox system consists of L-AA, monodehydroascorbate (MDHA) and dehydroascorbate (DHA) (Noctor and Foyer 1998). L-AA is stable when dry, but solutions readily oxidize, especially when it reacts with ROS. The first oxidation product of L-AA is the radical MDHA. *In vivo* MDHA is reduced back to L-AA by the activity of the NAD(P)-dependent enzyme, monodehydroascorbate reductase (MDHAR). However, MDHA is unstable and unless re-reduced, disproportionates to L-AA and DHA. DHA is also unstable in aqueous environments and reduced back to L-AA by the action of dehydroascorbate reductase (DHAR) which uses GSH as reductant (Conklin 2001). DHAR links L-AA to GSH and these reactions together constitute the ascorbate-glutathione (Halliwell-Asada) cycle (Foyer 1993), in which stabilization is achieved by three enzymes, namely, MDHAR, DHAR and glutathione reductase (GR) (Smirnoff and Wheeler 2000) (Fig. 5). As seen in Fig. 5, from glutathione disulphide (GSSG), GR regenerates reduced GSH linking it to the oxidation of NAD(P)H in ascorbate-glutathione cycle (Bray et al. 2000). The ascorbate-glutathione cycle is found in several compartments of the plant cell, such as chloroplasts, cytosol, mitochondria, apoplast and peroxisomes. The finding of this cycle in almost all cellular compartments tested suggests that this cycle plays a crucial role in controlling the level of ROS in these compartments (Mittler 2002).

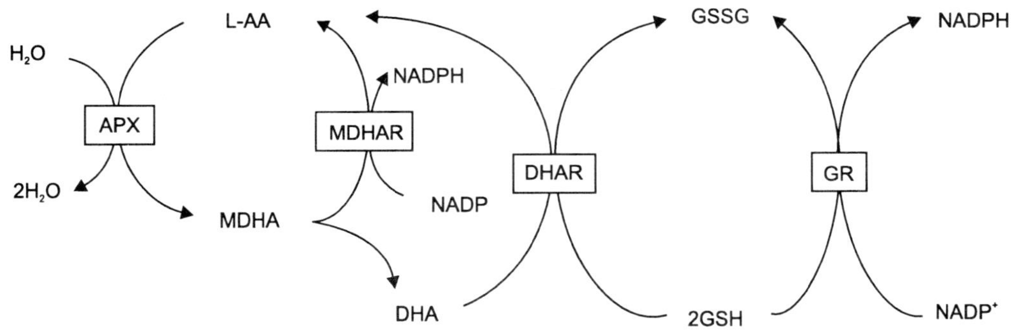

Fig. 5. Ascorbate-Glutathione (Halliwell-Asada) cycle (See text for the details).

3.2 Reduced Glutathione (GSH)

Glutathione (GSH), the tripeptide γ-glutamylcysteinylglycine (γ-Glu-Cys-Gly), is one of the crucial low molecular weight thiol compounds in most plants (Foyer et al., 1997; Arora et al., 2002). In the previous years, there are a number of reports about GSH in plants (Bergmann and Rennenberg 1993; Xiang et al., 2001; Noctor and Foyer 1998; Noctor et al., 2002a, b; Gomez et al., 2004a, b; Foyer and Noctor 2005). As the the major source of non-protein reduced sulphur in most plant cells (Bergmann and Rennenberg 1993; Noctor et al., 2000), GSH cannot be replaced functionally except by tripeptide homologs of GSH, such as homoglutathione, hydroxymethylglutathione and γ-glutamylcysteinylglutamic acid (Klapheck, 1988; Klapheck et al., 1992; Meuwly et al., 1993).

Unlike L-AA synthesis, the pathway of GSH synthesis is conserved in all organisms. GSH is synthesized from standard amino acids (glutamate, cysteine and glycine) in two steps. In those ATP-dependent reactions, first γ-glutamyl cysteine synthetase (γ-ECS) combines Glu and Cys to form γ-glutamyl-Cys (Hell and Bergmann 1990), then glutathione synthetase (GSase) catalyses the reaction between γ-EC and Gly to form GSH (Meister 1988). Biochemical studies of subcellular localization have found γ-ECS and GSase activities both inside and outside the chloroplast (reviewed by Noctor et al., 2002b).

GSH has been detected in plant tissues and localized virtually in all cell compartments, like cytosol, chloroplasts, endoplasmic reticulum, vacuoles, mitochondria, peroxisomes and apopolast (Millar et al., 2003; Shao et al., 2008; Ahmad et al., 2009). Synthesis of GSH in the root probably differs according to developmental stage as well as external factors, such as xenobiotics and heavy metals (like Cd). In several studies, foliar GSH contents have been found to be higher in the light. In seeds, in contrast to L-AA, high concentrations of GSH are found and GSase activity has also been notified in fruits (Noctor and Foyer 1998). However, it is clear that there are steep gradients in GSH concentration between different cellular compartments and in different cells and tissues (Fricker et al., 2000; Fricker and Meyer 2000; Gutierrez-Alcala et al., 2000). This difference in concentrations is due to an equilibrium between synthesis, degradation, usage and transport of GSH. Transport influences not only the individual amounts of GSH and GSSG in different organelles, but also the GSH/GSSG ratio (Foyer et al., 2001).

Due to the chemical reactivity of its thiol group, GSH can serve a wide range of functions in all organisms. Its physiological significance can be divided mainly into two categories, as regulation of sulphur metabolism and defence (Noctor and Foyer 1998), although its functions are not limited to these. GSH is a precursor of the phytochelatins (PCs, see Section 1.3.4) (Grill et al., 1989; Scheller et al., 1987; Clemens et al., 1999) and reported to be associated with several growth and development related events in plants, such as cell division (Sanchez-Fernandez et al., 1997; Shaul et al., 1996), cell differentiation, cell death, senesence, pathogen resistance and enzymatic regulation (Rausch and Wachter 2005). GSH functions also include signal transduction (Noctor et al., 2002a; Gomez et al., 2004a). Moreover, under stress conditions, GSH has been shown to stimulate the transcription of various genes, including hydroxyproline-rich glycoproteins, phenylalanineammonia lyase and chalcone synthase (Wingate et al., 1988).

Due to the nucleophilic property of its -SH group, GSH generates thiolate bonds and reacts with several electrophiles. Considering the relative stability and high water solubility of GSH, this property makes GSH a proper substance to protect organisms against stress conditions, such as oxidative stress, different intrinsic or extrinsic compounds, including heavy metals (Xiang et al., 2001). GHSs defence role is threefold. Firstly, GSH is a key antioxidant (Foyer et al., 2001) and redox buffer (Kunert and Foyer 1993; Meister 1994). As discussed above (Section 1.3), several environmental stress conditions cause an increase in ROS levels in plants and the response of GSH can be crucial for adaptive responses (Ahmad et al., 2009). Exposure to ROS can result in either decreased or enhanced antioxidant capacity. However, activation of GSH biosynthesis and its accumulation are a general feature of enhanced oxidation of the cytosol (Foyer and Noctor 2005; Foyer and Noctor 2001) accordingly increasing GSH biosynthetic capacity has been shown to enhance resistance to oxidative stress (Arisi et al., 1998; Zhu et al., 1999). GSH can scavenge H_2O_2, $^1O_2$, $O_2^{\bullet-}$, as well as other ROS like $OH^{\bullet}$ (Noctor and Foyer 1998; Smirnoff 1993; Larson 1988; Shao et al., 2005; Foyer and Noctor 2005). A central nucleophilic Cys residue is responsible for higher reductive potential of GSH (Ahmad et al., 2009). In metabolically active tissues, GSH acts as a key redox buffer, forming a barrier between protein Cys groups and ROS (Foyer and Noctor 2005). Under unstressed conditions, glutathione pool is generally > 95% in reduced form (Foyer et al., 2001). Upon the imposition of oxidative stress, the existing pool of reduced glutathione (GSH) is converted to oxidized glutathione (GSSG) (Grene 2002). It has been suggested that the GSH/GSSG ratio, indicative of the cellular redox balance, may act as a direct link between environmental cues and crucial molecular adaptive responses of plant cells (May et al., 1998a, b; Baginsky et al., 1999; Foyer and Noctor 2005; Millar et al., 2003). When the intensity of a stress increases, the rate of GSH oxidation exceeds GSSG reduction (Foyer et al., 2001). Therefore, the GSH/GSSG ratio decreases and leads to deterioration of the system (Tausz et al., 2004). It is generally believed that maintaining a high ratio of reduced to oxidized L-AA and GSH is essential for the proper scavenging of ROS in cells. This ratio is maintained by the three enzymes of ascorbate-glutathione cycle, namely, MDHAR, DHAR and GR, by using NAD(P)H as reducing power (see

Section 3.1) (Asada 1999; Noctor and Foyer 1998; Tsugane et al., 1999). However, the overall balance among different antioxidants has also to be tightly controlled. Because, enhanced glutathione biosynthesis in chloroplasts can result in oxidative damage to cells rather than their protection, possibly by altering the overall redox state of chloroplasts (Creissen et al., 1999). Secondly, GSH plays a protective role in detoxification reactions as the substrate of glutathione *S*-transferases (GSTs) (Edwards et al., 2000). GSTs (EC.2.5.1.18) generally constitute a dimeric enzyme and have been found in all living species, including plants. They are a superfamily of multifunctional enzymes which detoxify endobiotic and xenobiotic compounds by conjugating GSH to a hydrophobic substrate with electrophilic centres. As a result of the conjugation reaction between electrophiles and GSH, a conjugate that is generally less reactive than the parental compound is formed and the solubility of hydrophobic xenobiotics is increased. In plants, the *S*-glutathionylated conjugates are tagged and transported from the cytosol into the vacuole via the ATP-binding cassette (ABC) transporters (called glutathione pump) for storage excretion (Coleman et al., 1997). Besides their well investigated conjugation reaction for the detoxification of endogenous and xenobiotic compounds, plant GSTs can also be involved in both GSH dependent peroxidation (Cummins et al., 1999; Roxas et al., 1997) to counteract oxidative stress or isomerization reactions (Thom et al., 2001). (For a more detailed discussion for GST protein families readers are referred to recent reviews by Frova, 2006; Oztetik, 2008). And finally, GSH participates in the regeneration of L-AA from DHAR and indirectly essential in protecting membranes by maintaining α-tocopherol (see Section 3.3) and zeaxanthin in the reduced state (Arora et al., 2002; Xiang et al., 2001).

3.3 α-Tocopherol (vitamin E)

Tocopherols are a group of lipid soluble antioxidants collectively known as vitamin E and are essential components of human health due to their requirement in maintaning immune, neuronal and muscular functions (Pryor, 2000; Brigelius-Flohe et al., 2002). However, tocopherols are synthesized only by photosynthetic organisms (Abbasi et al., 2007; Li et al., 2008; Sattler et al., 2003). There are many reports present about different features of the tocopherols in the literature (Norris et al., 1995; Grusak and DellaPenna 1999; Porfirova et al., 2002; Cahoon et al., 2003; Falk et al., 2003; Qi et al., 2003). Tocopherols are essential components of biological membranes, exclusively located in the plastid or thylakoid membranes (Munne-Bosch 2005), where they have both antioxidant and non-antioxidant functions (Kagan 1989). They occur in four different types or isomers, namely, α-, β-, γ- and δ-tocopherols. All four types of tocopherols structurally consist of a chromanol head group attached to the phytyl tail, which together giving vitamin E compounds amphipathic character (Kamal-Eldin and Appelqvist 1996). Among others, α-tocopherol with its three methyl substituents has the highest antioxidant activity of tocopherols (Kamal-Eldin and Appelqvist 1996) and relative antioxidant activity of the tocopherol isomers *in vivo* is $\alpha > \beta > \gamma > \delta$, due to the methylation pattern and the amount of methyl groups attached to the phenolic ring (Blokhina et al., 2003).

The biosynthesis of tocopherols takes place in chloroplasts (Semchuk et al. 2009) and are synthesized from precursors derived from two metabolic pathways: homogentisic acid (HGA), an intermediate in aromatic amino acid degradation, and phytyl diphosphate (PDP), which arises from methylerythritol phosphate pathway. Homogentisate phytyltransferase (HPT) is a membrane-bound chloroplast enzyme, which catalyses the committed step of tocopherol biosynthesis, the condensation of HGA and PDP (Collakova and DellaPenna 2001; Savidge et al., 2002). δ- and γ-tocopherols are produced from their respective intermediates, 2- methyl-6-phytyl-1,4-benzoquinol (MPBQ) and 2,3-dimethyl-6- phytyl-1,4-benzoquinol (DMPBQ), via reactions catalyzed by a tocopherol cyclase (TC) (Arango and Heise 1998; Porfirova et al., 2002). A γ-tocopherol methyltransferase (γ-TMT) converts γ- and δ-tocopherols to α- and β-tocopherols, respectively (Bergmuller et al., 2003; Shintani and DellaPenna 1998).

Tocopherols are synthesized only in plants, algae and some cyanobacteria (Munne-Bosch 2005) and are found in all parts of plants (Blokhina et al., 2003). Like other low molecular weight antioxidants in the cell, tocopherol levels and composition vary in different tissues and fluctuate during plant growth and development and also in response to abiotic stress conditions (Collakova and DellaPena 2003). α-tocopherol, as the major tocopherol isomer, found in leaf chloroplasts, where it is located in the chloroplast envelope, thylakoid membranes and plastoglobuli (Munne-Bosch 2005; Mullineaux et al., 2006; Fey et al., 2005). Therefore, chloroplast membranes are thought to be well protected against photooxidative damage (Fryer 1992). The changes in the α-tocopherol levels are sensitive to environmental stress agents (such as light, drought, salt and cold), magnitude of the stress itself, species of the plant and degradation/recycling of α-tocopherol in plants (Munne-Bosch 2005; Shao and Chu 2005; Wu et al., 2007; Shao et al., 2007a, b). In general, while enhanced α-tocopherol accumulation may provide an additional protection from oxidative damage through stress tolerance, whereas decreased levels favour oxidative damage (Havaux et al., 2000; Munne-Bosch and Alegre 2002a). Studies have shown that oxidative stress activates the expression of genes responsible for the synthesis of tocopherols in higher plants (Shao et al., 2005; Wu et al., 2007).

To date, several lines of evidence indicate that tocopherols perform different functions in plants, as either non-antioxidants or antioxidants (Munne-Bosch and Falk 2004). For example, tocopherols have been suggested to participate in stabilization of membrane structures and in cyclic electron transport around photosystem II (reviewed by Munne-Bosch and Alegre 2002a; Munne-Bosch and Falk 2004; Fryer 1992). Earlier studies have shown that α-tocopherol modulates membrane fluidity in a similar manner to cholesterol and also membrane permeability to small ions and molecules (Fryer, 1992). The study of Kruk et al. (2000) has shown that there is an interaction between PSII with α-tocopherol. Complexation of tocopherol with free fatty acids and lysophospholipids protects membrane structures against their deleterious effects (Blockina et al., 2003).

On the other hand, most of the presented functions of tocopherol are related to their antioxidant properties both in animals and plants. As an antioxidant, α-tocopherol can react not only with singlet oxygen and superoxide radicals, but also with alkoxy radicals, lipid peroxy radicals, alkyl radicals and alkyl peroxy radicals (Buettner 1993, Fukuzawa and Gebicky 1983; Munne-Bosch and Alegre 2002a). Therefore, it deactivates ROS in photosynthetic membranes and prevents polyunsaturated fatty acids from lipid peroxidation by reducing lipid peroxyl radicals to the corresponding hydroperoxides. α-Tocopherol can physically quench the $^1O_2$ molecules and chemically scavenge both $^1O_2$ and lipid peroxyl radicals. Scavenging of lipid peroxyl radicals results in the formation of tocopheroxyl radicals (TOH·). Regeneration of TOH• back to its reduced form can be achieved by L-AA and GSH (Fryer, 1992 and see also review Munne-Bosch and Alegre 2002a). This situation shows the contribution of α-tocopherol to preservation of an adequate redox state in chloroplasts (Munne-Bosch and Alegre 2002a) and further opens the understanding of compensatory mechanism or interdependence between different antioxidants in plant responses to stress (Caretto et al., 2002; Munne-Bosch and Alegre 2003; Kanwischer et al., 2005).

3.4 Carotenoids

Among the various natural pigments, carotenoids comprise a large family of more than 700 structures (Britton et al., 2004), and are synthesized in plants and other photosynthetic organisms, as well as in some non-photosynthetic bacteria, fungi, algae, yeasts and moulds (Stahl and Sies 2003, 2005). Since animals and humans are unable to synthesize carotenoids de novo, they rely upon the diet as the source of these compounds (Fraser and Bramley 2004). During the previous years, carotenoids attracted attention, because a number of epidemiological studies have revealed that an increased consumption of a diet rich in carotenoids is correlated with a diminished risk for cardiovascular, ophthalmological, gastrointestinal or neurodegenerative disorders and some types of cancer (Mayne 1996; Van Duyn and Pivonka 2000). Therefore, since this discovery, the number of research has been increased in the literature about the way to enhancing carotenoid levels in crop plants and foods to achieve better health conditions for human.

Most carotenoids can be derived from a 40-carbon basal structure, which includes a polyene chain contains 3 to 15 conjugated double bonds. The pattern of conjugated double bonds in the polyene backbone of carotenoids determines their light absorbing properties and influences the antioxidant activity (see below) of carotenoids (Goodwin 1980; Britton et al., 1998). According to their chemical composition carotenoids are categorized as either carotenes or xanthophylls (oxocarotenoids) (Olson and Krinsky 1995). While the carotene group, such as β-carotene, α-carotene and lycopene, composed only of carbon and hydrogen atoms, xanthophylls, such as zeaxanthin, lutein, α- and β-cryptoxanthin, carry at least one oxygen atom (Stahl and Sies 2005; Chaudhary et al., 2010). Carotenoids within lipid bilayers may modify membrane properties such as fluidity and permeability Gruszecki (1999).

In higher plants, carotenoids are biosynthesized from the central isoprenoid pathway, in which a C5 building block isopentenyl pyrophosphate (IPP) is the common precursor of all isoprenoid compounds. IPP isomerises to dimethylallyl pyrophosphate (DMAPP) and geranyl pyrophosphate (GPP) forms by the condensation of DMAPP with a molecule of IPP. Addition of two more IPP units produces the C20 geranyl geranyl pyrophosphate (GGPP). Tail-to-tail reaction of two molecules of GGPP gives the compound called phytoene (Naik et al., 2003). From this point, the biosynthetic pathway continues as:

phytoene → phytofluene → ζ-carotene → neurosporene → lycopene → α-carotene and β-carotene. Each enzymatic step from phytoene to lycopene adds one double bound to the molecule. The biosynthetic step after lycopene involves enzymatic cyclization of the end groups, which results in α-carotene (one β-ring and one ε-ring), is the immediate precursor of lutein and β-carotene (two β-rings), is a precursor for several other carotenoids in plants like xanthophylls.

Carotenoids are found in most plant organs and tissues and accumulate in nearly all types of plastids. From those above-mentioned and commonly occurring carotenes, β-carotene and lycopene are found in chloroplasts and chromoplasts of some fruits and flowers, respectively. However, several xanthophylls, such as lutein, violaxanthin and neoxanthin are found as the main constituents of the photosynthetic plants in their light harvesting complexes (Woitsch and Romer 2003; Dall'Osto et al., 2006).

Besides their antioxidant activities (see below), carotenoids have many and diverse functions which influence some mechanisms in the plant cells. These functions range from plant development (Beveridge et al., 2003; McSteen and Leyser 2005), to the serving as precursors for biosynthesis of plant-growth regulator abscisic acid (Nambara and Marion-Poll 2005) and from an important role in photosystem assembly (Kulheim et al., 2002; Dall'Osto et al., 2007) to the protection of retinal degeneration of the eye. These and some other functions of carotenoids in plant cells are well summarized recently in Chaundry et al. (2010).

The best documented antioxidant action of carotenoids is their ability to quench singlet oxygen (Paiva and Russell 1999) via physical or chemical quenching (Stahl and Sies 2003). Physical quenching involves the transfer of excitation energy from $^1O_2$ to the carotenoid which leads to the formation of ground state oxygen and an excited triplet state carotenoid. Instead of further chemical reactions, the carotenoid returns to ground state dissipating its energy by interaction with the surrounding solvent. In the process of physical quenching the carotenoid remains intact and can be reused for further cycles of singlet oxygen quenching. The physical quenching activity of a carotenoid mainly depends on the number of conjugated double bonds of the molecule and is influenced to a lesser extent by carotenoid end groups (cyclic or acyclic) (Paiva and Russell 1999). Among different carotenoids, xanthophylls as well as carotenes (β-carotene) have shown to be efficient quenchers of singlet oxygen (Baltschun et al., 1997; Conn et al., 1991). However, lycopene, with eleven conjugated and two non-conjugated double bonds, is among the most efficient singlet oxygen quenchers of the natural carotenoids (Krinsky 1998). On the other hand, carotenoids efficiently scavenge peroxyl radicals, which are formed in the process of lipid peroxidation and scavenging of this species interrupts the reaction sequence that leads to damage in lipophilic compartments (Sies and Stahl 1995).

By the studies of Truscott (1996), Bohm et al. (1997) and Mortensen and Skibsted (1997), it has been shown that, similar to L-AA and GSH, β-carotene regenerates tocopherol from the tocopheroxyl radical. The resulting carotenoid radical cation may subsequently be repaired by vitamin C. Bohm et al. (1998a, b) demonstrated that the synergistic interaction of carotenoid with both vitamins E and C is possible *in vitro*. However, the synergistic protection afforded by these antioxidants is dependent upon a balance between them.

3.5 Flavonoids

The most important groups of secondary metabolites are phenolic compounds (phenols), which are characterized by at least one aromatic ring (C6) bearing one or more hydroxyl groups. Among these compounds, flavonoids constitute one of the most ubiquitous groups of plant phenolics. Together with bryophytes and pteridophytes, higher plants are the only natural source (Rausher 2006) and flavonoids cannot be synthesized by humans (Peterson and Dwyer 1998). Their isolation from oranges dates back to 1930s and they are thought to be a member of a new class of vitamins, so called vitamin P by that time (Tapas et al., 2008). While in 1980s, more than 4000 flavonoids were identified in plants (Kandaswami and Middleton 1994), today this number has increased up to over 9000 (Hernandez et al., 2009).

Although flavonoids are generally considered non-nutrients, in contrast to previous studies, as naturally occurring substances in plants that are thought to have positive effects on human health (Montoro et al., 2005). Today, due to the variety of pharmacological activities in the mammalian body (see below), flavonoids are more correctly referred as nutraceuticals (Tapas et al., 2008; Butelli et al., 2008). There are many studies related to different features of flavonoids in the literature. However, the number is still increasing today, because research on flavonoids gained more attention with the discovery of the French paradox, which defines the low cardiovascular mortality rate observed in Mediterranean populations in association with red wine consumption and a high saturated fat intake. This effect, at least in part, is due to the flavonoids in red wine (Formica and Regelson 1995; Ranaud and de Lorgeril 1992). Because, the concentration of flavonoids is higher in red than in white wines (Harborne and Williams 2000).

In plants, flavonoids are synthesized via the phenylpropanoid pathway. The precursors of the synthesis of most flavonoids are malonyl-CoA and p-coumaroyl-CoA, which are derived from carbohydrate metabolism and phenylpropanoid pathway, respectively (Forkmann and Heller 1999). The biosynthetic pathway starts with the condensation of one molecule of p-coumaroyl-CoA and three molecules of malonyl-CoA, yielding naringenin chalcone. This reaction is carried out by the enzyme chalcone synthase (CHS). Chalcone is isomerized to a flavanone by the enzyme chalcone flavanone isomerase (CHI). From these central intermediates, the pathway diverges into several side branches, each resulting in a different class of flavonoids, such as isoflavonoids, aurones, flavones, pro-anthocyanidins (PA) and flavonols. Chemically, the basic flavonoid structure is the flavan nucleus, that consists of 15 carbon atoms arranged in three rings (C6-C3-C6 configuration), which are labelled as A, B and C for convenience. The different classes of flavonoids differ in the level of oxidation and pattern of substitution of the C ring (Schijlen et al., 2004).

Flavonoids are generally present in a wide variety of edible plant sources (Marin et al. 2002) and in some beverages, like wine (especially red wine), tea and beer (Kuhnau 1976). In addition, they are also found in barks, roots, stems, leaves and flowers of plants (Middleton 1998). The synthesis of flavonoids fundamentally occurs in the cytosol and in the endoplasmic reticulum (ER)-linked multi-enzymatic complexes. After synthesis, they are transported to their subcellular targets (Hrazdina and Wagner 1985). Moreover, there is some research about the presence of flavonoids in other cell compartments, such as membranes, chloroplasts, vacuoles, nuclei and cuticle of leaves (Hernandez et al., 2009) or in the epidermal layers (Treutter 2006). However, as there are also some controversery results available, the relevance of flavonoids within these compartments still remains to be determined.

Flavonoids are involved in an array of processes and play different roles in the ecology and physiology of plants (Pietta 2000). Due to their attractive colours, they may act as visual signals for pollinating insects or because of their astringency, they can represent a defence system against herbivores harmful to the plant (Mazza and Miniati (1993). They are responsible for the characteristic red and blue colours of berries, wines and many vegetables (Heim et al., 2002) and involved in the seed development and allellopathy (Winkel-Shirley 2001). There is evidence that flavonoids may function in plants to screen harmful radiation, bind phytotoxins and help to regulate the stress response by controlling auxin transport (Winkel-Shirley 2002). Flavonoids are also important components in the human diet (De Groot and Rauen 1998) and it was suggested that flavonoid intake is inversely correlated with mortality due to coronary heart disease (Knekt et al., 1996). They have also been suggested to play a protective role in a wide range of biological actions, such as antimicrobial (antibacterial, antifungal, antiviral), antiulcer, hepatoprotective, anti-inflammatory, antidiabetic, vasorelaxant agent, antiatherosclerotic, antithrombogenic, cardioprotective, antineoplastic, anticarcinogenic and antiosteoporotic activities (For the details, see a recent review of Tapas et al., 2008).

All above-mentioned effects of flavonoids are believed to be related to the antioxidant properties and today there is a growing interest in the study of the antioxidant activity of flavonoids consequently. Flavonoid levels increase during exposure to biotic and abiotic stresses, such as wounding, drought, metal toxicity and nutrient deprivation (Dixon and Paiva 1995) and they have been shown to be highly effective antioxidants and protecting plants from oxidative stress (Bravo 1998; Barcelo and Poschenrieder 2002). The basic mechanism of antioxidant action which is applicable to flavonoids include ROS scavenging activity, containing quenching of singlet oxygen, chelation of transition metals and inhibition of enzymes, such as lipoxygenase, cyclooxygenase, monooxygenase, xanthine oxidase, phospholipase A_2, protein kinase C, mitochondrial succinoxidase and NADH oxidase, those are all involved in ROS generation (Cos et al., 2000; Pietta 2000). The completed studies have shown that flavonoids have fulfilled these criteria and *in vitro* antioxidant activity of flavonoids has been recognized for a considerable time. During the last 20 years, SAR studies of flavonoids have generated several consistent lines of evidence supporting the role of specific structural components as requisites for radical scavenging and chelating properties. According to Bors and colleagues, three structural elements are required for optimal radical scavenging activity. These are: (i) An ortho-dihydroxy structure (catechol moiety) in the B-ring, which is the radical target site for flavonoids with a saturated C-2 and C-3 double bond, confers a high stability to the flavonoid aroxyl radical and participates in electron delocalization, (ii) A 2,3-double bond in conjugation with a 4-oxo function is responsible for electron delocalization from the B-ring and, (iii) The 3- and 5-OH groups are, together with the 4-oxo function, essential for maximal radical scavenging activity. In the absence of a C2-C3 double bond, the presence of an aliphatic 3-OH group does not contribute to the radical scavenging activity (Bors et al., 1990a, b, 1996, 1992). As clear from the previous studies, the *in vitro* antioxidant capacities of flavonoids are several fold higher than those of L-AA, GSH or α-tocopherol (Rice-Evans et al., 1997; Heijnen et al., 2001; Chun et al., 2003). However, even with very high flavonoid intakes, the intracellular flavonoid concentrations in humans are likely to be 100-1000 times lower than concentrations of other antioxidants (Abd El Mohsen et al., 2002; Halliwell et al., 2000). Secondly, most circulating flavonoids are actually flavonoid metabolites and as metabolic modifications change their classical antioxidant nature, some circulating metabolites have lower antioxidant activity by comparing to their parent forms (Spencer et al., 2001; Miyake et al., 2000; da Silva et al., 1998). Therefore, the relative contribution of dietary flavonoids to plasma and tissue antioxidant function *in vivo* is likely to be very small or negligible (Williams et al., 2004; Frei and Higdon 2003; Lotito and Frei 2006). However, available evidence from cell culture experiments suggests that many of the biological effects of flavonoids are related to their ability to modulate cell-signalling pathways (Williams et al., 2004), as intracellular concentrations of flavonoids encountered *in vivo* are sufficiently high to have pharmacological activity at receptors, enzymes and transcription factors (Spencer et al., 2003; Hou et al., 2004).

Today, it has been known that a recycling pathway exist between GSH, L-AA and α-tocopherol and carotenoids, in which flavonoids might be involved. Bor and colleagues have been shown the interaction between L-AA and flavonoids with the formation of both L-AA and flavonoid radicals (Bors et al., 1995). Afterwards, Cossin et al. (1998) reported that some flavonoids compete with ascorbic acid for the ascorbate oxidase and therefore are preserving in a double way the ascorbic acid from oxidation. Vice versa, L-AA has been proposed to be involved in recycling oxidized flavonoids (Yamasaki et al., 1997). Another evidence come from the study in which, L-AA is oxidized by certain peroxidases in vine leaves, but only in the presence of phenolics (Perez et al., 2002). Moreover, some recent studies have shown that, under oxidative stress conditions, the pools of different antioxidants that establish recycling pathways with each other are oxidized in parallel (Munne-Bosch and Alegre 2002b; Caretto et al., 2002).

3.6 Other Low Molecular Weight Compounds

As tried to be explained in this review with details, several environmental stress conditions may stimulate oxidative stress, osmotic stress and lead damage to proteins, lipids and DNA in different extents. These stress conditions not only cause induction of ROS detoxification systems (enzymatic and non-enzymatic) and several stress proteins (i.e., heat shock proteins), but also accumulation of some compatible solutes, especially under abiotic stress, as part of the cellular adaptive responses in plants (Zhu 2002; Ashraf and Foolad 2007). The compatible organic solutes, generally known as osmolytes, are not highly charged, but are polar, highly soluble and low molecular weight compounds which are usually non-toxic at high cellular concentrations (Alonso et al., 2001). These are referred to as compatible metabolites, because they do not apparently interfere with the normal biochemical reactions and accumulate to high levels in the cytosol with increasing stress conditions, like drought, salinity or extreme temperatures (Hasegawa et al., 2000; Zhifang and Loescher 2003). The synthesis and accumulation of osmolytes varies among plant species as well as among different cultivars of the same species. Osmolytes are found in chloroplasts and cytoplasmic compartments in plant cells, in which they occupy 20% or less of the volume of mature cells (Ain-Lhout et al., 2001) and natural osmolyte concentrations can reach to 200 mM or more (Reddy et al., 2004). However, most probably, these solutes have another role in stabilizing the tertiary structure of proteins and function as osmoprotectants at low concentrations (Rhodes et al., 2002). Osmolytes protect plants from adverse effects of stress through different processes, which include cellular osmotic adjustment, detoxification of ROS, protection of membrane integrity and stabilization of enzymes and proteins (Ashraf and Foolad, 2007). A major category of organic osmotic solutes consists of simple sugars (mainly fructose and glucose), sugar alcohols (glycerol and methylated inositols) and complex sugars (trehalose, raffinose and fructans) (Kerepesi and Galiba 2000; Bohnert and Jensen 1996). Others include quaternary amino acid derivatives (proline, glycine betaine, β-alanine betaine, proline betaine, tertiary amines 1,4,5,6-tetrahydro-2-mehyl-4-carboxyl pyrimidine) (Khatkar and Kuhad 2000; Singh et al., 2000; Rhodes and Hanson 1993; Khan et al., 2000; Wang and Nil 2000) and sulphonium compounds (choline *O*-sulphate, dimethyl sulphonium propironate) (Yokoi et al., 2002). For today, even most of the effort has been devoted to engineer plants for overproduction of various osmoprotectants genetically, there has been little success in achieving the desired protective levels of these osmolytes. However, in some plants increased resistance to abiotic stresses has been achieved by exogenous application of various organic solutes, therefore, possibly this way may contribute to increased osmoprotectant levels under certain stress conditions (For a detailed review, see Ashraf and Foolad 2007 and Parvaiz and Satyawati 2008).

4. CONCLUSION AND PERSPECTIVES

ROS mediated oxidative stress/damage is controlled by a delicate balance between production and scavenging. A complex network of enzymatic and non-enzymatic antioxidants controls the concentration of ROS and repairs oxidative damage. Accumulated researches are revealing the complex and fine-tuned interplay between ROS and antioxidants in controlling growth, development and response to the abiotic and biotic stresses in plants. In this relation, antioxidants are not passive bystanders, but rather they function as key signalling components which constitute a dynamic interface inbetween the stress perception and physiological responses of plant cell. However, there is no such a compound as a universal antioxidant present. In fact, different compounds act as different lines of defence against ROS in the cell. This is most probably because of their mode of action as an antioxidant or due to the differential antioxidant concentrations between different compartments within the cell. Today, it is known that several antioxidants are working synergistically, as either part of the compensatory mechanism or interdependence between each other, to provide more effective barrier against to stress conditions. Although the highly compartmentalized nature of antioxidants involved in ROS scavenging is well defined, we have much to discover about the sensing and response mechanisms and how the delicate balance between production and scavenging is controlled. Many other questions are, how toxic are ROS to plant cells, what interactions exist between ROS and and calcium signalling, how antioxidants coordinate ROS metabolism, growth and development during a combination of abiotic and biotic stresses and during multiple abiotic stresses. To answer this and other possible questions, analysis can be performed on plants responding to abiotic or biotic stresses or combinations of both and can be complemented by using mutants with altered ability to produce or scavenge ROS. Advanced imaging methods, like the markers used for Ca^{2+} imaging, will also enhance the understanding of ROS metabolism. Even if that ROS scavenging mechanism is very complex, there can be no doubt that transgenic plants will be invaluable in assessing the precise role that main antioxidants and ROS play in the functional network which controls stress tolerance. As overexpression of antioxidant genes provides the opportunity to develop plants with enhanced tolerance to abiotic as well as biotic stress. In the near future, advanced functional genomics coupled with proteomics and metabolomics and some other new techniques will definitely offer a detailed insights into the networks of ROS and its related plant responses.

REFERENCES

Abbasi, A.R., Hajirezaei, M., Hofius, D., Sonnewald, U. and Voll, L.M. (2007). Specific roles of α- and γ-tocopherol in abiotic stress responses of transgenic tobacco. *Plant Physiol.* **143:** 1720-38.

Abd El Mohsen, M.M., Kuhnle, G., Rechner, A.R., Schroeter, H., Rose, S., Jenner, P. and Rice-Evans, C.A. (2002). Uptake and metabolism of epicatechin and its access to the brain after oral ingestion. *Free Radic. Biol. Med.* **33:** 1693-702.

Adam, A., Farkas, T., Somlayai, G., Hevesi, M. and Kiraly, Z. (1989). Consequence of O_2^- generation during bacterially induced hypersensitive reaction in tobacco: deterioration of membrane lipids. *Physiol. Mol. Plant Pathol.* **34:** 13-26.

Adam, Z., Rudella, A. and van Wijk, K.J. (2006). Recent advances in the study of Clp, FtsH and other proteases located in chloroplasts. *Curr. Opin. Plant Biol.* **9:** 234-40.

Agarwal, S. (2007). Increased antioxidant activity in *Cassia* seedlings under UV-B radiation. *Biol. Plant.* **51:** 157-60.

Ahmad, P., Jaleel, A., Azooz, M.M. and Nabi, G. (2009). Generation of ROS and non-enzymatic antioxidants during abiotic stress in plants. *Botany Research International.* **2:** 11-20.

Ahn,Y-J. and Zimmerman, J.L. (2006). Introduction of the carrot HSP17.7 into potato (*Solanum tuberosum* L.) enhances cellular membrane stability and tuberization *in vitro. Plant Cell Environ.* **29:** 95-104.

Ain-Lhout, F., Zunzunegui, M., Diaz Barradas, M.C., Tirado, R., Clavijo, A. and Gracia Novo, F. (2001). Comparision of proline accumulation in two Mediterranean shrubs subjected to natural and experimental water deficit. *Plant Soil.* **230:** 175-180.

Allakhverdiev, S.I., Sakamoto, A., Nishiyama, Y., Inaba, M. and Murata, N. (2000). Ionic and osmotic effects of NaCl-induced inactivation of photosystems-I and II in *Synechococcus* sp. *Plant Physiol.* **123:** 1047-1056.

Alonso, R., Elvira, S., Castillo, F.J. and Gimeno, B.S. (2001). Interactive effects of ozone and drought stress on pigments and activities of antioxidative enzymes in Pinis halpensis. *Plant Cell Environ.* **24:** 905-916.

Anderson, J.W., Foyer, C.H. and Walker, D.A. (1983). Light-dependent reduction of dehydroascorbate and uptake of exogenous ascorbate byspinach chloroplasts. *Planta.* **158:** 442-450.

Apel, K. and Hirt, H. (2004). Reactive Oxygen Species: Metabolism, Oxidative Stress, and Signal Transduction. *Annu. Rev. Plant Biol.* **55:** 373-399.

Aposto1, I., Heinstein, P.F. and Low, P.S. (1989). Rapid stimulation of an oxidative burst during elicitation of cultured plant cells. Role in defense and signal transduction. *Plant Physiol.* **90:** 109-116.

Arango, Y. and Heise, K-P. (1998). Tocopherol synthesis from homogentisate in *Capsicum anuum* L. (yellow pepper) chromoplast membranes: evidence for tocopherol cyclase. *Biochem. J.* **336:** 531-533.

Arisi, A.M., Cornic, G., Jouanin, L. and Foyer, C.H. (1998). Overexpression of iron superoxide dismutase in transformed poplar modifies the regulation of photosynthesis at low CO_2 partial pressures or following exposure to the prooxidant herbicide methyl viologen. *Plant Physiol.* **117:** 565-574.

Arora, A., Sairam, R.K. and Srivastava, G.C. (2002). Oxidative stress and antioxidative system in plants. *Current Science.* **82:** 1227-1238.

Arrigoni, O. (1994). Ascorbate system in plant development. *J. Bioenerget Biomemb.* **26:** 407-419.

Arrigoni, O. and De Tullio, M.C. (2002). Ascorbic acid: much more than just an antioxidant. *Biochimica et Biophysica Acta.* **1569:** 1-9.

Asada, K. (1999). The water–water cycle in chloroplasts: scavenging of active oxygen and dissipation of excess photons. *Annu. Rev. Plant Physiol. Plant Mol. Biol.* **50:** 601-639.

Asada, K. (2006). Production and scavenging of reactive oxygen species in chloroplasts and their functions. *Plant Physiol.* **141:** 391-396.

Ashraf, M. and Foolad, M.R. (2007). Roles of glycine betaine and proline in improving plant abiotic stress resistance. *Environmental and Experimental Botany.* **59:** 206-216.

Auh, C-K. and Murphy, T.M. (1995). Plasma membrane redox enzyme is involved in the synthesis of O_2- and H_2O_2 by Phytophthora elicitor-stimulated rose cells. *Plant Physiol.* **107:** 1241-1247.

Aziz, A. and Larher, F. (1998). Osmotic stress induced changes in lipid composition and peroxidation in leaf discs of *Brassica napus* L. *J. Plant Physiol.* **153:** 754-762.

Badiani, M., Paolacci, A.R., Fusari, A., D'Ovidio, R. and Scandalios, J.G. (1997). Non-optimal growth temperatures and antioxidants in the leaves of Sorghum bicolor (L.) Moench. II. Short-term acclimation. *J. Plant Physiol.* **151:** 409-421.

Baginsky, S., Tiller, K., Pfannschmidt, T. and Link, G. (1999). PTK, the chloroplast RNA polymerase-associated protein kinase from mustard (*Sinapis alba*), mediates redox control of plastid *in vitro* transcription. *Plant Mol. Biol.* **39:** 1013-1023.

Baker, C.J., Orlandi, E.W. and Mock, N.M. (1993). Harpin, an elicitor of the hypersensitive response in tobacco caused by *Enoinin* nmylouorn, elicits active oxygen production is suspension cells. *Plant Physiol.* **102:** 1341-1344.

Baltschun, D., Beutner, S., Briviba, K., Martin, H.D., Paust, J., Peters, M., Rover, S., Sies, H., Stahl, W., Steigel, A. and Stenhorst, F. (1997). Singlet oxygen quenching abilities of carotenoids. *Liebigs Ann.* 1887-1893.

Banhegyi, G., Braun, L., Csala, M., Puskas, F. and Mandl, J. (1997). Ascorbate metabolism and its regulation in animals. *Free Rad Biol. and Medicine.* **23:** 793-803.

Baniwal, S.K., Bharti, K., Chan, K.Y., Fauth, M., Ganguli, A., Kotak, S., Mishra, S.K., Nover, L., Port, M., Scharf, K-D., Tripp, J., Weber, C., Zielinski, D. and von Koskull-Doring, P. (2004). Heat stress response in plants: a complex game with chaperones and more than twenty heat stress transcription factors. *J. Biosci.* **29:** 471-487.

Barcelo, J. and Poschenrieder, C. (2002). Fast root growth responses, root exudates, and internal detoxification as clues to the mechanism of aluminium toxicity and resistance: a review. *Environ. Exp. Bot.* **48:** 75-92.

Bartoli, C.G., Pastori, G.M. and Foyer, C.H. (2000). Ascorbate Biosynthesis in Mitochondria is Linked to the Electron Transport Chain between Complexes III and IV. *Plant Physiol.* **123:** 335-344.

Bartosz, G. (1996). Peroxynitrite: mediator of the toxic action of nitric oxide. *Acta Biochem. Poi.* **43:** 645-660.

Bartosz, G. (1997). Oxidative stress in plants. *Acta Physiologiae Plantarum.* **19:** 47-64.

Bechtold, U., Murphy, D.J. and Mullineaux, P.M. (2004). Arabidopsis peptide methionine sulfoxide reductase2 prevents cellular oxidative damage in long nights. *Plant Cell.* **16:** 908-919.

Beligni, M.V. and Lamattina, L. (1999). Nitric oxide counteracts cytotoxic processes mediated by reactive oxygen species in plants. *Planta.* **208:** 337-344.

Bergmann, L. amd Rennenberg, H. (1993). Glutathione metabolism in plants. In: De Kok, L.J. (Ed.). Sulfur nutrition and assimilation in higher plants. SPB Academic Publishing, The Hague, The Netherlands, pp. 109-123.

Bergmuller, E., Porfirova, S. and Dormann, P. (2003). Characterization of an Arabidopsis mutant deficient in γ-tocopherol methyltransferase. *Plant Mol. Biol.* **52:** 1181-1190.

Beveridge, C.A., Gresshov, P.M., Rameau, C. and Turnbull, C.G.N. (2003). Additional signaling compounds are required to orchestrate plant development. *J. Plant Growth Regul.* **22:** 15-24.

Blokhina, O., Virolainen, E. and Fagerstedt, K.V. (2003). Antioxidants, oxidative damage and oxygen deprivition stress: a review. *Annals of Botany.* **91:** 179-194.

Bohm, F., Edge, R., Land, E.J., McGarvey, D.J. and Truscott, T.G. (1997). Carotenoids enhance vitamin E antioxidant efficiency. *J. Am. Chem. Soc.* **119:** 621-622.

Bohm, F., Edge, R., McGarvey, D.J. and Truscott, T.G. (1998a). Beta-carotene with vitamins E and C offers synergistic cell protection against NOx. *FEBS Lett.* **436:** 387-389.

Bohm, F., Edge, R., Lange, L. and Truscott, T.G. (1998b). Enhanced protection of human cells against ultraviolet light by antioxidant combinations involving dietary carotenoids. *J. Photochem. Photobiol. B: Biol.* **44:** 211-215.

Bohnert, H.J. and Jensen, R.G. (1996). Strategies for engineering waterstress tolerance in plants. *Trends Biotechnol.* **14:** 89-97.

Boo, Y.C. and Jung, J. (1999). Water deficit-induced oxidative stress and antioxidative defenses in rice plants. *J. Plant Physiol.* **155:** 255-261.

Bors, W., Heller, W., Michel, C. and Saran, M. (1990a). In: Packer, L. and Glazer, A.N. (Eds.). Methods in enzymology, Academic Press, New York, Vol. 186, pp. 343-355.

Bors, W., Heller, W., Michel, C. and Saran, M. (1990b). In: Emerit, I. (Ed.). Antioxidants in therapy and preventive medicine, Plenum Press, New York, pp. 165-170.

Bors, W., Heller, W., Michel, C. and Saran, M. (1992). In: Csomos, Q. and Feher, J. (Eds.). Free radicals and the liver, Springer, Berlin, pp. 77-95.

Bors, W., Michel, C. and Schikora, S. (1995). Interaction of flavonoids with ascorbate and determination of their univalent redox potentials: A pulse radiolysis study. *Free Rad Biol. Med.* **19:** 45-52.

Bors, W., Heller, W., Michel, C. and Stettmaier, K. (1996). In: Cadenas, E. and Packer, L. (Eds.). Handbook of antioxidants, Marcel Dekker, New York, pp. 409-466.

Boston, R.S., Viitanen, P.V. and Vierling, E. (1996). Molecular chaperones and protein folding in plants. *Plant Mol. Biol.* **32:** 191-222.

Braun, L., Garzo, T., Mandl, J. and Banhegyi, G. (1994). Ascorbic acid synthesis is stimulated by enhanced glycogenolysis in murine liver. *FEBS Lett.* **352:** 4-6.

Bravo, L. (1998). Polyphenols: Chemistry, dietary sources, metabolism and nutritional significance. *Nutrition Reviews.* **56:** 317-333.

Bray, E.A. (1997). Plant responses to water deficit. *Trends Plant Sci.* **2:** 48-54.

Bray, E.A., Bailey-Serres, J. and Weretilnyk, E. (2000). Responses to abiotic stresses. In: Buchanan, B.B., Gruissem, W. and Jones, R.L. (Eds.). Biochemistry and molecular biology of plants. ASPP, Rockville, pp. 1158.

Brigelius-Flohe, R., Kelly, F.J., Salonen, J.T., Neuzil, J., Zingg, J-M. and Azzi, A. (2002). The European perspective on vitamin E: current knowledge and future research. *Am. J. Clin. Nutr.* **76:** 703-716.

Britton, G., Liaaen-Jemsen, S. and Pfander, H. (1998). Carotenoids, biosynthesis and metabolism, Vol. 3, Birkauser Verlag, Basel, Switzerland.

Britton, G., Jensen, S.L. and Pfander, H. (2004). Carotenoids handbook. Birkhäuser Verlag, Basel.

Buettner, G.R. (1993). The pecking order of free radicals and antioxidants: lipid peroxidation, α-tocopherol, and ascorbate. *Archieves of Biochemistry and Biophysics.* **300:** 535-543.

Buettner, G.R. and Jurkiewicz, B.A. (1996). Chemistry and biochemistry of ascorbic acid. In: Cadenas, E. and Packer, L. (Eds.). Handbook of antioxidants, Dekker, New York, pp. 91-115.

Burke, J.J., Gamble, P.E., Hatfield, J.L. and Quisenberry, J.E. (1985). Plant Morphological and Biochemical Responses to Field Water Deficits: I. Responses of Glutathione Reductase Activity and Paraquat Sensitivity. *Plant Physiol.* **79:** 415-419.

Burns, J.J. (1960). Ascorbic acid. In: Greenberg, D.M. (Ed.) Metabolic Pathways, Academic Press, New York, pp. 341-356.

Butelli, E., Titta, L., Giorgio, M., Mock, H-P., Matros, A., Peterek, S., Schijlen, E.G.W.M., Hall, R.D., Bovy, A.G., Luo, J. and Martin, C. (2008). Enrichment of tomato fruit with healthpromoting anthocyanins by expression of select transcription factors. *Nat. Biotechnol.* **26:** 1301-1308.

Cahoon, E.B., Hall, S.E., Ripp, K.G., Ganzke, T.S., Hitz, W.D. and Coughlan, S.J. (2003). Metabolic redesign of vitamin E biosynthesis in plants for tocotrienol production and increased antioxidant content. *Nat. Biotech.* **21:** 1082-1087.

Caretto, S., Paradiso, A., D'Amico, L. and De Gara, L. (2002). Ascorbate and glutathione metabolism in two sunflower cell lines of differing α-tocopherol biosynthetic capability. *Plant Physiol. Biochem.* **40:** 509-513.

Caro, A. and Puntarulo, S. (1996). Effect of *in vivo* iron supplementation on oxygen radical production by soybean roots. *Biochim. Biophys. Acta.* **1291:** 245-251.

Casolo, V., Braidot, E., Chiandussi, E., Macrý, F. and Vianello, A. (2000). The role of mild uncoupling and non-coupled respiration in the regulation of hydrogen peroxide generation by plant mitochondria. *FEBS Lett.* **474:** 53-57.

Chaitanya, K.V., Sundar, D., Masilamani, S. and Reddy, A.R. (2002). Variation in heat stress-induced antioxidant enzyme activities among three mulberry cultivars. *Plant Growth Regul.* **36:** 175-180.

Chatterjee, I.B. (1973). Evolution and the biosynthesis of ascorbic acid. *Science.* **182:** 1271-1272.

Chaudhary, N., Nijhawan, A., Khurana, J.P. and Khurana, P. (2010). Carotenoid biosynthesis genes in rice: structural analysis, genome-wide expression profiling and phylogenetic analysis. *Mol. Genet Genomics.* **283:** 13-33.

Chaudiere, J. and Ferrari-Iliou, R. (1999). Intracellular antioxidants: from chemical to biochemical mechanisms. *Food Chem. Toxicol.* **37:** 949-962.

Cheeseman, J.M. (2006). Hydrogen peroxide concentrations in leaves under natural conditions. *J. Exp. Bot.* **57:** 2435-2444.

Christopher, S.C. (2000). Pytochelatins and their roles in heavy metal detoxification. *Plant Physiol.* **123:** 825-832.

Chun, O.K., Kim, D.O. and Lee, C.Y. (2003). Superoxide radical scavenging activity of the major polyphenols in fresh plums. *J. Agric. Food Chem.* **51:** 8067-8072.

Clemens, S., Kim, E.J., Neumann, D. and Schroeder, J.I. (1999). Tolerance to toxic metals by a gene family of phytochelatin synthases from plants and yeast. *EMBO J.* **15:** 3325-3333.

Clemens, S. (2001). Molecular mechanisms of plant metal tolerance and homeostasis. *Planta* **212:** 475-486.

Cobbett, C.S. (2000). Phytochelatins and their roles in heavy metal detoxification. *Plant Physiol.* **123:** 825-832.

Cobbet, C. and Goldsbrough, P. (2002). Phytochelatins and metallothioneins: roles in heavy metal detoxification and homeostasis. *Annu. Rev. Plant Biol.* **53:** 159-182.

Coleman, J.O.D., Blake-Kalff, M.M.A. and Davies, T.G.E. (1997). Detoxification of xenobiotics by plants: chemical modification and vascular compartmentation. *Trends Plant Sci.* **2:** 144-151.

Collakova, E. and DellaPenna, D. (2001). Isolation and functional analysis of homogentisate phytyltransferase from *Synechocystis* sp. PCC 6803 and Arabidopsis. *Plant Physiol.* **127:** 1113-1124.

Collakova, E. and DellaPenna, D. (2003). The role of homogentisate phytyltransferase and other tocopherol pathway enzymes in the regulation of tocopherol synthesis during abiotic stress plant physiology. *Plant Physiology* **133:** 930-940.

Collins, N.C., Thordal-Christensen, H., Lipka, V., Bau, S., Kombrink, E., Qiu, J.L., Huckelhoven, R., Stein, M., Freialdenhoven, A. and Somerville, S.C. (2003). SNARE-protein-mediated disease resistance at the plant cell wall. *Nature* **425:** 973-977.

Comba, M.E., Benavides, M.P. and Tomaro, M.L. (1998). Effect of salt stress on antioxidant defence system in soybean root nodules. *Aust. J. Plant Physiol.* **25:** 665-671.

Conklin, P.L. (2001). Recent advances in the role and biosynthesis of ascorbic acid in plants. *Plant, Cell and Environment.* **24:** 383-394.

Conn, P.F., Schalch, W. and Truscott, T.G. (1991). The singlet oxygen carotenoid interaction. *J. Photochem. Photobiol. B: Biol.* **11:** 41-47.

Cook-Johnson, R.J., Zhang, Q., Wiskich, J.T. and Soole, K.L. (2000). The nuclear origin of the non-phosphorylating NADH dehydrogenase of plant mitochondria. *FEBS Lett.* **454:** 37-41.

Cos, P., Calomme, M., Pieters, L., Vlietinck, A.J. and Vanden Berghe, D. (2000). Structure-activity relations of flavonoids as antioxidant and pro-oxidant compounds. In:Atta-ur-Rahman (Ed.). *Studies in Natural Products Chemistry.* **22:** 307-341.

Cossins, E., Lee, R. and Packer, L. (1998). ESR studies of vitamin C regeneration, order of reactivity of natural source phytochemical preparations. *Biochem. Mol. Biol. Int.* **45:** 583-597.

Creissen, G., Firmin, J., Fryer, M., Kular, B. and Leyland, N. (1999). Elevated glutathione biosynthetic capacity in the chloroplasts of transgenic tobacco plants paradoxically causes increased oxidative stress. *Plant Cell.* **11:** 1277-1292.

Cummins, I., Cole, D.J. and Edwards, R. (1999). A role for glutathione S-transferases functioning as glutathione peroxidases in resistance to multiple herbicides in black-grass. *Plant J.* **18:** 285-292.

da Silva, E.L., Piskula, M.K., Yamamoto, N., Moon, J.H. and Terao, J. (1998). Quercetin metabolites inhibit copper ion-induced lipid peroxidation in rat plasma. *FEBS Lett.* **430:** 405-408.

Dabrowski, K. (1994). Primitive actinopterygian fishes can synthesise ascorbic acid. *Experientia.* **50:** 745-748.

Dall'Osto, L., Lico, C., Alric, J., Giuliano, G., Havaux, M. and Bassi, R. (2006). Lutein is needed for efficient chlorophyll triplet quenching in the major LHCII antenna complex of higher plants and effective photoprotection *in vivo* under strong light. *BMC Plant Biol.* **6:** 32.

Dall'Osto, L., Fiore, A., Cazzaniga, S., Giuliano, G. and Bassi, R. (2007). Different roles of alpha- and beta-branch xanthophylls in photosystem assembly and photoprotection. *J. Biol. Chem.* **282:** 35056-35068.

Dangl, J.L. and Jones, J.D.G. (2001). Plant pathogens and integrated defense responses to infection. *Nature.* **411:** 826-833.

Dat, J.F., Foyer, C.H. and Scott, I.M. (1998a). Changes in salicylic acid and antioxidants during induced thermotolerance in mustard seedlings. *Plant Physiol.* **118:** 1455-1461.

Dat, J.F., Lopez-Delgado, H., Foyer, C.H. and Scott, I.M. (1998b). Parallel changes in H_2O_2 and catalase during thermotolerance induced by salicylic acid or heat acclimation in mustard seedlings. *Plant Physiol.* **116:** 1351-1357.

Dat, J., Vandenabeele, S., Vranova, E., Van Montagu, M., Inze, D. and Van Breusegem, F. (2000a). Dual action of the active oxygen species during plant stress responses. CMLS, Cell. *Mol. Life Sci.* **57:** 779-795.

Dat, J.F., Lopez-Delgado, H., Foyer, C.H. and Scott, I.M. (2000b). Effects of salicylic acid on oxidative stress and thermotolerance in tobacco. *J. Plant Physiol.* **156:** 659-665.

Davey, M.W., Van Montagu, M., Inze, D., Sanmartin, M., Kanellis, A., Smirnoff, N., Benzie, I.J.J., Strain, J.J., Favell, D. and John Fletcher, (2000). Plant L-ascorbic acid: chemistry, function, metabolism, bioavailability and effects of processing. *J. Sci. Food Agric.* **80:** 825-860.

Davies, M.J. and Shringarpure, R. (2006). Preferential degradation of oxidized proteins by the 20S proteasome may be inhibited in aging and in inflammatory neuromuscular diseases. *Neurology.* **66:** S93-S96.

Davletova, S., Rizhsky, L., Liang, H., Shengqiang, Z., Oliver, D.J., Coutu, J., Shulaev, V., Schlauch, K. and Mittler, R. (2005). Cytosolic ascorbate peroxidase 1 is a central component of the reactive oxygen gene network of Arabidopsis. *Plant Cell.* **17:** 268-281.

De Gara, L. and Tommasi, F. (1999). Ascorbate redox enzymes: a network of reactions involved in plant development. *Recent Dev. Phytochem.* **3:** 1-15.

De Gara, L., de Pinto, M.C. and Tommasi, F. (2003). The antioxidant systems vis-à-vis reactive oxygen species during plant–pathogen interaction. *Plant Physiology and Biochemistry.* **41:** 863-870.

De Groot, H. and Rauen, U. (1998). Tissue injury by reactive oxygen species and the protective effects of flavonoids. *Fundam. Clin. Pharma Col.* **12:** 249-255.

De Santis, A., Landi, P. and Genchi, G. (1999). Changes in mitochondrial properties in maize seedlings associated with selection for germination at low temperature: fatty acid composition, cytochrome *c* oxidase, and adenine nucleotide translocase activities. *Plant Physiol.* **119:** 743-754.

De Vos, R.C.H., Vonk, M.J., Voojis, R. and Schat, H. (1992). Gluthatione depletion due to copper-induced phytochelatin synthesis causes oxidative stress in *Silene cucubalus. Plant Physiol.* **98:** 853-858.

del Rio, L.A., Sandalio, L.M., Corpas, F.J., Palma, J.M. and Barroso, J.B. (2006). Reactive oxygen species and reactive nitrogen species in peroxisomes. Production, scavenging, and role in cell signaling. *Plant Physiol.* **141:** 330-335.

Dellapenna, D. and Pogson, B.J. (2006). Vitamin synthesis in plants: tocopherols and carotenoids. *Annu. Rev. Plant Biol.* **57:** 711-738.

Delledonne, M., Zeier, J., Marocco, A. and Lamb, C.J. (2001). Signal interactions between nitric oxide and reactive oxygen intermediates in the plant hypersensitive disease resistance response. *Proc. Natl. Acad. Sci.* **98:** 13454-13459.

Desikan, R., A-H Mackerness, S., Hancock, J.T. and Neill, S.J. (2001). Regulation of the *Arabidopsis* transcriptome by oxidative stress. *Plant Physiol.* **127:** 159-172.

Di Toppi, L.S., Lambardi, M., Pazzagli, L., Cappugi, G., Durante, M. and Gabbrielli, R. (1999). Response to cadmium in carrot *in vitro* plants and cell suspension cultures. *Plant Science.* **137:** 119-129.

Dixit, V., Pandey, V. and Shyam, R. (2001). Differential antioxidative responses to cadmium in roots and leaves of pea (*Pisum sativum* L. cv. Azad). *J. Exp. Bot.* **52:** 1101-1109.

Dixon, R.A. and Paiva, N.L. (1995). Stress induced phenylpropanoid metabolism. *Plant Cell.* **7:** 1085-1097.

Doke, N. (1983a). Involvement of superoxide anion generation in the hypersensitive response of potato tuber tissues to infection with an incompatible race of *Phytophthora infestans* and to the hyphal wall components. *Physiol. Plant Pathol.* **23:** 345-357.

Doke, N. (1983b). Generation of superoxide anion by potato tuber protoplasts during the hypersensitive response to hyphal cell wall components of Phytophthora *infestam* and specific inhibition of the reaction by suppressors of hypersensitivity. *Physiol. Plant Pathol.* **23:** 359-367.

Doke, N. and Ohashi, Y. (1988). Involvement of an 0_2^- generating system in the induction of necrotic lesions on tobacco leaves infected with tobacco mosaic virus. *Physiol. Mol. Plant Pathol.* **32:** 163-175.

Edwards, R., Dixon, D.P. and Walbot, V. (2000). Plant glutathione *S*-transferases: enzymes with multiple functions insickness and in health. *Trends Plant Sci.* **5:** 193-198.

Ehret, D.L. and Plant, A.L. (1999). Salt tolerance in crop plants. In: Dhaliwal, G.S. and Arora, R. (Eds.). Environmental stress in crop plants, Commonwealth Publishers, New Delhi, India, pp. 69-120.

Elstner, E.F., Wagner, G.A. and Schutz, W. (1988). Activated oxygen in green plants in relation to stress situations. *Curr. Topics Plant Biochem. Physiol.* **7:** 159-187.

Fadzillah, N.M., Finch, R.P. and Burdon, R.H. (1997). Salinity, oxidative stress and antioxidant responses in shoot cultures of rice. *J. Exp. Bot.* **48:** 325-331.

Falk, J., Andersen, G., Kernebeck, B. and Krupinska, K. (2003). Constitutive overexpression of barley 4-hydroxyphenylpyruvate dioxygenase in tobacco results in elevation of the vitamin E contentin seeds but not in leaves. *FEBS Lett.* **540:** 35-40.

Fenton, H.J.H. (1894). Oxidation of tartaric acid in presence of iron. *J. Chem. Soc. Trans.* **65:** 899-911.

Fenton, H.J.H. (1899). Oxidation of certain organic acids in the presence of ferrous salts. *Proc. Chem. Soc.* **25:** 224-228.

Fey, V., Wagner, R. and Brautigam, K. (2005). Photosynthetic redox control of nuclear gene expression. *J. Exp. Bot.* **56:** 1491-1498.

Foreman, J., Demidchik, V., Bothwell, J.H.F., Mylona, P., Miedema, H., Torres, M.A., Linstead, P., Costa, S., Brownlee, C., Jones, J.D.G., Davies, J.M. and Dolan, L. (2003). Reactive oxygen species produced by NADPH oxidase regulate plant cell growth. *Nature.* **422:** 442-446.

Forkmann, G. and Heller, W. (1999). Biosynthesis of flavonoids. In: Comprehensive natural products chemistry, Amsterdam, Elsevier, pp. 713-748.

Formica, J.V. and Regelson, W. (1995). Review of the biology of quercetin and related bioflavonoids. *Food Chem. Toxicol.* **33:** 1061-1080.

Fornazier, R.F., Ferreira, R.R., Pereira, G.J.G., Molina, S.M.G., Smith, R.J., Lea, P.J. and Azevedo, R.A. (2002) Cadmium stress in sugar cane callus cultures: effect on antioxidant enzymes. *Plant Cell Tissue Organ Cult.* **71:** 125-31.

Forti, G. and Elli, G. (1995). The function of ascorbic acid in photosynthetic phosphorylation. *Plant Physiol.* **109:** 1207-1211.

Foyer, C.H. (1993). Ascorbic acid. In: Alscher, R.G. and Hess, J.L. (Eds.). Antioxidants in higher plants, CRC Press, Boca Raton, pp. 32-57.

Foyer, C.H. and Harbinson, J. (1994). Oxygen metabolism and the regulation of photosynthetic electron transport. In: Foyer, C.H. and Mullineaux, P. (Eds.). Causes of photooxidative stresses and amelioration of defense systems in plants, CRC Press, Boca Raton, pp. 1-42.

Foyer, C. and Lelandais, M. (1996). A comparison of the relative rates of transport of ascorbate and glucose across the thylakoid, chloroplast and plasmalemma membranes of pea leaf mesophyll cells. *J. Plant Physiol.* **148:** 391-398.

Foyer, C.H., Delgado, H.L., Dat, J.F. and Scott, I.M. (1997). Hydrogen peroxide- and glutathione-associated mechanisms of acclimatory stress tolerance and signalling. *Physiol. Plant.* **100:** 241-254.

Foyer, C.H. and Noctor, G. (2000). Oxygen processing in photosynthesis: regulation and signaling. *New Phytol.* **146:** 359-388.

Foyer, C.H. and Noctor, G. (2001). The molecular biology and metabolism of glutathione. In: Grill, D. *et al.* (Eds.). Significance of glutathione in plant adaptation to the environment, Kluwer Academic Publishers, pp. 27-56.

Foyer, C.H., Theodoulou, F.L. and Delrot, S. (2001). The functions of inter- and intracellular glutathione transport systems in plants. *Trends in Plant Science.* **6:** 486-492.

Foyer, C.H. and Noctor, G. (2003). Redox sensing and signalling associated with reactive oxygen in chloroplasts, peroxisomes and mitochondria. *Physiol. Plant.* **119:** 355-364.

Foyer, C.H. and Noctor, G. (2005). Oxidant and antioxidant signaling in plants: a re-evaluation of the concept of oxidative stress in a physiological context. *Plant Cell and Environ.* **28:** 1056-1071.

Fraser, P.D. and Bramley, P.M. (2004). The biosynthesis and nutritional uses of carotenoids. *Progress in Lipid Research.* **43:** 228-265.

Frei, B. and Higdon, J.V. (2003). Antioxidant activity of tea polyphenols *in vivo*: evidence from animal studies. *J. Nutr.* **133:** 3275S-3284S.

Fricker, M.D. and Meyer, A.J. (2000). Direct measurement of glutathione in epidermal cells of intact *Arabidopsis* roots by two-photo laser scanning microscopy. *J. Microsc.* **198:** 174-181.

Fricker, M.D., May, M., Meyer, A.J., Sheard, N. and White, N.S. (2000). Measurement of glutathione levels in intact roots of *Arabidopsis. J. Microsc.* **198:** 162-173.

Frova, C. (2006). Glutathione transferases in the genomic era: New insights and perspectives. *Biomolecular Engineering.* **23:** 149-169.

Fry, S.C. (2004). Primary cell wall metabolism: Tracking the careers of wall polymers in living plant cells. *New Phytol.* **161:** 641-675.

Fryer, M.J. (1992). The antioxidants effects of thylakoid vitamin E (α-tocopherol). *Plant Cell and Environment.* **15:** 391-392.

Fukuzawa, K. and Gebicky, J.M. (1983). Oxidation of α-tocopherol in micelles and liposomes by the hydroxyl, perhydroxyl, and superoxide free radicals. *Arch. Biochem. Biophys.* **226:** 242-251.

Garcia-Hernandez, M., Murphy, A. and Taiz, L. (1998). Metallionein I and II have distinct but overlapping expression patterns in *Arabidopsis*. **118:** 387-397.

Gechev, T.S., Van Breusegem, F., Stone, J.M., Denev, I. and Laloi, C. (2006). Reactive oxygen species as signals that modulate plant stress responses and programmed cell death. *Bioessays.* **28:** 1091-1101.

Ghezzi, P. and Bonetto, V. (2003). Redox proteomics: Identification of oxidatively, modified proteins. *Proteomics.* **3:** 1145-1153.

Gogorcena, Y., Iturbe-Ormaetxe, I., Escuredo, P.R. and Becana, M. (1995). Antioxidant defenses against activate oxygen in pea nodules subjected to water stress. *Plant Physiol.* **108:** 753-759.

Gomes, E., Jakobsen, M.K., Axelsen, K.B., Geisler, M. and Palmgren, M.G. (2000). Chilling tolerance in Arabidopsis involves ALA1, a member of a new family of putative aminophospholipid translocases. *Plant Cell.* **12:** 2441-2453.

Gomez, L.D., Noctor, G., Knight, M. and Foyer, C.H. (2004a). Regulation of calcium signaling and gene expression by glutathione. *J. Experimental Botany.* **55:** 1851-1859.

Gomez, L., Vanacker, H., Buchner, P., Noctor, G. and Foyer, C.H. (2004b). The intercellular distribution of glutathione synthesis and its response to chilling in maize. *Plant Physiology.* **134:** 1662-1671.

Gonzalez-Meler, M.A., Ribas-Carbo, M., Giles, L. and Siedow, J.N. (1999). The effect of growth and measurement temperature on the activity of the alternative respiratory pathway. *Plant Physiol.* **120:** 765-772.

Goodwin, T.W. (1980). Carotenoids. In: Bell, E.A. and Charlwood, B.V. (Eds.). Encyclopedia of plant physiology, Springer-Verlag, Berlin, Vol. 8, pp. 257-281.

Govrin, E. and Levine, A. (2000). The hypersensitive response facilitates plant infection by the necrotrophic pathogen *Botrytis cinerea. Curr. Biol.* **10:** 751-757.

Grant, M., Brown, I., Adams, S., Knight, M., Ainslie, A. and Mansfield, J. (2000b). The RPM1 plant disease resistance gene facilitates a rapid and sustained increase in cytosolic calcium that is necessary for the oxidative burst and hypersensitive cell death. *Plant J.* **23:** 441-450.

Greenway, H. and Munns, R. (1980). Mechanisms of salt tolerance in nonhalophytes. *Annu. Rev. Plant Physiol.* **31:** 149-190.

Grene, R. (2002). Oxidative Stress and Acclimation Mechanisms in Plants. The arabidopsis book, *Am. Soc. Plant Biologists,* pp. 1-20.

Grill, E., Loffler, S., Winnacker, E.L. and Zenk, M.H. (1989). Phytochelatins, the heavymetal binding peptides of plants, are synthesized from glutathione by a specific γ-glutamylcysteine dipeptidyl transpeptidase (phytochelatin synthase). *Proc. Natl. Acad. Sci.* **86:** 6838-6842.

Grusak, M.A. and DellaPenna, D. (1999). Improving the nutrient composition of plants to enhance human nutrition and health. *Annu. Rev. Plant Physiol. Plant Mol. Biol.* **50:** 133-161.

Gruszecki, W.I. (1999). Carotenoids in membranes. In: Frank, H.A., Young, A.J., Britton, G. and Cogdell, R.J. (Eds.). The photochemistry of carotenoids, Kluwer Academic Publishers, Dordrecht, pp. 363-379.

Gutierrez-Alcala, G., Gotor, C., Meyer, A.J., Fricker, M., Vega, J.M. and Romero, L.C. (2000). Glutathion biosynthesis in *Arabidopsis* trichome cells. *Proc. Natl. Acad. Sci.* **97:** 11108-11113.

Haber, F. and Weiss, J. (1934). The catalytic decomposition of hydrogen peroxide by iron salts. *Proc. R. Soc. Lond. A.* **147:** 332-351.

Halliwell, B. and Gutteridge, J.M.C. (1989). Free Radicals in Biology and Medicine, Clarendon Press, Oxford.

Halliwell, B. and Gutteridge, J.M.C. (1999). Free Radicals in Biology and Medicine, Oxford Univ. Press. 3rd ed., Oxford.

Halliwell, B., Zhao, K. and Whiteman, M. (2000). The gastrointestinal tract: a major site of antioxidant action? *Free Radic. Res.* **33:** 819-830.

Halliwell, B. (2006). Reactive species and antioxidants. Redox biology is a fundamental theme of aerobic life. *Plant Physiol.* **141:** 312-322.

Hammond-Kosack, K. and Jones, J.D.G. (2000). Responses to plant pathogens. In: Buchanan, B.B., Gruissem, W. and Jones, R.L. (Eds.). Biochemistry and Molecular Biology of Plants, Rockville, MD: *Am. Soc. Plant Physiol.* pp. 1102-1156.

Harborne, J.B. and Williams, C.A. (2000). Advances in flavonoid research since 1992. *Phytochemistry.* **55:** 481-504.

Hartley-Whitaker, J., Ainsworth, G. and Mehrang, A.A. (2001). Copper- and arsenate- induced oxidative stress in *Holcus lanatus* L. clones with differential sensitivity. *Plant Cell Env.* **24:** 713-722.

Hasegawa, P.M., Bressan, R.A., Zhu, J-K. and Bohnert, H.J. (2000). Plant cellular and molecular responses to high salinity. *Annu. Rev. Plant Physiol. Plant Mol. Biol.* **51:** 463-499.

Havaux, M., Bonfils, J., Lutz, C. and Niyogi, K. (2000). Photodamage of the photosynthetic apparatus and its dependence on the leaf development stage in the *npq1 Arabidopsis* mutant deficient in the xanthophyll cycle enzyme violaxanthin de-epoxidase. *Plant Physiol.* **124:** 273-284.

Heijnen, C.G., Haenen, G.R., van Acker, F.A., van der Vijgh, W.J. and Bast, A. (2001). Flavonoids as peroxynitrite scavengers: the role of the hydroxyl groups. *Toxicol in Vitro.* **15:** 3-6.

Heim, K.E., Tagliaferro, A.R. and Bobilya, D.J. (2002). Flavonoid antioxidants: chemistry, metabolism and structure-activity. *Relationships J. Nutritional Biochemistry* **13:** 572-584.

Hell, R. and Bergmann, L. (1990). γ-Glutamylcysteine synthetase in higher plants: catalytic properties and subcellular localization. *Planta.* **180:** 603-612.

Hernandez, J.A., Corpas, F.J., Gomez, M., del Rio, L.A. and Sevilla, F. (1993). Salt-induced oxidative stress mediated by activated oxygen species in pea leaf mitochondria. *Physiol. Plant.* **89:** 103-110.

Hernandez, J.A., Olmos, E., Corpas, F.J., Sevilla, F. and del Rio, L.A. (1995). Salt-induced oxidative stress in chloroplasts of pea plants. *Plant Sci.* **105:** 151-167.

Hernandez, I., Alegre, L., Van Breusegem, F. and Munne-Bosch, S. (2009). How relevant are flavonoids as antioxidants in plants? *Trends in Plant Science.* **14:** 125-132.

Horemans, N., Foyer, C.H., Potters, G. and Asard, H. (2000). Ascorbate function and associated transport systems in plants. *Plant Physiol. Biochem.* **38:** 531-540.

Hou, Z., Lambert, J.D., Chin, K.V. and Yang, C.S. (2004). Effects of tea polyphenols on signal transduction pathways related to cancer chemoprevention. *Mutat Res.* **555:** 3-19.

Hrazdina, G. and Wagner, G.J. (1985). Metabolic pathways as enzyme complexes: evidence for the synthesis of phenylpropanoids and flavonoids on membrane associated enzyme complexes. *Arch. Biochem. Biophys.* **237:** 88-100.

Huang, M. and Guo, Z. (2005). Responses of antioxidative system to chilling stress in two rice cultivars differing in sensitivity. *Biol. Plant.* **49:** 81-84.

Iba, K. (2002). Acclimative response to temperature stress in higher plants: approaches of gene engineering for temperature tolerance. *Annu. Rev. Plant Biol.* **53:** 225-245.

Igamberdiev, A.U., Bykova, N.V. and Kleczkowski, L.A. (1999). Origins and metabolism of formate in higher plants. *Plant Physiol. Biochem.* **37:** 503-513.

Isbell, H.S., Frush, H.L. and Martin, E.T. (1973). Reactions of carbohydrates with hydroperoxides. 1. Oxidation of aldoses with sodium peroxide. *Carbohydr. Res.* **26:** 287-295.

Iturbe-Ormaetxe, I., Escuredo, P.R., Arrese-Igor, C. and Becana, M. (1998). Oxidative damage in pea plants exposed to water deficit or paraquat. *Plant Physiol.* **116:** 173-181.

Janska, H. (2005). ATP-dependent proteases in plant mitochondria: What do we know about them today? *Physiol. Plant.* **123:** 399-405.

Jeong, Y-C., Nakamura, J., Upton, P.B. and Swenberg, J.A. (2005). Pyrimido[1,2-α]-purin-10(3H)-one, M1G, is less prone to artifact than base oxidation. *Nucleic Acids Res*. **33:** 6426-6434.

Jiang, Y. and Huang, B. (2001). Drought and heat stress injury to two cool season turfgrasses in relation to antioxidant metabolism and lipid peroxidation. *Crop Sci*. **41:** 436-442.

Jimenez, A., Hernandez, J.A., del Rio, L.A. and Sevilla, F. (1997). Evidence for the presence of the ascorbate-glutathione cycle in mitochondria and peroxisomes of pea leaves. *Plant Physiol*. **114:** 275-284.

Juszczuk, I.M., Bykova, N.V. and Moller, I.M. (2007). Protein phosphorylation in plant mitochondria. *Physiol. Plant*. **129:** 90-113.

Kagan, V.E. (1989). Tocopherol stabilizes membarne against phospholipase A, free fatty acids, and lysophospholipids. In: Diplock, A.T., Machlin, J., Packer, L. and Pryor, W.A. (Eds.). Vitamin E:biochemistry and health implications, Annals of the New York Academy of Sciences, Vol: 570 New York, pp. 121-135.

Kamal-Eldin, A. and Appelqvist, L-A. (1996). The chemistry and antioxidant properties of tocopherol and tocotrienols. *Lipids*. **31:** 671-701.

Kandaswami, C. and Middleton, E. (1994). Free radical scavenging and antioxidant activity of plant flavonoids. *Adv. Exp. Med. Biol*. **366:** 351-376.

Kang, H-M. and Saltveit, M.E. (2002). Reduced chilling tolerance in elongating cucumber seedling radicles is related to their reduced antioxidant enzyme and DPPH-radical scavenging activity. *Physiol. Plant*. **115:** 244-250.

Kanwischer, M., Porfirova, S., Bergmuller, E. and Dormann, P. (2005). Alterations in tocopherol cyclase activity in transgenic and mutant plants of Arabidopsis affect tocopherol content, tocopherol composition, and oxidative stress. *Plant Physiol*. **137:** 713-723.

Kehrer, J.P. (2000). The Haber-Weiss reaction and mechanisms of toxicity. *Toxicology*. **149:** 43-50.

Keppler, L.D., Baker, C.J. and Atkinson, M.M. (1989). Active oxygen production during a bacteria induced hypersensitive reaction in tobacco suspension cells. *Phytopathology*. **79:** 974-978.

Kerepesi, I. and Galiba, G. (2000). Osmotic and salt stress-induced alteration in soluble carbohydrate content in wheat seedlings. *Crop Sci*. **40:** 482-487.

Kerkeb, L. and Kramer, U. (2003). The role of free histidine in xylem loading of nickel in *Alyssum lesbiacum* and *Brassica juncea*. *Plant Physiol*. **131:** 716-724.

Khan, M.A., Ungar, I.A. and Showalter, A.M. (2000). Effects of sodium chloride treatments on growth and ion accumulation of the halophyte Haloxylon recurvum. *Commun. Soil Sci. Plant Anal*. **31:** 2763-2774.

Khatkar, D. and Kuhad, M.S. (2000). Short-term salinity induced changes in two wheat cultivars at different growth stages. *Biol. Plant*. **43:** 629-632.

Kirst, G.O. (1989). Salinity tolerance of eukaryotic marine algae. *Annu. Rev. Plant Physiol. Plant Mol. Biol*. **40:** 21-53.

Klapheck, S. (1988). Homoglutathione: isolation, quantification and occurrence in legumes. *Physiol. Plant*. **74:** 727-732.

Klapheck, S., Chrost, B., Starke, J. and Zimmerman, H. (1992). γ-Glutamylcysteinylserine—a new homologue of glutathione in plants of the family *Poaceae*. *Botanica Acta*. **105:** 174-179.

Klessig, D.F., Durner, J., Noad, R., Navarre, D.A., Wendehenne, D., Kumar, D., Zhou, J.M., Shah, J., Zhang, S. and Kachroo, P. (2000). Nitric oxide and salicylic acid signaling in plant defense. *Proc. Natl. Acad. Sci*. **97:** 8849-8855.

Knekt, P., Jarvinen, R., Reunanen, A. and Maatela, J. (1996). Flavonoid intake and coronary mortality in Finland: a cohort study. *BMJ*. **312:** 478-481.

Korshunov, S.S., Korkina, O.V., Ruuge, E.K., Skulachev, V.P. and Starkov, A.A. (1998). Fatty acids as natural uncouplers preventing generation of O_2-and H_2O_2 by mitochondriain the resting state. *FEBS Lett*. **435:** 215-218.

Kraus, T.E. and Fletcher, R.A. (1994). Paclobutrazol protects wheat seedlings from heat and paraquat injury: is detoxification of active oxygen involved? *Plant Cell Physiol.* **35:** 45-52.

Krinsky, N.I. (1998). Overview of lycopene, carotenoids, and disease prevention. *Proc. Soc. Exp. Biol. Med.* **218:** 95-97.

Kruk, J., Schmid, G.H. and Strzalka, K. (2000). Interaction of α-tocopherol quinone, α-tocopherol and other prenyl lipids with photosystem II. *Plant Physiology and Biochemistry.* **38:** 271-277.

Kuhnau, J. (1976). The flavonoids. A class of semi-essential food components: their role in human nutrition. *World Rev. Nutr. Diet.* **24:** 117-191.

Kulheim, C., Agren, J. and Jansson, S. (2002). Rapid regulation of light harvesting and plant fitness in the field. *Science.* **308:** 267-269.

Kunert, K.J. and Foyer, C. (1993). Thiol/disulfide exchange in plants. In: de Kok, L.J., Stulen, I., Rennenberg, H., Brunold, C. and Rauser, W.E. (Eds.). Sulfur nutrition and sulfur assimilation in higher plants, The Hague: SPB Acad., pp. 139-151.

Kurepa, J., Herouart, D., Van Montagu, M. and Inze, D. (1997). Differential expression of CuZn- and Fe-superoxide dismutase genes of tobacco during development, oxidative stress and hormonal treatments. *Plant Cell Physiol.* **38:** 463-470.

Kwak, J.M., Mori, I.C., Pei, Z.M., Leonhardt, N., Torres, M.A., Dangl, J.L., Bloom, R.E., Bodde, S., Jones, J.D.G. and Schroder, J.I. (2003). NADPH oxidase AtrbohD and AtrbohF genes function in ROS-dependent ABA signaling in Arabidopsis. *EMBO J.* **22:** 2623-2633.

Laloi, M., Klein, M., Riesmeier, J.W., Muller-Rober, B. and Fleury, C. (1997). A plant cold-induced uncoupling protein. *Nature.* **389:** 135-136.

Lamb, C. and Dixon, R.A. (1997). The oxidative burst in plant disease resistance. *Annu. Rev. Plant Physiol. Mol. Biol.* **48:** 251-275.

Lane, B.R., Kajoika, R. and Kennedy, R. (1987). The wheat germ Ec protein is zinc-containing metallothionein. *Biochem. Cell Biol.* **65:** 1001-1005.

Langenkaneper, G., Manach, N., Broin, M., Cuine, S., Becuwe, N., Kuntz, M. and Rey, P. (2001). Accumulation of plastid lipid associated proteins (fibrillin/CDSP 34) upon oxidative stress, ageing and biotic stress in solanaceae and in response to drought in other species. *J. Exp. Bot.* **52:** 1545-1554.

Larkindale, J. and Knight, M.R. (2002). Protection against heat stress-induced oxidative damage in Arabidopsis involves calcium, abscisic acid, ethylene, and salicylic acid. *Plant Physiol.* **128:** 682-695.

Larson, R.A. (1988). The antioxidants of higher plants. *Phytochemistry.* **27:** 969-978.

Lee, B.R., Li, L.S., Jung, W.J., Jin, Y.L.J.C., Avice, J.C., Ourry, A. and Kim, T.H. (2009). Water deficit-induced oxidative stress and the activation of antioxidant enzymes in white clover leaves. *Biologia Plantarum.* **53:** 505-510.

Levine, A., Tenhaken, R., Dixon, R. and Lamb, C.J. (1994). H_2O_2 from the oxidative burst orchestrates the plant hypersensitive disease resistance response. *Cell.* **79:** 583-593.

Li, X.F., Ma, J.F., Matsumoto, H. (2000). Pattern of Al-induced secretion of organic acids differ between rye and wheat. *Plant Physiol.* **123:** 1537-1544.

Li, M., Hu, C.W., Zhu, Q., Chen, L., Kong, Z.M. and Liu, Z.L. (2006). Copper and zinc induction of lipid peroxidation and effects on antioxidant enzyme activities in the microalga *Pavlova viridis* (Prymnesiophyceae). *Chemosphere.* **62:** 565-572.

Li, Y., Wang, Z., Sun, X. and Tang, K. (2008). Current opinions on the function of tocopherol based on the genetic manipulation of tocopherol biosynthesis in plants. *J. Integr. Plant Biol.* **50:** 1057-1069.

Loewus, F.A. (1988). Ascorbic acid and its metabolic products. In: Preiss, J. (Ed.). The biochemistry of plants, Vol. 14, Academic Press, New York. pp. 85-107.

Loewus, F.A. and Loewus, M.W. (1987). Biosynthesis and metabolism of ascorbate in plants. *Crit. Rev. Plant Sci.* **5:** 101-119.

Lotito, S.B. and Frei, B. (2006). Consumption of flavonoid-rich foods and increased plasma antioxidant capacity in humans: cause, consequence, or epiphenomenon? *Free Radic. Biol. Med.* **41:** 1727-1746.

Low, P.S. and Merida, J.R. (1996). The oxidative burst in plant defense: function and signal transduction. *Physiol. Plant.* **96:** 533-542.

Luwe, M.W.F., Takahama, U. and Heber, U. (1993). Role of ascorbate in detoxifying ozone in the apoplast of spinach (*Spinacia oleracea* L.) leaves. *Plant Physiol.* **101:** 969-976.

Ma, J.F., Taketa, S. and Yang, Z.M. (2000). Aluminum tolerance genes on the short arm of chromosome 3R are linked to organic acid release in triticale. *Plant Physiol.* **122:** 687-694.

Mano, J. (2002). Early events in environmental stresses in plants-induction mechanisms of oxidative stress. In: Inze, D. and Montago, M.V. (Eds.). Oxidative stress in plants, Taylor and Francis Publishers, New York, USA, pp. 217-245.

Marin, F.R., Frutos, M.J., Perez-Alvarez, J.A., Martinez-Sanchez, F. and Del Rio, J.A. (2002). Flavonoids as nutraceuticals: Structural related antioxidant properties and their role on ascorbic acid preservation. In: Atta-ur-Rahman, (Ed.). *Studies in Natural Products Chemistry.* **26:** 741-778.

Martin, J. and Hartl, F.U. (1997). Chaperone assisted protein folding. *Curr. Opin. Struct. Biol.* **7:** 41-52.

Maxwell, D.P., Wang, Y. and McIntosh, L. (1999). The alternative oxidase lowers mitochondrial reactive oxygen production in plant cells. *Proc. Natl. Acad. Sci.* **96:** 8271-8276.

May, M.J., Vernoux, T., Leaver, C., Van Montagu, M. And Inze, D. (1998a). Glutathione homeostasis in plants: implications for environmental sensing and plant development. *J. Exp. Bot.* **49:** 649-667.

May, M.J., Vernoux, T., Sanchez-Fernandez, R., Van Montagu, M. and Inze, D. (1998b). Evidence for posttranscriptional activation of gamma-glutamylcysteine synthetase during plant stress responses. *Proc. Natl. Acad. Sci.* **95:** 12049-12054.

Mayne, S.T. (1996). Beta-carotene, carotenoids, and disease prevention in humans. *FASEB J.* **10:** 690-701.

Mazza, G. and Miniati, E. (1993). Anthocyanins in Fruits, Vegetables and Grains. CRC Press: Boca Raton.

McSteen, P. and Leyser, O. (2005). Shoot branching. *Annu. Rev. Plant Biol.* **56:** 353-374.

Meister, A. (1988). Glutathione metabolism and its selective modification. *J. Biol. Chem.* **263:** 17205-17208.

Meister, A. (1994). Glutathione-ascorbic acid antioxidant system in animals. *J. Biol. Chem.* **269:** 9397-9400.

Meuwly, P., Thibault, P. and Rauser, W.E. (1993). γ-Glutamylcysteinylglutamic acid – a new homologue of glutathione in maize seedlings exposed to cadmium. *FEBS Lett.* **336:** 472-476.

Middleton, E.J. (1998). Effect of plant flavonoids on immune and inflammatory cell function. *Adv. Exp. Med. Biol.* **439:** 175-182.

Millar, A.H., Mittova, V. and Kiddle, G. (2003). Control of ascorbate synthesis by respiration and its implications for stress responses. *Plant Physiol.* **133:** 443-447.

Mittler, R. (2002). Oxidative stress, antioxidants and stress tolerance. *Trends in Plant Science.* **7:** 405-410.

Mittler, R. and Zilinskas, B.A. (1992). Molecular cloning and characterization of a gene encoding pea cytosolic ascorbate peroxidase. *J. Biol. Chem.* **267:** 21802-21807.

Mittler, R. and Zilinskas, B.A. (1994). Regulation of pea cytosolic ascorbate peroxidase and other antioxidant enzymes during the progression of drought stress and following recovery from drought. *Plant J.* **5:** 397-405.

Mittler, R., Vanderauwera, S., Gollery, M. and Van Breusegem, F. (2004). Reactive oxygen gene network of plants. *Trends in Plant Science.* **9:** 490-498.

Mittova, V., Tal, M., Volokita, M. and Guy, M. (2002). Salt stress induces upregulation of an efficient chloroplast antioxidant system in the salt-tolerant wild tomato species *Lycopersicon pennellii* but not in the cultivated species. *Physiol. Plant.* **115:** 393-400.

Mittova, V., Tal, M., Volokita, M. and Guy, M. (2003). Upregulation of the leaf mitochondrial and peroxisomal antioxidative systems in response to salt-induced oxidative stress in the wild salt-tolerant tomato species *Lycopersicon pennellii. Plant Cell Environ.* **26:** 845-856.

Miyake, Y., Shimoi, K., Kumazawa, S., Yamamoto, K., Kinae, N. and Osawa, T. (2000). Identification and antioxidant activity of flavonoid metabolites in plasma and urine of eriocitrin-treated rats. *J. Agri. Food Chem.* **48:** 3217-3224.

Moller, I.M. (2001). Plant mitochondria and oxidative stress. Electron transport, NADPH turnover and metabolism of reactive oxygen species. *Annu. Rev. Plant Physiol. Plant Mol. Biol.* **52:** 561-591.

Moller, I.M. and Palmer, J.M. (1982). Direct evidence for the presence of a rotenoneresistant NADH dehydrogenase on the inner surface of the inner membrane of plant mitochondria. *Physiol. Plant.* **54:** 267-274.

Moller, I.M. and Kristensen, B.K. (2004). Protein oxidation in plant mitochondria as a stres indicator. *Photochem. Photobiol. Sci.* **3:** 730-735.

Moller, I.M., Jensen, P.E. and Hansson, A. (2007). Oxidative modifications to cellular components in plants. *Annu. Rev. Plant Biol.* **58:** 459-481.

Montoro, P., Braca, A., Pizza, C. and De Tommasi, N. (2005). Structure–antioxidant activity relationships of flavonoids isolated from different plant species. *Food Chemistry.* **92:** 349-355.

Mortensen, A. and Skibsted, L.H. (1997). Real time detection of reactions between radicals of lycopene and tocopherol homologues. *Free Radic. Res.* **27:** 229-234.

Mueller, M.J. (2004). Archetype signals in plants: The phytoprostanes. *Curr. Opin. Plant Biol.* **7:** 441-448.

Mullineaux, P. and Karpinski, S. (2002). Signal transduction in response to excess light: getting out of the chloroplast. *Curr. Opin. Plant Biol.* **5:** 43-48.

Mullineaux, P.M., Karpiniski, S. and Baker, N.R. (2006). Spatial dependence for hydrogen peroxide-directed signaling in light-stressed plants. *Plant Physiol.* **141:** 346-350.

Munne-Bosch, S. (2005). The role of a-tocopherol in plant stress tolerance. *J. Plant Physiology.* **162:** 743-748.

Munne-Bosch, S. and Alegre, L. (2002a). The function of tocopherols and tocotrienols in plants. *Crit. Rev. Plant Sci.* **21:** 31-57.

Munne-Bosch, S. and Alegre, L. (2002b). Interplay between ascorbic acid and lipophilic antioxidant defences in chloroplasts of water-stressed Arabidopsis plants. *FEBS Lett.* **524:** 145-148.

Munne-Bosch, S. and Alegre, L. (2003). Drought-induced changes in the redox state of a-tocopherol, ascorbate and the diterpene carnosic acid in chloroplasts of Labiatae species differing in carnosic acid contents. *Plant Physiol.* **131:** 1816-1825.

Munne-Bosch, S. Falk, J. (2004). New insights into the function of tocopherols in plants. *Planta.* **218:** 323-326.

Munns, R. (2002). Comparative physiology of salt and water stress. *Plant Cell Environ.* **25:** 239-250.

Munns, R. and Tester, M. (2008). Mechanisms of Salinity Tolerance. *Annu. Rev. Plant Biol.* **59:** 651-681.

Naik, P.S., Chanemougasoundharam, A., Khurana, S.M.P. and Kalloo, G. (2003). Genetic manipulation of carotenoid pathway in higher plants. *Current Science.* **85:** 1423-1430.

Nambara, E. and Marion-Poll, A. (2005). Abscisic acid biosynthesis and catabolism. *Annu. Rev. Plant Biol.* **56:** 165-185.

Nies, D.H. (1999). Microbial heavy metal resistance. *Applied Microbiology Biotechnology.* **51:** 730-750.

Nijs, D. and Kelley, P.M. (1991). Vitamins C and E donate single hydrogen atoms *in vivo. FEBS Lett.* **284:** 147-151.

Nishikimi, M. and Yagi, K. (1991). Molecular basis for the deficiency in humans of gulonolactone oxidase, a key enzyme for ascorbic acid biosynthesis *Am. J. Clin. Nutr.* **54:** 1203S-1208S.

Nishikimi, M., Yagi, K. (1996). Biochemistry and molecular biology of ascorbic acid biosynthesis. In: Harris, J. (Ed.). Subcellular biochemistry ascorbic acid: Biochemistry and biomedical cell biology, Plenum, New York. **25:** 17-39.

Niyogi, K.K. (1999). Photoprotection revised: genetic and molecular approaches. *Annu. Rev. Plant Physiol. Plant Mol. Biol.* **50:** 333-359.

Nobuhiro, S. and Mittler, R. (2006). Reactive oxygen species and temperature stresses: A delicate balance between signaling and destruction. *Physiol. Plant.* **126:** 45-51.

Noctor, G. and Foyer, C. (1998). Ascorbate and glutathione: keeping active oxygen under control. *Annu. Rev. Plant Physiol. Plant Mol. Biol.* **49:** 249-279.

Noctor, G., Veljovic-Jovanovic, S. and Foyer, C.H. (2000). Peroxide processing in photosynthesis: antioxidant coupling and redox signalling. *Phil. Trans R. Soc. Lond B.* **355:** 1465-1475.

Noctor, G., Veljovic-Jovanovic, S., Driscoll, S., Novitskaya, L. and Foyer, C.H. (2002a). Drought and oxidative load in wheat leaves: a predominant role for photorespiration? *Annals of Botany.* **89:** 841-850.

Noctor, G., Gomez, L., Vanacker, H. and Foyer, C.H. (2002b). Glutathione homeostasis and signaling: the influence of biosynthesis, compartmentation and transport. *J. Experimental Botany.* **53:** 1283-1304.

Norris, S.R., Barette, T.R. and DellaPenna, D. (1995). Genetic dissection of carotenoid synthesis in Arabidopsis defines plastoquinone as an essential component of phytoene desaturation. *Plant Cell.* **7:** 2139-2149.

O'Kane, D., Gill, V., Boyd, P. and Burdon, R. (1996). Chilling, oxidative stress and antioxidant responses in *Arabidopsis thaliana* callus. *Planta.* **198:** 371-377.

Olson, J.A. and Krinsky, N.I. (1995). Introduction: the colorful fascinating world of the carotenoids: important physiologic modulators. *FASEB J.* **9:** 1547-1550.

Overmyer, K., Brosche, M. and Kangasjarvi, J. (2003). Reactive oxygen species and hormonal control of cell death. *Trends Plant Sci.* **8:** 335-342.

Oztetik, E. (2008). A Tale of Plant Glutathione *S*-Transferases: Since 1970. *Bot. Rev.* **74:** 419-437.

Padh, H. (1990). Cellular functions of ascorbic acid. *Biochem. Cell Biol.* **68:** 1166-1173.

Paiva, S.A.R. and Russell, R.M. (1999). â-carotene and other carotenoids as antioxidants. *J. Am. College of Nutrition.* **18:** 426-433.

Panchuk, I.I., Volkov, R.A. and Schoffl, F. (2002). Heat stress- and heat shock transcription factor-dependent expression and activity of ascorbate peroxidase in Arabidopsis. *Plant Physiol.* **129:** 838-853.

Parida, A.K. and Das, A.B. (2005). Salt tolerance and salinity effects on plants: a review. *Ecotoxicology and Environmental Safety.* **60:** 324-349.

Parvaiz, A. and Satyawati, S. (2008). Salt stress and phyto-biochemical responses of plants – a review. *Plant Soil Environ.* **54:** 89-99.

Pastori, G.M. and Trippi, V.S. (1992). Oxidative stress induces high rate of glutathione reductase synthesis in a drought-resistant maize strahl. *Plant Cell Physiol.* **33:** 957-961.

Peltzer, D., Dreyer, E. and Polle, A. (2002). Temperature dependencies of antioxidative enzymes in two contrasting species. *Plant Physiol. Biochem.* **40:** 141-150.

Pereira, G.J.G., Molian, S.M.G., Lea, P.J. and Azevedo, R.A. (2002). Activity of antioxidant enzymes in response to cadmium in Crotalaria juncea. *Plant Soil.* **239:** 123-132.

Perez, F.J., Villegas, D. and Mejia, N. (2002). Ascorbic acid and flavonoid-peroxidase reaction as a detoxifying system of H_2O_2 in grapevine leaves. *Phytochemistry.* **60:** 573-580.

Peterson, J. and Dwyer, J. (1998). Flavonoids: Dietary occurrence and biochemical activity. *Nutrition Research.* **18:** 1995-2018.

Pietta, P-G. (2000). Flavonoids as antioxidants *J. Nat. Prod.* **63:** 1035-1042.

Pnueli, L., Liang, H., Rozenberg, M. and Mittler, R. (2003). Growth suppression, altered stomatal responses, and augmented induction of heat shock proteins in cytosolic ascorbate peroxidase (Apx1)-deficient Arabidopsis plants. *Plant J.* **34:** 187-203.

Polle, A. (2001). Dissecting the superoxide dismutase–ascorbate peroxidase–glutathione pathway in chloroplasts by metabolic modeling. Computer simulations as a step towards flux analysis. *Plant Physiol.* **126:** 445-462.

Porfirova, S., Bergmuller, E., Tropf, S., Lemke, R. and Dormann, P. (2002). Isolation of an Arabidopsis mutant lacking vitamin E and identification of a cyclase essential for all tocopherol biosynthesis. *Proc. Natl. Acad. Sci.* **99:** 12495-12500.

Poschenrieder, C., Tolra, R. and Barcelo, J. (2006). Can metals defend plants against biotic stress? *Trends in Plant Science.* **11:** 288-295.

Pospisil, P., Arato, A., Krieger-Liszkay, A. and Rutherford, A.W. (2004). Hydroxyl radical generation by photosystem II. *Biochemistry.* **43:** 6783-6792.

Prasad, T.K., Anderson, M.D., Martin, B.A. and Stewart, C.R. (1994). Evidence for chilling-induced oxidative stress in Maize seedlings and a regulatory role for hydrogen peroxide. *Plant Cell.* **6:** 65-74.

Price, A.H., Atherton, N.M. and Hendry, G.A.F. (1989). Plants under drought-stress generate activated oxygen. *Free Rad. Res. Comm.* **8:** 61-66.

Pryor, W.A. (1986). Oxy-radicals and related species: Their fornaation, lifetimes, and reactions. *Ann. Rev. Physiol.* **48:** 657-667.

Pryor, W.A. (2000). Vitamin E and heart disease: basic science to clinical intervention trials. *Free Radic. Biol. Med.* **28:** 141-164.

Puntorula, A., Sanchez, R.A. and Boveris, A. (1988). Hydrogen peroxide metabolism in soybean embryonic axes at the onset of germination. *Plant Physiol.* **86:** 626-630.

Purvis, A.C. and Shewfelt, R.L. (1993). Does the alternative pathway ameliorate chilling injury in sensitive plant tissues? *Physiol. Plant.* **88:** 712-718.

Qi, Q., Karunanandaa, B., Hao, M., Baszis, S., Wong, Y., Jiang, J., Jensen, P., Warren, A., Moshiri, F., Gruys, K.J., Venkatramesh, M. and Valentin, H.E. (2003). Metabolic engineering of tocopherol biosynthesis for increased tocopherol content in *Arabidopsis* and *Synechocystis* sp. PCC6803. *Plant Physiol.* [Suppl] Abstract No. 1074.

Rai, V.K. (2002) Role of amino acids in plant responses to stresses. *Biol. Plant.* **45:** 481-487.

Rains, D.W. (1972). Salt transport by plants in relation to salinity. *Annu. Rev. Plant Physiol.* **23:** 367-388.

Ranaud, S. and de Lorgeril, M. (1992). Wine, alcohol, platelets, and the french paradox for coronary heart disease. *Lancet.* **339:** 1523-1526.

Rao, K.V.M. and Sresty, T.V.S. (2000). Antioxidative parameters in the seedlings of pigeonpea (*Cajanus cajan* (L.) Millspaugh) in response to Zn and Ni stresses. *Plant Science.* **157:** 113-128.

Rausch, T. and Wachter, A. (2005). Sulfur metabolism: a versatile platform for launching defence operations. *Trends Plant Sci.* **10:** 503-509.

Rauser, W.E. (1990). Phytochelatins. *Annu. Rev. Biochem.* **59:** 61-86.

Rauser, W.E. (1999). Structure and function of metal chelators produced by plants; the case for organic acids, amino acids, phytin and metallothioneins. *Cell Biochem. Biophys.* **31:** 19-48.

Rausher, M.D. (2006). The evolution of flavonoids and their genes. In: Grotewold, E. (Ed). The science of flavonoids, Springer, pp. 175-211.

Rautenkranz, A.A.F., Li, L., Machler, F., Martinoia, E. and Oertli, J.J. (1994). Transport of ascorbic and dehydroascorbic acids across protoplast and vacuole membranes isolated from barley (*Hordeum vulgare* cv. Gerbil) leaves. *Plant Physiol.* **106:** 187-193.

Reddy, A.R., Chaitanya, K.V. and Sundar, D. (2000). Water stress mediated changes in antioxidant enzyme activities of mulberry (*Morus alba.* L). *J. Seric. Sci. Japan.* **69:** 169-175.

Reddy, A.R., Chaitanya, K.V. and Vivekanandan, M. (2004). Drought-induced responses of photosynthesis and antioxidant metabolism in higher plants. *J. Plant Physiol.* **161:** 1189-1202.

Rhodes, D. and Hanson, A.D. (1993). Quaternary ammonium and tertiary sulphonium compounds in higher plants. *Annu. Rev. Plant Physiol. Plant Mol. Biol.* **44:** 357-384.

Rhodes, D., Nadolska-Orczyk, A. and Rich, P.J. (2002). Salinity, osmolytes and compatible solutes. In: Lauchli, A. and Luttge, U. (Eds.). Salinity: environment-plants-molecules, Dordrecht, Netherlands: Kluwer, pp. 181-204.

Rice-Evans, C., Miller, N. and Paganga, G. (1997). Antioxidant properties of phenolic compounds. *Trends Plant Sci.* **2:** 152-159.

Richards, K.D., Schott, E.J., Sharma, Y.K., Davis, K.R. and Gardner, R.C. (1998). Aluminum induces oxidative stress genes in *Arabidopsis thaliana. Plant Physiol.* **116:** 409-418.

Robinson, M. and Bunce, J.A. (2000). Influence of drought-induced water stress on soybean and spinach leaf ascorbatedehydroascorbate level and redox status. *Int. J. Plant Sci.* **161:** 271-279.

Robinson, N.J., Tommey, A.M., Kuske, C. and Jackson, P.J. (1993). Plant metallothioneins. *Biochem. J.* **295:** 1-10.

Robson, C.A. and Vanlerberghe, G.C. (2002). Transgenic plant cells lacking mitochondrial alternative oxidase have increased susceptibility to mitochondria-dependent and independent pathways of programmed cell death. *Plant Physiol.* **129:** 1908-1920.

Ros Barcelo, A. (1997). Lignification in plant cell walls. *Int. Rev. Citol.* **176:** 87-132.

Rose, R.C. and Bode, A.M. (1993). Biology of free radical scavengers: an evaluation of ascorbate. *FASEB J.* **7:** 1135-1142.

Rouhier, N., Vieira Dos Santos, C., Tarrago, L. and Rey, P. (2006). Plant methionine sulfoxide reductase A and B multigenic families. *Photosynthesis Research.* **89:** 247-262.

Roxas, V.P., Smith, R.K., Allen, E.R. and Allen, R.D. (1997). Overexpression of Glutathione *S*-transferase/Glutathione peroxidase enhances the growth of transgenic tobacco seedlings during stress. *Nat. Biotechnol.* **15:** 988-991.

Sahw, B.P., Sahu, S.K. and Mishra, R.K. (2004). Heavy metal induced oxidative damage in terrestrial plants. In: Presad, M.N.V. (Ed.). Heavy metal stress in plants: From biomolecules to ecosystems, second ed. Springer, Berlin, pp. 84-126.

Sairam, R.K., Deshmuhh, P.S. and Saxena, D.C. (1998). Role of antioxidant systems in wheat genotypes tolerance to water stress. *Biol. Plant.* **41:** 387-394.

Sairam, R.K. and Tyagi, A. (2004). Physiology and molecular biology of salinity stress tolerance in plants. *Current Science.* **86:** 407-421.

Sakamoto, W. (2006). Protein degradation machineries in plastids. *Annu. Rev. Plant Biol.* **57:** 599-621.

Salt, D.E. and Rauser, W.E. (1995). MgATP-dependent transport of phytochelatins across the tonoplast of oat roots. *Plant Physiology.* **107:** 1293-1301.

Salt, D.E., Smith, R.D. and Raskin, I. (1998). Phytoremediation. *Ann. Rev. Plant Physiol. Plant Mol. Biol.* **49:** 643-668.

Sanchez-Fernandez, R., Fricker, M., Corben, L.B., White, N.S. and Sheard, N. (1997). Cell proliferation and hair tip growth in the *Arabidopsis* root are under mechanistically different forms of redox control. *Proc. Natl. Acad. Sci.* **94:** 2745-2750.

Saruyama, H. and Tanida, M. (1995). Effect of chilling on activated oxygen-scavenging enzymes in low temperature-sensitive and - tolerant cultivars of rice (*Oryza sativa* L.). *Plant Sci.* **109:** 105-113.

Sato, P. and Udenfriend, S. (1978). Scurvy-prone animals, including man, monkey, and guinea pig do not express the gene for gulonolactone oxidase. *Arch. Biochem. Biophys.* **71:** 293-299.

Sato, Y., Murakami, T., Funatsuki, H., Matsuba, S., Saruyama, H. and Tanida, M. (2001). Heat shock-mediated APX gene expression and protection against chilling injury in rice seedlings. *J. Exp. Bot.* **52:** 145-151.

Sattler, S.E., Cahoon, E.B., Coughlan, S.J., D. DellaPenna, D. (2003). Characterization of tocopherol cyclases from higher plants and cyanobacteria. Evolutionary implications for tocopherol synthesis and function. *Plant Physiol.* **132:** 2184-2195.

Savchenko, G.E., Klyuchareva, E.A., Abrabchik, L.M. and Serdyuchenko, E.V. (2002). Effect of periodic heat shock on the membrane system of etioplasts. *Russ. J. Plant Physiol.* **49:** 349-359.

Savidge, B., Weiss, J.D., Wong, Y-H.H., Lassner, M.W., Mitsky, T.A., Shewmaker, C.K., Post-Beittenmiller, D. and Valentin, H.E. (2002). Isolation and characterization of homogentisate phytyltransferase genes from *Synechocystis* sp. PCC 6803 and Arabidopsis. *Plant Physiol.* **129:** 321-332.

Scheller, H.V., Huang, B., Hatch, E. and Goldsbrough, P.B. (1987). Phytochelatin synthesis and glutathione levels in response to heavy metals in tomato cells. *Plant Physiol.* **85:** 1031-1035

Schijlen, E.G.W.M., Ric de Vos, C.H., van Tunen, A.J. and Bovy, A.G. (2004). Modification of flavonoid biosynthesis in crop plants. *Phytochemistry.* **65:** 2631-2648.

Schützendübel, A., Schwanz, P., Teichmann, T., Gross, K., Langenfeld-Heyser, R., Godbold, D.L. and Polle, A. (2001). Cadmium-induced changes in antioxidative systems, H_2O_2 contents and differentiation in pine (*Pinus sylvestris*) roots. *Plant Physiology.* **127:** 887-892.

Semchuk, N.M., Lushchak, O.V., Falk, J., Krupinska, K. and Lushchak, V.I. (2009). Inactivation of genes, encoding tocopherol biosynthetic pathway enzymes, results in oxidative stress in outdoor grown *Arabidopsis thaliana. Plant Physiology and Biochemistry.* **47:** 384-390.

Sgherri, C.L.M., Pinzino, C. and Navari-Izzo, F. (1993). Chemical changes and $O_2^{\cdot-}$ production in thylakoid membranes under water stress. *Physiol. Plant.* **87:** 211-216.

Sgherri, C.L.M., Pinzino, C. and Navari-Izzo, F. (1996). Sunflower seedlings subjected to increasing stress by water deficit: changes in $O_2^{\cdot-}$ production related to the compositionof thylakoid membranes. *Physiol. Plant.* **96:** 446-452.

Shao, H-B. and Chu, L.Y. (2005). Plant molecular biology in China: Opportunities and Challenges. *Plant Mol. Biol. Rep.* **23:** 345-358.

Shao, H-B., Chu, L-Y., Lu, Z-H. and Kang, C-M. (2008). Primary antioxidant free radical scavenging and redox signaling pathways in higher plant cells. *Int. J. Biol. Sci.* **4:** 8-14.

Shao, H.B., Chu, L.Y., Wu, G., Zhang, J.H., Lu, Z.H. and Hu, Y.C. (2007a). Changes of some anti-oxidative physiological indices under soil water deficits among 10 wheat (*Triticum aestivum* L.) genotypes at tillering stage. *Biointerfaces.* **54:** 143-149.

Shao, H.B., Guo, Q.J., Chu, L.Y., Zhao, X.N., Su, Z..L, Hu, Y.C. and Cheng, J.F. (2007b). Understanding molecular mechanism of higher plant plasticity under abiotic stress. *Biointerfaces.* **54:** 37-45.

Shao, H.B., Liang, Z.S., Shao, M.A., Sun, Q. and Hu, Z.M. (2005). Investigation on dynamic changes of photosynthetic characteristics of 10 wheat (*Triticum aestivum L.*) genotypes during two vegetative-growthstages at water deficits. *Biointerfaces.* **43:** 221-227.

Shaul, O., Mironov, V., Burssens, S., Van Montagu, M. and Inze, D. (1996). Two *Arabidopsis* cyclin promoters mediate distinctive transcriptional oscillation on synchronised tobacco3Y-2 cells. *Proc. Natl. Acad. Sci.* **93:** 4868-4872.

Shintani, D. and DellaPenna, D. (1998). Elevating the vitamin E content of plants through metabolic engineering. *Science.* **282:** 2098-2100.

Siedlecka, A., Tukendorf, A., Skorzanska-Polit, E., Maksymiec, W., Wojcik, M., Baszynski, T. and Krupa, Z. (2001). Angiosperms. In: Prasad, M.N.V. (Ed.). Metals in the environment. Analysis of biodiveristy. Marcel Dekker, Inc. New York, Hyderbad India, pp. 171-217.

Sies, H. and Stahl, W. (1995). Vitamins E and C, beta-carotene, and other carotenoids as antioxidants. *Am. J. Clin. Nutr.* **62:** 1315S-1321S.

Singh, S.K., Sharma, H.C., Goswami, A.M., Datta, S.P. and Singh, S.P. (2000). *In vitro* growth and leaf composition of grapevine cultivars as affected by sodium chloride. *Biol. Plant.* **43:** 283-286.

Skorzynska-polit, E., Drazkiewicz, M. and Krupa, Z. (2003–2004). The activity of the antioxidantive system in cadmium-treated *Arabidopsis thaliana. Biol. Plant.* **47:** 71-78.

Smirnoff, N. (1993). The role of active oxygen in the response of plants to water deficit and desiccation. *New Phytol.* **125:** 27-58.

Smirnoff, N. (1996). The function and metabolism of ascorbic acid in plants. *Ann. Bot.* **78:** 661-669.

Smirnoff, N. (1998). Plant resistance to environmental stress. *Curr. Opin. Biotechnol.* **9:** 214-219.

Smirnoff, N., Colombe, S.V. (1988). Drought Influences the Activity of Enzymes of the Chloroplast Hydrogen Peroxide Scavenging System. *J. Exp. Bot.* **39:** 1097-1108.

Smirnoff, N. and Cumbes, Q.J. (1989). Hydroxyl radical scavenging activity of compatible solutes. *Phytochemistry.* **28:** 1057-1060.

Smirnoff, N. and Wheeler, G.L. (2000). Ascorbic acid in plants: Biosynthesis and function. Critical Reviews in Plant Sciences. **19:** 267-290.

Somashekaraiah, B.V., Padmaja, K. and Prasad, A.R.K. (1992). Phytotoxicity of cadmium ions on germinating seedlings of mung bean *(Phaseolus vulgaris):* involvement of lipid peroxides in chlorophyill degradation. *Physiologia Plantarum.* **85:** 85-89.

Spencer, J.P., Rice-Evans, C. and Williams, R.J. (2003). Modulation of pro-survival Akt/protein kinase B and ERK1/2 signaling cascades by quercetin and its *in vivo* metabolites underlie their action on neuronal viability. *J. Biol. Chem.* **278:** 34783-34793.

Spencer, J.P.E., Schroeter, H., Crossthwaithe, A.J., Kuhnle, G., Williams, R.J. and Rice-Evans, C. (2001). Contrasting influences of glucuronidation and O-methylation of epicatechin on hydrogen peroxide-induced cell death in neurons and fibroblasts. *Free Radic. Biol. Med.* **31:** 1139-1146.

Sridevi, V., Satyanarayana, N.V. and Madhava Rao, K.V. (1999). Induction of heat shock proteins and acquisition of thermotolerance in germinating pigeonpea seeds. *Biol. Plant.* **42:** 589-597.

Srivalli, B., Chinnusamy, V. and Khanna-Chopra, R. (2003). Antioxidant defense in response to abiotic stresses in plants. *J. Plant Biol.* **30:** 121-139.

Stahl, W. and Sies, H. (2003). Antioxidant activity of carotenoids, *Mol. Aspects Med.* **24:** 345-351.

Stahl, W. and Sies, H. (2005). Bioactivity and protective effects of natural carotenoids. *Biochimica et Biophysica Acta.* **1740:** 101-107.

Steponkus, P.L., Uemura, M. and Webb, M.S. (1993). A contrast of the cryostability of the plasma membrane of winter rye and spring oat. Two species that widely differ in their freezing tolerance and plasma membrane lipid composition. In: Steponkus, P.L. (Ed.). Advances in low temperature biology, JAI Press, London, pp. 211-312.

Stone, I. (1972). The natural history of ascorbic acid in the evolution of the mammals and primates and its significance for present day man. *J. Orthomolecular Psychiatry.* **1:** 82-89.

Sumithra, K., Jutur, P.P., Dalton, C.B. and Reddy, A.R. (2006). Salinity-induced changes in two cultivars of *Vigna radiata*: responses of antioxidative and proline metabolism. *Plant Growth Regul.* **50:** 11-22.

Sweetlove, L.J. and Moller, I.M. (2009). Oxidation of Proteins in Plants-Mechanisms and Consequences. In: Jacquot, J-P. (Eds.). Advances in Botanical Research, Vol. 52, Burlington: Academic Press. pp. 1-23.

Tapas, A.R., Sakarkar, D.M. and Kakde, R.B. (2008). Flavonoids as Nutraceuticals: A Review *Tropical J. Pharmaceutical Research.* **7:** 1089-1099.

Tausz, M., Ircelj, H. and Grill, D. (2004). The glutathione system as a stress marker in plant ecophysiology: is a stress-response concept valid? *J. Exp. Bot.* **55:** 1955-1962.

Tester, M. and Davenport, R.J. (2003). Na^+ transport and Na^+ tolerance in higher plants. *Ann. Bot.* **91:** 503-527.

Thom, R., Dixon, D.P., Edwards, R., Cole, D.J. and Lapthorn, A. (2001). The structure of a zeta class glutathione S-transferase from A.thaliana: caharacterization of a GST with novel active site architecture and a putative role in tyrosine catabolism. *J. Mol. Biol.* **308:** 949-962.

Thomas, C.E., McLean, L.R., Parker, R.A. and Ohlweiler, D.F. (1992). Ascorbate and phenolic abtioxidant interactions in prevention of lipospmal oxidation. *Lipids.* **27:** 534-550.

Torres, M.A., Jones, J.D.G. and Dangl, J.L. (2006). Reactive Oxygen Species Signaling in Response to Pathogens. *Plant Physiology.* **141:** 373-378.

Treutter, D. (2006). Significance of flavonoids in plant resistance: a review. *Environ. Chem. Lett.* **4:** 147-157.

Truscott, T.G. (1996). Beta-carotene and disease: a suggested pro-oxidant and anti-oxidant mechanism and speculations concerning its role in cigarette smoking. *J. Photochem. Photobiol. B: Biol.* **35:** 233-235.

Tsugane, K., Kobayashi, K., Niwa, Y., Ohba, Y., Wada, K. and Kobayashi, H. (1999). A recessive *Arabidopsis* mutant that grows enhanced active oxygen detoxification. *Plant Cell.* **11:** 1195-1206.

Uemura, M., Joseph, R.A. and Steponkus, P.L. (1995). Cold acclimation of *Arabidopsis thaliana*: Effect on plasma membrane lipid composition and freeze-induced lesions. *Plant Physiol.* **109:** 15-30.

Umbach, A.L. and Siedow, J.N. (1993). Covalent and noncovalent dimers of the cyanide resistant alternative oxidase protein in higher plant mitochondria and their relationship to enzyme activity. *Plant Physiol.* **103:** 845-854.

Van Duyn, M.A. and Pivonka, E. (2000). Overview of the health benefits of fruit and vegetable consumption for the dietetics professional: selected literature. *J. Am. Diet. Assoc.* **100:** 1511-1521.

Vera-Estrella, R., Blumwald, E. and Higgins, V.J. (1992). Effect of specific elicitoirs of *Cladosporium fulvum* on tomato suspension cells. *Plant Physiol.* **99:** 1208-1215.

Vercesi, A.E., Martins, I.S., Silva, M.A.P. and Leite, H.M.F. (1995). Pumping plants. *Nature.* **375:** 24.

Vierstra, R.D. (2003). The ubiquitin/26S proteasome pathway, the complex last chapter in the life of many plant proteins. *Trends Plant Sci.* **8:** 135-142.

Wahid, A., Gelani, S., Ashraf, M. and Foolad, M.R. (2007). Heat tolerance in plants: *An Overview Environmental and Experimental Botany.* **61:** 199-223.

Wang, W.X., Vinocur, B., Shoseyov, O. and Altman, A. (2001). Biotechnology of plant osmotic stress tolerance: physiological and molecular considerations. *Acta. Hort.* **560:** 285-292.

Wang, Y. and Nil, N. (2000). Changes in chlorophyll, ribulose biphosphate carboxylase-oxygenase, glycine betaine content, photosynthesis and transpiration in Amaranthus tricolor leaves during salt stress. *J. Hortic. Sci. Biotechnol.* **75:** 623-627.

Waters, E.R., Lee, G.J. and Vierling, E. (1996). Evolution, structure and function of the small heat shock proteins in plants. *J. Exp. Bot.* **47:** 325-338.

Weast, R.C. (1984). CRC Handbook of chemistry and physics, 64th ed. CRC Press, Boca Raton.

Weckx, J.E.J. and Clijsters, H.M.M. (1996). Oxidative damage and defense mechanisms in primary leaves of *Phaseolus vulgaris* as a result of root assimilation of toxic amounts of copper. *Physiol. Plant.* **96:** 506-512.

Weckx, J.E.J. and Clijsters, H.M.M. (1997). Zn phytotoxicity induces oxidative stress in primary leaves of *Phaseolus vulgaris. Plant Physiol. Biochem.* **35:** 405-410.

Wheeler, G.L., Jones, M.A. and Smirnoff, N. (1998). The biosynthetic pathway of vitamin C in higher plants. *Nature.* **393:** 365-369.

Williams, R.J., Spencer, J.P. and Rice-Evans, C. (2004). Flavonoids: antioxidants or signalling molecules? *Free Radic. Biol. Med.* **36:** 838-849.

Wingate, V.P.M., Lawton, M.A. and Lamb, C.J. (1988). Glutathione causes a massive and selective induction of plant defense genes. *Plant Physiol.* **87:** 206-210.

Winkel-Shirley, B. (2001). It takes a garden. How work on diverse plant species has contributed to an understanding of flavonoid metabolism. *Plant Physiol.* **127:** 1399-1404.

Winkel-Shirley, B. (2002). Biosynthesis of flavonoids and effects of stress. *Current Opinion in Plant Biology.* **5:** 218-223.

Winkler, B.S., Orselli, S.M. and Rex, T.S. (1994). The Redox Couple between Glutathione and Ascorbic Acid: A Chemical and Physiological Perspective. *Free Radic. Biol. Med.* **17:** 333-349.

Woitsch, S. and Romer, S. (2003). Expression of xanthophyll biosynthetic genes during light-dependent chloroplast diVerentiation. *Plant Physiol.* **132:** 1508-1517.

Wojtaszek, P. (1997). Oxidative burst: an early plant response to pathogen infection. *Biochem. J.* **332:** 681-692.

Wu, G., Wei, Z.K. and Shao, H.B. (2007). The mutual responses of higher plants to environment: physiological and microbiological aspects. *Biointerfaces.* **59:** 113-119.

Xiang, C., Werner, B.L., Christensen, E.M. and Oliver, D.J. (2001). The Biological Functions of Glutathione Revisited in Arabidopsis Transgenic Plants with Altered Glutathione Levels. *Plant Physiology.* **126:** 564-574.

Xiong, L. and Zhu, J-K. (2002). Molecular and genetic aspects of plant responses to osmotic stress. *Plant Cell Environ.* **25:** 131-139.

Xiong, Y., Contento, A.L., Nguyen, P.Q. and Bassham, D.C. (2007). Degradation of oxidized proteins by autophagy during oxidative stress in Arabidopsis. *Plant Physiol.* **143:** 291-299.

Xu, S., Li, J., Zhang, X., Wei, H. and Cui, L. (2006). Effects of heat acclimation pretreatment on changes of membrane lipid peroxidation, antioxidant metabolites, and ultrastructure of chloroplasts in two cool-season turfgrass species under heat stress. *Environ. Exp. Bot.* **56:** 274-285.

Yamasaki, H., Sakihama, Y. and Ikehara, N. (1997). Flavonoid-peroxidase reaction as a detoxification mechanism of plant cells against H_2O_2. *Plant Physiol.* **115:** 1405-1412.

Yang, Z.M., Sivaguru, M., Horst, W.J. and Matsumoto, H. (2001). Aluminum tolerance is achieved by exudation of citric acid from roots of soybean (*Glycine max*). *Physiol. Plant.* **110:** 72-74.

Yokoi, S., Quintero, F.J., Cubero, B., Ruiz, M.T., Bressan, R.A., Hasegawa, P.M. and Pardo, J.M. (2002). Differential expression and function of *Arabidopsis thaliana* NHX Na^+/H^+ antiporters in the salt stress response. *Plant J.* **30:** 529-539.

Zhang, J. and Kirkham, M.B. (1994). Drought stress-induced changes in activities of superoxide dismutase, catalase and peroxidase in wheat species. *Plant Cett. Physiol.* **35:** 785-791.

Zhifang, G. and Loescher, W.H. (2003). Expression of a celery mannose 6- phosphate reductase in *Arabidopsis thaliana* enhances salt tolerance and induces biosynthesis of both mannitol and a glucosyl-mannitol dimmer. *Plant Cell Environ.* **26:** 275-283.

Zhu, D. and Scandalios, J.G. (1994). Differential accumulation of manganese-superoxide dismutase transcripts in maize in response to abscisic acid and high osmoticum. *Plant Physiol.* **106:** 173-178.

Zhu, J-K. (2002). Salt and drought stress signal transduction in plants. *Annu. Rev. Plant Physiol. Plant Mol. Biol.* **53:** 247-273.

Zhu, Y.L., Philon-Smits, E.A.H., Tarun, A.S., Weber, S.U., Jouanin, L. and Terry, N. (1999). Cadmium tolerance and accumulation in Indian mustard is enhanced by overexpressing gammaglutamylcysteine synthetase. *Plant Physiol.* **121:** 1169-1178.

Zornoza, P., Vazquez, S., Estebane, E., Fernandez-Pascual, M. and Carpena, R. (2002). Cadmium-stress in nodulated white lupin: strategies to avoid toxicity. *Plant Physiol. Biochem.* **40:** 1003-1009.

2

Environmental Factors Affecting Components of Ascorbate-Glutathione Pathway in Crop Plants

PAVLINA APOSTOLOVA[1], **GABRIELLA SZALAI**[2], **GÁBOR KOCSY**[2], **TIBOR JANDA**[2], **AND LOSANKA P. POPOVA**[1*]

[1]*Acad. M. Popov Institute of Plant Physiology, Bulgarian Academy of Sciences, Acad. G. Bonchev Street, Bldg. 21, 1113, Sofia, Bulgaria*

[2]*Agricultural Research Institute of the Hungarian Academy of Sciences, H-2462, Martonvásár, POB. 19, Hungary*

[*]***Corresponding author:*** *lpopova@bio21.bas.bg*

ABSTRACT

Both plant breeders and crop producers have an interest in finding crops capable of tolerating environmental changes with as little damage as possible. In order to develop such crops, knowledge of plant defence mechanisms and regulatory processes is essential. Ascorbate and glutathione are involved in both the direct and indirect control of the concentration of reactive oxygen species. As components of the ascorbate-glutathione pathway, they take part in the removal of excess H_2O_2. Besides its role in the antioxidative defence line, glutathione also takes part in the detoxifying of harmful compounds generated in or taken up by the cells. The present chapter focuses on the regulation of ascorbate and glutathione synthesis under stress conditions, such as low temperature, heat or drought. Since glutathione is the precursor of phytochelatins, compounds responsible for detoxifying heavy metals, this role will also be discussed.

Keywords: Acclimation, ascorbate, environment, glutathione, oxidative stress.

1. INTRODUCTION

The ascorbate-glutathione cycle plays an important role in the control of concentration of reactive oxygen species (ROS) produced as a consequence of the aerobic life form. As components of the ascorbate-glutathione pathway, glutathione (GSH) and ascorbate (vitamin C) (AsA) take part in the removal of excess H_2O_2 (Noctor and Foyer 1998), in a reaction in which GSH is oxidized. The high ratio of GSH to GSSG and AsA to dehydroascorbate (DHA) occurring under optimal growth conditions can be restored by means of higher activity of enzymes of AsA-GSH cycle, increased GSH or AsA synthesis, decreased degradation or the transport of these molecules. Four enzymes—ascorbate peroxidase, dehydroascorbate reductase, monodehydroascorbate reductase and glutathione reductase are involved in the ascorbate-glutathione cycle. Most of these enzymes are located in the cytosol, stroma of chloroplasts, mitochondria and peroxisomes (Jiménez et al., 1998). Ascorbate peroxidase (APX) is a key antioxidant enzyme in plants (Orvar and Ellis 1997) whilst glutathione reductase has a central role in maintaining the reduced glutathione pool during stress (Pastori et al., 2000). Two glutathione reductase complementary deoxyribonucleic acids have been isolated; one type encoding the cytosolic isoforms (Stevens et al., 2000) and the other encoding glutathione reductase proteins dual-targeted to both chloroplasts and mitochondria in different plants (Chew et al., 2003).

AsA is a metabolite with strong antioxidant activity and a cofactor for enzymes catalysing numerous biochemical reactions, including those neutralizing the effects of reactive oxygen species. It is also a necessary nutrient for a limited number of animals, including humans, that are incapable of its synthesis and that must secure vitamin C by means of dietary uptake. Ascorbic acid takes part both directly (e.g., H_2O_2 degradation) and indirectly (e.g., reduction of oxidised alpha-tocopherol) in the removal of ROS accumulating after adverse environmental effects (Noctor and Foyer 1998).

Besides the AsA-GSH cycle, GSH may also participate in H_2O_2 degradation in a reaction catalysed by glutathione peroxidase. However, the existence of GSH-specific peroxidases in plants is questionable since plant glutathione peroxidases do not react with GSH, but only with thioredoxins (Navrot et al., 2006). Besides H_2O_2, GSH removes lipid peroxides, methylglyoxal and herbicides (Moons 2005; Rausch et al., 2007; Yadav et al., 2008). In these processes GSH is a substrate for the glutathione S-transferases (GSTs), which catalyse the conjugation of GSH with potentially dangerous xenobiotics such as herbicides (Marrs 1996), and the conjugates are transported to the vacuoles. GSH not only participates in the direct detoxification of ROS, but may also protect cells against unfavourable stress effects through the activation of various defence mechanisms, due to its involvement in redox signalling (Foyer et al., 1997; Apel and Hirt 2004; Mittler et al., 2004; Mullineaux and Rausch 2005; Pitzschke et al., 2006; Foyer and Noctor 2009). In this signalling pathway GSH interacts with ROS, redox molecules such as thioredoxins or glutaredoxins, and plant hormones (Szalai et al., 2009). Besides the control of ROS levels, numerous physiological functions have been attributed to GSH in plants (Foyer et al., 1997; Szalai et al., 2009). GSH takes part in the regulation of growth, development, the cell cycle, gene expression and protein activity due to its effect on the redox state of the cells (Noctor et al., 1998; Aslund and Beckwitt 1999; Ogawa 2005; Shao et al., 2008). Important in the transfer and storage of sulphur, GSH also regulates inter-organ sulphur allocation (Herschbach and Rennenberg 2001), and it also serves as precursor of phytochelatins, which bind supra-optimal concentrations of heavy metals (Blum et al., 2007; Pál et al., 2006).

2. SYNTHESIS OF ASCORBATE AND GLUTATHIONE IN PLANT CELLS

2.1 Ascorbate Biosynthesis

Plants produce large amounts of AsA to facilitate resistance to the oxidative stresses associated with myriad biotic and abiotic challenges and inherent to photosynthesis. The study of plant ascorbate synthesis is highly relevant to both plant biology and human nutrition, with eventual engineering of plant ascorbate metabolism a logical and important target for crop quality improvement (Giovannoni 2007). A major component of plant ascorbate synthesis is through the L-galactose pathway, in which GDP-D-mannose is converted to L-ascorbate by four successive intermediates. Although additional pathways for ascorbate synthesis may contribute to the accumulation of this metabolite in plants, the L-galactose pathway is clearly a major contributor. The gene residing at the *vtc1* locus encodes GTP-mannose pyrophosphorolase, which catalyses the synthesis of GDP-mannose from mannose-1-P (Conklin et al., 1999), which is converted to GDP-l-galactose by GDP-mannose-3,5-epierimase. L-galactose guanyltransferase, encoded by the *vtc2* gene, catalyses the conversion of GDP-l-galactose to l-galactose-1-P. *Arabidopsis* lines harbouring mutant *vtc2* alleles have been shown to have reduced ascorbate levels (Conklin et al., 2000), impaired stress response (Conklin et al., 2000; Barth et al., 2004), and altered levels of stress response proteins (Giacomelli et al., 2006). The next step is catalysed by gene called *vtc4*, which encodes L-galactose-1-P phosphatase creating L-galactose, followed by L-galactose dehydrogenase, which catalyses the conversion of L-galactose to L-galactono-1,4-lactone. The final step is the synthesis of L-ascorbate by L-galactono-1,4-lactone dehydrogenase (Giovannoni 2007).

2.2 Glutathione Biosynthesis

Glutathione is found in the vast majority of prokaryotic and eukaryotic cells, where it often represents the major pool of non-protein reduced sulphur. The reduced form of glutathione, GSH is a tripeptide thiol (γ-glutamylcysteinyl glycine), which is synthesized in two sequential ATP-dependent reactions. First, the formation of γ-glutamylcysteine (γEC) is catalysed by γEC synthetase (γECS), then a glycine is added to the dipeptide by GSH synthetase (GS). The regulatory enzyme of GSH synthesis is γECS (Rüegsegger and Brunold 1992; Rennenberg et al., 2007). Several plant species produce analogous tripeptide derivatives of γEC in addition to, or in place of, GSH. Homoglutathione (hGSH: γ-glutamylcysteinyl-β-alanine) is present in the *Fabaceae* family (Klapheck 1988) and hydroxymethyl-GSH (hmGSH: γ-glutamylcysteinyl-serine) or γ-EC-glu in the *Gramineae* family (Klapheck et al., 1991; Meuwly et al., 1993). Since all functions thus far described for GSH in plants are related to the cysteine moiety of the tripeptide, these homologues may exercise similar biochemical roles to GSH.

3. CHANGES IN THE SIZE OF THE ASCORBATE AND GLUTATHIONE POOLS UNDER ABIOTIC STRESS CONDITIONS

The measurement of abiotic stress-induced changes in the glutathione levels and their comparison in several genotypes with different stress tolerance gave the first indication about the participation of GSH in the stress response. However, a better understanding of the role of GSH in stressed plants was only possible when GSH and GSSG contents were measured, since even at constant glutathione levels changes in the ratio of the two forms influence the redox state of the GSH/GSSG couple.

The redox state of the GSH pool was initially determined by the simultaneous spectrophotometric detection of GSH and GSSG (Mergel et al., 1979). Later on more sensitive high-performance liquid chromatography methods were introduced for monitoring the concentrations of GSH and GSSG and their precursors and homologues (Kranner and Grill 1996; Potesil et al., 2005; Rellán-Alvarez et al., 2006). In one HPLC method that detects the fluorescent monobromobimane derivatives of thiols, the levels of total and reduced thiols (Cys, γEC, GSH, hmGSH and hGSH) levels can be determined separately after reduction (Kranner and Grill 1996). The combination of the HPLC separation of thiols with electrochemical detection allowed the simultaneous determination of Cys, GSH, GSSG and phytochelatins (Potesil et al., 2005) and excludes possible mistakes originating from their separate detection. An HPLC electrospray/mass spectrometry method made the simultaneous measurement of GSH, GSSG, hGSH and homoglutathione disulphide (hGSSG) possible (Rellán-Alvarez et al., 2006) and a capillary zone electrophoresis approach ensured the simultaneous analysis of GSH and GSSG (Mendoza et al., 2004). One disadvantage of the latter three methods compared to fluorescent detection, is that the reduced and oxidized forms of the GSH precursors could not be determined.

The effect of abiotic stress on GSH synthesis can be monitored in various organs and cell compartments using selective antibodies against GSH and its precursors, Cys, glutamate and glycine (Zechmann et al., 2006; Zechmann and Müller 2008). Using this technique, 50% of glutathione was found in the mitochondria under control conditions and its relative level varied between 7 and 20 per cent in the other organelles in the leaves and roots of *Cucurbita pepo*, as calculated from the data of Zechmann and Müller (2008). Virus infection resulted in a 2-3-fold higher increase in Cys and GSH levels in plastids and nuclei compared to mitochondria, which also had altered ratios in the individual organelles (Zechmann and Müller 2008). Based on this observation, it can be anticipated that abiotic stresses also induce different changes in the concentration of GSH and its precursors in the various organelles. The current methods make it possible to detect GSH and changes in the redox state of GSH, even at subcellular levels, thus promoting a better understanding of its participation in stress responses. Experiments investigating the effects of abiotic stress on GSH synthesis in plant organs should be complemented in the future with studies at the cellular and subcellular levels.

3.1 Temperature Stresses

The role of GSH at low temperature is indicated by the greater glutathione contents observed in spruce during the winter (Anderson et al., 1992) and in chilling-tolerant maize genotypes compared to sensitive ones during cool spring periods in the field (Leipner et al., 1999) indicated a possible protective role of GSH during low temperature stress. This assumption was corroborated in growth chambers by comparing maize and rice genotypes with different levels of stress tolerance (Kocsy et al., 2001a; Guo et al., 2006). The importance of GSH was also shown in the case of heat stress which resulted in higher glutathione content in wheat and maize (Nieto-Sotelo and Ho 1986; Dash and Mohanty 2002).

The cold-induced increase in the glutathione concentration in maize was the result of a greater synthesis rate, as demonstrated by the incorporation of ^{35}S from sulphate into GSH and the higher activity of the two enzymes involved in GSH synthesis (Kocsy et al., 2001a). In addition, chilling increased the γECS activity and γEC content in the bundle sheath cells of maize leaves (Gómez et al., 2004b). Either high temperature or cold treatments induced a greater increase in GSH and hmGSH synthesis in tolerant wheat genotypes than in sensitive ones, as shown in ^{35}S-labelling experiments (Kocsy et al., 2000; 2004c). Chemical manipulation of the GSH metabolism and mutants with altered GSH levels were suitable for proving the protective role of GSH and of the corresponding metabolic enzymes in the response to abiotic stresses. In a chilling-sensitive maize genotype the GSH content and chilling tolerance increased when herbicide safeners were added (Kocsy et al., 2001b), but when increasing concentrations of buthionine sulphoximine, a specific inhibitor of γECS, were simultaneously applied the plants gradually became sensitive again. These results were confirmed by the inhibition of GSH synthesis in a tolerant maize genotype, which became sensitive following buthionine sulphoximine treatment (Kocsy et al., 2001a). Chilling tolerance could be restored by the addition of exogenous γEC or GSH. In these studies a correlation was found between GSH level, GR activity and chilling tolerance in maize. The glutathione level is also affected by the transport processes. GSH is a long distance transport metabolite for transporting reduced sulphur between shoots and roots and into developing seeds (Herschbach and Rennenberg 2001; Cairms et al., 2006). The uptake of GSH into the cells was studied in rice, where a putative GSH transporter was cloned, the function of which may be the retrieval of GSSG and GS-conjugates from the apoplast into the cytosol under stress conditions (Zhang et al., 2004).

By comparison of 5 wheat genotypes after 3 weeks cold hardening a correlation was found between the level of freezing tolerance and components of the ascorbate-glutathione cycle, thus H_2O_2 concentration, ascorbate/dehydroascorbate and GSH/GSSG ratios in the crowns (Fig. 1). This observation indicates the involvement of this cycle in the cold acclimation process.

The accumulation of glutathione under environmental stress conditions could be due not only to increased synthesis, but also to less intense degradation. Whereas GSH synthesis is performed in the chloroplast and cytosol, GSH and GS-conjugate degradation is restricted to the vacuole and perhaps to the apoplast (Foyer et al., 2001). γ-Glutamyltranspeptidases (GGT) are essential for the degradation of GSH and GS-conjugates. GGT can also catalyse the degradation of GSSG (Ohkama-Ohtsu et al., 2007). The effect of abiotic stress on GGT activity was described in pine, where it decreased

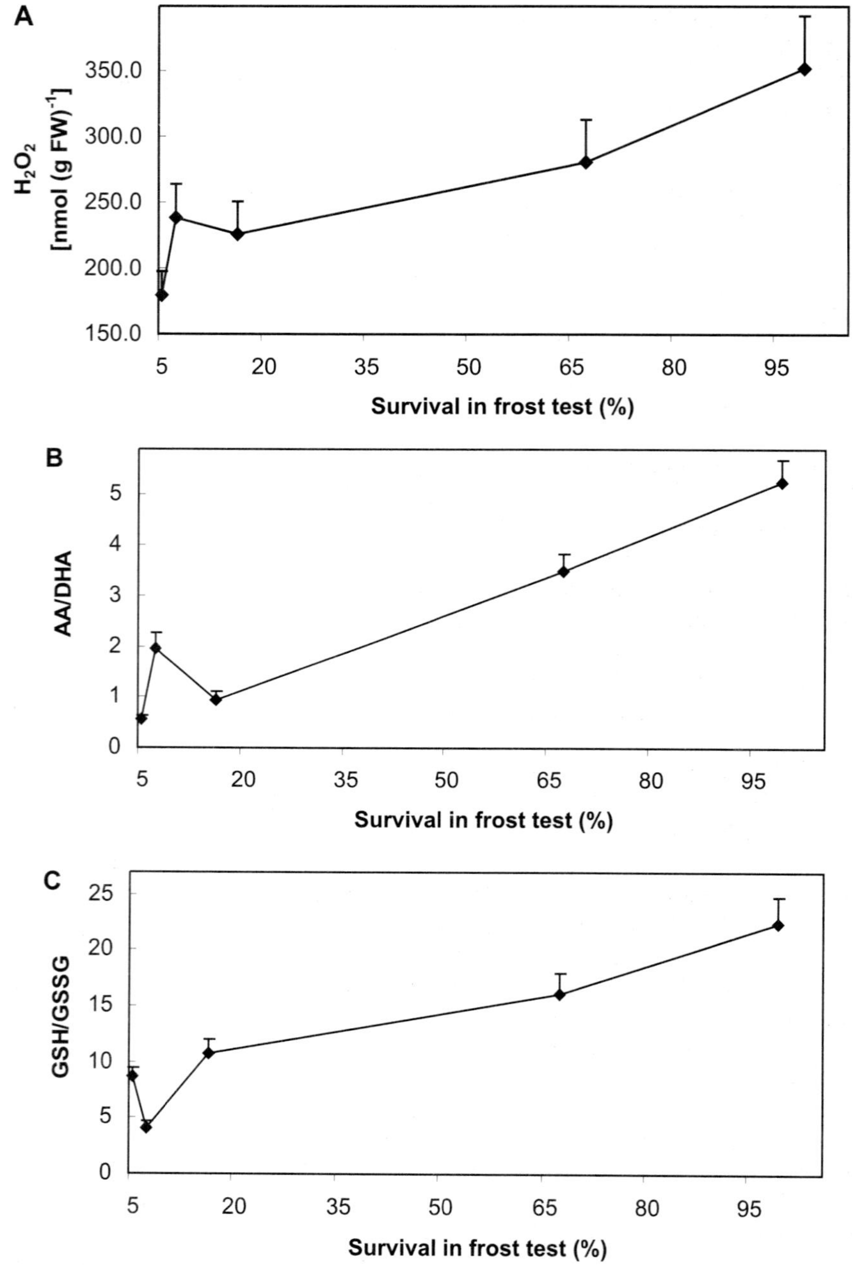

Fig. 1. Freezing-tolerance dependent differences in H_2O_2 level (A), ascorbic acid/dehydroascobate (AsA/DHA; B) and glutathione/glutathione disulphide (GSH/GSSG; C) ratios after 3 weeks cold-hardening at 2°C in wheat crowns. The examined genotypes from the left to the right: *Titicum spelta*, *T. aestivum* cv. Chinese Spring (*Triticum spelta* 5A), *T. ae.* cv. Chinese Spring, *T. ae.* cv. Chinese Spring (*T. ae.* cv. Cheyenne 5A), *T. ae.* cv. *Cheyenne*. Chromosome 5A substitution lines have been included into this study because of the large effect of this chromosome on freezing tolerance. The correlation coefficients were 0.95, 0.96 and 0.95, respectively.

during autumn, allowing the accumulation of GSH, which may be involved in the natural cold hardening process (Taulavuori et al., 1999). Recently a GGT-independent pathway of GSH catabolism to glutamate via 5-oxoproline was described in *Arabidopsis* (Ohkama-Ohtsu et al., 2008).

Besides synthesis and degradation, the conjugation of GSH with lipid peroxides, toxic metabolic products or xenobiotics also influences its concentration, as shown in various plant species (Dixon et al., 2002; Anderson and Davis 2004). This reaction is catalysed by GST, which is induced by abiotic stresses (Coleman et al., 1997). Conjugation takes place in the cytoplasm and the conjugates are transferred to the vacuole for further processing (Dixon et al., 2002).

Stress-induced changes in the GSH metabolism can be regulated at the transcriptional, translational and post-translational levels. The expression of the gene coding for ECS, the rate-limiting enzyme in GSH synthesis, was induced by chilling and ozone fumigation in maize and *Arabidopsis*, respectively (Gómez et al., 2004b; Sasaki-Sekimoto et al., 2005). Following various abiotic stress treatments the expression of *γECS* and *GR* genes increased in maize, and corresponding changes could also be detected at the level of GSH concentration and GR activity (Kellõs et al., 2008).

The post-transcriptional regulatory role of H_2O_2 was shown in maize subjected to various abiotic stresses, since the γECS and GR transcript levels either remained constant or decreased parallel to changes in the GSH synthesis, GSH/GSSG ratio and GR activity (Kellõs et al., 2008). The interaction between H_2O_2 and GSH in stress signalling was suggested in mungbean, where exogenous H_2O_2 increased both GSH levels and chilling tolerance (Yu et al., 2003).

The translational control of γECS activity was observed in *Arabidopsis* following H_2O_2 treatment and in maize subjected to chilling, since the amount of enzyme protein increased (Xiang and Bertrand 2000; Gómez et al., 2004b). The binding of the complex controlling the translation of γECS was induced by GSH and inhibited by GSSG under *in vitro* conditions (Xiang and Bertrand 2000). In pea higher GR activity was not accompanied by increased expression of the *GR* gene after heat stress (Kurganova et al., 1999; Escaler et al., 2000), suggesting the existence of translational or post-translational regulation.

Co-ordinated changes in the synthesis, degradation, transport and conjugation of GSH adjust its level and the GSH/GSSG ratio to stress conditions, allowing the effective participation of GSH in defence mechanisms.

Using HPLC separation and fluorescent detection, the importance of the high reduced/oxidised thiol ratio in the response to low temperature stress was shown in wheat, where higher GSH/GSSG and reduced/oxidized hydroxymethylglutathione (hmGSH/hmGSSG) ratios were found in freezing-tolerant genotypes compared to sensitive ones (Kocsy et al., 2001a). In addition, higher GR activity was detected in tolerant wheat and maize genotypes at suboptimal temperatures and in *Picea abies* during the winter indicating the involvement of GR in the maintenance of a high GSH/GSSG ratio in stressed plants (Esterbauer and Grill 1978; Leipner et al., 1999; Kocsy et al., 2001a). Besides the increase in total GR activity, the appearance of new, cold-specific GR-isoenzymes having high activity at low temperature is necessary for the efficient reduction of GR, as described for cold-hardened spruce (Hausladen and Alscher 1994). As in the case of cold-hardened wheat, the GSH/GSSG ratio at high temperatures correlated with heat tolerance in wheat and maize (Kocsy et al., 2004a, 2004c). The effective removal of H_2O_2 by GSH during heat stress is also facilitated by the increased GR activity observed in mustard seedlings (Dat et al., 1998).

For the accurate description of the involvement of GSH in stress responses it is necessary to determine the temporal and spatial changes in its concentration and in the GSH/GSSG ratio. The rapid initial decrease in the GSH and hmGSH contents during the first day of cold hardening coincided with a similar change in the GSH/GSSG and hmGSH/hmGSSG ratios in shoots and crowns of cold-hardened wheat (Kocsy G., unpublished results). Later on, during the first week of the 21-day treatment there was a transient increase in both the parameters of the two thiols in the shoots, while there was no significant change in their levels in the crowns. When investigating the spatial changes in GSH content in maize and wheat at low temperature, similar changes were found in shoots and roots (Kocsy et al., 2001a). Chilling increased γECS transcript and protein levels in the bundle sheath but not in the mesophyll cells, which could be the reason for the different GSH levels in the two cell types (Gómez et al., 2004b).

Studies on the subcellular localization of GSH it showed that GSH synthesis is possible in the chloroplasts and cytosol, and that the degradation of GSH and GS-conjugates occurs in the vacuoles and perhaps in the apoplast. Therefore, large GSH redox gradients may exist between the various subcellular compartments (Foyer et al., 2001). Thus, it is important to monitor the GSH and GSSG levels in the individual organelles (Schafer and Buettner 2001). The fluorescent dyes monochlorobimane and mercury orange were successfully used for the detection of GSH in living cells of *Allium cepa* (Müller et al., 1999). A specific antibody raised against GSH was appropriate for monitoring its intracellular distribution in various cell compartments (Müller et al., 2005). However, these methods did not allow the detection of GSSG or of changes in the GSH/GSSG ratio occurring during the response to abiotic stress. In addition, the fluorescent dyes may not be able to penetrate all the organelles. These problems could be solved using redox-sensitive green fluorescent protein, which is able to sense the redox potential of the cellular GSH buffer via Grx as a mediator (Meyer et al., 2007). The usefulness of this system was demonstrated in *Arabidopsis* roots, where the GSH/GSSG ratio was modified using exogenous H_2O_2 or dithiothreitol, and in the detection of wounding-induced redox changes in *Arabidopsis*. The complete oxidation of GSH, typical of the endoplasmatic reticulum, could also be detected via the transient expression of redox-sensitive green fluorescent protein in tobacco (Meyer et al., 2007).

The tracking of compartment-specific changes in the redox state of GSH is very important in relation to the redox regulation of the proteins. For example, γECS is localised exclusively in the cytosol (Pasternak et al., 2008) and is active in the oxidised state (Jez et al., 2004), while the active form of the non-expressor of pathogen-related genes (NPR1) is reduced and localized in the nucleus (Mou et al., 2003). Since both proteins are involved in the stress response, their simultaneous activation may be necessary. This would require different redox environments, which could be ensured by the localization of γECS and NPR1 in different cell compartments. If proteins differing in their redox activation mechanism need to be activated successively during the stress response, their induction could be ensured by changes in the redox state of the relevant compartment.

Extreme temperatures also influenced the ascorbic acid (AsA) concentration. After heat stress the DHA and AsA concentrations increased in *T. durum and T. aestivum, respectively* (Paolacci et al., 1997; Dash and Mohanty 2002). High temperature resulted in transient increase of AsA concentration and AsA /DHA ratio in mustard (Dat et al., 1998). The role of AsA in the adaptation to low temperature is shown by the fact that its concentration was much higher in the cold-tolerant *Soldanella* species growing in the high mountains compared to pea (Streb et al., 2003). In addition, the cold tolerance of cotton transformed with the APX gene also increased (Payton et al., 2001). In cereals APX activity was greater during cold hardening and there was a relationship between the frost tolerance and APX activity indicating the role of the enzyme in the frost tolerance (Janda et al., 2003).

3.2 Osmotic Stress

Besides extreme temperatures, the maintenance of a high GSH/GSSG ratio and GR activity also play an important role in salt, desiccation and drought tolerance as found in tomato, *Myrothamnus flabellifolia* and wheat, respectively (Shalata et al., 2001; Kranner et al., 2002; Kocsy et al., 2004b). A greater increase in GSH and hmGSH synthesis in tolerant wheat genotypes than in sensitive ones was observed not only during temperature stresses, but also under osmotic stress conditions (Kocsy et al., 2004b).

After NaCl treatment the γECS, GS and GST activity was much greater in a tolerant cotton cell line than in a sensitive one, indicating adaptation at the level of GSH synthesis, GSSG reduction and GS-conjugate formation (Gossett et al., 1996).

The glutathione content was increased not only by extreme temperatures but also by water deficit in sunflower seedlings (Sgherri and Navari-Izzo, 1995) and detached poplar leaves (Morabito and Guerrier 2000), and by salt treatment in groundnut cell lines (Jain et al., 2002). In addition, a comparison of various plant species revealed that salt tolerance was greater for those which had higher glutathione content (Tepe and Harms 1995). The accumulation of homoglutathione, was induced by drought in soybean (Fig. 2). This homologue of GSH may replace it in the control of the cellular redox state under stress conditions.

The increased level of GSH pool is generally regarded as a protective response against oxidative stress, although defence against stress situations sometimes occurs irrespective of the GSH concentration (Potters et al., 2004). It was also shown that the presence of NaCl in cell cultured medium may decrease the GSH content and the GR activity. This decreased level of GSH may be a reason for an increased activity of DHAR, which utilizes GSH as an electron donor in the reduction of DHA (Hoque et al., 2007). These results are also supported by the result of de Gara et al. (2000), who found that the conversion of DHA to AsA requires extensive utilization of GSH. In the AsA-GSH cycle, increased activity of DHAR leads to increased GSSG content in plant cells, which is regarded as oxidative stress. Drought increased the expression of the *GR*, *GST* and *GPx* genes,

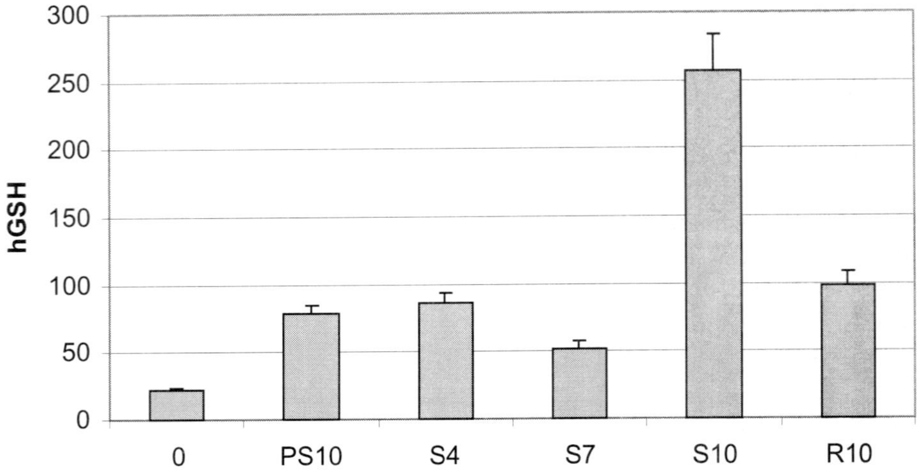

Fig. 2. Effect of drought on homoglutathione (hGSH) accumulation in soybean. The 6 weeks old plants were subjected to two subsequent 10 d long drought stress with one irrigation between them. Sampling was done before (0) and at the end of the first drought period (PS10), and after 4 (S4), 7 (S7) and 10 (S10) days during the second drought period and after 10 d recovery (R10). The youngest fully developed leaves were used for the biochemical analysis.

resulting in increases in the activities of the corresponding enzymes (Anderson and Davis 2004). Results obtained for pea show that the expression of the *GR* gene may be regulated by GSSG through its interaction with a possible GSSG binding site in the promoter region (Creissen et al., 1992).

Similarly to the low temperature, drought also resulted in an initial decrease in the GSH/GSSG ratio, but this change was followed by another decrease during the second half of the 23-day stress period (Tausz et al., 2004). After salt stress the GSH level and the GSH/GSSG ratio were lower after one week compared to the 3 d treatment in the maize inbred line Z7 (Kellõs et al., 2008). These results indicate that GSH may have an important role in ROS detoxification during the initial phase of various abiotic stresses. The decrease in the GSH/GSSG ratio could be due to the removal of ROS in the form of glutathione-conjugates (GS-conjugates) or due to GSH degradation. Following osmotic stress there was a similar increase in glutathione percentage in the shoots and roots in both drought-tolerant and sensitive wheat genotypes. However, the GSH/GSSG ratio was higher in the tolerant genotype under both control and stress conditions (Kocsy et al., 2004b). The GSH/GSSG ratio was similar in the shoots and roots under control and osmotic stress conditions. The changes in the size and redox state of the GSH pool at the cellular level could be even more important for the response to abiotic stress than the alteration of these parameters in various organs or tissues (Meyer 2008). Under control conditions different GSH levels were found in various cell types (Rennenberg et al., 2007), which may be the result of the compartmentation of the GSH metabolism (Kopriva and Koprivova 2005).

Besides GSH, AsA concentration and its redox state, is also important in the response to osmotic stress. In detached poplar leaves the amount of total ascorbic acid also increased parallel to the water loss compared to the control leaves (Morabito and Guerrier, 2000). Drought stress resulted an increased AsA content and AsA/DHA ratio in soybean leaves, which changes were affected by proline level (Kocsy et al., 2005). The role of AsA during drought stress was also demonstrated in *Myrothamnus flabellifolia,* in which it was oxidized during dehydration, and it was reduced again during rehydration (Kranner et al., 2002). Following water loss the activity of APX and DHAR was greater in the drought-tolerant *Triticum durum* cultivar compared to the sensitive one (Sgherri et al., 2000). These results indicate that a coordinated regulation of the AsA concentration and the enzymes affecting its redox state is necessary in order to reduce the damaging effect of water shortage.

Besides drought, high salt concentration also results in changes of the AsA concentration. After salt stress the amount of TAA and the activity of APX increased in *T. durum* protoplasts (Meneguzzo et al., 1998). Salt treatment also resulted in higher APX activity in radish while the level of the corresponding mRNA remained unchanged (Lopez et al., 1996). Following salt stress the APX activity increased in seedlings of the salt tolerant *Setaria italica* genotype while it decreased in the sensitive ones (Sreenivasulu et al., 2000). In cotton genotypes there was also a relationship between salt tolerance and APX activity (Gossett et al., 1996).

An increase in the activity of antioxidant enzymes, superoxide dismutase, peroxidase, catalase and polyphenol oxidase was also demonstrated in water-stressed maize plants. Foliar application of ascorbic acid reduced the activities of stress-induced enzymes, and also the proline and malondialdehyde levels in water-deficit stressed plants (Dolatabadian et al., 2009). These results show that ascorbic acid application helps the plants for better resistance under the stress by inactivation and scavenging of free radicals; thereupon it may be effective for the improvement of stressed plants in arid and semi-arid regions.

3.3 Environmental Pollutants, Flooding

Contrary to the other abiotic stresses, toxic concentrations of copper shifted the GSH/GSSG couple to a more oxidized state in *Silene cucubalus,* which could be the result of the use of GSH for phytochelatin synthesis (de Vos et al., 1992). This shift was prevented in cadmium-treated soybean roots due to the induction of GR (Ferreira et al., 2002).

The refining process in the petrochemical industry generates oil refinery sludges, a potentially contaminating waste product, with a high content of hydrocarbons and heavy metals. Organic contaminants can be stabilized within a soil matrix, taken up by plants and transformed or stored in a non-phytotoxic form. The presence of sludge in soil induces in plants an oxidative stress, which could be provoked by the direct toxicity of the petroleum sludge in the soil and/or a result of the contaminant properties that alter the physical and chemical properties of the soil, potentially affecting oxygen transfer, available water uptake, and nutrient mobility. An increase in the AsA/DHA ratio was shown in alfalfa plants grown in soils contaminated with sludge, due to a decrease in their DHA mutant, which agrees well with the enhancement of MDHAR, DHAR, and GR activities found in plants (Marti et al., 2009). It was also shown that total glutathione level was similar in both plants and that the GSH/GSSG ratio was not affected. On the other hand, together with GSH, some legume species contain a homolog thiol tripeptide, the homoglutathione (hGSH), and these compounds share an equivalent chemical structure, which points to similar functions in plants (Innocenti et al., 2007). The total concentration of hGSH, which is usually more abundant in alfalfa leaves than GSH was higher in plant leaves grown in presence of sludge showing that the synthesis of homoglutathione was induced in the presence of the contaminant (Marti et al., 2009) together with a non-significant change in the hGSH/hGSSG ratio, thus the plants grown in contaminated soils showed a lower ratio than plants grown in control soils. These results indicate that in spite of the activation of hGSH synthesis, the accumulation of hGSSG reflects its higher sensitivity to be oxidized under these conditions. The preferential oxidation of GSH, in comparison to AsA is in line with the GSH redox couple having a lower redox potential than the AsA (Foyer and Nocotr, 2000). The highly reduced glutathione pool maintained by GR is necessary for active protein function and avoids unspecific formation of mixed disulphide bonds that cause protein inactivation or aggregation. Although the hGSH level increased in alfalfa plants, a rapid ROS mediated oxidation of hGSH could take place in leaves during sludge growth conditions, so the oxidation observed in leaves proteins under sludge might be favoured by the low hGSH/hGSSG found in these plants. Results also show that the ascorbate-glutathione cycle plays an important role in the elimination of H_2O_2 in plants, which may favour plants grown on sludge not presenting a higher oxidative stress.

Flooding may also exert a detrimental effect through impairment of the photosynthetic machinery that can promote the overproduction of ROS. In addition, stress release may reinforce oxidative pressure due to the restoration of normal oxygen tension after ananoxic period (Yordanova et al., 2004; Lopez-Climent et al., 2008).

It was demonstrated that short-term limited oxygen and light availability did not cause irreversible modifications in barley leaf or root mitochondria functioning, but organ-specific modes of response and specific engagement of cell antioxidants were observed (Skutnik and Rychter 2009). In leaves, low-mass antioxidants, such as ascorbate and glutathione seem to play the major role in ROS defence. The extremely rapid oxidation of these compounds in barley leaves shows their engagement in detoxification of increasing ROS production by chloroplasts and mitochondria during post-anoxia. The light-dependency of leaf alternative oxidase (AOX) protein has been considered an evidence of an engagement of mitochondrial metabolism in the photosynthetic function of chloroplasts (Feng et al., 2007). In root cells, the capacity of ascorbate and glutathione is lower than in leaves. However, the root antioxidant system is supported by increased mitochondrial AOX capacity, which prevents elevated ROS production.

Within plant cells, SOD constitutes the first line of defence against ROS. It plays a crucial role in cellular defence against oxidative stress, as its activity directly modulates the amount of $O_2^{\cdot-}$ and H_2O_2, the two important Haber-Weiss reaction substrates. The excess $O_2^{\cdot-}$ generated under waterlogging stress could disproportionate into H_2O_2 by the action of SOD, which is then metabolized by the components of the ascorbate-glutathione cycle. The early activation in SOD and CAT enzymes observed in continuously flooded plants seems to be an active and efficient response of plants to cope with waterlogging stress. This was confirmed using tobacco plants over-expressing SOD isoforms, which could tolerate flooding better than the wild type plants (Yu and Rengel 1999). Continuously flooded plants also showed an increase in the activity of APX (Lin et al., 2004; Yordanova et al., 2004, Hossain et al., 2009). Unaltered or decreased MDHAR and DHAR activities during the entire anoxic and post-anoxic periods, together with the lack of positive correlation between DHAR activity and AsA/DHA ratio suggest an increase in de novo synthesis of AsA under waterlogged conditions. Alternatively, non-enzymatic disproportionation of MDHA might be responsible for such high levels of AsA accumulation (Hossain et al., 2009). Studies with waterlogged citrus plants show that elevated APX activity might be insufficient to scavenge all H_2O_2 produced in excess underwaterlogging stress, and the recycling activity for AsA did not seem to be essential in the maintenance of a balanced redox status for AsA. Isolated increases in AsA/DHA or GSH/GSSG ratios do not seem to prevent oxidative damage, and release from waterlogging conditions does not seem to improve plant performance; rather, probably due to a sudden oxygen burst soon after release of water, it may enhance the incidence of oxidative damage (Hoaasin et al., 2009).

Treatment of green gram (*Vigna radiata* L.) plants with chromium may also cause antioxidative responses, such as temporary increase in the content of hydrogen peroxide and superoxide dismutase activity (Karuppanapandian et al., 2006). While catalase activities and the contents of reduced AsA and DHA were significantly decreased in Cr-treated plants, ascorbate peroxidase and guaiacol peroxidase activities showed a later increase suggesting that increased enzyme activities would be responsible for the removal of H_2O_2. The reduced glutathione content decreased at early stages of Cr-treatment; However, it was restored to the normal level as in controls thereafter. In contrast, the glutathione disulphide content showed a progressive increase during the initial hours of Cr-treatment. The non-protein thiol content was shown to increase during the first several hours, but it declines at later stages. These results demonstrate that Cr-induced oxidative stress is an important component of the plant's reaction to toxic levels of Cr (Karuppanapandian et al., 2006).

The involvement of GSH in the response to heavy metal stress was also demonstrated, since the glutathione content was reduced by cadmium in bread wheat (Lin et al., 2007) and by copper in *Silene cucubalus* due to increased phytochelatin synthesis (de Vos et al., 1992), which detoxifies heavy metals by forming complexes. Correspondingly, in the *Arabidopsis* mutants *rax1-1* and *cad2-1* (cadmium-sensitive) (mutations in the γECS gene resulting in decreased GSH content) it was shown that GSH affects the expression of several genes involved in protection against environmental stresses (Ball et al., 2004). Exposure of durum wheat plants to Cd led to an increase in the activity enzymes of the AsA-GSH cycle in the leaves without any symptom of the oxidative damage, suggesting that the whole plant improved its antioxidant defence, even in those parts which had not yet been reached by Cd. (Paradiso et al., 2008). When ascorbate biosynthesis was enhanced, by feeding plants with its last precursor, L-galactono-g-lactone, the oxidative stress induced in the roots by the heavy metal was alleviated. Results suggest that different strategies can successfully cope with heavy metal toxicity. This precocious increase in the enzymes of the ascorbate-glutathione cycle further highlight the tight regulation and the relevance of this cycle in the defence against heavy metals, and highlights the interest in obtaining plants in which the ascorbate biosynthetic route has been enhanced, both for improving ROS scavenger systems and for removing heavy metals from metabolic reactions (Paradiso et al., 2008).

3.3.1 Phytochelatins

Following cadmium stress the enhanced synthesis of GSH and phytochelatins (PCs) was observed in various plant species (Mendoza-Cózati et al., 2005). Phytochelatins were first isolated by Grill et al. (1985) from a cell suspension culture of *Rauvolfia serpentina*. They have the structure [(γ-Glu - Cys)$_n$ - Gly], where n is the number of replications of the (γ-Glu - Cys) units, which is generally in the range 2–11. The primary structure of the cadmium phytochelatin complex is $[Cd_3(PC_3)_4]$, which contains discrete $Cd(SCys)_4$ units (Strasdeit et al., 1991). The peptide SH : Cd mol ratio is 3.78, declining to 1.01 as the complex becomes saturated with cadmium (Rauser and Meuwly 1995). This latter ratio was also recorded in maize roots, indicating that the PC formed in the roots is capable of maximum Cd binding with a very small quantity of SH, while in maize leaves this ratio was 3.41–4.73, suggesting the over-production of phytochelatins (Drazkiewicz et al., 2003). The enzyme responsible for their synthesis is β-glutamyl cysteine dipeptidyl transpeptidase (phytochelatin synthase: PCS), the substrate of which is glutathione (Grill et al., 1989). The enzyme is also expressed without heavy metal stress, but is primarily activated by the presence of heavy metals (Cobbett, 2000).

Even a very low (0.05 μM) concentration of cadmium stimulated the synthesis of PCs in maize roots after a period of 24 hours; in the presence of 3 μM Cd phytochelatins (PC_2–PC_4) appear after 2 h, while after 6 h PC_2 is largely replaced by PC_3 (Tukendorf and Rauser 1990). The quantities induced are closely correlated with the cadmium concentration and the treatment duration, but also depend on the age of the leaf (Wójcik and Tukendorf 1999; Drazkiewicz et al., 2003). Low molecular weight phytochelatins were dominant in maize root tips one day after treatment with 3 μM Cd, while in the older parts of the root high molecular weight phytochelatins were dominant after two days (Rauser 2003). Cadmium stress reduced the accumulation of glutathione but increase the amount of PC in *Arabidopsis* seedlings. Supplementation with salt increased the glutathione and PC levels under Cd stress markedly. However, no significant difference was detected in the accumulation of transcripts of the phytochelatin synthase gene, AtPCS1 regardless of the salt/Cd treatment used (Xu et al., 2010).

Little is known about the expression of phytochelatin synthase or the tissue-specificity of phytochelatin biosynthesis. In tomato plants phytochelatin synthase activity was recorded in the roots and stems, but not in the leaves and fruit (Chen et al., 1997). The PCS activity measured in maize under *in vitro* conditions increased in response to cadmium treatment in the leaves, but declined in the roots (Szalai et al., 2002; Pál et al., 2005). Since the biosynthesis of PCs is self-regulated (Loeffler et al., 1989), it can be assumed that a sufficient quantity of PCs was synthesized in maize roots to bind the cadmium, so there were no heavy metal ions to activate the enzyme (Pál et al., 2005). It is possible, however, that the synthesis of PCs is differently regulated under *in vivo* conditions.

Toxic metals (such as As or Cd) cause oxidative stress, and can deplete reduced glutathione, through the formation of As(III)-glutathione or Cd–glutathione complexes [As(III)-GS3 or Cd(II)-GS2] and As(III)-induced or Cd-induced PC synthesis. A major strategy to detoxify non-essential trace metal(-loid)s is the synthesis of specific low-molecular-weight chelators to avoid binding to physiologically important proteins and to facilitate their transport into the vacuoles. The favoured ligands of metal ions are thiols, present in glutathione and PC. GSH can bind to several metals and metalloids and is also a key metabolite in cellular redox balance, which is usually one of the targets of metal(loid) toxicity. Increasing GSH synthesis is considered a means of increasing metal(loid) binding capacity as well as a way to increase cellular defence against oxidative stress. Since glutathione is the precursor of PC, overexpression of γECS or GS usually also leads to higher rates of PC accumulation under metal exposure (Verbruggen et al., 2008). PCS is constitutively expressed, but requires post-translational activation by metal(loid)s, of which As and Cd belong to the most effective activators. The formation of As–GS3 for Cd–GS2 thiolates, which act as high-affinity substrates for the enzyme, seems to be sufficient for its activation (Vatamaniuk et al., 2000). PC synthesis seems to be the main factor for basal Cd and As tolerance (Cobbett and Goldsbrough 2002), but not in hypertolerant plants or hyperaccumulators (Ernst et al., 2008).

The constitutive presence of phytochelatin synthase is a further confirmation of the much more general role of phytochelatins. Phytochelatin synthase was found to play a role in the plant-specific degradation of GS-conjugates (Blum et al., 2007). It can be assumed that phytochelatins not only represent a cadmium detoxification system, but, at low concentrations, are also the key molecules in cellular homeostasis and are required for the transportation of metals to the relevant apoenzymes (Thumann et al., 1991). An other possible role of PCS that its peptidase activity might be involved in plant innate immunity (Clemens 2009). According to a hypothesis Cd and As are essential elements and the phytochelatin pathway is part of the homeostatic network for these elements (Clemens 2006). This hypothesis is so far not supported by experimental data for higher plants, but in the diatom *Thalassiosira weissflogii* a carboanhydrase was described that is expressed under conditions of low CO_2 and Zn deficiency. This enzyme uses Cd as cofactor in place of Zn (Xu et al., 2008).

4. CROSSTALKING OF GLUTATHIONE-MEDIATED STRESS RESPONSES WITH OTHER SIGNALLING PATHWAYS

4.1 Participation of Glutathione in Redox Signalling

The antioxidative system, including the AsA-GSH cycle may integrate signals originating from metabolic pathways and the environment, thus regulating the induction of adaptive or death processes (Foyer and Noctor 2005). Several regulatory and structural genes controlled by the thiol-disulphide status and ROS signalling have been identified in mutant and transgenic *Arabidopsis* and in wild-type plants treated with dithiothreitol or ROS-generating agents using transcript profiling, which could clarify the function of the redox network (Gadjev et al., 2006; Kolbe et al., 2006). This network controls the level of ROS by integrating signals from different cell compartments during abiotic stress, and the GSH/GSSG couple participates in its fine tuning (Meyer 2008).

In general, the protective and regulatory roles of GSH are based on changes in its redox state, which is defined by the reducing capacity of GSH (GSH concentration) and the half-cell reduction potential of the GSH/GSSG couple (Schafer and Buettner 2001). It differs in various organs, tissues, cells and compartments and also changes during the growth and development of the plants. While glutathione reductase (GR) uses NADPH to reduce GSSG to GSH, various free radicals and oxidants are able to oxidize GSH to GSSG. The proportion of glutathione in the reduced form reflects the relative rates of reduction and oxidation and is always greater than 0.9 under non-stress conditions. Since the concentration of GSH in the chloroplast stroma is thought to be close to 5 mM, the reduced form of glutathione may act as an important redox buffer, preventing enzyme inactivation by protecting potentially susceptible protein thiol groups (Noctor et al., 1998).

The stress-induced changes in GSH level and the GSH/GSSG ratio may derive from a higher rate of GSH synthesis. Abiotic stress-induced changes in the redox state of GSH precursors may also influence the synthesis and redox state of GSH, since there were similar differences in the ratios of GSH/GSSG, hmGSH/hmGSSG and their reduced/oxidized precursors between tolerant and sensitive wheat varieties subjected to osmotic or heat stress (Kocsy et al., 2004b; 2004c). Cystine reductase was described in pea (Romano and Nickerson 1954), but stress-induced changes in its activity were not studied. The accumulation of γ-glutamylcystine (ESSE), the oxidised form of the other GSH precursor in tobacco overexpressing γECS (Creissen et al., 1999), indicates that no enzyme exists for its efficient reduction in plants; therefore, it can only be removed by degradation or sequestration to the vacuole. The involvement of GSH in redox signalling is confirmed by the observation that inter- and intracellular GSH pools are linked by transport across the membranes, the rate of which could be similar to that of synthesis, as is the case for the chloroplast envelope (Noctor et al., 2002). The GSH transport in plants may also be regulated by antioxidants, since the promoter of the gene coding for a GSH transporter in mammals contains a functional antioxidant-responsive element (Wasserman and Fahl 1997).

The GSH/GSSG couple is able to modify the activity of various compounds (enzymes, regulatory proteins, etc.) directly through the reduction/oxidation of their disulphide bridges/sulfhydryl groups and through the (de)glutathionylation of sulfhydryl groups. The indirect regulation of proteins by the GSH/GSSG couple may occur due to crosstalk between GSH/GSSG and other redox systems through glutathionylation or thiol-disulphide transition, which may have a role in signalling and responses to abiotic stress (Rausch et al., 2007; Ying et al., 2007).

4.2 Interaction of Glutathione with Hormonal Systems

Both acorbate and the GSH/GSSG couple may interact with certain signalling pathways during the stress response. The effect of abiotic stresses on the H_2O_2, GSH and GSSG concentrations may be transmitted by various plant hormones. For example, the general regulatory role of NO, an important regulatory molecule, in stressed plants was demonstrated in several studies (for review, see Arasimowicz and Floriszak-Wieczorek 2007). NO affected H_2O_2 concentration due to the inhibition of Cat and ascorbate peroxidase (Clark et al., 2000) while exogenous H_2O_2 activated NO synthesis in tobacco (de Pinto et al., 2006), suggesting a bidirectional interaction between the two compounds. NO may also influence the GSH synthesis, as demonstrated in *Medicago truncula*ta roots, where the GSH level and gECS and GS gene expressions were increased by NO (Innocenti et al., 2007).

Components of the AsA-GSH cycle may interact with Ca^{2+}. H_2O_2 treatment alone or combined with low temperature increased the Ca^{2+} concentration in tobacco (Price et al., 1994), which could have a role in the Ca^{2+}-dependent regulation of the enzymes. In maize the interaction of Ca^{2+} and ROS was observed during the induction of the antioxidant system by ABA, and it was concluded that Ca^{2+} can be found both before and after ROS in the signalling pathway related to oxidative stress (Jiang and Zhang 2003). Yang and Poovaiah (2002) postulated a dual role for Ca^{2+} in the regulation of H_2O_2 homeostasis: (a) during positive regulation H_2O_2 will be produced due to the activation of NADPH oxidase; (b) during negative regulation the H_2O_2 concentration will decrease due to the activation of Cat. Interestingly, Ca^{2+} enhanced both the GSH concentration and the stress tolerance in rice (Lu et al., 1999). In tobacco, however, GSH and GSSG treatment resulted in a rapid, transient increase in the Ca^{2+} level, suggesting that GSH may be involved in the activation of Ca^{2+}-dependent protein kinases and in the early part of stress-induced signalling pathways (Gómez et al., 2004a). In plants, Ca^{2+} may interact not only with H_2O_2 but with other ROS, too. This assumption is based on observations on trout hepatoma cells, where the mobilization of Ca^{2+} was induced by ·OH (Burlando and Viarengo 2005). A review about interactions between ROS, Ca^{2+} and antioxidants were published recently (Noctor 2006).

Salicylic acid (SA) has long been known as a signal molecule in the induction of defence mechanisms in plants (Raskin 1992; Horváth et al., 2007). Although SA may also cause oxidative stress to plants, partially through the accumulation of hydrogen peroxide, the results published so far show that the preliminary treatment of plants with low concentrations of SA might have an acclimation like effect, causing enhanced tolerance toward most kinds of abiotic stresses due primarily to enhanced antioxidative capacity. The effect of exogenous SA depends on numerous factors such as the species and developmental stage of the plant, the mode of application, and the concentration of SA and its endogenous level in the given plant. SA increased the chilling tolerance of maize by inhibiting catalase, thus increasing the H_2O_2 concentration, which led to an increase in several protective mechanisms, including increase in the GR activity (Janda et al., 1999; Horváth et al., 2002). As also observed in chilled maize, SA stimulated the formation of ROS in *Arabidopsis* subjected to salt or osmotic stress (Borsani et al., 2001). GSH and GR were also affected by SA in a soybean cell suspension (Knörzer et al., 1999) and SA increased the GR activity in rice leaves (Ganesan and Thomas 2001). Pretreatment of barley seedlings with SA in the dark via the transpiration stream fully blocked the subsequent light-induced inhibitory effect of paraquat on photosynthesis and also decreased the paraquat-induced production of H_2O_2, lipid peroxidation, and electrolyte leakage (Ananieva et al., 2002), while inducing the activities of certain antioxidant enzymes (Ananieva et al., 2004). These results were also confirmed using SA-deficient NahG transgenic rice plants, which showed great sensitivity to paraquat. These plants also had lower glutathione content, which may explain the greater sensitivity of these plants to oxidative damage and also suggests that SA may regulate glutathione synthesis in rice (Kusumi et al., 2006).

Various stress hormones and abiotic stresses increased the GSH/GSSG ratio in maize leaves after one week, which shows the importance of the redox changes during the adaptation to the adverse environmental effects (Fig. 3).

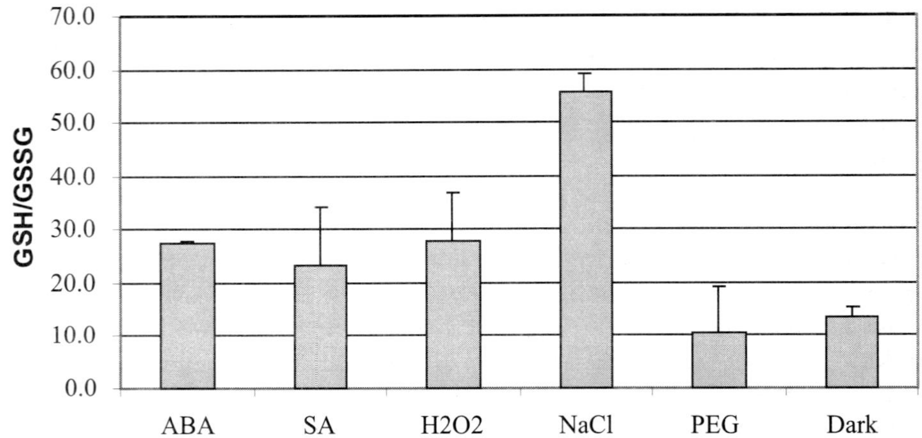

Fig. 3. Effect of various treatments on the relative GSH/GSSG ratio in maize. The GSH/GSSG ratios in treated plants are compared to the control plants the value in which (52.1) was taken as one. Eleven days old seedlings grown in hydroponics were subjected to the following treatments for one week: ABA – 0.1 mM abscisic acid, SA – 0.5 mM salicylic acid, 1 mM H_2O_2; 200 mM NaCl, PEG – 15% polyethylene glycol 4000, dark – without illumination. The youngest fully developed leaves were used for the biochemical analysis.

SA induced various alterations in the γEC and GSH contents, GR activity, and γECS and GR transcript levels in two maize genotypes with different levels of stress tolerance (Kellõs et al., 2008); the γEC and GSH concentrations were increased in both genotypes by SA. SA induces defence gene activation by NPR1, the cellular localization of which is regulated in turn by GSH; reduced NPR1 is transported to the nucleus, where it regulates gene expression (Mou et al., 2003). SA-induced NO production was found in *Arabidopsis,* and Ca^{2+} accumulation was a component of the signalling cascade (Zottini et al., 2007). Another direct evidence for the connection between SA and GSH signalling is that application of exogenous SA or H_2O_2 resulted in a greater reduced thiol content in cold tolerant Z7 than in cold sensitive Penjalinan maize inbred line. SA also increased the levels of precursors of GSH in young maize plants (Kellõs et al., 2008).

ABA treatment resulted in changes in ROS concentration in *Arabidopsis*, activating the Ca^{2+} channels of the cell membranes and increasing the Ca^{2+} concentration in the cytosol (Murata et al., 2001). The connection between the redox state of the cells and H_2O_2 and ABA was shown in the *Arabidopsis* mutant *glutathione peroxidase3* (*ATGPX3*), where the addition of oxidized ATGPX3 protein *in vitro* converted the protein phosphatase described in *ABA insensitive2* (*ABI2*) mutants to its oxidized form. *ABI2*, in turn, influences Ca^{2+} channels and stomatal closure (Miao et al., 2006). In addition, ABA influenced the GR activity in the cytosol of rice (Kaminaka et al., 1998). In two maize genotypes differing in their stress tolerance, ABA differentially affected the GSH content, GSH/GSSG ratio, GR activity and γECS transcript level (Kellõs et al., 2008). In *Vigna unguiculata* not only the GR activity, but also the expression of the corresponding gene was increased by ABA (Contour-Ansel et al., 2006). In summary, ABA was shown to affect the GSH/GSSG ratio and redox signalling in several plant species (Pastori and Foyer 2002).

Recent studies with maize plants differing in their stress tolerance indicate that SA and ABA may have different effects on GSH synthesis (Kellõs et al., 2008). Based on similarities in GSH synthesis following hormone and abiotic stress treatment, the possibility of a regulatory role for ABA during continuous darkness and salt stress, and for SA during H_2O_2 treatment and osmotic stress can be raised. These treatment- and genotype-specific differences in synthesis and reduction of GSH may be regulated at the post-transcriptional level.

Like SA and ABA, jasmonic acid (JA) also regulated gene expression through H_2O_2, as found in tobacco (Mur et al., 2006). In addition, JA influenced the GSH concentration and the genes involved in the GSH metabolism in *Arabidopsis* (Xiang and Oliver 1998; Sasaki-Sekimoto et al., 2005). Similarly to SA, ethylene and NO, JA also increased the transcript level of *GST* suggesting that the various plant growth regulators interact (Moons 2005).

The order of the components in the signalling pathway described above may vary, and some of them may be absent or additional ones may be present depending on environmental effects, plant species, organs and cell types. Multidirectional forward and backward interactions responsible for the regulation of metabolic pathways may exist between these compounds in order to ensure the most effective protection against environmental stress (Agarwal et al., 2005; Foyer and Noctor 2005; Noctor 2006; Dietz 2008; Miller et al., 2008).

CONCLUSION

In spite of the fact that the physiological role of the two main components of the AsA-GSH cycle, AsA and glutathione, have been extensively studied in several plant species, several questions remain to be clarified, for example the role of subcellular changes in the redox state of the redox pairs in stressed plants, and their interactions with other signalling molecules during the stress response (Szalai et al., 2009). Sequence comparisons of the related genes and proteins and the construction of phylogenetic trees could help to identify the evolutional events. Full knowledge of the genes involved in the L-galactose pathway of ascorbate synthesis in plants should facilitate a better understanding of the regulatory control of ascorbate accumulation and provide opportunities for increasing the vitamin C content of food crops. The majority of data on mechanisms related to these compounds were obtained using model plants. These results may provide a good background for strategies aimed at manipulating plants for elevated vitamin C and or altered glutathione levels in order to develop crops capable of tolerating environmental changes with as little damage as possible.

ACKNOWLEDGEMENTS

This work was supported by the Hungarian Scientific Research Fund and the National Office for Research and Technology (NKTH-OTKA K67906 and K68158) and the EU-FP7 Agrisafe Project No. 203288.

REFERENCES

Agarwal, S., Sairam, R.K., Srivastava, G.C., Tyagi, A. and Meena, R.C. (2005). Role of ABA, salicylic acid, calcium and hydrogen peroxide on antioxidant enzymes induction in wheat seedlings. *Plant Sci.* **169:** 559-570.

Ananieva, E.A., Alexieva, V.S. and Popova, L.P. (2002). Treatment with salicylic acid decreases the effects of paraquat on photosynthesis. *J. Plant Physiol.* **159:** 685-693.

Ananieva, E.A., Christov, K.N. and Popova, L.P. (2004). Exogenous treatment with salicylic acid leads to increased antioxidant capacity in leaves of barley plants exposed to paraquat. *J. Plant Physiol.* **161:** 319-328.

Anderson, J.V., Chevone, B.I. and Hess, J.L. (1992). Seasonal variation in the antioxidant system of eastern white pine needles. *Plant Physiol.* **98:** 501-508.

Anderson, J.V. and Davis, D.G. (2004). Abiotic stress alters transcript profiles and activity of glutathione S–transferase, glutathione peroxidase, and glutathione reductase in Euphorbia esula. *Physiol. Plant.* **120:** 421-433.

Apel, K. and Hirt, H. (2004). Reactive oxygen species: metabolism, oxidative stress, and signal transduction. *Ann. Rev. Plant Biol.* **55:** 373-379.

Arasimowicz, M. and Floryszak-Wieczorek, J. (2007). Nitric oxide as a bioactive signalling molecule in plant stress responses. *Plant Sci.* **172:** 876-887.

Aslund, F. and Beckwith, J. (1999). Bridge over troubled waters: sensing stress by disulphide bond formation. *Cell.* **96:** 751-753.

Ball, L., Accotto, G-P., Bechtold, U., Creissen, G., Funck, D., Jimenez, A., Kular, B., Leyland, N., Mejia-Carranza, J., Reynolds, H., Karpinski, S. and Mullineaux, P.M. (2004). Evidence for a direct link between glutathione synthesis and stress defence gene expression in *Arabidopsis. Plant Cell.* **16:** 2448-2462.

Barth, C., Moeder, W., Klessig, D.F. and Conklin, P.L. (2004). The timing of senescence and response to pathogens is altered in the ascorbate-deficient Arabidopsis mutant *vitamin c-1. Plant Physiol.* **134:** 1784-1792.

Blum, R., Beck, A., Korte, A., Stengel, A., Letzel, T., Lendzian, K. and Grill, E. (2007). Function of phytochelatin synthase in catabolism of glutathione-conjugates. *Plant J.* **49:** 740-749.

Borsani, O., Valpuesta, V. and Botella, M.A. (2001). Evidence for a role of salicylic acid in the oxidative damage generated by NaCl and osmotic stress in *Arabidopsis* seedlings. *Plant Physiol.* **126:** 1024-1030.

Burlando, B. and Viarengo, A. (2005). Ca^{2+} is mobilized by hydroxyl radical but not by superoxide in RTH-149 cells: The oxidative switching-on of Ca^{2+} signalling. *Cell Calcium.* **38:** 507-513.

Cairns, N.G., Pasternak, M., Wachter, A., Cobbett, C.S. and Meyer, A.J. (2006). Maturation of *Arabidopsis* seeds is dependent on glutathione biosynthesis within the embryo. *Plant Physiol.* **141:** 446-455.

Chen, J., Zhou, J. and Goldsbrough, P.B. (1997). Characterization of phytochelatin synthase from tomato. *Physiol. Plant.* **101:** 165-172.

Chew, O., Whelan, J. and Miller, A.H. (2003). Molecular definition of the ascorbate-glutathione cycle in Arabidopsis mitochondria reveals dual targeting of antioxidant defences in plants. *J. Biol. Chem.* **278:** 46869-46877.

Clark, D., Durner, J., Navarre, D.A. and Klessig, D.F. (2000). Nitric oxide inhibition of tobacco catalase and ascorbate peroxidase. *Mol. Plant-Microbe Interact.* **13:** 1350-1384.

Clemens, S. (2006). Evolution and function of phytochelatin synthases. *J. Plant Physiol.* **163:** 319-332.

Cobbet, C.S. (2000). Phytochelatins and their roles in heavy metal detoxification. *Plant Physiol.* **123:** 825-832.

Cobbett, C.S. and Goldsbrough, P. (2002). Phytochelatins and metallothioneins: roles in heavy metal detoxification and homeostasis. *Annu. Rev. Plant Physiol. Plant Mol. Biol.* **53:** 159-182.

Coleman, J.O.D., Randall, R. and Blake-Kalff, M.M.A. (1997). Detoxification of xenobiotics in plant cells by glutathione conjugation and vacuolar compartmentalization: a fluorescent assay using monochlorobimane. *Plant Cell Environ.* **20:** 449-460.

Conklin, P.L., Saracco, S.A., Norris, S.R. and Last, R.L. (2000). Identification of vitamin C-deficient Arabidopsis thaliana mutants. *Genetics.* **154:** 847-856.

Conklin, P.L., Norris, S.R., G.L., Wheeler, E.H., Smirnoff, W.N. and Last, R.L. (1999). Genetic evidence for the role of GDP-mannose in plant ascorbic acid (vitamin C) biosynthesis. *Proc. Natl. Acad. Sci. USA.* **96:** 4198-4203.

Contour-Ansel, D., Torres-Franklin, M.L., de Carvalho, M.H.C. and D'Arcy-Lameta, A. (2006). Glutathione reductase in leaves of cowpea: Cloning of two cDNAs, expression and enzymatic activity under progressive drought stress, desiccation and abscisic acid treatment. *Annals Bot.* **98:** 1279-1287.

Creissen, G., Firmin, J., Fryer, M., Kular, B., Leykand, N., Reynolds, H., Pastori, G., Wellburn, F., Baker, N., Wellburn, A. and Mullineaux, P. (1999). Elevated glutathione biosynthetic capacity in the chloroplasts of transgenic tobacco plants paradoxically causes increased oxidative stress. *Plant Cell.* **11:** 1277-1291.

Creissen, G.P., Edwards, E.A., Enard, C., Wellburn, A.R. and Mullineaux, P.M. (1992). Molecular characterization of glutathione reductase cDNAs from pea (*Pisum sativum* L.). *Plant J.* **2:** 129-131.

Dash, S. and Mohanty, N. (2002). Response of seedlings to heat-stress in cultivars of wheat: growth temperature-dependent differential modulation of photosystem 1 and 2 activity, and foliar anitoxidant defense capacity. *J. Plant Physiol.* **159:** 49-59.

Dat, J.F., Foyer, C.H. and Scott, I.M. (1998). Changes in salicylic acid and antioxidants during induced thermotolerance in mustard seedlings. *Plant Physiol.* **118:** 1455-1461.

de Gara, L., Paciolla, C., De Tullio, M.C., Motto, M. and Arrigioni, O. (2000). Ascorbate-dependent hydrogen peroxide detoxification and ascorbate regeneration during germination of a highly productive maize hybrid: evidence of an improved detoxification mechanism against reactive oxygen species. *Physiol. Plant.* **109:** 7-13.

de Pinto, M.C., Paradiso, A., Leonetti, P. and de Gara, L. (2006). Hydrogen peroxide, nitric oxide and cytosolic ascorbate peroxidase at the crossroad between defence and cell death. *Plant J.* **48:** 784-795.

de Vos, C.H.R., Vonk, M.J., Vooijs, R. and Schat, H. (1992). Glutathione depletion due to copper-induced phytochelatin synthesis causes oxidative stress in *Silene cucubalus. Plant Physiol.* **98:** 853-858.

Dietz, K.J. (2008). Redox signal integration: from stimulus to networks and genes. *Physiol. Plant.* **133:** 459-468.

Dixon, D.P., Lapthorn, A. and Edwards, R. (2002). Plant glutathione transferases. *Genome Biol.* **3:** 3004.1-3004.10.

Dolatabadian, A., Sanavy, S.A.M.M. and Sharifi, M. (2009). Alleviation of water deficit stress effects by foliar application of ascorbic acid on *Zea mays* L. *J. Agron. Crop Sci.* **195:** 347-355.

Dra¿kiewicz, A., Tukendorf, A. and Baszyñski, T. (2003). Age-dependent response of maize leaf segments to cadmium treatment: Effect on chlorophyll fluorescence and phytochelatin accumulation. *J. Plant Physiol.* **160:** 247-254.

Ernst, W.H., Krauss, G.J., Verkleij, J.A. and Wesenberg, D. (2008). Interaction of heavy metals with the sulphur metabolism in angiosperms from an ecological point of view. *Plant Cell Environ.* **31:** 123-143.

Escaler, M., Aranda, M.A., Roberts, I.M., Thomas, C.L. and Maule, A.J. (2000). A comparison between virus replication and abiotic stress (heat) as modifiers of host gene expression in pea. *Mol. Plant Pathol.* **1:** 159-167.

Esterbauer, H. and Grill, D. (1978). Seasonal variation of glutathione and glutathione reductase in needles of *Picea abies*. *Plant Physiol.* **61:** 119-121.

Feng, H., Li, H., Li, X., Duan, J., Liang, H., Zhi, D. and Ma, J. (2007). The flexible interrelation between AOX respiratory pathway and photosynthesis in rice leaves. *Plant Phys. Biochem.* **45:** 228-235.

Ferreira, R.R., Fornazier, R.F., Vitória, A.P., Lea, P.J. and Azevedo, R.A. (2002). Changes in antioxidant enzyme activities in soybean under cadmium stress. *J. Plant Nutr.* **25:** 327-342.

Foyer, C.H., Lopez-Delgado, H., Dat, J.F. and Scott, I.M. (1997). Hydrogen peroxide and glutathione-associated mechanisms of acclimatory stress tolerance and signalling. *Physiol. Plant.* **100:** 241-254.

Foyer, C.H. and Noctor, G. (2000). Oxygen processing in photosynthesis: regulation and signalling. *New Phytol.* **146:** 359-388.

Foyer, C.H. and Noctor, G. (2005). Redox homeostasis and antioxidant signalling: a metabolic link between stress perception and physiological responses. *Plant Cell.* **17:** 1866-1875.

Foyer, C.H., Theodoulou, F.L. and Delrot, S. (2001). The functions of inter- and intracellular glutathione transport systems in plants. *Trends Plant Sci.* **6:** 486-492.

Foyer, C.H. and Noctor, G. (2009). Redox regulation in photosynthetic organisms: signaling, acclimation, and practical implications. *Antiox. Redox Signal.* **11:** 861-905.

Gadjev, I., Vanderauwera, S., Gechev, T.S., Laloi, C., Minkov, I.N., Shulaev, V., Apel, K., Inzé, D., Mittler, R. and van Breusegem, F. (2006). Transcriptomic footprints disclose specificty of recative oxygen species signalling in *Arabidopsis*. *Plant Physiol.* **141:** 436-445.

Ganesan, V. and Thomas, G. (2001). Salicylic acid response in rice: influence of salicylic acid on H_2O_2 accumulation and oxidative stress. *Plant Sci.* **160:** 1095-1106.

Giacomelli, L., Rudella, A. and van Wijk, K.J. (2006). High Light Response of the Thylakoid Proteome in Arabidopsis Wild Type and the Ascorbate-Deficient Mutant vtc2-2. A Comparative Proteomics Study. *Plant Physiol.* **141:** 685-701.

Giovannoni, J.J. (2007). Completing a pathway to plant vitamin C synthesis. *PNAS.* **104:** 9109-9110.

Gómez, L.D., Noctor, G., Knight, M.R. and Foyer, C.H. (2004a). Regulation of calcium signalling and gene expression by glutathione. *J. Exp. Bot.* **55:** 1851-1859.

Gómez, L.D., Vanacker, H., Buchner, P., Noctor, G. and Foyer, C.H. (2004b). Intercellular distribution of glutathione synthesis in maize leaves and its response to short-term chilling. *Plant Physiol.* **134:** 1662-1671.

Gossett, D.R., Banks, S.W., Millhollon, E.P. and Lucas, C. (1996). Antioxidant response to NaCl stress in a control and an NaCl-tolerant cotton cell line grown in the presence of paraquat, buthionine sulfoximine, and exogenous glutathione. *Plant Physiol.* **112:** 803-809.

Grill, E., Loffler, S., Winnacker, E.L. and Zenk, M.H. (1989). Phytochelatins, the heavy-metal-binding peptides of plants, are synthesized from glutathione by a specific γ-glutamylcysteine dipeptidyl transpeptidase (phytochelatin synthase). *Proc. Natl. Acad. Sci. USA.* **86:** 6838-6842.

Grill, E., Winnacker, E.L. and Zenk, M.H. (1985). Phytochelatins: the principal heavy-metal complexing peptides of higher plants. *Science.* **230:** 674-676.

Guo, Z., Ou, W., Lu, S. and Zhong, Q. (2006). Differential responses of antioxidative system to chilling and drought in four rice cultivars differeing in sensitivity. *Plant Physiol. Biochem.* **44:** 828-836.

Hausladen, A. and Alscher, R.G. (1994). Cold-hardiness-specific glutathione reductase isozymes in red spruce. *Plant Physiol.* **105:** 215-223.

Herschbach, C. and Rennenberg, H. (2001). Significance of phloem-translocated organic sulphur compounds for the regulation of sulphur nutrition. *Progr. Bot.* **62:** 177-193.

Hoque, A., Banu, N.A., Okuma, E., Amako, K., Nakamura, Y., Shimoishi, Y. and Murata, Y. (2007). Exogenous proline and glycinebetaine increase NaCl-induced ascorbate–glutathione cycle enzyme activities, and proline improves salt tolerance more than glycinebetaine in tobacco Bright Yellow-2 suspension-cultured cells. *J. Plant Physiol.* **164:** 1457-1468.

Horváth, E., Janda, T., Szalai, G. and Páldi, E. (2002). *In vitro* salicylic acid inhibition of catalase activity in maize: differences between the isozymes and a possible role in the induction of chilling tolerance. *Plant Sci.* **163:** 1129-1135.

Horváth, E., Szalai, G. and Janda, T. (2007). Induction of abiotic stress tolerance by salicylic acid signalling. *J. Plant Growth Regul.* **26:** 290-300.

Hossain, Z., Lopez-Climent, M.F., Arbona, V., Perez-Clemente, R.M. and Gomez-Cadenas, A. (2009). Modulation of the antioxidant system in citrus under waterlogging and subsequent drainage. *J. Plant Physiol.* **166:** 1391-1404.

Innocenti, G., Pucciariello, C., Le Gleuher, M., Hopkins, J., Stefano, M., Delledonne, M. Puppo, A., Baudouin, E. and Frendo, P. (2007). Glutathione synthesis is regulated by nitric oxide in Medicago truncatula roots. *Planta.* **225:** 1597-1602.

Jain, M., Choudhary, D., Kale, R.K. and Bhalla-Sarin, N. (2002). Salt- and glyphosate-induced increase in glyoxalase I activity in cell lines of groundnut (*Arachis hypogaea*). *Physiol. Plant.* **114:** 499-505.

Janda, T., Szalai, G., Rios-Gonzalez, K., Veisz, O. and Páldi, E. (2003). Comparative study of frost tolerance and antioxidant activity in cereals. *Plant Sci.* **164:** 301-306.

Janda, T., Szalai, G., Tari, I. and Páldi, E. (1999). Hydroponic treatment with salicylic acid decreases the effects of chilling injury in maize (*Zea mays* L.) plants. *Planta.* **208:** 175-180.

Jez, J.M., Cahoon, R.E. and Chen, S. (2004). *Arabidopsis thaliana* glutamate-cysteine ligase. Functional properties, kinetic mechanism, and regulation of activity. *J. Biol. Chem.* **279:** 33463-33470.

Jiang, M. and Zhang, J. (2003). Crosstalk between calcium and reactive oxygen species originated from NADPH oxidase in abscisic acid-induced antioxidant defence in leaves of maize seedlings. *Plant Cell Environ.* **26:** 929-939.

Jiménez, A., Hernandez, J.A., Ros Barcelo, A., Sandalio, L.M., del Rio, L.A. and Sevilla, F. (1998). Mitochondrial and peroxisomal ascorbate peroxidase of pea leaves. *Physiol. Plant.* **104:** 687-692.

Kaminaka, H., Morita, S., Nakajima, M., Masumura, T. and Tanaka, K. (1998). Gene cloning and expression of cytosolic glutathione reductase in rice (*Oryza sativa* L.). *Plant Cell Physiol.* **39:** 1269-1280.

Karuppanapandian, T., Sinha, P.B., Haniya, A.M.K. and Manoharan, K. (2006). Differential antioxidative responses of ascorbate-glutathione cycle enzymes and metabolites to chromium stress in green gram (*Vigna radiata* L. Wilczek) leaves. *J. Plant Biol.* **49:** 440-447.

Kellõs, T., Tímár, I., Szilágyi, V., Szalai, G., Galiba, G. and Kocsy, G. (2008). Stress hormones and abiotic stresses have different effects on antioxidants in maize lines with different sensitivity. *Plant Biol.* **10:** 563-572.

Klapheck, S. (1988). Homoglutathione: isolation, quantification and occurrence in legumes. *Physiol. Plant.* **74:** 727-732.

Klapheck, S., Chrost, B., Starke, J. and Zimmermann, H. (1991). γ-Glutamylcysteinylserine—a new homologue of glutathione in plants of the family *Poaceae*. *Bot. Acta.* **105:** 174-179.

Knörzer, O.C., Lederer, B., Durner, J. and Böger, P. (1999). Antioxidative defense activation in soybean cells. *Physiol. Plant.* **107:** 294-302.

Kocsy, G., Galiba, G. and Brunold, C. (2001a). Role of glutathione in adaptation and signalling during chilling and cold acclimation in plants. *Physiol. Plant.* **113:** 158-164.

Kocsy, G., Kobrehel, K., Szalai, G., Duviau, M-P., Buzás, Z. and Galiba, G. (2004a). Thioredoxin h and glutathione as abiotic stress tolerance markers in maize. *Environ. Exp. Bot.* **52:** 101-112.

Kocsy, G., Laurie, R., Szalai, G., Szilágyi, V., Simon-Sarkadi, L., Galiba, G. and de Ronde, J.A. (2005). Genetic manipulation of proline levels affects antioxidants in soybean subjected to simultaneous drought and heat stresses. *Physiol. Plant.* **124:** 227-235.

Kocsy, G., Szalai, G. and Galiba, G. (2004b). Effect of osmotic stress on glutathione and hydroxymethylglutathione accumulation in wheat. *J. Plant Physiol.* **161:** 785-794.

Kocsy, G., Szalai, G., Sutka, J., Páldi, E. and Galiba, G. (2004c). Heat tolerance together with heat stress-induced changes in glutathione and hydroxymethylglutathione levels is affected by chromosome 5A of wheat. *Plant Sci.* **166:** 451-458.

Kocsy, G., Szalai, G., Vágújfalvi, A., Stéhli, L., Orosz, G. and Galiba, G. (2000). Genetic study of glutathione accumulation during cold hardening in wheat. *Planta.* **210:** 295-301.

Kocsy, G., von Ballmoos, P., Rüegsegger, A., Szalai, G., Galiba, G. and Brunold, C. (2001b). Increasing the glutathione content in a chilling-sensitive maize genotype using safeners increased protection against chilling-induced injury. *Plant Physiol.* **127:** 1147-1156.

Kolbe, A., Oliver, S.N., Fernie, A.R., Stitt, M., van Dongen, J.T. and Geigenberger, P. (2006). Combined transcript and metabolite profiling of *Arabidopsis* leaves reveals fundamental effects of the thiol-disuulphide status on plant metabolism. *Plant physiol.* **141:** 412-422.

Kopriva, S. and Koprivova, A. (2005). Sulphate assimilation and glutathione synthesis in C4 plants. *Photosynth. Res.* **86:** 363-372.

Kranner, I., Beckett, R.P., Wornik, S., Zorn, M. and Pfeifhofer, H.W. (2002). Revival of a resurrection plant correlates with its antioxidant status. *Plant J.* **31:** 13-24.

Kranner, I. and Grill, D. (1996). Determination of glutathione and glutathione disulphide in lichens: a comparison of frequently used methods. *Phytochem. Anal.* **7:** 24-28.

Kurganova, L.N., Veselov, A.P., Sinitsina, Y.V., Elikova, E.A. and Kulaeva, O.N. (1999). Lipid peroxidation products as possible mediators of heat stress response in plants. *Russ. J. Plant Physiol.* **46:** 181-185.

Kusumi, K., Yaeno, T., Kojo, K., Hirayama, M., Hirokawa, D., Yara, A. and Iba, K. (2006). The role of salicylic acid in the glutathione-mediated protection against photooxidative stress in rice. *Physiol. Plant.* **128:** 651-666.

Leipner, J., Fracheboud, Y. and Stamp, P. (1999). Effect of growing season on the photosynthetic apparatus and leaf antioxidative defenses in two maize genotypes of different chilling tolerance. *Environ. Exp. Bot.* **42:** 129-139.

Lin, R., Wang, X., Luo, Y., Du, W., Guo, H. and Yin, D. (2007). Effects of soil cadmium on growth, oxidative stress and antioxidant system in wheat seedlings (*Triticum aestivum* L.). *Chemosphere.* **69:** 89-98.

Loeffler, S., Hochberger, A., Grill, E., Winnacker, E.L. and Zenk, M.H. (1989). Termination of the phytochelatin synthase reaction through sequestration of heavy metals by the reaction product. *FEBS Lett.* **258:** 42-46.

Lopez, F., Vansuyt, G., Casse-Delbart, F. and Fourcroy, P. (1996). Ascorbate peroxidase activity, not the mRNA level is enhanced in salt-stressed *Raphanus sativus* plants. *Physiol. Plant.* **97:** 13-20.

Lopez-Climent, M.F., Arbona, A., Perez-Clemente, R.M. and Gomez-Cadenas, A. (2008). Relationship between salt tolerance and photosynthetic machinery performance in citrus. *Environ. Exp. Bot.* **62:** 176-184.

Lu, S.Y., Li, Y.C., Guo, Z.F., Li, B.S. and Li, M.Q. (1999). Enhancement of drought resistance of rice seedlings by calcium treatment. *Chin. J. Rice Sci.* **13:** 161-164.

Marrs, K.A. (1996). The function and regulation of glutathione S-transferases in plants. *Annu. Rev. Plant Physiol. Plant Mol. Biol.* **47:** 127-158.

Martí, M.C., Camejo, D., Fernández-García, N., Rellán-Álvarez, R., Marques, S., Sevilla, F. and Jiménez, A. (2009). Effect of oil refinery sludges on the growth and antioxidant system of alfalfa plants. *J. Hazard Mat.* **171:** 879-885.

May, M.J., Vernoux, T., Leaver, C., Van Montagu, M. and Inze, D. (1998). Glutathione homeostasis in plants: implications for environmental sensing and plant development. *J. Exp. Bot.* **49:** 649-667.

Mendoza, J., Soto, P., Ahumada, I. and Garrido, T. (2004). Determination of oxidised and reduced glutathione by capillary zone electrophoresis in *Brassica juncea* plants treated with copper and cadmium. *Electrophoresis.* **25:** 890-896.

Mendoza-Cózati, D., Loza-Tavera, H., Hernández-Navarro, A. and Moreno-Sánchez, R. (2005). Sulphur assimilation and glutathione metabolism under cadmium stress in yeast, protisits and plants. *FEMS Microbiol. Rev.* **29:** 653-671.

Meneguzzo, S., Sgherri, C.L.M., Navari-Izzo, F. and Izzo, N.R. (1998). Stromal and thylakoid-bound ascorbate preoxidases in NaCl-treated wheat. *Physiol. Plant.* **104:** 735-740.

Mergel, D., Andermann, G. and Andermann, C. (1979). Simultaneous spectrophotometric determination of oxidized and reduced glutathione in human and rabbit red cells. *Methods Find Exp. Clin. Pharmacol.* **1:** 277-283.

Meuwly, P., Thibault, P. and Rauser, W.E. (1993). γ-glutamylcysteinylglutamic acid—a new homologue of glutathione in maize seedlings exposed to calcium. *FEBS Lett.* **336:** 472-476.

Meyer, A.J. (2008). The integration of glutathione homeostasis and redox signalling. *J. Plant Physiol.* **165:** 1390-1403.

Meyer, A.J., Brach, T., Marty, L., Kreye, S., Rouhier, N., Jacquot, J-P. and Hell, R. (2007). Redox-sensitive GFP in *Arabidopsis thaliana* is a quantitative biosensor for the redox potential of the cellular glutathione redox buffer. *Plant J.* **52:** 973-986.

Miao, Y., Dong, L.V., Wang, P.C., Wang, X.C., Chen, J., Miao, C. and Song, C.P. (2006). An *Arabidopsis* glutathione peroxidase functions as both a redox transducer and a scavanger in abscisic acid and drought stress responses. *Plant Cell.* **18:** 2749-2766.

Miller, G., Shulaev, V. and Mittler, R. (2008). Reactive oxygen signalling and abiotic stress. *Physiol. Plant.* **133:** 481-489.

Mittler, R., Vanderauwera, S., Gollery, M. and van Breusegem, F. (2004). Reactive oxygen gene network of plants. *Trends Plant Sci.* **9:** 490-498.

Moons, A. (2005). Regulatory and functional interactions of plant growth regulators and plant glutathione S-transferases (GSTS). *Plant Hormones.* **72:** 155-202.

Morabito, D. and Guerrier, G. (2000). The free oxygen radical scavenging enzymes and redox status in roots and leaves of *Populus ×euramericana* in response to osmotic stress, desiccation and rehydration. *J. Plant Physiol.* **157:** 74-80.

Mou, Z., Fan, W. and Dong, X. (2003). Inducers of plant systemic acquired resistance regulate NPR1 function through redox changes. *Cell.* **113:** 935-944.

Müller, M., Tausz, M., Guttenberger, H. and Grill, D. (1999). Histochemical localization of glutathione in plant tissues: a comparison of two fluorescent staining methods. *Phyton (Austria).* **39:** 69-73.

Müller, M., Zellnig, G., Urbanek, A. and Zechmann, B. (2005). Recent developments in methods intracellulary localizing glutathione within plant tissues and cells (a minireview). *Phyton-Annales Rei. Bot.* **45:** 45-55.

Mullineaux, P.M. and Rausch, T. (2005). Glutathione, photosynthesis and the redox regulation of stress-responsive gene expression. *Photosynth Res.* **86:** 459-474 .

Mur, L.A.J., Kenton, P., Atzorn, R., Miersch, O. and Wasternack, C. (2006). The outcomes of concentration-specific interactions between salicylate and jasmonate signalling include synergy, antagonism, and oxidative stress leading to cell death. *Plant Physiol.* **140:** 249-262.

Murata, Y., Pei, Z-M., Mori, I.C. and Schroeder, J. (2001). Abscisic acid activation of plasma membrane Ca^{2+} channels in guard cells requires cytosolic NAD(P)H and is differentially disrupted upstream and downstream of reactive oxygen species production in *abi1-1* and *abi2-1* protein phosphatase 2C mutants. *Plant Cell.* **13:** 2513-2523.

Navrot, N., Collin, V., Gualberto, J., Gelhaye, E., Hirasawa, M., Rey, P., Knaff, D.B., Issakidis, E., Jacquot, J.P. and Rouhier, N. (2006). Plant glutathione peroxidases are functional peroxiredoxins distributed in several subcellular compartments and regulated during biotic and abiotic stresses. *Plant Physiol.* **142:** 1364-1379.

Nieto-Sotelo, J., Ho, T-H.D. (1986). Effect of heat shock on the metabolism of glutathione in maize roots. *Plant Physiol.* **82:** 1031-1035.

Noctor, G. (2006). Metabolic signalling in defence and stress: the central role of soluble redox couples. *Plant Cell Environ.* **29:** 409-425.

Noctor, G., Arisi, A-C.M., Jouanin, L., Kunert, K.J., Rennenberg, H. and Foyer, C.H. (1998). Glutathione: biosynthesis, metabolism and relationship to stress tolerance explored in transformed plants. *J. Exp. Bot.* **49:** 623-647.

Noctor, G. and Foyer, C.H. (1998). Ascorbate and glutathione: keeping active oxygen under control. *Annu. Rev. Plant Physiol. Plant Mol. Biol.* **49:** 249-279.

Noctor, G., Gómez, L., Vanacker, H. and Foyer, C.H. (2002). Interactions between biosynthesis, compartmentation and transport in the control of glutathione homeostasis and signalling. *J. Exp. Bot.* **53:** 1283-1304.

Ogawa, K. (2005). Glutathione-associated regulation of plant growth and stress responses. *Antiox Redox Sign.* **7:** 973-981.

Ohkama-Ohtsu, N., Oikawa, A., Zhao, P., Xiang, C.B., Saito, K. and Oliver, D.J. (2008). A γ-glutamyl transpeptidase-independent pathway of glutathione catabolism to glutamate via 5-oxoproline in *Arabidopsis*. *Plant Physiol.* **148:** 1603-1613.

Ohkama-Ohtsu, N., Zhao, P., Xiang, C.B. and David, O. (2007). Glutathione conjugates in the vacuole are degraded by γ-glutamyl transpeptidase GGT3 in *Arabidopsis*. *Plant J.* **49:** 878-888.

Orvar, B.L. and Ellis, B.E. (1997). Transgenic tobacco plants expressing antisense RNA for cytosolic ascorbate peroxidase show increased susceptibility to ozone injury. *Plant J.* **11:** 1297-1305.

Pál, M., Horváth, E., Janda, T., Páldi, E. and Szalai, G. (2005). Cadmium stimulate accumulation of salicylic acid and its putative precursors in maize (*Zea mays* L.) plants. *Physiol. Plant.* **125:** 356-364.

Pál, M., Horváth, E., Janda, T., Páldi, E. and Szalai, G. (2006). Physiological changes and defence mechanisms induced by cadmium stress in maize. *J. Plant Nutr. Soil Sci.* **169:** 239-246.

Paolacci, A.R., Badiani, M., D'Annibale, A., Fusari, A. and Matteucci, G. (1997). Antioxidants and photosynthesis in the leaves of *Triticum durum* Desf. seedlings acclimated to non-stressing high temeperature. *J. Plant Physiol.* **150:** 381-387.

Paradiso, A., Berardino, R., de Pinto, M.C., Sanitá di Toppi, L., Storelli, M.M., Tommasi, F. and De Gara, L. (2008). Increase in ascorbate-glutathione metabolism as local and precocious systemic responses induced by cadmium in durum wheat plants. *Plant Cell Physiol.* **49:** 362-374.

Pasternak, M., Lim, B., Wirtz, M., Hell, R., Cobbett, C.S. and Meyer, A.J. (2008). Restricting glutathione biosynthesis to the cytosol is sufficient for normal plant development. *Plant J.* **53:** 999-1012.

Pastori, G., Foyer, C.H. and Mullineaux, P. (2000). Low temperature-induced changes in the distribution of H_2O_2 and antioxidants between the bundle sheath and mesophyll cells of maize leaves. *J. Exp. Bot.* **51:** 107-113.

Pastori, G.M. and Foyer, C.H. (2002). Common components, networks, and pathways of cross-tolerance to stress. The central role of "redox" and abscisic acid-mediated controls. *Plant Physiol.* **129:** 460-468.

Payton, P., Webb, R., Kornyeyev, D., Allen, R. and Holaday, A.S. (2001). Protecting cotton photosynthesis during moderate chilling at high light intensity by increasing chloroplastic antioxidant enzyme activity. *J. Exp. Bot.* **52:** 2345-2354.

Pitzschke, A., Forzani, C. and Hirt, H. (2006). Reactive oxygen species signalling in plants. *Antiox Redox Sign.* **8:** 1757-1764.

Potesil, D., Petrlova, J., Adam, V., Vacek, J., Klejdus, B., Zehnalek, J., Trnkova, L., Havel, L. and Kizek, R. (2005). Simultaneous femtomole determination of cysteine, reduced and oxidised glutathione, and phytochelatin in maize (*Zea mays* L.) kernels using high-performance liquid chromatography with electrochemical detection. *J. Chromatogr.* **1084:** 134-144.

Potters, G., Horemans, N., Bellone, S., Caubergs, R.J., Trost, P., Guisez, Y. and Asard, H. (2004). Dehydroascorbate influences the plant cell cycle through a glutathione-independent reduction mechanism. *Plant Physiol.* **134:** 1479-1487.

Price, A.H., Taylor, A., Ripley, S.J., Griffiths, A., Trewavas, A.J. and Knight, M.R. (1994). Oxidative signals in tobacco increase cytosolic calcium. *Plant Cell.* **6:** 1301-1310.

Raskin, I. (1992). Role of salicylic acid in plants. *Annu. Rev. Plant Physiol. Plant Mol. Biol.* **43:** 439-463.

Rausch, T., Gromes, R., Liedschulle, V., Müller, I., Bogs, J., Galovic, V. and Wachter, A. (2007). Novel insight into the regulation of GSH biosynthesis in higher plants. *Plant Biol.* **9:** 565-572.

Rauser, W.E. (2003). Phytochelatin-based complexes bind various amounts of cadmium in maize seedlings depending on the time of exposure, the concentration of cadmium and the tissue. *New Phytol.* **158:** 269-278.

Rauser, W.E. and Meuwly, P. (1995). Retention of cadmium in roots of maize seedlings. *Plant Physiol.* **109:** 195-202.

Rellán-Alvarez, R., Hernández, L.E., Abadia, J. and Álvarez-Fernández, A. (2006). Direct and simultaneous determination of reduced and oxidised glutathione and homoglutathione by liquid chromatography-electrospray/mass spectrometry in plant tissue extracts. *Anal. Biochem.* **356:** 254-264.

Rennenberg, H., Herschbach, C., Haberer, K. and Kopriva, S. (2007). Sulfur metabolism in plants: Are trees different? *Plant Biol.* **9:** 620-637.

Romano, A.H. and Nickerson, W.J. (1954). Cystine reductase of pea seeds and yeasts. *J. Biol. Chem.* **208:** 409-416.

Rüegsegger, A. and Brunold, C. (1992). Effect of cadmium on g-glutamylcysteine synthesis in maize seedlings. *Plant Physiol.* **99:** 428-433.

Sasaki-Sekimoto, Y., Taki, N., Obayashi, T., Aono, M., Matsumoto, F., Sakurai, N., Suzuki, H., Hirai, M.Y., Noji, M., Saito, K., Masuda, T., Takamiya, K., Shibata, D. and Ohta, H. (2005). Coordinated activation of metabolic pathways for antioxidants and defence compounds by jasmonates and their roles in stress tolerance in *Arabidopsis*. *Plant J.* **44:** 653-668.

Schafer, F.O. and Buettner, G.R. (2001). Redox environment of the cell as viewed through the redox state of the glutathione disulphide/glutathione couple. *Free Rad. Biol. Med.* **30:** 1191-1212.

Sgherri, C.L.M., Maffei, M. and Navari-Izzo, F. (2000). Antioxidative enzymes in wheat subjected to increasing water deficit and rewatering. *J. Plant Physiol.* **157:** 273-293.

Sgherri, C.L.M. and Navari-Izzo, F. (1995). Sunflower seedlings subjected to increasing water deficit stress: oxidative stress and defence mechanisms. *Physiol. Plant.* **93:** 25-30.

Shalata, A., Mittova, V., Volokita, M., Guy, M. and Tal, M. (2001). Response of the cultivated tomato and its wild salt-tolerant relative *Lycopersicon pennellii* to salt-dependent oxidative stress: The root antioxidative system. *Physiol. Plant.* **112:** 487-494.

Shao, H.B., Chu, L.Y., Lu, Z.H. and Kang, C.M. (2008). Primary antioxidant free radical scavenging and redox signalling pathways in higher plant cells. *Int. J. Biol. Sci.* **4:** 8-14.

Skutnik, M. and Rychter, A.M. (2007). Differential response of antioxidant systems in leaves and roots of barley subjected to anoxia and post-anoxia. *J. Plant Physiol.* **166:** 926-937.

Sreenivasulu, N., Grimm, B., Wobus, U. and Weschke, W. (2000). Differential response of antioxidant compounds to salinity stress in salt-tolerant and salt-sensitive seedlings of foxtail millet (*Setaria italica*). *Physiol. Plant.* **109:** 435-442.

Stevens, R.G., Creissen, G.P. and Mullineaux, P.M. (2000). Characterization of pea cytosolic glutathione reductase expressed in transgenic tobacco. *Planta.* **211:** 537-545.

Strasdeit, H., Duhme, A.K., Kneer, R., Zenk, M.H., Hermes, C.H. and Nolting, H.F. (1991). Evidence for discrete $Cd(SCys)_4$ units in cadmium phytochelatin complexes from EXAFS Spectroscopy. *J. Chem. Soc. Chem. Commun.* **16:** 1129-1130.

Streb, P., Aubert, S., Gout, E. and Bligny, R. (2003). Reversibility of cold- and light-stress tolerance and accompanying changes of metabolite and antioxidant levels in the two high mountain plant species *Soldanella alpina* and *Ranunculus glacialis*. *J. Exp. Bot.* **54:** 405-418.

Szalai, G., Janda, T., Golan-Goldhirsh, A. and Páldi, E. (2002). Effect of Cd on phytochelatin synthesis in maize. *Acta. Biol. Szeg.* **46:** 121-122.

Szalai, G., Kellõs, T., Galiba, G. and Kocsy, G. (2009). Glutathione as an Antioxidant and regulatory molecule in plants under abiotic stress conditions. *J. Plant Growth Regul.* **28:** 66-80.

Taulavuori, E., Taulavuori, K., Sarjal, T. and Laine, K. (1999). Polyamines and glutathione metabolism in N-fertilized Scots pine seedlings during cold hardening. *J. Plant Physiol.* **154:** 179-184.

Tausz, M., Šircelj, H. and Grill, D. (2004). The glutathione system as a stress marker in plant ecophysiology: is a stress-response concept valid? *J. Exp. Bot.* **55:** 1955-1962.

Tepe, M. and Harms, H. (1995). Influence of abiotic stress on the GSH/GSSG system of plant cell cultures. *Z Pflanzenernähr Bodenkunde.* **158:** 75-78.

Thumann, J., Grill, E., Winnacker, E.L. and Zenk, M.H. (1991). Reactivation of metal-requiring apoenzymes by phytochelatin-metal complexes. *FEBS Lett.* **284:** 66-69.

Tukendorf, A. and Rauser, W.E. (1990). Changes in glutathione and phytochelatins in roots of maize seedlings exposed to cadmium. *Plant Sci.* **70:** 155-166.

Vatamaniuk, O.K., Mari, S., Lu, Y.P. and Rea, P.A. (2000). Mechanism of heavy metal ion activation of phytochelatin (PC) synthase: blocked thiols are sufficient for PC synthase-catalyzed transpeptidation of glutathione and related thiol peptides. *J. Biol. Chem.* **275:** 31451-31459.

Wasserman, W.W. and Fahl, W. (1997). Functional antioxidant responsive elements. *Proc. Natl. Acad. Sci. USA*. **94:** 5361-5366.

Wójcik, M. and Tukendorf, A. (1999). Cd-tolerance of maize, rye and wheat seedlings. *Acta Physiol. Plant.* **21:** 99-107.

Xiang, C. and Bertrand, D. (2000). Glutathione synthesis in *Arabidopsis*: Multilevel controls coordinate responses to stress. In: Brunold, C., Rennenberg, H., De Kok, L.J., Stulen, I. and Davidian, J-C. (Eds.). Sulfur Nutrition and Sulfur Assimilation in Higher Plants, Bern, Paul Haupt, pp. 409-412.

Xiang, C. and Oliver, D. (1998). Glutathione metabolic genes coordinately respond to heavy metals and jasmonic acid in *Arabidopsis*. *Plant Cell.* **10:** 1539-1550.

Xu, Y., Feng, L., Jeffrey, P.D., Shi, Y.G. and Morel, F.F.M. (2008). Structure and metal exchange in the cadmium carbonic anhydrase of marine diatoms. *Nature.* **452:** 56-61.

Xun, J., Yin, H., Liu, X. and Li, X. (2010). Salt effects plant Cd-stress responses by modulating growth and Cd accumulation. *Planta.* **231:** 449-459.

Yadav, S.K., Singla-Pareek, S.L. and Sopory, S.K. (2008). An overview on the role of methylglyoxal and glyoxalases in plants. *Drug Metabol. Drug Interact.* **23:** 51-68.

Yang, T. and Poovaiah, B.W. (2002). Hydrogen peroxide homeostasis: Activation of plant catalase by calcium/calmodulin. *Proc. Nat. Acad. Sci. USA*. **99:** 4097-4102.

Ying, J., Clavreul, N., Sethuraman, M., Adachi, T. and Cohen, R.A. (2007). Thiol oxidation in signalling and response to stress: detection and quantification of physiological and pathophysiological thiol modifications. *Free Rad. Bio. Med.* **43:** 1099-1108.

Yordanova, R.Y., Christov, K.N. and Popova, L.P. (2004). Antioxidative enzymes in barley plants subjected to soil flooding. *Environ. Exp. Bot.* **51:** 93-101.

Yu, C-W., Murphy, T.M. and Lin, C-H. (2003). Hydrogen-peroxide induced chilling tolerance in mung beans mediated through ABA-independent glutathione accumulation. *Funct. Plant Biol.* **30:** 955-963.

Yu, Q. and Rengel, Z. (1999). Waterlogging influences plant growth and activities of superoxide dismutases in narrow-leafed lupine and transgenic tobacco plants. *J. Plant Physiol.* **155:** 431-438.

Zechmann, B. and Müller, M. (2008). Effects of zucchini yellow mosaic virus infection on the subcellular distribution of glutathione and its precursors in a highly tolerant cultivar. *Botany.* **86:** 1092-1100.

Zechmann, B., Zellnig, G. and Müller, M. (2006). Immunocytochemical localization of glutathione precursors in plant cells. *J. Electron. Microsc.* **55:** 173-181.

Zhang, M.Y., Bourbouloux, A., Cagnac, O., Srikanth, C.V., Rentsch, D., Bachhawat, A.K. and Delrot, S. (2004). A novel family of transporter mediating the transport of glutathione derivatives in plants. *Plant Physiol.* **134:** 482-491.

Zottini, M., Costa, A., de Michele, R., Ruzzene, M., Carimi, F., lo Schiavo, F. (2007). Salicylic acid activates nitric oxide synthesis in *Arabidopsis*. *J. Exp. Bot.* **58:** 1397-1405.

3

Regulatory Role of Components of Ascorbate-Glutathione (AsA-GSH) Pathway in Plant Tolerance to Oxidative Stress

MOHAMMAD A. HOSSAIN[1,2] **AND MASAYUKI FUJITA**[1*]

[1]*Laboratory of Plant Stress Responses, Faculty of Agriculture, Kagawa University, Miki-cho, Kita-gun, Kagawa 761-0795, Japan*

[2]*Department of Genetics & Plant Breeding, Bangladesh Agricultural University, Mymensingh- 2202, Bangladesh*

[*]***Corresponding author:*** *fujita@ag.kagawa-u.ac.jp*

ABSTRACT

Due to their sessile nature, plants are constantly exposed to a multitude of environmental stresses to which they react with a battery of responses. The most important and best documented common response of plants to different abiotic and biotic stresses, such as heat, cold, high-light intensities, drought, salinity, heavy metals, wounding, UV-B radiation, ozone, and pathogens, is the accelerated production of reactive oxygen species (ROS). ROS are extremely reactive in nature because they can interact with a number of cellular molecules and metabolites thereby lead to irreparable metabolic dysfunction and death. Plants have well-developed enzymatic and non-enzymatic scavenging pathways or detoxification systems to counter the deleterious effects of ROS including ascorbate-glutathione (AsA–GSH) cycle. The AsA–GSH cycle involves ascorbate peroxidase (APX; EC 1.11.1.11), dehydroascorbate reductase (DHAR; EC 1.8.5.1), glutathione reductase (GR; EC 1.6.4.2), and monodehydroascorbate reductase (MDHAR; EC 1.6.5.4), reduced ascorbic acid (AsA), reduced glutathione (GSH) and NADPH in a series of cyclic reactions to detoxify H_2O_2 and also to regenerate AsA and GSH. In plant cells, this cycle has been shown to operate in different cell compartments in which ROS detoxification is needed, i.e. in chloroplasts, mitochondria, cytosol and peroxisomes. Apart from its function directly to detoxify ROS and thereby to combat the potentially harmful effects of environmental stresses, the AsA-GSH cycle is also involved in redox sensing and signalling. The specific interplay between ROS and the AsA-

GSH cycle components could generate compartment-specific changes in both the absolute concentrations of ROS and the antioxidant compounds, and in ascorbate and glutathione redox ratios. Under stress conditions these redox signals could interfere with signalling networks complementary to the antioxidant system and regulate defence gene expression, thus coordinating the necessary readjustments in the redox-regulated plant defence to overcome the oxidative stress. Although AsA and GSH function as cofactors of AsA-GSH cycle enzymes, both of them can also directly quench superoxide ($O_2^{•-}$), hydroxyl radicals ($^•OH$) and singlet oxygen ($^1O_2$) and provide essential information on cellular redox state, and regulate gene expression associated with biotic and abiotic stress responses to optimize defence and survival. ROS, more specifically, H_2O_2 plays versatile roles in normal plant physiological processes and in resistance to stresses. Recently, H_2O_2 has been perceived as a signalling molecule and regulator of the expression of genes in cells. In this chapter, we review the most recent information on the response and regulatory role of components of AsA-GSH cycle in plant tolerance to oxidative stress. Further, we discuss the progress made during the last few decades in improving oxidative stress tolerance of plants through genetic engineering by different components of AsA-GSH cycle.

Keywords: Reactive oxygen species, components of ascorbate-glutathione (AsA-GSH) cycle, oxidative stress tolerance.

1. OXIDATIVE STRESS IN PLANTS

As sessile organisms, plants live under a constant barrage of biotic and abiotic insults from which they cannot escape and, for which they have evolved a wide array of mechanisms for coping. The range of conditions to which different species have adapted to a greater or lesser extent, include extremes of temperature, salinity, drought, flooding, heavy metal toxicity, nutrient imbalances, ozone exposure, pathogens, wounding and harmful radiation (UV). The most important and best documented common feature of plant responses to the above-mentioned stimuli is the production of reactive oxygen species (ROS) such as singlet oxygen ($^1O_2$), superoxide radical ($O_2^{•-}$), hydrogen peroxide (H_2O_2), and hydroxyl radical ($^•OH$). ROS accumulation during stress greatly depends on the balance between ROS production and ROS scavenging (Mittler et al., 2004), that in turn depends on changes in growth conditions as well as the severity and the duration of the stress and the ability of the tissue to rapidly acclimate to the energy imbalance (Miller et al., 2010). Even under normal conditions, higher plants produce ROS during metabolic processes. However, during stress, their rate of production is dramatically elevated. Organelles with a highly oxidizing metabolic activity or with an intense rate of electron flow, such as chloroplast, mitochondria and microbodies, are major sources of ROS production in plant cells (Mittler et al., 2004). Chloroplast is the prime source of ROS having the capacity to produce high amounts of superoxide ($O_2^{•-}$) through Mehler reaction and H_2O_2, essentially during reduced rate of photosynthetic carbon fixation—a typical situation during abiotic stresses (Takashi and Murata 2008). In addition, chloroplasts can produce singlet oxygen ($^1O_2$) through excited chlorophyll molecule. In mitochondria, over-reduction of the electron transport chain (ETC) is the main source of $O_2^{•-}$ and H_2O_2 production under specific stress conditions (Moller 2001). Although, mitochondrial ROS production is much lower compared to chloroplasts, mitochondrial ROS are important regulators of a number of cellular processes, including stress adaptation and programmed cell death (PCD) (Robson and Vanlerberghe 2002). In peroxisomes, H_2O_2 is produced mainly during the photorespiration and also during β-oxidation of fatty acids, in the enzymatic reactions of flavin oxidases as well as in the dispropotionation of $O_2^{\bullet-}$. Photorespiration is a complex processes closely linked to photosynthesis. Under conditions that impair CO_2 fixation in chloroplasts, the oxygenase activity of ribulose-1,5-bisphosphate carboxylase/oxygenase increases and the produced glycolate moves to peroxisomes, where it is oxidized by glycolate oxidase forming H_2O_2. In glyoxysomes, acyl-CoA oxidase is the primary enzyme responsible for H_2O_2 generation. Plasma membrane-bound NADPH oxidases as well as cell-wall associated peroxidases are the main sources of $O_2^{\bullet-}$ and H_2O_2 producing apoplastic enzymes (Sagi and Fluhr 2006). Apoplastic ROS accumulation participates in the so-called oxidative burst observed as a part of the hypersensitive response (HR) to pathogens but also regulate cell growth, development and cell death (Gechev et al., 2006). $O_2^{\bullet-}$ and H_2O_2 are also produced by xanthine oxidase and various other oxidases induced by abiotic and biotic stresses (Dat et al., 2000). Additional sources of ROS in plant cells include the detoxifying reactions catalysed by cytochromes in both the cytoplasm and the endoplasmic reticulum (Urban et al., 1989). In addition to these metabolic ROS resources, in the presence of redox-active metals, $^{\bullet}OH$ can be formed from H_2O_2 through the Fenton reaction or from H_2O_2 and $O_2^{\bullet-}$ through the Haber-Weiss reaction. The extremely reactive $^{\bullet}OH$ in the cell causing extensive oxidative damage of membrane and other macromolecules including photosynthetic pigments, protein, DNA and lipids (Hernandez et al., 1995; Fadzilla et al., 1997; Gueta-Dahan et al., 1997; Foyer and Noctor 2005a; Gechev et al., 2006; Shulaev and Oliver 2006; Jaspers and Kangasjarvi 2010). ROS are normally scavenged immediately at the sites of their production by the locally present antioxidants. However, when this local antioxidant capacity cannot cope with ROS production (for example, during stress or temporary reduced antioxidant levels due to developmental signals), H_2O_2 can leak in the cytosol and diffuse to other compartments (Fig. 2).

2. ROS DETOXIFICATION IN RELATION TO AsA-GSH CYCLE

A strict control of ROS levels is essential to prevent their toxicity and to ensure an accurate execution of their signalling functions. Therefore, plants have evolved an elaborate enzymatic and non-enzymatic antioxidant system, which together with ROS producing enzymes maintains ROS homeostasis in all cellular compartments and regulates the adjustment of ROS levels according to the cellular needs at a particular time (Apel and Hirt 2004; Sairam and Tyagi 2004; Gechev et al., 2006; Moller et al., 2007). These antioxidants include the enzymes, superoxide dismutase (SOD; EC 1.15.1.1), catalase (CAT; EC 1.11.1.6), glutathione peroxidase (GPX, EC 1.11.1.9), glutathione S-transferases (GST; EC 2.5.1.18), ascorbate peroxidase (APX; EC 1.11.1.11), dehydroascorbate reductase (DHAR; EC 1.8.5.1), glutathione reductase (GR; EC 1.6.4.2), and monodehydroascorbate reductase (MDHAR; EC 1.6.5.4), the water-soluble compounds such as ascorbate, glutathione (GSH) and flavonoids, and the lipid-soluble compounds such as the carotenoids and tocopherols. The AsA–GSH cycle involves APX, MDHAR, DHAR, GR, reduced ascorbic acid (AsA), reduced glutathione (GSH) and NADPH in a series of cyclic reactions to detoxify H_2O_2 and also to regenerate AsA and GSH (Fig. 1). The presence of a complete AsA-GSH cycle has been identified in chloroplast (Asada 1999), mitochondria (Creissen et al., 1995; Jimenez et al., 1997, 1998), peroxisomes (Jimenez et al., 1997, 1998) and cytosol (Mittler et al., 2004), demonstrating the importance of ROS detoxification for cellular survival (Fig. 2). In the AsA-GSH cycle, APX catalyses the reduction of H_2O_2 into water with the concomitant generation of short-lived radical monodehydroascorbate (MDHA), which is converted to AsA by the action of NADPH-dependent MDHAR or disproportionates nonenzymatically to AsA and dehydroascorbate (DHA) (Asada 1992). DHA undergoes irreversible hydrolysis to 2, 3-diketogulonic acid or recycled to AsA by DHAR, which uses GSH as the reductant (Chen et al., 2003). GSH, which is oxidized in this process to glutathione disulphide (GSSG), is recycled by GR consuming NADPH. Apart from its function directly to detoxify ROS and thereby to combat the potentially harmful effects of environmental stresses, the AsA-GSH cycle is also involved in redox sensing and signalling. Hence, the specific interplay between ROS and the AsA-GSH cycle constituents could generate compartment-specific changes in both the absolute concentrations of ROS and the antioxidant compounds, and in ascorbate and glutathione redox ratios. Under stress conditions these redox signals could interfere with signalling networks complementary to the antioxidant system and regulate defence gene expression (May et al., 1998; Karpinski et al., 1999; Kuzniak and Sklodowska 2005a), thus coordinating the necessary readjustments in the redox-regulated plant defence to overcome the oxidative stress. Besides functioning as a cofactor of AsA-GSH cycle, both AsA and GSH can directly quench $O_2^{\bullet -}$, $^{\bullet}OH$ and $^1O_2$ and are associated with several growth- and development-related events in plants, including cell differentiation, cell death and senescence, pathogen resistance and enzymatic regulation. Most importantly, they provide essential information on cellular redox state, and regulate gene expression associated with biotic and abiotic stress responses to optimize defence and survival. ROS, more particularly, H_2O_2 plays versatile roles in normal plant physiological processes and in resistance to stresses. Recently, H_2O_2 has been regarded as a signalling molecule and regulator of the expression of genes in cells. Since the discovery of the AsA-GSH cycle in the mid-1970s, the enzyme-catalysed reactions of this pathway have attracted considerable interest and still are a matter of intensive research (Noctor and Foyer 1998; Noctor et al., 1998; Asada 1999). Studies with mutants or transgenic plants over- or underexpressing enzymes or metabolites of the AsA-GSH pathway established causal relationships between certain components of the cycle and oxidative stress tolerance (Scandalios 1993; Allen 1995; Miller et al., 2007; Lee et al., 2007; Lim et al., 2007; Gao and Zhang 2008; Li et al., 2009). Recent studies showed that coexpression of two or more antioxidant enzymes of the AsA-GSH pathway provides better antioxidant protection as compared to the single gene transformation (Lee et al., 2007). These results indicate that the balanced interaction of antioxidative enzymes may be necessary to obtain a substantial increase in oxidative stress tolerance.

The present review focuses on understanding the regulatory role of different components of AsA-GSH cycle in plants defence against oxidative stress caused by various abiotic and biotic stresses. The progress made during the last few decades in improving oxidative stress tolerance of plants through genetic engineering by different components of AsA-GSH cycle has also been discussed critically in this review.

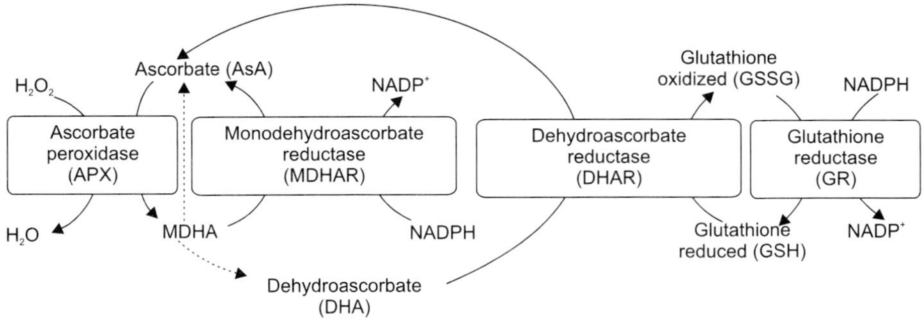

Fig. 1. The ascorbate-glutathione (AsA-GSH) cycle or Asada-Halliwell pathway. H_2O_2 is removed by APX and AsA is regenerated by the AsA-GSH cycle, involving MDHAR, DHAR and GR. AsA is first oxidized to MDHA. If MDHA is not rapidly reduced again to AsA by MDHAR, it will spontaneously disproportionate into AsA and DHA. DHAR recycles AsA from DHA using GSH that is regenerated through the action of GR in a NADPH-dependent reaction. The dotted line indicates non-enzymatic reactions.

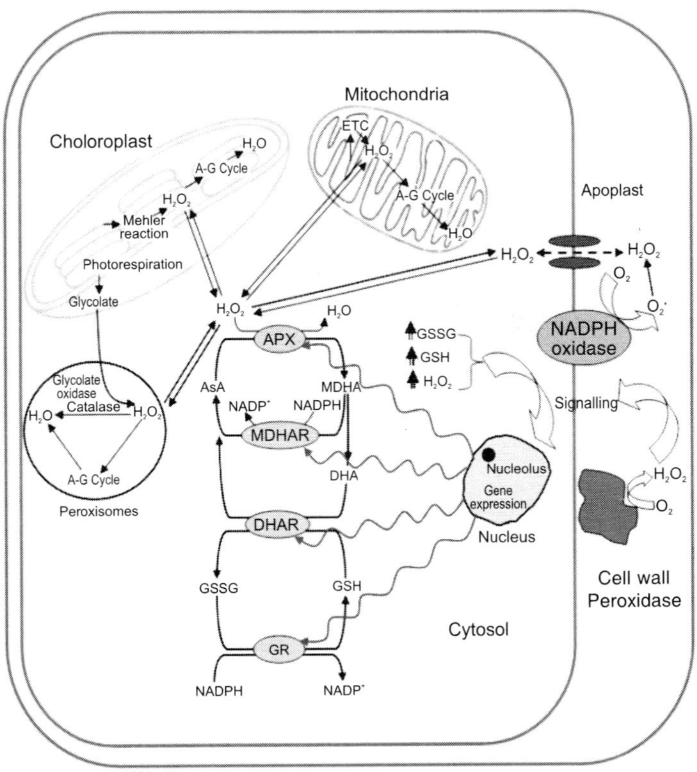

Fig. 2. Schematic representation of a generalized plant cell depicting major sources of ROS regeneration and scavenging through ascorbate-glutathione cycle. ETC, electron transport chain; A-G cycle, ascorbate-glutathione cycle.

3. COMPONENTS OF AsA-GSH CYCLE

The details of the AsA-GSH cycle may vary with the individual authors, but the basic framework is almost the same (Figs. 1 and 2). A typical AsA-GSH cycle has the following components which act in concert to detoxify H_2O_2 and plays diverse roles in plant tolerance to oxidative stress:

1. Ascorbate peroxidase (APX)
2. Monodehydroascorbate reductase (MDHAR)
3. Dehydroascorbate reductase (DHAR)
4. Glutathione reductase (GR)
5. Ascorbate (AsA)
6. Glutathione (GSH)
7. NADPH and
8. Hydrogen peroxide (H_2O_2)

3.1 Ascorbate Peroxidase (APX)

APX is a key enzyme in the AsA-GSH cycle and plays a vital role in plant defence against oxidative stress by catalysing the conversion of H_2O_2 to water. APX is a multi-gene family enzyme including isoforms in at least four distinct cellular compartments: stromal APX (sAPX), thylakoid-bound APX (tAPX) in chloroplasts, microbody (glyoxysome and peroxisome) membrane-bound APX (mAPX) and cytosolic APX (cAPX) (Shigeoka et al., 2002). Chloroplast-specific forms of APX can be targeted to mitochondria as well (Chew et al., 2003). Some of the above compartments, including the peroxisomes, glyoxysomes, and mitochondria, also contain catalase (CAT) but this enzyme has a low affinity for H_2O_2. In contrast, strong affinity to H_2O_2 makes APX a very versatile enzyme for a fine modulation of the levels of H_2O_2 in cells and/or within specific cellular compartments.

3.1.1 Regulation of APX activity during abiotic oxidative stress

In view of the physiological significance of APX against oxidative stress, recent studies have focused on the changes in activities of APX isoenzymes in higher plants subjected to various abiotic and biotic stresses that accelerate the production of ROS (Hernandez et al., 1995; Kuzniak and Sklodowska 2004a,b, 2005a,b; Perez-Lopez et al., 2009; Ozkur et al., 2009; Hossain and Fujita 2009b; Ryang et al., 2009; Kumutha et al., 2009; Song et al., 2009; Garcia-Limones et al., 2009). APX activity increased by 2.5 times when *Bruguiera parviflora* plants were exposed to two different salt stress levels (200 and 400 mM NaCl) for 5 d but a 4.5-fold increase was observed after 45 d of treatment (Parida et al., 2004). Salt tolerant *Lycopersicon pennellii* and *Plantago maritima* showed significant increase in APX activity but the activity decreased in the sensitive genotypes (Shalata et al., 2001; Mittova et al., 2002; Sekmen et al., 2007). APX activity in salt tolerant *Pisum sativum* was correlated to the inherent better protection from salt and oxidative stress (Hernandez et al., 2000). In addition, Mittova et al. (2003a) showed that APX activity increased in both mitochondrial and peroxisomal fraction of wild salt tolerant tomato species, however, the activity was unchanged in cultivated tomato. Recently, Perez-Lopez et al. (2009) reported that APX activity increased in both sensitive and tolerant genotypes of barley. APX activity was also increased in *Brassica juncea* seedlings subjected to salt stress with the increasing level of H_2O_2 and lipid peroxidation (Verma and Mishra 2005). It was hypothesized that the increase in activity of APX could be due to activation of pre-existing APX or due to synthesis of new APX upon salt exposure (Parida et al., 2004). In *B. parviflora*, $O_2^{\bullet -}$ formed owing to salt stress could be scavenged by increased activity of SOD. The end product of this reaction, H_2O_2, is partially dismutated by the increased activities of APX (Parida et al., 2004). The increase in APX activity and the concomitant increase in SOD activity suggest that this is an adaptation to remove the excess H_2O_2 generated due to increase in SOD activity. Drought-induced enhancement in APX activity resulted in lower H_2O_2 levels in the panicles of N22, a water stress tolerant genotype compared to N118, a susceptible genotype (Selote and Chopra 2004). Similar phenomenon was also observed in maize seedlings subjected to drought stress where the root of tolerant cultivar cv. Nova showed increase in APX activity after 4 h of exposure and a gradual increase was observed with prolonged stress (48 h). But the susceptible genotype cv. Ankora, showed decrease in APX activity within 24 h which indicates the higher ability of cv. Nova to tolerate drought stress and effectively cope with the oxidative damage (Kolarovic et al., 2009). Drought-resistant maize shows greater induction of APX activity than sensitive plants, in addition with higher GR activity (Pastori and Trippi 1992). Significant increase in APX activity was also observed in mung bean seedlings subjected to short-term drought stress with increased GST, GPX and Glyoxalase I activities (Hossain and Fujita 2009b). Drought stress significantly increased APX activity in poplar species (Yang et al., 2009); however, the increase was not enough to prevent the oxidation of lipid. APX activity was accelerated during drought stress in the 'Berangan' variety that has more tolerance to water deficit stress, but the activity was unaffected in 'Mas' that has less tolerance. APX activity increased 4 to 5-fold in perennial xerophyte *Capparis ovata* subjected to polyethylene glycol (PEG) induced drought stress. Higher activity in stressed leaves suggests a more effective H_2O_2 removal which might be produced by an enhanced activity of SOD in *C. ovata* under drought stress conditions (Ozkur et al., 2009). Therefore, higher APX activity was associated with greater protection against water-induced oxidative injury (Chai et al., 2005). In olive trees, the activities of SOD, APX, CAT, and peroxidase (POD) increased in relation to the severity of drought stress in both leaves and roots, and in particular, a marked increase in APX activity was found in leaves of plants subjected to severe drought stress (Sofo et al., 2005). APX activity increased in cold sensitive and tolerant variety of barley; however, cold acclimation increased the activity much higher in the tolerant variety as compared to sensitive variety indicating that APX is very important in detoxification of H_2O_2 for barely plants under freezing stress (Dai et al., 2009). Elevated level of APX and GR was also found to be associated with increasing freezing tolerance during cold acclimation in pines (Tao et al., 1998). Chilling tolerant rice cultivar (Xiangnuo-1) showed higher APX activity, however, the activity decreased in the chilling sensitive cultivars with higher level of MDA and electrolyte leakage (Huang and Guo 2005). Schützendübel et al. (2002) showed that short term (10 h) Cd stress significantly increased APX activity but the activity decreased with the increase to treatment duration. Accordingly, significant decrease in APX activity was observed in *B. juncea* and *B. napus* subjected to Cd stress (Nouairi et al., 2009). Differential biochemical response was observed in wheat shoots and roots subjected to nickel (Ni) stress. In root tissues, the APX activity decreased with the increasing level of Ni concentration but the activity increased significantly in shoot tissues (Gajewska et al., 2008). Susceptible pigeonpea genotypes (Pusa 207) showed decrease in APX activity with the increasing level of H_2O_2 and lipid peroxidation level subjected to waterlogging stress. In contrast, tolerant genotype (ICP 301) showed a consistent increase with the increase of the duration of the treatment in connection with the decreasing level of H_2O_2 and lipid peroxidation level (Kumutha et al., 2009). However, Wang and Jiang (2007) and Shi et al. (2008) reported that APX activity decreased significantly under flooding stress. Significant increase in APX activity was also observed in mulberry plant subjected to nitrogen (N), potassium (K) and phosphorus (P) deficiency stresses (Tewari et al., 2009). However, the opposite phenomenon was observed in sunflower plants subjected to iron deficiency, where the activity of APX decreased with decreasing level of AsA and increased NADPH oxidation and H_2O_2 level (Ranieri et al., 2003). Increase in APX activity was observed in *Liriodendron tulipifera* under elevated O_3 stress (Ryang et al., 2009). Most recently, Zhou et al. (2009) showed that lettuce leaves subjected to high light stress increases APX activity in addition with increased $O_2^{\bullet -}$ and lipid peroxidation. Numerous studies of ROS scavenging enzymes in plants have demonstrated that APX activity generally increases along with activities of other antioxidant enzymes like CAT, SOD, and GR in response to various environmental stress factors, suggesting that the components of ROS-scavenging systems are co-regulated. An *Arabidopsis* mutant (*pst1*) isolated based on its ability to grow photoautotrophically on a NaCl-containing medium showed significantly enhanced APX and SOD activities (Tsugane et al., 1999). The *pst1* also showed cross-resistance to other stress conditions, suggesting that the enhanced enzyme activities might be conferred by the mutation of a recessive nuclear gene and that regulation of these enzyme levels provides plants with additional stress-defence capabilities. Decrease in APX activity was observed in rice roots subjected to Al stress but CAT activity was increased with the decreasing level of H_2O_2 (Wang and Kao 2007). However, pumpkin seedlings subjected to Al-stress showed increase in H_2O_2 content after 24 h and reached maximum at 48 h and decreased at 72 h but the APX activity increased during the treatment period and was highest at 72 h indicating that APX plays a significant role in Al-induced ROS detoxification (Dipierro et al., 2005). From the foregoing account, it is increasingly evident that APX is a key enzyme for scavenging the ROS and maintenance of redox potential of the cell during oxidative stress.

3.1.2 Regulation of APX activity during biotic stress-induced ROS detoxification

One of the earliest detectable reactions of plant cells after pathogen attacks is oxidative burst (Baker and Orlandi 1995) and plant cells have to avoid excessive oxidative damage in non-invaded tissues around pathogen entry sites in order to exert effective defence response. Tomato leaf infected with cucumber mosaic virus (CMV) showed 2- and 3.65-

fold increase of APX activity in chloroplastic and mitochondrial fraction respectively, whereas 2-fold increase was observed in mitochondrial fraction of cucumber leaf (Song et al., 2009). When tomato leaves were infected with *Botrytis cinerea*, APX activity increased in chloroplastic and whole leaf fractions after 3 dpi (days post-inoculation), however, the activity decreased with the progression of disease symptom development in connection with the decreasing level of AsA and the GSH ratio which indicates breakdown of the plant defence mechanism through the AsA-GSH cycle (Kuzniak and Sklodowska 2001, 2005a). In contrast, APX activity decreased with the decreasing level of AsA after 1 d of *B. cinerea* infection in peroxisomal and mitochondrial fraction of tomato leaves (Kuzniak and Sklodowska 2004a, 2005b). In tolerant tomato genotype, a significant increase in APX activity was observed after 4 d of bacterial (*Pseudomonas syringae*) infection; however, the activity decreased in susceptive cultivars (Kuzniak and Sklodowska 2004b). Plum Pox Virus (PPV) infected peach leaves showed an 8.5-fold increase in APX activity due to long-term virus infection but this increase in APX activity could not stop the development of symptom resulting from infection by PPV, but may have helped to reduce the severity of the disease and possibly allow the plant to recover (Hernandez et al., 2004b). However, APX activity is decreased in both soluble and chloroplastic fraction of pea plant infected with PPV after 3 dpi and a greater decrease was observed in sensitive cultivar which was correlated with the increase in oxidative stress (increased level of H_2O_2 and oxidation of lipid and protein)(Diaz-Vivancos et al., 2008). Additionally, susceptible apricot cultivar infected with PPV showed decrease in APX activity whereas in resistant cultivar strong increases in both class I APX and class III APX activities were observed with the decreasing level of H_2O_2, indicating PPV resistant apricot cv. SEO seems to control apoplastic H_2O_2 level more properly than the susceptible cultivar (Diaz-Vivancos et al., 2006). Apoplastic APX activity increased in both sensitive and resistant lines of barley infected with *Blumeria graminis* but foliar APX activity decreased (Vanacker et al., 1998). In contrast, resistant chickpea cultivar roots infected with *Fusarium oxysporum* showed 2-fold increase in apoplastic APX activity after 15 d of inoculation whereas susceptible cultivar developed severe leaf chlorosis, flaccidity and wilting (Garcia-Limones et al., 2009). Thus, the result indicates that in the resistant interaction, APX participated in preventing disease development whereas in the susceptible one its activity decreases favouring the pro-oxidative state in the inoculated tissue could contribute to the ultimate success of the pathogen (Kuzniak and Sklodowska 2004b).

3.1.3 APX gene expression during oxidative stress condition

Progressive accumulation of cAPX and sAPX were observed in pea leaves subjected to high light stress and the increase in transcript levels were correlated with the increase of the foliar H_2O_2 content (Hernandez et al., 2004a). *Arabidopsis* mutants lacking cytosolic APX1, in which the entire chloroplastic H_2O_2 scavenging system collapses due to moderate light stress, H_2O_2 levels increased, and protein oxidation occurs (Davletova et al., 2005). In addition, Arabidopsis leaves subjected to high light intensity caused a rapid and selective increase in transcripts of APX1 and 2, which encode cytosolic APXs

(Karpinski et al., 1997). Upregulation of cAPX in response to high and low temperature in rice has also been reported (Sato et al., 2001). *Arabidopsis* fumigated by O_3 showed a large increase in the steady-state transcript level of cAPX (Conklin and Last 1995; Ovar and Ellis 1997). Tomato cultivar exposed to O_3 showed increase in APX gene expression with other antioxidant enzymes like GST, GPX, CAT and DHAR (Marco et al., 2008). In addition, Arabidopsis seedlings subjected to Cd stress (10 μM) for 3 weeks showed nearly 3-fold increase in APX1 gene expression (Smeets et al., 2008). *Hevea brasiliensis* subjected to chilling stress showed 4-fold increase in APX gene expression within 7 h of treatment (Mai et al., 2009). A remarkable increase in the transcript level of a pAPX isolated from barley (*HvAPX1*) in response to heat, salt and ABA treatment was seen and over-expression in Arabidopsis conferred tolerance to heat stress (Shi et al., 2001). In addition, peroxizomal ascorbate peroxidase from *Avicennia marina (Am-pAPX1)* upregulated in response to salinity, high light intensity, H_2O_2 and excess iron (Kavitha et al., 2008). The transcript level of pea cAPX also increases 4-fold in response to drought stress, but is even more dramatically enhanced (15-fold) following recovery from stress (Mittler and Zilinskas 1994). Two cytosolic (cAPX I, II) and a thylakoid-bound (tAPX) variants increased significantly in drought tolerant genotype whereas a cytosolic (cAPX I) and a stromal (sAPX II) transcriptional increase were observed in sensitive genotype (Secenji et al., 2010). Cytosolic APX transcripts increased after salt stress treatment in *Mesembryanthemum crystallinum* indicate that this enzyme plays a critical role in controlling increase of H_2O_2 concentration in plant cells during the initial salt-induced oxidative stress (Slesak et al., 2002). Expression of APX isoenzymes (chlAPX, two cAPXs, and mAPX) in response to several stress conditions were tested and the typical cAPX isoenzyme markedly increased in response to high light stress and paraquat treatment, but not in response to other stress treatments such as drought and salt (Yoshimura 2000). However, chlAPX and mAPX transcript levels did not change in response to any of the stress treatments indicating that the genes for chlAPX and mAPX are constitutively expressed. Panchuk et al. (2002) have reported that the transcript level of cAPX increased under heat stress conditions, while APX isoform localized in other compartments did not show any changes. Additionally, the transcript of cAPX is dramatically induced during the HR of tobacco plants infected with TMV (Mittler et al., 1999) and in chickpea roots infected with *F. oxysporum* (Garcia-Limones et al., 2009). Recently, Koussevitzky et al. (2008) reported that cytosolic ascorbate peroxidase 1(APX1) protein and mRNA accumulated during the drought and heat stress combination whereas the APX1-deficient mutant (*apx1*) accumulated more H_2O_2. Baek and Skinner (2003) have observed that expression level of tAPX was upregulated during cold acclimation of near-isogenic wheat lines. These findings clearly suggest the potential role of APX gene in oxidative stress tolerance.

3.1.4 Regulatory role of cAPX in redox signal modulation

In addition to H_2O_2 detoxification, cAPX isozymes have a dynamic function in redox signal modulation and gene expression. For example, upon pathogens attack accumulation of cellular ROS helps forward on induction of cell death in infected cells, or serves as a signal activating defence responses in distant uninfected cells. cAPX1 or tylAPX knockout mutant in *Arabidopsis* resulted in ROS accumulation that in turn generated a signal that enhanced the plant's tolerance to both osmotic and salt stresses (Miller et al., 2007). Likewise, transgenic tobacco cells with lowered cAPX activity due to co-suppression showed greater tolerance to environmental stress, such as heat and salinity, than wild-type cells due to activation of a specific protein kinase and upregulation of the expression of stress-responsive genes including heat shock protein (HSP) as well as increased cellular ROS accumulation (Ishikawa et al., 2005). cAPX expression in TMV-infected tobacco leaves is suppressed by inhibition of protein synthesis in the polysomes (Mittler et al., 1998), indicating that cAPX suppression contributes to endogenous ROS accumulation, which evokes redox-signalling. Pnueli et al. (2003) showed that cAPX (APX1) knockout *Arabidopsis* showed increased expression of various genes, including those for stress-related proteins and transcription factors under light stress conditions. Additionally, induction of HSP gene expression was much greater in APX1-knockout plants than in wild-type plants. These observations strongly support the idea that cAPX takes an active role in the regulation of redox gene expression by modulating the concentration of H_2O_2 (cellular redox status) to adjust its activity or expression to a level sufficient for second messenger activity (Fig. 2). In contrast, the increase in cAPX expression in response to H_2O_2 accumulation as described above is essential to reduce overproduction of ROS to toxic levels because prolonged accumulation of cellular ROS is eventually toxic and lethal (Ishikawa and Shigeoka 2008).

3.1.5 APX transgenic and oxidative stress tolerance

Transgenic plants overexpressing APX isoenzymes have been generated to improve oxidative stress induced by both biotic and abiotic stresses (Allen et al., 1997; Wang et al., 1999; Badawi et al., 2004; Murgia et al., 2004; Sarowar et al., 2005; Lim et al., 2007; Lee et al., 2007; Li et al., 2009; Wei-Feng et al., 2008; Xu et al., 2008a). Transgenic tobacco plant overexpressing either cAPX or chloroplast targeted cytosolic APX gene showed low membranes damage induced by methyl viologen (MV) and increased protection from photo-oxidative stress (Allen et al., 1997). Transgenic Arabidopsis plants overexpressing tAPX in chloroplasts showed remarkable tolerance against photo-oxidative stress following exposure to strong light and paraquat treatment (Murgia et al., 2004). Similarly, transgenic *A. thaliana* plants carrying a peroxisomal ascorbate peroxidase (pAPX) gene (*HvAPX1*) from barley (*Hordeum vulgare* L.) was found to be more tolerant to salt and heavy metal stresses due to higher activity of APX and lower level of H_2O_2 and lipid peroxidation (Wei-Feng et al., 2008; Xu et al., 2008a). Overexpression of cAPX gene in tomato could improve its resistance to heat and ultraviolet (UV-B) stress (Wang et al., 2006). Furthermore, overexpression of APX in tobacco chloroplasts enhanced plant tolerance to salt, water deficit, MV, paraquat and PEG induced oxidative stresses and increased net photosynthesis as compared to wild type plants (Badawi et al., 2004). In *Capsicum annum*, ascorbate peroxidase-like 1 gene (CAPOA1)-over expressed plants exhibited increased tolerance to MV-mediated oxidative stress, and also enhance resistance to *Phytopthora nicotianae* (Sarowar et al., 2005).

Transgenic sweet potato plants that express both CuZnSOD and APX in chloroplasts showed strong tolerance to MV-mediated oxidative stress, chilling, dehydration and air pollution stress, such as SO_2 (Lim et al., 2007). Transgenic tobacco plant overexpressing cAPX gene including CuZnSOD and DHAR showed enhance tolerance to oxidative stress induced by paraquat and salt (Lee et al., 2007). In addition, transgenic tobacco plants overexpressing tAPX gene showed increased tolerance to oxidative stress caused by MV, light intensity and by chilling stress combined with high light intensity. In wild-type plants, chlorophyll destruction increased but the redox status of ascorbate, and the activity of tAPX significantly decreased, while those in the transgenic plants were hardly changed. These observations suggest that tAPX is a limiting factor of antioxidative systems under photo-oxidative stress in chloroplasts, and that the enhanced activity of tAPX functions to maintain the AsA content and the redox status of AsA under stress conditions (Yabuta et al., 2002). A putative peroxisomal membrane-bound APX from Arabidopsis was expressed in tobacco plants, the transgenic plants had more protection against oxidative stress damage caused by aminotriazole which inhibits CAT activity that is found mainly in glyoxysomes and peroxisomes and leads to accumulation of H_2O_2 in those organelles (Wang et al., 1999). Most recently, Li et al. (2009) reported that transgenic tobacco plants overexpressing *Populus* peroxizomal ascorbate peroxidase (*PpAPX*) gene showed 80% higher APX activity in the leaves in response to drought or salt stress. The above evidence all points to APX having a key role in antioxidant and pathogen defence, with specialized and differentially regulated isoforms able to respond to salinity, drought, heavy metal, chilling, high light intensity, UV radiation and pathogen induced oxidative stress.

3.1.6 Conclusion

APX is one of the most important antioxidant enzymes of plant that detoxifies H_2O_2 using AsA for reduction. Considering the specific distribution, roles of the APX isoenzymes and the potential for active oxygen species production in each organelle, it seems likely that the APX isoenzymes are expressed by distinct regulatory mechanisms. APX activity and APX isoenzyme gene expression are variable even under the same stress conditions and some of them are constitutively expressed for the immediate and efficient detoxification of H_2O_2 under normal and oxidative stress conditions, while the gene expression for cAPX is responsive to environmental changes, resulting in the protection of important cellular compartments from oxidative stress and in strict control of the level of H_2O_2 in intracellular signalling.

3.2. Monodehydroascorbate Reductase (MDHAR)

MDHAR is one of the major components of the AsA–GSH cycle for the protection of plants against oxidative damage by ROS (Noctor and Foyer 1998). With its ability to directly regenerate AsA from MDHA, MDHAR plays an important role in maintaining reduced pool of AsA and ascorbate redox state. It has been suggested that regeneration of AsA is regulated in this cycle mainly by NADPH-dependent MDHAR activity (Shalata et al., 2001; Shalata and Tal 1998; Mittova et al., 2000). However, recent studies showed that both MDHAR and DHAR are equally important in regulating AsA level and its redox state under oxidative stress condition.

3.2.1 Regulation of MDHAR activity under abiotic oxidative stress

Increased MDHAR activity and total ascorbate content in response to a variety of stresses, which are believed to induce oxidative stress, have been documented in conifer needles (Mehlhorn et al., 1986), leaves of wheat (Mishra et al., 1993), spinach (Schoner and Krause 1990) and submerged rice seedlings (Ushimaru et al., 1992). Studies based on sensitive and tolerant cultivar of crop species also give us important clue to dissect the regulatory role of MDHAR during oxidative stress tolerance and acclimation. Shalata et al. (2001) showed that salt tolerant wild relative *L. pennellii* showed higher MDHAR activity in root tissues. However, in the case of sensitive cultivar the activity was decreased with the decreasing level of AsA and AsA redox state. In addition, chloroplastic MDHAR activity increased 3-fold in salt tolerant *L. pennellii* subjected to 100 mM NaCl stress as compared to cultivated species (Mittova et al., 2002). Similarly, Mittova et al. (2003a) showed that MDHAR activity increased in both mitochondrial and peroxisomal fraction in wild salt tolerant tomato species but the activity decreased in salt sensitive cultivar. The increase in MDHAR activity was accompanied by the increase in AsA content and AsA/DHA ratio. However, in sensitive cultivar, decrease in AsA content and increase in DHA content were observed with simultaneous decrease in AsA/DHA ratio, indicating that MDHAR activity plays a significant role in maintaining ascorbate redox state. Due to higher antioxidant capacity, salt tolerant *L. pennellii* maintained lower level of H_2O_2 and MDA, whereas cultivated tomato showed greater oxidative damage by increasing the H_2O_2 and MDA level. In another study, Hernandez et al. (2000) found that MDHAR activity increased in salt tolerant pea genotypes while the activity decreased in sensitive genotypes. MDHAR as well as DHAR activity decreased in ascorbate deficient *vtc* mutant subjected to short term (48 h) salt stress in addition with the decreasing level of AsA and AsA/DHA ratio, indicating the roles of both the enzymes in ascorbate recycling. The level of H_2O_2 was also increased in the mutant indicates the sensitivity of the mutant and inhibition of the AsA-GSH cycle (Huang et al., 2005). Sumithra et al. (2006) reported that the activity of MDHAR activity increased with the corresponding increased of salt stress level. Increase in MDHAR activity was observed in shoot tissues of rice seedlings under mild drought but the activity decreased under severe drought stress (Sharma and Dubey 2005). Similar phenomenon was also observed in mung bean seedlings subjected to drought stress where MDHAR activity increased under mild drought but decreased under severe drought stress with the increasing level of H_2O_2 and MDA (Hossain and Fujita 2009b). In another study, in mung bean seedlings MDHAR activity increased under moderate salt stress (150 mM NaCl), however, the activity declined under severe salt stress (300 mM NaCl) and the lower MDHAR activity was related to the lower AsA level (Hossain and Fujita unpublished data). However, drought acclimatized wheat seedlings maintained higher MDHAR activity in addition with other enzymes of AsA-GSH cycle and confer enhanced oxidative stress tolerance by well coordinated induction of antioxidant defence (Selote and Chopra 2006). Increase in MDHAR activity due to long-term drought stress was also observed in diploid hydrid *P. densata*. Hybrid plants showed higher MDHAR activity which was accompanied by higher AsA and lower DHA content as compared to their parents (Gao et al., 2009). In *Vinga unguiculata*, decrease in apoplastic AsA content was observed Mn-sensitive cultivar although the activity of MDHAR was increased in both the sensitive and resistant cultivars but sharp increase was observed in the resistant cultivars. But it is important to note that MDHAR plays a prominent role when DHAR is very low or absent (Fecht-Christoffers et al., 2003). MDHAR activity was unchanged in wheat roots subjected to Cd stress after 3 d of treatment but after 7 d of treatment the activity was increased with AsA level. However, the increase of MDHAR activity was not enough to prevent lipid peroxidation and protein oxidation (Paradiso et al., 2008). Significant increase in MDHAR activity was observed in scot pine seedlings subjected to 5 and 50 μM Cd concentrations after 96 h of treatment but the AsA content decreased with the decreasing level of APX activity (Schutzendubel et al., 2001). MDHAR activity increased in the rice seedlings subjected to Al stress (Sharma and Dubey 2007). Additionally, MDHAR and DHAR activity was found to increase in rice shoots and roots subjected to Ni stress suggesting that Ni activates the AsA regenerating system in order to maintain the elevated level of AsA (Maheshwari and Dubey 2009). In wheat plants, DHAR activity declined during the early stages of seedling growth under high-light regime and this implies that the regeneration of AsA for scavenging H_2O_2 is catalyzed mostly by MDHAR and not by DHAR (Mishra et al., 1995). Very recently, Zhou et al. (2009) showed that lettuce leaves subjected to excess light stress (1000-1200 μmol m^{-2} s^{-1}) nearly 2-fold increase in MDHAR activity was observed in comparison with low light stress. Increase in $O_2^{\bullet-}$ and MDA were accompanied by the increase of DHAR, GR, APX activity as well as non-enzymatic antioxidant like AsA, GSH level indicated that upon imposition of oxidative stress plants upregulate AsA-GSH cycle to reduce ROS mediated damage. In addition to increase of MDHAR activity, decrease in MDHAR activity was also observed in different plant species subjected to oxidative stress. Tobacco BY-2 cells subjected to salt and heavy metal stresses showed decrease in MDHAR activity with the decreasing level of AsA but H_2O_2 and lipid peroxidation level increased (Hoque et al., 2007; Islam et al., 2010). MDHAR activity remained unchanged in wheat leaves subjected to UV-B stress but subsequent recovery from freezing stress, sharp increase in MDHAR activity was observed. Enhanced H_2O_2 during UV-B irradiation was followed by the upregulation of MDHAR and other enzymes of AsA-GSH cycle, suggest that H_2O_2 acted as a signalling molecule than directly induced oxidative damage (Yang et al., 2007). Recently, Gao and Zhang (2008) reported that ascorbate deficient *Arabidopsis* mutant *vct1* showed increase in MDHAR activity due to exposure of UV-B radiation but the AsA level decreased. Although MDAHR and DHAR activities were increased, the increase is not sufficient to prevent oxidative damage. Reaction of O_3 with H_2O and solutes in the apoplast lead to the formation of ROS including H_2O_2, $O_2^{\bullet-}$, $^{\bullet}OH$, $^1O_2$ and apoplast AsA is most likely an important first line of defence against O_3, but it is essential that this antioxidant is to be maintained in a reduced state to effect this defensive action. In *Liriodendron tulipifera*, significant increase MDHAR activity was observed after 5 d of exposure to O_3 stress and showed continual increase up to 10 d of exposure and then decreased at 15 d of treatment. However, the AsA level was much higher after 5 d of O_3 exposure and the level reached the level of control plant after 15 d. This result implies that MDHAR plays a significant role in maintaining AsA level and its redox state during oxidative stress (Ryang et al., 2009).

Additionally, there are also few reports about MDHAR activity in other physiological processes that are related to oxidative stress. For instance, during senescence of pea leaves, a simultaneous decrease in MDHAR and APX activities was observed (Jimenez et al., 1998). During maturation of pepper (*Capsicum annuum*) fruits, an increase of AsA content has been described that can be attributed to an increase in L-galactono-γ-lactone dehydrogenase and MDHAR activities (Imahori et al., 1998). Recently, Stevens et al. (2008) found that the MDHAR activity and the AsA pool in tomato fruits are correlated with the reduced loss of firmness during chilling, and therefore, has implications for the post-harvest quality of tomatoes.

3.2.2 Regulation of MDHAR activity during biotic stress-induced ROS detoxification

ROS production occurs preferentially during early phases of beneficial plant/microbe interactions and it is generally believed that antioxidant enzymes are upregulated with the appearance of ROS during the infection to maintain a constant redox balance in the cytoplasm (Blilou et al., 2000; Lambais et al., 2003). Increased MDHAR activity was observed in chloroplastic, mitochondrial and soluble fractions of cucumber and tomato leaf infected with CMV (Song et al., 2009). MDHAR activity decreased in soluble fraction of pea plant infected with PPV after 3 dpi and increased in chloroplastic fraction. However, at 15 dpi the activity decreased in both soluble and chloroplastic fraction as compared to 3 dpi (Diaz-Vivancos et al., 2008). In peach, PPV infected leaves showed decreases of MDHAR activity in the chloroplastic fraction, suggesting that these plants have a lower capacity to AsA regeneration which favors the long-term PPV-induced oxidative damage as shown by the increase in lipid peroxidation and protein oxidation and the appearance of oxidative microbursts (Hernandez et al., 2004b). Infection of tomato leaves with the necrotrophic fungus *B. cinerea* resulted in substantial changes in enzymatic and non-enzymatic components of the AsA-GSH cycle, starting from second day of infection diminished the mitochondrial antioxidant capacity by decreasing activities of MDHAR as well as declining AsA and GSH contents. This was accompanied by DHA and GSSG accumulation and decreased AsA and GSH redox ratios (Kuzniak and Sklodowska 2004a). However, the enzymes responsible for AsA recycling showed a characteristic biphasic pattern of activity in tomato leaf infected with *B. cinerea* infection. Peroxisomal MDHAR activity was increased during the early plant-pathogen interaction whereas the appearance of the diseases symptoms was accompanied by a notable suppression of this enzyme (Kuzniak and Sklodowska 2005b). MDHAR activity increased in both sensitive and resistant lines in apoplastic fractions of barley leaves infected with *B. graminis* but decreased in foliar fractions (Vanacker et al., 1998). It appears, therefore, that the redox balance-related events during the infection process could be species specific or different compensatory mechanisms could be activated to cope with the changing ascorbate and glutathione redox status.

3.2.3 Expression of MDHAR gene during oxidative stress

Several expression studies showed that cytosolic, chloroplastic, mitochondrial and peroxizomal MDHAR genes are upregulated upon various stresses that accelerate the production of ROS (Ozturk et al., 2002; Chew et al., 2003; Yoon et al., 2004; Secenji et al., 2010). Rapid increase in *Brasscia campestries* MDHAR (*BcMdhar*) transcripts were observed in response to oxidative stress invoked by H_2O_2, SA, paraquat and O_3 indicating the importance of this gene in the ROS defence system in plants (Yoon et al., 2004). Expression of two MDHAR genes dramatically increased under osmotic stress in moss (*Physcomitrella patens* Hedw.) and upregulation of MDHAR gene expression in addition with MDHAR activity was also observed in leaves and roots of Kentucky bluegrass under drought stress (Lunde et al., 2006; Bian and Jiang 2009). Cold acclimation increased about 10-fold increase in MDHAR gene expression in near-isogenic wheat lines also suggest that increased expression for specific genes are responsible for oxidative stress tolerance (Baek and Skinner 2003). Pea peroxisomal MDHAR mRNA levels were found to increase in response to wounding, cold and 2, 4-D stresses (Leterrier et al., 2005). Similarly, Grantz et al. (1995) reported that mRNA levels of tomato MDHAR were increased by wounding. Upregulation of chloroplastic MDHAR mRNA were also reported by Chew et al. (2003) in *Arabidopsis* plant subjected to oxidative chemical stresses targeted to chloroplast and/or mitochondria. Hernandez et al. (2004a) reported that *chlMDHAR* increase in transcript level 2- to 7-fold in pea leaves subjected to high light conditions suggest that in chloroplast AsA is regenerated via MDHAR. Apart from chloroplastic and mitochondrial MDHAR gene expression, recent reports showed the upregulation of cytosolic MDHAR mRNA in both tolerant and sensitive wheat variety under drought stress conditions (Secenji et al., 2010). Additionally, sulphur starvation in rice seedlings showed upregulation of *OsMDHAR2* and protects the plant against oxidative stress due to lack of sufficient GSH pool (Lunde et al., 2008). Therefore, the enhanced gene expression along with increased or stable activities of MDHAR is beneficial to antioxidative protection.

3.2.4 Transgenic MDHAR plants and oxidative stress tolerance

Although MDHAR plays a significant role in maintaining AsA redox state, very few reports are available about oxidative stress tolerance of plants through overexpressing the MDHAR gene. Transgenic tobacco plant overexpressing *Arabidopsis thaliana* MDHAR gene (*AtMDR1*) in the cytosol showed enhanced stress tolerance in terms of significantly higher net photosynthesis rates under O_3, salt and PEG induced oxidative stresses and greater PSII effective quantum yield under O_3 and salt stresses. Transgenic plants exhibited up to 2.1-fold higher MDHAR activity and 2.2-fold higher level of AsA compared to non-transformed control plants. Transgenic plants also exhibited significantly lower H_2O_2 level under salt stress conditions (Eltayeb et al., 2007). Therefore, greater PSII photochemical activity in MDHAR transgenic plants might be conferred by efficient supply of the regenerated AsA maintained by MDHAR overexpressed transgenic tobacco. These observations clearly demonstrated an important function of MDHAR in leaves to maintain photosynthetic function by limiting ROS-mediated damage.

3.2.5 Conclusion

AsA is present in most cellular compartments and several pathways exist to ensure AsA recycling. MDHAR is an enzymatic component of the AsA-GSH cycle presents in plant cells and plays a dominant role in recycling of AsA to maintain ascorbate redox state. MDHAR gene expression, as well as MDHAR activity responds to oxidative stress conditions in different ways. Therefore, it is important to characterize which isoform is crucial for a particular stress and how the expression of that particular isoform is regulated. Transgenic plants with MDHAR gene targeted for different subcellular compartments further find their roles in ROS detoxification and regulation in the near future.

3.3 DEHYDROASCORBATE REDUCTASE (DHAR)

The univalent oxidation of AsA leads to the formation of MDHA. If MDHA is not rapidly reduced again to AsA by MDHAR, it will spontaneously disproportionate into AsA and DHA. DHA is then reduced to AsA by DHAR in a reaction requiring GSH (Chen et al., 2003). In the absence of sufficient DHAR, however, DHA undergoes irreversible hydrolysis to 2,3-diketogulonic acid. Rapid regeneration is necessary in order to maintain the antioxidative capacity of AsA. DHAR allows the plant to recycle DHA, thereby recapturing AsA before it is lost. Because AsA is the major reductant in plants, DHAR serves to regulate its redox state. Thus, DHAR is a physiologically important reducing enzyme in the AsA-GSH recycling reaction for higher plants.

3.3.1 Regulation of DHAR activity under abiotic oxidative stress

Increased DHAR activity was reported in response to various ROS-inducing stresses including H_2O_2, MV, O_3, salt, drought, and low temperature treatments (Mittova et al., 2003a; Eltayeb et al., 2006; Ushimaru et al., 2006; Lee et al., 2007). Mottova et al. (2003a) reported that salt stress significantly induced DHAR and MDHAR activity in mitochondrial function of salt tolerant *L. pennellii* whereas the activity unchanged in sensitive *L. esculentum* and tolerant genotypes showed higher AsA level and AsA/DHA ratio as compared to sensitive genotypes. In addition, tolerant *L. pennelli* showed lower oxidative damage. Likewise, Hernandez et al. (2000) found that DHAR activity increased to 2-fold in salt tolerant genotypes and the activity decreased in the salt sensitive genotypes. A sharp increase in DHAR activity was observed in shoot tissues of rice under mild drought stress but the activity decreased under severe drought stress (Sharma and Dubey 2005). In mung bean seedlings, mild drought stress slightly increased DHAR activity but the activity declined under severe drought stress with decreasing level of AsA and higher oxidative damage in terms of MDA (Hossain and Fujita 2009b). Ma et al. (2008) reported that heat stress induced DHAR activity in apple leaves. In addition, crofton weed subjected to heat and drought stress, the upregulation of DHAR activity was observed rather than MDHAR, suggesting that under these stress conditions, AsA is regenerated via GSH dependent DHAR (Lu et al., 2007). This is in contrast with the situation described by Moran et al. (1994) who proposed that MDHAR was the key detoxifying enzyme under drought stress conditions. It may be that DHAR activity could participate in AsA regeneration under conditions of severe stress when MDHAR activity is limited by the availability of NADPH (Asada and Takahashi 1987). UV-B radiation in ascorbate deficient mutant *vct1* showed increase in DHAR activity within 24 h but the activity decreased with the increase of treatment period. The decrease in DHAR activity was accompanied by the decrease of AsA level and increase of H_2O_2 and lipid peroxidation level (Gao and Zhang 2008). DHAR activity increased in wheat roots and leaves subjected to different concentration of Cd (10, 20, 40 μM) but the increase in AsA level is not enough to prevent protein or lipid oxidation (Paradiso et al., 2008). Schützendübel et al. (2002) showed that DHAR activity decreased in *Populus x Canescens* roots subjected to 50 μM Cd stress within 15 h of treatment but later the activity increased significantly with the progression of stress period. Additionally, DHAR activity was found to increase in rice shoots and roots subjected to Ni stress suggested that Ni activates the AsA regenerating system to maintain the elevated level of AsA (Maheshwari and Dubey 2009). *Liriodendron tulipifera* plant subjected to O_3 stress significant increase DHAR activity were observed after 5 d of exposure and but decreased after 10 d. Importantly, the activity of DHAR again increased when the MDHAR activity was decreased after 15 d of treatment. This result indicates that when one antioxidant system is inhibited then plant exhibits another antioxidant defence system to face the adverse challenge induced by oxidative stress (Ryang et al., 2009). Recently, Zhou et al. (2009) showed that lettuce leaves subjected to excess light stress nearly 2-fold increase in DHAR activity was observed in comparison with low light stress. Initial increase in DHAR activity was observed in *Hevea brasiliensis* subjected to chilling stress but the activity decreased with the increase of treatment duration (Mai et al., 2009). Very recently, Lu et al. (2009) reported that DHAR activity increased in Ginko biloba leaves subjected to both O_3 and elevated O_3 +CO_2 treatments on the 40th d and showed continual increase up to the 80th d but the H_2O_2 and lipid peroxidation level was significantly higher than the 40th d. It suggested that the increase in activity of antioxidant enzymes was an acclimation effect, but the increase was insufficient to reduce the ROS induced damage as evidenced by lower AsA content and higher MDA level. Rice seedlings subjected to heat stress showed significant increase in DHAR activity within 6 h of treatment (Urano et al., 2000). Heat induced alteration of DHAR activity in tobacco BY-2 cells showed an important phenomenon in maintaining redox balance when MDHAR activity significantly decreased. At 55°C significant increase in DHAR activity was observed in cytosolic and mitochondrial fraction. Therefore, DHAR may optimize the utilization of the still available AsA in a cellular organelle that is strongly subjected to oxidation by the ROS overproduced. The increase in DHAR activity could be a sort of feedback regulation mechanism occurring to improve AsA regeneration from DHA when AsA depletion occurs at its production sites due to enhanced ROS (Locato et al., 2009).

3.3.2 Regulation of DHAR under biotic stress induced ROS detoxification

DHAR activity deceased in tomato leaves infected with *B. cinerea* with an increase of DHA (Kuzniak and Sklodowska 2001). Later on, they showed that DHAR activity gradually decreased with the decreasing level of AsA in mitochondrial fraction of tomato leaves (Kuzniak and Sklodowska 2004a). Tomato plants infected with *B. cinerea*, the chloroplastic DHAR activity was not affected except for a marked increase after three days of infection, whereas mitochondrial and peroxisomal DHAR activities were distinctly decreased (Kuzniak and Sklodowska 2005a,b). This decrease of DHAR and other AsA-GSH cycle enzyme denotes a general shift of the cellular redox environment towards the oxidative state and *B. cinerea* was able to overcome protective antioxidant barrier related to the antioxidant AsA-GSH cycle capacity. The most dynamic changes in enzyme activities and redox potentials were correlated with the appearance of disease symptoms indicating the breakdown of an active plant defence response. DHAR activity did not change in the peach leaf infected with PPV (Hernandez et al., 2004b) but in apricot cultivars, foliar DHAR activity increased and the increase was much higher in resistant plants (300%), suggesting that resistant cultivar had a higher capacity than the susceptible plants for the regeneration of AsA (Hernandez et al., 2001). The higher DHAR induction in the PPV-resistant cultivar may contribute to an increased antioxidant capacity, which may be related, amongst other factors. Additionally, DHAR activity increased in chloroplastic, mitochondrial and soluble fraction of tomato leaf infected CMV but cucumber leaf infected with CMV showed increased DHAR activity only in mitochondrial fraction, decrease in chloroplast fraction and unchanged in soluble fraction (Song et al., 2009). DHAR activity increased in both sensitive and resistant lines in apoplastic and foliar fractions in barley leaves infected with *B. graminis* (Vanacker et al., 1998). Rice plants infected with endophytic fungus *Piriformospora indica* showed significant increase in DHAR activity after one day of inoculation which was accompanied by higher AsA level and lower DHA level, further demonstrating its role in regulating AsA level and AsA redox state under the condition of pathogen induced oxidative stress (Waller et al., 2005).

3.3.3 DHAR gene expression during oxidative stress

Chilling stress in *Hevea brasiliensis* showed 2-fold increase in DHAR gene expression after 96 h of treatment and the increase in gene expression was correlated with the decreased level of MDA (Mai et al., 2009). Likewise, Beak and Skinner (2003) reported an increased transcript level of the chloroplastic DHAR gene during cold acclimation of near-isogenic wheat lines. Ma et al. (2008) reported that DHAR gene expression enhanced in the leaves of apple plant subjected to short-term (4 h) heat stress but rice seedlings subjected to heat stress, the induction of the DHAR1 gene was recognized after 12 h, suggesting that this gene is not directly induced by heat shock. The DHAR1 gene might be rather induced by a second signal, such as an increase in DHA or in a ratio of DHA relative to ascorbate under stress conditions (Urano et al., 2000). Drought stress induced 1.5-fold cytosolic DHAR gene expressions in tolerant wheat genotype whereas 2.6-fold increase in chlDHAR in drought sensitive genotype (Secenji et al., 2010). Although chlDHAR gene expression did not showed sharp increase but higher basal expression of the genes result in more efficient ROS elimination due to more intense AsA recycling and thus contribute to the drought tolerance in resistant genotype. Salt stress induced a 2.2-fold in cytosolic DHAR mRNA expression in leaves and roots of model legume *Lotus japonicus*. The requirement for a high DHAR activity in the cytosol in connection with AsA regeneration in the apoplast would explain the reason for upregulation of cDHAR transcript under salt stress conditions (Rubio et al., 2009). Significant increase in DHAR transcript was also observed in *Arabidopsis* roots and leaves subjected to 10 μM $CdSO_4$ for 3 weeks (Smeets et al., 2008). Additionally, upregulation of DHAR were observed in tomato cultivar 'Valenciano' subjected to O_3 stress, but no increase in the transcript was observed in 'Nikita' and 'Alisacraig'. Variations in expression were also observed due to exposure time and also for plant growth conditions (Marco et al., 2008). Analysis of *Arabidopsis* mutant regarding cytosolic DHAR gene also dictates the role of DHAR gene in stress tolerance and response. Exposure of O_3 increased the expression of cytDHAR gene whereas cytDHAR deficient mutant was highly sensitive to O_3. Although, the amounts of total ascorbate and glutathione were similar in both lines, but the amount of apoplastic ascorbate in the mutant was 61.5% lower. These results indicate that the apoplastic AsA, which is generated through the reduction of DHA by cytDHAR, is important for O_3 tolerance (Yoshida et al., 2006).

3.3.4 Transgenic DHAR plants and oxidative stress tolerance

There is growing evidence that overexpression of DHAR gene in plants enhances tolerance to various oxidative stresses. Transgenic tobacco plants overexpressing the cytosolic DHAR gene exhibited 2.3- to 3.1-fold higher DHAR activity, 1.9-to 2.1-fold higher level of AsA and ascorbate redox state and exhibited an enhanced tolerance to O_3, drought, salt, and PEG induced oxidative stresses and showed higher net photosynthesis compared with non-transformed control plants (Eltayeb et al., 2006). Transgenic Arabidopsis that express the rice DHAR gene showed enhanced resistance to salt stress despite the slight increase of DHAR activity and total ascorbate (Ushimaru et al., 2006). Transgenic tobacco plants overexpressing the wheat DHAR gene showed enhanced AsA content and protection against O_3 (Chen et al., 2003; Chen and Gallie 2005). Furthermore, transgenic tobacco plants expressing the human DHAR gene showed increased tolerance to MV, H_2O_2, low temperature and salt induced oxidative stresses (Kwon et al., 2003). The cellular DHAR activity was drastically enhanced in the transgenic tobacco plants in both these cases. Lee et al. (2007) first reported that transgenic plant expressing three antioxidant enzymes including the DHAR gene enhanced tolerance to multiple stress tolerance by maintaining higher ratios of AsA/DHA and GSH/GSSG. Importantly, tobacco plants overexpressing DHAR gene showed increased susceptibility to drought stress because of a higher percentage of open stomata, while plants suppressed for DHAR expression exhibited increased drought tolerance (Chen and Gallie 2004). This suggests that excessive enhancement of the DHAR activity in transgenic plants results in decreased tolerance to some kind of stress and that an appropriate quantity of DHAR expression is important for improvement of stress tolerance. Recently, Yin et al.

(2010) showed that transgenic tobacco overexpressing DHAR gene confer tolerance to Al induced oxidative stress, due to efficient regeneration of DHA to AsA and higher APX activity and by reducing H_2O_2 and lower lipid peroxidation level. Although they previously reported that MDHAR transgenic plant enhanced tolerance to O_3, salt and PEG stress but MDHARs have little effect on AsA regeneration and redox state under Al-stress induced oxidative stress. Most recently, Wang et al. (2010) showed that DHAR overexpression was increased up to 1.5- to 5.4-fold, which enhanced foliar ascorbic acid levels 2- to 4.25-fold and ratio of AsA/DHA about 3- to 16-fold relative to wild type. In addition, the level of glutathione also increased as did its redox state. When whole plants were treated with high light and high temperature stress or *in vitro* leaf discs were subjected to 10 μM paraquat, transgenic plants showed a larger AsA pool size, lower membrane damage, a higher level of chlorophyll compared to control. Therefore, increasing AsA content through enhanced ascorbate recycling could limit the deleterious effects of environmental oxidative stress. Taken as a whole, it can be concluded that overexpressing DHAR in different subcellular compartments contributes significantly in enhancing plants tolerance to oxidative stress by efficient recycling of AsA.

3.3.5 Conclusion

Regulation of AsA levels in plants is tightly controlled by the level of its synthesis, recycling, degradation, and transportation. In addition to MDHAR, DHAR plays a critical role in the protection of cellular components by maintaining the proper level AsA and other cellular substrates. Elevated AsA levels through increase DHAR activity as well as overexpression of DHAR in different subcellular compartments contribute significantly in enhancing plants tolerance to oxidative stress. In addition, DHAR plays an important role in the regulation of growth, differentiation, and various metabolisms in higher plants. Further studies based on mutant and transgenic for different subcellular isoforms and determination of various ROS levels under stressing conditions are needed and will be valuable in answering any questions that might remain.

3.4 GLUTATHIONE REDUCTASE (GR)

GR belongs to a family of flavin-containing pyridine nucleotide-disulphide oxido-reductases which catalyses the reduction of GSSG to GSH with the accompanying oxidation of NADPH. GR plays an essential role in cell defence against reactive oxygen metabolites by sustaining the reduced status of glutathione and ascorbate pools which in turn maintain cellular redox state under stress (Lascano et al., 1999; Ansel et al., 2006; Romero-Puertas et al., 2006; Pastori et al., 2000; Hossain and Fujita 2010). Biochemical and molecular studies have shown that GR is an important enzyme for scavenging the ROS, which are continuously generated in different compartments during environmental and biotic stresses.

3.4.1 Regulation of GR activity under abiotic oxidative stress

Many studies have shown that stress-tolerant plants tend to have higher activities of GR. Hernandez et al. (2000) found that GR activity increased in salt tolerant pea genotype whereas the activity decreased in the salt sensitive genotype due to long-term salt stress (70 mM). Likewise, GR activity increased with the increasing level of salt stress in the salt tolerant *P. maritima* whereas a sharp decrease was observed in the sensitive genotype. This was accompanied by the higher lipid peroxidation level (Sekmen et al., 2007). In addition, Shalata et al. (2001) reported that salt tolerant wild relative of tomato, *L. pennellii*, showed significant increase in GR activity but in case of sensitive cultivar, the activity was as like as control plant with decreased GSH redox state. These adaptive behaviours of salt tolerant and sensitive genotypes clearly suggest that that GR plays a significant role in maintaining glutathione redox state under oxidative stress conditions. However, in most cases the activity of GR under stressful condition depends on the duration or intensity of stress imposed. In mung bean seedlings, GR activity increased significantly due to short term (24 h) of high salt stress (300 mM) but the activity declined after 48 h with the decreasing level of GSH and GSH redox state in connection with increased H_2O_2 and MDA level (Hossain and Fujita 2010). Similarly, moderate salt stress (150 mM NaCl) significantly increased GR activity; however, the activity decreased under severe salt stress (300 mM NaCl) with increased level of H_2O_2 and MDA (Hossain and Fujita unpublished data). GR activity increased in ascorbate deficient *vtc-1* mutant subjected to short term (48 h) salt stress and GSH is more accumulated in the mutant as compared to the wild type while the redox state of glutathione is constant both WT and *vct-1* (Huang et al., 2005). GR activity increased in *B. juncea* seedlings subjected to salt stress with increased level of H_2O_2 and lipid peroxidation (Verma and Mishra 2005). Increases of GR activity during salt stress were also reported in rice (Demiral and Türkan 2004), wheat (Sairam et al., 2005), melon (Yasar et al., 2006) and in mung bean (Sumithra et al., 2006). The increase of GR activity increased the ratio of NADP+/NADPH and ensure the availability of $NADP^+$ to accept electrons from the photosynthetic electron transport chain thus reducing the formation of $O_2^{\bullet -}$ which reduce the facilitation of metal-catalysed formation of the $^{\bullet}OH$ through the Haber-Weiss reaction (Sudhakar et al., 2001). Drought tolerant rice genotype (N22) significantly increased GR activity and antioxidant metabolites such as AsA and GSH in connection with decreased H_2O_2 level. However, in the susceptible genotype (N118) the activity of GR decreased with decreased level of GSH and AsA and increased level of H_2O_2. The author proposed that during water stress an integrated antioxidant defence system including AsA-GSH cycle in developing panicles are important factors related to enhanced spikelet fertility in the drought resistant rice genotype under upland conditions (Selote and Chopra 2004). High levels of GSH during drought stress are achieved through high activity of GR (Gamble and Burke 1984; Hossain and Fujita 2009b) but the GSH/GSSG ratio is likely to be influenced by the duration and severity of stress and the plant species (Smirnoff 1993; Hossain and Fujita 2009b, 2010). Short-term mild drought stress significantly increased in GR activity, but in case of severe drought stress the activity declined including glutathione redox state, indicating that GR activity plays a significant role in maintaining glutathione level in plants under stress (Hossain and Fujita 2009b). GR activity was increased in crofton weed subjected to cold and drought stress, however, the activity decreased under heat stress conditions (Lu et al., 2007). In contrast, Lascano et al. (2001) found no significant differences in the four drought-tolerant wheat varieties after one month of drought stress in the field but observed an increase in GSH in the leaves within 48 h of a short-term osmotic stress. Drought stress significantly increased GR activity in poplar species though the increase is not enough to prevent the oxidation of lipid (Yang et al., 2009). Short-term (24 h) drought stress treatment showed an increase in GSH and GR activity in wheat leaves (Bartoli et al., 1999). Progressive increase in GR activity was also observed in *P. densata* hybrid as compared to control when plants subjected to long-term drought stress (28 d) with concomitant increases in GSH and GSSG (Gao et al., 2009). In addition to the antioxidant enzymes described already, an increase in GR activity during stomatal closure in response to water stress has also been reported (Sairam et al., 1997/98). Therefore, a drought-tolerant genotype can not only retain sufficient water but also possess a highly active antioxidative system to guard against oxidative stress. Tobacco BY-2 cells subjected to heavy metal and salt stress showed decrease in GR activity with the decreasing level of GSH (Hoque et al., 2007; Islam et al., 2010) and with concomitant increase in H_2O_2 and lipid peroxidation. The breakdown of AsA and GSH regeneration system including GR activity is main reason behind the oxidative damage. Chilling tolerant rice cultivar (Xiangnuo-1) showed higher GR activity, whereas the activity decreased in the chilling sensitive cultivars in connection with higher level of MDA and electrolyte leakage (Huang and Guo et al., 2005). Zhou et al. (2009) showed that lettuce leaves subjected to high light stress had a significant increase in GR activity in addition to a higher GSH level. Increase in GR activity was observed in *Liriodendron tulipifera* under elevated O_3 stress (Ryang et al., 2009). GR activity increased in G. biloba leaves subjected to both O_3 and elevated O_3 +CO_2 treatments and increase in GR activity mitigated the oxidative damage to some extent (Lu et al., 2009). Likewise, increase in GR activity during O_3 stress was reported in pea (Edwards et al., 1994) and in *Arabidopsis* (Rao et al., 1996). Ascorbate deficient mutant *vct1* subjeted to UV-B radiation showed increase in GR activity within 8 h of treatment but the activity decreased with the increase of exposure time accompanied by the decrease of GSH level and increases of H_2O_2 level and lipid peroxidation (Gao and Zhang 2008). Twofold increase in GR activity was observed in both tolerant and susceptible genotype of pigeonpea up to 2 d of waterlogging. Further waterlogging resulted in decrease GR activity in susceptible genotypes whereas tolerant genotype showed 3-fold increase in GR activity up to 6th day of waterlogging (Kumutha et al., 2009). GR activity increased in scot pine seedlings subjected to Cd stress with the increasing level of GSH and decreasing the level of GSSG, indicating its role in maintaining glutathione level in plants (Schützendübel et al., 2001). It is now evident that GR responds differentially to different environmental and biotic stresses even within the same plant species, however, the increase of GR activity due to oxidative stress is a common phenomenon.

3.4.2 Regulation of GR activity during biotic stress induced ROS detoxification

In tomatoes, increased GR activity which is responsible for regeneration of GSH is seen in almost all the major cell organelles, showing the importance of GR (Kuzniak and Sklodowska 2001, 2004a,b, 2005a,b; Foyer and Noctor 2003). GR activity increased in apoplastic fraction of resistant barley lines infected with *B. graminis* but decreased in foliar fractions. Conversely, GR activity decreased in both fractions of sensitive cultivars (Vanacker et al., 1998). Additionally, GR activity increased following inoculation and persisted for whole course of experiment reaching maximum at 4 d in tomato plant infected with *B. cinerea* (Kuzniak and Sklodowska 1999, 2001). Substantial increase in GR activity was observed in leaf chloroplastic fraction but the peroxisomal and mitochondrial GR activities decreased (Kuzniak and Sklodowska 2005a). Similarly, GR activity was significantly decreased with decreasing level of GSH in peach leaf infected with PPV (Hernandez et al., 2004b), suggesting that the plants have a lower capacity of GSH regeneration during pathogen attack. In another study, Kuzniak and Sklodowska (2004a) showed that GR activity decreased with other AsA-GSH cycle enzymes in mitochondrial fraction of tomato plant infected with *B. cinerea* and it is found that GSH and AsA pools as well as ascorbate related enzymes were heavily suppressed when the spreading lesions started to develop. Song et al. (2009) showed that GR activity increased significantly due to CMV infection in tomato and cucumber. In the cucumber leaf, a significant increase in GR activity was observed in mitochondrial fraction but the activity was decreased in soluble fractions. In contrast, a significant increase in GR activity was observed in soluble fraction of tomato leaves. Tomato plants infected with *P. syringae* pv. tomato showed an increase in GR activity in both tolerant and susceptible cultivar, though, the increase in susceptible genotype was not high enough to prevent the oxidation of glutathione pool but its activity in the resistant one was high enough to account for an adequate glutathione turnover (Kuzniak and Sklodowska 2004b). GR activity was increased in soluble fraction after 3 dpi and decreased in chloroplastic fraction of pea plant infected with PPV. At 15 dpi, the GR activity was slightly decreased in soluble fraction whereas, significant decrease was observed in chloroplastic fraction (Hernandez et al., 2004b). Rice leaves infected with rice blast pathogen *Magnaporthe oryzae*, showed 90% increase in GR activity after one day of infection at seedling stage but infection in the neck region at the preliminary heading stage the GR activity was as like as control. Therefore, the less oxidative damage at seedlings stage was due to higher ROS scavenging activity indicating the role of GR in pathogen induced ROS detoxification (Hao et al., 2009). TMV inoculation and salicylic acid treatment showed significant increase in GR activity in inoculated leaves and non-inoculated leaves as well having contributes to the appearance of SAR in tobacco (Fodor et al., 1997).

3.4.3 Expression of GR gene during oxidative stress condition

Several expression studies showed that both cytosolic and chloroplastic GR genes are upregulated due to salt and drought stresses. Tolerant cultivar of wheat showed increased expression of both cytosolic *cytGR* and chloroplastic *chlGR* whereas in sensitive genotype the transcript level it remains unchanged (Secenji et al., 2010). In *Euphorbia esula*, 1.75- and 2-fold increases in GR transcript were observed due to drought and cold stress, respectively (Anderson and Davis 2004). In another study, Hong et al. (2009) showed that both cytosolic and chloroplastic isoforms of GR are upregulated in rice roots due to salt stress. Additionally, significant increases in both *chlGR* and *cytGR* transcript were observed in pea leaves subjected to excess light stress with the increasing level of H_2O_2 in the chloroplast (Hernandez et al., 2004a). A sharp increase in *cytGR* transcript was observed and this suggests that post-stress increase in *cytGR* transcript levels could be a common feature of different plant species. The author suggests that GSH levels or the redox status of glutathione could be involved in rapid signalling from the chloroplast to the nucleus during excess light stress (Karpinski et al., 1997). Besides *chlGR* transcript, *cytAPX* and *stAPX*, *chlMDHAR* transcripts were also increased following accumulation of H_2O_2, suggesting the role of H_2O_2 in their induction under high light stress. *Hevea brasiliensis* subjected to chilling stress, showed a 3-fold increase in GR gene expression after 96 h of treatment (Mai et al., 2009) and the increase in gene expression was correlated with DHAR and APX expression. The increase in activities and transcripts strongly suggest a response to cope with an accumulation of ROS. Similarly, GR gene expression was found to increase in apple leaves subjected to high temperature stress including APX and DHAR. In *Arabidopsis*, GR1 gene expression was significantly increased in both roots and leaves subjected to Cd stress although GR activity decreased. It is possible that Cd interacts with the translation mechanisms, disturbs the activity of GR. Indeed, an oxidative modification of this protein, which can result in a higher proteolytic degradation, has been reported after Cd exposure (Smeets et al., 2008; Romero-Puertas et al., 2002). Recently, Rubio et al. (2009) showed that both cytosolic and plastidal GR gene expression increased in *L. japonicus* due to salinity stress, indicating its role during salinity induced oxidative damage.

3.4.4 GR transgenic and oxidative stress tolerance

GR is a key enzyme for scavenging ROS and maintenance of redox potential of the cell during oxidative stress. Initial attempts to manipulate the levels of GR in transgenic plants were carried out with the GR gene from *Escherichia coli gor* gene in *N. tobaccum*. Transgenic plants showed 2-10 times increase in GR activity, but the cellular glutathione pools were not affected under both normal and MV treatments (Foyer et al., 1991). In addition, Aono et al. (1991, 1993) transformed bacterial *gor* gene and targeted to both cytosolic and chloroplastic compartments in *N. tabaccum* cv. SR1. Both the transgenic plants showed 3-fold increase in GR activity, but there was no effect on the glutathione pool. The transgenic plants showed less sensitivity to paraquat but not to the O_3 stress. Chloroplastic GR-transgenic plants showed lower susceptibility to SO_2 in light, suggesting that increasing GR activity both in chloroplasts and cytosol is effective in protecting cells from photo-oxidative damage. Additionally, transformed bacterial *gor* gene, in *B. juncea* showed 50-fold higher GR activity when targeted to chloroplast and a 2-fold increase when targeted to the cytosol. But only chloroplast GR transgenic showed Cd tolerance due to its higher glutathione level than the wild plants (Pilon-Smith et al., 2000). Transgenic poplar plants that expresses chloroplast-targeted *E. coli* GR were found to have GR activities up to 500-fold higher than untransformed plants. Leaves of these plants had approximately 2-fold higher levels of both glutathione and ascorbate relative to control plants or plants that expressed the GR gene without the chloroplast targeting sequence. Although these plants did not show increased protection from MV-induced inhibition of CO_2 assimilation, they were found to be more resistant to photoinhibition caused by high light intensity and chilling temperature (Foyer et al., 1995). These authors suggest that the protection from cold-induced photoinhibition in chloroplastic GR expressing plants could result from the increased pools of GSH and AsA and the more rapid cycling of these antioxidants to their reduced forms. GR transgenic cotton plants with increased GR activity have shown increased resistance to short-term chilling stress (Payton et al., 2001; Kornyeyev et al., 2001, 2003; Logan et al., 2003). Even though GR transgenic plants exhibited higher GR activity, there is no difference between the transgenic and wild-type plants when exposed to long-term chilling, indicating the need for other enzymes to combat the long-term chilling stress (Logan et al., 2003). Co-expression of cytosolic forms of both GR and SOD in tobacco plants provided more substantial protection from MV treatment than either enzyme alone indicating that the expression of combinations of antioxidant enzymes in transgenic plants may have synergistic effects on stress tolerance (Aono et al., 1995). Transgenic tobacco plants developed from bacterial GR gene suffered lipid peroxidation and under moderate light intensities in contrast to the wild type but these transgenic plants were more resistant towards oxidative stress induced by paraquat or H_2O_2 (Lederer and Boger 2003). Sometimes contrasting observation was observed in the transgenic plant transformed with the same construct and the transgenic showed nearly five times variation in GR activity (Foyer et al., 1995; Broadbent et al., 1995; Creissen et al., 1996). These differences in GR activity in different lines developed with the same construct might be due to position effect, cosuppression, or gene dosage effect.

3.4.5 Conclusion

In most of the biological functions, GSH is oxidized to GSSG, which should be converted back to GSH in plant cells to perform normal physiological functions. Hence, rapid recycling of GSH is more essential than synthesis of GSH, which is highly regulated and ATP-requiring process. GR is the only enzyme responsible for recycling of GSSG to GSH and control three important redox couples in plant cells. However, a large number of reports showed that GR activity along with other antioxidant enzymes showed better performance under stress condition, indicating the importance of coregulation of other enzymes along with the GR in limiting the concentration of ROS. Functional characterization of different subcellular isoform of GR and its interaction with plant hormone will clarify which isoform is important for a particular stress and how a particular isoform is regulated under oxidative stress condition.

3.5 Ascorbate (AsA)

In plant cells, ascorbic acid (AsA) is the most abundant antioxidant and serves as a major contributor to the cellular redox state (Smirnoff 2000). It has been detected in the majority of plant cell types, organelles and apoplast (Aro and Ohad 2003; Borland et al., 2006) and protects plants against oxidative damage resulting from aerobic metabolism, photosynthesis, and a range of biotic and abiotic stresses (Smirnoff 1996). AsA is synthesized in the mitochondria and is transported to the other cell components through a proton–electrochemical gradient or through facilitated diffusion (Shao et al., 2008). AsA protects the chloroplasts from photoinhibition (Asada 1999) and suppresses the accumulation of photoproduced H_2O_2 in the stroma by acting as the electron donor of APX to protect components of the Calvin cycle from inactivation (Asada 1999). Due to the ability of AsA to donate electrons in a number of enzymatic and non-enzymatic reactions, it is considered to be the most popular and powerful ROS detoxifying compound. It is the substrate of cAPX and the corresponding organellar isoforms, which are critical components of the AsA-GSH cycle for H_2O_2 detoxification (Nakano and Asada 1981; Dalton et al., 1986). In AsA-GSH cycle, two molecules of AsA is utilized by APX to reduce H_2O_2 to water with concomitant generation of MDHA (Fig. 3). AsA can directly quench $^1O_2$, $O_2^{\bullet-}$ and $^{\bullet}OH$ and regenerate α-tocopherol from α-chromanoxyl radical thereby providing protection to membranes (Thomas et al., 1992). AsA also acts as a co-factor of violaxanthin de-epoxidase, thus sustaining dissipation of excess excitation energy (Smirnoff 2000). Therefore, elevated levels of endogenous AsA in plants are necessary to offset oxidative stress in addition to regulating other plant metabolic processes (Smirnoff 2000; Athar et al., 2008). AsA influences many enzyme activities and minimizes the damage caused by oxidative process through synergic function with

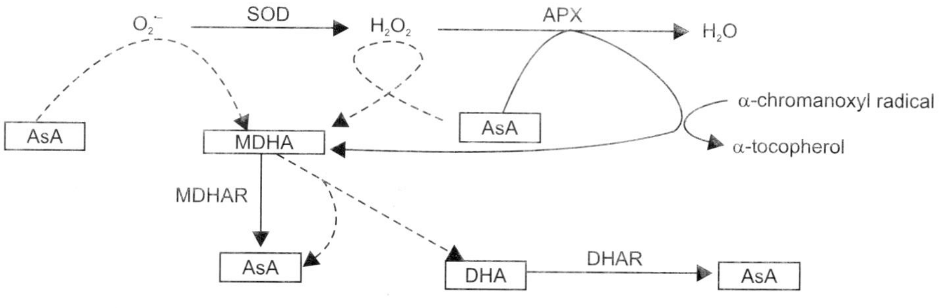

Fig. 3. The role of AsA in ROS detoxification. The dotted lines indicate non-enzymatic reactions.

other antioxidants (Foyer et al., 1994; Asada 1999). This multiplicity of functions has led some researchers to perceive that AsA may be, in addition to a powerful antioxidant and redox buffer, a signalling molecule involved in the regulation of complex processes such as the senescence of plants and their response to O_3, photo-oxidative conditions, or pathogen attack (Smirnoff 2000; Pastori et al., 2003; Conklin and Barth 2004; Puppo et al., 2005). In addition, AsA carries out a number of non-antioxidant functions in the cell. It has been implicated in the regulation of the cell division, cell cycle progression from G1 to S phase (Smirnoff 1996, Liso et al., 1998). More recently, it has been involved in some cell specific events such as stomatal closure (Chen and Gallie 2004). Moreover, the multiple analysis of gene expression in *Arabidopsis* mutants deficient in L-ascorbic acid has revealed that this compound can affect the expression of hundreds of genes in response to biotic stresses (Pastori et al., 2003). All these findings point to L-ascorbic acid being a signal molecule that is involved in many developmental and defence responses to oxidative stress in plants.

3.5.1 AsA metabolism in plants during abiotic oxidative stress

A continuous oxidative assault on plants during biotic and abiotic stresses has led to the presence of an arsenal of enzymatic and non-enzymatic antioxidant defenses to counter the phenomenon of oxidative stress (Shalata et al., 2001; Shao et al., 2007; Hossain and Fujita 2009b, 2010). Salt-tolerant *L. pennellii* showed increase in both AsA and AsA redox state in root tissues due to salt stress but the level of DHA and MDA was significantly higher in the sensitive genotypes (Shalata et al., 2001). Similarly, AsA content and its redox state increased significantly in the leaf mitochondrial fraction of tolerant genotype whereas they decreased in the cultivated *L. esculentum* (Mittova et al., 2003a). This phenomenon indicates that stress tolerant genotype responded defensively to protect cells against oxidative stress by upregulation of AsA level. Cold stress increased the AsA level in both tolerant and sensitive variety of barley but greater increase was observed in tolerant genotype in accordance with increases of APX and GR activities (Dai et al., 2009). Additionally, the AsA level increased significantly in response to short-term (24 h) mild salt stress (50 mM) in rice roots but the levels were significantly decreased under higher levels of salt stress (150 mM) (Khan and Panda 2007). In another study, Yasar et al. (2006) found that the AsA content increased significantly due to mild salt (100 mM NaCl) stress in melon cultivars. Total ascorbate content was increased in tomato and eggplants subjected to waterlogging stress (Lin et al., 2004) and mulberry plants subjected to N and P deficiencies (Tewari et al., 2009). Chilling tolerant cultivar (Xiangnuo-1) had high content of AsA under chilling, while sensitive cultivar (IR-50) had lower AsA content (Huang and Guo 2005). Increased AsA contents were also reported in marigold due to salt stress (Chaparzadeh et al., 2004), in scot pine roots due to heavy metal stress (Schützendübel et al., 2001), and in wheat due to low level of UV-B irradiation stress (Yang et al., 2007). In contrast, a decrease in AsA level was also observed in salt tolerant genotypes subjected to long-term salt stress but a greater decrease was observed in the sensitive genotypes (Hernandez et al., 2000). Kukreja et al. (2005) also noted significant decrease in AsA content under salinity stress in *C. arietinum* roots. AsA content increased in the drought tolerant rice genotype during panicle development stage while the H_2O_2 level decreased. However, in susceptible genotypes, AsA content decreased with the increased H_2O_2 level (Selote and Chopra 2004). Yang et al. (2008) reported that high light regime and drought stress significantly increased the AsA content in *P. asperata* seedlings. However, AsA content decreased due to drought stress in mungbean (Hossain and Fujita 2009b), in rice (Sharma and Dubey 2005), in wheat (Seole and Chopra 2006; Secenji et al., 2010), and in sunflower (Sgherri et al., 1995), and due to heat stress in tobacco Bright Yellow-2 cells (Locato et al., 2009). Although drought stress significantly decreases the AsA level but hybrid plant showed higher AsA content as compared to its parents, also indicating adaptive response of plants under stress (Gao et al., 2009). Agarwal (2007) reported that the AsA and DHA content as well as the GSH/GSSH ratio were significantly increased by the UV-B stress in *Cassia auriculata* seedlings. *N. tabacum* and *Populus x Canescens* plants have higher foliar AsA contents and improved tolerance to oxidative stress compared with wild type (Foyer et al., 1995). However, decrease in AsA content under Cd stress was found in *C. sativus* chloroplast and in the leaves of *A. thaliana, P. sativum* and *B. campestires* (Zhang et al., 2003; Romero-Puertas et al., 2007; Anjum et al., 2008). Most recently, Chao et al. (2010) reported that decline in AsA content is associated with Cd toxicity in rice seedlings.

Ascorbic acid is an integral weapon in the defence against ROS generated by O_3 and total ascorbate correlates with O_3 resistance across a wide variety of plant species. Apoplastic AsA is most likely an important first line of defence against O_3 (Conklin and Barth 2004). The ascorbate pool was approximately two times higher in the O_3 resistant Bel B tobacco relative to the O_3 sensitive Bel W3, in both control and O_3 treated leaves (Pasqualini et al., 2002). Furthermore, soybean leaflets of an O_3 tolerant cultivar (Essex) maintained higher total ascorbate levels than the O_3 sensitive (Forrest) under both O_3 treatment and control conditions (Robinson and Britz 2000). The ascorbate-deficient *A. thaliana* mutant *vtc1* contains approximately one-third AsA level of the wild type and was isolated by virtue of its sensitivity to O_3. Restoration of O_3 resistance in this mutant was achieved by artificial elevation of AsA by pretreatment with the AsA biosynthetic precursor, L-galactono-1,4-lactone (Conklin et al., 1997). Similarly, elevation of total ascorbate in radish with this biosynthetic precursor (Maddison et al., 2002) and in barley by direct ascorbate feeding was found to alleviate O_3 induced injury (Machler et al., 1995). Application of L-galactono-γ-lactone also increases the AsA level of water-stressed leaves of wheat, enhances photochemical and non-photochemical quenching of chlorophyll fluorescence, and thus reduces oxidative damage to their photosynthetic apparatus (Tambussi et al., 2003). Recently, it was found that application of L-galactono-γ-lactone reverses the effects of oxidative stress induced by heavy metal (Cd) by maintaining redox homeostasis and prevents the oxidative damage (lipid peroxidation and protein oxidation) in wheat and rice (Paredisо et al., 2008; Chao et al., 2010). Besides the application of AsA precursor, direct application of AsA also improves plant stress tolerance. Exogenous application of ascorbic acid in wheat and chickpea improves salt tolerance by improving photosynthetic capacity of plants against salt-induced oxidative stress and maintaining ion homeostasis and inducing protein stability and by reducing lipid peroxidation (Athar et al., 2008; Beltagi et al., 2008; Dolatabadian et al., 2008). Consistent findings reported on the beneficial effects of the exogenous application of AsA in mitigating partially the adverse effects of salt stress on growth, like cell division and cell enlargement. Moreover, in tomato seedlings, the exogenous supply of AsA increased tissue levels (Arrigoni et al., 1997) and percentage for surviving the toxic effects of NaCl (Shalata and Neumann 2001). Alternative organic carbon sources without direct antioxidant activity did not provide equivalent protection. These remarkable protective effects appeared to be specifically related to the antioxidant activity of ascorbic acid, rather than its possible utility as an organic substrate for respiratory energy metabolism (Shalata and Neumann 2001). Exogenous application of AsA improves the tolerance of *Cassia angustifolia* (a medical plant) to water stress (Singh et al., 2001). In rice, enhanced AsA level alleviated Al-induced inhibition of root growth (Guo et al., 2005), and exogenous AsA treatment recovered root elongation under Al stress (Wang and Kao 2007). These findings suggest that maintaining a high AsA level is important in achieving Al tolerance. Additionally, AsA content increased in pumpkin seedlings subjected to

50 μM Al stress within 24 h of treatment and due to higher level of AsA and APX activity the H_2O_2 level decreased (Dipierro et al., 2005). Based on the data accumulated so far, it is clear that a high level of endogenous AsA is essential for effectively maintaining the antioxidant system that protects plants from oxidative damage due to biotic and abiotic stresses

3.5.2 Role of AsA during biotic stress-induced ROS detoxification

The AsA pool in the apoplast is a key determinant of oxidant signal duration and strength as well overall redox status of the apoplast (Foyer and Noctor 2000, 2005b). In barley plants infected with *P. indica* significant increase in AsA content was observed after one week of infection with the decreasing level of DHA (Waller et al., 2005). In contrast, gradual decrease in AsA content was also found in tobacco plant infected with TMV (Fodor et al., 1997). AsA content and AsA/total ascorbate ratio was found to decrease in chloroplastic fraction of tomato leaves infected with *B. cinerea* with significant increase in DHA content (Kuzniak and Sklodowska 2001). Likewise, peroxisomal ascorbate pool was notably diminished in response to *B. cinerea* infection from 1st day onward with the decreasing level of total ascorbate and AsA/DHA ratios (Kuzniak and Sklodowska 2005b). *Cocksfoot mottle virus* infected *Dactylis glomerata* L. plants showed decrease in AsA content and increase in the DHA and H_2O_2 content. This would have caused a change in the redox status of infected leaves, with the cellular environment becoming more oxidized (Li and Burritt 2003). In *B. cinerea*-infected *A. thaliana* leaves a massive depletion of AsA level occurred before visible infection (Muckenschnabel et al., 2002). Barley plant infected with *B. graminis*, the apoplastic AsA level is unchanged in both resistant and susceptive lines but DHA level significantly lower in the resistant line. This may suggest that more rapid import of apoplastic DHA into the cytosol occurred after inoculation in resistant lines compared with sensitive one. In contrast, the net loss of total ascorbate observed in resistant leaves after inoculation is consistent with several previous experience of oxidative stress during HR (Vanacker et al., 1998). Tomato plant infected with *P. syringae* showed that AsA content increased in both tolerant and susceptible cultivars whereas strongest elevation of AsA content was evident in the susceptible cultivar just before the infection symptom development but the GSH level decreased in both the lines (Kuzniak and Sklodowska 2004b). This could be interpreted as a compensation of GSH depletion by AsA increase to maintain the overall cellular redox balance under stress condition, but an inverse phenomenon was found in mung bean seedlings subjected to short term drought stress (Hossain and Fujita 2009b) where GSH concentration increased with decreasing level of AsA. It has been reported that a decreased level of AsA could act as metabolic signal leading to increased GSH synthesis in both animals (Meister 1992) and plants (May and Leaver 1993).

3.5.3 AsA content modulates stress perception and signalling

The role that AsA plays in responding to oxidative stress has been shown with *Arabidopsis vtc* mutant which are impaired in AsA biosynthesis (Conklin et al., 1999, 2000). The *Arabidopsis* mutant *vtc1* (which is deficient in AsA) exhibits a moderately slower growth rate under normal growth conditions and is hypersensitive to O_3, SO_2, or UV-B light (Conklin et al., 1996, 1997; Veljovic-Jovanovic et al., 2001), suggesting that much of the AsA in plants is essential to respond to conditions of oxidative stress. Ascorbate deficient mutant *vtc1*, which has 30-60% of the AsA content of wild type plants, accumulated a much higher level of H_2O_2 than wild type coincides with a greater decrease in the ratio of reduced AsA to total AsA with reduced activity of the AsA-GSH cycle enzymes during salt stress (Huang et al., 2005). Similarly, *vtc2* mutants experience chronic photo-oxidative stress in high light and photobleach when transferred from low to high light accompanied by increased lipid peroxidation and photoinhibition (Muller-Moule et al., 2003, 2004). Additionally, the *Arabidopsis* mutant, *vtc1* and *vtc2*, which have constitutively low AsA, show constitutive expression of pathogenesis-related (PR) transcripts and the abundance of leaf AsA can modulate PR gene expression and SAR (Pastori et al., 2003; Kiddle et al., 2003; Barth et al., 2004; Foyer and Noctor 2005b). It has been suggested that the induction of PR proteins and enhanced pathogen resistance in *vtc1* are related to SA accumulation linked to earlier senescence in the mutants than the wild type (Barth et al., 2004; Foyer and Noctor 2005b). Moreover, the *vtc1* mutant shows micro-lesions made up of individual dead cells. It may, therefore, be that the induction of PR proteins and enhanced pathogen resistance observed in *vtc1* is due to enhanced local PCD and SAR rather than an earlier developmental senescence program. These results suggest that low AsA decreases the threshold for sensing stress and triggers PCD and SAR responses even under optimal growth conditions (Foyer and Noctor 2005b; Kangasjarvi et al., 2005).

3.5.4 Conclusion

AsA is a major component in plants defence against free redicals, H_2O_2 and oxidative stress. AsA plays a pivotal role in the destruction of active oxygen species, particularly H_2O_2 through the AsA-GSH cycle in different subcellular compartments. Apoplastic ascorbate could act both as a direct link between biotic and abiotic stress and several key adaptive responses in plants. The pool size and redox status of AsA are thought to affect signalling due to its quantitative predominance in the cell. Signalling should closely interlink with the phytohormone network because AsA is an essential cofactor of the biosynthesis of phytohormone such as ABA and gibberellins. Recent elucidation of the AsA biosynthesis in plants should provide a clue as to how it is produced and how its pool size is regulated. AsA deficient mutants are now available and transgenic plants with altered expression biosynthetic genes will soon be produced. The developments will enable production of plant with reduced or increased AsA content, which will be useful to probe the functions of this simple but multifaceted molecule, perhaps even revealing functions that are completely unsuspected.

3.6 Glutathione (GSH)

The cellular tri-peptide GSH is one of the major non-protein thiols and crucial metabolites found abundantly in all cell compartments of plants. The chemical reactivity and high water solubility of the thiol group of GSH makes it particularly suitable to serve a broad range of biochemical functions to protect plants against (stress induced) oxidative stress and endogenous organic chemicals (Foyer and Noctor 2005a). GSH takes part in the control of H_2O_2 levels through AsA-GSH cycle (Foyer and Noctor 2005a). It can also function directly as a free radical scavenger by reacting with $^1O_2$, $O_2^{\bullet -}$, and $^{\bullet}OH$ (Larson 1988). In addition, GSH plays a key role in the antioxidative defence system by regenerating AsA via the AsA-GSH cycle (Fig. 1) (Foyer and Halliwell 1976; Noctor and Foyer 1998). GSH protects proteins against denaturation caused by the oxidation of protein thiol groups under stress. It also plays an indirect role in protecting membranes by maintaining α-tocopherol and zeaxanthin in the reduced state. GSH is a substrate for GPX and GST, which are also involved in the removal of ROS and endogenous toxic compounds (Foyer and Noctor 2005a). It is also involved in the transfer and storage of sulphur (Herschbach and Rennenberg 2001) and in the detoxification of heavy metals where phytochelation (PC) derived from GSH forms heavy metal complexes (Blum et al., 2007). Recently, it has also been reported that GSH is involved in methylglyoxal (MG) detoxification, a toxic metabolite produced in ample amounts under stressful condition of plants (Yadav et al., 2005a, b; Hossain et al., 2009; Hossain and Fujita 2009b, 2010). The property of GSH is of great biological importance since it allows fine-tuning of the cellular redox environment under normal conditions and upon the onset of stress (Fig. 4). Along with its oxidized form (GSSG), it acts as a redox couple important for maintaining the cellular homeostasis, playing a key role in diverse signaling systems in plants (Noctor 2006). GSH accumulates in response to increased ROS, or to compensate for decrease in the defence capability of other antioxidants and GSH levels are constitutively higher in plants adapted to stress conditions (Gupta et al., 1991; May and

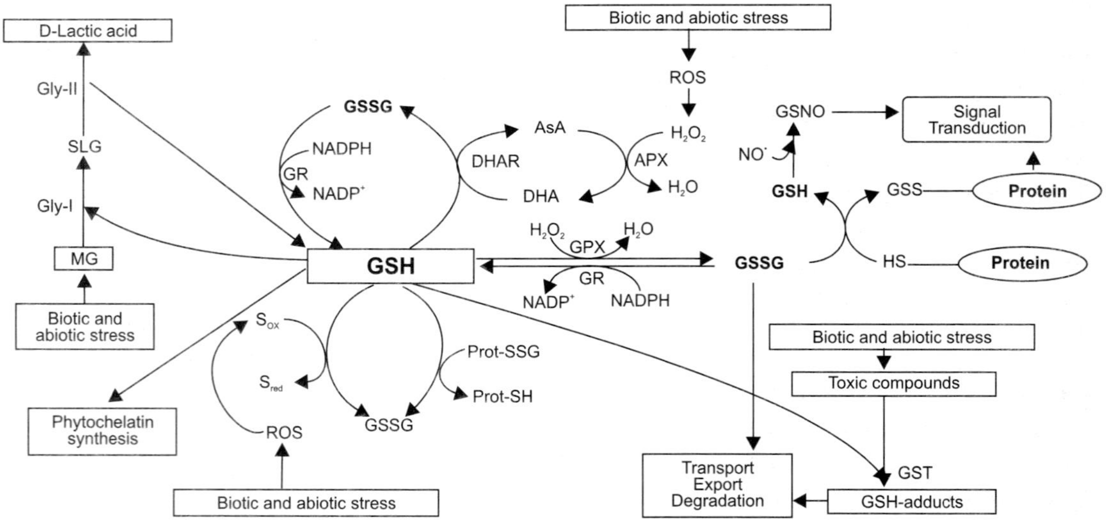

Fig. 4. Comprehensive scheme of possible role of GSH and GSH related enzyme in protection against ROS and MG. GSH can directly scavenge free radicals, act as a substrate or cofactor of GSTs, GPX and DHAR during the detoxification of H_2O_2, lipid hydroperoxides and electrophilic compounds and regeneration of AsA. The production of GSSG can lead to the formation of mixed disulphides in cellular proteins. Finally, the GSSG is reduced again to GSH with electrons derived from NADPH in a reaction catalysed by GR. Sox, oxidized substance; Sred, reduced substance; MG, methylglyoxal; Gly I, glyoxalase I; Gly II, glyoxalase II; SLG, S-D lactoylglutathione; GSNO- S-nitroso glutathione.

Leaver 1993; Marrs 1996). The regeneration from GSSG to GSH is catalysed by GR and NADPH is used as the reducing power (Fig. 1, 4). Recent experiments indicate that GSH may function as a cellular sensor to ensure maintenance of the NADPH pool (May et al., 1998) further illustrating the intertwining of the metabolism of different redox compounds. Additionally, GSH has been associated with several growth- and development-related events in plants, including cell differentiation, cell death and senescence, pathogen resistance and enzymatic regulation (Ogawa 2005).

3.6.1 Interaction between ROS and GSH during abiotic oxidative stress

The response of glutathione has been studied in various abiotic stresses induced oxidative damage due to its role as an antioxidant. GSH level and GSH/GSSG ratio are the decisive factors affecting the performance of the AsA-GSH cycle (Jimenez et al., 1998). In a range of plant species, glutathione levels and its redox state have been observed to alter in response to both long and short term exposure to a range of abiotic stresses (Hernandez et al., 2001; Shalata et al., 2001; Mittova et al., 2003a, b; Ogawa et al., 2004; Hossain and Fujita 2009b, 2010). The role of GSH in the antioxidant defence systems provides a strong basis for its use as a stress marker. Increasing glutathione biosynthetic capacity has been shown to enhance resistance to oxidative stress (Arisi et al., 1998; Zhu et al., 1999a). Salt tolerant *L. pennellii* showed higher GSH, GSH redox state and γ-Glutamyl-Cysteine synthetase (γ-ECS) activity than the salt sensitive *L. esculentum* (Shalata et al., 2001; Mittova et al., 2003a). This phenomenon clearly indicates plants response defensively to protect cells from oxidative damage by upregulation of GSH level. GSH level was found to increase significantly in response short-term (24 h) salt stress in rice and mung bean seedlings (Khan and Panda 2007; Hossain and Fujita 2010). Sumithra et al. (2006) reported that the activity of ROS scavenging enzyme and GSH concentration were found to be higher in the leaves of Pusa Bold than CO 4 cvs. of *Vigna radiata*, whereas, GSSG concentration was found to be higher in the leaves of CO 4 compared to in those of Pusa Bold. It indicates that Pusa Bold has efficient antioxidative characteristics which could prove better protection against oxidative damage in leaves under salt stressed conditions. However, a decrease in GSH level was also observed in salt tolerant genotypes subjected to long-term salt stress but a greater decrease was observed in the sensitive genotypes (Hernandez et al., 2000). Decrease in GSH content and increase in GSSG content was observed due to salt stress in strawberry leaves and tobacco BY-2 cells (Hoque et al., 2007; Tanou et al., 2009). In *Boea hygroscopica*, a dessication tolerant plant showed by almost 50 times increase in GSH content (Sgherri et al., 1994). Additionally, short-term (24 h) drought stress showed increases in GSH and GR activity in wheat leaves (Bartoli et al., 1999). The total GSH content was increased during water deficit in sunflower seedlings (Sgherri et al., 1995), and in a salt-treated groundnut cell lines (Jain et al., 2002). Maintaining a high ratio of GSH/GSSG that functions as redox couple was shown to play an important role in salt and drought tolerance as observed in tomato, wheat and mung bean (Shalata and Neumann 2001; Kocsy et al., 2002; Kranner et al., 2002; Hossain and Fujita 2009b, 2010). GSH content increased in the drought tolerant rice genotype during panicle development stage and with the decreased level of H_2O_2

while in susceptible genotype, GSH content decreased with an increased level of H_2O_2 indicating that GSH plays a role in regulating H_2O_2 level during drought stress (Selote and Chopra 2004). The increase of GSH content in response to drought stress was also observed in rice shoots and roots (Sharma and Dubey 2005). Selote and Chopra (2006) found that the acclimatized drought stressed seedlings showed higher GSH content and glutathione redox state. Hybrid plants subjected to long term drought stress showed high level of GSH content, GSH/GSSG ratio and higher activity of GSH related enzyme as compared to their parents (Gao et al., 2009). Treatment of maize roots to heat shock temperatures of 40°C resulted in decrease of cysteine levels and increase in GSH levels (Nieto-Sotelo and Ho 1986). There was an increase in the activity of the GSH synthesizing enzymes in maize root cells, which was related to the cells capacity to cope with heat stress conditions. Accumulation of GSH has also been observed in heat-stressed tomato seedlings (Rivero et al., 2004). In wheat, it was established that heat stress induced accumulation of GSH levels and increased the activity of the enzymes involved in GSH syntheses and the GSH/GSSG ratio (Kocsy et al., 2002). Chilling tolerant plant showed increased cysteine and GSH synthesis over chilling sensitive type (Kocsy et al., 1996). GSH induced chilling tolerance in maize was also shown by using herbicide in chilling-sensitive lines (Kocsy et al., 2001). This increased the pools of GSH and its precursors, cysteine and γ-glutamyl-cysteine, and there was an increase in GR activity as well, which increased the relative protection considerably. Cold stress as well as acclimation increased the GSH level in both tolerant and sensitive varieties of barley but greater increase was observed in tolerant genotype (Dai et al., 2009). Proteome analysis of chilling stress in rice showed the upregulation of cysteine synthase (Yan et al., 2006). This enzyme is responsible for the final step in cysteine biosynthesis, a key limiting step in GSH production. Thus, GSH and GR are important for resistance to chilling stress. However, the concentration of cellular GSH has a major effect on its antioxidant function and it varies considerably under Cd stress. An increase of GSH concentration has been observed with increasing Cd concentration in *B. juncea* and *B. campestires* (Qadir et al., 2004; Anjum et al. 2008) and *P. sativum* (Metwally et al., 2005). Schützendübel et al. (2002) showed that low level short-term Cd stress significantly induced GSH level while its level decreased with the increased duration and the level of Cd concentration. However, Cd-induced depletion of GSH has been mainly attributed to phytochelatin (PC) synthesis (Grill et al., 1985). The decline in the levels of GSH might also be attributed to an increased utilization from regeneration of AsA from DHA or for direct interaction with Cd (Pietrini et al., 2003). Recently, Gao and Zhang (2008) reported that ascorbate deficient mutant *vct1* showed increase in GSH level within 24 h of UV-B radiation treatment while its levels were deceased at 32 h. Acute O_3 resistant *Medicago truncatula* (accession, JE154) showed 2.5-fold increase in reduced GSH content (Puckette et al., 2009). Mulberry plants subjected to N, K and P deficiencies showed decrease in GSH content with the increasing level of H_2O_2 content (Tewari et al., 2009). Additionally sulphur starvation in rice seedlings reduced photosynthetic capacity and decreased GSH level (Lunde et al., 2008). Eggplant and tomato roots subjected to waterlogging stress (72 h) showed an increased level of total glutathione and as well as GSH content (Lin et al., 2004). Tobacco plants subjected to light stress also showed significant increase in

GSH and GSSG content (Gechev et al., 2003). It has been observed that there is a change in the redox state of glutathione under stress followed by increase in the synthesis and a highly reduced redox state of GSH under acclimation. During severe stress, this ratio becomes more in the oxidized state with loss and degradation of GSH. However, the response should be seen in totality along with the other components of the antioxidant system. The second point to consider would be that GSH levels are affected by the sulphur nutritional status of the plant as well. The variety of responses to different oxidative stress is probably related to the plant species, the age of the plant, and duration of the treatment.

3.6.2 Role of GSH during biotic stress-induced ROS detoxification

The role of GSH in response to biotic stresses has also widely been studied in plants and in spite of some controversial reports, it is now widely accepted that GSH participates in plant defence reactions against viral, fungal, and bacterial infections which induce oxidative damage. In *Curcurbita pepo*, GSH concentrations strongly increased in different subcellular fractions of leaves and roots during zucchini yellow mosaic virus (ZYMV) infection and a strong correlation was observed between elevated GSH concentrations and increased stress tolerance against oxidative stress induced by pathogen attack (Zechmann and Muller 2008). In tobacco plant infected with TMV and *C. pepo* infected with ZYMV, elevated GSH concentrations were found responsible for the suppression of symptom development and for the reduction of virus multiplication (Gullner et al., 1999; Zechmann et al., 2007). Elevated GSH levels are also involved in the development of SAR during TMV infection in tobacco plants (Fodor et al., 1997). The increased levels of GSH and GSH-related enzymes in TMV infected tobacco plants and reduced necrotizations of infected tissues were reported (Gullner et al., 1999). Barley and tomato plants infected with fungal pathogen showed increased GSH content in the resistant line but differential response in the susceptible line (Vanacker et al., 1998, 2000; Williams et al., 2002). The increase of GSH pool in the resistant lines and decreasing in the sensitive lines perhaps imply that increased synthesis and export of GSH have a crucial relationship with resistant response against oxidative stress (Vanacker et al., 1998; Komives et al., 1998). Notably, a marked accumulation of GSH occurred in tomato cells carrying resistance genes (Cf-9 or Cf-2) in response to specific fungal elicitors of the fungus *Cladosporium fulvum* (May et al., 1996a). These results indicate a positive role of the thiol antioxidant in pathogenesis-induced responses of plants. Adversely chloroplastic, mitochondrial and peroxisomal GSH pools were notably diminished in response to *B. cinerea* infection from 2 d onward with the decreasing level of total glutathione and GSH/GSSG ratios and increasing level of GSSG (Kuzniak and Sklodowska 2005a). A decrease in the GSH pool has been reported in oat and tomato leaves infected with necrotrophic fungal pathogens (Gönner and Schlösser 1993). GSH contents decreased in chloroplastic fraction of tomato leaves infected with *B. cinerea* with significant decrease in total glutathione content after 3 d of infection but the significant decrease in GSH/total GSH was observed after 5 d of infection (Kuzniak and Sklodowska 2001). In another study, they found that GSH content decreased progressively after 1 d of *B. cinerea* infected tomato leaves although GR activity increased (Kuzniak and Sklodowska 1999). However, in *Arabidopsis* mutant the depletion of GSH level by 70% did not change its response to fungal and bacterial pathogens, suggesting that changes in the GSH pool are inconsistent with plant accessibility to infection (May et al., 1996b). Moreover, both increased and decreased GSH concentration has been reported to signal phytoalexin accumulation following pathogen attack (Guo et al., 1993; Vanacker et al., 2000). The accumulation of phytoalexins in soybean and clover treated with sulphydryl reagents is considered to be mediated by the cellular redox potential (Gustine 1987; Stossel 1984). GSH-responsive elements have been identified on the promoters of genes involved in phytoalexin synthesis (Dron et al., 1988). Net increases in the levels of GSH have been measured in bean and alfalfa cell suspensions treated with fungal elicitors (Edwards et al., 1991) and in elicitor-treated liverwort cells (Nakagawara et al., 1993) but these were not necessarily directly related to the pathogen response. Since H_2O_2 has been found to mimic this effect, GSH and H_2O_2 must interact in the cellular redox balance that regulates these processes (Guo et al., 1993). Decreased GSH but elevated ascorbate levels were found in oats after infection with virulent *Drechslera* species (Gönner and Schlösser 1993). In incompatible *barley-barley* powdery mildew interactions the AsA-GSH cycle and other antioxidative enzymes (e.g., GST) were inactivated, and these processes might diminish the damaging effects of oxidative stress. However, in compatible interactions these antioxidative reactions were not or were only slightly activated (El-Zahaby et al., 1995). TMV inoculation led to a considerable increase in GSH level both in lower, inoculated and upper, non-inoculated leaves of *Nicotiana tabacum* L. cv. Xanthi-ne (Fodor et al., 1997). Therefore, it seems that elevated GSH concentration and the activation of AsA-GSH cycle in general are directly involved in the development of resistance during plant-microbe interactions (De Gara et al., 2003; Kuzniak and Sklodowska 2005a,b; Hernandez et al., 2006; Zechmann et al., 2007).

3.6.3 Glutathione in redox homeostasis and signalling

GSH redox couple is an information-rich redox buffer that interacts with numerous cellular components (Foyer and Noctor 2005a). The interaction between ROS and antioxidants provide the metabolic contact point between signals originating from metabolic pathways and the environment, thus regulating the induction of adaptive or death processes (Foyer and Noctor 2005a). Glutathione acts as a redox sensor of environmental cues and forms part of the multiple regulatory circuitry coordinating defence gene expression (Fig. 2). The redox state of the GSH/GSSH couples is altered under stress conditions because GSH is oxidized during the removal of the accumulating H_2O_2. Many stresses cause an initial in GSSG, which in turn stimulates GSH synthesis to maintain cellular homeostasis, often leading to an increase in total glutathione (Srivalli and Khanna-Chopra 2008). Stress induced changes in the H_2O_2 content and subsequently in the GSH/GSSG ratio, has a central role in signalling due to their effects on transcription, translation, post-translational modification of proteins and metabolic processes (Foyer et al., 1997; Neill et al., 2002; Dietz 2008; Quan et al., 2008; Szalai et al., 2009). The interaction between H_2O_2 and GSH in stress signalling was suggested in mungbean in which exogenous H_2O_2 increased both GSH levels and chilling tolerance (Yu et al., 2003). The indirect regulation of proteins by GSH/GSSG couple may occur due to crosstalk between GSH/GSSG and other redox systems through glutathionylation or thiol-disulphide transition, which may have a role in signalling and responses to biotic and abiotic stresses (Rausch et al., 2007; Ying et al., 2007). NO, an important regulatory molecule, affected H_2O_2 concentration due to the inhibition of CAT and APX (Clark et al., 2000), whereas exogenous H_2O_2 activated NO synthesis in tobacco (de Pinto et al., 2006), suggesting a bidirectional interaction between the two compounds. NO also influenced GSH synthesis, as demonstrated in *M. truncatula* roots in which the levels of GSH and γ-EC, and glutathione synthetase (GS) gene expressions were increased by NO (Innocenti et al., 2007; Xu et al., 2010). During the interaction of GSH with NO, S-nitrosoglutathione (GSNO) is formed in a reaction that may interconnect the ROS- and reactive nitrogen-based signalling pathways (Neill et al., 2002). The participation of GSNO in the stress response was shown in Cd-treated plants (Barroso et al., 2006). The general regulatory role of NO in stressed plants has been demonstrated in several studies. Moreover, GSH also participates in calcium signalling in plants by stimulating calcium release into cytosol (Gómez et al., 2004).

3.6.4 Role of GSH during GST, GPX, DHAR, Gly I and Gly II mediated ROS and MG detoxification

GSTs constitute a family of multifunctional enzymes present in both plants and animals. These dimeric enzymes catalyse the conjugation of GSH to a variety of electrophilic, hydrophobic, and often toxic substrates, thereby reducing their toxicity (Marrs et al., 1996; Hossain and Fujita 2002; Hossain et al., 2006a, b; Fujita and Hossain 2003a, b). GSTs have been shown to confer tolerance to various plants against oxidative and salt stress (Hossain and Fujita 2010; Katsuhara et al., 2005; Zhao and Zhang 2006). Induction of pumpkin GSTs by different abiotic stresses and with exogenous aldehydes and alcohols showed their importance in protection against oxidative damage (Hossain and Fujita 2002; Hossain et al., 2006b; Fujita and Hossain 2003a, b). It has also been reported that some isoenzymes have GSH dependent DHAR activity and GSTs are components of UV-inducible cell signalling pathways and potential regulators of apoptosis (Edwards et al., 2000; Kampranis et al., 2000; Dixon et al., 2002). GST activity increased in wheat, pea and maize tissues treated with herbicide (glyphosate). The elevation of thiol content in glyphosate-treated plants was shown to be primarily due to increase of GSH level (Uotila et al., 1995). These results suggest that the enhanced glutathione metabolism may have a role in the mode of action or degradation of this herbicide. In barley leaves, GST induced senescence have been described with both GSH conjugating and peroxidase activities and was suggested that it is induced as both an enzyme involved in secondary metabolism and an antioxidant protecting senescencing cells from ROS damage. In addition to GSH conjugation, GSTs may also exhibit GPX or isomerase activities, or function as binding proteins known as ligandins (Edwards et al., 2000). Recently, it is well known that GPXs are ubiquitously occurring enzymes in plant cells including cytosol, chloroplast, mitochondria, peroxisome, and apoplast that use GSH to reduce H_2O_2 and organic and lipid hydroperoxides (Eshdat et al., 1997; Navrot et al., 2006). Transgenic tobacco plants overexpressing both GST and GPX displayed improved seed germination and seedling growth under stress (Roxas et al., 1997). In addition to higher activities of GST/GPX, the transgenic seedlings also showed higher levels of AsA and GSH than wild-type plants (Roxas et al., 2000) indicating that there is close relation between GST, GPX and GSH in the cells. Although plant GPXs make a very little contribution of H_2O_2 detoxification as compared to APX or CAT, overexeperession of phospholipid hydrogenperoxide glutathione peroxidase (PHGPX) in transgenic tobacco plants has been reported to enhance stress tolerance (Ursini et al., 1995) and transgenic tobacco plant expressing a PHGPX-like protein either in cytosol (TcGPX) or chloroplasts (TpGPX) showed suppressed stress-induced production of MDA. Although GSH level did not increased, the GSSG levels were significantly lower than the wild type plants. DHAR exerts the regeneration of AsA by AsA-GSH cycle utilizing GSH as an electron donor. The deceased level of GSH under stressful conditions is due to the extensive utilization of GSH for the conversion DHA to AsA (De Gara et al., 2000; Hoque et al., 2007). The decrease in DHAR activity, due to decreased GSH content, was also observed in mung bean seedlings under severe salt stress (Hossain and Fujita unpublished data). Methylglyoxal (MG) is a highly reactive cytotoxic compound and produced in plants during normal physiological process as well. Recently, it has been reported that MG levels were increased significantly in plants in response to various stresses (Yadav et al., 2005a, b; Singla-Pareek et al., 2006; Hossain et al., 2009). In plants, MG is detoxified via the glyoxalase system that is comprised two enzymes, glyoxalase I (Gly I) and glyoxalase II (Gly II). Gly I convert MG to S-D-lactotylglutathione by utilizing GSH, while Gly II converts S-D-lactoylglutathione to D-lactate, and GSH is recycled back in the system (Thornalley 1990; Hossain and Fujita 2009a). Overexpression of the glyoxalase pathway in transgenic tobacco plants has been found to check an increase of ROS and MG under stress conditions by maintaining glutathione homoeostasis and antioxidant enzyme levels (Yadav et al., 2005b). Under stress condition GSH level is decreased and a smaller GSH pool is available for detoxification of ROS and MG. Exogenous applications of GSH reduced the MG levels in plants and enhanced salinity tolerance in both wild type and transgenic plants (Yadav et al., 2005a). Recent studies in mung bean seedlings showed decreased Gly I activity as well as lower GSH level under salinity stress (300 mM, 48 h) with the increasing level of H_2O_2 and lipid peroxidation. However, proline or betaine treated seedlings maintained higher Gly I activity and higher level of GSH content with lower level of H_2O_2 and MDA. Therefore, it could be concluded that endogenous GSH level directly regulates the activity of GSH utilizing and regenerating enzymes and plays significant role in ROS and MG detoxification (Hossain and Fujita 2010).

3.6.5 Transgenic for higher GSH synthesis and oxidative stress tolerance

Since GSH metabolism plays a crucial role in the ROS detoxification, transgenic poplar plants overexpressing γ-ECS gene showed enhanced tolerance to herbicide induced oxidative stress due to higher GSH level and GST activity (Gullner et al., 2001). Two *E. coli* genes involved in GSH biosynthesis, GS or γ-ECS, were constitutively expressed in *B. juncea* (Zhu et al., 1999a, 1999b). The expression of either GS or γ-ECS in the plant species resulted in higher accumulation of GSH and PCs as well as enhanced tolerance to Cd induced oxidative stress. Additionally, transgenic poplar plants overexpressing a cytosol targeted γ-ECS gene showed 3- to 4-fold increase in GSH levels (Noctor et al., 1996). However, transgenic tobacco overexpressing chloroplast targeted γ-ECS contained elevated glutathione levels but exhibited light-dependent chlorosis and necrosis (Creissen et al., 1999). This was attributed to the increased levels of GSSG and a failure of the redox sensing mechanism in chloroplasts of the transgenic plants. Recently, Reisinger et al. (2008) showed that *B. juncea* overexpressing GS or γ-ECS showed enhanced tolerance to a variety of heavy metals (particularly As, Cd, Zn and Pb) due to higher capacity of GSH synthesis as well as PC synthesis.

3.6.6 Conclusion

The activation of GSH-related antioxidant defence systems in plants in response to oxidative stress is a general phenomenon and GSH plays a central role in ROS detoxification process. Besides its multiple functions, especially as an antioxidant, it functions as a key signalling compound for stress perception, adaptation and cellular homeostasis. When the intensity of a stress increases, glutathione concentration usually declines leading to the decrease of GSH/GSSG ratio although differences exist for different abiotic or biotic stresses, plant species and developmental stages. The ability of the plant cells to restore this ratio may be accompanied by increase in the total pool of glutathione and would be subject to regulation by multiple controls such as a high GR activity and the role of phytohormone in the regulation of GSH needs to be clarified. Identification of other factors could provide us with tools for manipulation of plant to induce oxidative stress tolerance. Additionally, plant responses in glutathione levels under acclimation conditions of various environmental stresses would give us more information regarding its regulation and function during oxidative stress tolerance.

3.7 NADPH

NADPH, the reduced form of nicotinamide-adeninde dinucleotide phosphate, plays an important role in antioxidation and many other cellular processes. Besides functioning as a coenzyme, the reducing feature of cellular NADPH makes it a major resource of electrons for cellular antioxidative processes. It is well understood that NADPH acts as an indirect co-antioxidant by helping with the re-reduction of oxidized antioxidants, which is crucial for the cell to defend against oxidative stress and maintain the normal redox environment (Schafer and Buettner 2001). However, a recent research also suggests that NADPH has the potential to be a direct antioxidant by directly scavenging free radicals and repair damaged biomolecules (Kirsch and de Groot 2001). NADPH helps in maintaining glutathione and ascorbate redox state by recycling of GSSG to GSH and MDHA to AsA. During the conversion of GSSG to GSH by GR and regeneration of AsA from MDHA by MDHAR, NADPH serves as the electron donor. Besides the ascorbate and glutathione system, NADPH is also known to provide electrons to the glutathione/glutaredoxin and thioredoxin systems which are major systems controlling cellular redox homeostasis (Holmgren 1989). Recently, Marty et al. (2009) reported that *Arabidopsis* mutant deficient in GR (*gr1* and *gr2*) gene, the NADPH-dependent thioredoxin system constitutes a functional backup for cytosolic GR indicating further roles of NADPH in redox system regulation and oxidative stress tolerance.

3.8 Hydrogen Peroxide (H_2O_2)

The role of ROS, especially H_2O_2, in plant response to stresses has been the focus of much attention. It is the most stable of the so-called ROS, and plays a central role in response to both abiotic and biotic stresses in plants. This molecule seems to be a "master hormone" that controls a variety of stress responses and physiological adjustments, including the ROS/hormonal homeostasis in the cell. Intra- and intercellular levels of H_2O_2 increase during environmental stresses and the rate of H_2O_2 production depends on the strength and duration of the imposed stress. H_2O_2 plays a dual role in plants: at low concentration, it acts as a signal that induces the expression of numerous defence genes and activates multiple defence responses to abiotic and biotic stresses, while excessive accumulation leads to cellular oxidative damage and even programmed death (Levine et al., 1994; Prasad et al., 1994). H_2O_2 induces the expression of genes potentially involved in its synthesis, such as NADPH oxidase (Neill et al., 2002) and also of those encoding proteins involved in its degradation, implying a complex mechanism for cellular regulation of oxidative status. H_2O_2 interacts with thiol-containing proteins and activates different signalling pathways as well as transcription factors, which in turn regulate gene expression and cell-cycle processes. Using cDNA microarray technology, Desikan et al. (2001) have showed that 113 Arabidopsis transcripts were induced and 62 were repressed by H_2O_2. H_2O_2 is continuously generated from various sources during normal metabolism (Fig. 2) and a wide range of steady-state H_2O_2 concentrations (μM-mM) has been reported, although such variation may reflect technical difficulties in quantifying H_2O_2. Moreover, differences in the H_2O_2 content of different subcellular compartments are not yet known (Neill et al., 2002). Although the precise subcellular concentrations of H_2O_2 that are likely to be toxic will vary, high rates of H_2O_2 production are normally balanced by very efficient antioxidant systems and AsA-GSH cycle is the most important one (Smirnoff 1995; Asada 1999).

3.8.1 Regulatory role of H_2O_2 during abiotic oxidative stress

ROS, particularly H_2O_2, leading to the severe oxidative tissue damage was reported in response to various environmental constraints such as drought, flooding, salt, high light, chilling, heat, UV radiation, O_3, herbicides (Prasad et al., 1994; Selote and Chopra 2004; Gao and Zhang 2008; Kumutha et al., 2009; Zhou et al., 2009; Lu et al., 2009; Locato et al., 2009; Tanou et al., 2009; Liu et al., 2009; Hossain and Fujita 2009b, 2010; Islam et al., 2010). In contrast, mild oxidative stress induced by H_2O_2 treatments appears to be beneficial to subsequently stressed seedlings (Gong et al., 2001; Uchida et al., 2002). Exogenous uses of H_2O_2 have been employed to reduce the adverse effects of oxidative stresses and the appropriate level of H_2O_2 pretreatment can enhance tolerance to abiotic stresses such as chilling (Prasad et al., 1994; Yu et al., 2002, 2003,), salinity (Uchida et al., 2002; Azevedo-Neto et al., 2005; Wahid et al., 2007; Xu et al., 2008b), heat (Lopez-Delgado et al., 1998; Gong et al., 2001; Uchida et al., 2002; Larkindale and Huang 2004; Kang et al., 2009), osmotic (Liu et al., 2009), heavy metals (Hsu and Kao 2007; Chao et al., 2009), paraquat (Moskova et al., 2009) and multi-adversity (Gong et al., 2001). Prasad et al. (1994) showed that application of H_2O_2 rendered maize seedlings more tolerant to chilling stress probably because the transient increase in H_2O_2 activated the acclimatory mechanisms. The authors suggest that H_2O_2 has dual effect in maize: during chilling acclimation it serves as a signal to induce the synthesis of ROS-scavenging enzymes while in non-acclimated seedlings it accumulates to higher levels and acts as a destructive agent. In rice seedlings, H_2O_2 pretreatment induced CAT, APX, GPX activities and reduces heavy-metal induced oxidative stress (Hu et al., 2009). Similarly, Chao et al. (2009) reported that H_2O_2 pretreatment increased the GSH level in rice seedlings and showed better tolerance to Cd induced oxidative stress. Recently, Kang et al. (2009) reported that H_2O_2 pretreatment induces APX and glucose-6-phosphated dehydrogenase (G6PDH) activities in cucumber and tomato seedlings and induce tolerance to heat stress. Exogenous H_2O_2 increases antioxidant activity in cucumber leaves and thereby decreases lipid peroxidation and enhanced oxidative stress tolerance (Liu et al., 2009). Tolerance against oxidative stress generated by high light intensities or the catalase inhibitor, aminotriazole, was induced by spraying with H_2O_2 in tobacco (Gechev et al., 2002). Moreover, it was reported that both H_2O_2 and salicylic acid (SA) could mediate the induction of protective mechanisms to abiotic stresses. SA pretreatment induced an increase in H_2O_2 concentration which in turn increased antioxidant enzyme activity finally leading to enhanced chilling tolerance in maize seedlings (Janda et al., 1999). Similarly, in mustard seedlings both SA and H_2O_2 were involved in signal transduction leading to acclimation during heat stress (Dat et al., 1998). The induction of tolerance to specific environmental stresses involving oxidative stress also increases tolerance to other abiotic stresses as well as to biotic stress. This so-called cross adaptation was evidenced in numerous plants. Yang et al. (2007) reported that UV-B irradiation induces freezing tolerance in winter wheat due to irradiation induced H_2O_2 accumulation. Gong et al. (2001) reported that heat shock which induced increase of H_2O_2 induces the tolerance of maize seedlings for heat, chilling, drought and salt stresses. Water deficit induced paraquat tolerance in cotton plants whereas exposure to UV irradiation enhanced the resistance in tobacco to TMV infection (Burke et al., 1985; Yalpani et al., 1994). O_3 exposure activated a set of pathogenesis-related proteins and induced resistance to *P. syringae* in *Arabidopsis thaliana* (Sharma et al., 1996). The heat treatment which resulted in H_2O_2 generation induced resistance to powdery mildew in barley (Vallelian-Bindschedler et al., 1998). Additionally, Hsu and Kao (2007) showed that heat shock-mediated H_2O_2 accumulation and protection against Cd toxicity in rice seedlings. These results strongly support the hypothesis that H_2O_2 plays a role of a regulatory signal in activation of genes encoding enzymes and proteins involved in the general oxidative stress defence, regardless of its biotic or abiotic origin.

3.8.2 Regulatory role of H_2O_2 during biotic stress response

Plant cells have defensive responses to pathogen attack associated with changes in oxidative metabolism. H_2O_2 is believed to play two distinct roles in pathogenesis. One involves the oxidative burst in the HR, which restricts pathogen growth and the other activates plant defence responses including induction of phytoalexins, second messengers or signalling intermediates, PR proteins, antioxidant enzymes and cell wall reinforcement (Romero-Romero and Lopez-Delgado 2009). A rapid and transient production of ROS terms as oxidative burst is a hallmark of successful recognition of plant pathogens. The rapidity of its production and the potential for H_2O_2 to free diffuse across membranes suggested that ROS could exert this function either directly through redox regulation of transcription factors or indirectly by interacting with other signalling components like phosphorylation cascade (Torres 2010). H_2O_2 was shown to activate mitogen activated protein (MAP) kinase in Arabidopsis suspension (Desikan et al., 1999) and in response to a number of infectious conditions (Hirt 1997) although it remains unclear whether the activities of protein kinase are essential for H_2O_2-mediated PCD and gene expression. H_2O_2 production in barley infected with *B. graminis* led to cell wall cross-linking (Thordal-Christensen et al., 1997). Feruloyl-arabinoxylan trisaccharide accumulation in the cell wall was shown to oxidatively coupled by H_2O_2, creating a barrier against pathogens (Encina and Fry 2005). H_2O_2 may also be involved in the induction of SAR by SA (Chen et al., 1993). In addition, direct injection of H_2O_2 enhanced the expression of PR genes (Chen et al., 1993), antioxidant enzymes (Ivenish et al., 1995), and phytoalexins (Degousie et al., 1994) and enhanced accumulation of signalling component, SA (Leon et al., 1995). Production of an oxidative burst is also observed in response to pest insects, both in response of sap-feeding and chewing herbivores. These ROS will be produced not only by the interaction with the insect but also by the mechanical wounding associated with feeding. ROS could mediate the establishment of defensive mechanism, local and systemic, including the production of volatile products that could act not only against the insect but also against the subsequent invasion of microbial pathogens (Torres 2010).

3.8.3 H_2O_2: a part of signalling network

In order to exert such a profound effect on plant metabolism, ROS must utilize and/or interfere with other signalling molecules. The connection between H_2O_2 and signalling networks has been extensively documented for a number of stress responses, including pathogen elictors, insect feeding, wounding, high temperature, high light and ABA associated stomatal closure (reviewed by Quan et al., 2008). An increasing number of reports suggest that H_2O_2 emerges following synthesis of NO and that NO collaborates with H_2O_2 in plant disease resistance. NO is indispensable to SA function as a SAR inducer (Song and Goodman 2001). Tobacco plants with reduced peroxisomal CAT activity produce ethylene as an earliest response of high light irradiance (Chamnongpol et al., 1998). In addition, exogenous application of H_2O_2 increases ethylene production in pine needles in a concentration dependent manner (Ivenish and Tillberg 1995). H_2O_2 is an essential signal in mediating stomatal closure induced by ABA via the activation of calcium-permeable channels in the plasma membrane (Neill et al., 2002). Methyl jasmonate is a well-established signal molecule in plant defence responses and an effective inducer of accumulation of secondary metabolites such as H_2O_2 in plant cell cultures (Wang and Wu 2005). Therefore, plant hormones are not only located downstream of ROS signal, but ROS themselves are also secondary messengers in many hormone signalling pathways. In addition, feedback and feedforward interactions may conceivably occur between different plant hormones and H_2O_2 (Bhattacharjee 2005).

3.8.4 Transgenic plant and H_2O_2 mediated stress tolerance

The expression of H_2O_2-inductive enzymes in transgenic plants has provided an innovative approach to study the plant defence resistance. Overexpression of apoplastic *swpa4* gene in transgenic tobacco plants showed tolerance to a variety of abiotic and biotic stresses due to increase in H_2O_2 production followed by the induction of defence-related genes, including apoplastic acidic PR genes (Kim et al., 2008). H_2O_2 produced by the overexpressed oxalate oxidase in transgenic tobacco plants triggers the signalling pathways to upregulate expression of antioxidant enzymes genes which in turn result in elevated tolerance to MV and high light induced oxidative stresses (Wan et al., 2009). Similarly, Wu et al. (1995) observed that transgenic potato plants expressing a fungal gene encoding glucose oxidase which generates H_2O_2, exhibited strong resistance to *Erwinia carotovora* subsp *carotovora*, and to *Phytophthora infestans*. This resistance to soft rot and to potato late blight was apparently mediated by elevated levels of H_2O_2. Transgenic soybean, tobacco and sunflower (*Helianthus annuus*) expressing a wheat oxalate oxidase gene have increased resistance to *Sclerotinia sclerotiorum* (Donaldson et al., 2001; Hu et al., 2003). Additionally, transgenic poplar expressing the same gene also exhibited resistance to *Septoria musiva* (Liang et al., 2001).

3.8.5 Conclusion

H_2O_2 production is inevitable during plant growth, development and resistance responses. As an element of the oxidative stress H_2O_2 has a deleterious impact on cell components when accumulated in excess, and simultaneously it acts as an inducer of protective mechanisms, especially at the early stages of plant stress response. Evidence to date suggests a central role of H_2O_2 in intracellular and systemic signalling processes giving rise to successful acclimation and tolerance to biotic and abiotic stresses. It is likely that redox poise of the cell may control cellular events through interactions with H_2O_2 and redox-sensitive molecules such as GSH, AsA, thioredoxins and peroxiredoxins. Several other signalling molecules e.g. SA, ABA, jasmonate, NO and Ca^{2+} were proposed to interact with H_2O_2 in regulation of plant defence responses at both the gene and protein levels (Van Camp et al., 1998; Kuzniak and Urbanek 2000). The use of transgenic plants that are impaired in H_2O_2 generation or increased H_2O_2 production and the isolation of H_2O_2-signalling mutants will be invaluable in elucidating further biological roles of H_2O_2 in specific cells and in response to various oxidative stresses. Post-genomic developments in transcriptomics and proteomics will facilitate further insights into cellular responses to H_2O_2. Future work will no doubt reveal novel signalling roles for H_2O_2 and its interaction with ROS scavenging system including AsA-GSH cycle.

4. SUMMARY AND FUTURE PERSPECTIVES

Oxidative stress is one of the major limiting factors of plant productivity. In plants, ROS are continuously produced as by-products of various metabolic reactions. However, imbalances in metabolic processes due to abiotic or biotic stresses may lead to increased accumulation of ROS, forming a potential threat of oxidative damage to cellular components. Usually, a dynamic balance has to be maintained between ROS generation and scavenging in different subcellular compartments in order to guarantee normal plant growth. There are many antioxidant pathways in plants but AsA-GSH pathway is the most vital one (Smirnoff 1995). The AsA-GSH cycle present in different subcellular compartments plays the central role in regulating ROS levels (in particular H_2O_2 level) by regulating three interdependent redox couples, i.e., AsA/DHA, GSH/GSSG and, NADPH/NADP$^+$. It is becoming increasingly evident that ROS regulate a complex signal transduction network within the plant development and its response and adaptation to both biotic and abiotic stressors. Under stress conditions, ROS may play two very different roles: damaging the cellular components or signalling the activation of defence responses. Such a dual function was first described in pathogenesis but has recently been demonstrated during abiotic stress responses. Extensive research findings support that coordinated induction and regulation of the component of AsA-GSH is necessary in order to obtain substantial tolerance against oxidative stress and inhibition of one component of (either enzyme or metabolite) AsA-GSH cycle limit the activity of other enzymes or metabolites and thereby leading to deterioration of the system. We have highlighted the major sources of ROS and sites of production in plant cells, together with the regulatory role of the key components of AsA-GSH cycle, i.e., APX, DHAR, MDHAR, GR, AsA, GSH, NADPH and H_2O_2 and their interaction, consequence and adaptation of plants under oxidative stress conditions. Nevertheless, a major gap exists in our understanding about how ROS induces large-scale and coordinated expression of different enzymes and metabolites of AsA-GSH cycle. Additionally, ascorbate and glutathione pools within the cellular compartments are connected by transport across membranes. Little is known how the transporters are involved in ROS metabolism and signalling. Understanding the interaction between ROS and plant hormone and components of AsA-GSH cycle in different subcellular compartments will reveal the more subtle regulatory role of this pathway in oxidative stress tolerance.

Physiological and genetic evidence clearly indicate that the ROS scavenging system of plants is an important component of the stress protective mechanism. During the last few years, considerable progress has been made in understanding how plants protect themselves against ROS. Several genes of encoding for antioxidant enzymes and metabolites of AsA-GSH cycle have been cloned, characterized and used in the construction of transgenic lines and in few cases promising results were obtained (Eltayeb et al., 2006, 2007; Lim et al., 2007; Li et al., 2009; Yin et al., 2010; Wang et al., 2010). Although manipulation of antioxidant gene seems to be a sound approach to counteract oxidative stress, attempts to improve oxidative stress tolerance particularly by manipulation of single antioxidant have limited success presumably because of the need for a balanced interaction of protective enzymes and other metabolites (Lee et al., 2007;

Lim et al., 2007; Tseng et al., 2007). Therefore, future studies should focus on the clarification of the bottleneck factors affecting the performance of AsA-GSH cycle under oxidative stress as well as further probe into an effective pathway to mediate the cycle. Ascorbate deficient mutant or gene-knockout mutants are now available. Genomics tools are accelerating the discovery of ROS-responsive genes on a global scale and are expanding our understanding of the oxidative stress response and the pleiotropic roles of ROS in signalling and gene expression. In the long term, we need to understand how plants sense oxidative stress in different subcellular compartments and how this stress signal is transduced to turn on the activation of different components of AsA-GSH cycle. It is likely that identification of master regulators that control stress response activation will provide promising avenues to enhancing crop tolerance to oxidative stress. Additionally, field tests of transgenic plants that overexpress different components of AsA-GSH cycle and other antioxidant enzymes will surely provide the answer to these and other questions about the utility of enhancing the ROS scavenging systems of crop plants.

ACKNOWLEDGEMENTS

The authors wish to thank Peter Lutes for his valuable review of this manuscript and Mr. Mirza Hasanuzzaman for the artwork.

REFERENCES

Agarwal, S. (2007). Increased antioxidant activity in *Cassia* seedlings under UV-B radiation. *Biol. Plant.* **51:** 157-160.

Allen, R.D. (1995). Dissection of oxidative stress tolerance using transgenic plants. *Plant Physiol.* **107:** 1049-1054.

Allen, R.D., Webb, R.P. and Schake, S.A. (1997). Use of transgenic plants to study antioxidant defences. *Free Rad. Biol. Med.* **23**: 473-479.

Anderson, J.V. and Davis, D.G. (2004). Abiotic stress alters transcript profiles and activity of glutathione S-transferase, glutathione peroxidase, and glutathione reductase in *Euphorbia esula. Physiol. Plant.* **120:** 421-433.

Anjum, N.A., Umar, S., Ahmed, A., Iqbal, M. and Khan, N.A. (2008). Sulfur protects mustard (*Brassica campestries* L.) from cadmium toxicity by improving leaf ascorbate and glutathione. *Plant Growth Regul.* **54:** 271-279.

Ansel, D.C., Franklin, M.L.T., De Carvalho, M.H.C., Lameta, A.D.A. and Fodil, Y.Z. (2006). Glutathione reductase in leaves of cowpea: cloning of two cDNAs, expression and enzymatic activity under progressive drought stress desiccation and abscisic acid treatment. *Ann. Bot.* **98:** 1279-1287.

Aono, J., Saji, H., Sakamoto, A., Tanaka, K., Kondo, N. and Tanaka, K. (1995). Paraquat tolerance of transgenic *Nicotiana tabacum* with enhanced activities of glutathione reductase and superoxide dismutasae. *Plant Cell Physiol.* **36:** 1687-1691.

Aono, M., Kubo, A., Saji, H., Natori, T., Tanaka, K. and Kondo, N. (1991). Resistance to active oxygen toxicity of transgenic *Nicotiana tabacum* that expresses the gene for glutathione reductase from Escherichia coli. *Plant Cell Physiol.* **32:** 691-698.

Aono, M., Kubo, A., Saji, H., Tanaka, K. and Kondo, N. (1993). Enhanced tolerance to photooxidative stress of transgenic *Nicotiana tabacum* with high chloroplastic glutathione reductase activity. *Plant Cell Physiol.* **34:** 129-136.

Apel, K. and Hirt, H. (2004). Reactive oxygen species: metabolism, oxidative stress, and signal transduction. *Annu. Rev. Plant Biol.* **55:** 373-399.

Arisi, A.C.M., Cornic, G., Jouanin, L. and Foyer, C.H. (1998). Overexpression of iron superoxide dismutase in transformed poplar modifies the regulation of photosynthesis at low CO_2 partial pressures or following exposure to the prooxidant herbicide methyl viologen. *Plant Physiol.* **117:** 565-574.

Aro, E.M. and Ohad, I. (2003). Redox regulation of thylakoid protein phosphorylation. *Antiox. Redox Signal.* **5:** 56-67.

Arrigoni, O., Calabrese, G., De Gara, L., Bitonti, M.B. and Liso, R. (1997). Correlation between changes in ascorbate and growth of *Lupinus albus* seedlings. *J. Plant Physiol.* **150:** 302-338.

Asada, K. (1992). Ascorbate peroxidase—a hydrogen peroxide-scavenging enzyme in plants. *Physiol. Plant.* **85:** 235-241.

Asada, K. (1999). The water-water cycle in chloroplasts: scavenging of active oxygen and dissipation of excess photons. *Annu. Rev. Plant Physiol. Plant Mol. Biol.* **50:** 601-639.

Asada, K. and Takahashi, M. (1987). Production and scavenging of active oxygen in photosynthesis. In *Photoinhibition*. pp. 227-287. Eds. Kyle, D.J., Osmond, C.B. and Amtzen, C.J. (Elsevier Sci. Pub.: Amsterdam).

Athar, H.R., Khanb, A. and Ashraf, M. (2008). Exogenously applied ascorbic acid alleviates salt-induced oxidative stress in wheat. *Environ. Exp. Bot.* **63:** 224-231.

Azevedo-Neto, A.D., Prisco, J.T., Eneas-Filho, J., Medeiros, J.V.R. and Gomes-Filho, E. (2005). Hydrogen peroxide pretreatment induces salt stress acclimation in maize plants. *J. Plant Physiol.* **162:** 1114-1122.

Badawi, G.H., Kawano, N., Yamauchi, Y., Shimada, E., Sasaki, R., Kubo, A. and Tanaka, K. (2004). Over-expression of ascorbate peroxidase in tobacco chloroplasts enhances the tolerance to salt stress and water deficit. *Physiol. Plant.* **121:** 231-238.

Baek, K.H. and Skinner, D.Z. (2003). Alteration of antioxidant enzyme gene expression during cold acclimation of near-isogenic wheat lines. *Plant Sci.* **165:** 1221-1227.

Baker, C.J. and Orlandi, E.W. (1995). Active oxygen in plant pathogenesis. *Anuu. Rev. Phytopath.* **33:** 299-321.

Barroso, J.B., Corpa, F.J., Carreras, A., Rodriguez-Serrano, M., Esteban, F.J., Fernandez-Ocana, A., Chaki, M., Romero-Puertas, M.C., Valderrama, R., Sandalio, L.M. and del Rio, L.A. (2006). Localization of S-nitrosoglutathione and expression of S-nitrosoglutathione reductase in pea plants under cadmium stress. *J. Exp. Bot.* **57:** 1785-1793.

Barth, C., Moeder, W., Klessig, D.F. and Conklin, P.L. (2004). The timing of senescence and response to pathogens is altered in the ascorbate-deficient mutant vitamin *c-1*. *Plant Physiol.* **134:** 178-192.

Bartoli, C.G., Simontacchi, M., Tambussi, E., Beltrano, J., Montaldi, E. and Puntarulo, S. (1999). Drought and watering-dependent oxidative stress: effect on antioxidant content in *Triticum aestivum* L. leaves. *J. Exp. Bot.* **50:** 375-383.

Beltagi, M.S. (2008). Exogenous ascorbic acid (vitamin C) induced anabolic enhances for salt tolerance in chickpea (*Cicer arietinum* L.) plants. African *J. Plant Sci.* **2:** 118-123.

Bhattacharjee, S. (2005). Reactive oxygen species and oxidative burst: Roles in stress, senescence and signal transduction in plants. *Current Sci.* **89:** 1113-1121.

Bian, S. and Jiang, Y. (2009). Reactive oxygen species, antioxidant enzyme activities and gene expression patterns in leaves and roots to Kentucky bluegrass in response to drought stress and recovery. *Scientia. Hort.* **120:** 264-270.

Blilou, I., Bueno, P., Ocampo, J.A. and Garcia-Garrido, J.M. (2000). Induction of catalase and ascorbate peroxidase activities in tobacco roots inoculated with the arbuscular mycorrhizal fungus *Glomus mosseae*. *Mycol. Res.* **104:** 722-725.

Blum, R., Beck, A., Korte, A., Stengel, A., Letzel, A., Lendzian, K. and Grill, E. (2007). Function of phytochelatin synthase in catabolism of glutathione-conjugates. *Plant J.* **49:** 740-749.

Borland, A., Elliott, S. and Patterson, S. (2006). Are the metabolic components of crassulacean acid metabolism upregulated in responses to an increase in oxidative burden? *J. Exp. Bot.* **57:** 319-328.

Broadbent, P., Creissen, G.P., Kular, B., Wellburn, A.R. and Mullineaux, P.M. (1995). Oxidative stress responses in transgenic tobacco containing altered levels of glutathione reductase activity. *Plant J.* **8:** 247-255.

Burke, J.J., Gamble, P.E., Hartfield, J.L. and Quinsberry, J.E. (1985). Plant morphological and biochemical responses to field water deficit. I. Responses of glutathione reductase activity and paraquat sensitivity. *Plant Physiol.* **79:** 415-419.

Chai, T.T., Fadzillah, N.M., Kusnan, M. and Mahmood, M. (2005). Water stress-induced oxidative damage and antioxidant responses in micropropagated banana plantlets. *Biol. Plant.* **49:** 153-156.

Chamnongpol, S., Willekens, H., Moeder, W., Langebartels, C., Sandermann, H., Montagu, M.V., Inze, D. and Camp, W.V. (1998). Defense activation and enhanced pathogen tolerance induced by H_2O_2 in transgenic plants. *Proc. Natl. Acad. Sci. USA.* **95:** 5818-5823.

Chao, Y.Y., Hong, C.Y. and Kao, C.H. (2010). The decline in ascorbic acid content is associated with cadmium toxicity in rice seedlings. *Plant Physiol. Biochem.* (in press), doi:10.1016/j.plaphy. 2010.01.009.

Chao, Y.Y., Hsu, Y.T. and Kao, C.H. (2009). Involvement of glutathione in heat shock- and hydrogen peroxide-induced cadmium tolerance of rice (*Oryza sativa* L.) seedlings. *Plant Soil.* **318:** 37-45.

Chaparzadeh, N., D'Amico, M.L., Khavari-Nejad, R.A., Izzo, R. and Navari-Izzo, F. (2004). Antioxidative responses of *Calendula officinalis* under salinity conditions. *Plant Physiol. Biochem.* **42:** 695-701.

Chen, Z. and Gallie, D.R. (2005). Increasing tolerance to ozone by elevating foliar ascorbic acid confers greater protection against ozone than increasing avoidance. *Plant Physiol.* **138:** 1673-1689.

Chen, Z. and Gallie, D.R. (2004). The ascorbic acid redox state controls guard cell signalling and stomatal movement. *Plant Cell.* **16:** 1143-1162.

Chen, Z., Shilva, H. and Klessig, R.F. (1993). Active oxygen species in the induction of plant systemic acquired resistance by SA. *Science.* **262:** 1883-1886.

Chen, Z., Young, T.E., Ling, J, Chang, S.C. and Gallie, D.R. (2003). Increasing vitamin C content of plants through enhanced ascorbate recycling. *Proc. Natl.Acad. Sci. USA.* **100:** 3525-3530.

May, M., Vernoux, T., Leaver, C., van Montagu, M. and Inze, D. (1998). Glutathione homeostasis in plants: implications for environmental sensing and plant development. *J. Exp. Bot.* **49:** 649-667.

Chew, O., Whelan, J. and Millar, A.H. (2003). Molecular definition of the ascorbate-glutathione cycle in *Arabidopsis* mitochondria reveals dual targeting of antioxidant defenses in plants. *J. Biol .Chem.* **278:** 46869-46877.

Clark, D., Durner, J., Navarre, D.A. and Klessig, D.F. (2000). Nitric oxide inhibition of tobacco catalase and ascorbate peroxidase. *Mol. Plant Microbe Interact.* **13:** 1350-1384.

Conklin, P.L. and Barth, C. (2004). Ascorbic acid, a familiar small molecule intertwined in the response of plants to ozone, pathogens, and the onset of senescence. *Plant Cell Environ.* **27:** 959-970.

Conklin, P.L. and Last, R.L. (1995). Differential accumulation of antioxidant mRNA in *Arabidopsis thaliana* exposed to ozone. *Plant Physiol.* **109:** 203-212.

Conklin, P.L., Norris, S.N., Wheeler, G.L., Williams, E.H., Smirnoff, N. and Last, R.L. (1999). Genetic evidence for the role of GDP-mannose in plant ascorbic acid (vitamin C) biosynthesis. *Proc. Natl. Acad. Sci. USA.* **96:** 4198-4203.

Conklin, P.L., Pallanca, J.E., Last, R.L. and Smirnoff, N. (1997). L-Ascorbic Acid metabolism in the Ascorbate-Deficient Arabidopsis Mutant vtc1. *Plant Physiol.* **115:** 1277-1285.

Conklin, P.L., Saracco, S.A., Norris, S.R. and Last, R.L. (2000). Identification of Ascorbic Acid-Deficient *Arabidopsis thaliana Mutants. Genetics.* **154:** 847-856.

Conklin, P.L., Williams, E.H. and Last, R.L. (1996). Environmental stress sensitivity of an ascorbic acid-deficient Arabidopsis mutant. *Proc. Natl. Acad. Sci. USA.* **93:** 9970-9974.

Creissen, G., Firmin, J., Fryer, M., Kular, B., Leyland, N., Reynolds, H., Pastori, G., Wellburn, F., Baker, N., Wellburn, A. and Mullineaux, P. (1999) Elevated glutathione biosynthetic capacity in the chloroplasts of transgenic tobacco plants paradoxically causes increased oxidative stress. *Plant Cell.* **11:** 1277-1291.

Creissen, G., Reynolds, H., Xue, Y. and Mullineaux, P. (1995). Simultaneous targeting of pea glutathione reductase and of a bacterial fusion protein to chloroplasts and mitochondria in transgenic tobacco. *Plant J.* **8:** 167-175.

Creissen, G.P., Broadbent, P., Stevens R., Wellburn, A.R. and Mullineaux, P.M. (1996). Manipulation of glutathione metabolism in transgenic plants. *Biochem. Soc. Trans.* **24:** 465-469.

Dai, F., Huang, Y., Zhou, M. and Zhang, G. (2009). The influence of cold acclimation on antioxidative enzymes and antioxidants in sensitive and tolerant barley cultivars. *Biol. Plant.* **53:** 257-262.

Dalton, D.A., Russell, S.A., Hanus, F.J., Pascoe, G.A. and Evans, H.J. (1986). Enzymatic reactions of ascorbate and glutathione that prevent peroxide damage in soybean root nodules. *Proc. Natl. Acad. Sci. USA.* **83:** 3811-3815.

Dat, J., Vandenabeele, S., Vranová, E., Van Montagu, M., Inze D and Van Breusegem, F. (2000). Dual action of the active oxygen species during plant stress responses. *Cell Mol. Life Sci.* **57:** 779-795.

Dat, J.F., Foyer, C.H. and Scott, I.M. (1998). Changes in salicylic acid and antioxidants during induced thermotolerance in mustard seedlings. *Plant Physiol.* **118:** 1455-1461.

Davletova, S., Rizhsky, L., Liang, H., Shengqiang, Z., Oliver, D.J., Coutu, J., Shulaev, V., Schlauch, K. and Mittler, R. (2005). Cytosolic ascorbate peroxidase 1 is a central component of the reactive oxygen gene network of Arabidopsis. *Plant Cell.* **17:** 268-281.

De Gara, L., de Pinto, M.C. and Tomasi, F. (2003). The antioxidant systems vis-a-vis reactive oxygen species during plant-pathogen interaction. *Plant Physiol. Biochem.* **41:** 863-870.

De Gara, L., Paciolla, C., De Tullio, M.C., Motto, M. and Arrigioni, O. (2000). Ascorbate-dependent hydrogen peroxide detoxification and ascorbate regeneration during germination of a highly productive maize hybrid: evidence of an improved detoxification mechanism against reactive oxygen species. *Physiol. Plant.* **109:** 7-13.

de Pinto, M.C., Paradiso, A., Leonetti, P. and de Gara, L. (2006). Hydrogen peroxide, nitric oxide and cytosolic ascorbate peroxidase at the crossroad between defence and cell death. *Plant J.* **48:** 784-795.

Degousie, H., Triantaphilides, C. and Montillet, J.L. (1994). Involvement of oxidative process in the signaling mechanism leading to the activation of glyceollin synthesis in soybean (*Glycine max*). *Plant Physiol.* **104:** 945-952.

Demiral, T. and Türkan, I. (2004). Does exogenous glycinebetaine affect antioxidative system of rice seedlings under NaCl treatment? *J. Plant Physiol.* **161:** 1089-1100.

Desikan, R., Clarke, A., Hancock, J.T. and Neill, S.J. (1999). H_2O_2 activates a MAP kinase-like enzymes in *Arabidopsis thaliana* suspension cultures. *J. Exp. Bot.* **50:** 1863-1866.

Desikan, R.S., Mackerness, A.H., Hancock, J.T. and Neill, S.J. (2001). Regulation of the *Arabidopsis* transcriptome by oxidative stress. *Plant Physiol.* **127:** 159-172.

Diaz-Vivancos, P., Clemente-Moreno, M.J., Rubio, M., Olmos, E., Garcia, J.A., Martinez-Gomez, P. and Hernandez, J.A. (2008). Alteration in the chloroplastic metabolism leads to ROS accumulation in pea plants in response to plum pox virus. *J. Exp. Bot.* **59:** 2147-2160.

Diaz-Vivancos, P., Rubio, M., Mesonero, V., Periago, P.M., Barceló, A.R., Martínez-Gómez, P. and Hernandez, J.A. (2006). The apoplastic antioxidant system in *Prunus*: response to long-term plum pox virus infection. *J. Exp. Bot.* **57:** 3813-3824.

Dietz, K.J. (2008). Redox signal integration: from stimulus to networks and genes. *Physiol. Plant.* **133:** 459-468.

Dipierro, N., Mondelli, D., Paciolla, C., Brunetti, G. and Dipierro, S. (2005). Changes in the ascorbate system in the response of pumpkin (*Cucurbita pepo* L.) roots to aluminum stress. *J. Plant Physiol.* **162:** 529-536.

Dixon, D.P., Lapthorn, A. and Edwards, R. (2002). Plant glutathione transferase: protein family review. *Genome Biol.* **3:** reviews 3004.1-3004.10.

Dolatabadian, A., Sanavy, S.A.M.M. and Chashmi, N.A. (2008). The Effects of Foliar Application of Ascorbic Acid (Vitamin C) on Antioxidant Enzymes Activities, Lipid Peroxidation and Proline Accumulation of Canola (*Brassica napus* L.) under Conditions of Salt Stress. *J. Agron. Crop Sci.* **194:** 206-213.

Donaldson, P.A., Anderson, T., Lane, B.G., Davidson, A.L. and Simmonds, D.H. (2001). Soybean plants expressing an active oligomeric oxalate oxidase from the wheat gf-2.8 (germin) gene are resistant to the oxalate-secreting pathogen *Sclerotinia sclerotiorum. Physiol. Mol. Plant. Pathol.* **59:** 97-307.

Dron, M., Clouse, S.D., Dixon, R.A., Lawton, M. A. and Lamb, C. J. (1988). Glutathione and fungal elicitor regulation of a plant defense gene promoter in electroporated protoplasts. *Proc. Natl. Acad. Sci. USA.* **85:** 6738-6742.

Edwards, E.A., Enard, C., Creissen, G.P. and Mullineaux, P.M. (1994). Synthesis and properties of glutathione reductase in stressed peas. *Planta.* **192:** 137-143.

Edwards, R., and Dixon, D.P. and Walbot, V. (2000). Plant glutathione S-transferases: enzymes with multiple functions in sickness and health. *Trends Plant Sci.* **5:** 193-198.

Edwards, R., Blount, J.W. and Dixon, R.A. (1991). Glutathione and elicitation of the phytoalexin response in legume cell cultures. *Planta.* **184:** 403-409.

El-Zahaby, H.M., Gullner, G. and Kiraly, Z. (1995). Effect of powdery mildew infection of barley on the ascorbate-glutathione cycle and other antioxidants in different host-pathogen interactions. *Phytopath.* **85:** 1225-1230.

Eltayeb, A.E., Kawano, N., Badawi, G., Kaminaka, H., Sanekata, T. and Morishima, I. (2006). Enhanced tolerance to ozone and drought stresses in transgenic tobacco overexpressing dehydroascorbate reductase in cytosol. *Physiol. Plant.* **127:** 57-65.

Eltayeb, A.L., Kawano, N., Badawi, G.H., Kaminaka, H., Sanekata, T., Shibahar, T., Inanaga, S. and Tanaka, K. (2007). Overexpression of monodehydroascorbate reductase in transgenic tobacco confers enhanced tolerance to ozone, salt and polyethylene glycol stresses. *Planta.* **225:** 1255-1264.

Encina, A. and Fry, S.C. (2005). Oxidative coupling of feruloylarabinoxylan trisaccharide (FAXX) in the walls of living maize cells requires endogenous hydrogen peroxide and is controlled by a low-Mr apoplastic inhibitor. *Planta.* **223:** 77-89.

Eshdat, Y., Holland, D., Faltin, Z. and Ben-Hayyim, G. (1997). Plant glutathione peroxidases. *Physiol. Plant.* **100:** 234-240.

Fadzilla, N.M., Robert, P., Finch, R.P. and Burdon, R.H. (1997). Salinity, oxidative stress and antioxidant response in shoot cultures of rice. *J. Exp. Bot.* **48:** 325-31.

Fecht-Christoffers, M.M., Maier, P. and Horst, W.J. (2003). Apoplastic peroxidase and ascorbate are involved in manganese toxicity and tolerance of *Vigna unguiculata. Physiol. Plant.* **117:** 237-244.

Fodor, J., Gullner, G., Adam, A.L., Barna, B., Komives, T. and Kiraly, Z. (1997). Local and Systemic Responses of Antioxidants to Tobacco Mosaic Virus Infection and to Salicylic Acid in Tobacco (Role in Systemic Acquired Resistance). *Plant Physiol.* **114:** 443-1451.

Foyer, C.H. and Halliwell, B. (1976). The presence of glutathione and glutathione reductase in chloroplasts: a proposed role in ascorbic acid metabolism. *Planta.* **133:** 21-25.

Foyer, C.H. and Noctor, G. (2000). Oxygen processing in photosynthesis: regulation and signalling. *New Phytol.* **146:** 359-388.

Foyer, C.H. and Noctor, G. (2003). Redox sensing and signalling associated with reactive oxygen in chloroplasts, peroxisomes and mitochondria. *Physiol. Plant.* **119:** 355-364.

Foyer, C.H. and Noctor, G. (2005a). Redox homeostasis and antioxidant signaling: A metabolic interface between stress perception and physiological responses. *Plant Cell.* **17:** 1866-1875.

Foyer, C.H. and Noctor, G. (2005b) Oxidant and antioxidant signalling in plants: a re-evaluation of the concept of oxidative stress in a physiological context. *Plant Cell Environ.* **28:** 1056-1071.

Foyer, C.H., Descourvières, P. and Kunert, K.J. (1994). Protection against oxygen radicals: an important defence mechanism studied in transgenic plants. *Plant Cell Environ.* **17:** 507-523.

Foyer, C.H., Lelandais, M., Galap, C. and Kunert, K.J. (1991). Effects of elevated cytosolic glutathione reductase activity on the cellular glutathione pool and photosynthesis in leaves under normal and stress conditions. *Plant Physiol.* **97:** 863-872.

Foyer, C.H., Lopez-Delgado, H., Dat, J.F. and Scott, I.M. (1997). Hydrogen peroxide and glutathione-associated mechanisms of acclimatory stress tolerance and signalling. *Physiol. Plant.* **100:** 241-254.

Foyer, C.H., Souriau, N., Perret, S., Lelandais, M., Kunert, K.J., Pruvost, C. and Jouanin, L. (1995). Overexpression of glutathione reductase but not glutathione synthetase leads to increases in antioxidant capacity and resistance to photoinhibition in poplar trees. *Plant Physiol.* **109:** 1047-1057.

Fujita, M. and Hossain, M.Z. (2003a). Molecular cloning of three tau-type glutathione *S*-transferases in pumpkin (*Cucurbita maxima*) and their expression. *Physiol. Plant.* **117:** 85-92.

Fujita, M. and Hossain, M.Z. (2003b). Modulation of Pumpkin Glutathione *S*-Transferases by Aldehydes and Related Compounds. *Plant Cell Physiol.* **44:** 481-490.

Gajewska, E. and Sklodowska, M. (2008). Differential biochemical responses of wheat shoots and roots to nickel stress: antioxidative reactions and proline accumulation. *Plant Growth Regul.* **54:** 179-188.

Gamble, P.E. and Burke, J.J. (1984). Effect of water stress on the chloroplastic antioxidant system. I. Alterations in glutathione reductase activity. *Plant Physiol.* **76:** 615-621.

Gao, D., Gao, Q., Xu, HY., Ma, F., Zhao, C.M. and Liu, J.Q.(2009). Physiological responses to gradual drought stress in the diploid hybrid *Pinus densata* and its two parental species. *Trees.* **23:** 717-728.

Gao, Q. and Zhang, L. (2008). Ultraviolet-B-induced oxidative stress and antioxidant defense system responses in ascorbate-deficient *vtc1* mutants of *Arabidopsis thaliana. J. Plant Physiol.* **165:** 138-148.

Garcia-Limones, C., Dorado, G., Navas-Cortes, J.A., Jimenez-Diaz, R.M. and Tena, M. (2009). Changes in the redox status of chickpea roots in response to infection by *Fusarium oxysporum*

f. sp *ciceris*: apoplastic antioxidant enzyme activities and expression of oxidative stress-related genes. *Plant Biol.* **11:** 194-203.

Gechev, T., Gadjev, I., Breusegem, F.V., Inze, D., Dukiandjiev, S., Toneva, V. and Minkov, I. (2002). Hydrogen peroxide protects tobacco from oxidative stress by inducing a set of antioxidant enzymes. *Cell Mol. Life Sci.* **59:** 708-714.

Gechev, T., Willekens, H., Van Montagu, M., Inze,´ D., Van Camp, W., Toneva, V. and Minkov, I. (2003). Different responses of tobacco antioxidant enzymes to light and chilling stress. *J. Plant Physiol.* **160:** 509-515.

Gechev, T.S., Van Breusegem, F., Stone, J.M., Denev, L. and Laloi, C. (2006). Reactive oxygen species as signals that modulate plant stress responses and programmed cell death. *Bioessays.* **28:** 1091-1101.

Gong, M., Chen, B., Li, Z.G. and Guo, L.H. (2001). Heat-shock-induced cross adaptation to heat, chilling, drought and salt stress in maize seedlings and involvement of H_2O_2. *J. Plant Physiol.* **158:** 1125-1130.

Grantz, A.A., Brummell, D.A. and Bennett, A.B. (1995). Ascorbate free radical reductase mRNA levels are induced by wounding. *Plant Physiol.* **108:** 411-418.

Grill, E., Winnacker, E.L. and Zenk, M.H. (1985). Phytochelatins: the principal heavy metal complexing peptide of higher plants. *Science.* **230:** 674-676.

Gueta-Dahan, Y., Yaniv Z., Zilinkas, B.A. and Ben-Hayyim, G. (1997) Salt and oxidative stress: similar and specific responses and their relation to salt tolerance in citrus. *Planta.* **203:** 460-469.

Gullner, G., Kömives, T. and Rennenberg, H. (2001). Enhanced tolerance of transgenic poplar plants overexpressing γ-glutamylcysteine synthetase towards chloroacetanilide herbicides. *J. Exp. Bot.* **52:** 971-979.

Gullner, G., Tóbiás, I., Fodor, J. and Kömives, T. (1999). Elevation of glutathione level and activation of glutathione-related enzymes affect virus infection in tobacco. *Free Redic. Res.* **31:** 155-161.

Guo, Z., Tan, H., Zhu, Z., Lu, S. and Zhou, B. (2005). Effect of intermediates on ascorbic acid and oxalate biosynthesis of rice and in relation to its stress resistance. *Plant Physiol. Biochem.* **43:** 955-962.

Guo, Z.J., Nakagawara, S., Sumitani, K. and Ohta, Y. (1993). Effect of Intracellular Glutathione Level on the Production of 6-Methoxymellein in Cultured Carrot (*Daucus carota*) Cells. *Plant Physiol.* **102:** 45-51.

Gupta, A.S., Alscher, R.G. and MeCume, D. (1991). Response of photosynthesis and cellular antioxidants to ozone in *Polulus* leaves. *Plant Physiol.* **96:** 650-655.

Gustine, D.L. (1987). Induction of Medicarpin Biosynthesis in Ladino Clover Callus by p-Chloromercuribenzoic Acid is Reversed by Dithiothreitol. *Plant Physiol.* **84:** 3-6.

Gómez, L.D., Noctor, G., Knight, M.R .and Foyer, C.H. (2004). Regulation of calcium signalling and gene expression by glutathione. *J. Exp. Bot.* **55:** 1851-1859.

Gönner, M.V. and Schlösser, E. (1993). Oxidative stress in interactions between *Avena sativa* L. and *Drechslera* spp. *Physiol. Mol. Plant. Pathol.* **42:** 221-234.

Hao, Z.N., Wang, L.P. and Tao, R.X. (2009). Expression patterns of defence genes and antioxidant defence responses in a rice variety that is resistant to leaf blast but susceptible to neck blast. *Physiol. Mol. Plant. Path.* (in press), doi:10.1016/j.pmpp.2009.11.003.

Hassan, N.M. and Nemat Alla, M.M. (2005). Oxidative stress in herbicide and broad bean and maize plants. *Acta Physiol. Plant.* **27:** 429-438.

Hernandez, J., Jimenez, A., Millineaus, P. and Sevilla, F. (2000). Tolerance to pea (*Pisum sativum* L.) to long-term salt stress is associated with induction of antioxidant defenses. *Plant Cell Environ.* **23:** 853-862.

Hernandez, J.A., Diaz-Vuvancos, P., Rubio, P., Olmos, M., Ros-Barcelo, E., Martinez-Gomez, P. (2006). Long-term plum pox virus infection produces an oxidative stress in a susceptible apricot, *Prunus armeniaca,* cultivar but not in a resistant cultivar. *Physiol. Plant.* **126:** 140-152.

Hernandez, J.A., Escobar, C., Crissen, G. and Mullineaux, P.M. (2004a). Role of hydrogen peroxide and the redox state of ascorbate in the induction of antioxidant enzymes in pea leaves under excess light stress. *Fun. Plant Biol.* **31:** 359-368.

Hernandez, J.A., Olmos, E., Corpas, F.J., Sevilla, F. and del Rio, L.A. (1995). Salt-induced oxidative stress in chloroplasts of pea plants. *Plant Sci.* **105:** 151-67.

Hernandez, J.A., Rubio, M., Olmos, E., Ros-Barcelo, A. and Martinez-Gomez, P. (2004b). Oxidative stress induced by long-term plum pox virus infection in peach (*Prunus persica*). *Physiol. Plant.* **122:** 486-495.

Hernandez, J.A., Tlavera, J.M., Martinez-Gomez, P., Dicenta, F. and Sevilla, F. (2001). Response of antioxidative enzymes to plum pox virus in two apricot cultivars. *Plant Physiol.* **111:** 313-321.

Herschbach, C. and Rennenberg, H. (2001). Significance of phloem-translocated organic sulfur compounds for the regulation of sulfur nutrition. *Prog. Bot.* **62:** 177-193.

Hirt, H. (1997). Multiple roles of MAP kinases in plant signal transduction. *Trends Plant Sci.* **2:** 11-15.

Holmgren, A. (1989). Thioredoxin and glutaredoxin systems. *J. Biol. Chem.* **264:** 13963-13966.

Hong, C.Y., Chao, Y.Y., Yang, M.Y., Cheng, S.Y., Cho, S.C. and Kao, C.H. (2009). NaCl-induced expression of glutathione reductase in roots of rice (*Oryza sativa* L.) seedlings is mediated through hydrogen peroxide but not abscisic acid. *Plant Soil.* **320:** 103-115.

Hoque, M.A., Banu, M.N.A., Okuma, E., Amako, K., Nakamura, Y., Shimoishi, Y. and Murata, Y. (2007). Exogenous proline and glycinebetaine ingresses NaCl-induced ascorbate-glutathione cycle enzyme activities and proline improves salt tolerance more than glycinebetaine in tobacco Bright yellow-2 suspension- cultured cells. *J. Plant Physiol.* **164:** 553-561.

Hossain, M.A. and Fujita, M. (2009a). Purification of Glyoxalase I from Onion Bulbs and Molecular Cloning of Its cDNA. *Biosci. Biotechnol. Biochem.* **73:** 2007-2013.

Hossain, M.A. and Fujita, M. (2009b). Exogenous glycinebetaine and proline modulate antioxidant defense and methylglyoxal detoxification systems and reduce drought-induced damage in mung bean seedlings (*Vigna radiata* L.). Poster presented in The 3rd International Conference on Integrated Approaches to Improve Crop Production Under Drought-Prone Environments, Shanghai, China. pp. 138.

Hossain, M.A. and Fujita, M. (2010). Evidence for a role of exogenous glycinebetaine and proline in antioxidant defense and methylglyoxal detoxification systems in mung bean seedlings under salt stress. *Physiol. Mol. Biol. Plants.* (in press).

Hossain, M.A., Hossain, M.Z. and Fujita, M. (2009). Stress-induced changes of methylglyoxal level and glyoxalase I activity in pumpkin seedlings and cDNA cloning of glyoxalase I gene. *Aust. J. Crop Sci.* **3:** 53-64

Hossain, M.Z. and Fujita, M. (2002). Purification of a phi-type glutathione S-transferase from pumpkin flowers and molecular cloning of the cDNA. *Biosci. Biotechnol. Biochem.* **66:** 2068-2076.

Hossain, M.Z., Hossain, M.D. and Fujita, M. (2006b). Induction of pumpkin glutathione *S*-transferase by different stresses and its possible mechanisms. *Biol. Plant.* **50:** 210-218.

Hossain, M.Z., da Silva, J.A.T. and Fujita, M. (2006a) Differential roles of glutathione *S*-transferase in oxidative stress modulation. In *Floriculture, Ornamental and Plant Biotechnology.* Vol. 3. pp. 108-116 (Global Science Books: UK).

Hsu, Y.T. and Kao, C.H. (2007). Heat shock-mediated H_2O_2 accumulation and protection against Cd toxicity in rice seedlings. *Plant Soil.* **300:** 137-147.

Hu, X., Bidney, D.L., Yalpani, N., Duvick, J.P., Crasta, O., Folkerts, O. and Lu, G. (2003). Overexpression of a gene encoding hydrogen peroxide-generating oxalate oxidase evokes defense responses in sunflower. *Plant Physiol.* **133:** 170-181.

Hu, Y., Ge, Y., Zhang C., Ju, T. and Cheng, W. (2009). Cadmium toxicity and translocation in rice seedlings are induced by hydrogen peroxide pretreatment. *Plant Growth Regul.* **59:** 51-61.

Huang, C., He W., Guo, J., Chang, X., Su, P. and Zhang, L. (2005) Increased sensitivity to salt stress in an ascorbate-deficient *Arabidopsis* mutant. *J. Exp. Bot.* **56:** 3041-3049.

Huang, M. and Guo, Z. (2005). Responses of antioxidative system to chilling stress in two rice cultivars differing in sensitivity. *Biol. Plant.* **49:** 81-84.

Imahori, Y., Zhou, Ueda, Y.F., and Chachin, K. (1998). Ascorbate metabolism during maturation of sweet pepper (*Capsicum annuum* L.) fruit. *J. Jpn. Soc. Hort. Sci.* **67:** 798-804.

Innocenti, G., Pucciariello, C., Le Gleuher, M., Hopkins, J., Stefano, M., Delledonne, M., Puppo, A., Baudouin, E. and Frendo, P. (2007). Glutathione synthesis is regulated by nitric oxide in *Medicago truncatula* roots. *Planta.* **225:** 1597-1602.

Ishikawa, T. and Shigeoka, S. (2008). Recent advances in ascorbate biosynthesis and the physiological significance of ascorbate peroxidase in phtotosynthesizing organisms. *Biosci. Biotechnol. Biochem.* **72:** 1143-1154.

Ishikawa, T., Morimoto, Y., Rapolu, M., Sawa, Y., Shibata, H., Yabuta, Y., Nishizawa, A. and Shigeoka, S.(2005). Acclimation to diverse environmental stresses caused by a suppression of cytosolic ascorbate peroxidase in tobacco BY-2 cells. *Plant Cell Physiol.* **46:** 1264-1271.

Islam, M.M., Hoque, M.A., Okuma, E., Banu, M.N.A., Shimoishi, Y., Nakamura, Y. and Murata, Y. (2010). Exogenous proline and glycinebetaine increase antioxidant enzyme activities and confer tolerance to cadmium stress in cultured tobacco cells. *J. Plant Physiol.* **166:** 1587-1597.

Ivenish, G. and Tillberg, E.J. (1995). Stress-induced ethylene biosynthesis in pine needles: A search for putative ACC-independent pathway. *Plant Physiol.* **145:** 308-317.

Ivenish, G, Valeina, A. and Ozol, D. (1995). Induction of ascorbate peroxidase activity in stressed pine (*Pinus sylvestsis* L.) needles – A putative role for ethylene. *Plant Sci.* **112:** 167-173.

Jain, M., Choudhary, D., Kale, R.K. and Bhalla-Sarin, N. (2002). Salt and glyphosphate-induced increase in glyoxalase I activity in cell lines of groundnut (*Arachis hypogaea*). *Physiol. Plant.* **114:** 499-505.

Janda, T., Szalai, G. and Paldi, E. (1999). Hydroponic treatment with salicylic acid decreases the effects of chilling injury in maize (*Zea mays* L.) plants. *Planta.* **208:** 175-180.

Jaspers, P. and Kangasjarvi, J. (2010). Reactive oxygen species in abiotic stress signalling. *Phyiol. Plant.* **138:** 405-413.

Jimenez, A., Hernandez, J.A., Pastori, G., del Rio, L.A. and Sevilla, F. (1998). Role of the ascorbate-glutathione cycle of mitochondria and peroxisomes in the senescence of pea leaves. *Plant Physiol.* **118:** 1327-1335.

Jimenez, A., Hernandez, J.A., del Rio, L.A. and Sevilla, F. (1997). Evidence for the Presence of the Ascorbate-Glutathione Cycle in Mitochondria and Peroxisomes of Pea Leaves. *Plant Physiol.* **114:** 275-284.

Kampranis, S.C., Damianova, R., Atallah, M., Toby G., Kondi, G., Tsichlisi, P.N. and Makris, A.M. (2000). A Novel Plant Glutathione S-Transferase/Peroxidase Suppresses Bax Lethality in Yeast. *J. Biol. Chem.* **275:** 29207-29216.

Kang, N.J., Kang, Y.I., Kang, K.H. and Jeong, B.R. (2009). Induction of thermotolerance and activation of antioxidant enzymes in H_2O_2 pre-applied leaves of cucumber and tomato seedlings. *J. Jpn. Soc. Hort. Sci.* **78:** 320-329.

Kangasjarvi, J., Jaspers, P. and Kollist, H. (2005). Signalling and cell death in ozone-exposed plants. *Plant Cell Environ.* **28:** 1021-1036.

Karpinski, S., Escorbar, C., Karpinska, B., Creissen, G. and Mullineaux, P.M. (1997). Photosynthetic electron transport regulates the expression of cytosolic ascorbate peroxidase genes in *Arabidopsis* during excess light stress. *Plant Cell.* **9:** 627-640.

Karpinski, S., Reynolds, H., Karpinska, B., Wingsle, G., Creissen, G. and Mullineaux, P. (1999). Systemic signalling and acclimation in response to excess excitation energy in *Arabidopsis. Science.* **284:** 654-657.

Katsuhara, M., Otsuka, T. and Ezaki, B. (2005). Salt stress-induced lipid peroxidation is reduced by glutathione *S*-transferase, but this reduction of lipid peroxides is not enough for a recovery of root growth in *Arabidopsis. Plant Sci.* **169:** 369-373.

Kavitha, K., Venkataraman, G. and Parida, A. (2008). An oxidative and salinity stress induced peroxisomal ascorbate peroxidase from *Avicennia marina*: Molecular and functional characterization. *Plant Physiol. Biochem.* **46:** 794-804.

Khan, M.H. and Panda, S.K. (2007). Alterations in root lipid peroxidation and antioxidative responses in two rice cultivars under NaCl-salinity stress. *Acta Physiol. Plant.* **30:** 81-89.

Kiddle, G., Pastori, G.M., Bernard, S., Pignocchi, C., Antoniw, J., Verrier, P.J. and Foyer, C.H. (2003). Effects of ascorbate signaling on defense and photosynthesis genes. *Antiox. Redox Signal.* **5:** 23-32.

Kim, Y.H., Kim, C.Y., Song, W.K., Park, D.S., Kwon, S.Y., Lee, H.S., Bang, J.W. and Kwak, S.S. (2008). Overexpression of sweet potato *swpa4* peroxidase results in increased hydrogen peroxide production and enhances stress tolerance in tobacco. *Planta.* **227:** 867-881.

Kirsch, M. and de Groot, H. (2001). NAD (P)H, a directly operating antioxidant? *FASEB J.* **15:** 1569-1574.

Kocsy, G., von Ballmoos, P., Rüegsegger, A., Szalai, G., Galiba, G. and Brunold, C. (2001). Increasing the glutathione content in a chilling-sensitive maize genotype using safeners increased protection against chilling-induced injury. *Plant Physiol.* **127:** 1147-1456.

Kocsy, G., Brunner, M., Rüegsegger, A., Stamp, P. and Brunold, C. (1996). Glutathione synthesis in maize genotypes with different sensitivity to chilling. *Planta.* **198:** 365-370.

Kocsy, G., Szalai, G. and Galiba, G. (2002). Effect of heat stress on glutathione biosynthesis in wheat. *Acta Biol. Szeged.* **46:** 71-72.

Kolarovic, L., Valentovic, P., Luxova, M. and Gasparikova, O. (2009). Changes in antioxidants and cell damage in heterotrophic maize seedlings differing in drought sensitivity after exposure to short-term osmotic stress. *Plant Growth Regul.* **59:** 21-26.

Komives, T., Guller, G. and Kiraly, Z. (1998). Role of glutathione and glutathione-related enzymes in response of plants to environmental stress. *Ann. N.Y. Acad. Sci.* **851:** 251-258.

Kornyeyev, D., Logan, B.A., Payton, P., Allen, R.D. and Holaday, A.S. (2001). Enhanced photochemical light utilization and decreased chilling induced photoinhibition of photosystem II in cotton over expressing genes encoding chloroplast targeted antioxidant enzymes. *Physiol Plant.* **113:** 323-331.

Kornyeyev, D., Logan, B.A., Payton, P., Allen, R.D. and Holaday, A.S. (2003). Elevated chloroplast glutathione reductase activities decrease chilling induced photoinhibition by increasing rates of photochemistry, but not thermal energy dissipation in transgenic cotton. *Func. Plant Bio.* **30:** 101-110.

Koussevitzky, S., Suzuki, N., Huntington, S., Armijo, L., Sha, W., Cortes, D., Shulaev, V. and Mittler, R. (2008). Ascorbate Peroxidase 1 Plays a Key Role in the Response of *Arabidopsis thaliana* to Stress Combination. *J. Biol. Chem.* **283:** 34197-34203.

Kranner, I., Beckett, R.P., Wornik, S., Zorn, M. and Pfeifhofer, H.W. (2002). Revival of a resurrection plant correlates with its antioxidant status. *Plant J.* **31:** 13-24.

Kukreja, S., Nandwal, A., Kumar, N., Sharma, S., Unvi, V. and Sharma, P. (2005). Plant water status, H_2O_2 scavenging enzymes, ethylene evolution and membrane integrity of *Cicer arietinum* roots as affected by salinity. *Biol. Plant.* **49:** 305-308.

Kumutha, D., Ezhilmathi, K., Sairam, R.K., Srivastava, G.C., Deshmukh, P.S. and Meena, R.C. (2009). Waterlogging induced oxidative stress and antioxidant activity in pigeonpea genotype. *Biol. Plant.* **53:** 75-84.

Kuzniak, E. and Sklodowska, M. (1999). The effect of *Botrytis cinerea* infection on ascorbate-glutathione cycle in tomato leaves. *Plant Sci.* **148:** 69-76.

Kuzniak, E. and Sklodowska, M. (2001). Ascorbate, glutathione and related enzymes in chloroplasts of tomato leaves infected by *Botrytis cinerea. Plant Sci.* **160:** 723-731.

Kuzniak, E. and Sklodowska, M. (2004a). The effect of *Botrytis cinerea* infection on the antioxidant profile of mitochondria from tomato leaves. *J. Exp. Bot.* **397:** 605-612.

Kuzniak, E. and Sklodowska, M. (2004b). Differential implication of glutathione, Glutathione–Metabolizing enzymes and ascorbate in tomato resistance to *Pseudomonas syringae. J. Phytopath.* **152:** 529-536.

Kuzniak, E. and Sklodowska, M. (2005a). Compartment-specific role of the ascorbate-glutathione cycle in the response to tomato leaf cells to *Botrytis cinerea* infection. *J. Exp. Bot.* **413:** 921-933.

Kuzniak, E. and Sklodowska, M. (2005b). Fungal pathogen-induced changes in the antioxidant systems of leaf peroxisomes from infected tomato plants. *Planta.* **222:** 192-200.

Kuzniak, E. and Urbanek, H. (2000). The involvement of hydrogen peroxide in plant responses to stresses. *Acta Physiol. Plant.* **22:** 195-203.

Kwon, S.Y., Choi, S.M., Ahn, Y.O., Lee, H.S., Lee, H.B. and Park, Y.M. (2003). Enhanced stress-tolerance of transgenic tobacco plants expressing a human dehydroascorbate reductase gene. *J. Plant Physiol.* **160:** 347-353.

Lambais, M.R., Rios-Ruiz, W.F. and Andrade, R.M. (2003). Antioxidant responses in bean (*Phaseolus vulgaris*) roots colonized by arbuscular mycorrhizal fungi. *New Phytol.* **160:** 421-428.

Larkindale, J. and Huang, B. (2004). Thermotolerance and antioxidant systems in *Agrostis stolonifera*: involvement of salicylic acid, abscisic acid, calcium, hydrogen peroxide, and ethylene. *J. Plant Physiol.* **161:** 405-413.

Larson, R.A. (1988). The antioxidants of higher plants. *Phytochem.* **27:** 969-978.

Lascano, H.R., Casano, L.M., Melchiorre, M.N., Trippi, V.S. (2001). Biochemical and molecular characterization of wheat chloroplastic glutathione reductase. *Biol. Plant.* **44:** 509-516.

Lascano, H.R., Gomez, L.D., Casano, L.M. and Trippi, V.S. (1999). Wheat chloroplastic glutathione reductase activity is regulated by the combined effect of pH, NADPH and GSSG. *Plant Cell Physiol.* **40:** 683-690.

Lederer, B. and Boger, P. (2003.) Antioxidant responses to tobacco expressing a bacterial glutathione reductase. *Z. Naturforsch.* **58:** 843-849.

Lee, Y.P., Kim, S.H., Bang, J.W., Lee, H.S., Kwak, S.S. and Kwon, S.Y. (2007). Enhanced tolerance to oxidative stress in transgenic tobacco plants expressing three antioxidant enzymes in chloroplasts. *Plant Cell Rep.* **26:** 591-598.

Leon, J., Lawton, M.A. and Raskin, I. (1995). H_2O_2 stimulates salicylic acid biosynthesis in tobacco. *Plant Physiol.* **108:** 1673-1678.

Leterrier, M., Corpas, F.J., Barroso, J.B., Sandalio, L.M. and del Rio, L.A. (2005). Peroxisomal monodehydroascorbate reductase. Genomic clone characterization and functional analysis under environmental stress conditions. *Plant Physiol.* **138:** 2111-2123.

Levine, A., Tenhaken, R., Dixon, R. and Lamb, C. (1994). H_2O_2 from the oxidative burst orchestrates the plant hypersensitive disease resistance response. *Cell.* **79:** 583-593.

Li, Y.J., Hai, R.L., Du, X.H., Jiang, X.N. and Lu, H. (2009). Over-expression of a populus peroxisomal ascorbate peroxidase (PpAPX) gene in tobacco plants enhances stress tolerance. *Plant Breed.* **128:** 404-410.

Li, Z. and Burritt, D.J. (2003). The influence of Cocksfoot mottle virus on antioxidant metabolism in the leaves of *Dactylis glomerata* L. *Physiol. Mol. Plant Path.* **62:** 285-295.

Liang, H., Maynard, C.A., Allen, R.D., Powell, W.A. (2001). Increased *Septoria musiva* resistance in transgenic hybrid poplar leaves expressing a wheat oxalate oxidase gene. *Plant Mol. Biol.* **45:** 619-629.

Lim, S., Kim, Y.H., Kim, S.H., Kwon, S.Y., Lee, H.S., Kim, J.S., Cho, K.Y., Paek, K.Y. and Kwak, S.S. (2007). Enhanced tolerance of transgenic sweetpotato plants that express both CuZnSOD and APX in chloroplasts to methyl viologen-mediated oxidative stress and chilling. *Mol. Breed.* **19:** 227-239.

Lin, K.H.R., Weng, C.C., Lo, H.F. and Chenc, J.T. (2004). Study of the root antioxidative system of tomatoes and eggplants under waterlogged conditions. *Plant Sci.* **167:** 355-365.

Liso, R., Innocenti, A.M., Bitonti, M.B. and Arrigoni, O. (1998). Ascorbic acid induced progression of quiescent centre cells from G1 to S phase. *New Phytol.* **110:** 469-471.

Liu, Z.J., Guo, Y.K. and Bai, J.G. (2009). Exogenous Hydrogen Peroxide Changes Antioxidant Enzyme Activity and Protects Ultrastructure in Leaves of Two Cucumber Ecotypes under Osmotic Stress. *J. Plant Growth Regul.* (in press), doi: 10.1007/s00344-009-9121-8.

Locato, V., de Pinto, M.C. and De Gara, L. (2009). Different involvement of the mitochondrial, plastidal and cytosolic ascorbate-glutathione redox enzymes in heat shock responses. *Phyiol. Plant.* **135:** 296-306.

Logan, B.A., Monteiro, G., Kornyeyev, D., Payton, P., Allen, R.D. and Holaday, A.S. (2003). Transgenic overproduction of glutathione reductase does not protect cotton, *Gossypium hirsutum* (Malvaceae), from photoinhibition during growth under chilling conditions. *Am. J. Bot.* **90:** 1400-1403.

Lopez-Delgado, H., Dat, J.F., Foyer, C.H. and Scott, I.M. (1998). Induction of thermotolerance in potato microplants by acetylsalicylic acid and H_2O_2. *J. Exp. Bot.* **321:** 713-720.

Lu, P., Sang, W.G. and Ma, K.P. (2007). Activity of stress-related antioxidant enzymes in the invasive plant crofton weed (*Eupatorium adenophorum*). *J. Integr. Plant Biol.* **49:** 1555-1564.

Lu, Tao., He, X., Chen, W., Yan, K. and Zhao, T. (2009). Effects of Elevated O_3 and/or Elevated CO_2 on Lipid Peroxidation and Antioxidant Systems in *Ginkgo biloba* Leaves. *Bull. Environ. Contam. Toxicol.* **83:** 92-96.

Lunde, C., Baumann, U., Shirley, N.J., Drew, D.P. and Fincher, G.B. (2006). Gene structure and expression pattern analysis of three monodehydroascorbate reductase (*Mdhar*) genes in *Physcomitrella patens*: Implications for the evolution of the MDHAR family in plants. *Plant Mol. Biol.* **60:** 259-275.

Lunde, C., Zygadlo, A., Simonsen, H. T., Nielsen, P. L., Blennow, A. and Haldrup, A. (2008). Sulfur starvation in rice: the effect on photosynthesis, carbohydrate metabolism, and oxidative stress protective pathways. *Physiol. Plant.* **134:** 508-521.

Ma, Y.H., Ma, F.W., Zhang, J.K., Li, M.J., Wang, Y.H. and Liang, D. (2008). Effects of high temperature on activities and gene expression of enzymes involved in ascorbate-glutathione cycle in apple leaves. *Plant Sci.* **175:** 761-766.

Machler, F., Wasescha, M.R., Krieg, F. and Oertli, J.J. (1995). Damage by ozone and protection by ascorbic acid in barley leaves. *J. Plant Physiol.* **147:** 469-473.

Maddison, J., Lyons, T., Plochl, M. and Barnes, J. (2002). Hydroponically cultivated radish fed 1-galactono-1, 4-lactone exhibits increased tolerance to ozone. *Planta* **214:** 383-391.

Maheshwari, R. and Dubey, R.S. (2009). Nickel-induced oxidative stress and the role of antioxidant defence in rice seedlings. *Plant Growth Regul.* **59:** 37-49.

Mai, J., Herbette, S., Vandame, M., Kositsup, B., Kasemsap, P., Cavaloc, E., Julien, J.L., Ameglio, T. and Roeckel-Drevet, P. (2009). Effect of chilling on photosynthesis and antioxidant enzymes in *Hevea brasiliensis* Muell. Arg. *Trees.* **23:** 863-874.

Marco, F., Calvo, E., Carrasco, P. and Sanz, M.J. (2008). Analysis of molecular markers in three different tomato cultivars exposed to ozone stress. *Plant Cell Rep.* **27:** 197-207.

Marrs, K. (1996). The function of regulation of glutathione S-transferases in plants. *Annu. Rev. Plant. Physiol. Plant. Mol. Biol.* **47:** 127-158.

Marty, L., Siala, W., Schwarzlander, M., Fricker, M.D., Wirtz, M., Sweetlove, L.J., Meyer, Y., Meyer, A.J., Reichheld, J.P. and Hell, R. (2009). The NADPH-dependent thioredoxin system constitutes a functional backup for cytosolic glutathione reductase in Arabidopsis. *Proc. Natl. Acad. Sci. USA.* **106:** 9109-9114.

May, M.J. and Leaver, C.J. (1993). Oxidative stimulation of glutathione synthesis in *Arabidopsis thaliana* suspension culture. *Plant Physiol.* **103:** 157-166.

May, M.J., Parker, J.E., Daniels, M.J., Leaver, C.J. and Cobbett, C.S. (1996a). Involvement of reactive oxygen species, glutathione metabolism and lipid peroxidation in the *Cf*-gene-dependent defense response of tomato cotyledons induced by race-specific elicitors of *Cladosporium fulvum. Plant Physiol.* **110:** 1367-1379.

May, M.J., Parker, J.E., Daniels, M.J., Leaver, C.J. and Cobbett, C.S. (1996b). An *Arabidopsis* mutant depleted in glutathione shows unaltered responses to fungal and bacterial pathogens. *Mol. Plant Microbe Interact.* **9:** 349-356.

May, M., Vernoux, T., Leaver, C., van Montagu, M. and Inze, D. (1998). Glutathione homeostasis in plants: implications for environmental sensing and plant development. *J. Exp. Bot.* **49:** 649 -667.

Mehlhorn, H., Seufert, G., Semidt, A. and Kunert, K.J. (1986). Effect of SO_2 and O_3 on production of antioxidant in conifers. *Plant Physiol.* **82:** 336-338.

Meister, A. (1992). On the antioxidant effects of ascorbic acid and glutathione. *Biochem. Pharmacol.* **44:** 1905-1915

Metwally, A., Safronova, V.I., Belimov, A.A. and Dietz, K.J. (2005). Genotypic variation of the response to cadmium toxicity in *Pisum sativum* L. *J. Exp. Bot.* **56:** 167-178.

Miller, G., Sujuki, N., Ciftci-Yilmaz, S. and Mittler, R. (2010). Reactive oxygen species homeostasis and signalling during drought and salinity. *Plant Cell Environ.* **33:** 453-467.

Miller, G., Suzuki, N., Rizhsky, L., Hegie, A., Koussevitzky, S. and Mittler, R. (2007). Double deficient in cytosolic and thylakoid ascorbate peroxidase reveal a complex mode of interaction between reactive oxygen species, plant development, and response to abiotic stresses. *Plant Physiol.* **144:** 1777-1785.

Mishra, N.P., Fatma, T. and Singhal, G.S. (1995). Development of antioxidative defense system of wheat seedlings in response to high light. *Physiol. Plant.* **95:** 77-82.

Mishra, N.P., Mishra, R.K. and Singhal, G.S. (1993). Changes in the activities of antioxidant enzymes during exposure of intact wheat leaves to strong visible light at different temperatures in the presence of protein inhibitors. *Plant Physiol.* **102**: 903-910.

Mittler, M and Zilinskas, BA (1994). Regulation of pea cytosolic ascorbate peroxidase and other antioxidant enzymes during the progression of drought stress and following recovery from drought. *Plant J.* **5:** 397-405.

Mittler, R., Feng, X. and Cohen, M. (1998). Post-transcriptional suppression of cytosolic ascorbate peroxidase during pathogen-induced programmed cell death in tobacco. *Plant Cell.* **10:** 461-473.

Mittler, R., Herr, E.H., Orvar , B.L., van Camp, W., Willekens, H., Inzé, D. and Ellis, B.E. (1999). Transgenic tobacco plants with reduced capability to detoxify reactive oxygen intermediates are hyperresponsive to pathogen infection. *Proc. Nat. Acad. Sci. USA.* **96:** 14165-14170.

Mittler, R., Vanderauwera, S., Gollery, M. and Breusegem, F.V. (2004). Reactive oxygen gene network of plants. *Trends Plant Sci.* **9:** 490-498.

Mittova, V., Tal, M., Volokita, M. and Guy, M. (2002). Salt stress induces upregulation of an efficient chloroplast antioxidant system in the salt-tolerant wild tomato species *Lycopersicon pennellii* but not in the cultivated species. *Physiol. Plant.* **115:** 393-400.

Mittova, V., Tal, M., Volokita, M. and Guy, M. (2003a). Upregulation of the leaf mitochondrial and peroxisomal antioxidative systems in response to salt-induced oxidative stress in the wild salt-tolerant tomato species *Lycopersicon pennellii. Plant Cell Environ.* **26:** 845-856.

Mittova, V., Theodoulou, F.L., Kiddle, G., Gomez, L., Volokita, M. and Tal, M. (2003b). Co-ordinate induction of glutathione biosynthesis and glutathione metabolizing enzymes is correlated with salt tolerance. *FEBS Lett.* **554:** 417-421.

Mittova, V., Volokita, M., Guy, M. and Tal, M. (2000). Activities of SOD and the ascorbate-glutathione cycle enzymes in subcellular compartments in leaves and roots of the cultivated tomato and its wild salt-tolerant relative *Lycopersicon pennellii. Physiol. Plant.* **110:** 42-51.

Moller, I.M. (2001). Plant mitochondria and oxidative stress: Electron transport, NADPH turnover, and metabolism of reactive oxygen species. *Annu. Rev. Plant. Physiol. Plant Mol. Biol.* **52:** 561-591.

Moller, I.M., Jensen, P.E. and Hansson, A. (2007). Oxidative modifications to cellular components in plants. *Annu. Rev. Plant Biol.* **58:** 459-481.

Moran, J.F., Becana, M., Iturbe-Ormaetxe, I., Frechilla, S., Klucas, R.V. and Aparicio-Tejo, P. (1994). Drought induced oxidative stress in pea plants. *Planta.* **194:** 346-352.

Moskova, I., Todorova, D., Alexieva, V., Ivanov, S. and Sergiev, I. (2009). Effect of exogenous hydrogen peroxide on enzymatic and nonenzymatic antioxidants in leaves of young pea plants treated with paraquat. *Plant Grwoth Regul.* **57:** 193-202.

Muckenschnabel, I., Goodman, B.A., Williamson, B., Lyon, G.D. and Deighton, N. (2002). Infection of leaves of *Arabidopsis thaliana* by *Botrytis cinerea*: changes in ascorbic acid, free radicals and lipid peroxidation products. *J. Exp. Bot.* **53:** 207-214.

Muller-Moule, P., Golan, T. and Niyogi, K.K. (2003). Zeaxanthin deficiency enhances the high light sensitivity of an ascorbate-deficient mutant of Arabidopsis. *Plant Physiol.* **133:** 748-760.

Muller-Moule, P., Golan, T. and Niyogi, K.K. (2004). Ascorbate-deficient mutants of *Arabidopsis* grow in high light despite chronic photooxidative stress. *Plant Physiol.* **134:** 1163-1172.

Murgia, I., Tarantino, D., Vannini, C., Bracale, M., Carravieri, S. and Soave, C. (2004). *Arabidopsis thaliana* plants overexpressing thylakoidal ascorbate peroxidase show increased resistance to Paraquat-induced photooxidative stress and to nitric oxide-induced cell death. *Plant J.* **38:** 940-953.

Nakagawara, S., Nakamura, N., Guo, Z.J., Sumitani, K., Katoh, K. and Ohta, Y. (1993). Enhanced formation of a constitutive sesquiterpenoid in cultured cells of a liverwort, *Calypogeia granulata* Inoue during elicitation: Effects of a vanadata. *Plant Cell Physiol.* **34:** 421-429.

Nakano, Y. and Asada, K. (1981). Hydrogen peroxide is scavenged by ascorbate specific peroxidase in spinach chloroplast. *Plant Cell Physiol.* **22:** 867-880.

Navrot, N., Collin, V., Gualberto, J., Gelhaye, E. and Hirasawa, M. (2006). Plant glutathione peroxidases are functional peroxiredoxins distributed in several subcellular compartments and regulated during biotic and abiotic stresses. *Plant Physiol.* **142:** 1364-1379.

Neill, S., Desikan, R. and Hancock, J. (2002). Hydrogen peroxide signalling. *Curr. Opin. Plant. Biol.* **5:** 388-395.

Nieto-Sotelo, J. and Ho, T.H.D. (1986). Effect of heat shock on the metabolism of glutathione in maize roots. *Plant Physiol.* **82:** 1030-1035.

Noctor, G. (2006). Metabolic signalling in defence and stress: the central roles of soluble redox couples. *Plant Cell Environ.* **29:** 409-425.

Noctor, G. and Foyer, C.H. (1998). Ascorbate and glutathione: keeping active oxygen under control. *Annu. Rev. Plant Physiol. Plant Mol. Biol.* **49:** 249-279.

Noctor, G., Arisi, A., Jouanin, L., Kunert, K.J., Rennenberg, H. and Foyer, C. (1998). Glutathione: biosynthesis, metabolism and relationship to stress tolerance explored in transformed plants. *J. Exp. Bot.* **49:** 623-647.

Noctor, G., Strohm, M., Jouanin, L., Kunert, K.J., Foyer, C.H., Rennenberg, H. (1996). Synthesis of glutathione in leaves of transgenic poplar overexpressing gamma- glutamylcysteine synthetase. *Plant Physiol.* **112:** 1071-1078.

Nouairi, I., Ammar, W.B., Youssef, N.B., Miled, D.D.B., Ghorbal, M.H. and Zarrouk, M. (2009). Antioxidant defense system in leaves of Indian mustard (*Brassica juncea*) and rape (*Brassica napus*) under cadmium stress. *Acta Physiol. Plant.* **31:** 237-247.

Ogawa, K. (2005). Glutathione-associated regulation of plant growth and stress responses. *Antiox. Red. Signal.* **7:** 973-981.

Ogawa, K., Hatano-Iwasaki, A., Yanagida, M. and Iwabuchi, M. (2004). Level of glutathione is regulated by ATP-dependent ligation of glutamate and cysteine through photosynthesis in *Arabidopsis thaliana*: mechanism of strong interaction of light intensity with flowering. *Plant Cell Physiol.* **45:** 1-8.

Ovar, B.L. and Ellis, B.E. (1997). Transgenic tobacco plants expressing antisense RNA for cytosolic ascorbate peroxidase show increased susceptibility to ozone injury. *Plant J.* **11:** 1297-1305.

Ozkur, O., Ozdemir, F., Bor M., and Turkan, I. (2009). Physiochemical and antioxidant responses of the perennial xerophyte *Capparis ovata* Desf. to drought. *Environ. Exp. Bot.* **66:** 487-492.

Ozturk, Z.N., Talame, V., Deyholos, M., Michalowski, C.B., Galbraith, D.W., Gozukirmizi, N., Tuberosa, R. and Bohnert, H.J. (2002). Monitoring large-scale changes in transcript abundance in drought- and salt-stressed barley. *Plant Mol. Biol.* **48:** 551-573.

Panchuk, I.I., Volkov, R.A. and Schöffl, F. (2002). Heat stress- and heat shock transcription factor-dependent expression and activity of ascorbate peroxidase in *Arabidopsis. Plant Physiol.* **129:** 838-853.

Paradiso, A., Berardino, R., de Pinto, M., di Toppi, L.S., Storelli, F.T. and de Gara, L. (2008). Increase in ascorbate-glutathione metabolism as local and precocious systemic responses induced by cadmium in durum wheat plants. *Plant Cell Physiol.* **49:** 362-374.

Parida, A.K., Das, A.B. and Mohanthy, P. (2004). Defense potentials to NaCl in mangrove, *Bruguiera parviflora*: differential changes of isoforms of some antioxidative enzymes. *J. Plant Physiol.* **161:** 531-542.

Pasqualini, S., Della Torre, G., Ferranti, F., Ederli, L., Piccioni, C., Reale, L. and Antonielli, M. (2002). Salicylic acid modulates ozone induced hypersensitive cell death in tobacco plants. *Physiol. Plant.* **115:** 204-212.

Pastori, G., Foyer, C.H. and Mullineaux, P. (2000). Low temperature induced changes in the distribution of H_2O_2 and antioxidants between the bundle sheath and mesophyll cells of maize leaves. *J. Exp. Bot.* **51:** 107-113.

Pastori, G., Kiddle, G., Antoniw, J., Bernard, S., Veljovic-Jovanovic, S., Verrier, P.J., Noctor, G. and Foyer, C.H. (2003). Leaf vitamin C contents modulate plant defense transcripts and regulate genes that control development through hormone signaling. *Plant Cell.* **15:** 939-951.

Pastori, G.M. and Trippi, V.S. (1992). Oxidative stress induces high rate of glutathione reductase synthesis in a drought resistant maize strain. *Plant Cell Physiol.* **33:** 957-961.

Payton, P., Webb, R., Kornyeyev, D., Allen, R. and Holaday, A.S. (2001). Protecting cotton photosynthesis during moderate chilling at high light intensity by increasing chloroplastic antioxidant enzyme activity. *J. Exp. Bot.* **52:** 2345-2354.

Perez-Lopez, U., Robredo, A., Lacuesta, M., Sgherri, C., Munoz-Rueda, A., Navari-Izzo, F. and Mena-Petite, A. (2009). The oxidative stress caused by salinity in two barley cultivars is mitigated by elevated CO_2. *Physiol. Plant.* **135:** 29-42.

Pietrini, F., Innelli, M.A., Pasqualini, S. and Massacci, A. (2003). Introduction of cadmium with glutathione and photosynthesis in developing leaves and chloroplast of phragmites australis (cav.) Trin. Ex Steudel. *Plant Phsyiol.* **133:** 829-837.

Pilon-Smith, E., Zhu, Y.L., Sears, T. and Terry, N. (2000). Over-expression of glutathione reductase in *Brassica juncea*: effects on cadmium accumulation and tolerance. *Physiol. Plant.* **110:** 455-460.

Pnueli, L., Liang, H., Rozenberg, M. and Mittler, R. (2003). Growth suppression, altered stomatal responses, and augmented induction of heat shock proteins in cytosolic ascorbate peroxidase (Apx1)-deficient Arabidopsis plants. *Plant J.* **34:** 187-203.

Prasad, T.K., Anderson, M.D., Martin, B.A. and Stewart, C.R. (1994). Evidence for chilling-induced oxidative stress in maize seedlings and a regulatory role for hydrogen peroxide. *Plant Cell.* **6:** 65-74.

Puckette, M., Lila, P., Jarrod,S., Tang, Y. and Ramamurthy, M. (2009). Ozone responsive genes in *Medicago truncatula*: Analysis by suppression subtraction hybridization. *J. Plant Physiol.* **166:** 1284-1295.

Puppo, A., Groten, K., Bastian, F., Carzaniga, R., Soussi, M., Lucas, M.M., De Felipe, M.R., Harrison, J., Vanacker, H. and Foyer, C.H. (2005). Legume nodule senescence: roles for redox and hormone signalling in the orchestration of the natural aging process. *New Phytol.* **165:** 683-701.

Qadir, S., Qureshi, M.I., Javed, S. and Abdin, M.Z. (2004). Genotypic variation in phytoremediation potential of *Brassica juncea* cultivars exposed to Cd stress. *Plant Sci.* **167:** 1171-1181.

Quan, L.J., Zhang, B., Shi, W.W. and Li, H.Y. (2008). Hydrogen peroxide in plants: a versatile molecule of the reactive species network. *J. Integr. Plant Biol.* **50:** 2-18.

Ranieri, A., Castagna, A., Baldan, B., Sebastiani, L. and Soldatini, G.F. (2003). H_2O_2 accumulation in sunflower leaves as a consequence of iron deprivation. *J. Plant Nutr.* **26:** 2187-2196.

Rao, M.V., Paliyath, G. and Ormrod, D.P. (1996). Ultraviolet-B and ozone-induced biochemical changes in antioxidant enzymes of *Arabidopsis thaliana*. *Plant Physiol.* **110:** 125-136.

Rausch, T., Gromes, R., Liedschulle, V., Müller, I., Bogs, J., Galovic, V. and Wachter, A. (2007). Novel insight into the regulation of GSH biosynthesis in higher plants. *Plant Biol.* **9:** 565-572.

Reisinger, S., Schiavon, M., Norman, T. and Pilon-Smits, E.A.H. (2008). Heavy metal tolerance and accumulation in Indian mustard (*Brassica juncea* L.) expressing bacterial gamma-glutamylcysteine synthetase or glutathione synthetase. *Int. J. Phytoremediation.* **10:** 1-15.

Rivero, R.M., Ruiz, J.M. and Romero, L. (2004). Oxidative metabolism in tomato plants subjected to heat stress. *J. Hort. Sci. Biotech.* **79:** 560-564.

Robinson, J.M. and Britz, S.J. (2000). Tolerance of a field grown soybean cultivar to elevated ozone level is concurrent with higher leaflet ascorbic acid level, higher ascorbate-dehydroascorbate redox status, and long-term photosynthetic productivity. *Photosynthesis Res.* **64:** 77-87.

Robson, C.A. and Vanlerberghe, G.C. (2002). Transgenic plant cells lacking mitochondrial alternative oxidase have increased susceptibility to mitochondria-dependent and independent pathway of programmed cell death. *Plant Physiol.* **129:** 1908-1920.

Romero-Puertas, M., Palma, J.M., Gomez, M., Dong-Hee, L. and Sandalio, L.M. (2002). Cadmium causes the oxidative modification of proteins in pea plants. *Plant Cell Environ.* **25:** 677-686.

Romero-Puertas, M.C., Corpas, F.J., Rodriguez-Serrano, M., Gomez, M., del Rio, L.A. and Sandalio, L.M. (2007). Differential expression and regulation of antioxidative enzymes by cadmium in pea plants. *J. Plant Physiol.* **164:** 1346-1357.

Romero-Puertas, M.C., Corpas, F.J., Sandalio, L.M., Leterrier, M., Rodriguez-Serrano, M., del Rio, L.A. and Palma, J.M. (2006). Glutathione reductase from pea leaves: response to abiotic stress and characterization of the peroxisomal isozyme. *New Phytol.* **170:** 43-52.

Romero-Romero, M.T. and Lopez-Delgado, H.A. (2009). Ameliorative Effects of Hydrogen Peroxide, Ascorbate and Dehydroascorbate in *Solanum Tuberosum* Infected by Phytoplasma. *Am. J. Pot. Res.* **86:** 218-226.

Roxas, V.P., Lodhi, S.A., Garrett, D.K., Mahan, J.R. and Allen, R.D. (2000). Stress tolerance in transgenic tobacco seedlings that overexpress glutathione S-transferase/glutathione peroxidase. *Plant Cell Physiol.* **41:** 1229-1234.

Roxas, V.P., Smith, R.K., Allen, E.R. and Allen, R.D. (1997). Overexpression of glutathione S-transferase/glutathione peroxidase enhances the growth of transgenic tobacco seedlings during stress. *Nat. Biotechnol.* **15:** 988-991.

Rubio, M.C., Busots-Sanmamed, P., Clemente, M.R. and Becana, M. (2009). Effect of salt stress on the expression of antioxidant genes and proteins in the model legume *Lotus japonicus. New Phytol.* **181:** 851-859.

Ryang, S.Z., Woo, S.Y., Kwon, S.Y., Kim, S.H., Lee, S.H., Kim, K.N. and Lee, D.K. (2009). Changes of net photosynthesis, antioxidant enzyme activities, and antioxidant contents of *Liriodendron tulipifera* under elevated ozone. *Photosynthetica.* **47:** 19-25.

Sagi, M. and Fluhr, R. (2006). Production of reactive oxygen species by plant NADPH oxidases. *Plant Physiol.* **141:** 336-340.

Sairam, R.K., Shukla, D.S. and Saxena, D.C. (1997/98). Stress-induced injury and antioxidant enzymes in relation to drought tolerance in wheat genotypes. *Biol. Plant.* **40:** 357-364.

Sairam, R.K., Srivastava, G.C., Agarwal, S. and Meena, R.C. (2005). Differences in antioxidant activity in response to salinity stress in tolerant and susceptible wheat genotypes. *Biol. Plant* **49:** 85-91.

Sairam, R.K. and Tyagi, A. (2004). Physiology and molecular biology of salinity stress tolerance in plants. *Curr. Sci.* **86:** 407-421.

Sarowar, S., Kim, E.N., Kim, Y.J., Ok, S.H., Kim, K.D., Hwang, B.K. and Shin, J.S. (2005). Overexpression of a pepper ascorbate peroxidase-like I gene in tobacco plants enhances tolerance to oxidative stress and pathogens. *Plant Sci.* **169:** 55-63.

Sato, Y., Murakami, T., Funatsuki, H., Matsuba, S., Saruyama, H. and Tanida, M. (2001). Heat shock-mediated APX gene expression and protection against chilling injury in rice seedlings. *J. Exp. Bot.* **52:** 145-151.

Scandalios, J.G. (1993). Oxygen stress and superoxide dismutases. *Plant Physiol.* **101:** 7-12.

Schafer, F.Q. and Buettner, G.R. (2001). Redox environment of the cell as viewed through the redox state of the glutathione disulfide/glutathione couple. *Free Radic. Biol. Med.* **30:** 1191-1212.

Schoner, S. and Krause, G.H. (1990). Protective systems against active oxygen species and spinach: Response to cold acclimation in excess light. *Planta.* **180:** 383-389.

Schützendübel, A., Schwanz, P., Teichmann, T., Gross, K. and Langenfeld-Heyser, R. (2001). Cadmium-induced changes in antioxidative systems, hydrogen peroxide content, and differentiation in Scots pine roots. *Plant Physiol.* **127:** 887-898.

Schützendübel, A., Nikolova, P., Rudolf, C. and Polle, A. (2002). Cadmium and H_2O_2-induced oxidative stress in *Populus canescens* roots. *Plant Physiol. Biochem.* **40:** 577-584.

Secenji, M., Hideg, E., Babes, A. and Gyorgyey, J. (2010). Transcriptional differences in gene families of the ascorbate-glutathione cycle in wheat during mild water deficit. *Plant Cell Rep.* **29:** 37-50.

Sekmen, A.H., Türkan, I. and Takio, S. (2007). Differential responses of antioxidative enzymes and lipid peroxidation to salt stress in salt-tolerant *Plantago maritima* and salt-sensitive *Plantago media*. *Physiol. Plant.* **131:** 399-411.

Selote, D.S. and Chopra, R.K. (2004). Drought-induced spikelet sterility is associated with an inefficient antioxidant defense in rice panicles. *Physiol. Plant.* **121:** 462-471.

Selote, D.S. and Chopra, R.K. (2006). Drought acclimation confers oxidative stress tolerance by inducing co-ordinated antioxidant defense at cellular and subcellular level in leaves of wheat seedlings. *Physiol. Plant.* **127:** 494-506.

Sgherri, C.L.M. and Navari-Izzo, F. (1995). Sunflower seedlings subjected to increasing water deficit stress: oxidative stress and defence mechanisms. *Physiol. Plant.* **93:** 25-30.

Sgherri, C.L.M, Loggini, B., Bochicchio, A. and Navari-Izzo, F. (1994). Antioxidant system in Boea hygroscopica: changes in response to desiccation and rehydration. *Phytochem.* **37:** 377-381.

Shalata, A. and Neumann, P.M. (2001). Exogenous ascorbic acid (vitamin C) increases resistance to salt stress and reduces lipid peroxidation. *J. Exp. Bot.* **52:** 2207-2211.

Shalata, A. and Tal, M. (1998). The effect of salt stress on lipid peroxidation and antioxidants in the leaf of the cultivated tomato and its wild salt-tolerant relative *Lycopersicon pennellii. Physiol. Plant.* **104**: 169-174.

Shalata, A., Mittova, V., Volokita, M., Guy, M. and Tal, M. (2001). Response of the cultivated tomato and its wild salt-tolerant relative *Lycopersicon pennellii* to salt-dependent oxidative stress: the root antioxidative system. *Physiol. Plant.* **112:** 487-494.

Shao, H.B., Chu, L.Y., Shao, M.A., Cheruth, A.J. and Mi, H.M. (2008). Higher plants antioxidant and redox signalling under environmental stresses. *Comptes. Rendus. Biologies.* **331:** 433-441.

Shao, H.B., Chu, L.Y., Wu, G., Zhang, J.H., Lu, Z.H. and Hu, Y.C. (2007). Changes of some anti-oxidative physiological indices under soil water deficits among 10 wheat (*Triticum aestivum* L.) genotypes at tillering stage. *Biointerfaces.* **54:** 143-149.

Sharma, P. and Dubey, R.S. (2005). Drought induces oxidative stress and enhances the activities of antioxidant enzymes in growing rice seedlings. *Plant Growth Regul.* **46:** 209-221.

Sharma, P. and Dubey, R.S. (2007). Involvement of oxidative stress and role of antioxidative defense system in growing rice seedlings exposed to toxic concentrations of aluminum. *Plant Cell Rep.* **26:** 2027-2038.

Sharma, Y.K., Leon, J., Raskin, I. and Devis, K.R. (1996). Ozone-induced responses in *Arabidopsis thaliana*—the role of salicylic acid in the accumulation of defence related transcripts and induced resistance. *Proc. Natl. Acad. Sci. USA.* **93:** 5099-5104.

Shi, F., Yamamoto, R., Shimamura, S., Hiraga, S., Nakayama, N., Nakamura, T., Yukawa, K., Achinohe, M., Matsumoto, H. and Komatsu, S. (2008). Cytosolic ascorbate peroxidase 2 (cAPX2) is involved in the soybean response to flooding. *Phytochem.* **69:** 1295-1303.

Shi, W.M., Muramoto, Y., Ueda, A. and Takabe, T. (2001). Cloning of peroxizomal ascorbate peroxidase gene from barley and enhanced thermotolerance by overexpressing in *Arabidopsis thaliana. Gene.* **273:** 23-27.

Shigeoka, S., Ishikawa, T., Tamoi, M., Miyagawa, Y., Takeda, T., Yabuta, Y. and Yoshimura, K. (2002). Regulation and function of ascorbate peroxidase isoenzymes. *J. Exp. Bot.* **53:** 1305-1319.

Shulaev, V. and Oliver, D.J. (2006). Metabolic and Proteomic Markers for Oxidative Stress: New tools for reactive oxygen species research. *Plant Physiol.* **141:** 367-72.

Singh, D.V., Srivastava, G.C. and Abdin, M.Z. (2001). Amelioration of negative effect of water stress in *Cassia angustifolia* by benzyladenine and/or ascorbic acid. *Biol. Plant.* **41:** 141-143.

Singla-Pareek, S.L., Yadav, S.K., Pareek, A., Reddy, M.K. and Sopory, S.K. (2006). Transgenic tobacco overexpressing glyoxalase pathway enzymes grow and set viable seeds in zinc-spiked soils. *Plant Physiol.* **140:** 613-623.

Slesak, I., Miszalski, Z., Karpinska, B., Niewiadomska, E., Ratajczak, R. and Karpinski, S. (2002). Redox control of oxidative stress responses in the C_3-CAM intermediate plant *Mesembryanthemum crystallinum. Plant Physiol. Biochem.* **40:** 669-677.

Smeets, K., Ruytinx, J., Semane, B., Van Belleghem, F., Remans, T., Van Sanden, S., Vangronsveld, J. and Cuypers, A. (2008). Cadmium-induced transcriptional and enzymatic alterations related to oxidative stress. *Environ. Exp. Bot.* **63:** 1-8.

Smirnoff, N. (1993). The role of active oxygen in the response of plants to water deficit and desiccation. *New Phytol.* **125:** 27-58.

Smirnoff, N. (1995). Antioxidant systems and plant response to the environment. In. *Environment and Plant Metabolism: Flexibility and Acclimation.* pp. 217–243. Eds. Smirnoff, N. (Oxford Bio. Sci. Pub.: UK).

Smirnoff, N. (1996). The function and metabolism of ascorbic acid in plants. *Ann. Bot.* **78:** 661-669.

Smirnoff, N. (2000). Ascorbic acid: metabolism and functions of a multifaceted molecule. *Curr. Opin. Plant. Biol.* **3:** 229-235.

Sofo, A., Dichio, B., Xiloyannis, C. and Masia, A. (2005). Antioxidant defences in olive trees during drought stress: changes in activity of some antioxidant enzymes. *Funct. Plant Biol.* **32:** 45-53.

Song, F. and Goodman, R.M. (2001). Activity of nitric oxide is dependent on, but is particularly required for function of salicylic acid in the signaling pathway in tobacco systemic acquired resistance. *Mol. Plant Microbe. Int.* **14:** 1458-1462.

Song, X.S., Wang, Y.J., Mao, W.H., Shi, K., Zhou, Y.H., Nogues S. and Yu, J.Q. (2009). Effects of cucumber mosaic virus infection on electron transport and antioxidant system in chloroplasts and mitochondria of cucumber and tomato leaves. *Physiol. Plant.* **135:** 246-257.

Srivalli, S. and Khanna-Chopra, R. (2008). Role of glutathione in abiotic stress tolerance. In *sulfur Assimilaiton and abiotic stress in plants.* pp. 207-225. Eds. Khan, Nafees A., Sarvajeet, S. and Shahid, U. (Springer-Verlag: Berlin).

Stevens, R., Page. D., Gouble, B., Garchery, C., Zamir, D. and Causse. M. (2008). Tomato fruit ascorbic acid content is linked with monohydroascorbate reductase activity and tolerance to chilling stress. *Plant Cell Environ.* **31:** 1086-1096.

Stossel, P. (1984). Regulation by sulphydryl groups of glyceollin accumulation in soybean hypocotyls. *Planta.* **160:** 314-319

Sudhakar, C.L., Akshm, A. and Giridarakumar, S. (2001). Changes in the antioxidant enzyme efficacy in two high yielding genotypes of mulberry (*Morus alba* L.) under NaCl salinity. *Plant Sci.* **161:** 613-619.

Sumithra, K., Jutur, P.P., Carmel, B.D. and Reddy, A.R. (2006). Salinity-induced changes in two cultivars of *Vigna rediata*: responses of antioxidative and proline metabolism. *Plant Growth Regul.* **50:** 11-22.

Szalai, G., Kellos, T., Galiba, G. and Kocsy, G. (2009). Glutathione as an antioxidant and regulatory molecule in plants under abiotic stress conditions. *J. Plant Grwoth Regul.* **28:** 66-80.

Takashi, S. and Murata, N. (2008). How do environmental stresses accelerate photoinhibition? *Trends Plant Sci.* **13:** 178-182.

Tambussi, E.A., Beltrano, B.J., Guiamet, J.J. and Araus, J.L. (2003). Oxidative damage to thylakoid proteins in water-stressed leaves of wheat (*Triticum aestivum*). *Physiol. Plant.* **108:** 398-404.

Tanou, G., Molassiotis, A. and Diamantidis, G. (2009). Induction of reactive oxygen species and necrotic death-like destruction in strawberry leaves by salinity. *Environ. Exp. Bot.* **65:** 270-281.

Tao, D.L., Oquist, G. and Wingle, G. (1998). Active oxygen scavengers during cold acclimation of Scots pine seedlings in relation to freezing tolerance. *Cryobiol.* **37:** 38-45.

Tewari, R.K., Kumar, P. and Sharma, P.N. (2009). Oxidative Stress and Antioxidant Responses in Young Leaves of Mulberry Plants Grown Under Nitrogen, Phosphorus or Potassium Deficiency. *J. Integr. Plant. Biol.* **49:** 313-322

Thomas, C.E., MeLean, L.R., Parker, R.A. and Ohlweiler, D.F. (1992). Ascorbate and phenolic antioxidant interactions in prevention of liposomal oxidation. *Lipids.* **27:** 543-550.

Thordal-Christensen, H., Zhang, Z., Wei, Y. and Collinge, D.B. (1997). Subcellular localization of H_2O_2 in plants. H_2O_2 accumulation in papillae and hypersensitive response during the barley–powdery mildew interaction. *Plant J.* **11:**1187-1194.

Thornalley, P.J. (1990). The glyoxalase system: new developments towards functional characterization of a metabolic pathway fundamental to biological life. *Biochem. J.* **269:** 1-11.

Torres, M.A. (2010). ROS in biotic interactions. *Phyisol. Plant.* **138:** 414-429.

Tseng, M.J., Liu, C.W. and Yiu, J.C. (2007). Enhanced tolerance to sulfur dioxide and salt stress of transgenic Chinese cabbage plants expressing both superoxide dismutase and catalase in chloroplasts. *Plant Physiol. Biochem.* **45:** 822-833.

Tsugane, K., Kobayashi, K., Niwa, Y., Ohba, Y., Wada, K. and Kobayashi, H. (1999). A recessive *Arabidopsis* mutant that grows photoautotrophically under salt stress shows enhanced active oxygen detoxication. *Plant Cell.* **11:** 1195-1206.

Uchida, A., Jagendorf, A.T., Hibino, T., Takabe, T. and Takabe, T. (2002). Effects of hydrogen peroxide and nitric oxide on both salt and heat stress tolerance in rice. *Plant Sci.* **163:** 515-523.

Uotila, M., Gullner, G. and Komives, T. (1995). Induction of glutathione S-transferase activity and glutathione level in plants exposed to glyphosate. *Physiol. Plant.* **93:** 689-696.

Urano, J., Nakagawa, T., Maki, Y., Masumura, T., Tanaka, K. and Murata, N. (2000). Molecular cloning and characterization of a rice dehydroascorbate reductase. *FEBS Lett.* **466:** 107-111.

Urban, P., Mignotte, C., Kazmaier, M., Delorme, F. and Pompon, D. (1989). Cloning, Yeast Expression and characterization of the coupling of two distantly related *Arabidopsis thaliana* NADPH-Cytochrome P450 reductases with P450 CYP73A5. *J. Biol. Chem.* **272:** 19176-19186.

Ursini, F., Mariorino, M., Berigelius-Flohe, R., Aumann, K.D., Roveri, A., Schomburg, D. and Flohe, L. (1995). Diversity of glutathione peroxidase. *Methods Enzymol.* **252:** 38-53.

Ushimaru, T., Nakagawa, T., Fujioka, Y., Daicho, K. , Naito, M., Yamauchi, Y., Nonaka, H., Amako, K., Yamawaki, K. and Murata, N. (2006). Transgenic *Arabidopsis* plants expressing the rice dehydroascorbate reductase gene are resistant to salt stress. *J. Plant Physiol.* **163:** 1179-1184.

Ushimaru, T., Shibasaka, M., Tsuji, H. (1992). Development of the O^-_2-detoxification system during adaptation to air of submerged rice seedlings. *Plant Cell Physiol.* **33**: 1065-1071.

Vallelian-Bindschedler, L., Schweizer, P., Mosinger, E. and Metraux, J.P. (1998). Heat-induced resistance in barley to powdery mildew (*Blumeria graminis* f.sp. *hordei*) is associated with a burst of active oxygen species. *Physiol. Mol. Plant Pathol.* **52:** 185-199.

Van Camp, W., Van Montagu, M. and Inze, D. (1998). H_2O_2 and NO: redox signals in disease resistance. *Trends Plant Sci.* **3:** 330-334.

Vanacker, H., Carver, T.L.W. and Foyer, C.H. (1998). Pathogen induced changes in the antioxidant status of the apoplast in Barley leaves. *Plant Physiol.* **117:** 1103-1114.

Vanacker, H., Carver, T.L.W. and Foyer, C.H. (2000). Early H_2O_2 hydrogen peroxide in mesophyll cells leads to induction of glutathione during the hypersensitive response in the Barley-Powdery Mildew Interaction. *Plant Physiol.* **123:** 1289-1300.

Veljovic-Jovanovic, S.D., Pignocchi, C., Noctor, G. and Foyer, C.H. (2001). Low ascorbic acid in the vtc-1 mutant of Arabidopsis is associated with decreased growth and intracellular redistribution of the antioxidant system. *Plant Physiol.* **127:** 426-435.

Verma, S. and Mishra, S.N. (2005). Putrescine alleviation of growth in salt stressed *Brassica juncea* by inducing antioxidative defense system. *J. Plant Phsysiol.* **162:** 669-677.

Wahid, A., Perveen, M., Gelani, S. and Basra, S.M.A. (2007). Pretreatment of seed with H_2O_2 improves salt tolerance of wheat seedlings by alleviation of oxidative damage and expression of stress proteins. *J. Plant Physiol.* **164:** 283-294.

Waller, F., Achatz, B., Baltruschat, H., Fodor, J., Becker, K., Fischer, M., Heier, T., Huckelhoven, R., Neumann, C., von Wettstein, D., Franken, P. and Kogel, K.H. (2005). The endophytic fungus *Piriformospora indica* reprograms barley to salt-stress tolerance, disease resistance, and higher yield. *Proc. Natl. Acad. Sci. USA.* **102:** 13386-13391.

Wan, X., Tan, J., Lu, S., Lin, C., Hu, Y. and Guo, Z. (2009). Increased tolerance to oxidative stress in transgenic tobacco expressing a wheat oxalate oxidase gene via induction of antioxidant enzymes is mediated by H_2O_2. *Physiol. Plant.* **136:** 30-44.

Wang, J., Zhang, H., and Allen, R.D. (1999). Overexpression of an *Arabidopsis* peroxisomal ascorbate peroxidase gene in tobacco increase protection against oxidative stress. *Plant Cell Physiol.* **40:** 725-732.

Wang, J.W. and Kao, C.H. (2007). Protective effect of ascorbic acid and glutathione on $AlCl_3$-inhibited growth of rice roots. *Biol. Plant.* **51:** 493-500.

Wang, J.W. and Wu, J.Y. (2005). Nitric oxide is involved in methyl jasmonate-induced defense responses and secondary metabolism activities of Taxus cells. *Plant Cell Physiol.* **6:** 923-930.

Wang, K. and Jiang, Y. (2007). Antioxidant responses of creeping bent-grass roots to waterlogging. *Crop Sci.* **47:** 232-238.

Wang, Y., Wisniewski, M., Meilan, R., Cui, M. and Fuchigami, L. (2006). Transgenic tomato (*Lycopersicon esculentum*) overexpressing cAPX exhibits enhanced tolerance to UV-B and heat stress. *J. Applied Hort.* **8:** 87-90.

Wang, Z., Zhang, L., Xiao, Y., Chen, W. and Tang, K. (2010). Increased Vitamin C Content Accompanied by an Enhanced Recycling Pathway Confers Oxidative Stress Tolerance in *Arabidopsis. J. Integr. Plant Biol.* (in press).

Wei-Feng, X.U., Wei-Ming, S.H.I., Ueda, A. and Takabe, T. (2008). Mechanism of salt tolerance in transgenic *Arabidopsis thaliana* carrying a peroxisomal ascorbate peroxidase gene from barley. *Pedosphere.* **18:** 486-495.

Williams, J.S., Hall, S.A., Hawkesford, M.J., Beale, M.H. and Cooper, R.M. (2002). Elemental Sulfur and Thiol Accumulation in Tomato and Defense against a Fungal Vascular Pathogen. *Plant Physiol.* **128:** 150-159.

Wu, G., Shortt, B.J., Lawrence, E.B., Levine, E.B., Fitzssimmons, K.C. and Shah, D.M. (1995). Disease resistance conferred by expression of a gene encoding H_2O_2-generating glucose oxidase in transgenic potato plants. *Plant Cell.* **7:** 1357-1368.

Xu, J., Wang, W., Yin, H., Liu, X., Sun, H. and Mi, Q. (2010). Exogenous nitric oxide improves antioxidative capacity and reduces auxin degradation in roots of *Medicago truncatula* seedlings under cadmium stress. *Plant Soil.* **326:** 321-330.

Xu, Q., Xu, X., Zhao, Y., Jiao, K., Herbert, S.J. and Hao, L. (2008b). Salicylic acid, hydrogen peroxide and calcium-induced saline tolerance associated with endogenous hydrogen peroxide homeostasis in naked oat seedlings. *Plant Growth Regul.* **54:** 249-259.

Xu, W., Shi, W., Liu, F., Ueda, A. and Takabe, T. (2008a). Enhanced zinc and cadmium tolerance and accumulation in transgenic *Arabidopsis* plants constitutively overexpressing a barley gene (*HvAPX1*) that encodes a peroxisomal ascorbate peroxidase. *Botany.* **86:** 567-575.

Yabuta, Y., Motoki, T., Yoshimura, K., Takeda, T., Ishikawa, T. and Shigeoka, S. (2002). Thylakoid membrane-bound ascorbate peroxidase is a limiting factor of antioxidative systems under photo-oxidative stress. *Plant J.* **32:** 915-925.

Yadav, S.K., Singla-Pareek, S.L., Reddy, M.K. and Sopory, S.K. (2005a). Methylglyoxal levels in plants under salinity stress are dependent on glyoxalase I and glutathione. *Biochem. Biophys. Res. Commun.* **337:** 61-67.

Yadav, S.K., Singla-Pareek, S.L., Reddy, M.K. and Sopory, S.K. (2005b). Transgenic tobacco plants overexpressing glyoxalase enzymes resist an increase in methylglyoxal and maintain higher reduced glutathione levels under salinity stress. *FEBS Lett.* **579:** 6265-6271.

Yalpani, N., Enyedi, A.J., Leon, J. and Raskin, I. (1994). Ultraviolet light and ozone stimulate accumulation of salicylic acid, pathogenesis-related proteins and virus resistance in tobacco. *Planta.* **193:** 372-376.

Yan, S.P., Zhang, Q.Y., Tang, Z.C., Su, W.A. and Sun, W.N. (2006). Comparative proteomic analysis provides new insights into chilling stress responses in rice. *Mol. Cell Proteomics.* **5:** 484-496.

Yang, F., Xu, X., Xiao, X. and Li, C. (2009). Responses to drought stress in two poplar species originating from different altitudes. *Bio. Plant.* **53:** 511-516.

Yang, S.H., Wang, L.J. and Li, S.H. (2007). Ultraviolet-B irradiation-induced freezing tolerance in relation to antioxidant system in winter wheat (*Tiriticum aestivum* L.) leaves. *Environ. Expt. Bot.* **60:** 307.

Yang, Y., Han, C., Liu, Q., Lin, B. and Wang, J. (2008). Effect of drought and low light on growth and enzymatic antioxidant system of *Picea asperata* seedlings. *Acta Physiol. Plant.* **30:** 433-440.

Yasar, F., Kusvuran, S. and Ellialtioglu, S. (2006). Determination of antioxidant activities in some melon (*Cucumis melo* L.) varieties and cultivars under salt stress. *J. Hort. Sci. Biotech.* **81:** 627-630.

Yin, L., Wang, S., Eltayeb, A.E., Uddin, M.I., Yamamoto, Y., Tsuji, W., Takeuchi, Y. and Tanaka, K. (2010). Overexpression of dehydroascorbate reductase, but not monodehydroascorbate reductase, confers tolerance to aluminum stress in transgenic tobacco. *Planta.* **231:** 609-621.

Ying, J., Clavreul, N., Sethuraman, M., Adachi, T. and Cohen, R.A. (2007). Thiol oxidation in signalling and response to stress: detection and quantification of physiological and pathophysiological thiol modifications. *Free Radic. Bio. Med.* **43:** 1099-1108.

Yoon, H.S., Lee, H., Lee, I.A., Kim, K.Y. and Jo, .J (2004). Molecular cloning of the monodehydro ascorbate reductase gene from *Brassica campestris* and analysis of its mRNA level in response to oxidative stress analysis of its mRNA level in response to oxidative stress. *Biochem. Biophys. Acta.* **1658:** 181-186.

Yoshida, S., Tamaoki, M., Shikano, T., Nakajima, N., Ogawa, D., Ioki, M., Aono, M., Kubo, A., Kamada, H., Inoue, Y. and Saji, H. (2006). Cytosolic Dehydroascorbate Reductase is Important for Ozone Tolerance in *Arabidopsis thaliana. Plant Cell Physiol.* **47:** 304-308.

Yoshimura, K., Yabuta, Y., Ishikawa, T. and Shigeoka, S. (2000). Expression of spinach ascorbate peroxidase isoenzymes in response to oxidative stresses. *Plant Physiol.* **123:** 223-234.

Yu, C.W., Murphy, T. and Lin, C.H. (2003). Hydrogen peroxide-induced chilling tolerance in mung bean mediated through ABA-independent glutathione accumulation. *Funct. Plant Biol.* **30:** 955-963.

Yu, C.W., Murphy, T.M., Sung, W.W. and Lin, C.H. (2002). H_2O_2 treatment induces glutathione accumulation and chilling tolerance in mung bean. *Funct. Plant Biol.* **29:** 1081-1087.

Zechmann, B. and Muller, M. (2008). Effects of zucchini yellow mosaic virus infection on the subcellular distribution of glutathione and its precursors in a highly tolerant *Cucurbita pepo* cultivar. *Can. J. Bot.* **86:** 1092-1100.

Zechmann, B., Zellnig, G., Urbanek-Krajnc, A. and Müller, M. (2007). Artificial elevation of glutathione affects symptom development in ZYMV-infected *Cucurbita pepo* L. plants. *Archives Virol.* **152:** 747-762.

Zhang, F.Q., Shi, W.Y., Jin, Z.X. and Shen, Z.G. (2003). Response of antioxidative enzymes in cucumber choloplast to cadmium toxicity. *J. Plant Nutr.* **26:** 1779-1788.

Zhao, F. and Zhang, H. (2006). Salt and paraquat stress tolerance results from co-expression of the *Suaeda salsa* glutathione S-transferase and catalase in transgenic rice. *Plant Cell Tiss. Organ Cult.* **86:** 349-358.

Zhou, Y.H., Zhang Y.Y., Zhao, X., Yu, H.J., Shi, K. and, Yu, J.Q. (2009). Impact of Light Variation on Development of Photoprotection, Antioxidants, and Nutritional Value in *Lactuca sativa* L. *J. Agric. Food Chem.* **57**: 5494-5500

Zhu, Y., Pilon-Smits, E.A.H., Jouanin, L. and Terry, N. (1999a). Overexpression of glutathione synthetase in *Brassica juncea* enhances cadmium tolerance and accumulation. *Plant Physiol.* **119**: 73-79.

Zhu, Y.L., Pilon-Smits, E.A.H., Tarun, A., Weber, S.U., Jouanin, L., Terry, N. (1999b). Cadmium tolerance and accumulation in Indian mustard is enhanced by overexpressing γ-glutamylcysteine synthetase. *Plant Physiol.* **121:** 1169-1177.

4

Ascorbate and Glutathione: Saviours Against Oxidative Stress

VISHAL PRASAD AND R.S. UPADHYAY*

Department of Botany, Banaras Hindu University, Varanasi – 221 005, India
***Corresponding author:** upadhyay_bhu@yahoo.co.uk*

ABSTRACT

Oxidative stress may arise from any abiotic and biotic stress that promotes the formation of reactive oxygen species (ROS) having potential to damage or kill the cell. These ROS can accumulate in plant tissues, resulting in oxidative damage of cellular components like lipids, proteins and nucleic acids leading to an overall reduction in plant growth and yield. To cope with such an oxidative injury caused by ROS the plant system possesses a very well-defined and fine-tuned antioxidative system. The antioxidative system comprises several enzymic and non-enzymic antioxidants which maintain the cellular level of ROS under control. The present review highlights the role of ascorbate and glutathione in plant tolerance against oxidative stress.

Keywords: Ascorbate, glutathione, oxidative stress, reactive oxygen species.

1. INTRODUCTION

Due to photosynthetic activities on the earth, oxygen concentrations in the atmosphere are kept at ca. 21%, and provide a basis for our continued existence. A paradox in metabolism is that while the vast majority of complex life on the earth requires oxygen for its existence, oxygen is a highly reactive molecule and has all the potential to damage living organisms. It is absolutely required for aerobic metabolism, but reduction of molecular di-oxygen is a risky process and frequently results in the formation of reactive oxygen species (ROS) that can cause cellular damage. All forms of life maintain a reducing environment within their cells. Disturbances in this normal redox state can cause toxic effects within a cell.

Biochemically, stress situations can be characterized as a transition from mainly two-electron (ionic) reactions to one-electron reactions (radical reactions) within the cells (Elstner and Osswald 1994). Whereas the ionic reactions are tightly controlled, radical reactions easily escape metabolic controls and proliferate in chain reaction pathways. An increased production of ROS is typical for the effect of most environmental stresses on plant cells.

Since a certain amount of ROS is produced under normal metabolic conditions (Foyer and Noctor 2000), plants have developed effective defence systems to survive. These systems include a number of different functions, such as protection against excess energy absorption, alternative regeneration of the electron acceptor, the scavenging of ROS, the repair of oxidatively damaged structures, and regulation processes.

The use of oxygen as part of the process for generating metabolic energy produces ROS. Oxidative stress is caused by an imbalance between the production of ROS and a biological system's ability to readily detoxify the reactive intermediates or easily repair the resulting damage in any cell compartment (Foyer and Noctor 2000; Ahmad et al., 2008; Tuteja et al., 2009).

In chemical terms, oxidative stress is a large rise (becoming less negative) in the cellular reduction potential, or a large decrease in the reducing capacity of the cellular redox couples. The effects of oxidative stress depend upon the size of these changes, with a cell being able to overcome small perturbations and regain its original state. ROS such as superoxide (O_2^-), hydrogen peroxide (H_2O_2), hydroxyl radicals (HO^-) and singlet oxygen ($^1O_2$) are an unavoidable consequence of aerobic metabolism (Foyer and Noctor 2000). Most of these oxygen-derived species are produced at a low level by normal aerobic metabolism and the damage they cause to cells is constantly repaired, but when excess ROS are produced they can damage proteins, lipids, carbohydrates and DNA ultimately resulting in cell death (Mittler et al., 2004; Foyer and Noctor 2005; Ahmad et al., 2008; Tuteja et al., 2009).

2. REACTIVE OXYGEN SPECIES

About two billion years ago molecular oxygen (O_2) entered into our atmosphere, by oxygen-evolving photosynthetic organisms, since then ROS have been the unwelcome companions of aerobic metabolism. ROS are partially reduced forms of atmospheric oxygen. They typically result from the excitation of O_2 to form singlet oxygen ($^1O_2$) or from the transfer of one, two or three electrons to O_2 to form, respectively, a superoxide radical (O_2^-), hydrogen peroxide (H_2O_2) or a hydroxyl radical (HO^-). Singlet oxygen ($^1O_2$) can be formed in reaction with excited chlorophyll (^{3}Chl) and has a specific electron spin combination of two electrons on the outer suborbitals that allow for acceptance of new electron and it is therefore usually considered a ROS. ROS are highly reactive molecules and may react with a wide variety of other molecules such as DNA, pigments, lipids, proteins and other essential cellular molecules which lead to a destructive chain of events (Asada and Takahashi 1987; Hammond and Jones 1996; Lamb and Dixon 1997; Asada 1999; Dat et al., 2000).

2.1 Production of ROS in Cells

The steady state level of ROS in the different cellular compartments is determined by interplay between multiple ROS-producing pathways, and ROS-scavenging mechanisms. These are controlled by the ROS-signal transduction pathway and constitute the "basic ROS cycle". There are many potential sources of ROS in plants during normal growth and development (Table 1).

Table 1: Production of ROS in plants

Production Source/Process	Primary ROS	Localization
Excited Chlorophyll	$^1O_2$	Chlorophyll
Photosynthesis Electron transport and PSI or PSII	O_2^-	Chlorophyll
Respiration Electron transport	O_2^-	Mitochondria
Glycolate oxidase	H_2O_2	Peroxisome
Fatty acid β-oxidation	H_2O_2	Peroxisome
Xanthine oxidase	O_2^-	Peroxisome
Amine oxidase	H_2O_2	Apoplast
Oxalate oxidase	H_2O_2	Apoplast
NADPH oxidase	O_2^-	Plasma membrane
Peroxidase	O_2^-, H_2O_2	Cell wall

ROS are formed as a direct consequence of diverse biochemical processes in many subcellular compartments (Mittler et al., 2004; Gechev et al., 2006). The largest producer of ROS during the photoperiod in the plant cells is the chloroplast together with the peroxisome (Kozaki and Takeba 1996; Asada 1999; Foyer and Noctor 2003; Asada 2006). Chloroplast is the major producer of O_2^- and H_2O_2 in plants. In chloroplast thylakoids, the reaction centres of PSI and PSII are the major generation sites of ROS (Asada 2006). The peroxisomes produce O_2^- and H_2O_2 in several key metabolic reactions.

Overreduction of the respiratory chain leads to ROS production in the mitochondria (Maxwell 1999). The mitochondria produce O_2^- at complex I and III, as by-products (Moller 2001). The NADPH oxidase in the plasma membrane produces O_2^-, which participates in several physiological processes (Torres and Dangl 2005). Amine oxidases, cell wall peroxidases and extracellular peroxidases are apoplastic ROS producers and are involved in several stress responses (Allan and Fluhr 1997).

Hydrogen peroxide is an oxidizing agent and is also reactive but its main significance lies in it being a source of hydroxyl radicals. The hydroxyl radical is an extremely reactive oxidizing radical that will react with most biomolecules at diffusion controlled rates. Hydroxyl radicals are known to be produced by Haber-Weiss reaction (Haber and Weiss 1934).

The Haber-Weiss cycle consists of the following two reactions:

$$H_2O_2 + OH\cdot \rightarrow H_2O + O_2^- + H^+ \text{ and}$$

$$H_2O_2 + O_2^- \rightarrow O_2 + OH^- + OH\cdot$$

Although the second reaction has a negligible rate constant it acts as a possible source of hydroxyl radicals.

H_2O_2 can give rise to $HO^\bullet$ through the Fenton reaction, which is catalyzed mainly by free transition metal ions.

$$Fe_2^+ + H_2O_2 \rightarrow Fe_3^+ + OH^\bullet + OH^-$$

This is the iron-salt-dependent decomposition of H_2O_2, generating the highly reactive hydroxyl radical, possibly via an oxoiron (IV) intermediate.

Hydroxyl radicals are also formed during exposure of high energy radiations like X-rays or gamma-rays to the living tissues. Most of the energy is absorbed by cell sap having very high water content. It may result in splitting of one of the convent bonds of water. The hydroxyl radicals are most reactive free radicals in the living system known so far. They can damage almost every type of molecule found in cell. Although it is highly reactive, it has very short life of about a micro-second.

Under normal growth conditions low amounts of ROS production in cells have been found (240μMs^{-1} O_2^- and a steady-state level of 0.5μM H_2O_2 in chloroplasts) (Polle 2001). However, under stressful conditions like drought, desiccation, salinity, chilling, heat shock, heavy metals, UV radiation, air pollutants such as ozone and SO_2, mechanical stress, nutrient deprivation, pathogen attack and high light stress that disrupt the cellular homeostasis an enhanced production of ROS has been observed (upto 240–720μMs^{-1} O_2^- and a steady-state level of 5–15μM H_2O_2) (Bowler et al., 1992; Allen 1995; Noctor and Foyer 1998; Dat et al., 2000; Desikan et al., 2001; Polle 2001). The production of ROSs during these stresses results from pathways such as photorespiration, from the photosynthetic apparatus and from mitochondrial respiration. In addition, pathogens and wounding or environmental stresses (e.g., drought or osmotic stress) also trigger the active production of ROSs by NADPH oxidases (Hammond and Jones 1996; Orozco and Ryan 1999; Cazale et al., 1999; Pei et al., 2000). Such an enhanced production of ROSs during stress can pose a threat to cells.

The most common mechanism of stress induced ROS production in the chloroplasts is based on an imbalance between the consumption of reductant (NADPH) in carbon fixation, and the need of the electron transport chain for the regenerated electron acceptor at the PS I site (NADP). Most environmental stresses impair the function of the Calvin cycle (i.e., the fixation and the NADPH and ATP consumption). At the same time, light driven electron transport remains fully active, which leads to an overreduction of the electron transport chains and forces electrons to leak to alternate acceptors, predominantly to molecular oxygen The univalent reduction of O_2 by leaking electrons yields the O_2^-.

2.2 Toxic Effects of ROS

At low levels ROS perform important housekeeping functions (Foreman et al., 2003; Lam 2004; Mori and Schroeder 2004) but at elevated levels ROS are extremely toxic. The steady-state level of ROS can be used by plants to monitor their intracellular level of stress; however, this level has to be kept under tight control because over-accumulation of ROS can result in cell death (Asada and Takahashi 1987; Hammond and Jones 1996; Asada 1999; Dat et al., 2000). They are the major players in processes ranging from photobleaching and necrosis to defence responses and programmed cell death (Van Breusegem et al., 2001; Neill et al., 2002; Blokhina et al., 2003; Dat et al., 2003; Laloi et al., 2004; Gechev et al., 2006). Most environmental stresses may provoke a temporary increase in ROS production (Karpinski et al., 1999; Desikan et al., 2001; Van Breusegem et al., 2001; Pastori and Foyer 2002; Blokhina et al., 2003; Karpinski et al., 2003; Laloi et al., 2004; Mori and Schroeder 2004).

ROS-induced cell death can result from oxidative processes such as membrane lipid peroxidation, protein oxidation, enzyme inhibition and DNA and RNA damage. Superoxide (O_2^-) is a free radical with potentially destructive effects on proteins, lipids, nucleic acids, and pigments. For example, it may initiate free radical chain reactions in membrane lipids, which produce lipid peroxides and lipid hydroperoxides. Superoxide and H_2O_2 can inactivate various macromolecules directly, but it is their conversion to OH^-, catalysed by transition metals (i.e., the Haber-Weiss reaction), that accounts for their main toxicity. Hydroxyl radicals react instantaneously with proteins, lipids, and DNA, causing rapid cell damage. The polyunsaturated fatty acids (PUFA) are particularly susceptible to attack by $^1O_2$ and HO^-, giving rise to complex mixtures of lipid hydroperoxides (Mueller 2004). Extensive PUFA peroxidation decreases the fluidity of the membrane, increases leakiness and causes secondary damage to membrane proteins (Halliwell 2006). DNA can be modified by ROS in many different ways. The most reactive HO^- and $^1O_2$ primarily attacks guanine and 8-hydroxyguanine which are the most commonly observed modifications (Wiseman and Halliwell 1996). ROS can cause damage to both mitochondrial and nuclear DNA (Tuteja et al., 2009). In addition to direct DNA oxidation, ROS can also indirectly modify DNA. A common type of damage involves conjugation of PUFA breakdown product MDA with guanine (Jeong et al., 2005). In addition to mutations, oxidative DNA modifications can lead to changes in the methylation of cytosine, which is important for regulating gene expression. Protein oxidation occurs in terms of covalent modification of protein induced by ROS or byproducts of oxidative stress. Most types of protein oxidations are essentially irreversible, whereas, a few involving sulphur containing amino acids are reversible (Ghezzi and Bonetto 2003). Protein oxidation is widespread and is often used as a diagnostic marker for oxidative stress.

3. ANTIOXIDANTS

An antioxidant is a molecule capable of slowing or preventing the oxidation of other molecules. Oxidation is a chemical reaction that transfers electrons from a substance to an oxidizing agent. Oxidation reactions can produce free radicals, which start chain reactions that damage cells. Antioxidants terminate these chain reactions by removing free radical intermediates, and inhibit other oxidation reactions by being oxidized themselves.

The ability of the plant to tolerate different levels of ROS is dependent on the efficiency of plants antioxidant systems and ultimately on metabolism (Wingsle and Karpinski 1996; Yu 1999; Blokhina et al., 2003; Couee et al., 2006). Antioxidants can be generally divided into enzymatic, non-enzymatic and water or lipid soluble. Glutathione (GSH) and Ascorbate (ASC) are the major low molecular weight water-soluble non-enzymatic antioxidants and make up the foundation of the redox control within the cellular compartments, since they provide the ability to scavenge ROS (Foyer and Halliwell 1976; Asada 1999; Asada 2006). Carotenoids, xanthophylls and tocopherols are examples of low molecular weight lipid soluble non-enzymatic antioxidants.

The non-enzymatic antioxidants, such as ASC, GSH, α-tocopherols, and carotenoids, occur in high concentrations in plants, especially the concentrations of ASC and GSH are high in chloroplasts and other cellular compartments (5–20 mM ASC and 1–5 mM GSH) as they play crucial role in plants defence against oxidative stress caused by both biotic and abiotic sources (Noctor and Foyer 1998). Mutants with suppressed ASC levels (Conklin et al., 1996) and transgenic plants with altered content of GSH (Creissen et al., 1999) exhibit hypersensitive response during biotic and abiotic stress conditions. Enzymes like glutathione reductase (GR), monodehydroascorbate reductase (MDAR) and dehydroascorbate reductase (DHAR) help in maintaining a high reduced per oxidized ratio of ASC and GSH by using NADPH as reducing power and help in proper scavenging of ROSs in cells (Noctor and Foyer 1998; Asada 1999).

The toxic effects of ROS are also counteracted by enzymatic antioxidative system such as: superoxide dismutase (SOD), catalase (CAT), ascorbate peroxidase (APX) and glutathione reductase (GR) (Ahmad et al., 2008). Normally, each cellular compartment contains more than one enzymatic activity that detoxifies a particular ROS. Hydroxyl radicals are too reactive to be eliminated enzymatically, but formation is limited by scavenging of O_2^- and H_2O_2.

Superoxide dismutase (SOD) is a metal-containing enzyme that eliminates O_2^- radicals. Different isoforms of SOD have been identified on the basis of their metal co-factor. In general, plants contain a mitochondrial MnSOD, as well as a cytosolic and a chloroplastic Cu/ZnSOD (Bowler 1994). Besides the ubiquitous chloroplastic Cu/ZnSOD, many plants also contain chloroplastic FeSOD (Bowler 1994; Van Camp et al., 1990).

Hydrogen peroxide is eliminated by catalases and peroxidases. Catalases are peroxisomal enzymes that do not require a reducing substrate for their activity. Ascorbate peroxidases (APXs) are also important H_2O_2 scavengers operating both in the cytosol and chloroplasts. They use ASC as a reducing substrate and form part of a cycle, known as the Ascorbate-glutathione or Halliwell-Asada cycle. Other enzymes involved in this oxidation-reduction cycle are monodehydroascorbate reductase (MDAR), dehydroascorbate reductase (DHAR), and glutathione reductase (GR) (Bowler et al., 1992; Willekens et al., 1997).

The balance between SOD and APX or CAT activities in cells is crucial for determining the steady-state level of O_2^- and H_2O_2 (Bowler et al., 1991). This balance, together with sequestering of metal ions, is thought to be important to prevent the formation of the highly toxic hydroxyl radical via the metal-dependent Haber–Weiss or the Fenton reactions (Asada and Takahashi 1987). Both APX and CAT have different affinities for H_2O_2. APX having affinity in μM range might be responsible for the fine modulation of ROS for signalling, whereas CAT showing affinity in mM range might be responsible for the removal of excess ROSs during stress. The major ROS-scavenging pathways of plants shown in (Fig. 1) include SOD which is found in almost all cellular compartments, the water–water cycle in chloroplasts, the Ascorbate–glutathione cycle

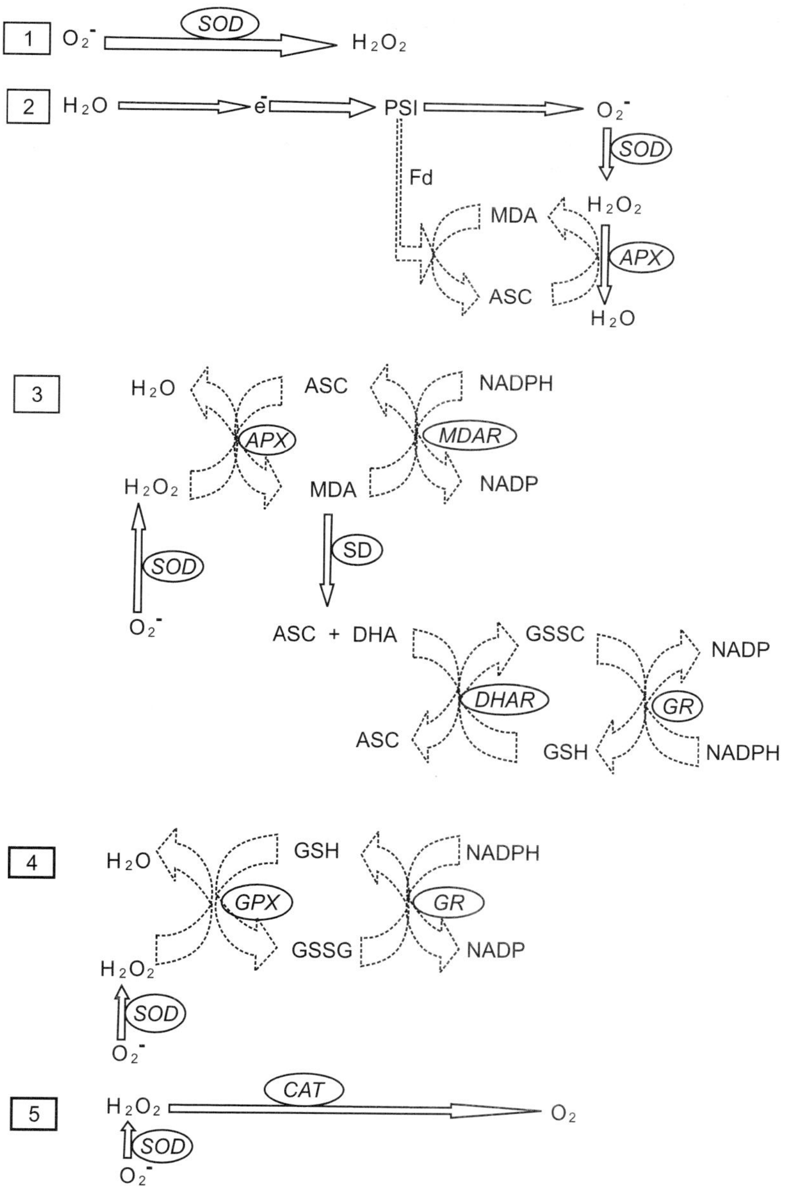

Fig. 1. Various antioxidants and different ROS removal pathways in plants. 1. Superoxide dismutase cycle; 2. Water-water cycle; 3. Ascorbate-glutathione cycle; 4. Glutathione peroxidase cycle; 5. Catalase cycle.

in chloroplasts, cytosol, mitochondria, apoplast and peroxisomes, glutathione peroxidase (GPX), and CAT in peroxisomes. The finding of the ascorbate–glutathione cycle in almost all cellular compartments tested to date, as well as the high affinity of APX for H_2O_2, suggests that this cycle plays a crucial role in controlling the level of ROSs in these compartments.

ROS production can also be decreased in cells by the alternative channelling of electrons in the electron transport chains of the chloroplasts and mitochondria by a group of enzymes called alternative oxidases.

4. ASCORBATE

L-Ascorbate (ASC) or L-Ascorbic acid is also commonly known as "vitamin C". It is a monosaccharide antioxidant found both in animals as well as plants. Ascorbate (ASC) is one of the best characterized components contributing to the nutritive value of green vegetables and fruits. During human evolution one of the enzymes needed to make ASC has been lost by mutation, hence, humans are devoid of ASC and must obtain it from the diet as a vitamin. ASC is a reducing agent and can reduce, and thereby neutralize, ROS such as hydrogen peroxide. In addition to its direct antioxidant effects, ASC is also a substrate for the antioxidant enzyme ascorbate peroxidase (APX), a function that is particularly important for stress resistance in plants. ASC also plays a pivotal role in the synthesis of hydroxyproline-rich proteins (analogous to roles in collagen formation) which are components of the plant cell wall. In addition, it also plays a role in plant cell growth and division. ASC plays a key role in the prevention of scurvy and also fulfils many other essential functions in animals and plants.

The organic structure of ASC is L-threo-hexenon-1,4-lactone (Fig. 2). ASC is one of the predominant weak acids in the plant cell which dissociates into the ascorbate anion the predominant form of the molecule at a physiological pH (Arrigoni et al., 1992; Asada 1994; Arrigoni and De Tullio 2000). The enediol structure of ASC at C-2 and C-3 provides it the acidic properties (pKa1 4.2) and renders it a reducing behaviour. ASC is utilized in the cell as an electron donor, and the first ASC oxidation product is the semiquinone-like free radical monodehydroascorbate (MDA) (Fig. 2). Although MDA is a free radical with highly delocalized free electron, it does not react readily with other molecules and is therefore generally considered to be less destructive (Navas et al., 1994). At pH 7

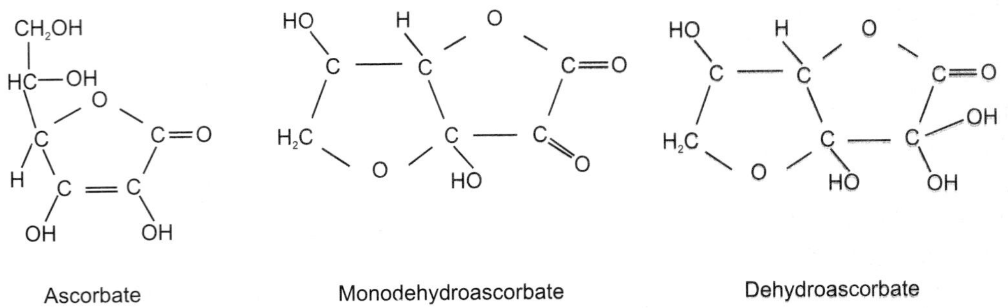

Fig. 2. Structure of ascorbate and its metabolites.

MDA disproportionates spontaneously to ASC and dehydroascorbate (DHA) with a rate constant between 10^5 and 2.8×10^6 M^{-1} s^{-1}(at pH 7).

The oxidized form, DHA (Fig. 2), which exists as a bicyclic monomeric hydrate in aqueous solution, does not have acid characteristics. It is a rather unstable molecule, which undergoes spontaneous and irreversible hydrolysis to 2,3-diketogulonic acid (Deutsch 2000), which breaks down further to yield approximately 50 different compounds, including oxalate and tartrate (Keats et al., 2000).

ASC is involved in the protection of different cell compartments against oxidative damage. Different isozymes of APX are found in the cell wall, in the cytosol, the mitochondria and in the peroxisomes (Vanacker et al., 1998), and assist in the detoxification of ROS at these locations. ASC is one of the most important factors in the scavenging of ozone and ozone-derived ROS in the apoplast, either through direct (chemical) interaction, or by detoxifying ozone breakdown products (Chameides 1989; Luwe et al., 1993). ROS are generated during aerobic metabolism in the chloroplast and during exposure to stress conditions, and are bound to inflict serious damage upon the cell, if not readily detoxified (Di Cagno et al., 2001). ASC reacts rapidly with O_2^- and $^1O_2$, chemically with ozone, and enzymatically with H_2O_2 (through APX), and thus assists in neutralizing these ROS. Additionally, ASC regenerates the chloroplastic lipophilic antioxidant α-tocopherol (vitamin E) from the α-chromanoxyl radical (Asada 1994). Carotenoid pigments like carotenes and xanthophylls also depend upon ASC for their regeneration. At a physiological level, the best-studied phenomena involving ASC is its participation in an oxygen scavenging pathway in the chloroplast known as the ascorbate-glutathione cycle (Fig. 1).

4.1 Biosynthesis and Metabolism of Ascorbate

In general ASC is present in mM concentrations in various parts of the plant tissue. In higher plants variable concentrations of ASC is present in different cell compartments, like cytosol, chloroplast, vacuoles, mitochondria and extracellular matrix (Horemans et al., 2000). Chloroplastic and cytosolic ASC concentrations can be high (10-20 and 20 mM, respectively) (Asada 1999). Such higher concentrations in these compartments reflect the high need for scavenging the toxic ROS. However, ASC contents vary considerably between tissues and depend on the physiological status of the plant as well as on environmental factors (Loewus and Loewus 1987; Smirnoff 1996). Photosynthetic tissues have high concentrations of ASC as do fruits and other storage organs (Fukuda 1995; Loewus and Loewus 1987). ASC contents are generally higher in younger tissues than in older ones and ASC accumulates in actively growing tissues such as meristems (Degara and Tommasi 1999; Luwe 1993).

The onset of senescence triggers loss of tissue ASC while ASC loss is retarded in stay-green mutants (Borraccino 1995; Kerk and Feldman 1995). Dry seeds and dormant *Malus* sp. buds have little or no ASC (Arrigoni 1992; Wang and Faust 1992; Tommasi 1999). During germination and bud breakage a sharp increase is observed in both tissue ASC and ASC-metabolizing enzymes. Cells in the quiescent centre of *Zea mays* roots have negligible ASC and can be induced to divide upon the addition of exogenous ASC (Kerka and Feldman 1995). ASC is present in most, if not all, cell compartments. For example, ASC accumulates in chloroplasts to high concentrations, where it can amount from 12 to 30% of the total ASC content of the leaves (Foyer and Lelandais 1996; Foyer et al., 1993). Substantial amounts of ASC are also found in the apoplast (Vanacker et al., 1998) cytosol and vacuole (Rautenkranz et al., 1994). The cytosolic ASC pool has been calculated to be approximately 20 mM (Foyer and Lelandais 1996). The ultimate step of ASC synthesis catalysed by L-galactono-γ-lactone dehydrogenase occurs in the mitochondria and isolated mitochondria produce large amounts of ASC in the presence of the precursor, L-galactono-γ-lactone. The ASC pool of the intracellular compartments is largely reduced except under conditions of oxidative stress (de Pinto et al., 1999). In marked contrast, in the extracellular matrix of plant cells, the redox status of ASC (i.e., the relative occurrence of reduced and oxidized ASC forms) depends on the age of the tissue, the degree of stress to which a plant is subjected and appears also to be species-dependent (de Pinto et al., 1999).

Advance in the understanding of the biosynthetic pathway of ASC in plants has been made studying *Pisum sativum* and *Arabidopsis* plants (Wheeler 1998). The biosynthetic pathways of ASC in plants is little different from that in animals (Fig. 3) (Wheeler 1998).

One way that plants synthesise ASC involves D-glucose-6-P, D-fructose- 6-P, D-mannose-6-P, D-mannose-1-P, GDP-Dmannose, GDP-L-galactose, L-galactose-1-P, and L-galactose as intermediates in the production of L-galactono-γ-lactone, the immediate precursor of ASC. Apparently, ozone sensitive *Arabidopsis* mutant *vtc1* (formerly *soz1*) is deficient in the biosynthesis of ASC (Conklin et al., 1999). Since ASC levels in this mutant still amount up to 30% of wild type, there are possibilities of other ASC biosynthetic pathways in addition to the above pathway.

The subcellular localisation of the different steps in the ASC biosynthesis remains to be elucidated but the enzyme L-galactono-γ-lactone-dehydrogenase catalysing the final conversion of L-galactono-γ-lactone to ASC is situated on the inner mitochondrial membrane (Mutsuda et al., 1995; Siendones et al., 1999). The pathway has a number of key roles in plant metabolism because it provides not only intermediates for ASC biosynthesis but also supplies intermediates for cell wall biosynthesis and protein glycosylation (Conklin et al., 1997).

ASC is generally involved as an electron donor transferring electrons in two sequential steps. The first step results in the formation of ASC free radical monodehydroascorbate (MDA) and the second in the fully oxidized dehydroascorbate (DHA). Both oxidation products are relatively unstable in aqueous solution. The MDA concentration of cells is generally low because of the rapid disproportionation of the radicals to ASC and DHA (Skotland and Ljones 1980). MDA does not readily interact with other molecules such as lipids and is therefore relatively harmless. DHA undergoes spontaneous and irreversible hydrolysis to 2,3-diketogulonic acid (Washko et al., 1992). ASC turnover is remarkably slow, in the order of 2% turnover of the total pool per hour in darkness (Imai et al., 1999). Efficient cycling between oxidized and reduced forms of ASC thus minimises DHA degradation and decreases the requirement for high rates of de novo synthesis. Feedback inhibition of ASC on the pathway of its own biosynthesis is strongly indicated (Mozafer and Oertli 1993; Imai et al., 1999).

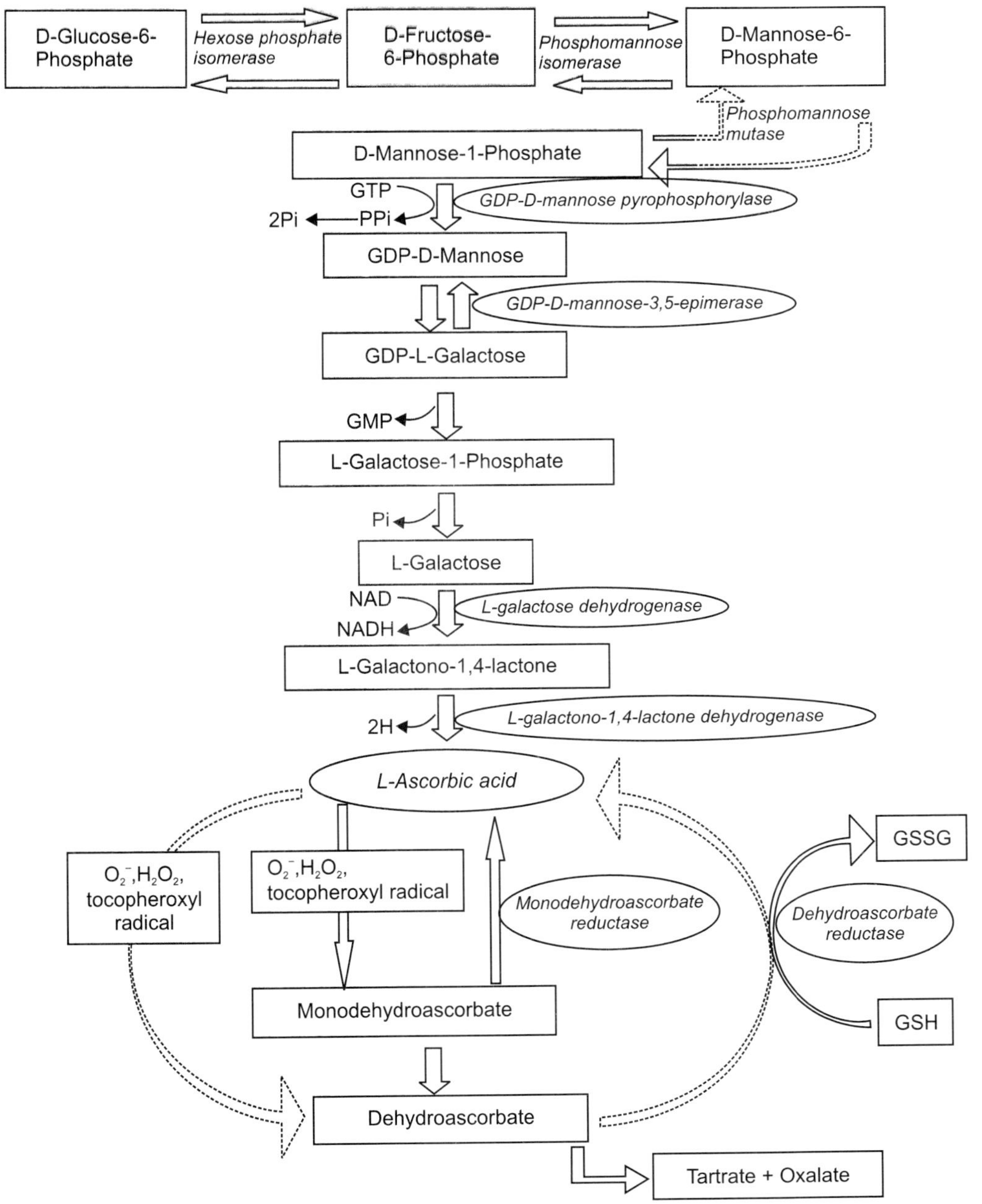

Fig. 3. Biosynthesis, functions and degradation of L-Ascorbate in plants.

4.2 Functions of Ascorbate

The biochemical functions of ASC can be divided into four categories:

(i) Antioxidant. (ii) Enzyme cofactor (iii) Electron transport (iv) Oxalate and tartrate synthesis. By far the best studied function of ASC in plant cells is its role in the protection of the chloroplast against oxidative damage (Fig. 1; Fig. 3). ROS are inevitably formed during photosynthesis and photorespiration (Foyer et al., 1991; Noctor and Foyer 1998). Under normal conditions, ROS are generated during photosynthesis by singlet oxygen formation as well as oxygen photoreduction. Photooxidative damage can occur when ROS production exceeds that of the antioxidant capacity. Such conditions occur when high light is combined with other environmental conditions such as drought, temperature extremes, or nutrient deprivation. Other abiotic factors in the plant's environment also lead to increased ROS including UV-B, air pollutants (e.g., ozone, SO_2), redox-active herbicides (e.g., paraquat), and phytotoxic metals (e.g., Zn, Cu, Cd). Plants also generate ROS *in planta* in oxidative bursts that occur during pathogen infection (Lamb and Dixon 1997). H_2O_2 generated during the oxidative burst is thought to play an important role in initiation of the hypersensitive response (Chen, Silva and Klessig 1993; Levine et al., 1994), although given the toxicity of ROS such as the hydroxyl radical, ROS levels must be tightly controlled. This control is accomplished through an interconnected network of antioxidants. Within this network, ASC eliminates ROS through multiple mechanisms. ASC has the capacity to directly eliminate several different ROS including $^1O_2$, O_2^-, and OH^- (Padh 1990; Athar et al., 2008). The scavenging of O_2^- and H_2O_2 produced in the chloroplast is catalysed by stromal and thylakoid-bound SOD and APX (Miyake et al., 1993; Foyer et al., 1994). MDA formed by the action of the latter enzyme can be directly reduced by ferredoxin (Miyake et al., 1993; Foyer et al., 1994; Eltayeb et al., 2007). This series of reactions occurs close to the thylakoid membrane and is known as the Mehler peroxidase reaction sequence. In addition, a second enzymatic cascade (commonly known as ascorbate-glutathione cycle) (Foyer and Halliwell 1976) is important in the regeneration of ASC from DHA. The presence of two ASC regeneration systems guarantees the complete regeneration of reduced ASC at the expense of glutathione and NAD(P)H (Foyer and Halliwell 1976). While the Mehler peroxidase sequence is restricted to the chloroplast, the ASC-glutathione cycle is found in several compartments of the plant cell. ASC not only functions as a major antioxidant in the soluble phase of plant cells but also supports essential membrane-bound antioxidants in their role as cellular protectants. For example, carotenoid pigments (carotenes and xanthophylls) and α-tocopherol present in the chloroplast membranes quench triplet chlorophyll and singlet oxygen and remove ROS as they are formed within the membrane environment. These membrane-associated antioxidants protect against lipid peroxidation, and depend on ASC in two ways. First, the de-epoxidation of violaxanthin, which produces zeaxanthin, an essential component of the protective xanthophyll cycle, is dependent on ASC (Neubauer and Yamamoto 1994). Second, ASC acts as a secondary antioxidant involved in the direct regeneration of reduced α-tocopherol from its oxidized form. ASC is also involved in the protection of other cell compartments against oxidative damage. APX are found in the cytosol and in the peroxisomes. In these peroxisomes, APX may support catalase in the removal of the large amounts of H_2O_2 produced in the photorespiratory glycollate pathway. There is apparently an inverse relationship between total foliar ASC concentration and sensitivity to ozone (Kelly et al., 1995; Conklin et al., 1996; Dietz 1997). Exposure to oxidative stress increases tissue ASC accumulation and results in an enhancement of ASC dependent detoxification processes.

Many studies over the years have investigated the consequences of an oxidative stress on the ASC status in plant cells. Often, the pool of ASC is found to become increasingly oxidized due to a combination of increased oxidation and/or inefficient regeneration. In *Triticum durum* seedlings grown at an elevated temperature (~30°C) for 34d the ASC:DHA ratio declined significantly (Paolacci et al., 1997). Treatment of *Nicotiana plumbaginifolia* with toxic concentrations of either NaCl or $CuSO_4$ for 120h resulted in complete oxidation of ASC (Savouré et al., 1999). MDA radicals increase in several plant species upon exposure to a variety of ROS-generating stresses including the redox-active herbicide paraquat, cyanide, high light, drought and senescence (Heber et al., 1996). The redox status of ASC is at least partially dependent on the size of the ASC pool. The *Arabidopsis thaliana* mutant, *vtc1*, which has only 30% of the ASC present in the wild type, shows increased sensitivity to ozone and to oxidative stress caused by exposure to UV-B irradiation or SO_2 pollution (Conklin et al., 1997).

4.3 Ascorbate Peroxidase

Ascorbate peroxidases (APXs) are the major consumers of ASC, generating MDA in the reduction of H_2O_2 to water. APX has a key role in removal of H_2O_2, and APX isoenzymes are located in the chloroplast, cytosol, peroxisome, and mitochondria (Mittler and Zilinskas 1991; Asada 1992; Jimenez et al., 1997; Panchuk et al., 2005; Teixeira et al., 2006). There are evidences for regulation of APX at both the level of transcription as well as at the level of translation. An increased ROS triggers an increase in the steady-state level of one or more APX mRNAs (Conklin and Last 1995; Kubo et al., 1995; Donahue et al., 1997; Morita et al., 1999). Photooxidative stress caused by exposure to excess light causes rapid elevation of two cytosolic APX mRNAs in Arabidopsis (Karpinski et al., 1997). APX is also subject to post-transcriptional regulation. Stress imposed by both salinity and heavy metal cause an increase in APX activity whereas the level of the mRNA remains constant (Lopez et al., 1996) or even declines (Savouré et al., 1999). Such results could be obtained if APX is under post-transcriptional control. Post-transcriptional control of APX has been shown to occur during induction of programmed cell death in tobacco mosaic virus-infected tobacco (Mittler et al., 1998). Although transcripts encoding a specific cytosolic APX accumulated to high levels, the level of the corresponding APX protein declined. The inhibition of APX expression was shown to occur post-transcriptionally after binding of the APX mRNAs to ribosomes.

In addition to its role in protection against photo-oxidative stress through the activity of APX, ASC also has a key role in photoprotection. ASC levels increase in plants grown under high light, and ASC is a cofactor utilized in the xanthophylls cycle (Eskling et al., 1997). This cycle (conversion of violaxanthin to zeaxanthin across the thylakoid membrane) is thought to be involved in non-photochemical quenching of excess light energy in photosystem II (Demmig-Adams and Adams 1990). ASC is required *in vitro* as a cofactor for the enzyme violaxanthin de-epoxidase in the conversion of violaxanthin to zeaxanthin. The ASC-deficient Arabidopsis mutants (Conklin et al., 2000) have lower levels of non-photochemical quenching due to a decrease in this de-epoxidation reaction (Smirnoff 2000).

As an enzyme cofactor, ASC plays significant parts in photoprotection, the wounding response, and insect herbivory as well as cell expansion and division. ASC is also a cofactor for some hydroxylase enzymes (e.g., prolyl hydroxylase) involved in hydroxyproline (HP) and hydroxylysine synthesis. HP-rich glycoproteins, such as extensin, are cell wall structural proteins (Carpita and Gibeaut, 1993). Post-translational formation of prolyl residues in these proteins is an important use for intracellular ASC in growing cells (Liso et al., 1985). ASC is also a cofactor for violaxanthin de-epoxidase. Ascorbate is also required, at least *in vitro*, by ethylene-forming enzyme. This enzyme has sequence homology with other Ascorbate dependent hydroxylases (Smith, Ververidis and John, 1992). Cell wall ASC and cell wall-localized ascorbate oxidase (AO) have been implicated in control of growth. High AO activity is associated with rapidly expanding cells. ASC has also been implicated in regulation of cell division by influencing progression from G1 to S phase of the cell cycle.

4.4 The Occurrence of Ascorbate

Ascorbate and ascorbate peroxidase (APX) are universal in photosynthetic eukaryotes including algae and bryophytes (Loewus, 1980; Miyake et al., 1991). Amongst prokaryotes, APX occurs in some cyanobacteria (Miyake et al., 1991). It is not known if cyanobacteria which lack APX also lack ASC. ASC occur in the cytosol, chloroplasts, vacuoles, mitochondria and cell wall (Anderson, Foyer and Walker, 1983; Rautenkranz et al., 1994). The concentration in chloroplasts can be high (up to 50 mm in spinach) and is probably related to its central role in photosynthesis (Foyer, 1993). Occurrence in the cell wall is not accidental since high affinity carriers for both ASC (Km = 90 μm) and DHA (Km = 20 μm) occur on the plasma membrane of barley leaf protoplasts. DHA, the oxidized form of ASC, is taken up more rapidly than ASC. The effects of ionophores and channel blockers on ASC uptake suggest a requirement for a proton electrochemical gradient across the plasma membrane. In contrast, transport into isolated vacuoles is slow and not carrier-mediated (Rautenkranz et al., 1994). Isolated spinach chloroplasts take up ASC with a suprisingly low affinity (Km = 18-40 mm) via a saturable carrier (Anderson et al., 1983; Beck et al., 1983). Uptake into chloroplasts is inhibited by DHA (Anderson et al., 1983; Beck et al., 1983). High affinity carriers on the plasma membrane, if generally occurring, would facilitate movement of ASC into the cell wall and allow movement from cell to cell via the apoplast.

ASC is also known as an *in vitro* electron donor for photosynthetic and mitochondrial electron transport. MDA can act as an electron acceptor from PSII *in vivo* (Miyake and Asada, 1992) and that it could act as both electron donor and acceptor in transmembrane electron transport (Asard et al., 1995).

ASC is also used for synthesis. ASC can be cleaved to form oxalate and tartrate (Loewus, 1980; 1988; Saito, 1996). Other pathways could contribute to oxalate synthesis but ASC is probably the main source of tartrate in grapes.

5. GLUTATHIONE

Glutathione (GSH) is a cysteine-containing tripeptide found in most forms of aerobic life. It is not required in the diet and is instead synthesized in cells from its constituent amino acids. GSH is an abundant and ubiquitous low molecular weight thiol in fungal, plant and animal tissues and occurs in two distinct redox forms. GSH has antioxidant properties since the thiol group in its cysteine moiety is a reducing agent and can be reversibly oxidized and reduced. In cells, glutathione is maintained in the reduced form by the enzyme glutathione reductase (GR). The reduced form GSH, is a tripeptide (γ-Glu–Cys–Gly). The chemical reactivity of its thiol group, its relative stability and high solubility in water make GSH a particularly adequate electron acceptor or donor in physiological reactions. GSH reduces other metabolites and enzyme systems, such as ASC in the glutathione-ascorbate cycle, glutathione peroxidases and glutaredoxins, as well as reacting directly with oxidants. Oxidation causes two reduced molecules to be linked by way of a cysteine disulphide bond, designated GSSG. Due to its high concentration and its central role in maintaining the cell's redox state, GSH is one of the most important cellular antioxidants.

Due to the particular properties of the molecule it plays multiple roles in cellular metabolism. It is a central compound in sulphur metabolism and is considered the main transport form of reduced sulphur (Rennenberg and Lamoureux 1990). It links the sulphur reduction pathways to the protein synthesis and functions as a buffer for reduced sulphur. GSH also plays an important role in the scavenging of toxic ROS. The capacity of the GSH redox system to detoxify dangerous ROS is potentially dependent on the pool size of total GSH, on the redox ratio GSH/GSSG and on the activity of the regenerating enzyme system, the NADPH-dependent glutathione reductase. Chloroplastic GSH concentrations are estimated to be 1–4.5 mM (Noctor and Foyer 1998). Elevated levels of GSH have also been found in plants under stress (Alscher 1989).

5.1 Biosynthesis and Metabolism of Glutathione

The synthesis and degradation of GSH occurs continuously through the glutamyl cycle. The pathway of glutathione synthesis appears to be common to all organisms that contain GSH. Two ATP-dependent steps, catalysed by γ-glutamylcysteine synthetase (γ-ECS) and glutathione synthetase (GS), lead to the sequential formation of γ-glutamylcysteine (γ-EC) and glutathione (GSH) (Meister 1988). The second step involves the addition of glycine by the enzyme GS. Species that produce γ-glu-cys-β-ala have homoglutathione synthetase activity in the second step, whose affinity for β-alanine is much higher than for glycine (Klapheck et al., 1988; Macnicol 1987).

GSH accumulates in response to increased ROS generation or is constitutively higher in plants adapted to stress conditions. Under stress conditions, oxidation of GSH is accompanied by net glutathione degradation.The degradation of GSH involves first the cleavage of the bond between glutamate and cysteine by glutamyl transpeptidase and the transfer of the glutamate residues to an acceptor amino acid. Subsequently the Cys-Gly dipeptide is degraded by dipeptidases and Glu-aa by glutamylcyclotransferase.

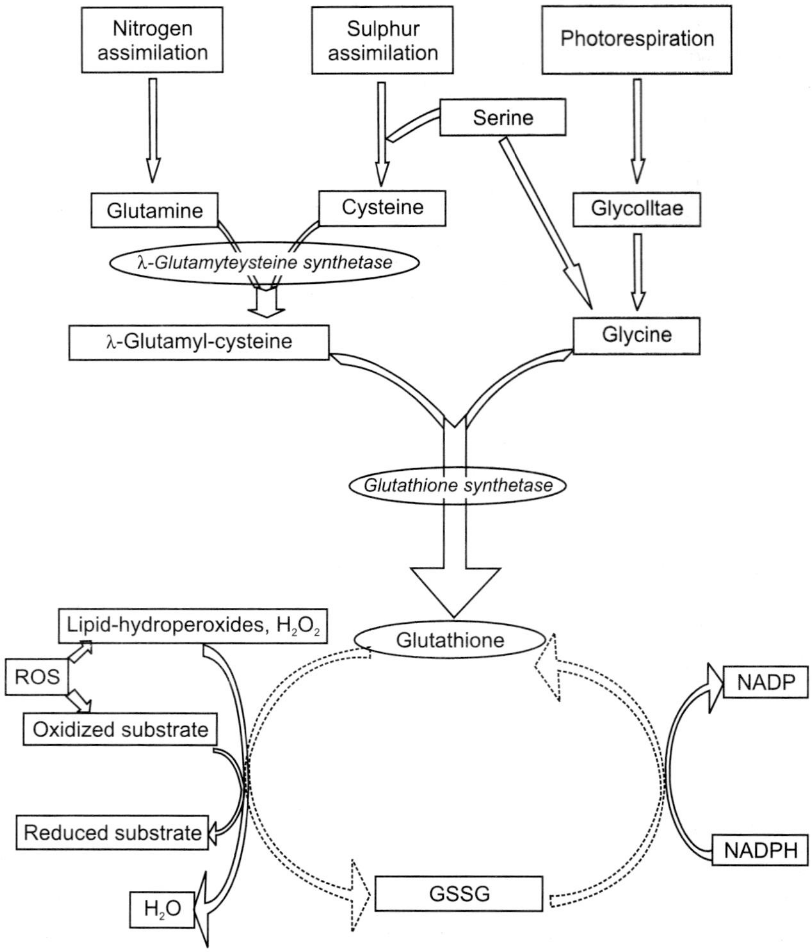

Fig. 4. Biosynthesis and functions of glutathione.

5.2 Functions of Glutathione

In general, the physiological significance of GSH in plant cells covers three categories: (i) defence against oxidative stress (ii) sulphur metabolism (iii) detoxification of xenobiotic compounds. The fact that GSH is involved in defence reactions against oxidative stress, as an antioxidant, is widely acknowledged. Its importance is illustrated by the fact that GSH biosynthesis is stimulated when the cell encounters stress conditions, as if the cell builds up its defence capability. GSH accumulates in response to increased ROS generation, or to compensate for decreases in the defence capacity of other antioxidants, and GSH levels are constitutively higher in plants adapted to stress conditions (Marrs 1996; May and Leaver 1993; Sen Gupta et al., 1991). Supplementary evidence for the role of GSH in cell defence comes from by transgenic *Arabidopsis* plants with a limited biosynthetic capacity for GSH. The decreased GSH pool apparently renders them substantially more sensitive to different stresses (such as Cu, Cd, photo-oxidative or ozone stress) (Xiang et al., 2001). Furthermore, the reactive cystein residue enables GSH to keep thiol group-containing proteins in their native state during stress conditions. For example, under water stress conditions, GSH protects specific protein thiol groups from irreversible formation of intramolecular disulphide bonds.

GSH metabolism is also tied into other defence pathways. For example, GSH plays an important role in ASC regeneration, both in the chloroplast and in the cytosol (Noctor and Foyer 1998). GSSG reduction itself requires NADPH as a source of electrons for its reduction. The GSH–GSSG redox pair can therefore only function with an adequate supply of NADPH. GSH may also function as a cellular sensor to ensure maintenance of the NADPH pool, which further illustrate the intertwining of the metabolism of different redox compounds (May et al., 1998). These connections undoubtedly increase the chances of the plant cell for displaying an optimal defence capacity against stressful conditions.

Due to its sulfhydryl group, the GSH molecule has strong antioxidative properties. It may react directly with ROS or with oxidized substrates reducing them. By doing so GSH is eventually transferred to its oxidized form, GSSG. The rate constants of the direct reaction of GSH with O_2^- and $^1O_2$ are also of comparable magnitude as compared to that of ASC. Due to its antioxidative properties GSH may stabilize membrane structures. Membrane lipids are a prominent target for ROS reactions and lipid peroxidation forms acyl peroxides, which can be removed by GSH (McKersie and Leshem 1994). GSH may reduce disulphide bonds of proteins and thus counteract the detrimental effects of ROS because the formation of protein disulphides is one of the consequences of oxidative attack. Furthermore, the thioldisulphide exchange is of more general importance in enzyme activations, protein protection, and cellular regulation processes (Kunert and Foyer 1993). A non-enzymatic reduction of dehydroascorbate by GSH is possible at a pH greater than 7 and GSH concentrations above 1mM. Within the living plant cell such a milieu exists in the stroma of illuminated chloroplasts with a pH about 8 and sufficient GSH concentrations (McKersie and Leshem 1994).

GSH may serve as a substrate for peroxidase reactions. Selenium containing glutathione peroxidase (GPX) has been demonstrated in animals. In plant systems, increased GPX activities were measured upon stress impacts. A protein of strong homology to animal GPX was characterized unequivocally in *Citrus sinensis*. The best substrate for this enzyme is not but phospholipids hydroperoxides, which are toxic products in ROS-induced lipid peroxidation reaction chains. That means that the *Citrus* enzyme is a phospholipidhydroperoxide glutathione peroxidase. A role in removing toxic products of oxidative stress was ascribed to this enzyme. In contrast to the animal enzyme, the plant GPX has a cysteine in the active centre instead of a Se-cysteine and the specific activity of the plant enzyme is much lower (Eshdat et al., 1997). In pea and *Arabidopsis* it has been observed that genes for a chloroplast targeted GPX exist and respond to stress (Mullineaux et al., 1998).

GSH plays an important role during drought stress (Secenji et al., 2008). To minimize water loss during drought stress homoiohydric plants respond to water deficiency with stomatal closure. Although this stomatal closure restricts gas exchange and is efficient in preventing water loss, but it causes a lack of CO_2 in the chloroplasts which then leads to the situation of increased oxidative stress in illuminated chloroplast (Smirnoff 1993). To dissipate the excess accumulated light energy photo-protective changes in xanthophylls occurs which is followed by a drop in GSH/GSSG ratio (Tausz et al., 2001).

Algae, lichens and many mosses are able to endure significant water losses by means of poikilohydry, while only few higher plants have the ability to show poikilohydry. Complete metabolic activity is quickly regained upon wetting these poikilohydric plants. An important trait connected with poikilohydry is the capacity to bear the high oxidation of the GSH pool and to quickly reduce it upon re-hydration. An increase in activities of GR, GPX and GST and also an oxidized GSH pool is observed in a poikilohydric moss *Tortula ruralis*, during its slow dehydration, but upon rapid dehydration no such changes are observed (Dhindsa 1987, 1991). The slowly dehydrated moss shows a quick reduction of their glutathione pool upon wetting and the activities of the enzymes are resumed to control levels. On the other hand, a transient increase in oxidized glutathione and the enzyme activities and a subsequent return to control levels within the next hours after wetting are observed in quickly dehydrated plants. Some lichen species also show a quick reduction of highly oxidized glutathione pools (70 to 90% of total) upon wetting (Kranner and Grill 1997). The resurrection plant *Boea hygroscopica* which is one amongst the few higher plants that show poikilohydry shows an increase in GSH contents as a possible protection of proteins from oxidation upon dehydration (Navari-Izzo et al., 1997).

Many plants have poikilohydric resting stages in form of dormant seeds or spores. The dormancy of seeds and the loss of viability of recalcitrant seeds are accompanied by the action of ROS. The antioxidative defence of recalcitrant seeds is rapidly lost (Hendry et al., 1992). Highly oxidized GSH pools are present in the resting seeds which are rapidly reduced upon imbibition. Spores of *Neurospora crassa* and dormant seeds of *Triticum*, *Hordeum*, *Pisum*, and *Lycopersicum* all are shown to reactivate their glutathione system when they resume physiological activity upon wetting reported the same response (Kranner and Grill 1997).

GSH also protects from excessive light energy induced oxidative stress in the chloroplasts. Higher GSH concentrations are found in sun-exposed spruce needles than shade needles of the same tree (Grill et al., 1987). The evergreen oak tree *Quercus ilex* also exhibits similar results in their sun and shade grown leaves (García-Plazaola et al., 1999). Higher GR activities are observed in chloroplasts of pea leaves grown under high light than those grown under low light (Gillham and Dodge 1987). The etiolated wheat seedlings show a rapid increase in GSH level within hours upon illumination (Mattagajasingh and Kar 1989).

A protective role of GSH is also observed against low temperature induced stress. Low temperature imposes various adverse effects on plants: (i) Photo-chilling is an ill effect of low temperature where the Calvin cycle is slowed down or inhibited, but the absorbed light energy still drives the electron transport which leads to a situation of overproduction of ROS in the chloroplasts; (ii) Low temperature also results in frost and frost is very much detrimental to plant tissues mainly due the formation of ice crystals, as it removes water from the liquid phase resulting in damage to proteins and the ultrastructure of cells. The level of damage is dependent on the plant species, the developmental and phenological stage, the organ and tissues, and the degree of stress. ROS are also involved in freezing stress, participating in lipid peroxidation, protein destruction, and collapses of the antioxidant defence systems (Polle 1997).

Chilling has a strong impact on the antioxidative systems, especially in plants sensitive to chilling. Maize shows symptoms of chilling stress at temperatures as high as 11°C with an increased production of GSH and other antioxidants. Young plants upon short-term chilling are shown to drop down their GSH/GSSG ratio, whereas chilling-sensitive cultivars show an increase in their GSH concentrations upon long-term chilling and their overall GSH/GSSG ratios are lower than those of more tolerant lines (Hodges et al., 1996). Increased levels of GSH are observed during winter in needles of spruce (Esterbauer and Grill 1978; Doulis et al., 1993) and pine (Anderson et al., 1992). The winter wheat (Sagisaka et al., 1991) and the living bark of apple trees (Kuroda and Sagisaka 1998) also show an increase in GSH level during winter.

GSH also gives protection against high temperature. ROS are one of the major players involved in high temperature damage to plants. Upon a heat shock of 40°C the GSH synthesis in excised roots of maize is strongly strengthened (Nieto-Sotelo and Tuan-Hua 1986).

The non-enzymatic antioxidant GSH also shows protection against harmful UV radiations. UV radiations due to their shorter wavelength transmit more energy per quantum than visible light and are potentially dangerous to organisms. Three forms of UV radiations exist in the environment UV-A (315-400 nm), UV-B (290-315 nm) and UV-C (220-290 nm) and the most important stress factor amongst these is UV-B because the UV-B doses are expected to mount up in the environment due to the depletion of the protective stratospheric ozone layer. UV-A is only a minimal stress factor for plants, while UV-C is highly destructive it is mostly screened by the atmosphere. Stress symptoms are observed in plants exposed to elevated UV-B doses (Polle 1997). Direct production of ROS at the cell surfaces may occur due to UV-B absorption, apart from increasing the ROS production in chloroplasts. In maize UV-B does not change the pool size or redox state of GSH, but accelerates the turnover of this pool along with increased GR activities (Masi et al., 1995).

Higher constitutive glutathione concentrations are found in plants growing at high elevations like spruce (Polle and Rennenberg 1992; Tausz et al., 1997), pine (Tausz et al., 1999), and some herbaceous alpine species (Wildi and Lütz 1996) to cope up with a combination of stress factors which are imposed by high elevation and others such as high irradiation and lower temperatures, which are sometimes combined with higher atmospheric ozone concentrations.

Upon exposure to salt stress GSH and cysteine levels increase in *Arabidopsis thaliana* (Barroso et al., 1999), indicating a strongly induced GSH synthesis. Salinity is a common stress that can act in multiple ways in plants: (i) by inhibiting enzymes of the Calvin cycle, salinity induces oxidative stress in plant cells; (ii) high salinity can cause water loss in cells and induce drought stress in the plant tissues due to changes in the osmotic balance. Salinity causes an induction of GR in rice (Kaminaka et al., 1998). Salt stress induces the activity of GPX in less tolerant cells of *Citrus* as a result of the stress-induced formation of phospholipidhydroperoxides (Avsiankretchmer et al., 1999).

Role of GSH in sulphur metabolism is evident from the fact that GSH is the predominant non-protein thiol. Furthermore, GSH regulates sulphur uptake at root level and the pathways for sulphur assimilation and GSH synthesis are intertwined at a molecular level (Herschbach and Rennenberg 1994; Anjum et al., 2008).

As a substrate of glutathione *S*-transferases, GSH is responsible for the removal of xenobiotics or other adverse organic compounds from the cytoplasm (Mendoza-Cozatl and Moreno-Sanchez 2006; Sun et al., 2007). Some GSTs are also able to catalyse the reaction of GSH with products of lipid peroxidation chains and so function as repair enzymes in oxidative stress situations. If the peroxidation products are lipid hydroperoxides GSTs mimic a peroxidase activity.

6. CONCLUDING REMARKS

The ROS generated in plants by various biotic and abiotic factors results in an oxidative stress. Most of the ROS contain an unpaired electron; and thus, are highly reactive and damaging to plants at the molecular level. This is due to their interaction with nucleic acid, proteins, and lipids. ASC acts as an antioxidant by being available for energetically favourable oxidation. The ROS oxidizes ASC first to MDA and then to DHA. In this cycle of reaction the ROS are reduced to water, while the oxidized forms of ASC are relatively stable and unreactive, and do not cause any harm to the cells. GSH also participates in removal of these toxic ROS by either neutralizing them directly or by curtailing them in conjugation with ASC. Thus, these two non-enzymatic antioxidants of the plants defence system are very important in maintaining the overall plant health under stressful conditions. An improvement in our understanding of ASC and GSH in plants will lead to the possibility of increasing their concentration in plants by genetic manipulation. This will be helpful in increasing tolerance of plants to various biotic and abiotic factors causing oxidative stress.

REFERENCES

Ahmed, P., Sarwat, M. and Sharma, S. (2008). Reactive oxygen species, antioxidant and signaling in plants. *J. Plant Biol.* **51:** 167-173.

Allan, A.C. and Fluhr, R. (1997). Two distinct sources of elicited reactive oxygen species in tobacco epidermal cells. *Plant Cell.* **9:** 1559-1572.

Allen, R. (1995). Dissection of oxidative stress tolerance using transgenic plants. *Plant Physiol.* **107:** 1049-1054.

Alscher, R.G. (1989). Biosynthesis and antioxidant function of glutathione in plants. *Physiol. Plant.* **77:** 457-464.

Anderson, J.W., Foyer, C.H. and Walker, D.A. (1983). Light-dependent reduction of dehydroascorbate and uptake of exogenous ascorbate by spinach chloroplasts. *Planta.* **158:** 442-450.

Anderson, J., Chevone, B. and Hess, J. (1992). Seasonal variation in the antioxidant system of eastern white pine needles. *Plant Physiol.* **98:** 501-508.

Anjum, N.A., Umar, S., Ahmad, A., Iqbal, M. and Khan, N.A. (2008). Sulphur protects mustard (*Brassica campestris* L.) from cadmium toxicity by improving leaf ascorbate and glutathione. *Plant Growth Regul.* **54:** 271-279.

Arrigoni, O., De Gara, L., Tommasi, F. and Liso, R. (1992). Changes in the ascorbate system during seed development of *Vicia faba* L. *Plant Physiol.* **99:** 235-238.

Arrigoni, O. and De Tullio, M.C. (2000). The role of ascorbic acid in cell metabolism: between gene-directed functions and unpredictable chemical reactions. *J. Plant Physiol.* **157:** 481-488.

Asada, K. and Takahashi, M. (1987). Production and scavenging of active oxygen in photosynthesis. In: *Photoinhibition: Topics in Photosynthesis* (Eds. Kyle, D.J., Osmond, C.B. and Arntzen, C.J.) pp. 227-287, Elsevier, Amsterdam.

Asada, K. (1992). Ascorbate peroxidase–a hydrogen peroxide scavenging enzyme in plants. *Physiologia Plantarum.* **85:** 235-241.

Asada, K. (1994). Mechanisms for scavenging reactive molecules generated in chloroplasts under light stress. In: *Photoinhibition of photosynthesis. From molecular mechanisms to the field* (Eds. Baker, N.R. and Bowyer, J.R.) pp. 129-142, Bios Scientific Publishers, Oxford.

Asada, K. (1999). The water–water cycle in chloroplasts: scavenging of active oxygen and dissipation of excess photons. *Annu. Rev. Plant Physiol. Plant Mol. Biol.* **50:** 601-639.

Asada, K. (2006). Production and scavenging of reactive oxygen species in chloroplast and their functions. *Plant Physiol.* **141:** 391-396.

Asard, H., Horemans, N. and Caubergs, R.J. (1995). Involvement of ascorbic acid and a b-type cytochrome in plant plasma membrane redox reactions. *Protoplasma.* **184:** 36-41.

Athar, H.R., Khan, A. and Asraf, M. (2008). Exogenously applied ascorbic acid alleviates salt-induced oxidative stress in wheat. *Env. Exp. Bot.* **63:** 224-231.

Avsiankretchmer, O., Eshdat, Y., Guetadahan, Y. and Benhayyim, G. (1999). Regulation of stress induced phospholipid hydroperoxide glutathione-peroxidase expression in *Citrus. Planta.* **209:** 469-477.

Barroso, C., Romero, L.C., Cejudo, F.J., Vega, J.M. and Gotor, C. (1999). Salt-specific regulation of the cytosolic O-acetylserine(thiol)lyase gene from *Arabidopsis thaliana* is dependent on abscisic acid. *Plant Mol. Biol.* **40:** 729-736.

Beck, E., Burkert, A. and Hofman, M. (1983). Uptake of l-ascorbate by intact spinach chloroplasts. *Plant Physiology.* **73:** 41-45.

Borraccino, G., Mastropasqua, L., De Leonardis, S. and Dipierro, S. (1994). The role of the ascorbic acid system in delaying the senescence of oat (*Avena sativa* L.) leaf segments. J. *Plant Physiol.* **144:** 161-166.

Bowler, C., Slooten, L., Vandenbranden, S., De Rycke, R., Botterman, J., Sybesma, C., Van Montagu, M. and Inzé, D. (1991). Manganese superoxide dismutase can reduce cellular damage mediated by oxygen radicals in transgenic plants. *EMBO J.* **10:** 1723-1732.

Bowler, C., Van Montagu, M. and Inzé, D. et al. (1992). Superoxide dismutases and stress tolerance. *Annu. Rev. Plant Physiol. Plant Mol. Biol.* **43:** 83-116

Bowler, C., Van Camp, W., Van Montagu, M. and Inzé, D. (1994). Superoxide dismutase in plants. *Crit. Rev. Plant Sci.* **13:** 199-218.

Carpita, N.C. and Gibeaut, D.M. (1993). Structural models of primary cell walls in flowering plants: consistency of molecular structure with the physical properties of the walls during growth. *The Plant J.* **3:** 1-30.

Cazale, A.C., Droillard, M.J., Wilson, C., Heberle-Bors, E., Barbier-Brygoo, H. and Lauriere, C. (1999). MAP kinase activation by hypoosmotic stress of tobacco cell suspensions: towards the oxidative burst response? *Plant J.* **19:** 297-307.

Chameides, W.L. (1989). The chemistry of ozone deposition to plant leaves: Role of ascorbic acid. *Environ. Sci. Technol.* **23:** 595-600.

Chen, Z., Silva, H. and Klessig, D.F. (1993). Active oxygen species in the induction of plant systemic acquired resistance by salicylic acid. *Science.* **262:**1883-1886.

Conklin, P.L., Williams, E.H. and Last, R.L. (1996). Environmental stress sensitivity of an ascorbic acid-deficient *Arabidopsis* mutant. *Proc. Natl. Acad. Sci. USA.* **93:** 9970-9974.

Conklin, P.L., Pallanca, J.E., Last, L. and Smirnoff, N. (1997). L-Ascorbic acid metabolism in the ascorbate-deficient *Arabidopsis* mutant *vtc1. Plant Physiol.* **115:** 1277-1285.

Conklin, P.L., Norris, S.R., Wheeler, G.L., Williams, E.H., Smirnoff, N. and Last, R.L. (1999). Genetic evidence for the role of GDP-mannose in plant ascorbic acid (vitamin C) biosynthesis. *Proc. Natl. Acad. Sci. USA.* **96:** 4189-4203.

Conklin, P.L., Saracco, S.A., Norris, S.R. and Last, R.L. (2000). Identification of ascorbic acid-deficient *Arabidopsis thaliana* mutants. *Genetics.* **154:** 847-856.

Creissen, G., Firmin, J., Fryer, M., Kular, B., Leyland, N., Reynolds, H., Pastori, G., Wellburn, F., Baker, N., Wellburn, A. and Mullineaux, P. (1999). Elevated glutathione biosynthetic capacity in the chloroplasts of transgenic tobacco plants paradoxically causes increased oxidative stress. *Plant Cell.* **11:** 1277-1292.

Dat, J., Vandenabeele, S., Vranovà, E., Van Montagu, M., Inzé, D. and Van Breusegem, F. (2000). Dual action of the active oxygen species during plant stress responses. *Cell. Mol. Life Sci.* **57:** 779-795.

De Gara, L. and Tommasi, F. (1999). Ascorbate redox enzymes: a network of reactions involved in plant development. *Recent Dev. Phytochem.* **3:** 1-15.

de Pinto, M.C., Francis, D. and De Gara, L. (1999). The redox state of the ascorbate-dehydroascorbate pair as a specific sensor of cell division in tobacco BY-2 cells. *Protoplasma.* **209:** 90-97.

Demmig-Adams, B. and Adams, W.W.I. (1990). The carotenoid zeaxanthin and high-energy-state quenching of chlorophyll fluorescence. *Photosynthesis Research.* **25:** 187-198.

Desikin, R., Mackerness, S.A.H., Hancock, J.T. and Neill, S.J. (2001). Regulation of the *Arabidopsis* transcriptosome by oxidative stress. *Plant Physiol.* **127:** 159-172.

Deutsch, J.C. (2000). Dehydroascorbic acid. *J. Chromat.* **881:** 301-309.

Dhindsa, R.S. (1987). Glutathione status and protein synthesis during drought and subsequent rehydration in *Tortula ruralis. Plant Physiol.* **83:** 816-819.

Dhindsa, R.S. (1991). Drought stress, enzymes of glutathione metabolism, oxygen injury, and protein synthesis in *Tortula ruralis. Plant Physiol.* **95:** 648-651.

Di Cagno, R., Guidi, L., De Gara, L. and Soldatini, G.F. (2001). Combined cadmium and ozone treatments affects photosynthesis and ascorbate-dependent defences in sunflower. *New Phytol.* **151:** 627-636.

Dietz, K.J. (1997). Functions and responses of the leaf apoplast under stress. *Prog. Bot.* **58:** 221-254.

Donahue, J.L., Okpodu, C.M., Cramer, C.L., Grabau, E.A. and Alscher, R.G. (1997). Responses of antioxidants to paraquat in pea leaves. *Plant Physiol.* **113:** 249-257.

Doulis, A.G., Hausladen, A., Mondy, B., Alscher, R.G., Chevone, B.I., Hess, J.L. and Weiser, R.L. (1993). Antioxidant response and winter hardiness in red spruce (*Picea rubens* Sarg.). *New Phytol.* **123:** 365-374.

Elstner, F.E. and Osswald, W. (1994). Mechanisms of oxygen activation during plant stress. *Proc. Royal Soc. Edinb.* **102**: 131-154.

Eltayeb, A.E., Kawano, N., Badawi, G.H., Kaminak, H., Saekata, T., Shibahara, T., Inanaga, S. and Tanaka, K. (2007) Overexpression of monodehydroascorbate reductase in transgenic tobacco confers enhanced tolerance to ozone, salt and polyethylene glycol stress. *Planta.* **225:** 1255-1264.

Eshdat, Y., Holland, D., Faltin, Z. and Ben-Hayyim, G. (1997). Plant glutathione peroxidases. *Physiol. Plant.* **100:** 234-240.

Eskling, M., Arvidsson, P.O. and Åkerlund, H.E. (1997). The xanthophylls cycle, its regulation and components. *Physiologia Plantarum.* **100:** 806-816.

Esterbauer, H. and Grill, D. (1978). Seasonal variation of glutathione and glutathione reductase in needles of *Picea abies. Plant Physiol.* **61:** 119-121.

Foyer, C.H. (1993). Ascorbic acid. In: *Antioxidants in higher plants.* (Eds. Alscher, R.G. and Hess, J.L.) pp. 31-58, CRC Press, Boca Raton.

Foyer, C. and Halliwell, B. (1976). Presence of glutathione and glutathione reductase in chloroplasts: a proposed role in ascorbic acid metabolism. *Planta.* **133:** 21-25.

Foyer, C.H., Lelandais, M., Edwards, E.A. and Mullineaux, P.M. (1991). The role of ascorbate in plants, interactions with photosynthesis, and regulatory significance. In: *Active Oxygen, Oxidative Stress and Plant Metabolism* (Eds. Pell, E.J. and Steffen, K.L.) pp. 131-144, *American Society of Plant Physiologists,* USA.

Foyer, C.H., Lelandais, M. and Kunert, K.J. (1994). Photooxidative stress in plants. *Physiol. Plant.* **92:** 696-717.

Foyer, C.H. and Lelandais, M. (1996). A comparison of the relative rates of transport of ascorbate and glucose across the thylakoid, chloroplast and plasmalemma membranes of pea leaf mesophyll cells. *J. Plant Physiol.* **148:** 391-398.

Foyer, C.H. and Noctor, G. (2000). Oxygen processing in photosynthesis. Regulation and signaling. *New Phytol.* **146:** 359-388.

Foyer, C.H. and Noctor, G. (2003). Redox sensing and signaling associated with reactive oxygen in cloroplasts, peroxisomes and mitochondria. *Physiol. Plant.* **119:** 355-64.

Foyer, C.H. and Noctor, G. (2005). Oxidant and antioxidant signaling in planta: A re-evaluation of the concept of oxidative stress in a physiological context. *Plant Cell Environ.* **28:** 1056-1071.

Foyer, C.H., Rowell, J. and Walker, D.A. (1983). Measurements of the ascorbate content of spinach leaf protoplasts and chloroplasts during illumination. *Planta.* **157:** 239-244.

Fukuda, M., Kunisada, Y., Noda, H., Tagaya, S., Yamamoto, Y. and Kida, Y. (1995). Effect of storage time of potatoes after harvest on increase in the ascorbic acid content by wounding. *Nippon Shokuhin Kagaku Kogaku Kaishi.* **42:** 1031-1034.

García-Plazaola, J.I., Artetxe, U. and Becerril, J.M. (1999). Diurnal changes in antioxidant and carotenoid composition in the mediterranean schlerophyll tree *Quercus ilex* (L.) during winter. *Plant Sci.* **143:** 125-133.

Ghezzi, P. and Bonetto, V. (2003). Redox proteomics: Identification of oxidatively, modified proteins. *Proteomics.* **3:** 1145-1153.

Gillham, D. and Dodge, A. (1987). Chloroplast superoxide and hydrogen peroxide scavenging systems from pea leaves: seasonal variations. *Plant Sci.* **50:** 105-109.

Grill, D., Pfeifhofer, H. and Esterbauer, H. (1987) Further investigations on the thiol content of Norway spruce needles. *Phyton.* **27:** 311-317.

Haber, F. and Weiss, J. (1934). The catalytic decomposition of hydrogen peroxide by iron salt, *Proc. R. Soc. Lond. A. Math. Phys. Sci.* **147:** 337-351.

Hammond-Kosack, K.E. and Jones, J.D.G. (1996). Resistance gene-dependent plant defense responses. *Plant Cell.* **8:** 1773-1791.

Halliwell, B. (2006). Reactive species and antioxidant. Redox biology is fundamental theme of aerobic life. *Plant Physiol.* **141:** 312-322.

Heber, U., Miyake, C., Mano, J., Ohno, C. and Asada, K. (1996). Monodehydroascorbate radical detected by electron paramagnetic resonance spectrometry is a sensitive probe of oxidative stress in intact leaves. *Plant Cell Physiol.* **37:** 1066-1072.

Hendry, G., Finch-Savage, W.E., Thorpe, P.C., Atherton, N.M., Buckland, S.M., Nilsson, K.A. and Seel, W.E. (1992). Free radical processes and loss of seed viability during desiccation in the recalcitrant species *Quercus robur* L. *New Phytol.* **122:** 273-279.

Herschbach, C. and Rennenberg, H. (1994). Influence of glutathione on net uptake of sulphate and sulphate transport in tobacco plants. *J. Exp. Bot.* **45:** 1069-1076.

Hodges, M.D., Andrews, C.J., Johnson, D.A. and Hamilton, R.I. (1996). Antioxidant compound responses to chilling stress in differentially sensitive inbred maize lines. *Physiol. Plant.* **98:** 685-692.

Horemans, N., Foyer, C.H., Potters, G. and Asard, H. (2000). Ascorbate function and associated transport systems in plants. *Plant Physiol. Biochem.* **38:** 531-540.

Imai, T., Kingston-Smith, A.H. and Foyer, C.H. (1999). Inhibition of endogenous ascorbate synthesis in potato leaves supplied with exogenous ascorbate. *Free Radical Res.* **31:** 171-179.

Jeong, Y.C., Nakamura, J., Upton, P.B. and Swenberg, J.A. (2005). Pyrimido [1,2-a]-10(3H)one, M1G, is less prone to artifact then base oxidation. *Nucleic Acids Res.* **33:** 6426-6434.

Jimenez, A., Hernandez, J.A., del-Rio, L.A. and Sevilla, F. (1997). Evidence for the presence of the ascorbate-glutatione cycle in mitochondria and peroxisomes of pea leaves. *Plant Physiology.* **114:** 275-284.

Kaminaka, H., Morita, S., Nakajima, M., Masumura, T. and Tanaka, K. (1998). Gene cloning and expression of cytosolic glutathione-reductase in rice (*Oryza sativa* L.). *Plant Cell Physiol.* **39:** 1269-1280.

Karpinski, S., Escobar, C., Karpinska, B., Creissen, G. and Mullineaux, P.M. (1997). Photosynthetic electron transport regulates the expression of cytosolic ascorbate peroxidase genes in Arabidopsis during excess light stress. *Plant Cell.* **9:** 627-640.

Keates, S.E., Tarlyn, N.M., Loewus, F.A. and Franceschi, V.R. (2000). L-ascorbic acid and L-galactose are sources for oxalic acid and calicium oxalate in *Pistia stratiotes. Phytochemistry.* **53:** 433-440.

Kelly, F.J., Mudway, I., Krishna, M.T. and Holgate, S.T. (1995). The free radical basis of air pollution: focus on ozone. *Respir. Med.* **89:** 647-656.

Kerk, N.M. and Feldman, L.J. (1995). A biochemical model for the initiation and maintenance of the quiescent center: implications for organization of root meristems. *Development.* **121:** 2825-2833.

Klapheck, S., Zopes, H., Levels, H.G. and Bergmann, L. (1988). Properties and localization of the homoglutathione synthetase from *Phaseolus coccineus* leaves. *Physiol. Plant.* **74:** 733-39.

Kubo, A., Saji, H., Tanaka, K. and Kondo, N. (1995). Expression of Arabidopsis cytosolic ascorbate peroxidase in response to ozone or sulfur dioxide. *Plant Molecular Biology.* **29:** 479-489.

Kranner, I. and Grill, D. (1997). Desiccation and the subsequent recovery of cryptogamics that are resistant to drought. *Phyton.* **37:** 139-150.

Kunert, K.J. and Foyer, C.H. (1993). Thiol/disulfide exchange in plants. In: *Sulfur nutrition and assimilation in higher plants; regulatory, agricultural, and environmental aspects.* (Eds. De Kok, L.J., Stulen, I., Rennenberg, H., Brunold, C. and Rauser, W.E.) pp. 139-152, SPB Academic Publishing, The Hague, The Netherlands.

Kuroda, H. and Sagisaka, S. (1998). Metabolic and enzymatic responses associated with oxidative stress in plants acclimatized to cold environments. *Rec. Res. Devel. Agric. Biol. Chem.* **2:** 395-410.

Lamb, C. and Dixon, R.A. (1997). The oxidative burst in plant disease response. *Ann. Rev. Plant Physio. Plant Mole. Bio.* **48:** 251-275.

Levine, A., Tenhaken, R., Dixon, R. and Lamb, C. (1994). H_2O_2 from the oxidative burst orchestrates the plant hypersensitive disease resistance response. *Cell.* **79:** 583-593.

Liso, R., De Gara, L., Tommasi, F. and Arrigoni, O. (1985). Ascorbic acid requirement for increased peroxidase activity during potato tuber slice aging. *FEBS Lett.* **187:** 141-145.

Loewus, F.A. (1980). l-Ascorbic acid: metabolism, biosynthesis, function. In: *The Biochemistry of plants.* (Ed. Preiss, J.) Vol. 3. pp. 77-99, Academic Press, New York.

Loewus, F.A. and Loewus, M.W. (1987). Biosynthesis and metabolism of ascorbic acid in plants. *Crit. Rev. Plant Sci.* **5:** 101-119.

Loewus, F.A. (1988). Ascorbic acid and its metabolic products. In: *The biochemistry of plants.* (Ed. Preiss, J.) Vol. 14 pp. 85-107, Academic Press, New York.

Lopez, F., Vansuyt, G., Casse, D.F. and Fourcroy, P. (1996). Ascorbate peroxidase activity, not the mRNA level, is enhanced in salt stressed *Raphanus sativus* plants. *Physiologia Plantarum.* **97:** 13-20.

Luwe, M.W.F., Takahama, U. and Heber, U. (1993). Role of ascorbate in detoxifying ozone in the apoplast of spinach (*Spinacia oleracea* L.) leaves. *Plant Physiol.* **101:** 969-976.

Macnicol, P.K. (1987). Homoglutathione and glutathione synthetases of legume seedlings: partial purification and substrate specificity. *Plant Sci.* **53:** 229-35.

Marrs, K. (1996). The functions and regulation of glutathione S-transferases in plants. *Annu. Rev. Plant Physiol. Plant Mol. Biol.* **47:** 127-158.

Masi, A., Ferretti, M., Passera, C. and Ghisi, R. (1995). Antioxidative responses in *Zea mays* seedlings grown under UV-B radiation. *Plant Physiol. Biochem.* **Special Issue:** 264-265.

Mattagajasingh, S.N. and Kar, M. (1989). Changes in the antioxidant system during greening of etiolated wheat leaves. *J. Plant Physiol.* **134:** 656-660.

May, M.J. and Leaver, C.J. (1993). Oxidative stimulation of glutathione synthesis in *Arabidopsis thaliana* suspension cultures. *Plant Physiol.* **103:** 621-627.

May, M.J., Vernoux, T., Leaver, C., Van Montagu, M. and Inzé, D. (1998). Glutathione homeostasis in plants: implications for environmental sensing and plant development. *J. Exp. Bot.* **49:** 649-667.

McKersie, B.D. and Leshem, Y.Y. (1994). Stress and stress coping in cultivated plants. pp. 256, Kluwer Publishers, Dordrecht, The Netherlands.

Meister, A. (1988). Glutathione metabolism and its selective modification. *J. Biol. Chem.* **263:** 17205-17208.

Mendoza-Cozatl, D.G. and Moreno-Sanchez, R. (2006). Control of glutathione and phytochelatin synthesis under cadmium stress. Pathways modeling for plants. *J. Theor. Biol.* **238:** 919-936.

Mittler, R. and Zilinskas, B.A. (1991). Purification and characterization of pea cytosolic ascorbate peroxidase. *Plant Physiology.* **97:** 962-968.

Mittler, R.S., Vanderauwera, M. and Van Breusegem, F. (2004). Reactive oxygen gene network of plant. *Trends Plant Sci.* **9:** 490-98.

Miyake, C., Michihata, F. and Asada, K. (1991). Scavenging of hydrogen peroxide in prokaryotic and eukaryotic algae: acquisition of ascorbate peroxidase during the evolution of cyanobacteria. *Plant and Cell Physiology.* **32:** 33-43.

Miyake, C., Cao, W.H. and Asada, K. (1993). Purification and molecular properties of the thylakoid-bound ascorbate peroxidase in spinach chloroplasts. *Plant Cell Physiol.* **34:** 881-889.

Moller, I.M. (2001). Plant mitochondria and oxidative stress. Electron transport, NADPH tumover and metabolism of reactive oxygen species. *Annu. Rev. Plant Physiol. Plant Mol. Biol.* **52:** 561-591.

Morita, S., Kaminaka, H., Masumura, T. and Tanaka, K. (1999). Induction of rice cytosolic ascorbate peroxidase mRNA by oxidative stress; the involvement of hydrogen peroxide in oxidative stress signalling. *Plant Cell Physiol.* **40:** 417-422.

Mozafar, A. and Oertli, J.J. (1993). Vitamin C, (ascorbic acid): Uptake and metabolism by soybean. *J. Plant Physiol.* **141:** 316-321.

Mueller, M.J. (2004). Archetype signals in plants. The phytoprostanes. *Curr. Opin. Plant Biol.* **7:** 441-448.

Mullineaux, P.M., Karpinski, S., Jiménez, A., Cleary, S.P., Robinson, C. and Creissen, G.P. (1998). Identification of cDNAs encoding plastid-targeted glutathione peroxidase. *Plant J.* **13:** 375-379.

Mutsuda, M., Ishikawa, T., Takeda, T. and Shigeoka, S. (1995). Subcellular localization and properties of L-galactono-ã-lactone dehydrogenase in spinach leaves. *Biosci. Biotech. Biochem.* **59:** 1983-1984.

Navari-Izzo, F., Meneguzzo, S., Loggini, B., Vazzana, C. and Sgherri, C.L.M. (1997). The role of the glutathione system during dehydration of *Boea hygroscopica. Physiol. Plant.* **99:** 23-30.

Navas, P., Villalba, J.M. and Córdoba, F. (1994). Ascorbate function at the plasma membrane. *Biochim. Biophys. Acta.* **1197:** 1-13.

Neubauer, C. and Yamamoto, H.Y. (1994). Membrane barriers and Mehler-peroxidase reaction limit the ascorbate available for violaxanthin de-epoxidase activity in intact chloroplasts. *Photosynth. Res.* **39:** 137-147.

Nieto-Sotelo, J. and Tuan-Hua, D.H. (1986). Effect of heat shock on the metabolism of glutathione in maize roots. *Plant Physiol.* **82:** 1031-1035.

Noctor, G. and Foyer, C.H. (1998). Ascorbate and glutathione: keeping active oxygen under control. *Annu. Rev. Plant Physiol., Plant Mol. Biol.* **49:** 249-279.

Orozco-Cardenas, M. and Ryan, C.A. (1999). Hydrogen peroxide is generated systemically in plant leaves by wounding and systemin via the octadecanoid pathway. *Proc. Natl. Acad. Sci. USA.* **96:** 6553-6557

Padh, H. (1990). Cellular functions of ascorbic acid. *Biochem. Cell Biol.* **68:** 1166-1173.

Panchuk, I.I., Zentgraf, U. and Volkov, R.A. (2005). Expression of the *Apx* gene family during leaf senescence of *Arabidopsis thaliana. Planta.* **222:** 926-932.

Paolacci, A.R., Badiani, M., D'Annibale, A., Fusari, A. and Matteucci, G. (1997). Antioxidants and photosynthesis in the leaves of *Triticum durum* Desf. seedlings acclimated to non-stressing high temperature. *J. Plant Physiol.* **150:** 381-387.

Pei, Z.M., Murata, Y., Benning, G., Thomine, S., Klusner, B., Allen, G.J., Grill, E. and Schroeder, J.I. (2000). Calcium channels activated by hydrogen peroxide mediate abscisic acid signaling in guard cells. *Nature.* **406:** 731-734.

Polle, A. and Rennenberg, H. (1992). Field studies on Norway spruce trees at high altitudes: II. Defence systems against oxidative stress in needles. *New Phytol.* **121:** 635-642.

Polle, A. (1997). Defense against photoxidative damage in plants. In: *Oxidative stress and the molecular biology of antioxidant defenses* (Ed. Scandalios, J.G.) pp. 623-666, Cold Spring Harbor Laboratory Press, New York, USA.

Polle, A. (2001). Dissecting the superoxide dismutase–ascorbate peroxidase–glutathione pathway in chloroplasts by metabolic modeling. Computer simulations as a step towards flux analysis. *Plant Physiol.* **126:** 445-462.

Rautenkranz, A.A.F., Li, L., Mächler F, Märtinoia, E., Oertli, J.J. (1994). Transport of ascorbic and dehydroascorbic acid across protoplast and vacuole membranes isolated from barley (*Hordeum vulgare* L. cv Gerbel) leaves. *Plant Physiol.* **106:** 187-193.

Rennenberg, H. Lamoureux, G.L. (1990). Physiological processes that modulate the concentration of glutathione in plant cells. In: *Sulfur nutrition and sulfur assimilation in higher plants; fundamental, environmental, and agricultural aspects.* (Eds. Rennenberg, H., Brunold, C., De Kok, L.J. and Stulen, I.) pp. 53-65, SPB Academic Publishing, The Hague, The Netherlands.

Sagisaka, S., Matsuda, Y., Okuda, T. and Ozeki, S. (1991). Comparative studies of the changes in enzymatic activities in hardy and less hardy cultivars of winter wheat in late fall and in winter under snow. *Soil Sci. Plant Nutr.* **37:** 543-550.

Saito, K. (1996). Formation of l-ascorbic acid and oxalic acid from d-glucosone in *Lemna minor*. *Phytochemistry*. **41:** 145-149.

Savouré, A., Thorin, D., Davey, M., Hua, X.J., Mauro, S., Van Montagu, M., Inzé, D. and Verbruggen, N. (1999). NaCl and $CuSO_4$ treatments trigger distinct oxidative defence mechanisms in *Nicotiana plumbaginifolia* L. *Plant Cell Environ.* **22:** 387-396.

Sen Gupta, A., Alscher, R.G. and McCume, D. (1991.) Response of photosynthesis and cellular antioxidants to ozone in *Populus* leaves. *Plant Physiol.* **96:** 650-655.

Secenji, M., Bebes, A., Hideg, É. and Györgyey, J. (2008). Transcriptional changes in ascorbate-glutathione cycle under drought conditions. *Acta Biol. Szeged.* **52:** 93-94.

Siendones, E., González-Reyes, J.A., Santos-Ocana, C., Navas, P. and Cordoba, F. (1999). Biosynthesis of ascorbic acid in kindney beans. L-Galactono-γ-lactone dehydrogenase is an intrinsic protein located at the mitochondrial inner membrane. *Plant Physiol.* **120:** 907-912.

Skotland, T. and Ljones, T. (1980). Direct spectrophotometric detection of ascorbate free radical formed by dopaminemonooxygenase and ascorbate reductase. *Biochim. Biophys. Acta.* **630:** 30-35.

Smirnoff, N. (1993). The role of active oxygen in the response of plants to water deficit and desiccation. *New Phytol.* **125:** 27-58.

Smirnoff, N. (1996). The function and metabolism of ascorbic acid in plants. *Ann. Bot.* **78:** 661-669.

Smirnoff, N. (2000). Ascorbate biosynthesis and function in photoprotection. *Philo. Trans. R. Soc. Biol. Sci.* **355:** 1455-1464.

Smith, J.J., Ververidis, P. and John, P. (1992). Characterization of the ethylene forming enzyme partially purified from melon. *Phytochemistry.* **31:** 1485-1494.

Sun, Q., Ye, Z.H., Wang, X.R. and Wong, M.H. (2007). Cadmium hyperaccumulation leads to an increase of glutathione rather than phytochelatins in the cadmium hyperaccumulator *Sedum alfredii*. *J. Plant Physiol.* **164:** 1489-1498.

Tausz, M., Grill, D., Jiménez, M.S. and Morales, D. (1997). A physiological comparison between environmental effects on Canary pine in Tenerife and Norway spruce in Austria. In: *Impacts of Global Change on Tree Physiology and Forest Ecosystems.* (Eds. Mohren, F., Kramer, K. and Sabaté, S.) pp. 103-110, Kluwer Publishers, Amsterdam, The Netherlands.

Tausz, M., Bytnerowicz, A., Arbaugh, M.J., Weidner, W. and Grill, D. (1999). Antioxidants and protective pigments of *Pinus ponderosa* needles at gradients of natural stresses and ozone in the San Bernardino Mountains in California. *Free Rad. Res.* **31:** 113-120.

Tausz, M., Wonisch, A., Peters, J., Jiménez, M.S., Morales, D. and Grill, D. (2001). Short-term changes in free-radical scavengers and chloroplast pigments in *Pinus canariensis* needles as affected by mild drought stress. *J. Plant Physiol.* **158**: 213-219.

Teixeira, F.K., Menezes-Benavente, L., Galvao, V.C., Margis, R. and Margis-Pinhero, M. (2006). Rice ascorbate peroxidase gene family encodes functionally diverse isoforms localized in different subcellular compartments. *Planta.* **224:** 300-324.

Tommasi, F., Paciolla, C. and Arrigoni, O. (1999). The ascorbate system in recalcitrant and orthodox seeds. *Physiol. Plant.* **105:** 193-198.

Torres, M.A. and Dangl, J.L. (2005). Functions of the respiratory burst oxidase in biotic interactions, abiotic stress and development. *Curr. Opin. Plant Biol.* **9:** 397-403.

Tuteja, N., Ahmed, P., Panda, B.B. and Tuteja, R. (2009). Genotoxic Stress in plants: Shedding light on DNA damage, repair and DNA repair helicases. *Mutat Res.Rev Mutat. Res.* **681:** 134-149.

Van Camp, W., Bowler, C., Villarroel, R., Tsang, E.W.T., Van Montagu, M. and Inzé, D. (1990). Characterization of iron superoxide dismutase cDNAs from plants obtained by genetic complementation in *Escherichia coli*. *Proc. Natl. Acad. Sci. USA.* **87:** 9903-9907.

Vanacker, H., Carter, T.L.W. and Foyer, C.H. (1998). Pathogen induced changes in the antioxidant status of the apoplast in barley leaves. *Plant Physiol.* **117:** 1103-1114.

Wang, S.Y. and Faust, M. (1992). Ascorbic acid oxidase activity in apple buds: relation to thidiazuron-induced lateral budbreak. *Hort. Sci.* **27:** 1102-1105.

Washko, P.W., Welch, R.W., Dhariwal, K.R., Wang, Y. and Levine, M. (1992). Ascorbic acid and dehydroascorbic acid analysis in biological samples. *Anal. Biochem.* **204:** 1-14.

Wheeler, G.L., Jones, M.A. and Smirnoff, N. (1998). The biosynthetic pathway of vitamin C in higher plants. *Nature.* **393:** 365-369.

Wildi, B. and Lütz, C. (1996). Antioxidant composition of selected high alpine plant species from different altitudes. *Plant Cell Environ.* **19:** 138-146.

Willekens, H., Chamnongpol, S., Davey, M., Schraudner, M., Langebartels, C., Van Montagu, M., Inzé, D. and Van Camp, W. (1997). Catalase is a sink for H_2O_2 and is indispensable for stress defence in C-3 plants. *EMBO J.* **16:** 4806-4816.

Wiseman, H. and Halliwell, B. (1996). Damage to DNA by reactive oxygen and nitrogen species: Role in inflammatory disease and progression to cancer. *Biochem. J.* **313:** 17-29.

Xiang, C., Werner, B.L., Christensen, E.M. and Olver, D.J. (2001). The biological functions of glutathione revisited in *Arabidopsis* transgenic plants with altered glutathione levels. *Plant Physiol.* **126:** 564-574.

5

Ascorbate-Glutathione and Plant Tolerance to Various Abiotic Stresses

Aryadeep Roychoudhury[1*] and Supratim Basu[2]

[1]*Post Graduate Department of Biotechnology, St. Xavier's College, 30, Mother Teresa Sarani, Park Street, Kolkata–700 016, West Bengal, India*

[2]*Department of Botany, Bose Institute, 93/1, A. P. C. Road, Kolkata–700 009, West Bengal, India*

[*]***Corresponding author:*** *aryadeep19@rediffmail.com*

ABSTRACT

Oxygen is essential for the existence of aerobic life, but toxic reactive oxygen species (ROS), including the superoxide anion radicals, hydroxyl radicals, hydrogen peroxide and singlet oxygen tend to increase in plants exposed to a variety of stressful conditions. The injuries associated with ROS, collectively referred to as oxidative stresses, are among the most pronounced damaging factors in plants. Usually oxidative stress occurs if the generation rates or titre of free radicals exceeds the natural capacity of their scavenging or degradation, creating an imbalance between ROS production and degradation, ultimately resulting in premature senescence, necrosis or apoptosis. The enzymatic sources of ROS include lipoxygenases (LOXs), peroxidases (POXs), NADPH oxidase and xanthine oxidase (XO). The plants have developed a number of antioxidant defence mechanisms, employing antioxidant enzymes including superoxide dismutase (SOD), POXs and catalases (CATs), as well as low molecular weight antioxidants of paramount importance like ascorbic acid (AA), reduced glutathione (GSH) and phenolic compounds. They are the major assimilate sinks present in many tissues at millimolar concentration and acting in synchrony. Since AA and GSH are considered the main information-rich redox cell buffers and redox sensors, we shall emphasize on the role of these two antioxidants in abiotic stress tolerance in several plant species. The AA, synthesized in mitochondria, and transported to the other cellular compartments, is an important water-soluble antioxidant-reductant, which can directly neutralize secondary products of ROS reaction. They also play role in regenerating tocopherols from tocopheroxyl radicals, thus providing membrane stabilization, protection of oxidized carotenoids and synthesis of zeaxanthin in the xanthophyll cycle. The non-protein, sulphur-containing tripeptide, glutathione, a low molecular weight thiol, is also a widely used marker of oxidative stress and a universal redox sensing component at the cellular level. It is essential for the detoxification of xenobiotics and being the precursor of heavy metal-binding phytochelatins, also sequesters heavy metals. Glutathione plays antioxidant roles in cellular compartments other than chlorophyll, like mitochondria, cytosol, peroxisomes and nuclei. The ascorbate-glutathione cycle (AGC), which utilizes GSH as an electron donor to regenerate ascorbate (ASC) from its oxidized form dehydroascorbate (DHA) is considered the main pathway of free radical removal in the chloroplast (stroma) by equilibrating the redox status. This cycle is catalysed by a set of four enzymes, ascorbate peroxidase (APX), monodehydroascorbate reductase (MDHAR), glutathione-dependent dehydroascorbate reductase (DHAR) and glutathione reductase (GR). The ASC and glutathione participate in a cyclic transfer of reducing equivalents, permitting the reduction of H_2O_2, using electron derived from NADPH. The progressive oxidation and degradation of glutathione and ASC pools eventually leads to senescence and plant death follows. The present review will focus on the functional aspects of the different components of the AGC in scavenging the ROS, with special emphasis on the antioxidative role of ASC and glutathione, in maintaining the redox homeostasis in several plant species during salinity, drought, oxygen scarcity (anoxia or flooding), strong light, photo-chilling, temperature extremes, high concentration of air pollutants such as ozone, UV-B or sulphur dioxide, redox-active herbicides and heavy metal phytotoxicity, thereby outlining the foundations of ASC glutathione-dependent signalling events during multifaceted oxidative stress reactions.

Keywords: Abiotic stress, ascorbate, glutathione, ascorbate-glutathione cycle, reative oxygen species, tolerance.

1. INTRODUCTION

Life in aerobic conditions requires the acquisition of efficient strategies allowing the cells to cope with the unavoidable presence of reactive oxygen species (ROS), the transient species generated as a general physiological response due to autoxidation, enzymatic oxidation, photosensitization, cellular energy transfer reactions and the reaction of metalloenzymes with oxygen, ozone (O_3), water, unsaturated fatty acids, and many other materials *in vivo*. Molecular oxygen is relatively non-reactive due to its electronic configuration. However, four-electron reduction of oxygen in the respiratory electron transport chain (ETC) is always accompanied with a partial one- to three-electron reduction, yielding the formation of ROS. The ROS are free radicals that are atoms or groups of atoms having at least one unpaired electron. This is a highly unstable configuration; so the radicals promptly react with other molecules to generate more free radicals. However, the term ROS not only includes free radicals (superoxide anion radicals and hydroxyl radicals), but also molecules such as H_2O_2, singlet oxygen and O_3. Both superoxide and hydroperoxyl radicals undergo spontaneous dismutation to produce H_2O_2. Although H_2O_2 is less reactive than superoxide, in the presence of reduced transition metals such as Fe^{2+} in the chelated form, the formation of potentially dangerous hydroxyl radicals can occur in the iron-catalysed Fenton reaction or Cu^{2+}-catalysed Haber-Weiss reaction (Blokhina et al., 2003). All these different forms of ROS are created to a large extent by a number of environmental or abiotic stresses such as salinity, drought, nutrient deprivation, UV-B irradiation, flooding or waterlogging stress, anoxia, strong light, temperature extremes, redox-active herbicides (like paraquat), phytotoxic heavy metals and air pollutants such as sulphur dioxide and O_3 systems (Tambussi et al., 2000; Vacca et al., 2004 a, 2004b). The ROS generated upon pathogen attack have been proposed to play two different roles: exacerbating the harmful oxidative effect of infection or participating in the defence response by being toxic to the invading pathogen, contributing to the programmed cell death during the hypersensitive response, driving the cell wall reinforcement processes as well as serving as signal molecules for the activation of local and systemic resistance (Grant and Loake 2000; Corpas et al., 2001). Thus, production of ROS by the Mehler reaction in chloroplasts, the glycolate oxidase reaction in peroxisomes and the electron transfer chain in mitochondria is enhanced by conditions limiting CO_2 fixation, such as drought, salt, heat and cold stress as well as the combination of these conditions with high light (Foyer and Noctor 2003; Mittler 2002; Moller 2001; Noctor et al., 2002a). Due to their metabolically active nature and short living period, ROS can act as signalling molecules and their damaging effects are usually restricted at the sites of their production (Mittler 2002).

2. SITE OF ROS PRODUCTION IN PLANTS AND OXIDATIVE STRESS INITIATION

In green tissues of plants and eukaryotic algae, ROS are largely produced in chloroplasts due to an abundance of O_2 produced by the primary photochemical reactions of photosynthesis (Asada and Takahashi 1987), whereas in non-green tissues or dark-growing plants, mitochondrial respiration and plasma membrane-linked electron transport systems represents the principal ROS source, such as in all other aerobic organisms (Foyer and Noctor 2003; Ishikawa and Shigeoka 2008) even under optimal conditions. Plants also produce significant amounts of ROS in cell walls, especially under adverse environmental conditions and in the microbodies, in particular, during photorespiration or germination of oilseeds (del Rio et al., 2003). In natural environments, the excessive ROS production, leading to an imbalance between ROS production and degradation, results in a toxic state called oxidative stress (Dat et al., 2000; Bolwell et al., 2002). This phenomenon being a common trait of all environmental stresses is regarded as the major cause of the oxidative damage found in cells under stressful conditions. If the generation rates or titre of free radicals exceeds the natural capacity of their scavenging or degradation under stress conditions, cells suffer oxidative stress, which can result in premature senescence, necrosis, or apoptosis (Foyer and Noctor 2000). The consequences of oxidative stress depend on tissue and/or species (i.e., their tolerance), on membrane properties, on endogenous antioxidant content and on the ability to induce the response in the antioxidant system. Oxidative injury to intracellular constituents follows through the propagation of ROS-fuelled chain reactions associated with the induction of a hypersensitive-like reaction, i.e., activation of programmed cell death pathways (Rao et al., 2000; Langebartels et al., 2002). The ROS generated in excess unbalance the cellular redox system in favour of oxidized forms (Halliwell and Gutteridge 1984). The main cellular components susceptible to damage by free radicals are lipids (peroxidation of unsaturated fatty acids in the membrane), proteins (denaturation), carbohydrates and nucleic acids. The ROS can react with membrane lipids to generate lipid peroxides that can initiate a series of reactions producing damaging reactive oxygen intermediates (Hernández-Jiménez et al., 2002; Puppo et al., 2005). These damaging free radicals and their products react with proteins, DNA and membrane lipids to cause the reduced photosynthesis, electrolyte leakage and accelerated senescence (Halliwell and Gutteridge 1984). Peroxidation of polyunsaturated fatty acids by a ROS attack can lead to chain breakage and shortening, which will increase membrane fluidity and permeability. Breakdown products of lipid peroxidation (LP), notably 4-hydroxy-2-nonenal (HNE), affect several mitochondrial processes. Proteins can be damaged by ROS either through direct chemical interaction or indirectly, involving end products of LP. A number of amino acids can be modified; for example, cysteine can be oxidized to cystine, and both proline and arginine are converted to glutamyl semialdehyde.

3. MECHANISM OF ROS PRODUCTION AND OXIDATIVE INJURY

The mechanisms for the generation of ROS in biological systems are represented by both enzymatic and non-enzymatic reactions. Among the enzymatic sources of ROS are LOXs (EC 1.13.11.12), POXs, NADPH oxidase (EC 1.6.99.6) and XO (EC 1.1.3.22). The XO is responsible for the initial activation of dioxygen, using xanthine, hypoxanthine or acetaldehyde as electron donors. The next enzymatic step generates H_2O_2 through the dismutation of the superoxide anion by the SOD (EC 1.15.1.1). Due to its relative stability, the level of H_2O_2 is regulated enzymatically by an array of CATs (EC 1.11.1.6) and POXs, localized in almost all compartments of the plant cell. The POXs, besides their main function in H_2O_2 elimination, can also catalyse superoxide radical and H_2O_2 formation by a complex reaction in which NADH is oxidized, using trace amounts of H_2O_2. The NAD radical, thus formed, reduces O_2 to superoxide, some of which dismutates to H_2O_2 and O_2. Thus, POX and CAT play an important role in the fine regulation of ROS concentration in the cell through activation and deactivation of H_2O_2. The LOX reaction is another possible source of ROS, catalysing the hydroperoxidation of polyunsaturated fatty acids (PUFA). The hydroperoxidative derivatives of PUFA undergo autocatalytic degradation initiating the chain reaction of LP. The LOX-mediated formation of singlet oxygen or superoxide has also been shown. The hydroxyl radicals and singlet oxygen can react with the methylene groups of PUFA forming conjugated dienes, lipid peroxy radicals and hydroperoxides. The lipid alkoxyl and peroxyl radical formed is highly reactive and is able to initiate and propagate further chain reactions. The formation of conjugated dienes occurs when free radicals attach the hydrogens of methylene groups separating double bonds and leading to a rearrangement of the bonds. The lipid hydroperoxides produced can undergo reductive cleavage by reduced metals such as Fe^{2+}. Apart from LOX, several apoplastic enzymes may also lead to ROS production under normal and stress conditions. Other oxidases, responsible for the two-electron transfer to dioxygen (amino acid oxidases and glucose oxidase) can also contribute to H_2O_2 accumulation. The amine oxidases catalyse the oxidation of biogenic amines to the corresponding aldehyde with the release of NH_3 and H_2O_2 (Blokhina et al., 2003). The non-enzymatic mechanism of ROS generation involves its formation as by-products in the ETC of chloroplasts, mitochondria and plasma membrane (cytochrome b-mediated electron transfer). It is now well evident that some of the electrons flowing through the respiratory chain may leak away from the primary route thereby directly reducing dioxygen to superoxide anion (Mittler 2002). Plant mitochondrial ETC, with its redox-active electron carriers, is considered the most probable candidate for the intracellular ROS formation. Mitochondria have been shown to produce ROS (superoxide anion and the succeeding H_2O_2) due to electron leakage at the ubiquinone site—the ubiquinone: cytochrome b region and at the matrix side of complex 1 (NADH dehydrogenase). The H_2O_2 generation by higher plant mitochondria and its regulation by uncoupling of ETC and oxidative phosphorylation have been demonstrated earlier. For example, chloroplasts generate H_2O_2 in the range of 150-250 pmoles mg^{-1} chlorophyll h^{-1} during photosynthesis (Asada 1994) through the Mehler reaction (Mehler 1951). H_2O_2 is produced through dismutation of two superoxide anions, or from the β-oxidation of fatty acids and peroxisomal photorespiration reactions (Kuzniak and Sklodowska 2005). Singlet oxygen is produced through the energetic activation of ground state oxygen. Under normal growth conditions, the production of $O_2^{\bullet -}$ is as low as 240 μM s^{-1} and the steady state level of H_2O_2 in chloroplasts is 0.5 μM. On the contrary, the production of $O_2^{\bullet -}$ and H_2O_2 touches the figures of 720 μM s^{-1} and 15 μM respectively under stress conditions like salinity (Dat et al., 2000).

4. DEFENCE STRATEGY: ENZYMATIC AND NON-ENZYMATIC ANTIOXIDANT SYSTEMS IN PLANTS

There is a growing body of evidence that either ROS or changes in the redox balance generated at the sites of primary action of stressors could activate a coordinated response in other compartments to ensure a successful defence strategy (Foyer and Noctor 2003). Thus, the tightly regulated balance between ROS production and removal at both the cellular and subcellular levels seems to be of primary importance for fulfilling the multiple functions of ROS and controlling redox homeostasis. The formation of ROS is prevented by an elaborate antioxidant system, which acts as a cooperative network employing a series of redox reactions. The term antioxidant can be considered to describe any compound capable of quenching ROS without itself undergoing conversion to a destructive radical. The determinants of the competence of the antioxidant system include several aspects like compartmentalization of ROS formation and antioxidant localization, synthesis and transport of antioxidants, the ability to induce the antioxidant defence and cooperation (and/or compensation) between different antioxidant systems (Blokhina et al., 2003). Subtle changes in the activity of antioxidant isozymes in different cell compartments are considered more important than overall enzymatic activities. The delicate balance between ROS production and scavenging that allows this duality in function to exist in plants is thought to be orchestrated by a large network of genes termed the 'ROS gene network', which includes more than 152 genes in *Arabidopsis*, tightly regulating ROS production and scavenging. The ROS signalling is therefore predominantly controlled by production and scavenging (Mittler et al., 2004). The common ROS-scavenging enzymes are the SOD, CAT, POXs and glutathione peroxidases

(Glu-POX, EC 1.11.1.9). The multiple (three) isoforms of the upstream enzyme SOD (mitochondrial Mn-, chloroplastic Fe- and cytosolic/chloroplastic/peroxisomal-Cu/Zn-SODs) catalyzes the dismutation or disproportionation of two superoxide radicals to H_2O_2 and oxygen, thus maintaining superoxide radicals to steady state levels (Mittova et al., 2004; del Rio et al., 2006). The CAT, that does not require an additional substrate, can decompose the H_2O_2 formed into H_2O and O_2 within the peroxisome. The scavenging of H_2O_2 in other cell compartments (cytosol, vacuole, cell wall and extracellular space) depends on distinct POXs such as guaiacol peroxidases (GPX, EC 1.11.1.7) and ascorbate peroxidases (APX, EC 1.11.1.11) that use a substrate for their activity (Inze and Van Montagu 1995; Noctor and Foyer 1998). While CAT is primarily localized in peroxisomes, isoforms of POXs and SODs are distributed throughout the cell and can be found in cytosolic, mitochondrial and chloroplastic compartments. The CAT functions through an intermediate CAT-H_2O_2 complex (Compound I) and produces water and dioxygen or can decay to the inactive Compound II. In the presence of an appropriate substrate, Compound I drives the peroxidatic reaction. Compound I is a much more effective antioxidant than H_2O_2 itself, thus the reaction of Compound I with another H_2O_2 molecule (CAT action) represents a one-electron transfer, which splits peroxide and produces another strong antioxidant, the hydroxyl radical. The hydroxyl radical is a very strong oxidant and can initiate radical chain reactions with organic molecules, particularly with PUFA in membrane lipids. Apart from ROS-scavenging enzymes, there are enzymes like glutathione S-transferases (GST, EC 2.5.1.18), phospholipid hydroperoxide Glu-POX and APX, detoxifying the LP products and other electrophilic xenobiotics. The GST catalyses the conjugation of electrophilic toxic molecules with GSH, which are then transported out of the cytosol by glutathione pumps. In addition, a whole array of enzymes is needed for the regeneration of the reduced forms of the antioxidants, like monodehydroascorbate reductase (MDHAR, EC 1.6.5.4), dehydroascorbate reductase (DHAR, EC 1.8.5.1) and glutathione reductase (GR, EC 1.6.4.2). The non-enzymatic antioxidant components involve a network of low molecular mass antioxidants with high reducing potentials like ascorbic acid (AA), cysteine, non-protein thiols, glutathione, tocopherols, β-carotene, polyamines, phenolic compounds, flavonoids, tannins and lignin precursors. Interactions between AA and glutathione or AA and phenolic compounds are well known (Blokhina et al., 2003). The AA and glutathione are considered the main information-rich redox cell buffers and redox sensors (Pastori et al., 2003; Foyer and Noctor 2005a, 2005b). A decrease in their redox status leads to a loss of cell redox homeostasis (Foyer and Noctor 2003). They together are coupled or work in concert with such enzymes as APX, DHAR and GR to scavenge the H_2O_2 generated in the chloroplasts and modulate the oxidation state of the cell (Creissen et al., 1994), the corresponding biochemical pathway is termed the ascorbate-glutathione cycle (AGC) or Halliwell–Asada pathway (Noctor and Foyer 1998; Chew et al., 2003; Liu et al., 2007). Figure 1 depicts the free radical formation and their scavenging in plant cells. In this review, we limit our discussion to AA and glutathione, together with the enzymatic components of AGC, with particular emphasis on the recent perspectives of their regulatory expression, activity and protective role in various common abiotic oxidative stress responses. We review the biochemical and molecular evidence showing their vital

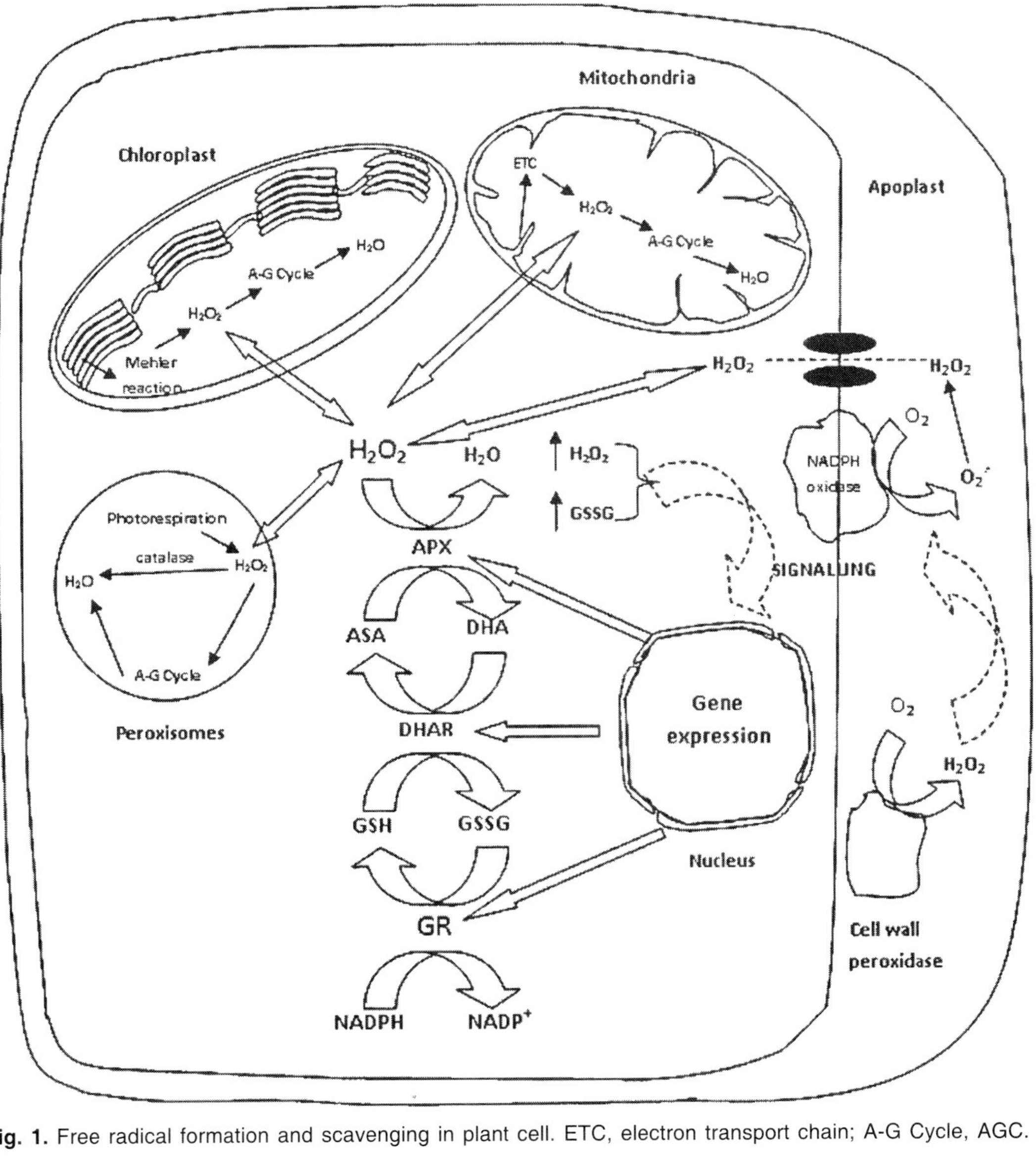

Fig. 1. Free radical formation and scavenging in plant cell. ETC, electron transport chain; A-G Cycle, AGC.

importance in conferring stress tolerance and plant survival under adverse or challenged environmental conditions.

5. ANTIOXIDANT ROLE OF ASCORBATE AND GLUTATHIONE

5.1 Ascorbate

Ascorbate (AA) is the most abundant, small, water-soluble (hydrophilic), powerful antioxidant, which acts as the primary substrate in the cyclic pathway for enzymatic detoxification of H_2O_2 (Anjum et al. 2008, 2011). In addition, it acts directly to neutralize superoxide or hydroxyl radicals (Padh 1990; Smirnoff and Wheeler 2000; Noctor and Foyer 1998; Asada 1999) and singlet oxygen and as a secondary antioxidant during non-enzymatic reductive recycling of the oxidized form (tocopheroxyl radical) of α-tocopherol, another lipophilic antioxidant molecule (Noctor and Foyer 1998). In plants, AA concentration ranges from 1 to 10 mmol g^{-1} FW (fresh weight) in photosynthetic tissues (Noctor 2006) to over 150 mmol g^{-1} FW (i.e., over 3% on a fresh weight) such as in the leaves of *Diospyros oleifera* and the fruit of *Terminalia ferdiandiana* (Li et al., 2009). Generally, the concentration of AA is higher in leaves than that in other plant parts and it is 5–10 times higher than that of glutathione (Smirnoff 2005). The H_2O_2-induced stomatal closure was reversed by exogenous application of AA, which is consistent with its role as a scavenger of H_2O_2 (Zhang et al., 2001; Schroeder et al., 2001a, 2001b; Chen et al., 2003). The stomatal guard cells with increased AA redox state exhibited reduced levels of H_2O_2 and were less responsive to abscisic acid (ABA) signalling, whereas those with a decreased AA redox state had an elevated level of H_2O_2. These observations suggested that the redox state of AA plays an important role in controlling H_2O_2-mediated stomatal closure (Chen and Gallie 2004). Consequently, plants with increased AA might be predicted to exhibit reduced responsiveness to ABA or H_2O_2 signalling. Within the chloroplasts, AA acts as cofactor of violaxanthin de-epoxidase, an enzyme that converts violaxanthin to zeaxanthin under excess light (the xanthophyll cycle), thus sustaining dissipation and non-photochemical quenching of excess excitation energy in photosystem II (PSII) (Demmig-Adams and Adams 1990; Eskling et al., 1997; Niyogi 1999). Due to its ability to donate electrons in a wide range of enzymatic and non-enzymatic reactions, AA constitutes the most important ROS-detoxifying compound. Under physiological conditions, AA exists mostly in the reduced form (90% of the ASC pool) in leaves and chloroplasts, though it occurs in cytosol, vacuoles, mitochondria and cell wall as well. In its intracellular concentration, AA can build up to millimolar range (e.g., 20 mM in cytosol and 20-300 mM in chloroplast stroma). About 4 to 10% of the leaf ASC pool resides in the apoplast (Noctor and Foyer 1998; Veljovic-Jovanovic et al., 2001), giving an ASC concentration in the low millimolar range. The redox state of the apoplastic ASC pool (which can be defined as % total ASC in the reduced form), i.e., 100 [ASC]/([MDHA]+[DHA]+[ASC]), may be a key regulator of apoplastic defence and redox-linked signalling. The redox potential of AA at pH 7.0 is 58 mV. The redox status of ASC is at least partially dependent on the size of the ASC pool. In an ASC-deficient mutant of *Arabidopsis* (*vtc1*), grown under non-stress conditions, the ASC/DHA ratio was nearly two-times lower than in the wild-type progenitor (Conklin et al., 1996).

5.2 AA-Mediated Gene Expression

The *Arabidopsis vtc1* (ASC-deficient) mutant, which was isolated via its sensitivity to O_3, showed increased sensitivity to other abiotic stresses such as freezing and UV-B irradiation (Conklin et al., 1996). Microarray analysis was used to identify genes that were differentially expressed in and also following feeding the *vtc1* leaf discs with ASC (Kiddle et al., 2003; Pastori et al., 2003). In addition to the dramatic changes in the abundance of transcripts encoding pathogenesis related (PR) proteins and similar proteins associated with plant defence responses against biotic stress, effects of ASC on the abundance of transcripts encoding chloroplast proteins were observed (Kiddle et al., 2003). Low ASC in *vtc1* led to the activation of a suite of genes that might be considered to provide a molecular signature of ASC deficiency in plants (Pastori et al., 2003), while high ASC not only led to the repression of these transcripts but resulted also in changes in other transcripts, at least in the short term (Kiddle et al., 2003; Pastori et al., 2003).

5.3 Glutathione

This is another major low-molecular weight thiol tripeptide in plant tissues involved in the cellular defence against the toxic action of xenobiotics, oxyradicals as well as the metal cations (Anjum et al., 2008, 2011). It has been found virtually in all cellular compartments in millimolar concentrations, viz., cytosol, chloroplasts, endoplasmic reticulum, vacuole and mitochondria, where it executes multiple functions (Maughan and Foyer 2006). The concentration of glutathione is the highest in the chloroplasts (1-4 mM). Glutathione is the main storage pool of non-protein sulphur. It is an important sulphur sink, a predominant transported or storage form of reduced sulphur, and a regulator of sulphur assimilation (Hell 1997; May et al., 1998). Moreover, glutathione also participates in calcium signalling in plants by stimulating calcium release into cytosol. It also plays an indirect role in protecting membranes by maintaining α-tocopherol and zeaxanthin in the reduced state. Glutathione protects proteins against denaturation caused by the oxidation of protein thiol groups under stress. It acts as a potent detoxifier of xenobiotics through conjugation reactions catalysed by GST (Marrs 1996). It can serve as a precursor or substrate of phytochelatins (PCs), which have an affinity for heavy metals and are transported as complexes into the vacuole, thus allowing the plants to have some level of resistance or tolerance to heavy metals in plants. The nucleophilic nature of the thiol group also is important in the formation of mercaptide bonds with metals and for reacting with selected electrophiles. This reactivity, along with the relative stability and high water solubility, makes glutathione an ideal biochemical to protect plants against stress, including oxidative stress, heavy metals and certain exogenous and endogenous organic chemicals. Glutathione exists with two different forms, i.e., the reduced form (γ-Glu-Cys-Gly, GSH) and the oxidized form (glutathione disulphide, GSSG), the occurrence or maintenance of GSH form is more predominating. The proportion of glutathione in the reduced form is normally greater than 0.9 under non-stress conditions (Noctor et al., 1998). The physiological functions of glutathione have been mainly attributed to its reduced form in plants. Thus, it is necessary that high proportion of GSH should be maintained in plants. The GSH participates in signalling processes by itself, and also as nitrosoglutathione (GSNO) after its reaction with nitric oxide (Noctor et al., 2002 a, 2002b; del Río et al., 2003; Díaz et al., 2003). While acting as a major antioxidant in plant cells, GSH is oxidized to GSSG, a reaction termed thiolation or protein glutathionylation. Reversible protein thiolation protects essential thiol groups on key proteins from irreversible inactivation during oxidative stress and also plays an important regulatory role in controlling metabolism, protein turnover and gene transcription. Together with GSSG, GSH maintains a redox couple in the cellular compartments for maintaining cellular homeostasis (Noctor and

Foyer 1998). The latter property is of great biological significance since it allows fine-tuning of the cellular redox environment under normal conditions and upon the onset of stress, and provides the basis of GSH stress signalling. The plants have evolved the mechanisms to maintain a high proportion of GSH. Indeed, the role for GSH in redox regulation of gene expression has been described earlier. The reduction of GSSG to GSH can be catalysed by GR with the accompanying oxidation of NADPH (Noctor and Foyer 1998; Asada 1999). The glutathione pool is maintained largely in the reduced state by GR, utilizing NADPH as a reductant. This reaction affords further protection against photoinhibition by forming $NADP^+$, the preferred electron acceptor for photosynthetic electron transport. Due to redox properties of GSH/GSSG pair and reduced SH group of GSH, it can participate in the regulation of cell cycle.

5.4 Role of GSH as Antioxidant

Functioning of GSH as antioxidant under oxidative stress has received enough attention during the last decade. A central nucleophilic cysteine residue in GSH is responsible for high reductive potential of GSH. This reactive cysteine residue enables it to keep thiol group-containing proteins in their native state during stress condition. It scavenges cytotoxic H_2O_2, and reacts non-enzymatically with other ROS like singlet oxygen, superoxide and hydroxyl radicals. The GSH also takes part in thioredoxin-related regulation of many enzymes of photosynthetic metabolism. It is essential that sufficient amounts of GSH and high GSH/GSSG ratios are present for glutathione to fulfil its roles in metabolism and defence. In the absence of GR, the GSH pool is lost (Kunert and Foyer 1993). In the chloroplasts, which are supposed to contain the most GSH within the cell, their levels were estimated by biochemical methods, and found to lie between 2 and 5 mM. Cytosolic concentrations were calculated to be between 0.1 and 0.2 mM (Rennenberg and Brunold 1994; Noctor et al., 1998; Foyer and Rennenberg 2000). The GSH has an oxidation reduction potential of –0.23 V that allows it to act as an effective electron acceptor and donor for numerous biological reactions (Xiang et al., 2001). It is also known to regulate expression of certain stress defence genes during environmental stresses (Ball et al., 2004; Wingate et al., 1988), whose products are involved in redox regulation and/or in the enhancement of stress tolerance. The targets might be transcription factors altering the expression of certain stress-related genes, or metabolic enzymes, which might allow for alterations in metabolic flux rates after post-translational modification. The GSH is also known to protect –SH groups of some enzymes and structural proteins against oxidation, either by acting as scavenger for oxidizing substances or by repairing the –SH groups through the GSH-disulphide exchange reaction (Mayer and Hell 2005). Further, GSH is an effective donor of reducing equivalents to ASC in the ROS scavenging system. In almost all of its metabolic functions, oxidation of GSH and its further reduction by GR are extremely efficient reactions to dissipate energy and to modulate the ATP/NADPH ratios, at times when CO_2 fixation is limited in plants under unfavourable growth conditions (Chen et al., 2004).

6. ASCORBATE-GLUTATHIONE CYCLE (AGC)

6.1 Components

This cycle (Fig. 2A) is very efficient in regenerating the reduced forms of ASC. The H_2O_2 is eliminated by AGC involving successive redox reactions of ASC, glutathione and NADPH, which are catalysed by four enzymes, namely, APX, MDHAR, DHAR and GR, acting in an orchestrated manner (Noctor and Foyer 1998). The APX, because of its presence in all cell compartments and its high affinity to H_2O_2, detoxifies H_2O_2 using ASC as electron donor and concomitantly oxidizing ASC to the monodehydroascorbate (MDHA) radicals, some of which disproportionates and spontaneously converted to ASC and dehydroascorbate (DHA) non-enzymatically. The ability of ASC to lose or donate electrons to produce MDHA is the basis of its biologically useful antioxidant capacity (Buettner and Schafer 2004). The MDHAR re-reduces the remainder of the MDHA radicals and regenerates ASC at the expense of nicotinamide adenine dinucleotide 2′-phosphate (NADPH). A previous kinetic study suggested that NADPH reduces the enzyme's FAD, and the reduced FAD donates electrons to MDHA radicals through two successive transfers of electrons. The DHAR catalyses the re-reduction of DHA to ASC, thus regenerating ASC using GSH as an electron donor, converting the latter to the disulphide, i.e., GSSG. The DHA undergoes easy irreversible hydrolysis to 2, 3-L-diketogulonate, which is very unstable and degrades further or converted to non-renewable oxalic acid and/or tartaric acid (Debolt et al., 2007). Therefore, rapid reduction of DHA to ASC by DHAR is important for maintenance of the total ASC level (ASC plus DHA). Thus, MDHAR and DHAR allow the plant to

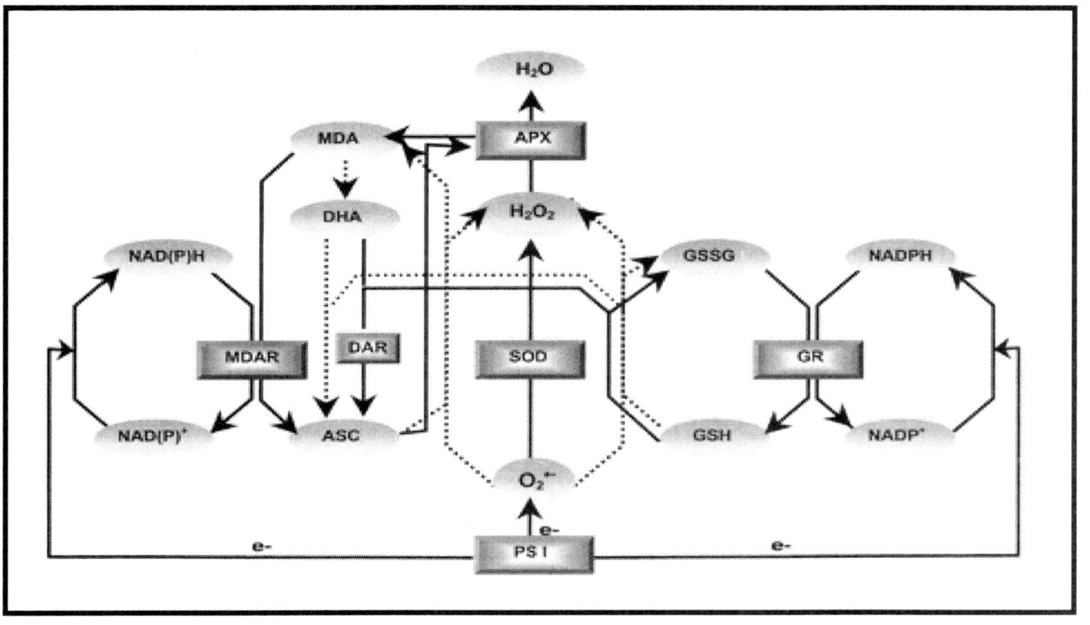

Fig. 2A. Schematic representation of the AGC (thick arrows depict enzymatic reactions, the broken arrows show non-enzymatic redox reactions).

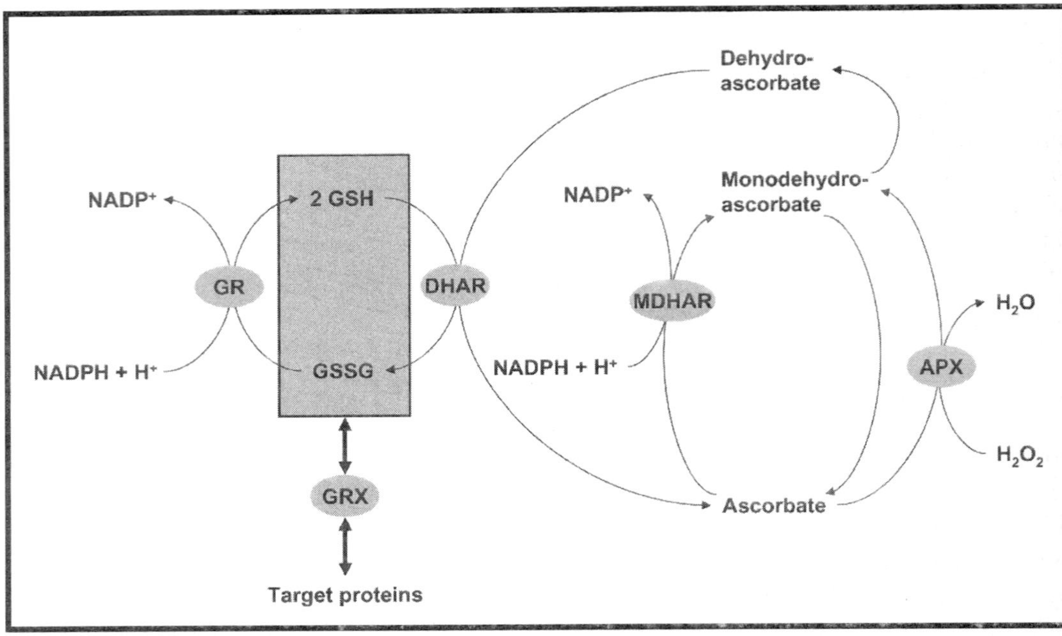

Fig. 2B. The AGC with different enzymes, GR: glutathione reductase; DHAR: dehydroascorbate reductase; MDHAR: monodehydroascorbate reductase; APX: ascorbate peroxidase; GRX: glutaredoxin. The spontaneous disproportionation of monodehydroascorbate has been shown. The ASC part of the cycle is not shown stoichiometrically.

recycle oxidized ASC thereby recapturing it before it is lost. The activities of both these ASC-reducing enzymes are crucial to regenerate ASC. The enzyme GR plays a key role in reducing GSSG and regenerating GSH at the expense of NADPH, thus allowing a high GSH/GSSG ratio to be maintained (Fig. 2B). This is considered to be the rate-limiting step of the cycle (Apel and Hirt 2004; Palma 2006; Liu et al., 2008 a, 2008b). The AGC comprises three interdependent redox couples: ASC/DHA, GSH/GSSG and NADPH/NADP. The specific interplay between ROS and the AGC constituents could generate compartment-specific changes in both the absolute concentrations of ROS and the antioxidant compounds, and in ASC and glutathione redox ratios. Under stress conditions, these redox signals could interfere with signalling networks, complementary to the antioxidant system, and regulate defence gene expression (Grant and Loake 2000; Vranova et al., 2002; Kiddle et al., 2003), thus coordinating the necessary re-adjustments in the redox-regulated plant defence to overcome the oxidative stress.

6.2 Cellular Site of Operation of AGC

The AGC enzymes are present in all cell organelles like cytosol, membrane, chloroplasts (stromal as well as thylakoid membrane-bound), plastids, mitochondria and microbodies (glyoxysome and peroxisomes) (Yoshimura et al., 2000; Mittova et al., 2002; Shigeoka et al., 2002; Madhusudhan et al., 2003). It is apparent that APX has numerous isoenzymes in various plant cell compartments and that it plays a central role in scavenging H_2O_2 in plants and eukaryotic algae. The APX also helps to dissipate excess photon energy through the water-to-water cycle in chloroplasts. In addition, peroxiredoxins were recently identified as members of the antioxidant defence system of chloroplast (Dietz et al., 2002). In plant cells, AGC has been shown to operate in cytosol and all organelles in which ROS detoxification is needed (Edjolo et al., 2001). This effective antioxidant defence system pathway prevents excessive build up of ROS under normal conditions. However, when a plant is stressed, the production of ROS can exceed the capacity of the scavenging systems, resulting in oxidative damage. Thus, the ability of a plant to improve its active-oxygen-scavenging capacity may be a key element in stress tolerance.

7. ROLE IN ABIOTIC STRESS

Abiotic stresses, in general, elicit an oxidative response via increases in ROS in plant cells. These include stresses such as drought stress, high salinity, temperature extremes, herbicide challenge, wounding, pollutants like SO_2 fumigation and O_3 or UV-B exposure, heavy metal toxicity, etc. (Inzé and Van Montagu 1995). These stresses are serious threats to agriculture, as they cause morphological, molecular and biochemical changes which affect plant growth and productivity. In fact, different abiotic stresses are often interconnected and cause similar cellular damages. The plant responses to these stresses are complex and multifold. The phenomenon of "cross tolerance" has emphasized the unity of diverse stress conditions and underlined the common feature of enhanced ROS production. To prevent oxidative damage, the antioxidant response can be crucial in plants. Recent years have witnessed a plethora of reports, correlating increases in one or more of the antioxidants/antioxidant enzymes with either stress conditions or ameliorated stress resistance. Whether improved stress tolerance is generated in the plants probably depends on a multitude of factors, including nature of the stress imposed, isoenzyme overexpressed, intracellular targeting of the antioxidants, strength of overexpression of the antioxidative enzymes, leaf age and growth conditions. The degrees to which the activities of AA and glutathione or the AGC-antioxidative enzymes are increased or decreased as a result of stress imposition is extremely variable and, in many cases, relatively minor. We present in this review a comprehensive and exhaustive overview of the importance of ASC-glutathione metabolites and AGC enzymes in abiotic stress response, taking each individual stress one by one, to see if any general consensus emerges from these studies and how valid it is to interpret stress levels through antioxidative responses of ASC and glutathione.

7.1 Salinity Stress

Soil salinity is a major abiotic stress factor affecting agricultural productivity. Increased salinization of arable land is expected to have devastating global effects, resulting in 30% land loss within the next 25 years, and up to 50% by the year 2050 (Wang et al., 2003). Salinity stress results in osmotic inhibition and ionic toxicity, which can affect the physiological and biochemical functions of the plant cell (Chinnusamy et al., 2005; Sumithra et al., 2006). Hence, increasing the yield of crop plants in salinized regions is essential for feeding the world. Several strategies have been evolved to improve salinity stress tolerance. Adaptation of the plants to high salinity involves osmotic adjustment and the compartmentation of toxic ions and tolerating oxidative stress. The correlation between antioxidant capacity and salt tolerance is well known (Hernandez et al., 2000; Sreenivasulu et al., 2000; Shalata et al., 2001; Sudhakar et al., 2001; Khan et al., 2006; Roychoudhury et al., 2008). Both ASC and glutathione, along with the enzymes of AGC, have been reported to regulate salt tolerance. The role of ASC in salt tolerance was established through experiments in ASC-deficient *Arabidopsis* mutant, *vtc-1*, having only 30% ASC contents of the wild type, which showed increased sensitivity to salt stress, accompanied with decrease in CO_2 assimilatory capacity, impaired PSII function, dramatic increase in H_2O_2 content and lowered MDHAR and DHAR activities, leading to lower intrinsic ASC and impaired AGC under salt stress, but higher induction in the reduced form of ASC, which resulted in enhanced ROS contents (Huang et al., 2005). The electron spin resonance (ESR) method directly showed that most (40-50%) of the superoxide radical scavenging activity in *Arabidopsis* was due to ASC (Nagata et al., 2003). In general, the positive effects of ASC in mitigating the adverse effects of salt stress have been ascribed to activation of some of the enzymatic reactions (Kefeli 1981) or attributed to the stabilization and protection of photosynthetic pigments and the photosynthetic apparatus from oxidative damages (Neubauer and Yamamoto 1992). Transgenic tobacco plants expressing the ascorbate oxidase (AAO, EC 1.10.3.3) gene in antisense orientation, so as to suppress the catalytic oxidation of ASC to MDHA, showed higher percentage germination, photosynthetic activity and seed yields with lower H_2O_2 levels under high salinity. On the contrary, AAO in sense orientation showed salt sensitivity with decreased redox state of symplastic and apoplastic ASC and higher H_2O_2 in the symplastic and apoplastic spaces. Thus, suppressed expression of apoplastic AAO or conservation of AA pool under salt stress conditions provided an index to salt tolerance (Yamamoto et al., 2005). The salt tolerant transgenic tobacco, overexpressing the group II *lea* gene *Rab16A* from rice, also showed enhanced AA level or its induction following salt treatment, compared to the wild type seedlings. The AAO activity in the transgenics remained unchanged or decreased following salt stress, whereas it increased in wild type plants (Roychoudhury et al., 2007), again clearly pinpointing to the importance of maintaining adequate AA pool within the cell.

Consistent findings reported on the beneficial effects of the exogenous application of AA in mitigating partially the adverse effects of salt stress on growth like cell division and cell enlargement (Mozafar and Oertli 1992). In chick pea, the significant synergistic effect between salt stress (40 mM) and the exogenous supply of AA (4 mM) improved the fresh and dry matter gain in roots, shoots and leaves, increased the contents of chlorophyll *a* and chlorophyll stability index (CSI %) in leaves and helped in the induced stability of protein synthesis for the sake of salt resistance (Beltagi 2008). The addition of AA as an antioxidant compound in the stressing medium increased the percentage of *Vicia faba* seedlings to survive the toxic effects of salinity by protecting the protein turnover machinery against stress damage and by upregulating stress protective proteins. It also partially alleviated the inhibitory effects of NaCl or mannitol on RNA and DNA contents (Younis et al., 2008, 2009; Hasaneen et al., 2008). Golan-Goldhirsh et al. (1995)

indicated that soybean plants treated with AA showed increase in photosynthetic processes. Singh et al. (2001) detected that foliar application of AA increased the content of macro nutrients (N, P and K) in *Cassia angustifolia*. Afzal et al. (2006) showed that hormonal priming with 50 ppm AA increased the ability of wheat to grow successfully under saline conditions. Foliar spray with exogenous AA in salt-treated (150 mM NaCl) wheat seedlings protected the photosynthetic machinery from salt injuries (by improving leaf "chlorophyll *a*" content) but could not alleviate the stress-induced reduction in seedling growth. Moreover, AA application enhanced the toxic Na^+ accumulation in the leaves of salt-stressed plants (Khan et al., 2006). The salt-stress-induced ultrastructural damages in sorghum seedlings were alleviated by pretreatment (pre-soaking or pre-soaking plus spraying) with AA at 100 ppm (Arafa et al., 2007). The salt-treated alfalfa (*Medicago sativa*) seedlings, co-treated with AA (1-2 mM) showed salt tolerance through improved seed germination or seedling survival, increased number of new roots and leaves through cell division and differentiation of meristem cells and also increased activity of acid phosphates, chlorophyll content and dry mass, with decreased sodium accumulation or balanced Na^+/K^+ levels (Arab and Ehsanpour 2006). In tomato seedlings, the exogenous supply of AA increased tissue levels (Arrigoni et al., 1997). Shalata and Neumann (2001) found that AA (0.5 mM), applied through the rooting medium, prior to and during salt treatment, counteracted the salt induced reduction in growth, partially inhibited progressive increase in LP and facilitated the subsequent recovery and long-term survival of wilted tomato seedlings; this was specifically related to its antioxidant activity, rather than its possible utility as an organic substrate for respiratory energy metabolism. The addition of AA increased the leakage of essential electrolytes following peroxidative damage to plasma membrane (Blokhina et al., 2003; Bourgeois-Chillou and Gurrier 1992). In case of wheat, exogenous application of AA through rooting medium could alleviate or counteract the adverse effects of salt stress. This was associated with enhanced endogenous AA level and CAT activity, increased stomatal conductance and higher photosynthetic capacity, and accumulation of K^+ and Ca^{2+} in the leaves; though this response was cultivar specific (Athar et al., 2008). The endogenous level of AA can be increased by exogenous application of AA through the rooting medium (Chen and Gallie 2004, 2005), as foliar spray or as seed priming. However, the effectiveness of exogenously applied AA depends on the mode of application. Though foliar application of AA had positive effects on various growth parameters, enabling to overcome destructive effect of salinity (El-Aziz et al., 2006), it was found to be less effective in improving the growth of salt stressed plants, as compared to AA application in the rooting medium.

Enhanced activities of antioxidant enzymes have been reported in salt-tolerant cultivars of pea (Hernandez et al., 2000) and foxtail millet (Sreenivasulu et al., 2000) under salt stress, whereas this increase was not observed in the sensitive cultivars. The significant increase in the activities of SOD, CAT, APX, POX and GR in the NaCl-stressed barley root was highly correlated with the temporal regulation of the constitutive isoforms as well as the induction of new isoforms (Kim et al., 2005). The antioxidant systems SOD, GSH, GR, APX and carotenoids were considered selection criteria for salt tolerance in *Sorghum*. Particularly the increase in GR and APX activity were highly pronounced in most tolerant genotypes at 50 mM NaCl than in sensitive genotypes. The increase in enzyme activity might be due to increasing the synthesis of the enzyme or an increased activation of constitutive enzyme pool (Hefny and Abdel-Kader 2009). de Azevedo Neto et al. (2006) reported that in the leaves of salt-stressed maize plants, the SOD, APX, GPX and GR activities increased with time; this increase was more pronounced in the salt-tolerant than in the salt-sensitive genotype. In salt-stressed roots of the salt-tolerant genotype, SOD and CAT activities decreased and APX, GPX and GR activities remained unchanged. In roots of the salt-sensitive genotype, salinity reduced the activity of all studied enzymes indicating a decreased GSH turnover rate and a less active AGC in roots. In French bean (*Phaseolus vulgaris*), whereas AA contents doubled during high temperature and salt stress, GSH content showed a twofold increase during salt stress and a marginal but significant increase during high temperature stress (Nagesh Babu and Devaraj Corr 2008). The chickpea (*Cicer arietinum* L. cv. Gokce) seedlings, subjected to salt stress regimes showed significant increase in APX and GR activities in the leaves (Kukreja et al., 2005; Eyidogan and Oz 2007). The enhanced activities of the tested enzymes like GST, Glu-POX, total SOD and Cu/Zn-SOD in NaCl treated-*Lycopersicon esculentum* Mill. (cv Perkoz) roots indicated their involvement in early and late defence systems against ROS under salinity stress (Gapinska et al., 2008). The higher ratio of SOD to APX activity in salt-tolerant *Lycopersicon pennellii* was correlated to the inherently better protection from salt and oxidative stress (Mittova et al., 2000). Increased GR activities have also been shown by Molina et al. (2002) and Mittova et al. (2003). In case of *Vigna radiata*, Sumithra et al. (2006) have shown that the Pusa Bold proved to be a superior cultivar for salinity tolerance than CO4 because of higher ROS-scavenging enzyme (MDHAR) activity and GSH concentration, indicating more ASC regeneration in the former, whereas CO4 leaves showed higher GSSG concentration. The APX, especially cytosolic APX (cAPX), is thought to play an essential role in protecting plants from such stress (Shigeoka et al., 2002). Transgenic *Arabidopsis* overexpressing two rice cytosolic APX (*OsAPXa* or *OsAPXb*) exhibited increased tolerance to salt stress, with higher tolerance conferred by *OsAPXb*. The overproduction of OsAPXb enhanced and maintained APX activity to a much higher degree than OsAPXa in transgenics during treatment with different concentrations of NaCl, enhanced the ROS scavenging system and protected plants from salt stress by equilibrating H_2O_2 metabolism (Lu et al., 2007). A similar improvement in salt stress tolerance was also observed in transgenic tomato overexpressing pea cAPX and transgenic tobacco overexpressing the *Arabidopsis* cAPX gene (Badawi et al., 2004; Wang et al., 2005). It has also been reported that, under normal growth conditions, very low expression of OsAPx2 (OsAPXb), which encodes a cytosolic isoform in rice, could be highly induced by NaCl treatment, while the accumulation of the rice cytosolic OsAPx2 transcript was more obvious than that of OsAPx1 (OsAPXa) in the presence of salt (Teixeira et al., 2006). A threefold increase in cysteine and glutathione was observed in wild-type plants of *Brassica napus* L. canola, but not in the salt-tolerant transgenic, overexpressing a vacuolar Na^+/H^+ antiporter (Ruiz and Blumwald 2002). This could be a possible protective mechanism against salt-induced oxidative damage. Different antioxidative capacities, both in the apoplast and in the symplast of pea leaves, contributed to a better protection against salt stress in relatively salt-tolerant cultivars. Salt-induced oxidative damage led to necrotic lesions in the minor veins of pea leaves, as oxidative stress was higher in the apoplasts (Hernandez et al., 2001). The GSH/GSSG ratio declined under salt stress, and there was no GR activity in the apoplasts; the sensitive cultivar Lincoln was more affected than the tolerant cv. Puget with reduction in growth response. Continuous exposure to salt stress in rice seedlings made them more tolerant when the GSH/GSSG levels returned to normal values after an initial decline (Fadzilla et al., 1997). Under short-term NaCl stress, the salt-tolerant rice Pokkali showed higher activity of CAT and enhanced levels of antioxidants like ASC and GSH than the sensitive cultivar Pusa Basmati 1 (Vaidyanathan et al., 2003). The results of Khan and Panda (2008) with two rice varieties Lunishree and Begunbitchi showed decreased AA content under salt stress except in lower concentrations of NaCl in Lunishree, allowing better antioxidant protection as reported for other plants. The increase in ASC levels in lower NaCl concentrations in Lunishree depended both on the rate of its synthesis and regeneration. Increased glutathione content in the stressed and recovered roots of Lunishree, whereas decreased content in Begunbitchi reflected, at least partially, the increased demand of glutathione in ROS detoxification in Lunishree compared to Begunbitchi. The GR activity showed a greater decline in Begunbitchi than in Lunishree under NaCl stress, while constitutive levels of GR were higher in Lunishree than in Begunbitchi. In *Calendula officinalis* (Chaparzadeh et al., 2004), a decrease in total glutathione and an increase in total ASC (ASC + DHA), accompanied with enhanced GR and APX activities, were observed in leaves under high salinity stress. The decrease in DHAR and MDHAR activities suggested that other mechanisms played a major role in the regeneration of reduced ASC. The higher ASC content observed at high salinity could suggest that ASC synthesis was stimulated or ASC catabolism was inhibited. The low ASC/DHA ratio in leaves at high salinity might be an indication of APX participation in ROS scavenging. A proper increase of ASC, during H_2O_2 increase in conditions of high salinity, may be important for maintaining APX activity; APX being inactivated when ASC concentration falls down. Among the 126 salinity-tolerant cDNAs identified and isolated by subtractive hybridization from the roots of a mangrove plant *Bruguiera cylindrica*, one was a putative GST (Wong et al., 2007). The Glu-POX is also a major enzyme, which can reduce alkyl and lipid hydroperoxides to protect the cells from LP. Transgenic tobacco plants overexpressing both GST and GPX displayed improved seed germination and seedling growth under stress (Roxas et al., 1997). In addition to higher activities of GST/GPX, the transgenic seedlings also showed higher levels of glutathione and ASC than wild-type plants (Roxas et al., 2000). The glyoxalase enzymes are important for the GSH-based detoxification of methylglyoxal, and their overexpession led to further enhanced levels of GSH. Under salinity stress, GSH levels and the GSH/GSSG ratio were higher in the transgenic plants overexpressing glyoxalase enzymes than wild-type plants (Singla-Pareek et al., 2003; Yadav et al., 2005). Several authors have suggested that the salt tolerance character is related to increased GR activity (Hernandez et al., 2000; Sudhakar et al., 2001; Meloni et al., 2003). Broadbent et al. (1995) found that transgenic tobacco plants overexpressing GR had both high levels of GSH and increased tolerance to oxidative stress. Increase in GR activity during salinity stress has been reported in several plants such as lentils (Bandeoglu et al., 2004), cantaloupe (Fahmy et al., 1998), citrus (Gueta-Dahan et al., 1997), soybean (Comba et al., 1998), rice (Demiral and Turkan 2005; Dionisio-Sese and Tobita 1998; Lin and Kao 2000a; Tsai et al., 2005), *Arabidopsis thaliana* (Huang et al., 2005), *Setaria italica* (Sreenivasulu et al., 2000), *Helianthus annuus* (Davenport et al., 2003) and a tolerant cultivar of wheat (Sairam et al., 2004). A NaCl-tolerant cell line of cotton exhibited significantly higher activities of GR (287%) and GST (500%) than the control plants (Gossett et al., 1996). Buthionine sulphoximine (BSO), an inhibitor of GSH, reduced growth of the control lines by almost 94%, which was less in the NaCl-tolerant cell line; exogenous GSH restored growth in both the cell lines. The DHAR activity is also increased in response to various environmental stresses that promote ROS generation (Urano et al., 2000; Ushimaru et al., 1992), suggesting that it is important for protection against such stresses. The plants with high DHAR activity were rich in ASC, suggesting that DHAR function is important for maintenance of the cellular ASC pool. Transgenic tobacco plants, overexpressing the wheat DHAR gene, showed enhanced ASC content and protection against ozone (Chen and Gallie 2005; Chen et al., 2003). Transgenic tobacco plants expressing the human DHAR gene showed increased tolerance to oxidative, low temperature and salt stresses (Kwon et al., 2003). Similarly, transgenic *Arabidopsis thaliana* expressing DHAR1 gene (encoding glutathione-dependent DHAR, from rice bran) showed dramatic improvement of salt-stress tolerance, despite slight increases in DHAR activity and total ASC (Ushimaru et al., 2006).

7.2 Hormonal Control by ABA

During environmental cues, ABA levels are increased in plants which are used for signal transduction pathways specifically for scavenging the excess ROS produced during oxidative stress (Anderson et al., 1994; Ansel et al., 2006; Bueno et al., 1998; Gong et al., 1998; Jiang and Zhang 2001, 2002; Roychoudhury et al., 2009). The pathway of ASC-mediated induction of PR genes has been suggested to involve both enhanced ABA (Pastori et al., 2003) and enhanced salicylic acid. It has been observed that GSH levels increased on treatment with ABA in maize (Jiang and Zhang 2001). However, it is not clear whether the increased GSH levels are in direct response to ABA or the result of enhanced oxidative stress (Guan et al., 2000). Increase in GR activity has been reported in maize seedlings and rice roots with increased ABA treatment (Jiang and Zhang 2001; 2002, 2003; Tsai and Kao 2004). H_2O_2 is known to be involved in ABA-induced GR activity in plant tissues (Anderson et al., 1994; Bueno et al., 1998; Gong et al., 1998; Jiang and Zhang 2001, 2002; Tsai and Kao 2004). Kaminaka et al. (1998) have shown that cytosolic GR in rice was modulated by two ABA-responsive elements, and the expression of rice cytosolic GR gene was regulated by ABA-mediated signal transduction pathway. Most recently, Ansel et al. (2006) stated that the ABA-mediated signal transduction pathway is involved in the expression of GR genes during drought and desiccation. However, in pea, there are no such ABA-responsive elements for GR cDNAs. In fact, the role of phytohormones in the regulation of GSH synthesis needs to be investigated in further details (Kopriva and Rennenberg 2004).

7.3 Drought Stress

During drought stress, the stomata are closed and the availability of the CO_2 for carbon assimilation is decreased, which in turn results in the non-availability of electron acceptor (NADP), leading to the generation of free radicals (Asada 2000; Reddy et al., 2004; Sairam and Saxena 2000; Smirnoff 1993). The alteration of antioxidant metabolism or changes in the activities of antioxidant enzyme under drought stress depends on plant species, cultivar and stress intensity or duration. The role of ASC in mitigating oxidative damage under drought stress has been shown in Okra (Amin et al., 2009). In case of cowpea, the concentration of AA was shown to increase in severe drought stress. The leaves of the tolerant variety Kanakamony showed higher AA levels than the susceptible variety Pusakomal (Nair et al., 2008). However, in wheat seedlings, no variety-specific differences in the response of low molecular antioxidants to drought were found. The levels of both ASC and thiols strongly decreased during drought and were restored during recovery from water deficit stress. The ratio between reduced and oxidised ASC was more or less conserved. These results strongly suggested participation of the low-molecular antioxidative compounds in the defence against ROS under severe drought and rather good functionality of the AGC (Simova-Stoilova et al., 2008). ASC may exert effects on water stress response by regulating stomatal opening in at least two ways. First, it is a cofactor in the synthesis of ABA, the phytohormone that regulates stomatal pore size and hence entry of CO_2 and loss of water from the leaf. Stomatal pores close in response to water stress through the action of ABA, which causes an increase in cytosolic calcium concentrations via H_2O_2-activated channels and release from the vacuole and other intracellular stores (Price et al., 1994; Kohler and Blatt 2002). The ASC-dependent dioxygenases are involved in the pathway of ABA biosynthesis. In particular, ASC is required for activity of 9-*cis*-epoxycarotenoid dioxygenase (NCED), an enzyme catalysing the formation of xanthoxin, a precursor of ABA. It is noteworthy that NCED expression is modulated by ASC, such that transcripts are increased when ASC is low and decreased when ASC is high (Pastori et al., 2003). The second interaction between ASC and stomata involves its role in regulating the redox state of the guard cell apoplast. H_2O_2 interacts strongly with ABA signalling. In particular, H_2O_2 is thought to fulfil a signalling role in guard cells, activating plasma membrane-localized anion channels, leading to guard cell depolarization, K^+ efflux and loss of turgor and volume, concluding in stomatal closure (Schroeder et al., 2001a, b). However, H_2O_2-induced stomatal closure was reversed by the application of exogenous ASC, presumably reflecting H_2O_2-scavenging (Zhang et al., 2001). Moreover, high constitutive expression of DHAR in transgenic plants increased the amount of ASC relative to DHA in leaves and guard cells (Chen et al., 2003) and significantly affected guard cell signalling and stomatal movement. The leaves of the DHAR-overexpressors contained less H_2O_2 in the guard cells and had a higher percentage of open stomata and increased stomatal conductance (Chen and Gallie 2004). These results suggested that plants with higher cellular ASC are better able to maintain open stomata and hence photosynthesis, at least under optimal conditions of water supply. The pH of the apoplast is about 5.5–6.3 in well-watered plants as opposed to about 7.2 in drought-stressed plants, favouring a preferential accumulation of ABA and AA in stomata under drought stress. The above evidence suggests that the distribution of ASC between the cytoplasm and the apoplast could also be influential in the regulation of stomatal opening. However, high apoplastic ASC concentrations would act antagonistically to ABA by limiting the extent of ABA-induced H_2O_2 accumulation. It is likely that the apoplastic ASC pool could be important in regulating stomatal opening by contributing to POX-catalysed detoxification of H_2O_2.

Dalmia and Savhney (2004) observed an involvement of antioxidant metabolites in ROS detoxification under drought with increased pools of ASC and glutathione at the beginning of the water stress and diminution when the stress became more severe. Lunde et al. (2006) reported that expression of two MDHAR genes dramatically increased under osmotic stress in moss (*Physcomitrella patens* Hedw.). In *Boea hygroscopica* (a dessication-tolerant plant species), the constitutive GSH increased by almost 50 times (Sgherri et al., 1994), protecting the plant through protection of sulfhydryl groups of proteins (Navari-Izzo et al., 1997). High levels of GSH during drought stress are achieved through high activity of GR. An increase in GR activity during stomatal closure in response to water stress has been reported (Sairam et al., 1997/98; Lascano et al., 2001; Ansel et al., 2006; Romero-Puertas et al., 2006). Increased GR activity during drought was reported in various plant species, including barley (Smirnoff and Colombe 1988), maize (Jiang and Zhang 2002; Pastori and Trippi 1992), tobacco (Van Rensburg and Kruger 1994), wheat (Chen et al., 2004; Sairam et al., 1997/8; Selote and Chopra 2006), rice (Lin and Kao 2000b; Selote and Chopra 2004; Sharma and Dubey 2005; Srivalli et al., 2003), *Tortula ruralis* (Dhindsa 1991) and pea (Gogorcena et al., 1995). Contour-Ansel et al. (2006) showed that drought stress upregulated leaf cytosolic GR gene, directly related to the intensity of the stress in both resistant and susceptible cowpeas (*Vigna unguiculata* L.). In maize, the changes in GSH content and GR activity during drought stress, as observed by Kocsy et al. (2004), are supported by the observations of Bartoli et al. (1999) and Loggini et al. (1999). In banana plantlets, the APX activity was accelerated during drought stress in the 'Berangan' variety that has more tolerance to water deficit stress, but was unaffected in 'Mas' that has less tolerance. Therefore, higher APX activity was associated with greater protection against drought-induced oxidative injury (Chai et al., 2005). In olive trees, the activities of SOD, APX, CAT and POX increased in relation to the severity of drought stress in both leaves and roots, and in particular, a marked increase in APX activity was found in leaves of plants subjected to severe drought stress (Sofo et al., 2005a, b). In cut rose (*Rosa hybrida* L.) cv. Samantha, water deficit caused inhibition of the flower opening process and a decrease in APX activity, which was prevented by AA. The AA, along with its relevant artificial substrates, cooperatively exerted powerful functions of enhancing APX activity, further improving the tolerance to water deficit (Jin et al., 2006). Selote and Chopra (2004) also reported drought-induced enhancement in APX activity in the water stress-tolerant upland rice genotype N22 compared to N118, a susceptible genotype. N22 was endowed with four isozymes of APX whereas N118 showed only one isozyme. In addition, all the APX isozymes in N22 showed enhancement under stress, whereas the N118 isoenzyme declined in water-stressed panicles. The expression patterns of genes encoding antioxidant enzymes are complex and may not be consistent with changes in protein expression and enzyme activity under drought stress. For instance, the transcript abundance of cytosolic APX and cytosolic Cu/Zn-SOD largely increased with increasing drought stress severity, but relatively small increases in APX protein and activity were observed in the leaves of peas (Mittler and Zilinskas 1994).

The GSH/GSSG ratio is likely to be influenced by the duration and severity of stress and the plant species (Smirnoff 1993). Basu et al. (2009) showed a remarkable decrease in leaf GSH/GSSG ratio during polyethylene glycol [20% (w/v), 48 h]-mediated water stress in the salt-sensitive (IR-29) and aromatic (Pusa Basmati) rice varieties, but unaffected in the salt-tolerant rice Pokkali, indicating greater dehydration resistance potential of Pokkali and hence lesser oxidative damages. When wheat plants (*T. durum* cv. Ofanto, drought tolerant) were subjected to two phases of 12 days water stress with rewatering after every phase, GSH/GSSG showed an increase as compared to the controls (Menconi et al., 1995). However, ASC/DHA ratio remained unchanged. The GR declined only in the second phase of stress by 27% of the controls. In another study, this cultivar was compared with a drought-sensitive cultivar, Adamello, and subjected to 35 days of water stress and then rewatered for 4 days (Loggini et al., 1999). Drought stress caused a 30% decline in total glutathione and GSH in both the cultivars, and full recovery was not possible after rewatering. In Adamello, GR increased to 136% of the controls during drought stress, whereas no change was observed in the cv. Ofanto. These differences were attributed to the ability of the tolerant cultivar to acclimate to drought stress, whereas Adamello, the sensitive cultivar, had to increase its antioxidative defence mechanism to prevent irreversible damage due to drought stress. In contrast, Lascano et al. (2001) found no significant differences in the four drought-tolerant wheat varieties, after one month of drought stress in the field, but observed an increase in glutathione in the leaves within 48 h of a short-term osmotic stress. Another study on a short-term drought stress treatment of 24 h showed an increase in GSH and GR activity in wheat leaves (Bartoli et al., 1999). Thus, no consistent picture emerges for GSH levels in drought-stressed wheat, as the water stress intensities have varied across these studies. On comparison of a drought-tolerant wheat cultivar C306 with a drought-susceptible one, Moti, it was observed that C306 had higher tolerance to oxidative stress with higher GR activity under stress (Chopra and Selote 2007). Moti showed a significant decline in the GSH/GSSG ratio as compared to C306 under water stress.

Another interesting feature is that the ROS level, not only during drought stress, but also during recovery from drought stress, may indicate the potential of oxidative stress or signalling in plants (Foyer and Noctor 2005a, 2005b). The differential responses of the enzyme activities to rewatering may depend on plant species, stress severity and intensity of ROS production. As mentioned by Sgherri et al. (1994), the hours following rehydration are also the most critical periods in drought tolerance mechanisms. Upadhyaya et al. (2008) found that lower ROS level and higher recovery antioxidant property of tea (*Camellia sinensis* L.), in response to rehydration after drought stress, showed a better recovery potential. Sofo et al. (2005a, 2005b) also showed that upon re-watering, the H_2O_2 content decreased in the leaves of interspecific *Prunus* hybrids and gas exchange of the plants recovered, suggesting a lower ROS production accompanied with recovered physiological activity. The increased activities of APX, MDHAR, DHAR and GR were found in the leaves subjected to the prolonged drought stress, but enzyme activities were downregulated during the rewatering phase. A strong accumulation of soluble GR mRNA isoform was also reported in pea plants (Stevens et al., 1997) after recovery from drought. Similarly, Mittler and Zilinskas (1994) reported a higher expression level of pea cytosolic APX during rehydration than during drought. Ratnayaka et al. (2003) reported an enhanced APX and GR activity in cotton during drought stress as well as 2 weeks after recovery, compared with well-watered plants. They suggested that drought stress might lead to acclimation tolerance to a subsequent more severe drought. The antioxidant enzymes and their gene expressions were also differentially or cooperatively involved in the defence mechanisms in the leaves and roots of Kentucky bluegrass (*Poa pratensis* L.) exposed to drought stress and recovery (Bian and Jiang 2009). The increased or stable activities of the four enzymes APX, MDHAR, DHAR and GR, observed in both leaves and roots of Kentucky bluegrass under drought stress (except for a reduced DHAR activity in the roots), could maintain H_2O_2 detoxification. The increased APX and DHAR activities in the leaves or increased GR and MDHAR activities and unchanged APX activities in the roots, under drought stress, helped to maintain levels of AA and GSH. The higher activities of APX, MDHAR and DHAR in the leaves and higher activities of MDHAR and DHAR in the roots, after rewatering, indicated the potential need for removal of ROS when plants were rehydrated. Strong up-regulation of GR expression in the leaves under drought stress and recovery, was similar to the trend in other studies where moderate drought stress induced gene expression of cytosolic GR in the susceptible cultivar but remained stable in the tolerant cultivar of cowpea leaves (Torres-Franklin et al., 2008), quite contrary to the results obtained in *V. unguiculata* (Contour-Ansel et al., 2006), where no significant differences in GR activity between the two cultivars differing in tolerance to drought, have been observed. Temporary stimulation of cytosolic GR gene expression during rehydration in bean and cowpea plants corresponded to a hardening process, preparing for a possible new drought episode. Thus, it can be suggested that the production of ROS or regulations of APX, DHAR and GR gene expression might occur at different levels under drought stress and recovery.

Studies performed on the C_4 plant sorghum (*Sorghum vulgare* L.), exposed to water stress, showed the differential distribution of antioxidant enzymes between the two cell compartments, namely, the bundle sheath cells (BSCs) and mesophyll cells (MCs). This is crucial to the susceptibility of each cell type to oxidative damage in sorghum leaves. Under water stress conditions, most of the SOD activity was found in the BSCs. Little or no activity of the enzymes CAT, APX or GPX was found in the MCs. The GR activity increased when exposed to low water regimes, allowing the MCs to maintain large pools of reduced ASC and GSH, despite a large increase in H_2O_2 contents. The MCs showed less damage from oxidative stress when compared to the BSCs, which had the capacity to detoxify superoxide and H_2O_2, but the regeneration of ASC from DHA and GSH from GSSG was restricted, justifying its lesser ability to withstand water stress (Sundar et al., 2004). Impairment in the transport of accumulated antioxidants from the BSCs to the MCs, caused by an abiotic factor like drought stress, would cause an over-accumulation of these in the BSCs, which would in turn inhibit re-reduction of ASC and oxidized glutathione by GR and DHAR, resulting in severe oxidative damages.

7.4 Temperature Stress

7.4.1 Chilling or low temperature

Chilling injury is a physiological disorder commonly found in crops indigenous to subtropical and temperate regions. Many important food crops such as rice and maize are very sensitive to chilling stress when air temperatures fall below a non-freezing critical threshold. The duration of chilling stress can have an important effect on chilling-sensitive plants. Because most of the enzymes do not operate optimally at low temperature, some of the antioxidant enzymes during chilling can be very labile and could become weak points in the antioxidant systems of chilling-sensitive plants (Hakam and Simon 1996; Peltzer et al., 2002). Chilling temperatures mostly limit the activity of the Calvin cycle enzymes by disrupting the sulfhydryl groups, thus reducing the utilization of absorbed light energy for CO_2 assimilation. Restricted carbon metabolism is the symptom of low temperature stress, which leads to an inadequate supply of natural electron acceptor, resulting in an overreduction of the photosystem reaction centres and leading to increased ROS in the cell. Increases in AA content and in AFR reductase activity have been noted during cold acclimation in spinach leaves (Schoner and Krause 1990). Acclimation to chilling temperatures generally led to increased GSH contents and GR activities (Anderson et al., 1992). Increased GR activity to cold stress were reported by Esterbauer and Grill (1978), Guy et al. (1984), Foyer et al. (1994) and Hodges et al. (1997a, 1997b). A positive correlation between resistance to chilling-induced photoinhibition and high GR activities has been shown in French bean seedlings (El-Saht 1998), rice (Huang and Guo 2005; Kuk et al., 2003; Oidaira et al., 2000), maize (Hodges et al., 1996), *Sorghum* (Badiani et al., 1997), wheat (Kocsy et al., 2000a, 2000b), tomato (Walker and McKersie 1993), jack pine (Zhao and Blumwald 1998), eastern white pine (Anderson et al., 1992) and poplars (Foyer et al., 1995). The activities of antioxidant enzymes SOD, CAT, APX or GR and contents of AA and GSH were enhanced greatly after chilling stress in chilling-tolerant rice cultivars, while they decreased in the chilling-sensitive cultivars (Guo et al., 2006). The chilling-tolerant cultivars have higher activities of antioxidant enzymes than susceptible cultivars in other crops as well like cucumber (Kang and Saltveit 2002a, 2002b; Shen et al., 1999) and maize (Hodges et al., 1997 a, 1997b). Chilling tolerance, induced by heat shock treatment in rice, was also correlated with elevated antioxidant enzyme activity (Kang and Saltveit 2002) and the induction of *APX* gene (Sato et al., 2001). Proteome analysis of chilling stress in rice showed the upregulation of cysteine synthase (Yan et al., 2006). This enzyme is responsible for the final step in cysteine biosynthesis, a key-limiting step in GSH production. In wheat, chromosome 5A is involved in the regulation of GSH accumulation during cold hardening (Kocsy et al., 2000a, 2000b). In maize, chilling stress increased the cysteine and GSH levels in the chilling-tolerant genotype as compared to the chilling-sensitive type. Increasing GSH synthesis by exogenous GSH treatment increased chilling tolerance of maize, while inhibiting GSH synthesis reduced tolerance (Kocsy et al., 2000a, 2000b). The importance of GSH in protection against chilling injury was shown by the use of herbicide safeners on chilling-sensitive lines (Kocsy et al., 2001). This increased the pools of GSH and its precursors, cysteine and γ-glutamyl-cysteine, and there was an increase in GR activity as well, which increased the relative protection considerably. Cold acclimation in maize was chiefly due to a 34%-47% increase in GR activity and induction in new isoforms of GR. Just as we discussed earlier for drought, previous reports from chilling stress in maize had also shown that compared to BSCs, the MCs suffer lesser damages from oxidative stress arising from H_2O_2 accumulation due to the localization of APX and GR enzymes exclusively in the MCs (Doulis et al., 1997; Pastori et al., 2000). Attempts to enhance chilling tolerance via transgenic overproduction of GR have met with some success (Kornyeyev et al., 2001, 2003; Payton et al., 2001). Transgenic poplar trees overexpressing chloroplast GR were less sensitive to cold-induced photoinhibition than control plants (Foyer et al., 1995). A 30- to 40-fold overproduction of chloroplastic GR in cotton decreased the levels of PSII and PSI photoinhibition during abruptly imposed, short-term exposure of leaf discs from warm-grown conditions to 10°C (Kornyeyev et al., 2001, 2003). The maintenance of greater rates of photochemistry, along with decreased PSII reduction states, partly explained the enhanced chilling tolerance exhibited by these transgenic plants under these conditions (Kornyeyev et al., 2001, 2003). The absence of an effect of GR overproduction and lack of protection against photoinhibition under longer-term chilling in transgenic cotton was explained, in part, by the fact that wild-type cotton acclimated to chilling by upregulating native GR activity (Logan et al., 2003). Genetic transformation studies in maize for freezing tolerance showed the upregulation of three genes, including GSTs, under both normal and cold-acclimated conditions (Wang et al., 2005). The expression of DHAR has been observed in response to low temperature in wheat (Baek and Skinner 2003). Gechev et al. (2003) has made observations on chilling stress with tobacco, where low temperature lability of CAT and DHAR provoked a significant loss in their specific activities, the effect of which was best visible after four days of chilling treatment. The low specific activity of DHAR led to impaired ASC recovery, with a lower ASC/DHA ratio and irreversible conversion of the accumulated DHA to the potentially toxic L-diketogulonic acid (Smirnoff and Wheeler 2000). The rest of the antioxidant enzymes of the AGC demonstrated sensitivity to low temperature to a lesser extent. The maintenance or unalteration of SOD activity suggested that this enzyme is hardly a rate-limiting step in the antioxidant machinery during chilling stress. The strong induction in Glu-POX levels could partially compensate the lack of CAT and AGC enzymes and resisted the accumulation of GSSG. The induction of Glu-POX protein levels is related with the responsiveness of this enzyme to various stress conditions and related to the membrane damage characteristic of the chilling injuries (Beeor-Tzahar et al., 1995; Chamnongpol et al., 1998). Its role in chilling tolerance has been established earlier (Roxas et al., 1997, 2000). The APX was suggested to be very important for adaptation to low temperatures because in *Fagus sylvatica*, its temperature optimum is below 10°C, which is unusually low (Peltzer et al., 2002). The importance of SOD for chilling tolerance was demonstrated by overexpression of this enzyme in tobacco (Sen Gupta et al., 1993a, 1993b).

7.4.2 Heat or high temperature

Heat stress damages cellular structure and metabolic pathways and contributes to secondary water stress. Grain yield reductions between 4%-10% have been estimated for every rise of 1°C in mean temperature, but these reductions could be greater, particularly in the grain-filling stages due to increased rate of leaf senescence. Many researches suggest that heat shock (HS) proteins and HS transcription factors are induced by H_2O_2 (Lee et al., 2000). Moreover, it has been demonstrated that, in the early phase of HS response, the binding of HS transcription factors with HS elements requires H_2O_2 and is inhibited by the ROS scavenging action of ASC (Volkov et al., 2006). These results are consistent with the evidence that many compounds inducing oxidative burst also increase thermotolerance (Dat et al., 1998). Several reports have highlighted the oxidative stress and the response of antioxidant defence mechanisms in heat-stressed plants (Dat et al., 1998; Anderson and Padhye 2004). Treatment of maize roots to HS temperatures of 40°C resulted in decrease of cysteine levels and increase in GSH levels (Nieto-Sotelo and Ho 1986). There was an increase in the activity of the GSH synthesizing capacity in maize root cells, which was related to the cell capacity to cope with heat stress conditions. In maize, the chilling-tolerant line Z7 suffered less damage during chilling compared to the chilling-sensitive line Penjalinan (Pen) during heat stress. A great increase in the GSH and thioredoxin (Trx) *h* levels were induced by chilling in Z7 and conversely by heat stress in Pen. Extreme temperatures had a similar effect on the level of cysteine and γ-glutamylcysteine (GSH precursors) and on the ratio of reduced to oxidized forms of the three low-molecular-weight thiols, as well as on the activity of GR. A combination of ABA or PEG treatment with heat stress induced similar changes in GSH and Trx *h* contents to those induced by heat stress alone (Kocsy et al., 2004). Accumulation of GSH was also observed in heat-stressed tomato seedlings (Rivero et al., 2004). In wheat, it was established that heat stress induced accumulation of GSH levels and increased the activity of the enzymes involved in GSH synthesis and the GSH/GSSG ratio (Kocsy et al., 2002). This was correlated to the frost sensitivity of the wheat genotypes, as well as with the higher accumulation in the sensitive genotypes. In fact, heat stress increased GSH levels in the flag leaf of two wheat genotypes, showing contrasting behaviour in heat tolerance, at all the stages during grain development. Overall, GSH levels and GSH/GSSG ratio declined during grain development in flag leaves of wheat in both control and heat-stressed plants. The rate of decline was, however, higher under a heat stress environment. Field studies were also conducted in cotton seedlings to observe the temperature variations on oxidative stress (Mahan and Mauget 2005). There was no alteration in the GSH levels and GR activity to low or high temperatures, which could have been an acclimatory response of the plants. Another recent study on apple (*Malus domestica* Borkh.) plants, subjected to high temperature (40°C), showed increased contents of total ASC, AA, total glutathione and GSH, reaching the peak at the 2 h high temperature duration, followed by a continuous decline to the lowest level with the duration of treat time. The ratios of AA/DHA and GSH/GSSG were decreased under the high temperature conditions. The reduction on GSH after 2 h, in spite of higher APX and GR activities, indicated that mechanism of antioxidant defence was by enhanced oxidation of GSH. The high thermal activation of both these enzymes suggested their protective role in heat tolerance. The higher activity of DHAR resulted from GSH utilization as the electron donor for DHA reduction. The activity of these protective enzymes (APX, GR and DHAR), and their gene expression reached the highest point at 4 h of high temperature treatment duration, and then decreased (Ma et al., 2008).

The APX and GR play important role in the protection of plants from high temperature stress by preventing the oxidation of enzymes and membranes (Sairam et al., 2000). It has been reported that transgenic tobacco cells underexpressing a cAPX (APX1) are more resistant to adverse environmental situations, including heat stress (Ishikawa et al., 2005), probably because in the APX1-depleted cells, alternative defence systems are already activated when the stress is imposed. Consistently, in the APX1 underexpressing plants, the expression of HS proteins was enhanced, and this reinforcement of the HS defence system was proposed to act as an alternative strategy controlling ROS accumulation (Pnueli et al., 2003). In *Arabidopsis thaliana*, a thermostable APX was constitutively expressed in transgenic plants overexpressing HS transcription factor 3 (Panchuk et al., 2002). This transcription factor has been recently characterized as a regulator of stress genes in *A. thaliana* (Lohmann et al., 2004). Moreover, this novel APX was also inducible in wild-type plants, as a consequence of moderate HS (Karpinski et al., 1999). The involvement of cAPX has been reported in the programmed cell death (PCD) induced by HS. In this case, HS determined a decrease in the activity and expression of cAPX, probably as part of the strategy aimed at generating the typical oxidative burst occurring during PCD (Vacca et al., 2004a, 2004b; Locato et al., 2008). Locato et al. (2009) observed that in case of tobacco TBY-2 cells, exposure to a moderate HS at 35°C allowed the cells to promptly recover their redox homeostasis, whereas 55°C HS induced a much more drastic oxidative stress against which PCD was activated. At 35°C, an increase in the cAPX and plastidial MDHAR were aimed to remove the H_2O_2, overproduced under this condition and to restore the cellular redox homeostasis. On the other hand, no alteration in the redox state of ASC and GSH occurred, in spite of an increase in the APX-dependent oxidation of ASC being observed. This suggested that the subcellular activities of the ASC recycling enzymes are probably redundant to better cope with moderate and commonly occurring oxidative stress. Under HS conditions that induce PCD (55°C), the activities of AGC enzymes strongly increased in the cytosolic fraction, with a strong decrease of ASC and GSH pools. The cytosolic increases in the ASC and GSH recycling enzymes were aimed to compensate for the depletion in these redox metabolites. On the other hand, in the cellular organelles, the activities of the AGC enzymes decreased, with the exception of that of the mitochondrial DHAR that showed increase. The activation of PCD induced the release, from mitochondria, of an apoptosis-inducing factor, homologous to a MDHAR isoenzyme, which could explain the particularly evident drop of this enzyme activity observed in the mitochondria of cells undergoing PCD. The increase in DHAR activity at 55°C could be a sort of feedback regulation mechanism occurring to improve ASC regeneration from DHA when ASC depletion occurred at its production site. The decrease in APX activity allowed oxidative burst to take place and, as a consequence, induced the activation of all the other metabolic events that orchestrate PCD. Increases in GR have also been reported in other species during heat stress (Almeselmani et al., 2006).

The AGC activity in stored broccoli florets was higher at 3°C than at 15°C and the advance of senescence was delayed by 3°C storage (Yamauchi and Kusabe 2001), suggesting that the cycle might be involved in the control of senescence by scavenging H_2O_2 produced in cell organelles and the cytosol. In heat treated broccoli (*Brassica oleracea* L. cv Erude) florets, the AGC in the cell, including the cytosol, also helped in scavenging ROS, associated with the enhancement of the cAPX activity. The contents of AA and GSH were higher in heat-treated broccoli florets than in control florets (Shigenaga et al., 2005). Sato et al. (2001) observed that APX activity was higher in heat-treated rice seedlings than in the control. The cAPX mRNA expression in heat-treated rice seedlings was also higher than without heat treatment under chilling stress. They suggested that the reduction of chilling injury could be because of heat induction of the cAPX gene.

7.5 Light Stress

Light stress is one of the most common sources of oxidative stress in plants (Karpinski et al., 1997, 1999; Dat et al., 2000). This stress is particularly damaging when combined with conditions limiting the CO_2 fixation like extreme temperatures, water deficit and Mg deficiency (Wise and Naylor 1987; Cakmak and Marschner 1992). Combined chilling and light stress is damaging to the photosynthetic apparatus of chilling-sensitive species, a process referred to as chilling-dependent photoinhibition of photosynthesis (Long and Humphries 1994). The activity level of CAT, GR and APX in light and low temperature conditions was significantly higher than that of dark and low temperature conditions in saffron (*Crocus sativus* L.). Increased activity of CAT, GR and APX alleviated the damaging effect of light at a low temperature on cell membranes (Esfandiari et al., 2009). Long-term exposures of chilling-sensitive plants to low temperature after an initial lag period resulted in photo-oxidation, an oxygen-dependent bleaching of the carotenoid and chlorophyll pigments (Powles 1984). The response of antioxidant enzymes to light stress is believed to be an answer to the increased generation of ROS and one of the protective mechanisms against oxidative stress (Cakmak and Marschner 1992; Mishra et al., 1993; Slooten et al., 1995). The upregulation of *APX* transcript levels at high light stress was reported earlier in *Arabidopsis* (Karpinski et al., 1997), a process related to the phenomenon of systemic acquired acclimation (Karpinski et al., 1999). It has been reported that APX is systemically induced in *Arabidopsis* when a single leaf, or part of it, undergoes high light stress (Mullineaux and Karpinski 2002). Likewise, the Glu-POX protein levels were induced in tobacco exposed to light stress (Willekens et al., 1997). A protective role against light stress was also attributed to the isoforms of GPX (Prasad et al., 1995). Gechev et al. (2003) also noted the induction of different classes of POXs in tobacco by light in a time-specific pattern; for instance, the transient elevation of Glu-POX or the observed light-dependent increase in one of the GPX isoforms.

7.6 Oxygen Deprivation Stress

7.6.1 Anoxia and post-anoxia

Under natural conditions, oxygen deprivation stress in plant cells is distinguished by various physiologically different transition states: transient hypoxia, anoxia or subsequent oxidative stress characterized by different oxygen concentrations as well as particular physiological states of the tissue and finally by reoxygenation (Gutteridge and Halliwell 1990, Garnczarska 2005). At the onset of anaerobiosis, plants are initially affected by decreased oxygen concentration – hypoxia. Anoxia occurs when the oxygen concentration in the environment falls below 2%. Generation of ROS is characteristic for hypoxia and especially for reoxygenation (Blokhina et al., 2003). Investigations on H_2O_2 accumulation in barley (Kalashnikov et al., 1994) and wheat (Biemelt et al., 2000; Blokhina et al., 2001) have shown its increase in the roots suffering from hypoxia. Prolonged anaerobiosis results in damage and eventually in death of plant cells. Mitochondria, being major oxygen consumers in the cell, are strongly affected by oxygen deficiency, which may change their structure and inhibit their functions. The acclimational changes in the structure and metabolism of plant tissues during the first hours of oxygen shortage enhance tolerance to anoxia. In case of barley leaves, immediately after the end of the dark anaerobic treatment, a rapid oxidation of ASC and glutathione pool occurred indicating that these antioxidants are the "first line" of ROS detoxification. In roots, under the same conditions, ASC and glutathione remained highly reduced, reflecting the more reduced cell environment during oxygen deprivation. Anoxia resulted in decreased APX activity, decreased DHAR activity and GSH concentration and hence restricted AA regeneration. Higher GSSG levels after anoxia was due to decreased activity of GR, declined GST activity and inhibited GSH biosynthesis. Re-aeration during the post-anoxic period, which occurs when water recedes, also resulted in serious cell damage (Blokhina et al., 2000). Depletion of ATP, increased reduction state of the cell environment and hyper-polarization of the inner mitochondrial membrane during anoxia created conditions of enhanced ROS production and oxidative stress after re-exposure to air, called "post-anoxic injury" (Blokhina et al., 2003). It has been suggested that perturbation of the cell structure and function during post-anoxia can be far more severe than during the period of uninterrupted anaerobiosis (Drew 1997). However, there are other reports suggesting that the post-anoxia resulted in only a moderate increase in H_2O_2 content with no signs of lipid destruction due to efficient action of various antioxidative enzymes. At least partial restoration of low-mass antioxidant redox state, with an overall induction of antioxidant systems was observed (Skutnik and Rychter 2009). The significant accumulation of reduced ASC during post anoxia could result from increased respiratory activity since ASC biosynthesis depends on the electron transfer through Complex I (Millar et al., 2003). Szal et al. (2004) showed that hypoxic treatment decreased superoxide anion production by the mitochondria isolated from barley roots. Immediately after hypoxia, a significant increase in the reduced forms of ASC and glutathione was found in lupine (*Lupinus luteus* L.) roots. However, hypoxia followed by reoxygenation through immediate re-aeration resulted in increased intensity of free radical signal in electron paramagnetic resonance (EPR) spectra (Garnczarska et al., 2004) along with increase in SOD activity, which could be due to increased formation of superoxide anion. Higher superoxide anion generation was also observed in post-anoxic soybean seedlings (VanToai and Bolles 1991). The decreased total glutathione level observed in lupine roots subjected to hypoxia followed by re-aeration might be due to its utilization as a reducing substrate in the synthesis of ASC, leading to the increased content of ASC. Re-aeration actually caused the decline of the ratios of reduced to oxidized forms of both the metabolites, indicating oxidative stress. The enhancement of APX activity during re-aeration appeared to be due to preferential induction of all isoforms and particularly to the synthesis of isoform APX-5, which was not observed during hypoxia. The depletion of GSH and total glutathione in spite of higher GR activity or expression of GR-1 isoform upon exposure to air indicated that the mechanism of antioxidant defence was through enhanced oxidation of GSH to GSSG by DHAR yielding AA (Garnczarska 2005). Wollenweber-Ratzer and Crawford (1994) have shown that in the intolerant plants, antioxidative enzyme activities were very low or without any changes. The GSH decreased significantly during the post-anoxic period, while AA showed increased values in the tolerant species. In roots of intact wheat seedlings, made hypoxic or anoxic by sparging the roots with N_2, there was an increase in reduced forms of ASC and glutathione. Nevertheless, a rapid decrease in the redox state of both the antioxidants was observed after return to air. The activities of MDHAR, DHAR and GR were diminished or only slightly influenced under hypoxia, but anoxia caused a significant inhibition of enzyme activities (Biemelt et al., 1998). In anaerobically germinated rice seedlings, the induction of AGC enzymes was observed after transfer to air (Ushimaru et al., 1997). The imposition of anoxia and subsequent re-oxygenation caused a decrease both in the content of ASC and in its reduction state in the roots of cereals and the rhizomes of *Iris* spp (Blokhina et al., 2000).

7.6.2 Flooding

Plants in the natural environment often experience limited oxygen availability due to transient or continuous flooding (Drew 1997). Progressive decreases in soil oxygen concentration and redox potential are the major physiological consequences of soil flooding (Syvertsen et al., 1983; Ruizsanchez et al., 1996), an environmental factor of seasonal occurrence, which ultimately contribute to the apparition of several reduced compounds of either chemical or biochemical origin. During prolonged periods of soil flooding, a decrease in root hydraulic conductance (Ruizsanchez et al., 1996) causes impairment of water uptake, which eventually leads to leaf wilting and chlorosis (Arbona et al., 2008). One of the early responses to waterlogging stress appears to involve closing of stomata to avoid water loss, with a subsequent downregulation or impairment of the photosynthetic machinery (Ahmed et al., 2002; Yetisir et al., 2006; Garcýa-Sanchez et al., 2007). As a consequence, the photosynthetic electron transport chain gets over-reduced, forming superoxide radicals and singlet oxygen species in chloroplasts. Flooding stress also contributes to the accumulation of acetaldehyde and other compounds derived from the anaerobic metabolism, which could be susceptible to degradation, yielding H_2O_2 as an end product (Blokhina et al., 2003). Flooding release by drainage can cause a sudden oxygen burst that reinforces oxidative pressure due to the restoration of normal oxygen tension after an anoxic period and further exacerbates oxidative damage (Yordanova et al., 2004). Waterlogging induced an over-reduction in the antioxidant machinery in *Vigna radiata* L. (Ahmed et al., 2002). The involvement of oxidative stress in the flooding-induced damages and antioxidant responses, as indicative of tolerance or sensitivity, have also been studied in sweet potato, *Crocus* and bentgrasses (Lin et al., 2006; Keyhani et al., 2006; Wang and Jiang 2007), showing a direct relationship between an increased antioxidant activity and stress tolerance. Increase in the activity of APX has been reported in barley, eggplants and citrus subjected to waterlogging stress (Lin et al., 2004; Yordanova et al., 2004; Hossain et al., 2009). However, increased APX activity was found to be insufficient to scavenge all H_2O_2 produced in excess under waterlogging stress. Unaltered or decreased MDHAR and DHAR activities during the entire anoxic and post-anoxic periods, together with the lack of positive correlation between DHAR activity and AA/DHA ratio, suggested an increase in de novo synthesis of AA. Elevated APX activity or discrete increases in AA or glutathione concentrations seemed inefficient in maintaining low levels of oxidative damage. Waterlogging stress release by soil drainage did not improve plant performance but, on the contrary, enhanced oxidative stress and even accelerated plant injury (Hossain et al., 2009). Differences in the relative tolerance of citrus genotypes to waterlogging have been reported (Arbona et al., 2008). The flooding-induced changes in GSH pools showed that the maintenance of a higher leaf GSH/GSSG ratio is related to genotype tolerance. An impairment of the collaborative action of GR and APX contributed to a lower ability for ROS detoxification and to the higher sensitivity of citrus genotypes to flooding. The effective activation of antioxidant defences in the flooding-tolerant genotypes, together with the increase in the ability to recycle active forms of AA and GSH, partially coped with oxidative damages, delaying the apparition of senescence.

7.6.3 Submergence

The adverse effect of submergence is a result of various interrelated factors. The limited gas diffusion of CO_2, O_2 and the accumulated ethylene (Setter et al., 1988), the decreased light availability under water (Setter et al., 1995), and the turbidity of flood water (Palada and Vergara 1972) are just a few factors that slow down photosynthesis during submergence. The visual damage caused by submergence generally develops soon after the water level had receded during recovery from complete submergence (Gutteridge and Halliwell 1990). At this point, plants kept in limited conditions for underwater photosynthesis are then exposed to a completely different environment, i.e., higher O_2 level and higher light intensity relative to submerged conditions. Submergence affects both the light intensity and spectral composition when it passes through the water. In the absence of light or at very low light intensity, mitochondria are the main source of ROS due to the activity of the mitochondrial electron transport chain (mETC). Excessive illumination during recovery of submerged rice seedlings may induce an oxidative stress because of abnormal amount of ROS. The peroxidation products of lipid breakdown such as malondialdehyde (MDA), ethylene and ethane occur to a much smaller extent in submergence tolerant than in intolerant species of wheat and potato (Albrecht and Wiedenroth 1994; Pfister-Sieber and Braendle 1994). The rhizomes of the two submergence-tolerant marsh plants *Acorus calamus* and *Schoenoplecus lacustris* showed very slow LP, with stable membranes, even after 50-70 days of submergence under anoxia. In contrast, the rhizomes of the less tolerant *Iris germanica* showed high peroxidation after only 7-10 days and the membranes were completely disintegrated within 2-3 weeks (Henzi and Braendle 1993). Increases in AA content and in AFR reductase activity have been noted in response to increased oxygen tension in soybean nodules (Dalton et al., 1991) and submerged rice seedlings (Ushimaru et al., 1992). The submergence-tolerant rice FR13A was reported to show less LP or reduction in chlorophyll content due to maintenance of considerably higher level of ASC and GR activity during recovery than the intolerant variety IR-42, where the AGC operated at much slower rate because of limited ASC and GR activity (Ella et al., 2003).

7.7 Herbicide Stress

The herbicide-induced growth reduction could result from alterations in certain metabolic processes (Farago et al., 1993; Chun Yan et al., 2000; Nemat Alla and Hassan 2006; Nemat Alla et al., 2008a) and production of oxidative stress agents (Aono et al., 1995, Foyer et al., 1995). Growth reduction of several plant species was reported following the application of triazine and triazinone herbicides (Nemat Alla et al., 2008a) and also of α-chloroacetanilides (Hassan and Nemat Alla 2005). The free radicals generated by the herbicides are effectively scavenged by efficient regeneration of ASC and glutathione (Chang and Kao 1997). In this context, Farago et al. (1993) found that the decreased GSH levels in maize shoots enhanced the susceptibility to metolachlor. Nemat Alla et al. (2007) concluded that plant tolerance to butachlor was related to GSH induction and that isoproturon-induced oxidative stress was accompanied with depletion of GSH. Not only that, the differential tolerance of wheat, maize and soybean to butachlor was related to the differential induction of GSH and GSH-associated enzymes. Moreover, Nemat Alla and Hassan (2007) affirmed that high doses of isoproturon induced the oxidative stress and caused inhibitions in activities of γ-glutamyl-cysteine synthetase (γ-GCS, EC 6.3.2.2), glutathione synthetase (GS, EC 6.3.2.3), GPX, GST and GR. Gehin et al. (2006) found that the glyphosate-caused depletion of GSH was accompanied with Glu-POX disorders. Pyon et al. (2004) also indicated that one of the paraquat-resistant mechanisms in *Erigeron canadensis* might be related to detoxicative enzymes and GSH content. Increased GR activity to paraquat treatment was reported in pea (Edwards et al., 1994; Donahue et al., 1997) and maize (Pastori and Trippi 1992). The upregulation of the MDHAR transcript was reported in *Conyza bonariensis* by paraquat (Ye and Gressel 2000). The increased ratio of GSH to GSSG was maintained by GR during herbicide treatment, indicating the key role of GR in herbicide stress tolerance in plants (Romero-Puertas et al., 2006). Zabalza et al. (2007) concluded that the enhancement of the GSH content, detected in imazethapyr-treated pea roots, can be related to the increase of glutathione-associated enzyme activity. Nemat Alla et al. (2008b) have also shown that differential tolerance of the two herbicides, metribuzin and pretilachlor, in maize seedlings is related to differential glutathione level or glutathione-associated enzymes. Pretilachlor led to far greater induction in thiols and GSH, with high values of GSH/GSSG ratio than metribuzin. It also elevated the activities of GSTs, γ-GCS, GS, Glu-POX and GR to a greater extent than metribuzin. The extended drop in GSH by metribuzin might indicate a deficiency in the detoxification of the herbicide and also of ROS. On the other hand, the conjugation of GSH with pretilachlor, through catalytic action of GSTs, might be considered one of their major mechanisms of ROS detoxification.

7.8 Mechanical Wounding

Tomato tissues, subjected to mechanical wounding showed increased abundance of ascorbate free radical (AFR) reductase mRNA and increased demand for AA. There also occurred induction in ASC-dependent prolyl hydroxylation required for the accumulation of mRNA species encoding hydroxyproline-rich glycoproteins. In addition, AFR reductase contributed to maintaining the levels of AA for protection against wound-induced free radical-mediated damages. Increased synthesis of ASC and induction of AFR reductase also were seen in the adjacent, non-wounded cells to prevent the spread of damaging free radicals and to localize necrosis to wound or infection sites (Grantz et al., 1995). The up-regulation of the MDHAR transcript has been reported in tomato after wounding and mechanical stimulation (Ben Rejeb et al., 2004), while in *Pisum sativum* leaves, the up-regulation of peroxisomal MDHAR enzyme activity was noted during mechanical wounding (Leterrier et al., 2005).

7.9 Environmental Pollutants

7.9.1 Ozone

Tropospheric ozone (O_3) is a widespread phytotoxic secondary air pollutant that is generated by photochemical oxidation of precursor gases such as NOx, CO, hydrocarbons and volatile organic compounds in the atmosphere. While the emission of these precursor gases and consequently, the O_3 concentration has decreased in recent years in developed regions such as the USA or the European Union, a rapid increase was seen in developing and emerging countries, especially in Asia (Jonson et al., 2006; Forster et al., 2007). For example, O_3 concentrations exceeding 80 nl l^{-1} occur regularly, even in rural areas of India or China (Cheung and Wang 2001). The current ambient O_3 concentrations in certain geographical locations have been shown to cause not only significant yield losses in agricultural crops but also to be linked with forest decline. The ingress of O_3 seems to depend on the number and size of stomata in different plant species (McLaughlin and Taylor 1981; Heath 1994; Conklin and Last 1995). O_3 effects on plants have been found dependent on leaf age, with young leaves hardly affected (Strohm et al., 1999). Transgenic tobacco plants, overexpressing wheat DHAR, and maintaining larger AA pool size, but reduced guard cell responsiveness, exhibited a lower oxidative load, reduced induction of oxidative-related enzyme activities, greater chlorophyll *a* content, and a higher level of photosynthetic activity, following an acute O_3 exposure than plants with a reduced AA pool size, but increased guard cell responsiveness. Following a chronic O_3 exposure, plants with a larger AA pool size exhibited a higher oxidative load, but retained a higher level of photosynthetic activity, despite a larger stomatal area than plants with a reduced AA pool size, which exhibited a lower oxidative load, but also a substantially lower level of photosynthetic activity. Together, these data indicated that, despite a reduced ability to respond to O_3 through stomatal closure, increasing the level of AA through enhanced AA recycling provided a greater degree of protection than reducing stomatal area or increasing guard cell responsiveness (Chen and Gallie 2005). Several workers have shown AA implication in plant detoxification of O_3 (Tanaka et al., 1985; Mehlhorn et al., 1986; Luwe et al., 1993; Barnes et al., 2002).

The increase in the level of ROS in guard cells limits further diffusion of O_3 into the leaf by promoting stomatal closure. Once O_3 enters the stomata, it reacts instantaneously with the apoplastic cell structures and undergoes breakdown to generate secondary oxyproducts including several forms of ROS, which alter plant metabolism by structurally modifying proteins and enhancing their susceptibility to proteolytic degradation (Pell and Dann 1991; Moldau and Bichele 2002). Such an O_3-induced oxidative burst (Wohlgemuth et al., 2002) results in tissue damage that produces visible leaf damage or 'leaf bronzing', possibly due to a ROS-induced cell death process (Baier et al., 2005; Fiscus et al., 2005; Rao and Davis 2001). In addition to its effect on guard cells, O_3 exposure can result in reduced photosynthesis. This has been suggested to occur, in part, from reduced stomatal conductance or accelerated loss of D1 protein affecting photosynthetic activity (Chen and Gallie 2005). In wheat and beans, O_3 exposure was shown to reduce the photosynthetic rate, mostly due to a loss of carboxylation efficiency (Farage et al., 1991; Morgan et al., 2003, 2004; Fiscus et al., 2005). O_3 concentrations around 120 nl l^{-1} have already been measured in rice-producing peri-urban and rural sites in China and India (Frei et al., 2008). The sensitivity of rice to O_3 was demonstrated through fumigation studies, which reduced biomass production and grain yield due to decreasing light use efficiency (Kobayashi and Okada 1995). Frei et al. (2008) have shown that rice seedlings exhibited substantial stress symptoms that included visible leaf symptoms and reduced biomass and tillering. Loss of photosynthetic capacity was particularly pronounced in older rice leaves exposed to O_3 for an extended period of time, similar to observations reported by Morgan et al. (2004) in soybeans. It is hypothesized that the reduction in photosynthetic capacity in O_3-stressed plants is due to a combination of accelerated leaf ageing and damage of the photosynthetic apparatus, and that these processes proceed more rapidly in intolerant varieties compared with tolerant ones. Significant natural genotypic variation may offer the possibility for breeding O_3-tolerant rice varieties. Likewise, *Arabidopsis thaliana* plants fumigated continuously for two days with 100–150 ppb O_3 developed accelerated yellowing of older leaves in addition to the downward leaf curling (Kubo et al., 1995). The O_3 pollution may cause rice yield losses of up to 16% with no change in agricultural practices (Ainsworth et al., 2008), which would put food security in Asia at substantial risk.

During O_3 stress, minimal protection would be at the stomatal level and is largely dependent on protective mechanisms operating at the cellular level (Rao 1992; Maddison et al., 2002). It has been suggested that differential sensitivity to O_3 exhibited by different varieties within a plant species is dependent on the relative amounts of ASC, GSH and levels of glutathione-ascorbate pathway enzymes. The most convincing evidence that ASC plays a critical role in plant response to O_3 stress was found in the *vtc1* mutant of *Arabidopsis* (Conklin et al., 1996), where a decrease in AA content was associated with increased O_3 sensitivity. While it is clear that minimal levels of AA are required, the concentration of AA that naturally occurs in leaf tissue is not always well correlated with O_3 tolerance (Burkey et al., 2000). Sensitivity to O_3 is inversely correlated with the level of foliar AA and the level of AA in the apoplast, and the rate of its regeneration contributes significantly to the detoxification of invading O_3. Apoplastic ASC has been proposed to be the first line of defence against O_3 pollution in a wide range of plant species (Turcsanyi et al., 2000; Conklin and Barth 2004). A low level of AA observed in O_3-intolerant rice varieties indicated that a lack of AA would lead to the formation of leaf symptoms (Frei et al., 2008). O_3 fumigation caused an increase in apoplastic ASC levels in *Sedum album* leaves (Castillo and Greppin 1988) and in parsley (Eckey-Kaltenbach et al., 1993). The exposure of wheat to twice ambient levels of O_3 caused an accumulation of reduced ASC in flag leaves (Fangmeier et al., 1994). These results, and others, indicated that ASC probably plays an important role in providing resistance to oxidative stress imposed by O_3 exposure. Potentially, AA could serve as either a direct chemical scavenger of O_3 in the apoplast, although this mechanism has been questioned, or as a substrate for extracellular enzymes like APX that attenuate ROS levels and thus affect the propagation of the initial O_3 signal. Changes in leaf extracellular ASC content and redox status have been observed in O_3-treated plants (Burkey 1999), evidence that extracellular AA is involved in detoxification process. Recent studies have found higher levels of AA in the leaf apoplast of O_3-tolerant genotypes of *Phaseolus vulgaris* (Burkey et al., 2003) and *Plantago major* (Zheng et al., 2000) compared with sensitive lines, suggesting that localization of AA in the cell wall is important. Another study (Burkey et al., 2006) identified three aspects of ASC metabolism that contributed to plant defence against O_3 stress: synthesis of sufficient quantities of AA, maintenance of a reduced ASC pool and the capacity to accumulate reduced ASC in the leaf apoplast. Cutleaf coneflower (*Rudbeckia laciniata* L.), crown-beard (*Verbesina occidentalis* Walt.) and tall milkweed (*Asclepias exaltata* L.) exhibited differences in leaf AA content and oxidation state, that corresponded with species differences in O_3-induced foliar injury. The injury was much greater in cutleaf coneflower than in crown-beard, confirming an earlier report (Chappelka et al., 1997). The DHA was elevated in crown-beard and cutleaf coneflower relative to tall milkweed suggesting a diminished capacity for converting DHA into AA. The apoplastic ascorbate oxidase (AO, EC 1.10.3.3), whose function is to produce MDHA and DHA, removes ASC (Pignocchi and Foyer 2003). The effect of AO on the redox state of the apoplastic ASC pool and O_3 tolerance was clearly demonstrated in transformed plants with an AO transgene. High constitutive overexpression of AO in tobacco decreased the apoplastic ASC redox state to only 3% without any appreciable effect on the total leaf ASC accumulation or the redox state of the whole leaf ASC pool (Pignocchi et al., 2003). The decrease in apoplastic ASC greatly enhanced leaf injury upon chronic O_3 exposure. Earlier, Sanmartin et al. (2003) also showed that overexpression of cucumber AAO gene in tobacco caused a decrease of AA/DHA ratios, that increased sensitivity to O_3.

The O_3 treatment of plants has also been shown to affect glutathione metabolism (Hausladen and Alschesr 1993). The O_3 caused an increase in the activity of components of the AGC in clonal poplars (*Populus deltoides* × *Populus* cv *caudina*). Clonal poplars, fumigated with 180 ppb O_3, exhibited an initial decline in GSH levels, after which, glutathione levels and SOD activity increased. Total glutathione (reduced plus oxidized) increased in fumigated leaves throughout the exposure period. The ratio of GSH/GSSG also decreased from 12.8 to 1.2 in O_3 exposed trees. The elevated antioxidant levels were maintained 21 h after O_3 exposure (Sen Gupta et al., 1991). In another study, the differential O_3 sensitivity of *Phaseolus vulgaris* cultivars was found to be correlated with higher GSH levels and GR activity in the tolerant cultivar (Guri 1983). Besides changes in the levels of GSH and GSSG, O_3 treatment of plants resulted in the increased expression of GST (Conklin and Last 1995). GST probably contributes to protection against O_3 and other oxidative stresses by detoxification of hydrophobic electrophilic substances associated with LP.

The O_3-induced alteration of antioxygenic activities and/or the gene expression of several antioxidant enzymes have been reported in various plants, indicating antioxidative systems linking to O_3 tolerance (Creissen et al., 1994; Kangasjarvi et al., 1994; Willekens et al., 1994). The exposure to 300 ppb O_3 exhibited increased accumulation of a number of mRNAs corresponding to antioxidant defence genes including GST, Cu/Zn-SOD and a neutral POX in *Arabidopsis*. The *A. thaliana* plants treated with 100–150 ppb O_3 revealed enhanced cAPX mRNA levels within 3–4 h that led to approximately 3–4 fold accumulation by day 4 of treatment (Kubo et al., 1995). Elevated APX mRNA levels in O_3-treated plants were correlated with higher levels of enzyme activity and protein accumulation; thus the ASC-mediated removal of ROS by APX contributed to tolerance against O_3. Kubo et al. (1995) did not observe any significant changes in the activities of enzymes of the AGC including MDHAR, DHAR and GR in plants exposed to 150 ppb O_3. In contrast, another study reported a 98% increase in total GR activity exposed to 200 ppb O_3 (Rao et al., 1996). Increased GR activity has been reported for plants in response to O_3 treatment (Mehlhorn et al., 1986, 1987; Castillo and Greppin 1988) as in pea (Mehlhorn et al., 1987; Madamanchi et al., 1992; Edwards et al., 1994), *Spinacia oleracea* (Tanaka et al., 1982), *Triticum aestivum* and *Arabidopsis thaliana* (Kubo et al., 1995). Edwards et al. (1994) reported that O_3 exposure induced two isoforms of GR in pea plants with no significant changes, either in the GR protein or in the mRNA transcripts encoding GR. Activities of GR and other enzymes of AGC were higher in sensitive rice cultivar (TN1) than in the tolerant cultivar (TNG67), suggesting that other physiological processes may confirm O_3 tolerance in these cultivars. Pretreatment of TN1 with ABA significantly facilitated stomatal closure and prevented the ingress of O_3. However, the actual role of ABA in O_3 detoxification in plants needs to be studied (Plochl et al., 2000). The presence of two ABA-responsive elements in the rice cDNA sequence of GR has been reported. It was suggested that increase in GSH synthesis or the GSH pool in cytosol or chloroplast may not be sufficient for O_3 stress tolerance, and enhanced tolerance may be achieved by increasing antioxidant capacity in the apoplast, which is likely to be the first target of O_3 stress (Polle 1998).

7.9.2 Ultraviolet-B

The overexpression of antioxidants might play an important role in detoxifying heavy loads of ROS during ultraviolet B (UV-B) stress to provide protection (Balakumar et al., 1997; Mazza et al., 1999). Earlier studies with *A. thaliana* exposed to UV-B showed that the exposure preferentially enhanced GPX, APX and other POXs, specific to coniferyl alcohol and modified the substrate affinity of APX. UV-B radiation, in contrast to O_3, enhanced ROS by increasing membrane-localized NADPH oxidase activity and decreasing CAT activities. While UV-B exposure preferentially induced the POX-related enzymes, O_3 exposure invoked the enzymes of AGC (Rao et al., 1996). It was suggested that the plants recognize UV-B radiation through mechanisms identical with those used to detect pathogen infection. The UV-B- and pathogen-induced signal transduction pathways were suggested to be similar (Grene and Fluhr 1995). A semi-dominant, monogenic and O_3-sensitive mutant (*soz1*) from *Arabidopsis*, accumulating only 30% of the normal ASC concentration, not only showed increased foliar susceptibility to O_3, but also appeared to be hypersensitive to both SO_2 and UV-B irradiation, thus implicating ASC in defence against varied environmental stresses (Conklin et al., 1996). The regulation of the antioxidant defence system by UV-B was also determined in a marine macroalga *Ulva fasciata* Delile (Chia-Tai and Tse-Min 2005). Under low UV-B doses, increases in CAT, POX, APX and GR activities and AA and glutathione pools, as well as AA regeneration ability, functioned to keep the balance of cellular H_2O_2. The DHAR and MDHAR were responsible for AA regeneration under low and medium UV-B radiations respectively. The appearance of oxidative damage in medium and high UV-B flux was attributable to a lower induction of the AGC as an antioxidant defence system. The SOD activity was also increased by UV-B. Overexpression of cAPX in transgenic tomato enhanced resistance of their leaf and fruit tissues to UV-B stress compared to wild-type plants. The APX activity in leaves of transgenic plants was several folds higher than in leaves of wild-type plants when exposed to UV-B (Wang et al., 2006). Wisniewski et al. (2002) reported earlier that transgenic apple plants that overexpressed cAPX also showed tolerance to UV-B stress.

7.9.3 Sulphur dioxide

SO_2 is an air pollutant phytotoxic at comparable concentrations, causing various types of damages such as leaf chlorosis and necrosis, inhibition of photosynthesis and growth, and reduced yield (Treshow and Anderson 1989). SO_2 toxicity may be classified into direct and indirect effects of the metabolic derivatives of SO_2, viz., sulphite and bisulphite (HSO_3^-) on plants. When plants are fumigated with SO_2, it gets entry into the leaf tissue through stomata and subsequently produces H^+, HSO_3^- and SO_3^{2-} in the cells. Most of the SO_3^{2-} and HSO_3^- get photo-oxidized to the less toxic SO_4^{2-} in chloroplasts and the conversion increases the formation of oxygen free radicals. Formation of these derivatives causes many toxic effects in plant tissues. The three enzymes APX, GR and DHAR were involved in scavenging of free radicals in SO_2 fumigated plants (Tanaka et al., 1982). The localization of APX, GR and DHAR activities in chloroplasts provided evidence for the photogeneration of ASC. The scavenging of H_2O_2 in the chloroplast, due to ASC regenerated from DHA by the system PS I $\rightarrow$ Fd (ferredoxin) $\rightarrow$ NADP $\rightarrow$ glutathione, can be considered a means for preliminary detoxification of SO_2 by chloroplasts (Renuga and Paliwal 1995). SO_2 has been proved to affect protein and mRNA levels of antioxidant enzymes (Madamanchi et al., 1994). In the leaves of *A. thaliana*, fumigated with 0.1-0.15 ppm O_3 or SO_2 up to about 1 week, only the activities of APX and GPX increased, but had little effect on the activities of SOD, CAT, MDHAR, DHAR or GR. The *APX1* mRNA level was also elevated. These results proved that O_3 and SO_2 have similarity of action, although O_3 elevated the mRNA and proteins more rapidly and extensively than SO_2 (Kubo et al., 1995). Similar result was found in case of *Nicotiana plumbaginifolia* L., where the effects of O_3, SO_2 and UV-B on the antioxidant genes were found to be similar, although the response to SO_2 was generally less pronounced and delayed. The effects of different stresses were characterized by a decline in catalase gene *Catl,* a moderate increase in *Cat3,* and a strong increase in *Cat2* and Glu-POX. Remarkably, SODs and cAPX were not affected. It was proposed that alterations in mRNA levels of CAT and Glu-POX, but not of SODs and cAPX formed part of the initial antioxidant response to O_3, SO_2 and UV-B in *Nicotiana* (Willekens et al., 1994). Increases in AA content and in AFR reductase activity have been noted in response to a variety of plant stresses and SO_2 or O_3 exposure in conifer needles (Mehlhorn et al., 1986, 1987). In the leaves of *Cassia siamea,* SO_2 elevated the SOD and GPX activities, but had little effect on GR activity (Rao 1992).

7.10 Heavy Metal Toxicity

Heavy metal contamination of agricultural soils is a major environmental problem that can reduce both the productivity of plants and the safety of plant products as foods and feeds. In recent years, research has been focused on accumulation of heavy metals in crop plants, aquatic plants and naturally growing weeds. Industrial mining and burning of fossil fuels have contaminated large areas of land in developed countries with high concentrations of heavy metals causing environmental pollution. Heavy metals, unlike organic pollutants, usually cannot be chemically degraded or biodegraded by microorganisms. Phytoremediation is an alternative cheap and efficient biological approach to deal with this problem (Peuke and Rennenberg 2005; Salt et al., 1995a, 1995b). The metal uptake process and accumulation by different plants depend on the concentration of available metals in soils, solubility sequences and the plant species growing on these soils. Some of the heavy metals are essential for the plant growth when they are present in normal levels; but when present in excess, they cause toxic effects on plant growth, ultimately resulting in decreased yield or death of the plant. Heavy metals are also known to induce free radicals in plants, and consequently, oxidative damages (Dietz et al., 2006; Anjum et al. 2008, 2011). As most studies are based on the work done in laboratory conditions, information is scanty on status of heavy metals in soils, aquatic plants and growing plants receiving industrial effluents (Srivastava et al., 2004).

7.10.1 Cadmium

Cadmium (Cd) is a widespread trace pollutant released into the soil by industrial processes or by the utilization of fertilizers and pesticides containing it. The hitherto studies on the effect of Cd on plants have shown that it intensifies two types of unfavourable processes in plants: inactivation of macromolecules and cellular structures and induction of oxidative stress. Cd strongly interferes with plant metabolism: it binds sulfhydryl groups, altering the activity of several enzymes; on the other hand, it causes deficiency of other essential metals, competing with them for the same carriers. Cd inhibits the activities of several groups of enzymes involved in the Calvin cycle, nitrogen metabolism, sugar metabolism and sulphate assimilation. The interaction of Cd^{2+} with sulfhydryl-containing amino acids and peptides/proteins plays a major role in their environmental and biochemical behaviour (Diaz-Cruz et al., 1998). Accordingly, the intracellular fate of Cd ions strongly depends on thiol-containing molecules, particularly sulfhydryl-containing amino acids, glutathione and PCs in plants (Zenk 1996). It has been shown that Cd induces the synthesis of PCs, which bind metals in the cytosol and sequester them in the vacuole. The PCs are the glutathione-derived thiol peptides having the general structure (γ-Glu-Cys) n-Gly, where n is the number of repetitions of the γ-Glu-Cys unit, which can vary from two to eleven (more commonly from two to five) (Grill et al., 1985). The roles of GSH and PCs in heavy metal tolerance were well illustrated in Cd-sensitive mutants of *Arabidopsis* (Howden et al., 1995). Although Cd is not able to trigger Fenton-type reactions and hence cannot directly produce ROS, it can indirectly activate the plasma membrane NADPH oxidase, thus inducing generation of ROS in the exposed tissues or it can suppress the antioxidant defences, thus increasing LP, protein oxidation and nucleic acid damages (Paradiso et al., 2008). The PCs form various complexes with Cd, due to the presence of the thiolic groups of cysteine, which chelate Cd and prevent it from circulating as a free ion in the cytosol. The observed preponderance of PC_3 may be ascribed to the prompt formation of a cytosolic 'low molecular weight' complex, produced by Cd bound to PC_3. The PC synthesis does not induce GSH depletion in the roots; in contrast, a net increase in GSH occurs in parallel with Cd accumulation in this organ.

It was earlier reported that one of the pathways through which Cd induces an increase in ROS and oxidative stress in plants is the inhibition of antioxidant enzymes or the depletion of antioxidant molecules (Schickler and Caspi 1999; Chien et al., 2001). In the leaves of Cd-exposed *Helianthus annuus* plants, the activities of ASC-glutathione related defence enzymes were decreased (Gallego et al., 1996). A decrease in the amount of GSH in 3-day-old corn seedlings and in the roots of spinach subjected to Cd was reported (Tukendorf and Rauser 1990; Tukendorf 1993). Anjum et al. (2008, 2011) have shown that the synthesis of glutathione is crucial for protection of *Brassica campestris* plants against Cd stress. They further noticed that increased content of GSH and AA with sulphur nutrition could alleviate Cd-induced effects in plants. In Cd-treated cucumber seedlings, APX activity was inhibited due to GSH depletion (Zhang et al., 2002; Goncalves et al., 2007). Such a decrease has also been reported in other Cd-treated plants (Gomes-Júnior et al., 2006). Studying the GSH level in the tubers of two potato cultivars differing in their resistance to Cd, it was found that the level was ten times higher over control in the tissues of the more resistant cultivar Bzura during the second phase of stress, while in the sensitive cultivar; the higher level was noted in the early phase. The GR activity was reduced due to Cd treatment, but the enzyme from the more resistant cultivar possessed lower affinity to the toxic metal. Similar "direct" inactivation of other enzymes such as SOD, CAT and APX by Cd was found in potato tuber tissues during the first phase of stress (Stroinski et al., 1999). Arisi et al. (2000) found that elevated glutathione concentrations in transgenic poplars overexpressing γ-EC synthetase (EC 6.3.2.2) activities did not protect from Cd toxicity. Perhaps the increases in thiols at stages when Cd did not increase any longer were caused by enhanced sulphate assimilation. Cd promoted oxidative stress in pea by reducing the activity of some antioxidative enzymes (Sandalio et al., 2001; Romero-Puertas et al., 1999, 2004, 2007). A reduction of the total glutathione content (GSH+GSSG) was noted, with GSH being the most affected. The ASC was also decreased. The total content of thiols in leaves was increased by Cd. A similar situation has been described in *Brassica juncea* (Lee and Leusteck 1999), in maize roots (Nocito et al., 2002) and *Arabidopsis* (Harada et al., 2002). The transcript levels of CAT and MDHAR showed a Cd-dependent increase, although CAT activity and its protein content were depressed, which suggested a post-translational modification of this protein. The GR and APX did not change significantly, either in activity or accumulation of transcript. However, Cd treatment provoked a strong reduction in mRNA, protein level and activity of Cu/Zn-SOD, being the most negatively affected antioxidative enzyme. Increase of peroxisomal GR and APX activity has been described in leaves of pea plants under Cd toxicity (Romero-Puertas et al., 1999). The MDHAR transcript level was also increased with Cd. In the tobacco leaves, the level of GSH dropped insignificantly in the first hours of the experiment; yet after 48 hours, it did not differ from the level reported in the control sample (Vogeli-Lange and Wagner 1990). In case of Cd-treated Scots pine (*Pinus sylvestris*) seedlings, the systems involved in H_2O_2 removal like glutathione, GR, CAT and APX were inhibited initially, but with prolonged treatment, the GR activities were recovered to control levels, and APX and CAT were also stimulated. A significant depletion of glutathione was found at earlier stages. This is a common response to Cd caused by an increased consumption of or reduction in glutathione for increased rate of PC production. Because the synthesis of glutathione is demand driven, the low glutathione concentration might have triggered increased sulphur uptake and its own synthesis, thus resulting in elevated glutathione concentrations at later stages. Likewise, the redox state of ASC initially increased, but with more Cd exposure, ASC was consumed and DHA accumulated (Schutzendubel et al., 2001, 2002). The toxic critical threshold value of soil Cd in inducing oxidative stress to wheat seedlings was found to lie between 3.3 mg kg^{-1} and 10 mg kg^{-1}. At lower Cd concentration, the activity levels of SOD, CAT, GPX, APX and GR did not change much, but fluctuated drastically at high Cd concentrations. The GSH contents and GSH/GSSG ratios decreased at low Cd concentrations, indicating excess consumption of GSH relative to the supplement, but increased sharply at high Cd concentrations. When organisms are exposed to low concentrations of Cd, their intrinsic GSH might be quickly consumed due to a high cellular requirement for SH compounds to resist stress by inducing PC synthesis or activating the GSH-involved ROS scavenging mechanism (Schutzendubel et al., 2002; Corticeiro et al., 2006) and decreasing the GSH/GSSG ratio. This excess response might cause organisms to keep a low redox status to maintain the high rate of growth. However, high levels of Cd must significantly induce GSH synthesis to resist aggravating stress, resulting in high GSH/GSSG ratios. The wheat seedlings might overcompensate for the stress at low Cd concentrations, resulting in an increased quenching rate, decreased accumulation of free radicals, lower oxidative stress and a positive effect on growth.

The changes in biochemical parameters would occur before any visible symptom of toxicity appeared, and the end point based on these parameters might be more sensitive or indicative than morphological observations in revealing the eco-toxicity of Cd. Under Cd stress, tomato plants (*Lycopersicon esculentum*) showed a rapid loss of chloroplastic APX activity, especially that of the thylakoid-bound chloroplastic APX isoenzymes, with the induction of thylakoid-bound and stromal Fe-SOD and Cu/Zn-SODs. The concurrently increased SOD activity and decreased APX activity favoured the accumulation of H_2O_2. The APX inactivation was mainly related to the decrease of ASC concentration at the site where APXs were compartmentalized, as supported by *in vitro* treatment of exogenous ASC, which retarded Cd-induced APX inactivation The APX from Cd-treated seedlings showed a hyperbolic dependence on ASC concentration (Liu et al., 2008b). Moreover, the decline of the antioxidant enzymes, associated with Cd toxicity, was noted in *Phaseolus vulgaris* (Chaoui et al., 1997), *P. aureus* (Shaw 1995), *Pisum sativum* (Dixit et al., 2001) and *Secale cereale* (Streb et al., 1993).

In contradiction to the above observations, Cd has been reported to activate the ROS producing enzymes in different plant species (Romero-Puertas et al., 2004; Ranieri et al., 2005; Rodrýguez-Serrano et al., 2006). The roots and leaves of *P. vulgaris, P. sativum* (Dixit et al., 2001) as well as suspension cultures of tobacco (*Nicotiana tabacum*) cells contained elevated APX activities after Cd exposure (Chaoui et al., 1997, Piqueras et al., 1999). An overall increase in the metabolite levels (stimulated AA synthesis concurrent with enhanced GSH pool) and APX or GR was observed in ramie (*Bechmeria nivea* L. Gaud) while prolongation of exposure resulted in decrease of GR activity and GSH pool in roots, or enhanced AA oxidation due to increased DHA level and DHA/AA ratios, which limited the operation of the whole AGC. This showed that the oxidative damages in ramie were closely associated with the efficiency of its intrinsic antioxidant mechanisms and the accelerated operation of AGC. It was suggested that ramie seedlings can survive moderate Cd toxicity and 7 mg l^{-1} was recommended as the threshold of Cd tolerance (Liu et al., 2007). Cd-treated aquatic plant *Bacopa monnieri* showed elevated APX activity, though the AA content of the roots and leaves were decreased because of its oxidation to DHA (Singh et al., 2006). Accumulation of GSH, coupled with activation of the enzyme GS, under the influence of Cd, was observed in the roots and leaves of pea and maize seedlings (Riiegsegger et al., 1990; Riiegsegger and Brunold 1992). It has been shown that exogenous GSH may decrease Cd uptake (Kang 1992). Gowrinathan and Rao (1992) attributed the reversal of heavy metal toxicity by AA to two possible mechanisms: AA may bind metals, thereby affecting their movement across biological membranes; or AA may act as a reducing agent, protecting the oxidation of the mercapto (-SH) group by contributing electron or reducing power for PSII. The formation and transitory elevation of GSH in barley after 5 d Cd exposure, was considered one of the "truly adaptive responses". The Cd-sensitive variety Wumaoliuling showed significant reduction in GSH concentration in both stems and roots, while no significant reduction was noted in the tolerant variety ZAU3 due to its higher GSH biosynthetic capacity. The Cd-induced AA reduction was more pronounced for Cd-sensitive genotype with relatively less reduction in Cd-tolerant genotype (Wu et al., 2004). The enhancement in AA and marked increase in non-protein thiols during Cd exposure reflected the ability of cucumber to tolerate the cellular metal load. Several other investigations have similarly reported increase in AA and non-protein thiols for other plant species (Tiryakioglu et al., 2006, Liu et al., 2007). Increases in GR activity as a consequence of Cd exposure have previously been observed in *Brassica juncea* (Qadir et al., 2004), *P. vulgaris* (Smeets et al., 2005) and *Bacopa monnieri* L. (Mishra et al., 2006). In Cd-treated *P. vulgaris*, significant increases of APX and GR activity and in total AA and GSH or GSSG contents were observed, along with an increase in the DHA/AA ratio, due to elevated DHA level, without a contemporaneous decrease in the amount of AA. The AA content remained stable due to an elevated AA synthesis (Smeets et al., 2005). The durum wheat plants (*Triticum durum* cv. Creso) showed enhancement in the metabolite level and enzymatic activity of AGC, in both roots and leaves, during Cd exposure. The whole plant improved its antioxidant defence even in those parts which had not yet been reached by Cd. The Cd is able to activate the onset of a signal that is transferred from the site of production to other parts of the plant (Ueki and Citovsky 2001). This local and precocious systemic response, mediated through the increase in the enzymes of the AGC, highlight the tight regulation and the relevance of this cycle in the defence against Cd. The increase in ASC and GSH, observed in durum wheat roots as a consequence of Cd exposure, is not in agreement with the behaviour observed in pea roots subjected to similar Cd concentrations (Rodrýguez-Serrano et al., 2006), where decreases in both the antioxidant metabolites and some ROS-scavenging enzymes (CAT, SOD and class III POX) were observed. Treatment with L-galactono-c-lactone (GL), the last ASC precursor, was able to completely reverse the effects of oxidative stress induced by Cd in durum wheat. GL treatment also induced a slight increase in the GSH pool, as has been reported to occur in tobacco cells treated with GL, and it might be due to a type of GSH protection by the presence of an increased amount of another antioxidant metabolite (de Pinto et al., 1999). The GL could probably substitute for PCs in protecting cellular metabolism with a similar mechanism. Consistently, the inhibition of the Cd-dependent H_2O_2 production, by increasing ASC biosynthesis, has been reported (Zhao et al., 2005). The increase in GSH biosynthesis, induced by γ-GCS overexpression, had a marginal effect on Cd tolerance (Arisia et al., 2000). The *A. thaliana* mutants, lacking γ-GCS, or plants treated with L-buthionine-[S,R] sulphoximine, a specific inhibitor of GSH biosynthesis, were more sensitive to Cd than wild-type or control plants respectively, as a consequence of the GSH depletion (Cobbet et al., 1998; Arisia et al., 2000). This is consistent with the requirement of GSH for PC biosynthesis, in addition to protecting tissues against ROS.

7.10.2 Lead

Lead (Pb) is one of the most abundant, ubiquitous toxic elements posing a critical concern to human and environmental health in that it is a persistent contaminant, has low solubility, and is classified as carcinogenic and mutagenic (Diels et al., 2002). Pb-contaminated soils cause sharp decreases in crop productivity and thus pose a serious problem for agriculture. The primary effect of Pb toxicity in plants is a rapid inhibition of root growth, probably due to inhibition of cell division in root tips (Eun et al., 2000). Secondarily, it may induce oxidative stress (Ruley et al., 2004; Reddy et al., 2005) with excessive ROS generation, that damages cell and its components such as chloroplasts, and alter the concentration of different metabolites including soluble protein, proline, ASC and glutathione, and the enzymatic antioxidants including SOD, APX, GR and CAT (Kastori et al., 1992). The oxidative stress induced by Pb treatment is directly proportional to its concentration. Exposure of 45-days old plants of Indian senna (*Cassia angustifolia* Vahl.) to lead acetate caused a reduction in the overall ASC and GSH pool, involving a rapid increase in DHA, indicating a change in redox balance in the antioxidant system. The stress was accompanied with the observed increase in GR and APX activities, along with increase in GSSG and total glutathione at all stages of growth (Qureshi et al., 2007). In case of lead-accumulating ecotypes of *Sedum alfredii*, biochemical responses to Pb included enhancement in the activity of antioxidative enzymes like SOD, GPX, CAT, APX and DHAR in both leaves and roots, but depressed GR activity. The accelerated GSH synthesis and increased ASC pools were also remarkable (Liu et al., 2008a). These results were in agreement with previous studies by Sun et al. (2005).

7.10.3 Copper

Cu concentrations in soil generally lie between 2 and 250 ppm and in healthy plant tissues range from 20–30 $\mu g\ g^{-1}$ dry weight. However, excess concentrations are said to generate oxidative stress within subcellular compartments. The reactivity of Cu not only makes it suitable for redox activities but also renders it more toxic causing destruction of thylakoid structure of chloroplasts and considerable modification of the lipid and protein composition of thylakoid membrane (Maksymiec 1997), even at mild excess. Free copper ions can readily oxidize the thiol bond present in the proteins, causing disruption of their structure and functions. The destruction of membrane lipids and leakage of electrolytes from root membranes was reported to result from enhanced generation of ROS in Cu-excess wheat plants (Quartacci et al., 2001). Many reports concerning Cu excess-induced oxidative stress and antioxidant responses are available (Gupta et al., 1999; Teisseire and Guy 2000; Nagalakshmi and Prasad 2001; Drazkiewicz et al., 2003; Raeymaekers et al., 2003; Morelli and Scarano 2004; Lombardi and Sebastiani 2005). Progressive depletion of GSH content in the cells of *Scenedesmus bijugatus* was observed with increasing concentrations of Cu. There was an increase in the protein thiol content while the non-protein thiol content decreased. There was also an initial elevation and later decrease in H_2O_2 level in the cells. Cu stress increased the activities of γ-GCS, GST and Glu-POX and decreased the activity of GSSG reductase (EC 1.6.4.2). These results suggested that Cu alters the equilibrium between synthesis and utilization of GSH, either due to its antioxidant role or by serving as a precursor in the synthesis of PCs. The AGC has a pivotal role in Cu detoxification in algal cells (Nagalakshmi and Prasad 2001). Investigation of the cellular defence mechanisms used by the marine diatom *Phaeodactylum tricornutum* to cope with short-term Cu toxicity showed that Cu induced a rapid synthesis of PCs of different degree of polymerization, initially formed at the expense of the cellular pool of glutathione. These complexes were detected as early as 1 h after Cu exposure, and increased with time. The GR activity, after an initial partial inhibition, also enhanced indicating the need to restore the oxidative balance of glutathione. The glutathione and PCs formed the first line of defence to combat against Cu-induced ROS formation. Thereafter, the activity of at least three antioxidant enzymes SOD, CAT and GR was enhanced to counteract the oxidative stress induced by Cu; a rapid oxidation of the ASC pool or a depletion of NADPH also occurred (Morelli and Scarano 2004). Other investigations concerned on the Cu influence on APX activity and GSH level in *Ceratophyllum* (Devi and Prasad 1998) or changes in the GSH pool and activities of APX and GR in cyanobacterium under Cu treatment (Mallick and Rai 1999). In the roots of *Silene cucubalus* treated with $CuSO_4$, the level of GSH was observed to decrease in less tolerant plants, while it remained the same in the tolerant cultivar (De Vos et al., 1991). Mallick (2004) reported a concentration-dependent decrease in APX and GR activities as well as AA and glutathione pools in *Chlorella vulgaris* following Cu exposure. In Cu-treated *A. thaliana* (Drazkiewicz et al., 2003), the enhanced GR activity could result from de novo GR protein synthesis, connected with Cu action on genes encoding this enzyme. The *A. thaliana* responded to Cu by increasing the transcription of glutathione metabolic genes as well as genes for GR (Xiang and Oliver 1998). The protective effect of the glutathione against free radical formation was attributed to its ability to stabilize copper in the Cu (I) oxidation state, preventing redox cycling and the generation of free radicals (Stohs and Bagchi 1995). Response of the AGC in the leaves of *A. thaliana* to excess Cu was first of all time-dependent, but also Cu concentration-dependent. The Cu-treated *A. thaliana* could counteract DHA accumulation mainly by elevating DHAR activity, accompanied by decrease of GSH level; the diminished GSH pool seemed to limit the operation of the AGC. The GR and MDHAR were especially active to maintain large pools of active form of glutathione and ASC, respectively in cells. Prolongation of exposition time to Cu resulted in decrease of APX activity (Drazkiewicz et al., 2003). Response of APX to excess Cu in *A. thaliana* was opposite to that found in leaves of *P. vulgaris* treated with this metal (Weckx and Clijsters 1996; Cuypers et al., 2000). Luna et al. (1994) demonstrated that Cu ions decreased the APX activity in oat leaf segments due to a direct effect of Cu-induced ROS on the enzyme protein. In case of Cu-treated *Withania somnifera*, APX, MDHAR, DHAR, GST and GPX activities of leaves were increased in the presence of Cu when compared to control plants. At least six clear APX and four GPX isoforms were detected in stressed tissues. The Cu treatment gradually activated GST with increasing concentration of Cu compared to the control indicating increased detoxification of endogenously produced electrophiles like 4-hydroxy alkenals and base propenals (Khatun et al., 2008). In Cu-excess mulberry (*Morus alba* L.) cv. Kanva 2 plants, the ratio of the redox couple (DHA/AA) increased in Cu-excess plants. The activities of APX and GR also increased in Cu-excess plants. Excess of Cu-damaged roots accelerated the rate of senescence in the older leaves, induced antioxidant responses and disturbed the cellular redox environment in the young leaves of mulberry plants. The DHA/AA ratio was found to be a better index of oxidative stress compared to the concentration of ASC (Tewari et al., 2008).

7.10.4 Mercury

The environmental mercury (Hg) pollution is becoming increasingly severe because of mineral exploitation and industrial waste emission, as well as improper application of

Hg^{2+}-containing pesticides and chemical fertilizers. Hg residues in soils and crops seriously endanger human health through the food chain. Hg can enhance the load of ROS in *Arrowhead* (Zeng et al., 2003), *Potamogeton* (Gu et al., 2002), *Azolla* (Shi et al., 2003) and other plant species, and conduce oxidative damage of macromolecules, resulting in plant injuries. There are several reports regarding the effect of Hg on antioxidant enzyme (Pang et al., 2001; Kenichi 2005). Rice is one of the crops that easily absorbs Hg from soils. The GSSG and DHA of the leaves were higher in the wild type than in the Hg^{2+}-tolerant mutant of the Hg^{2+}-sensitive rice cultivar Zhonghua 11 in response to 0.4 mmol L^{-1} Hg^{2+} treatments. As Hg^{2+} can easily bind to sulfhydryl groups, the ratio of GSH to GSSG and the ratio of AA to DHA were higher in the mutant than in the wild type, while the accumulation of Hg^{2+} in roots and stems of the mutant was more than that in the wild type. These results indicated that the AGC was less inhibited in the mutant than in the wild type, and that an effective AGC is important for the Hg resistance of the rice mutant (Bin et al., 2008). The GSH was also effective in reducing Hg uptake in *Chlorella vulgaris* and in *Nostoc calcicola*.

7.10.5 Iron

The high levels of free iron may cause damage to cells by enhancing the generation of ROS (Halliwell and Gutteridge 1984). Partially reduced forms of oxygen in the cell can react with free iron leading to the formation of ROS, such as the hydroxyl radicals, produced via the Fenton and the Haber-Weiss reactions. Recent studies suggested that excess of free iron can induce an antioxidative response in plants. In *Nicotiana plumbaginifolia,* it stimulated the oxidation of glutathione and ASC (Kampfenkel et al., 1995). In *P. vulgaris* and *N. plumbaginifolia,* it led to an increase in CAT and APX activities (Kampfenkel et al., 1995; Sheinberg et al., 1996). The demonstration of iron-mediated plant antioxidative gene expression was reported in *Brassica napus,* where iron triggered a rapid induction of cAPX expression (Vansuyt et al., 1997). Because the presence of free iron is thought to be tightly linked to free radical production and oxidative stress, this result demonstrated the unique connection between iron metabolism and oxidative stress. However, because cAPX is a haem peroxidase which contains iron as a cofactor, it was also possible that excess of free iron may have triggered the gene expression simply because the intracellular level of free iron is the limiting factor for cAPX biosynthesis. Pekker et al. (2002) observed that cAPX expression, i.e., mRNA and protein, was rapidly induced in response to iron overload in the leaves of de-rooted bean plants and this induction correlated with the increase in iron content in leaves. GSH was used as an intermediate signal to enhance the induction of APX mRNA by iron. Using cAPX-antisense transgenic plants, cAPX expression was shown to be essential to prevent iron-mediated tissue damage in tobacco. Increased GR activity was recorded in rice plants exposed to Fe stress (Fang et al., 2001).

7.10.6 Zinc

Zn is the second most abundant transition metal after iron and is involved in various biological processes in organisms (Broadley et al., 2007). Like other heavy metals, excess

Zn can have negative effects on plants. At organism level, excess Zn inhibits seed germination, plant growth (Mrozek and Funicelli 1982) and root development (Lingua et al., 2008) and causes leaf chlorosis (Ebbs and Kochian 1997). At the cellular level, excess Zn can significantly alter mitotic activity (Rout and Das 2003), affect membrane integrity or permeability (Stoyanova and Doncheva 2002) and even kill cells (Chang et al., 2005). Excess Zn exerts its toxicity partially through disturbing nutrient balance, inducing oxidative stress and LP in plants (Prasad et al., 1999; Madhava Rao and Sresty 2000). Several recent studies (Madhava Rao and Sresty 2000; Cuypers et al., 2001, 2002; Wójcik et al., 2006; Tewari et al., 2008) have demonstrated the effects of Zn stress on the activity of many antioxidative enzymes (APX, SOD, POD and CAT) and antioxidant contents (ASC and GSH) in plants. These data suggest there is crosstalk between Zn-induced differentially expressed genes and ROS-induced antioxidative defensive genes, and represents a complex mechanism developed to cope with Zn toxicity (Van de Mortel et al., 2006). The APX increased significantly in leaves of rapeseed seedlings treated with 0.56 mM Zn and decreased significantly at 1.12 mM. However, decreased APX activity was noted in roots (Wang et al., 2009). Increased APX activity was detected in leaves of *Brassica juncea* (Prasad et al., 1999) and pigeonpea (Madhava Rao and Sresty 2000), while decreased activity in *P. vulgaris* (Cuypers et al., 2001). The AA concentration in rapeseed was higher at 0.07–0.14 mM Zn, but lower at 0.28–1.12 mM, compared with controls; and for DHA, it was significantly greater at 0.14–1.12 mM (Wang et al., 2009). Similar results were reported by Cuypers et al. (2001) in *P. vulgaris* plants treated with 50 μM Zn. Decreased AA content was also detected in leaves of *B. juncea* (Prasad et al., 1999) and pigeonpea (Madhava Rao and Sresty 2000). Madhava Rao and Sresty (2000) found that GSH content in the roots and shoots decreased with increased concentrations of externally supplied Zn. Sun et al. (2005) suggested that GSH content increased more in leaves than in roots, and that GSH might play important roles in Zn transport, accumulation and tolerance in *Sedum alfredii*. Thus, increased GSH might enhance the ability of leaves to resist the increased Zn concentration. The reduction in GSH may be explained by its degradation, direct reaction with Zn as a chelator or consumption by the synthesis of protective materials such as PCs in roots (Pawlik-Skowronska 2003). The GSH content in Zn-treated rapeseed increased significantly in leaves, treated with 0.28 and 0.56 mM Zn; but in roots, it decreased with increased Zn concentration. The GST activity increased in all Zn treatments in leaves; in roots, it increased at 0.07–0.28 mM Zn but decreased significantly at 0.56 and 1.12 mM (Wang et al., 2009).

7.10.7 Nickel

At excess concentration, Ni becomes toxic for most plant species. High dose of Ni have a negative effect on photosynthesis, mineral nutrition, carbohydrate transport and water relations. The most common symptoms of Ni toxicity in plants are inhibition of growth, chlorosis, necrosis and wilting. The mechanisms of Ni toxicity and its metabolism in plants are not well understood. It has been shown that Ni is able to bind different organic macromolecules and damage their structure (Gajewska and Sklodowska 2005). Exposure of pea plants to Ni toxicity reduced the APX activity in roots (Gajewska and Skodowska

2005), either due to the deficiency of Fe in APX metalloprotein complex, since Ni is known to reduce Fe concentration in plant tissues (Pandey and Sharma 2002), or due to interaction of Ni with functional sulfhydryl groups, which have been suggested to participate in enzymatic activity of APX (Elia et al., 1992). Decline in APX activity in response to Ni stress was previously found in hairy roots of *Alyssum bertolonii* and *N. tabacum* (Boominathan and Doran 2002). In contrast, Baccouch et al. (2001) reported an increase in APX activity in roots of maize plants exposed to Ni. However, APX activity in leaves of pea plants increased in response to Ni treatment (Gajewska and Skodowska 2005). Induction of leaf APX activity, as a result of Ni application, was also earlier reported by Baccouch et al. (1998). The activity of GST increased in pea plants at high Ni concentration, more markedly in roots than in leaves, contributing to Ni tolerance (Gajewska and Sklodowska 2005).

7.10.8 Chromium

The Cr, in contrast to most trace metals like Cd, Pb, Hg and Al, with no known function in plants has got relatively less attention from plant scientists. The phenomenon of Cr-induced production of ROS has been described as oxidative stress response (Breusegem et al., 2001). However, the mechanisms in Cr-induced antioxidative responses are not yet fully understood. Differential toxicity of Cr speciation to plants is well documented (Cervantes et al., 2001; McGrath 1982) wherein Cr (VI) has been found to be more toxic than Cr (III). The toxic property of Cr (VI) originates from the action of this form itself as an oxidizing agent, as well as from the formation of free radicals during the reduction of Cr (VI) to Cr (III) occurring inside the cell. Cr (III) on the other hand, apart from generating ROS, if present in high concentrations, can cause toxic effects due to its ability to coordinate various organic compounds resulting in inhibition of some metallo-enzyme systems (Kotas and Stasicka 2000). Cr ion-induced inactivation of mitochondrial electron transport and superoxide generation has been demonstrated in higher plants (Dixit et al., 2002). Roots accumulate several magnitudes higher Cr under both speciations as compared to shoots (Zayed et al., 1998). The toxic effects of excess Cr in the leaves probably caused the decrease in biomass (Karuppanapandian et al., 2006a, 2006b). Evidence from several plant species revealed that Cr caused oxidative stress by mediating the activities of antioxidative enzymes (Karuppanapandian et al., 2006a, 2006b). Panda and Khan (2004) reported that Cr treatment triggered distinct oxidative defence mechanisms in *Hydrilla verticillata*. Cr-induced H_2O_2 accumulations in wheat and green gram seedlings significantly elevated the APX activity (Sharma and Sharma 1996; Karuppanapandian et al., 2006a). The response of the antioxidant enzymes and metabolites of the AGC to oxidative stress, caused by equal concentration (50 μM) of Cr (III) and Cr (VI), was studied in 15-day-old seedlings of green gram (*Vigna radiata* L. Wilczek) by Shanker et al. (2004). Under Cr (VI), the APX activity increased, together with a steep rise of the AA content between 5 and 24 h of treatment. In contrast to GSH content, which reduced after 24 h treatment, the GSSG increased steadily under both Cr (III) and Cr (VI) speciation. The rate of decline in the GSH/GSSG ratio was much faster in Cr (III) than Cr (VI). The results suggested differential response to AA and H_2O_2

signaling by Cr (III) and Cr (VI), and that AA in combination with APX was more effective in mitigating oxidative stress as against the role of GSH as an antioxidant. The Cr-induced antioxidative responses (in the form of potassium dichromate supplemented in the medium) of AGC enzymes and metabolites in green gram *(V. radiata* L.) leaves were also investigated in both dose and time-dependent manner by Karuppanapandian et al. (2006b). The APX activities were low in 300 μM Cr-treated leaves during the first 96 h, but significantly increased thereafter, suggesting that increased enzyme activities would be responsible for the removal of ROS. The low level of APX activity during the initial period of Cr treatment, suggested that plants require a time lag for induction of the enzyme. The contents of reduced ASC and DHA were significantly decreased under 300 μM Cr treatments, indicating their pronounced consumption during scavenging. The GSH content decreased at early stages of Cr treatment. However, it was restored to the normal level as in controls thereafter. In contrast, the GSSG content showed a progressive increase during the initial hours of Cr treatment. The non-protein thiol content was shown to increase during the first several hours, but it declined at later stages.

7.10.9 Aluminium

Al is a major factor reducing crop production in acid soils throughout the world. The most toxic Al species seems to be Al^{3+}, although a very toxic polynuclear complex, which is referred to as Al_{13}, has been shown to be formed in partially neutralized solutions (Kinraide 1991). The primary site of Al accumulation and toxicity is the root meristem, suggesting that Al interacts with actively dividing and elongating cells (Delhaize and Ryan 1995). Al preferentially accumulates in the root tips; it affects cell wall and plasma membrane characteristics, sensitizes membranes to an Fe-mediated free-radical chain reaction, leading to LP (Yamamoto et al., 2001), oxidation of proteins, interference with signal transduction and binding directly to DNA or RNA (Simonovicova et al., 2004). The root growth is one of the earliest and most dramatic symptoms exhibited by plants suffering from Al^{3+} toxicity. When roots of growing plants are exposed to Al^{3+}, an Al-specific signal transduction occurs between roots and shoots that lead to myriad of toxicity symptoms in shoots. Since Al can rapidly reach the cytosol, it probably affects physiological parameters such as membrane potential, ion fluxes and signal transduction pathways (Pineros and Kochian 2001). For instance, Al is able to block Ca^{2+} channels at the plasma membrane of cultured tobacco cells. Although Al itself is not a transition metal and cannot catalyse redox reactions, the involvement of oxidative stress in Al toxicity has been suggested in many plant species (Jones et al., 2006). As Al induces the expression of diverse genes in plant species like wheat, maize, sugar cane, tobacco, *Arabidopsis*, etc. and many of these genes encode antioxidant enzymes, a strong correlation appears between Al toxicity and oxidative stress in plants (Boscolo et al., 2003). The oxidative stress genes, including POX and GST were induced in *Arabidopsis* in the presence of Al. Further evidence corroborating the relation between Al stress and oxidative stress in plants has been obtained with transgenic *Arabidopsis* plants (Ezaki et al., 2000). Rice seedlings, exposed to toxic concentrations of Al, elevated the amount of oxidized glutathione and DHA, with decline in the concentrations of thiols (-SH) and AA (Sharma and Dubey 2007). Guo et al. (2005) observed enhanced Al resistance in rice roots, fed with AA, and suggested that this possibly might be the result of increased production of oxalate, which acts as a metal chelator. The activities of APX, MDHAR, DHAR and GR increased significantly in rice, while chloroplastic APX declined at higher Al concentration. The formation of new isoenzymes of APX was noted in shoots and roots. It was interpreted that the genes encoding different APX isoforms were induced differentially by Al that might act directly on trans-acting factors for specific genes or through signal transduction. The expression of the APX genes might also be regulated post transcriptionally, as has been reported for cAPX expression in pea plants subjected to drought stress (Mittler and Zilinskas 1994). In response to Al stress, the pumpkin (*Cucurbita pepo* L.) roots showed increased level of APX and ASC free radical reductase (AFRR, EC 1.6.5.4) activities, while DHAR and GR did not change. The levels of AA in the roots were also increased by the Al treatment. An oxidative burst was probably involved in the toxicity of Al in pumpkin roots and that plants reacted to the enhanced production of ROS by expressing higher levels of scavenging systems such as the AA–APX system (Dipierro et al., 2005).

7.10.10 Arsenic

Arsenic (As) is a toxic metalloid, which enters the environment mainly through anthropogenic activities (Abedin et al., 2002). As contamination in soil and groundwater is a serious health risk problem worldwide, especially in Bangladesh, China and India, resulting from natural geologic activity and manmade sources such as mining, heavy industry, semiconductor manufacturing, forest products, landfill leachates, fertilisers, pesticides and sewage (Francisco et al., 2002). Plants normally take up As, predominantly in trivalent (As III) and pentavalent (AsV) forms, which are known to interfere with various metabolic pathways in cells like interaction with sulfhydryl groups and replacement of phosphate from ATP. Hence, plants sensitive to As show toxic patterns such as decrease in plant growth and crop yield (Meharg and Hartley-Whitaker 2002). Arsenic has been reported to stimulate the formation of free radicals and ROS, leading to oxidative stress (Singh et al., 2007a, 2007b). Requejo and Tena (2005) reported the effect of As exposure on maize (*Zea mays* L.) root proteome and concluded that the induction of oxidative stress is the main process underlying As toxicity in plants. The widespread use of As contaminated groundwater for irrigation in rice field elevates its concentration in surface soil and eventually into rice plants and grains (Williams et al., 2007; Rahman et al., 2007). Numerous studies have been carried out in relation to uptake and translocation of As from soil to plants (Tripathi et al., 2007). The predominantly accepted model for plant detoxification of As in plants is complexation of As (III) with GSH and/or PCs through –SH coordination. For the detoxification purpose, As (III) must be compartmentalized in vacuoles that may be achieved by shuttling from the cytoplasm, probably as As-PC complexes. The resultant effect of As (V) and As (III) toxicity in rice seedlings, with the role of antioxidant enzyme activities and their isozymes, were investigated very recently by Shri et al.

(2009). The activities of the antioxidative enzymes in different tissues were related to As concentrations but in a differential manner. As (III) induced the enzymes more prominently in the roots, whereas As (V) enhanced their activity more in the shoot tissue, suggesting that the metabolism of As (III) and As (V) occurs involving different pathways in roots and shoots. As-generated oxidative stress was more pronounced in As (III) treatment. The differential role of antioxidant systems and PCs has also been proposed earlier for As (III) and As (V) toxicity (Srivastava et al., 2007). One APX isoenzyme was detected especially in the roots of 100 mM As (III) treated seedlings. The new isoforms of GR were induced during As (III) stress with elevated GSH turnover. Moreover, As (III) significantly decreased the GSH content in rice roots, due to its conversion to PCs. The GSH and Cys supplementation resulted in partial protection against As stress, reducing the MDA content and restoring seedling growth of As (V) exposed seedlings, preventing the toxic effects of As (III)/As (V). However, GSH had a negative effect on seedling growth, which may be due to feedback regulation of glutathione biosynthesis within plants (Richman and Meister 1975). In case of wheat seedlings, high concentration of As (5–20 mg kg^{-1}) affected seed germination, plant growth and their biomass with enhanced ROS generation and MDA accumulation. The APX activities decreased at low concentrations of As, but increased at high concentrations. The results indicated that As could exert harmfulness in the early development stage of wheat at inappropriate concentrations (Chun-xi et al., 2007). Khan et al. (2009) investigated with the Indian mustard (cv. Pusa Jai Kisan) against the As-induced toxicity in 20-day-old plants, exposed to 5 and 25 mM As for 96 h in hydroponic culture. Reduction in plant growth, measured in terms of root and shoot dry weights, was insignificant with 5 mM, but highly significant with 25 mM As treatment. Shoots accumulated more As than roots. The high level of endogenous As-induced production of superoxide radicals, led to increased LP. The plant was able to detoxify the low As level through induction of antioxidant defence mechanism, because the As treatments significantly increased the activities of APX or GR and the contents of glutathione and ASC; the increase being dependent on the duration of exposure. Large genotypic differences in As tolerance, concomitant with variations in antioxidative responses have been reported also within fern species (Meharg 2003). A previous study, comparing two fern species, *Ptreis vittata* and *Nephrolepis exaltata* (Boston fern, a non-As-hyperaccumulator), demonstrated that *P. vittata* displayed a greater As uptake influx rate than *N. exaltata*, when subjected to As (Tu and Ma 2004). The greater production of ROS and chlorophyll damages was noted in *P. ensiformis* (non As-hyperaccumulator) than *P. vittata* (As-hyperaccumulator) at 133 mM As stress. The levels of ASC and glutathione, and their reduced/oxidized ratios in the fronds of *P. vittata* of the control was much greater than *P. ensiformis*, indicating greater inherent antioxidant potential of *P. vittata* than *P. ensiformis*. The lower levels of ASC and glutathione in *P. ensiformis* than *P. vittata* are consistent with its greater exposure to ROS and lower scavenging ability. Thus, protection from oxidative damage by a greater level of ascorbate–glutathione pool was involved in the As-tolerance in As-hyperaccumulator *P. vittata* (Singh et al., 2006).

7.10.11 Uranium

Radioactive contamination of the environment where uranium (U) has been mined and processed has occurred in many countries. U is reported to be the most frequent radionuclide contaminant in ground and surface water soils. U in soil does not often present a radiological hazard to humans, but toxicity to plants could lead to prescribed clean-up and assessment criteria for industrial activities. U is a radiotoxic and chemotoxic heavy metal. There is a greater risk of chemical toxicity than of radiological toxicity when the decay half-life is very long (i.e., the radionuclide has a very low specific activity: decay half life of 238U is 4.47 milliard years). Only enriched U presents a radiotoxicological problem. General physiological phenomena occurring at the cellular level following radiation are direct effects on molecules or indirect effects through a water radiolytic reaction resulting in the production of ROS (Ribera et al., 1996). Vandenhove et al. (2006) showed the antioxidative mechanism with 10-day-old seedlings of beans (*P. vulgaris*), exposed to different concentrations of U toxicity. They observed that U can cause oxidative stress and cellular redox imbalance. The antioxidative enzyme capacities in roots were slightly stimulated with increasing contaminant concentrations. For roots exposed to 1000 μM U, enzyme capacities were significantly reduced. The enzyme capacities in leaves were not affected by U treatment. The U treatment caused an accumulation of total and reduced GSH in the primary leaves, except at 1000 μM, when the level dropped (almost to zero in case of roots), indicating a complete disruption of the cellular redox status. Total root GSH content slightly decreased by U treatment. The ratio of GSH/GSSG was higher in exposed plants than for the control plants. In leaves, the GSH metabolism seemed to be more sensitive to U exposure and preceded the antioxidative enzymes in maintaining normal cell functioning. The root DNA integrity was hampered at the highest external U concentration. For *P. vulgaris*, the U toxicity threshold is expected to be between 100 and 1000 μM U.

7.10.12 Combined metal stress

Cd is often associated with Zn as a contaminant, up to 5%, in the processed Zn ores of Zn mines and smelters. Cd and Zn have similar structural, geochemical and environmental properties and hence, their association leads to interactions, which are of considerable importance of study. The interaction between Zn, an essential micronutrient and Cd, a non-essential element, was earlier studied in *Ceratophyllum demersum* L. (Coontail), a free floating freshwater macrophyte (Aravind and Prasad 2005). A clear indication of oxidative stress with Cd was revealed with decreased thiol content and enhanced oxidation of AA and GSH, which were counteracted by Zn supplementation to Cd. The Cd slightly induced APX, but inhibited MDHAR, DHAR and GR enzymes. Zn supplementation restored and enhanced the functional activity of all these AGC enzymes. Zn supplements also increased GST activity to a greater extent than Cd alone and simultaneously restored Glu-POX activity impaired by Cd toxicity. Zn-alone treatments did not change the above investigated parameters. These results clearly indicated the protective role of Zn in modulating the redox status of the plant system through the AGC antioxidant pathway for combating Cd induced oxidative stress.

7.11 Mineral Nutrient Deficiency

Little is known about the relationship of mineral nutrient deficiency, as a factor of abiotic stress, with the antioxidative defence mechanisms in plants. It has been found that Mg deficiency enhances the concentrations of ASC and GSH and the activities of SOD, APX and GR in bean leaves (Cakmak and Marschner 1992). Plants respond to Fe deficiency stress by inducing a series of physiological and morphological changes in the roots to facilitate the mobilization of sparingly soluble Fe compounds in the root environment. The responses may also include a shift in the redox balance inside the root cell towards more reduced state, along with the increased capacity of the roots to reduce extracellular Fe. When grown under Fe-limiting conditions, plants can develop oxidative stress since Fe is a constituent or cofactor of some antioxidant enzymes, and therefore it may control their activity. Increased H_2O_2 accumulation has been found in sunflower leaves as a consequence of Fe deprivation and was related to the reduced APX, POX and SOD activities (Ranieri et al., 2002). The levels of ASC and GSH increased in Fe-deficient cucumber (Zaharieva et al., 1999) and sugar beet roots (Zaharieva and Abadia 2003). The dynamics of sugar beet (*Beta vulgaris* L.) root metabolic responses to Fe deficiency for four days was studied by Zaharieva et al. (2004). The time course curves of ASC and GSH concentrations in the roots showed gradual increases with the advancement of Fe deficiency, suggesting that plant responses to Fe deficiency stress may include a shift in the redox balance inside the root cells towards more reduced state. Considering the up-regulation of ASC and GSH levels by Fe deficiency, they could be related to plant adaptive responses to Fe deficiency stress. The time course of APX activity revealed that it could be a sensitive indicator of the initial changes in the plant Fe nutritional status. The root APX activities decreased with Fe deficiency, which is likely to reflect a depletion of the physiologically active Fe pool in the root cell, since no reduction of the substrate (ASC) was observed. It has been reported that APX activity and APX protein were significantly decreased in Fe-deficient sunflower leaves (Ranieri et al., 2002), which is consistent with the results reported by Vansuyt et al. (1997) and Pekker et al. (2002), who showed that APX mRNA abundance increased in response to Fe excess. Increased ASC concentrations, as well as enhanced activity of the plasma-membrane bound ascorbate free radical reductase (AFR-R) in the roots have been found in sugar beet (Zaharieva and Abadia 2003) and cucumber (Zaharieva et al., 1999), after two weeks of Fe deprivation. It has been proposed that an ASC-mediated system could be involved in Fe reduction, both at the root plasma membrane (along with the FC-R) (Zaharieva et al., 1999) and/or in the intracellular Fe reduction processes, as previously shown in other plant organs (Laulhère and Briat 1993). The fact that GSH concentrations are rapidly affected by Fe deficiency suggested its role in the nutrient stress response of the plant. Increased GR activity was reported in Fe-deficient barley, wheat and maize roots and Fe-deficient rice, barley and maize leaves. This proved that GR may play a role in internal Fe homeostasis and Fe mobilization in graminaceous plants and thereby allow plants to cope with Fe deficiency-induced oxidative stress through the Glu-POX in combination with SOD (Bashir et al., 2007).

8. CONCLUSION AND PERSPECTIVES

Redox signal transduction is a universal feature of aerobic life honed through evolution to balance information from metabolism and the environment. The antioxidants fulfil signalling roles to provide information on general plant health, particularly in terms of robustness for defence, using kinase-dependent and independent pathways that are initiated by redox-sensitive receptors modulated by thiol status. Antioxidants are not passive bystanders in this crosstalk, but rather function as key signalling compounds that constitute a dynamic metabolic interface between plant cell stress perception and physiological responses. All the above considerations presented in this review assign a central role to the antioxidant pair ASC-glutathione for maintaining cellular redox poise. The determination of total foliar activities may not adequately reflect the importance of compartment-specific changes. In particular, regulation and compartmentation of glutathione and ASC biosynthesis, degradation and the exchange of glutathione and ASC between different compartments via specific transporters, all impact on AGC function by affecting the pool sizes of both the antioxidants. Together, these two compounds fight the potentially villainous ROS in a unified fashion, limit ROS lifetime and accumulation, with a high degree of coupling that provides interdependency in the regenerative cycle and some compensation of function. Differential antioxidant concentrations between compartments permit antioxidant-driven vectorial signalling through processes such as ASC-driven electron transport or futile cycles. The current data suggest that glutathione is a key arbiter of the intracellular redox potential, controlling the thiol/disulphide status of proteins, while AA is particularly influential in setting thresholds for apoplastic and cytoplasmic signalling and influencing the overall availability of reductant in the hydrophilic phase. The effects of ASC deficiency and feeding on gene expression demonstrate the extensive metabolic crosstalk between the different defence processes in plants. Glutathione also has comparable roles in orchestration of stress-related gene expression. The future will determine more precisely how AA and glutathione are involved in initiating and controlling redox signal transduction and how they trigger defence-related responses in diverse plant species to optimize survival strategies. Although there are obviously still large gaps to fill, in order to elucidate the precise relationships of ASC-glutathione and AGC in redox signalling *in vivo,* it is likely that ROS production in cellular compartments, affecting the redox signalling during stress, are extensively scavenged by these two major antioxidants and the AGC, in which they are the key players. The detailed analysis of ROS accumulation, ASC-glutathione levels and the antioxidative enzyme activities in each compartment during individual as well as combined abiotic stress, coupled with comprehensive gene and proteomics assays in plants and overexpressing or suppressing an antioxidative enzyme-encoding target gene will further provide better insight in this respect. However, most of the literatures undoubtedly exaggerate the importance of antioxidative enzymes in many stress situations, since positive results find their way into press easier than negative ones. Consequently, the absence of evidence for a massive, concerted, potent upregulation of antioxidant enzyme activities under stress conditions is all the more conspicuous. Perhaps the fault lies not in the plant cell but in our presuppositions. Rather than a system designed to annihilate ROS, the antioxidative system may have evolved primarily to allow adjustment of the cellular redox state and to enable redox signalling.

ACKNOWLEDGEMENTS

The authors gratefully acknowledge the financial assistance from the Department of Biotechnology, Government of India, New Delhi and from the Bose Institute in the form of Institute Research Fellowship.

REFERENCES

Abedin, M.J., Cresser, M.S., Meharg, A.A., Feldmann, J. and Cotter-Howells, J. (2002). Arsenic accumulation and metabolism in rice (*Oryza sativa* L.). *Environ. Sci. Technol.* **36:** 962-968.

Afzal, I., Basra, S.M.A., Farooq, M. and Nawaz, A. (2006). Alleviation of salinity stress in spring wheat by hormonal priming with ABA, salicylic acid and ascorbic acid. *Int. J. Agri. Biol.* **8**.

Ahmed, S., Nawata, E., Hosokawa, M., Domae, Y. and Sakuratani, T. (2002). Alterations in photosynthesis and some antioxidant enzymatic activities of mungbean subjected to water-logging. *Plant Sci.* **163:** 117-123.

Ainsworth, E.A., Rogers, A. and Leakey, A.D.B. (2008). Targets for crop biotechnology in a future high-CO_2 and high-O_3 world. *Plant Physiol.* **147:** 13-19.

Albrecht, G. and Wiedenroth, E.M. (1994). Protection against activated oxygen following re-aeration of hypoxically pretreated wheat roots: the response of glutathione system. *J. Exp. Bot.* **45:** 449-455.

Almeselmani, M., Deshmukh, P.S., Sairam, R.K., Kushwaha, S.R. and Singh, T.P. (2006). Protective role of antioxidant enzymes under high temperature stress. *Plant Sci.* **171:** 382-388.

Amin, B., Mahleghah, G., Mahmood, H.M.R. and Hossein, M. (2009). Evaluation of interaction effect of drought stress with ascorbate and salicylic acid on some of physiological and biochemical parameters in Okra (*Hibiscus esculentus* L.). *Res. J. Biol. Sci.* **4:** 380-387.

Anderson, J.A. and Padhye, S.R. (2004). Protein aggregation, radical scavenging capacity, and stability of hydrogen peroxide defense systems in heat-stressed *Vinca* and sweet pea leaves. *J. Am. Soc. Hort. Sci.* **129:** 54-59.

Anderson, J.V., Chevone, B.I. and Hess, J.L. (1992). Seasonal variation in the antioxidant system of eastern white pine needles. *Plant Physiol.* **98:** 501-508.

Anderson, M.D., Prasad, T.K., Martin, B.A. and Stewart, C.R. (1994). Differential gene expression in chilling acclimated maize seedlings and evidence for the involvement of abscisic acid in chilling tolerance. *Plant Physiol.* **105:** 331-339.

Anjum, N.A., Umar, S., Iqbal, M. and Khan, N.A. (2011). Cadmium causes oxidative stress in moongbean [*Vigna radiata* (L.) Wilczek] by affecting antioxidant enzyme systems and ascorbate-glutathione cycle metabolism. *Russ. J. Plant Physiol.* **58:** 92-99.

Anjum, N.A., Umar, S., Ahmad, A., Iqbal, M. and Khan, N.A. (2008). Sulphur protects mustard (*Brasasica campestris* L.) from cadmium toxicity by improving leaf ascorbate and glutathione. *Plant Growth Regul.* **54:** 271-279

Ansel, D.C., Franklin, M.L.T., De Carvalho, M.H.C., Lameta, A.D.A. and Fodil, Y.Z. (2006). Glutathione reductase in leaves of cowpea: cloning of two cDNAs, expression and enzymatic activity under progressive drought stress desiccation and abscisic acid treatment. *Ann. Bot.* **98:** 1279-1287.

Aono, M., Saji, H., Fujiyama, K., Sugita, M., Kondo, N. and Tanaka, K. (1995). Decrease in activity of glutathione reductase enhances paraquat sensitivity in transgenic *Nicotiana tabacum*. *Plant Physiol.* **107:** 645-648.

Apel, K. and Hirt, H. (2004). Reactive oxygen species: metabolism, oxidative stress, and signal transduction. *Annu. Rev. Plant Biol.* **55:** 373-379.

Arab, L. and Ehsanpour, A.A. (2006). The effects of ascorbic acid on salt induced alfalfa (*Medicago sativa* L.) in *in vitro* culture. *Biokemistri.* **18:** 63-69.

Arafa, A.A., Khafagy, M.A. and El-Banna, M.F. (2007). The effect of glycinebetaine or ascorbic acid on the salt-stress induced damages in sorghum plant cells. *Int. J. Bot.* **3:** 251-259.

Aravind, P. and Prasad, M.N.V. (2005). Cadmium-induced toxicity reversal by zinc in *Ceratophyllum demersum* L. (a free floating aquatic macrophyte) together with exogenous supplements of amino- and organic acids. *Chemosphere.* **61:** 1720-1733.

Arbona, V., Hossain, Z., Lopez-Climent, M.F., Perez-Clemente, R.M. and Gomez-Cadenas, A. (2008). Antioxidant enzymatic activity is linked to waterlogging stress tolerance in citrus. *Physiol. Plant.* **132:** 452-466.

Arisi, A.C.M., Mocquot, B., Lagriffoul, A., Mench, M., Foyer, C.H. and Jouanin, L. (2000). Responses to cadmium in leaves of transformed poplars overexpressing gamma-glutamylcysteine synthetase. *Physiol. Plant.* **109:** 143-149.

Arrigoni, O., Calabrese, G., De Gara, L., Bitonti, M.B. and Liso, R. (1997). Correlation between changes in ascorbate and growth of *Lupinus albus* seedlings. *J. Plant Physiol.* **150:** 302-338.

Asada, K. (1994). Production and action of active oxygen species in photosynthetic tissues. In: Foyer, C.H. and Mullineaux, P.M. (Eds.). Causes of photooxidative stress and amelioration of defense systems in plants, CRC Press, Boca Raton, pp. 77-104.

Asada, K. (1999). The water-water cycle in chloroplasts: scavenging active oxygen species and dissipation of excess photons. *Annu. Rev. Plant Physiol. Mol. Biol.* **50:** 601-639.

Asada, K. (2000). The water-water cycle as alternative photon and electron sinks. *Phil. Trans. R. Soc. Lond.* **B 355:** 1419-1431.

Asada, K. and Takahashi, M. (1987). Production and scavenging of active oxygen in photosynthesis. In: Kyle, D.L., Osmond, C.B. and Arntzen, C.J. (Eds.). Photoinhibition. Elsevier Scientific Publishers, pp. 227-287.

Athar, H-u-R., Khan, A. and Ashraf, M. (2008). Exogenously applied ascorbic acid alleviates salt-induced oxidative stress in wheat. *Environ. Exp. Bot.* **63:** 224-231.

Baccouch, S., Chaoui, A. and El Ferjani, E. (1998). Nickel-induced oxidative damage and antioxidant responses in *Zea mays* shoots. *Plant Physiol. Biochem.* **36:** 689-694.

Baccouch, S., Chaoui, A. and El Ferjani, E. (2001). Nickel toxicity induces oxidative damage in *Zea mays* roots. *J. Plant Nutr.* **24:** 1085-1097.

Badawi, G.H., Kawano, N., Yamauchi, Y., Shimada, E., Sasaki, R., Kubo, A. and Tanaka, K. (2004). Over-expression of ascorbate peroxidase in tobacco chloroplasts enhances the tolerance to salt stress and water deficit. *Physiol. Plant.* **121:** 231-238.

Badiani, M., Paolacci, A.R., Fusari, A., D'ovidio, R., Scandalios, J.G., Porceddu, E. and Sermanni, G.G. (1997). Non-optimal growth temperatures and antioxidants in the leaves of *Sorghum bicolor* (L.) Moench. II. Short-term acclimation. *J. Plant Physiol.* **151:** 409-421.

Baek, K-H. and Skinner, D.Z. (2003). Alteration of antioxidant enzyme gene expression during cold acclimation of near-isogenic wheat lines. *Plant Sci.* **165:** 1221-1227.

Baier, M., Kandlbinder, A., Golldack, D. and Dietz, K-J. (2005). Oxidative stress and ozone: Perception, signalling and response. *Plant Cell Environ.* **28:** 1012-1020.

Balakumar, T., Garathri, B. and Anbudurai, P.R. (1997). Oxidative stress injury in tomato plants induced by supplemental UV-B radiation. *Biol. Plant.* **39:** 215-221.

Ball, L., Accotto, G.P., Bechtold, U., Creissen, G.P., Funck, D., Jimenez, A., Kular, B., Leyland, N., Mejia-Carranza, J., Reynolds, H., Karpinski, S. and Mullineaux, P.M. (2004). Evidence for a direct link between glutathione biosynthesis and stress defense gene expression in *Arabidopsis*. *Plant Cell.* **16:** 2448-2462.

Bandeoglu, E., Eyidogan, F., Yücel, M. and Öktem, H.A. (2004). Antioxidant responses of shoots and roots of lentil to NaCl-salinity stress. *Plant Growth Regul.* **42:** 69-77.

Barnes, J., Zheng, Y. and Lyons, T. (2002). Plant resistance to ozone: the role of ascorbate. In: Omasa, K., Saji, H., Youssefian, S. and Kondo, N. (Eds.). Air pollution and plant biochemistry: prospects for phytomonitoring and phytoremediation. Springer, Berlin Heidelberg New York, pp. 235-252.

Bartoli, C.G., Simontacchi, M., Tambussi, E., Beltrano, J., Montaldi, E. and Puntarulo, S. (1999). Drought and watering dependent oxidative stress: effect on antioxidant content in *Triticum aestivum* L. leaves. *J. Exp. Bot.* **50:** 375-383.

Bashir, K., Nagasaka, S., Itai, R.N., Kobayashi, T., Takahashi, M., Nakanishi, H., Mori, S. and Nishizawa, N.K. (2007). Expression and enzyme activity of glutathione reductase is upregulated by Fe-deficiency in graminaceous plants. *Plant Mol. Biol.* **65:** 277-284.

Basu, S., Roychoudhury, A., Saha, P.P. and Sengupta, D.N. (2010). Comparative analysis of some biochemical responses of three indica rice varieties during polyethylene glycol-mediated water stress exhibits distinct varietal differences. *Acta Physiol. Plant.* **32:** 551-563.

Beeor-Tzahar, T., Ben-Hayyim, G., Holland, D., Faltin, Z. and Eshdat, Y. (1995). A stress-associated citrus protein is a distinct plant phospholipid hydroperoxide glutathione peroxidase. *FEBS Lett.* **366:** 151-155.

Beltagi, M.S. (2008). Exogenous ascorbic acid (vitamin C) induced anabolic changes for salt tolerance in chick pea (*Cicer arietinum* L.) plants. *Afr. J. Plant Sci.* **2:** 118-123.

Ben Rejeb, I., Lenne, C., Leblanc, N., Julien, J., Ammar, S. and Bouzid, S. (2004). Iron-superoxide dismutase and monodehydroascorbate reductase transcripts accumulate in response to internode rubbing in tomato. *C.R. Biol.* **327:** 679-686.

Bian S. and Jiang, Y. (2009). Reactive oxygen species, antioxidant enzyme activities and gene expression patterns in leaves and roots of Kentucky bluegrass in response to drought stress and recovery. *Scientia Horticulturae.* **120:** 264-270.

Biemelt, S., Keetman, U. and Albrecht, G. (1998). Re-aeration following hypoxia or anoxia leads to activation of antioxidative defense system in roots of wheat seedlings. *Plant Physiol.* **116:** 651-658.

Biemelt, S., Keetman, U., Mock, H-P. and Grimm, B. (2000). Expression and activity of isoenzymes of superoxide dismutase in wheat roots in response to hypoxia and anoxia. *Plant Cell Environ.* **23:** 135-144.

Bin, Z., Fei-Juan, W., Cheng, Z. and Zong-Xiu, S. (2008). Effect of ascorbate-glutathione (AsA-GSH) cycle on Hg^{2+}-tolerance in rice mutant. *Acta Agron. Sin.* **34:** 823-830.

Blokhina, O., Chirkova, T.V. and Fagerstedt, K.V. (2001). Anoxic stress leads to hydrogen peroxide formation in plant cells. *J. Exp. Bot.* **52:** 1179-1190.

Blokhina, O.B., Virolainen, E. and Fagerstedt, K.V. (2003). Antioxidants, oxidative damage and oxygen deprivation stress: a review. *Ann. Bot.* **91:** 179-194.

Blokhina, O.B., Virolainen, E., Fagerstedt, K.V., Hoikkala, A., Wähälä, K. and Chirkova, T.V. (2000). Antioxidant status of anoxia-tolerant and -intolerant plant species under anoxia and reaeration. *Physiol. Plant.* **109:** 396-403.

Bolwell, G.P., Bindschedler, L.V., Blee, K.A., Butt, V.S., Davies, D.R., Gardner, S.L., Gerrish, C. and Minibayeva, F. (2002). The apoplastic oxidative burst in response to biotic stress in plants: a three-component system. *J. Exp. Bot.* **53:** 1367-1376.

Boominathan, R. and Doran, P.M. (2002). Ni-induced oxidative stress in roots of the Ni hyperaccumulator, *Alyssum bertolonii. New Phytol.* **156:** 205-215.

Boscolo, P.R.S., Menossi, M. and Jorge, R.A. (2003). Aluminum-induced oxidative stress in maize. *Phytochem.* **62:** 181-189.

Bourgeois-Chillou, P. and Gurrier, G. (1992). Salt responses in *Lycopersicon esculentum* call and whole plants. *J. Plant Physiol.* **14:** 494-501.

Breusegem, F.V., James, E.V., Dat, F. and Inze, D. (2001). The role of active oxygen species in plant signal transduction. *Plant Sci.* **161:** 423-431.

Broadbent, P., Creissen, G.P., Kular, B., Wellburn, A.R. and Mullineaux, P. (1995). Oxidative stress responses in transgenic tobacco containing altered levels of glutathione reductase activity. *Plant J.* **8:** 247-255.

Broadley, M.R., White, P.J., Hammond, J.P., Zelko, I. and Lux, A. (2007). Zinc in plants. *New Phytol.* **173:** 677-702.

Bueno, P., Piqueras, A., Kurepa, J., Savoure, A., Verbruggen, N., Van Montagu, M. and Inze, D. (1998). Expression of antioxidant enzymes in response to abscisic acid and high osmoticum in tobacco BY-2 cell cultures. *Plant Sci.* **138:** 27-34.

Buettner, G.R. and Schafer, F.Q. (2004). Ascorbate as an antioxidant in vitamin C. In: Asard, H., May, J.M. and Smirnoff, N. (Eds.), Functions and Biochemistry in Animals and Plants. Bios Scientific Publishers, Oxford, pp. 173-188.

Burkey, K.O. (1999). Effects of ozone on apoplast/cytoplasm partitioning of ascorbic acid in snap bean. *Physiol. Plant.* **107:** 188-193.

Burkey, K.O., Eason, G. and Fiscus, E.L. (2003). Factors that affect leaf extracellular ascorbic acid content and redox status. *Physiol. Plant.* **117:** 51-57.

Burkey, K.O., Neufeld, H.S., Souza, L., Chappelka, A.H. and Davison, A.W. (2006). Seasonal profiles of leaf ascorbic acid content and redox state in ozone-sensitive wild flowers. *Environmental Pollution.* **143:** 427-434.

Burkey, K.O., Wei, C., Palmer, G., Ghosh, P. and Fenner, G.P. (2000). Antioxidant metabolite levels in ozone-sensitive and tolerant genotypes of snap bean. *Physiol. Plant.* **110:** 195-200.

Cakmak, I. and Marschner, H. (1992). Magnesium deficiency and high light intensity enhance activities of superoxide dismutase, ascorbate peroxidase, and glutathione reductase in bean leaves. *Plant Physiol.* **98:** 1222-1227.

Castillo, F.J. and Greppin, H. (1988). Extracellular ascorbic acid and enzyme activities related ascorbic acid metabolism in *Sedum album* L. leaves after ozone exposure. Analysis by high-performance liquid chromatography. *Environ. Exp. Bot.* **28:** 231-238.

Cervantes, C., Garcia, J.C., Devars, S., Corona, F.G., Tavera, H.L., Carlos Torres-guzman, J. and Sanchez, R.M. (2001). Interactions of chromium with micro-organisms and plants. *FEMS Microb. Rev.* **25:** 335-347.

Chai, T.T., Fadzillah, N.M., Kusnan, M. and Mahmood, M. (2005). Water stress-induced oxidative damage and antioxidant responses in micropropagated banana plantlets. *Biol. Plant.* **49:** 153-156.

Chamnongpol, S., Willekens, H., Moeder, W., Langebartels, C., Sandermann, H., Van Montagu, M., Inzé, D. and Van Camp, W. (1998). Defense activation and enhanced pathogen tolerance induced by H_2O_2 in transgenic tobacco. *Proc. Natl. Acad. Sci. USA.* **95:** 5818-5823.

Chang, C.J. and Kao, C.H. (1997). Paraquat toxicity is reduced by metal chelators in rice leaves. *Physiol. Plant.* **101:** 471-476.

Chang, H.B., Lin, C.W. and Huang, H.J. (2005). Zinc-induced cell death in rice (*Oryza sativa* L.) roots. *Plant Growth Regul.* **46:** 261-266.

Chaoui, A., Mazhoudi, S., Ghorbal, M.H. and El Ferjani, E. (1997). Cadmium and zinc induction of lipid peroxidation and effects on antioxidant enzyme activities in bean (*Phaseolus vulgaris* L.). *Plant Sci.* **127:** 139-147.

Chaparzadeh, N., D'Amico, M.L., Khavari-Nejad, R-A., Izzo, R. and Navari-Izzo, F. (2004). Antioxidative responses of *Calendula offýcinalis* under salinity conditions. *Plant Physiol. Biochem.* **42:** 695-701.

Chappelka, A.H., Renfro, J.R., Somers, G. and Nash, B. (1997). Evaluation of ozone injury on foliage of black cherry (*Prunus serotina*) and tall milkweed (*Asclepias exaltata*) in Great Smoky Mountains National Park. *Environmental Pollution.* **95:** 13-18.

Chen, K.M., Gong, H.J., Chen, G.C., Wang, S.M. and Zhang, C.L. (2004). Gradual drought under field conditions influences the glutathione metabolism, redox balance and energy supply in spring wheat. *J. Plant Growth Regul.* **23:** 20-28.

Chen, Z. and Gallie, D.R. (2004). The ascorbic acid redox state controls guard cell signaling and stomatal movement. *Plant Cell.* **16:** 1143-1162.

Chen, Z. and Gallie, D.R. (2005). Increasing tolerance to ozone by elevating foliar ascorbic acid confers greater protection against ozone than increasing avoidance. Plant *Physiol.* **138:** 1673-1689.

Chen, Z., Young, T.E., Ling, J., Chang, S.C. and Gallie, D.R. (2003). Increasing vitamin C content of plants through enhanced ascorbate recycling. *Proc. Natl. Acad. Sci. USA.* **100:** 3525-3530.

Cheung, V.T.F. and Wang, T. (2001). Observational study of ozone pollution at a rural site in the Yangtze Delta of China. *Atmospheric Environment.* **35:** 4947–4958.

Chew, O., Whelan, J. and Millar, A.H. (2003). Molecular definition of the ascorbate–glutathione cycle in *Arabidopsis* mitochondria reveals dual targeting of antioxidant defenses in plants. *J. Biol. Chem.* **278:** 46869-46877.

Chia-Tai, S. and Tse-Min, L. (2005). Ultraviolet-B-induced oxidative stress and responses of the ascorbate-glutathione cycle in a marine macroalga *Ulva fasciata*. *J. Exp. Bot.* **56:** 2851-2865.

Chien, H-F., Wang, J-W., Lin, C.C.C. and Kao, C.H. (2001). Cadmium toxicity of rice leaves is mediated through lipid peroxidation. *Plant Growth Regul.* **33:** 205-213.

Chinnusamy, V., Jagendorf, A. and Zhu, J.K. (2005). Understanding and improving salt tolerance in plants. *Crop Sci.* **45:** 437-448.

Chopra, R.K. and Selote, D.S. (2007). Acclimation to drought stress generates oxidative stress tolerance in drought-resistant than–susceptible wheat cultivar under field conditions. *Environ. Exp. Bot.* **60:** 276-283.

Chun Yan, H., Tie Bao, C., Yu, W. and Bao Hong, S. (2000). Study on the effects of three herbicides on the growth of maize seedlings. *Plant Protect.* **26:** 17-19.

Chun-xi, L., Shu-li, F., Yun, S., Li-na, J., Xu-yang, L. and Xiao-li, H. (2007). Effects of arsenic on seed germination and physiological activities of wheat seedlings. *J. Environ. Sci.* **19:** 725-732.

Cobbet, C.S., May, M.J., Howden, R. and Rolls, B. (1998). The glutathione-deficient, cadmium sensitive mutant, cad-2, of *Arabidopsis thaliana* is deficient in γ-glutamylcysteine synthetase. *Plant J.* **16:** 73-78.

Comba, M.E., Benavides, M.P. and Tomaro, M.L. (1998). Effect of salt stress on antioxidant defense system in soybean root nodules. *Aus. J. Plant Physiol.* **25:** 665-671.

Conklin, P.L. and Barth, C. (2004). Ascorbic acid, a familiar small molecule intertwined in the response of plants to ozone, pathogens, and the onset of senescence. *Plant Cell Environ.* **27:** 959-970.

Conklin, P.L. and Last, R.L. (1995). Differential accumulation of antioxidant mRNAs in *Arabidopsis thaliana* exposed to ozone. *Plant Physiol.* **109:** 203-212.

Conklin, P.L., Williams, E.H. and Last, R.L. (1996). Environmental stress sensitivity of an ascorbic acid-deficient *Arabidopsis* mutant. *Proc. Natl. Acad. Sci. USA.* **93:** 9970-9974.

Contour-Ansel, D., Ttorres-franklin, M.L., Cruz de Carvalho, M.H., D'arcy-Lameta, A. and Zuily-Fodil, Y. (2006). Glutathione reductase in leaves of cowpea: cloning of two cDNAs, expression and enzymatic activity under progressive drought stress, desiccation and abscisic acid treatment. *Ann. Bot.* **98:** 1279-1287.

Corpas, F.J., Barroso, J.B. and del Rio, L.A. (2001). Peroxisomes as a source of reactive oxygen species and nitric oxide signal molecules in plant cells. *Trends Plant Sci.* **6:** 145-150.

Corticeiro, S.C., Lima, A.I.G. and de Almeida Paula Figueira, E.M. (2006). The importance of glutathione in oxidative status of *Rhizobium leguminosarum* biovar *viciae* under Cd exposure. *Enzyme and Microbial Technology.* **40:** 132-137.

Creissen, G.P., Edwards, E.A. and Mullineaux, P.M. (1994). Glutathione reductase and ascorbate peroxidase. In: Foyer, C.H. and Mullineaux, P.M. (Eds.). Causes of photoxidative stress and amelioration of defense systems in plants, Boca Raton: CRC Press, pp. 343-364.

Cuypers, A., Vangronsveld, J. and Clijsters, H. (2000). Biphasic effect of copper on the ascorbate-glutathione pathway in primary leaves of *Phaseolus vulgaris* seedlings during the early stages of metal assimilation. *Physiol. Plant.* **110:** 512-517.

Cuypers, A., Vangronsveld, J. and Clijsters, H. (2001). The redox status of plant cells (AsA and GSH) is sensitive to zinc imposed oxidative stress in roots and primary leaves of *Phaseolus vulgaris. Plant Physiol. Biochem.* **39:** 657-664.

Cuypers, A., Vangronsveld, J. and Clijsters, H. (2002). Peroxidases in roots and primary leaves of *Phaseolus vulgaris* copper and zinc phytotoxicity: a comparison. *J. Plant Physiol.* **159:** 869-876.

Dalmia, A. and Savhney, V. (2004). Antioxidant defense mechanism under drought stress in wheat seedlings. *Physiol. Mol. Biol. Plants.* **10:** 109-114.

Dalton, D.A., Post, C.J. and Langeberg, L. (1991). Effects of ambient oxygen and of fixed nitrogen on concentrations of glutathione, ascorbate, and associated enzymes in soybean root nodules. *Plant Physiol.* **96:** 812-818.

Dat, J.F., Lopez-Delgado, H., Foyer, C.H. and Scott, I.M. (1998). Parallel changes in H_2O_2 and catalase during thermotolerance induced by salicylic acid or heat acclimation in mustard seedlings. *Plant Physiol.* **116:** 1351-1357.

Dat, J., Vandenabeele, S., Vranová, E., Van Montagu, M., Inzé, D. and Van Breusegem, F. (2000). Dual action of active oxygen species during plant stress responses. *Cellular and Molecular Life Sciences.* **57:** 779-795.

Davenport, S.B., Gallego, S.M., Benavides, M.P. and Tomaro, M.L. (2003). Behaviour of antioxidant defense system in the adaptive response to salt stress in *Helianthus annuus* L. cells. *Plant Growth Regul.* **40:** 81-88.

de Azevedo Neto, A.D., Prisco, J.T., Eneas-Filho, J., de Abreu, C.E.B. and Gomes-Filho, E. (2006). Effect of salt stress on antioxidative enzymes and lipid peroxidation in leaves and roots of salt-tolerant and salt-sensitive maize genotypes. *Environ. Exp. Bot.* **56:** 87-94.

de Pinto, M.C., Francis, D. and De Gara, L. (1999). The redox state of the ascorbate-dehydroascorbate pair as a specific sensor of cell division in tobacco BY-2 cells. *Protoplasma.* **209:** 90-97.

De Vos, C.H., De Waal, M.A.M., Vooijs, R., Schat, H. and Ernst, W.H.O. (1991). Increased resistance to copper-induced damage of root cell plasmalemma in copper tolerant *Silene cucubalus. Physiol. Plant.* **82:** 523-528.

Debolt, S., Melino, V. and Ford, C.M. (2007). Ascorbate as a biosynthetic precursor in plants. *Ann. Bot.* **99:** 3-8.

del Río, L.A., Corpas, F.J., Sandalio, L.M., Palma, J.M. and Barroso, J.B. (2003). Plant peroxisomes, reactive oxygen metabolism and nitric oxide. *IUBMB Life.* **55:** 71-81.

del Rio, L.A., Sandalio, L.M., Corpas, F.J., Palma, J.M. and Barroso, J.B. (2006). Reactive oxygen species and reactive nitrogen species in peroxisomes production, scavenging, and role in cell signaling. *Plant Physiol.* **141:** 330-335.

Delhaize, E. and Ryan, P.R. (1995). Aluminum toxicity and tolerance in plants. *Plant Physiol.* **107:** 315-321.

Demiral, T. and Turkan, I. (2005). Comparative lipid peroxidation, antioxidant defense systems and proline content in roots of two rice cultivars differing in salt tolerance. *Environ. Exp. Bot.* **53:** 247-257.

Demmig-Adams, B. and Adams, W.W.I. (1990). The carotenoid zeaxanthin and high-energy-state quenching of chlorophyll fluorescence. *Photosyn. Res.* **25:** 187-198.

Devi, S.R. and Prasad, M.N.V. (1998). Copper toxicity in *Ceratophyllum demersum* L. (Coontail) a free floating macrophyte: response of antioxidant enzymes and antioxidants. *Plant Sci.* **138:** 157-165.

Dhindsa, R.S. (1991). Drought stress, enzymes of glutathione metabolism, oxidative injury and protein synthesis in *Tortula ruralis*. *Plant Physiol.* **95:** 648-651.

Díaz, M., Achkor, H., Titorenko, E. and Martínez, M.C. (2003). The gene encoding glutathione-dependent formaldehyde dehydrogenase/GSNO reductase is responsive to wounding, jasmonic acid and salicylic acid. *FEBS Lett.* **543:** 136-139.

Diaz-Cruz, M.S., Mendieta, J., Monjonell, A., Tauler, R. and Esteban, M. (1998). Study of the zinc-binding properties of glutathione by differential pulse polarography and multivariate curve resolution. *J. Inorg. Biochem.* **70:** 91-98.

Diels, L., van der Lelie, N. and Bastiaens, L. (2002). New developments in treatment of heavy metal contaminated soils. *Reviews in Environmental Science and Biotechnology.* **1:** 75-82

Dietz, K.J., Horling, F., Konig, J. and Baier, M. (2002). The function of the chloroplast 2-cysteine peroxiredoxin in peroxide detoxification and its regulation. *J. Exp. Bot.* **53:** 1321-1329.

Dietz, K-J., Jacob, S., Oelze, M-L., Laxa, M., Tognetti, V., de Miranda, M.S.N., Baier, M. and Finkemeier, I. (2006). The function of peroxiredoxins in plant organelle redox metabolism. *J. Exp. Bot.* **57:** 1697-1709.

Dionisio-Sese, M.L. and Tobita, S. (1998). Antioxidant responses of rice seedlings to salinity stress. *Plant Sci.* **135:** 1-9.

Dipierro, N., Mondelli, D., Paciolla, C., Brunetti, G. and Dipierro, S. (2005). Changes in the ascorbate system in the response of pumpkin (*Cucurbita pepo* L.) roots to aluminium stress. *J. Plant Physiol.* **162:** 529-536.

Dixit, V., Pandey, V. and Shyam, R. (2001). Differential antioxidative responses to cadmium in roots and leaves of pea (*Pisum sativum* L. cv. Azad). *J. Exp. Bot.* **52:** 1101-1109.

Dixit, V., Pandey, V. and Shyam, R. (2002). Chromium ions inactivate electron transport and enhance superoxide generation *in vivo* in pea (*Pisum sativum* L. cv. Azad) root mitochondria. *Plant Cell Environ.* **25:** 687-690.

Donahue, J.L., Okpodu, C.M., Cramer, C.L., Grabau, E.A. and Alscher, R.G. (1997). Responses of antioxidants to paraquat in pea leaves: relationships to resistance. *Plant Physiol.* **113:** 249-257.

Doulis, A., Debian, N., Smith, A.K. and Foyer, C.H. (1997). Characterization of chilling sensitivity in maize. I. Differential localization of antioxidants in maize leaves. *Plant Physiol.* **114:** 1031-1037.

Drazkiewicz, M., Skorzynska-Polit, E. and Krupa, Z. (2003). Response of the ascorbate-glutathione cycle to excess copper in *Arabidopsis thaliana* (L). *Plant Sci.* **164:**195-202.

Drew, M.C. (1997). Oxygen deficiency and root metabolism: injury and acclimation under hypoxia and anoxia. *Ann. Rev. Plant Physiol. Plant Mol. Biol.* **48:** 223-250.

Ebbs, S.D. and Kochian, L.V. (1997). Toxicity of zinc and copper to *Brassica* species: implications for phytoremediation. *J. Environ. Qual.* **26:** 776-781.

Eckey-Kaltenbach, H., Heller, W., Sonnebichler, J., Zetl, I., Schafer, W., Ernst, D. and Sandermann, H. (1993). Oxidative stress and plant secondary metabolism: 6″-*O*-malonylapiin in parsley. *Phytochem.* **34:** 687-691.

Edjolo, A., Laffray, D. and Guerrier, G. (2001). The ascorbate-glutathione cycle in the cytosolic and chloroplastic fractions of drought-tolerant and drought-sensitive poplars. *J. Plant Physiol.* **158:** 1511-1517.

Edwards, E.A., Enard, C., Creissen, G.P. and Mullineaux, P.M. (1994). Synthesis and properties of glutathione reductase in stressed peas. *Planta.* **192:** 137-143.

El-Aziz, N.G.A., Mazher, A.A.M. and El-Habba, E. (2006). Effect of foliar spraying with ascorbic acid on growth and chemical constituents of *Khaya senagalensis* grown under salt conditions. *American-Eurasian J. Agric. Environ. Sci.* **1:** 207-214.

Elia, M.R., Borraccino, G. and Dipierro, S. (1992). Soluble ascorbate peroxidase from potato tubers. *Plant Sci.* **85:** 17-21.

Ella, E.S., Kawano, N. and Ito, O. (2003). Importance of active oxygen-scavenging system in the recovery of rice seedlings after submergence. *Plant Sci.* **165:** 85-93.

El-Saht, H.M. (1998). Responses to chilling stress in French bean seedlings: antioxidant compounds. *Biol. Plant.* **41:** 395-402.

Esfandiari, E., Alavi-Kia, S.S., Bahmani, A. and Aazami, M.A. (2009). The effect of light on ROS-scavenging systems and lipid peroxidation under cold conditions in saffron (*Crocus sativus* L.). *Afr. J. Agricul. Res.* **4:** 378-382.

Eskling, M., Arvidsson, P. and Akerlund, H.E. (1997). The xanthophylls cycle, its regulation and components. *Physiol. Plant.* **100:** 806-816.

Esterbauer, H. and Grill, D. (1978). Seasonal variation of glutathione and glutathione reductase in needles of *Picea abies*. *Plant Physiol.* **61:** 119-121.

Eun, S.O., Youn, H.S. and Lee, Y. (2000). Lead disturbs microtubule organization in the root meristem of *Zea mays*. *Physiol. Plant.* **110:** 357-365.

Eyidogan, F. and Oz, M.T. (2007). Effect of salinity on antioxidant responses of chickpea seedlings. *Acta Physiol. Plant.* **29:** 485-493.

Ezaki, B., Gardner, R.C., Ezaki, Y. and Matsumoto, H. (2000). Expression of aluminum-induced genes in transgenic *Arabidopsis* plants can ameliorate aluminum stress and/or oxidative stress. *Plant Physiol.* **122:** 657-666.

Fadzilla, N.M., Finch, R.P. and Burdon, R.H. (1997). Salinity, oxidative stress and antioxidant responses in shoot cultures of rice. *J. Exp. Bot.* **48:** 325-331.

Fahmy, A.S., Mohamed, T.M., Mohamed, S.A. and Saker, M.M. (1998). Effect of salt stress on antioxidant activities in cell suspension cultures of cantaloupe (*Cucumis melo*). *Egyptian J. Physiol. Sci.* **22:** 315-326.

Fang, W.C., Wang, J.W., Lin, C.C. and Kao, C.H. (2001). Iron induction of lipid peroxidation and effect on antioxidative enzyme activities in rice leaves. *Plant Growth Regul.* **35:** 75-80.

Fangmeier, A., Brunschon, S. and Jager, H-J. (1994). Time course of oxidant stress biomarkers in flag leaves of wheat exposed to ozone and drought stress. *New Phytol.* **126:** 63-69.

Farage, P.K., Long, S.P., Lechner, E.G. and Baker, N.R. (1991). The sequence of change within the photosynthetic apparatus of wheat following short-term exposure to ozone. *Plant Physiol.* **95:** 529-535.

Farago, S., Kreuz, K. and Brunold, C. (1993). Decreased glutathione levels enhance the susceptibility of maize seedlings to metolachlor. *Pestic. Biochem. Physiol.* **47:** 199-205.

Fiscus, E.L., Booker, F.L. and Burkey, K.O. (2005). Crop responses to ozone: Uptake, modes of action, carbon assimilation and partitioning. *Plant Cell Environ.* **28:** 997-1011.

Forster, P., Ramaswamy, V., Artaxo, P., Berntsen, T., Betts, R., Fahey, D.W., Haywood, J., Lean, J., Lowe, D.C., Myhre, G., Nganga, J., Prinn, R., Raga. G.M.S. and Van Dorland, R. (2007). Changes in atmospheric constituents and in radiative forcing. In: Solomon, S. et al. (Eds.). Climate Change 2007: The Physical Science Basis. Contribution of Working Group I to the

Fourth Assessment Report of the Intergovernmental Panel on Climate Change. Cambridge University Press, Cambridge, UK.

Foyer, C.H., Descourvieres, P. and Kunert, K.J. (1994). Protection against oxygen radicals: an important defence mechanism studied in transgenic plants. *Plant Cell Environ.* **17:** 507-523.

Foyer, C.H. and Noctor, G. (2000). Oxygen processing in photosynthesis: regulation and signaling. *New Phytol.* **146:** 359-388.

Foyer, C.H. and Noctor, G. (2003). Redox sensing and signaling associated with reactive oxygen in chloroplasts, peroxisomes and mitochondria. *Physiol. Plant.* **119:** 355-364.

Foyer, C.H. and Noctor, G. (2005a). Oxidant and antioxidant signalling in plants: a reevaluation of the concept of oxidative stress in a physiological context. *Plant Cell Environ.* **28:** 1056-1071.

Foyer, C.H. and Noctor, G. (2005b). Redox homeostasis and antioxidant signaling: a metabolic interface between stress perception and physiological responses. *Plant Cell.* **17:** 1866-1875.

Foyer, C.H. and Rennenberg, H. (2000). Regulation of glutathione synthesis and its role in abiotic and biotic stress defence. In: Brunold, C., Rennenberg, H., De Kok, L.J., Stulen, I. and Davidian, J-C. (Eds.). Sulfur nutrition and sulfur assimilation in higher plants. Paul Haupt, Bern, pp. 127-153.

Foyer, C.H., Souruau, N., Perret, S., Lelandais, M., Kunert, K-J., Provust, C. and Jouanin, L. (1995). Overexpression of glutathione reductase but not glutathione synthetase leads to increases in antioxidant capacity and resistance to photoinhibition in poplar trees. *Plant Physiol.* **109:** 1047-1057.

Francisco, F.R., Joni, M.B. and Debby, F. (2002). Evaluation of a GFP reporter gene construct for environmental arsenic detection. *Talanta.* **58:** 181-188.

Frei, M., Tanaka, J.P. and Wissuwa, M. (2008). Genotypic variation in tolerance to elevated ozone in rice: dissection of distinct genetic factors linked to tolerance mechanisms. *J. Exp. Bot.* **59:** 3741-3752.

Gajewska, E. and Sklodowska, M. (2005). Antioxidative responses and proline level in leaves and roots of pea plants subjected to nickel stress. *Acta Physiol. Plant.* **27:** 329-339.

Gallego, S.M., Benavides, M.P. and Tomaro, M.L. (1996). Effect of heavy metal ion excess on sunflower leaves: evidence for involvement of oxidative stress. *Plant Sci.* **121:** 151-159.

Gapinska, M., Sklodowska, M. and Gabara, B. (2008). Effect of short- and long-term salinity on the activities of antioxidative enzymes and lipid peroxidation in tomato roots. *Acta Physiol. Plant.* **30:** 11-18.

Garcýa-Sanchez, F., Syvertsen, J.P., Gimeno, V., Botia, P. and Perez-Perez, J.G. (2007). Responses to flooding and drought stress by two citrus root stock seedlings with different water-use efficiency. *Physiol. Plant.* **130:** 532-542.

Garnczarska, M. (2005). Response of the ascorbate–glutathione cycle to re-aeration following hypoxia in lupine roots. *Plant Physiol. Biochem.* **43:** 583-590.

Garnczarska, M., Bednarski, W. and Morkunas, I. (2004). Re-aeration-induced oxidative stress and antioxidative defenses in hypoxically pretreated lupine roots. *J. Plant Physiol.* **161:** 415-422.

Gechev, T., Willekens, H., Van Montagu, M., Inzé, D., Camp, W.V., Toneva, V. and Minkov, I. (2003). Different responses of tobacco antioxidant enzymes to light and chilling stress. *J. Plant Physiol.* **160:** 509-515.

Gehin, A., Guyon, C. and Nicod, L. (2006). Glyphosate-induced antioxidant imbalance in HaCaT: the protective effect of vitamins C and E. *Environ. Toxic. Pharm.* **22:** 27-34.

Gogorcena, Y., Iturbe-Ormaetxe, I., Escuredo, P.R. and Becana, M. (1995). Antioxidant defenses against activated oxygen in pea nodules subjected to water stress. *Plant Physiol.* **108:** 753-759.

Golan-Goldhirsh, A., Mozafar, A. and Oerti, J.J. (1995). Effect of ascorbic acid on soybean seedlings grown on medium containing a high concentration of copper. *J. Plant Nutr.* **18:** 1735-1741.

Gomes-Júnior, R.A., Moldes, C.A., Delite, F.S., Pompeu, G.B., Gratão, P.L., Mazzafera, P., Lea, P.J. and Azevedo, R.A. (2006). Antioxidant metabolism of coffee cell suspension cultures in response to cadmium. *Chemosphere.* **65:** 1330-1337.

Gonçalves, J.F., Becker, A.G., Cargnelutti, D., Tabaldi, L.A., Pereira, L.B., Battisti, V., Spanevello, R.M., Morsch, V.M., Nicoloso, F.T. and Schetinger, M.R.C. (2007). Cadmium toxicity causes oxidative stress and induces response of the antioxidant system in cucumber seedlings. *Braz. J. Plant Physiol.* **19:** 223-232.

Gong, M., Li, Y.J. and Chen, S.Z. (1998). Abscisic acid-induced tolerance in maize seedlings is mediated by calcium and associated with antioxidant systems. *J. Plant Physiol.* **153:** 488-496.

Gossett, D.R., Banks, S.W., Millhollon, E.P. and Lucas, M.C. (1996). Antioxidant response to NaCl stress in a control and an NaCl-tolerant cell line grown in the presence of paraquat, buthionone sulfoximine, and exogenous glutathione. *Plant Physiol.* **112:** 803-809.

Gowrinathan, K.P. and Rao, V.N.R. (1992). Reversal of heavy metal toxicity by ascorbic acid in microalgae. *J. Swa. Bot.* **9:** 27-29.

Grant, J.J. and Loake, G.J. (2000). Role of reactive oxygen intermediates and cognate redox signalling in disease resistance. *Plant Physiol.* **124:** 21-29.

Grantz, A.A., Brummell, D.A. and Bennett, A.B. (1995). Ascorbate free radical reductase mRNA levels are induced by wounding. *Plant Physiol.* **108:** 411-418.

Grene, R. and Fluhr, R. (1995). UV-B-induced PR-1 accumulation is mediated by active oxygen species. *Plant Cell.* **7:** 203-212.

Grill, E., Winnacker, E-L. and Zenk, M.H. (1985). Phytochelatins: the principal heavy metal complexing peptides of higher plants. *Science.* **230:** 674-676.

Gu, W., Shi, G.X., Zhang, C.Y., Wang, W., Xu, Q.S., Xu, N., Zeng, X.M., Zhang, X.L. and Zhou, H.W. (2002). Toxic effects of Hg^{2+}, Cd^{2+} and Cu^{2+} on photosynthetic systems and protective enzyme systems of *Potamogeton crispus*. *J. Plant Physiol. Mol. Biol.* **28:** 69-74.

Guan, L.M., Zhao, J. and Scandalios, J.G. (2000). Cis-elements and trans-factors that regulate expression of the maize *Cat1* antioxidant gene in response to ABA and osmotic stress: H_2O_2 is the likely intermediary signaling molecule for the response. *Plant J.* **22:** 87-95.

Gueta-Dahan, Y., Yaniv, Z., Zilinkas, B.A. and Ben-Hayyim, G. (1997). Salt and oxidative stress: similar and specific responses and their relation to salt tolerance in citrus. *Planta.* **203:** 460-469.

Guo, Z., Ou, W., Lu, S. and Zhong, Q. (2006). Differential responses of antioxidative system to chilling and drought in four rice cultivars differing in sensitivity. *Plant Physiol. Biochem.* **44:** 828-836.

Guo, H-S., Xie, Q., Fei, J-F. and Chua, N-H. (2005). MicroRNA directs mRNA cleavage of the transcription factor NAC1 to downregulate auxin signals for *Arabidopsis* lateral root development. *Plant Cell.* **17:** 1376-1386.

Gupta, C.P., Sharma, A., Dubey, R.C. and Maheshwari, D.K. (1999). *Pseudomonas aeruginosa* as a strong antagonist of *Macrophomina phaseolina* and *Fusarium oxysporum*. *Cytobios.* **99:**183-189.

Guri, A. (1983). Variation in glutathione and ascorbic acid content among selected cultivars of *Phaseolus vulgaris* prior to and after exposure to ozone. *Can. J. Plant Sci.* **63:** 733-737.

Gutteridge, J.M.C. and Halliwell, B. (1990). Reoxygenation injury and antioxidant protection: a tale of two paradoxes. *Arch. Biochem. Biophys.* **283:** 223-226.

Guy, R.D., Warme, P.G. and Reid, D.M. (1984). Glycinebetaine content of halophytes: Improved analysis by liquid chromatography and interpretations of results. *Physiol. Plant.* **61:** 195-202.

Hakam, N. and Simon, J.P. (1996). Effects of low temperatures on the activity of oxygen-scavenging enzymes in two populations of the C_4 grass *Echinochloa crusgalli*. *Physiol. Plant.* **97:** 209-216.

Halliwell, B. and Gutteridge, J. (1984). Oxygen toxicity, oxygen radicals, transition metals and disease. *Biochem. J.* **219:** 1-14.

Harada, E., Yamaguchi, Y., Koizumi, N. and Sano, H. (2002). Cadmium stress induces production of thiol compounds and transcripts for enzymes involved in sulfur assimilation pathways in *Arabidopsis*. *J. Plant Physiol.* **159:** 445-448.

Hasaneen, M.N.A., Younis, M.E. and Tourky, S.M.N. (2008). Plant growth, metabolism and adaptation in relation to stress conditions. XXIII. Salinity biofertility interactive effects on growth, carbohydrates and photosynthetic efficiency of *Lactuca sativa*. *Plant Omics J.* **2:** 60-69.

Hassan, N.M. and Nemat Alla, M.M. (2005). Oxidative stress in herbicide treated broad bean and maize plants. *Acta Physiol. Plant.* **27:** 429-438.

Hausladen, A. and Alschesr, R.G. (1993). Glutathione. In: Alscher, R. and Hess, J. (Eds.). Antioxidants in Higher Plants. Boca Radon: CRC Press, pp. 1-30.

Heath, R.L. (1994). Alterations of plant metabolism by ozone exposure. In: Alscher, R.G. and Wellburn, A.R. (Eds.). Plant Responses to the Gaseous Environment. Chapman and Hall, London, pp. 121-145.

Hefny, M. and Abdel-Kader, D.Z. (2009). Antioxidant-enzyme system as selection criteria for salt tolerance in forage Sorghum genotypes (*Sorghum bicolor* L. Moench). In: Ashraf, M. et al. (Eds.). Salinity and Water Stress, pp. 25-36.

Hell, R. (1997). Molecular physiology of plant sulfur metabolism. *Planta.* **202:** 138-148.

Henzi, T. and Braendle, R. (1993). Long term survival of rhizomatous species under oxygen deprivation. In: Jackson, M.B. and Black, C.R. (Eds.). Interacting Stresses on Plants in a Changing Climate, NATO ASI Series 116, Springler, Berlin, Heidelberg, pp. 305-314.

Hernandez, J.A., Ferrer, M.A., Jimenez, A., Barcelo, A.R. and Sevilla, F. (2001). Antioxidant systems and $O_2^{\cdot-}/H_2O_2$ production in the apoplast of pea leaves: its relation with salt-induced necrotic lesions in minor veins. *Plant Physiol.* **127:** 817-831.

Hernandez, J.A., Jimenez, A., Mullineaux, P. and Sevilla, F. (2000). Tolerance of pea (*Pisum sativum* L.) to long-term salt stress is associated with induction of antioxidant defences. *Plant Cell Environ.* **23:** 853-862.

Hernández-Jiménez, M.J., Lucas, M.M. and de Felipe, M.R. (2002). Antioxidant defence and damage in senescencing lupin nodules. *Plant Physiol. Biochem.* **40:** 645-657.

Hodges, D.M., Andrews, C.J., Johnson, D.A. and Hamilton, R.I. (1997a). Antioxidant enzyme responses to chilling stress in differentially sensitive inbred maize lines. *J. Exp. Bot.* **48:** 1105-1113.

Hodges, D.M., Andrews, C.J., Johnson, D.A. and Hamilton, R.I. (1997b). Antioxidant enzyme and compound responses to chilling stress and their combining abilities in differentially sensitive maize hybrids. *Crop Sci.* **37:** 857-863.

Hodges, D.M., Andrews, C.J., Johnson, D.A. and Hamilton, R.I. (1996). Antioxidant compound responses to chilling stress in differentially sensitive inbred maize lines. *Physiol. Plant.* **98:** 685-692.

Hossain, Z., Lopez-Climent, M.F., Arbona, V., Perez-Clemente, R.M. and Gomez-Cadenas, A. (2009). Modulation of the antioxidant system in citrus under waterlogging and subsequent drainage. *J. Plant Physiol.* **166:** 1391-1404.

Howden, R., Goldsbrough, P.B., Andersen, C.R. and Cobbett, C.S. (1995). Cadmium-sensitive, *cad1* mutants of *Arabidopsis thaliana* are phytochelatin deficient. *Plant Physiol.* **107:** 1059-1066.

Huang, C., He, W., Guo, J., Chang, X., Su, P. and Lixin Zhang, L. (2005). Increased sensitivity to salt stress in an ascorbate-deficient *Arabidopsis* mutant. *J. Exp. Bot.* **56:** 3041-3049.

Huang, M. and Guo, Z. (2005). Responses of antioxidative system to chilling stress in two rice cultivars differing in sensitivity. *Biol. Plant.* **49:** 81-84.

Inze, D. and Van Montagu. M. (1995). Oxidative stress in plants. *Curr. Opinion. Biotech.* **6:** 153-158.

Ishikawa, T., Morimoto, Y., Madhusudhan, R., Sawa, Y., Shibata, H., Yabuta, Y., Nishizawa, A. and Shigeoka, S. (2005). Acclimation to diverse environmental stresses caused by a suppression of cytosolic ascorbate peroxidase in tobacco BY-2 cells. *Plant Cell Physiol.* **46:** 1264-1271.

Ishikawa, T. and Shigeoka, S. (2008). Recent advances in ascorbate biosynthesis and the physiological significance of ascorbate peroxidase in photosynthesizing organisms. *Biosci. Biotechnol. Biochem.* **72:** 1143-1154.

Jiang, M. and Zhang, J. (2001). Effect of abscisic acid on active oxygen species, antioxidative defence system and oxidative damage in leaves of maize seedlings. *Plant Cell Physiol.* **42:** 1265-1273.

Jiang, M. and Zhang, J. (2002). Water stress-induced abscisic acid accumulation triggers the increased generation of reactive oxygen species and up-regulates the activities of antioxidant enzymes in maize leaves. *J. Exp. Bot.* **53:** 2401-2410.

Jiang, M. and Zhang, J. (2003). Crosstalk between calcium and reactive oxygen species originated from NADPH oxidase in abscisic acid-induced antioxidant defense in leaves of maize seedlings. *Plant Cell Environ.* **26:** 929-939.

Jin, J., Shan, N., Maa, N., Bai, J. and Gao, J. (2006). Regulation of ascorbate peroxidase at the transcript level is involved in tolerance to postharvest water deficit stress in the cut rose (*Rosa hybrida* L.) cv. Samantha. *Postharvest Biology and Technology.* **40:** 236-243.

Jones, S., Gonzalez, D.O. and Vodkin, L.O. (2006). Global gene expression profiles of early soybean seed development using microarrays. *Am. Soc. Plant Biologists*, Boston, MA, pp. 255.

Jonson, G., Park, J.C., Noh, T.W., Kim, M.J., Hyun, J.N. and Kim, J.G. (2006). Development of an efficient mechanical inoculation technique to screen barley genotypes for resistance to barley mild mosaic virus disease and its comparison to natural infection. *Plant Pathol. J.* **22:** 348-352.

Kalashnikov, Y.E., Balakhnina, T.I. and Zakrzhevsky, D.A. (1994). Effect of soil hypoxia on activation of oxygen and the system of protection from oxidative destruction in roots and leaves of *Hordeum vulgare*. *Russ. J. Plant Physiol.* **41:** 583-588.

Kaminaka, H., Morita, S., Nakajima, M., Masumura, T. and Tanaka, K. (1998). Gene cloning and expression of cytosolic glutathione reductase in rice (*Oryza sativa* L.). *Plant Cell Physiol.* **39:** 1269-1280.

Kampfenkel, K., Van Montagu, M. and Inzé, D. (1995). Effects of iron excess on *Nicotiana plumbaginifolia* plants: implications to oxidative stress. *Plant Physiol.* **107:** 725-735.

Kang, H.M. and Saltveit, M.E. (2002a). Effect of chilling on antioxidant enzymes and DPPH-radical scavenging activity of high- and low-vigour cucumber seedling radicles. *Plant Cell Environ.* **25:** 1233-1238.

Kang, H.M. and Saltveit, M.E. (2002b). Antioxidant enzymes and DPPH-radical scavenging activity in chilled and heat shocked rice (*Oryza sativa* L.) seedlings radicles. *J. Agr. Food Chem.* **50:** 513-518.

Kang, Y.J. (1992). Exogenous glutathione decreases cellular cadmium uptake and toxicity. *Drug Metab. Dispos.* **20:** 714-718.

Kangasjarvi, J., Talvinen, J., Utriainen, M. and Karjalainen, R. (1994). Plant defense systems induced by ozone. *Plant Cell Environ.* **17:** 783-794.

Karpinski, S., Escobar, C., Karpinska, B., Creissen, G. and Millineaux, P. (1997). Photosynthetic electron transport regulates the expression of cytosolic ascorbate peroxidase genes in *Arabidopsis* during excess light stress. *The Plant Cell.* **9:** 627-640.

Karpinski, S., Reynolds, H., Karpinska, B., Wingsle, G., Creissen, G. and Millineaux, P. (1999). Systemic signaling in response to excess excitation energy in *Arabidopsis. Science.* **284:** 654-657.

Karuppanapandian, T., Sinha, P.B., Haniya, A.M.K. and Manoharan, K. (2006b). Differential antioxidative responses of ascorbate-glutathione cycle enzymes and metabolites to chromium stress in green gram *(Vigna radiata* L. Wilczek) leaves. *J. Plant Biol.* **49:** 440-447.

Karuppanapandian, T., Sinha, P.B., Premkumar, G. and Manoharan, K. (2006a). Chromium toxicity: Correlated with increased in degradation of photosynthetic pigments and total soluble protein and increased peroxidase activity in green gram *(Vigna radiata* L.) seedlings. *J. Swamy Bot-Cl.* **23:** 117-122.

Kastori, S., Petrovic, M. and Petrovic, N. (1992). Effect of excess lead, cadmium, copper and zinc on water relations in sunflower. *J. Plant Nutr.* **15:** 2527-2439.

Kefeli, V.I. (1981). Vitamins and some other representatives of non-hormonal plant growth regulators. *Prible Biochem. Microbiol.* **17:** 5-15.

Kenichi, O. (2005) Glutathione-associated regulation of plant growth and stress responses. *Antioxid. Redox Signal.* **7:** 973-981.

Keyhani, E., Ghamsari, L., Keyhani, J. and Hadizadeh, M. (2006). Antioxidant enzymes during hypoxia-anoxia signaling events in *Crocus sativus* L. corm. *Ann. N. Y. Acad. Sci.* **1091:** 65-75.

Khan, A., Ahmad, M.S.A., Athar, H-U. and Ashraf, M. (2006). Interactive effect of foliarly applied ascorbic acid and salt stress on wheat (T*riticum aestivum* L.) at the seedling stage. *Pak. J. Bot.* **38:** 1407-1414.

Khan, I., Ahmad, A. and Iqbal, M. (2009). Modulation of antioxidant defence system for arsenic detoxification in Indian mustard. *Ecotoxicol. Environ. Safety.* **72:** 626-634.

Khan, M.H. and Panda, S.K. (2008). Alterations in root lipid peroxidation and antioxidative responses in two rice cultivars under NaCl-salinity stress. *Acta Physiol. Plant.* **30:** 81-89.

Khatun, S., Ali, M.B., Hahn, E-J. and Paek, K-Y. (2008). Copper toxicity in *Withania somnifera*: Growth and antioxidant enzymes responses of *in vitro* grown plants. *Environ. Exp. Bot.* **64:** 279-285.

Kiddle, G., Pastori, G.M., Bernard, S., Pignocchi, C., Antoniw, J., Verrier, P.J. and Foyer, C.H. (2003). Effects of leaf ascorbate content on defence and photosynthesis gene expression in *Arabidopsis thaliana. Antioxiants and Redox Signalling.* **5:** 3-32.

Kim, S.Y., Lim, J-H., Park, M.R., Kim, Y.J., Park, T.I., Seo, Y.W., Choi, K.G. and Yun, S.J. (2005). Enhanced antioxidant enzymes are associated with reduced hydrogen peroxide in barley roots under saline stress. *J. Biochem. Mol. Biol.* **38:** 218-224.

Kinraide, T.B. (1991). Identity of the rhizotoxic aluminium species. *Plant Soil.* **134:** 167-178.

Kobayashi, K. and Okada, M. (1995). Effects of ozone on the light use of rice *(Oryza sativa* L.) plants. *Agric. Ecosys. Environ.* **53:** 1-12.

Kocsy, G., von Ballmoos, P., Rüegsegger, A., Szalai, G., Galiba, G. and Brunold, C. (2001). Increasing the glutathione content in a chilling-sensitive maize genotype using safeners increased protection against chilling-induced injury. *Plant Physiol.* **127:** 1147-1156.

Kocsy, G., Kobrehel, K., Szalai, G., Duviau, M-P., Buzás, Z. and Galiba, G. (2004). Abiotic stress-induced changes in glutathione and thioredoxin *h* levels in maize. *Environ. Exp. Bot.* **52:** 101-112.

Kocsy, G., Szalai, G. and Galiba, G. (2002). Effect of heat stress on glutathione biosynthesis in wheat. *Acta Biol. Szeged.* **46:** 71-72.

Kocsy, G., Szalai, G., Vagujfalvi, A., Stehli, L., Orosz, G. and Galiba, G. (2000a). Genetic study of glutathione accumulation during cold hardening in wheat. *Planta.* **210:** 295-301.

Kocsy, G., von Ballmoons, P., Suter, M., Ruegsegger, A., Galli, U., Szalai, G., Galiba, G. and Brunold, C. (2000b). Inhibition of glutathione synthesis reduces chilling tolerance in maize. *Planta.* **211:** 528-536.

Kohler, B. and Blatt, M.R. (2002). Protein phosphorylation activates the guard cell Ca^{2+} channel and is a prerequisite for gating by abscisic acid. *Plant J.* **32:** 185-194.

Kopriva, S. and Rennenberg, H. (2004). Control of sulphate assimilation and glutathione synthesis: interaction with N and C metabolism. *J. Exp. Bot.* **55:** 1831-1842.

Kornyeyev, D., Logan, B.A., Payton, P., Allen, R.D. and Holaday, A.S. (2001). Enhanced photochemical light utilization and decreased chilling-induced photoinhibition of photosystem II in cotton overexpressing genes encoding chloroplast-targeted antioxidant enzymes. *Physiol. Plant.* **113:** 323-331.

Kornyeyev, D., Logan, B.A., Payton, P., Allen, R.D. and Holaday, A.S. (2003). Elevated chloroplastic glutathione reductase activities decrease chilling-induced photoinhibition by increasing rates of photochemistry, but not thermal energy dissipation, in transgenic cotton. *Funct. Plant Biol.* **30:** 101-110.

Kotas, J. and Stasicka, Z. (2000). Commentary: chromium occurrence in the environment and methods of its speciation. *Environ. Poll.* **107:** 263-283.

Kubo, A., Saji, H., Tanaka, K. and Kondo, N. (1995). Expression of *Arabidopsis* cytosolic ascorbate peroxidase gene in response to ozone or sulfur dioxide. *Plant Mol. Biol.* **29:** 479-489.

Kuk, Y.I., Shin, J.S., Burgos, N.R., Hwang, T.E., Han, O., Cho, B.H., Jung, S. and Guh, J.O. (2003). Antioxidative enzymes offer protection from chilling damage in rice plants. *Crop Sci.* **43:** 2109-2117.

Kukreja, S., Nandwal, A.S., Kumar, N., Sharma, S.K., Sharma, S.K., Unvi, V. and Sharma, P.K. (2005). Plant water status, H_2O_2 scavenging enzymes, ethylene evolution and membrane integrity of *Cicer arietinum* roots as affected by salinity. *Biol. Plant.* **49:** 305-308.

Kunert, K.J. and Foyer, C.H. (1993). Thiol/disulphide exchange in plants In: De Kok, L.J., Stulen, I., Rennenberg, H., Brunhold, C. and Rausen, W. (Eds.). Sulfur nutrition and assimilation in higher plants. Regulatory Agricultural and environmental Aspects, SPB Acedemics, The Hague, pp. 139-151.

Kuzniak, E. and Sklodowska, M. (2005). Fungal pathogen-induced changes in the antioxidant systems of leaf peroxisomes from infected tomato plants. *Planta.* **222:** 192-200.

Kwon, S.Y., Choi, S.M., Ahn, Y.O., Lee, H.S., Lee, H.B. and Park, Y.M. (2003). Enhanced stress-tolerance of transgenic tobacco plants expressing a human dehydroascorbate reductase gene. *J. Plant Physiol.* **160:** 347-353.

Langebartels, C., Wohlgemuth, H., Kschieschan, S., Grun, S. and Sandermann, H. (2002). Oxidative burst and cell death in ozone exposed plants. *Plant Physiol. Biochem.* **4:** 567-575.

Lascano, H.R., Casano, L.M., Melchiorre, M.N. and Trippi, V.S. (2001). Biochemical and molecular characterization of wheat chloroplastic glutathione reductase. *Biol. Plant.* **44:** 509-516.

Laulhère, J.P. and Briat, J.F. (1993). Iron release and uptake by plant ferritin: effects of pH, reduction and chelation. *Biochem J.* **290:** 693–699.

Lee, B.H., Won, S.H., Lee, H.S., Miyao, M., Chung, W.I., Kim, I.J. and Jo, J. (2000). Expression of the chloroplast-localized small heat shock protein by oxidative stress in rice. *Gene.* **245:** 283-290.

Lee, S. and Leusteck, T. (1999). The effect of cadmium on sulfate assimilation enzymes in *Brassica juncea. Plant Sci.* **141:** 201-207.

Leterrier, M., Corpas, F.J., Barroso, J.B., Sandalio, L.M. and del Rio, L.A. (2005). Peroxisomal monodehydroascorbate reductase. Genomic clone characterization and functional analysis under environmental stress conditions. *Plant Physiol.* **138:** 2111-2123.

Li, M., Liang, D., Pu, F., Ma, F., Hou, C. and Lu, T. (2009). Ascorbate levels and the activity of key enzymes in ascorbate biosynthesis and recycling in the leaves of 22 Chinese persimmon cultivars. *Scientia Horticulturae.* **120:** 250-256.

Lin, C.C. and Kao, C.H. (2000a). Effect of NaCl stress on H_2O_2 metabolism in rice leaves. *Plant Growth Regul.* **30:** 151-155.

Lin, J.N. and Kao, C.H. (2000b). Involvement of lipid peroxidation in water stress promoted senescence of detached rice leaves. *Biol. Plant.* **43:** 141-145.

Lin, K.H., Chao, P.Y. and Yang, C.M. (2006). The effects of flooding and drought stresses on the antioxidant constituents in sweet potato leaves. *Bot. Stud.* **47:** 417-426.

Lin, K.H., Weng, C.C., Lo, H.F. and Chen, J.T. (2004). Study of the root antioxidative system of tomatoes and eggplant under waterlogged conditions. *Plant Sci.* **167:** 355-365.

Lingua, G., Franchin, C, Todeschini, V., Castiglione, S., Biondi, S., Burlando, B., Parravicini, V., Torrigiani, P. and Berta, G. (2008). Arbuscular mycorrhizal fungi differentially affect the response to high zinc concentrations of two registered poplar clones. *Environ. Pollut.* **153:** 137-147.

Liu D, Li, T-Q., Jin, X-F., Yang, X-E., Islam, E. and Mahmood, Q. (2008a). Lead induced changes in the growth and antioxidant metabolism of the lead accumulating and non-accumulating ecotypes of *Sedum alfredii. J. Integr. Plant Biol.* **50:** 129-140.

Liu, K-L., Shen, L., Wang, J-Q. and Sheng, J-P. (2008b). Rapid inactivation of chloroplastic ascorbate peroxidase is responsible for oxidative modification to Rubisco in tomato (*Lycopersicon esculentum*) under cadmium stress. *J. Integr. Plant Biol.* **50:** 415-426.

Liu, Y.G., Wang, X., Zeng, G.M., Qu, D., Gu, J.J. and Zhou, M. (2007). Cadmium-induced oxidative stress and response of the ascorbate glutathione cycle in *Bechmeria nivea* (L.) Gaud. *Chemosphere.* **69:** 99-107.

Locato, V., de Pinto, M.C. and De Gara, L. (2009). Different involvement of the mitochondrial, plastidial and cytosolic ascorbate-glutathione redox enzymes in heat shock responses. *Physiol. Plant.* **135:** 296-306.

Locato, V., Gadaleta, C., De Gara, L. and de Pinto, M.C. (2008). Production of reactive species and modulation of antioxidant network in response to heat shock: a critical balance for cell fate. *Plant Cell Environ.* **31:** 1608-1619.

Logan, B.A., Monteiro, G., Kornyeyev, D., Payton, P., Allen, R.D. and Scott Holaday, A. (2003). Transgenic overproduction of glutathione reductase does not protect cotton, *Gossypium hirsutum* (Malvaceae), from photoinhibition during growth under chilling conditions. *American J. Bot.* **90:** 1400-1403.

Loggini, B., Scartazza, A., Brugnoli, E. and Navari-Izzo, F. (1999). Antioxidative defence system, pigment composition, and photosynthtic efficiency in two wheat cultivars subjected to drought. *Plant Physiol.* **119:** 1091-1099.

Lohmann, C., Eggers-Schumacher, G., Wunderlich, M. and Schoffl, F. (2004). Two different heat shock transcription factors regulate immediate early expression of stress genes in *Arabidopsis. Mol. Genet. Genomics.* **271:** 11-21.

Lombardi, L. and Sebastiani, L. (2005). Copper toxicity in *Prunus cerasifera*: growth and antioxidant enzymes response of *in vitro* grown plants. *Plant Sci.* **168:** 797-802.

Long, S.P. and Humphries, S. (1994). Photoinhibition of photosynthesis in nature. *Annu. Rev. Plant Physiol. Plant Mol. Biol.* **45:** 633-662.

Lu, Z., Liu, D. and Liu, S. (2007). Two rice cytosolic ascorbate peroxidases differentially improve salt tolerance in transgenic *Arabidopsis. Plant Cell Rep.* **26:** 1909-1917.

Luna, C.M., Gonzalez, C.A. and Trippi, V.S. (1994). Oxidative damage caused by an excess of copper in oat leaves. *Plant Cell Physiol.* **35:** 11-15.

Lunde, C., Baumann, Y., Shirley, N.J., Drew, D.P. and Fincher, G.B. (2006). Gene structure and expression pattern analysis of three monodehydroascorbate reductase (Mdhar) genes in *Physcomitrella patens*: implications for evolution of the MDHAR family in plants. *Plant Mol. Biol.* **60:** 259-275.

Luwe, M.F.W., Takahama, U. and Heber, U. (1993). Role of ascorbate in detoxifying ozone in the apoplast of spinach (*Spinuciu oleraceu* L.) leaves. *Plant Physiol.* **101:** 969-976.

Ma, Y-H., Ma, F-W., Zhang, J-K., Li, M-J., Wang, Y-H. and Liang, D. (2008). Effects of high temperature on activities and gene expression of enzymes involved in ascorbate–glutathione cycle in apple leaves. *Plant Sci.* **175:** 761-766.

Madamanchi, N.R., Donahue, J.L., Cramer, C.L., Alscher, R.G. and Pedersen, K. (1994). Differential response of Cu, Zn superoxide dismutases in two pea cultivars during a short-term exposure to sulfur dioxide. *Plant Mol. Biol.* **26:** 95-103.

Maddison, J., Lyons, T., Plochl, M. and Barnes, J.D. (2002). Hydroponically cultivated radish fed L-galactono-1, 4-lactone exhibit increased tolerance to ozone. *Planta.* **214:** 383-391.

Madhava Rao, K.V. and Sresty, T.V.S. (2000). Antioxidative parameters in the seedlings of pigeonpea (*Cajanus cajan* (L.) Millspaugh) in response to Zn and Ni stresses. *Plant Sci.* **157:** 113-128.

Madhusudhan, R., Ishikawa, T., Sawa, Y., Shigeoka, S. and Shibata, H. (2003). Characterization of an ascorbate peroxidase in plastids of tobacco BY-2 cells. *Physiol. Plant.* **117:** 550-557.

Mahan, J.R. and Mauget, S.A. (2005). Antioxidant metabolism in cotton seedlings exposed to temperature stress in the field. *Crop Sci.* **45:** 2337-2345.

Maksymiec, W. (1997). Effect of copper on cellular processes in higher plants. *Photosynthetica.* **34:** 321-342.

Mallick, N. (2004). Copper-induced oxidative stress in the chlorophycean microalga *Chlorella vulgaris*: response of the antioxidant system. *J. Plant Physiol.* **161:** 591-597.

Mallick, N. and Rai, L.C. (1999). Response of the antioxidant systems of the nitrogen fixing cyanobacterium *Anabena doliolum* to copper. *J. Plant Physiol.* **155:** 146-149.

Marrs, K. (1996). The functions and regulation of gluthathione S-transferases in plants. *Annu. Rev. Plant Physiol. Plant Mol. Biol.* **48:** 127-158.

Maughan, S. and Foyer, C.H. (2006). Engineering and genetic approaches to modulating the glutathione network in plants. *Physiol. Plant.* **126:** 382-397.

May, M., Vernoux, T., Leaver, C., Van Montagu, M. and Inze, D. (1998). Glutathione homeostasis in plants: implications for environmental sensing and plant development. *J. Exp. Bot.* **49:** 649-667.

Mayer, A.J. and Hell, R. (2005). Glutathione homeostasis and redox-regulation by sulfhydryl groups. *Photosynthetic Res.* **86:** 435-457.

Mazza, C.A., Battista, D., Zima, A.M., Szwarcberg-Bracchitta, M., Giordano, C.V., Acevedo, A., Scopel, A.L. and Ballare, C.L. (1999). The effects of solar ultraviolet-B radiation on the growth and yield of barley are accompanied by increased DNA damage and antioxidant responses. *Plant Cell Environ.* **22:** 61-70.

McGrath, S.P. (1982). The uptake and translocation of tri- and hexavalent chromium and effects on the growth of oat in flowing nutrient solution and in soil. *New Phytol.* **92:** 381-390.

McLaughlin, S.B. and Taylor, G.E. (1981). Relative humidity: Important modifier of pollutant uptake by plants. *Science.* **211:** 167-169.

Meharg, A.A. (2003). Variation in arsenic accumulation-hyperaccumulation in ferns and their allies. *New Phytol.* **157:** 25-31.

Meharg, A.A. and Hartley-Whitaker, J. (2002). Arsenic uptake and metabolism in arsenic resistant and non-resistant plant species. *New Phytol.* **154:** 29-43.

Mehler, A.H. (1951). Studies on reactions of illuminated chloroplasts. I. Mechanisms of the reduction of oxygen and other Hill reagents. *Arch. Biochem. Biophys.* **33:** 65-77.

Mehlhorn, H., Cottam, D.A., Lucas, P.W. and Wellburn, A.R. (1987). Induction of ascorbate peroxidase and glutathione reductase activities by interactions of mixtures of air pollutants. *Free Rad. Res. Comm.* **3:** 1-5.

Mehlhorn, H., Seufert, G., Schmidt, A. and Kunert, K.J. (1986). Effect of SO_2 and O_3 on production of antioxidants in conifers. *Plant Physiol.* **82:** 336-338.

Meloni, D.A., Oliva, M.A., Martinez, C.A. and Cambraia, J. (2003). Photosynthesis and activity of superoxide dismutase, peroxidase and glutathione reductase in cotton under salt stress. *Environ. Exp. Bot.* **49:** 69-76.

Menconi, M., Sgherri, C.L.M., Pinzino, C. and Navari-Izzo, F. (1995). Activated oxygen production and detoxification in wheat plants subjected to a water deficit programme. *J. Exp. Bot.* **46:** 1123-1130.

Millar, A.H., Mittova, V., Kiddle, G., Heazlewood, J.L., Bartoli, C.G. and Theodoulou, F.L. (2003). Control of the ascorbate synthesis by respiration and its implication for stress responses. *Plant Physiol.* **133:** 443-447.

Mishra, N.P., Mishra, R.K. and Singhal, G.S. (1993). Changes in the activities of anti-oxidant enzymes during exposure of intact wheat leaves to strong visible light at different temperatures in the presence of protein synthesis inhibitors. *Plant Physiol.* **102:** 903-910.

Mishra, R.P.N., Singh, R.K., Jaiswal, H.K., Kumar, V. and Maurya, S. (2006). Rhizobium mediated induction of phenolics and plant growth promotion in rice (*Oryza sativa* L.). *Curr. Microbiol.* **52:** 383-389.

Mittler, R. (2002). Oxidative stress, antioxidants and stress tolerance. *Trends Plant Sci.* **7:** 405-410.

Mittler, R., Vanderauwera, S., Gollery, M. and Van Breusegem, F. (2004). Reactive oxygen gene network of plants. *Trends Plant Sci.* **9:** 490-498.

Mittler, M. and Zilinskas, B.A. (1994). Regulation of pea cytosolic peroxidase and other antioxidant enzymes during the progression of drought stress and following recovery from drought. *Plant J.* **5:** 397-405.

Mittova, V., Guy, M., Tal, M. and Volokita, M. (2004). Salinity upregulates the antioxidative system in root mitochondria and peroxisomes of the wild salt-tolerant tomato species *Lycopersicon pennellii*. *J. Exp. Bot.* **55:** 1105-1113.

Mittova, V., Tal, M., Volokita, M. and Guy, M. (2002). Salt stress induces upregulation of an efficient chloroplast antioxidant system in the salt tolerant wild tomato species *Lycopersicon pennellii* but not in the cultivated species. *Physiol. Plant.* **115:** 393-400.

Mittova, V., Tal, M., Volokita, M. and Guy, M. (2003). Upregulation of the leaf mitochondrial and peroxisomal antioxidative systems in response to salt-induced oxidative stress in the wild salt-tolerant tomato species *Lycopersicon pennellii*. *Plant Cell Environ.* **26:** 845-856.

Mittova, V., Volokita, M., Guy, M. and Tal, M. (2000). Activities of SOD and the ascorbate–glutathione cycle enzymes in subcellular compartments in leaves and roots of the cultivated tomato and its wild salt-tolerant relative *Lycopersicon pennellii*. *Physiol. Plant.* **110:** 42-51.

Moldau, H. and Bichele, I. (2002). Plasmalemma protection by the apoplast as assessed from above-zero ozone concentrations in leaf intercellular air spaces. *Planta.* **214:** 484-487.

Molina, A., Bueno, P., Marín, M.C., Rosales, M.P.R., Belver, A., Venema, K. and Donaire, J.P. (2002). Involvement of endogenous salicylic acid content, lipoxygenase and antioxidant enzyme activities in the response of tomato cell suspension cultures to NaCl. *New Phytol.* **156:** 409-415.

Moller, I.M. (2001). Plant mitochondria and oxidative stress: electron transport, NADPH turnover, and metabolism of reactive oxygen species. *Annu. Rev. Plant Physiol. Plant Mol. Biol.* **52:** 561-591.

Morelli, E. and Scarano, G. (2004). Copper-induced changes of nonprotein thiols and antioxidant enzymes in the marine microalga *Phaeodactylum tricornutum*. *Plant Sci.* **167:** 289-296.

Morgan, P.B., Bernacchi, C.J., Ort, D.R. and Long, S.P. (2004). An *in vivo* analysis of the effect of season-long open-air elevation of ozone to anticipated 2050 levels on photosynthesis in soybean. *Plant Physiol.* **135:** 2348-2357.

Morgan, C.L.S., Norman, J.M. and Lowery, B. (2003). Estimating plant-available water across a field with an inverse yield model. *Soil Sci. Soc. Am. J.* **67:** 620-629.

Mozafar, A. and Oertli, J.J. (1992). Uptake and transport of thiamin (vitamin B_1) by barley and soybean. *J. Plant Physiol.* **436:** 442.

Mrozek, J.E. and Funicelli, N.A. (1982). Effect of zinc and lead on germination of *Spartina alterniflora* Loisel seeds at various salinities. *Environ. Exp. Bot.* **22:** 23-32.

Mullineaux, P. and Karpinski, S. (2002). Signal transduction in response to excess light: getting out of the chloroplast. *Curr. Opin. Plant Biol.* **5:** 43-48.

Nagalakshmi, N. and Prasad, M.N.V. (2001). Responses of glutathione cycle enzymes and glutathione metabolism to copper stress in *Scenedesmus bijugatus*. *Plant Sci.* **160:** 291-299.

Nagata, T., Todoriki, S., Masumizu, T., Suda, I., Furuta, S., Du, Z. and Kikuchi, S. (2003). Levels of active oxygen species are controlled by ascorbic acid and anthocyanin in *Arabidopsis*. *J. Agric. Food Chem.* **51:** 2992-2999.

Nagesh Babu, R. and Devaraj, V.R. (2008). High temperature and salt stress response in French bean (*Phaseolus vulgaris*). *Austr. J. Crop Sci.* **2:** 40-48.

Nair, A.S., Abraham, T.K. and Jaya, D.S. (2008). Studies on the changes in lipid peroxidation and antioxidants in drought stress induced cowpea (*Vigna unguiculata* L.) varieties. *J. Environ. Biol.* **29:** 689-691.

Navari-Izzo, F., Meneguzzo, S., Loggini, B., Vazzana, C. and Sgherri, C.L.M. (1997). The role of the glutathione system during dehydration of *Boea hygroscopica*. *Physiol. Plant.* **99:** 23-30.

Nemat Alla, M.M., Badawi, A-H.M., Hassan, N.M., El-Bastawisy, Z.M. and Badran, E.G. (2008b). Herbicide tolerance in maize is related to increased levels of glutathione and glutathione-associated enzymes. *Acta Physiol. Plant.* **30:** 371-379.

Nemat Alla, M.M., Badawi, A.M., Hassan, N.M., El-Bastawisy, Z.M. and Badran, E.G. (2007). Induction of glutathione and glutathione associated enzymes in butachlor-tolerant plant species. *Am. J. Plant Physiol.* **2:** 195-205.

Nemat Alla, M.M., Badawi, A.M., Hassan, N.M., El-Bastawisy, Z.M. and Badran, E.G. (2008a). Effect of metribuzin, butachlor and chlorimuron-ethyl on amino acid and protein formation in wheat and maize seedlings. *Pestic. Biochem. Physiol.* **90:** 8-18.

Nemat Alla, M.M. and Hassan, N.M. (2006). Changes of antioxidants levels in two maize lines following atrazine treatments. *Plant Physiol. Biochem.* **44:** 202-210.

Nemat Alla, M.M. and Hassan, N.M. (2007). Changes of antioxidants and GSH-associated enzymes in isoproturon-treated maize. *Acta Physiol. Plant.* **29:** 247-258.

Neubauer, C. and Yamamoto, H. (1992). Mehler-peroxidase reaction mediates zeaxanthin-related fluorescence quenching in intact chloroplasts. *Plant Physiol.* **99:** 1354-1361.

Nieto-Sotelo, J. and Ho, T-H.D. (1986). Effect of heat shock on the metabolism of glutathione in maize roots. *Plant Physiol.* **82:** 1030-1035.

Niyogi, K.K. (1999). Photoprotection revisited: genetic and molecular approaches. *Annu. Rev. Plant Physiol. Plant Mol. Biol.* **50:** 333-359.

Nocito, F.F., Pirovano, L., Cocucci, M. and Sacchi, G.A. (2002). Cadmium induced sulfate uptake in maize roots. *Plant Physiol.* **129:** 1872-1879.

Noctor, G. (2006). Metabolic signalling in defence and stress: the central roles of soluble redox couples. *Plant Cell Environ.* **29:** 409-425.

Noctor, G., Arisi, A.M., Jouanin, L., Kunert, K.J., Rennenberg, H. and Foyer, C.H. (1998). Glutathione: biosynthesis, metabolism and relationship to stress tolerance explored in transformed plants. *J. Exp. Bot.* **49:** 623-647.

Noctor, G. and Foyer, C.H. (1998). Ascorbate and glutathione: keeping active oxygen under control. *Annu. Rev. Plant Physiol. Plant Mol. Biol.* **49:** 249-279.

Noctor, G., Gomez, L., Vanacker, H. and Foyer, C.H. (2002a). Interactions between biosynthesis, compartmentation and transport in the control of glutathione homeostasis and signalling. *J. Exp. Bot.* **53:** 1283-1304.

Noctor, G., Veljovic-Jovanovic, S., Driscoll, S., Novitskaya, L. and Foyer, C.H. (2002b). Drought and oxidative load in the leaves of C3 plants: a predominant role for photorespiration? *Ann. Bot.* **89:** 841-850.

Oidaira, H., Sano, S., Koshiba, T. and Ushimaru, T. (2000). Enhancement of antioxidative enzyme activities in chilled rice seedlings. *J. Plant Physiol.* **156:** 811-813.

Padh, H. (1990). Cellular function of ascorbic acid. *Biochem. Cell Biol.* **68:** 1270-1275.

Palada, M. and Vergara, B.S. (1972). Environmental effect on the resistance of rice seedlings to complete submergence. *Crop Sci.* **12:** 209-212.

Palma, J.M. (2006). Glutathione reductase from pea leaves: response to abiotic stress and characterization of the peroxisomal isozyme. *New Phytol.* **170:** 43-52.

Panchuk, I.I., Volkov, R.A. and Schoffl, F. (2002). Heat stress- and heat shock transcription factor-dependent expression and activity of ascorbate peroxidase in *Arabidopsis*. *Plant Physiol.* **129:** 838-853.

Panda, S.K. and Khan, M.H. (2004). Changes in growth and superoxide dismutase activity in *Hydrilla verticillata* L. under abiotic stress. *Braz. J. Plant Physiol.* **16:** 115-118.

Pandey, N. and Sharma, C.P. (2002). Effect of heavy metals Co^{2+}, Ni^{2+} and Cd^{2+} on growth and metabolism of cabbage. *Plant Sci.* **163:** 753-758.

Pang, X., Wang, D.H. and Peng, A. (2001). Effects of mercury stress on the activity of antioxidant enzymes. *Environ. Chem.* **20:** 351-355.

Paradiso, A., Berardino, R., de Pinto, M.C., Sanita' di Toppi, L., Storelli, M.M., Tommasi, F. and Gara, L.D. (2008). Increase in Ascorbate–Glutathione metabolism as local and precocious systemic responses induced by cadmium in Durum wheat plants. *Plant Cell Physiol.* **49:** 362-374.

Pastori, G., Foyer, C.H. and Mullineaux, P. (2000). Low temperature-induced changes in the distribution of H_2O_2 and antioxidants between the bundle sheath and mesophyll cells of maize leaves. *J. Exp. Bot.* **51:** 107-113.

Pastori, G.M., Kiddle, G., Antoniw, J., Bernard, S., Veljovic-Jovanovic, S., Verrier, P.J., Noctor, G. and Foyer, C.H. (2003). Leaf vitamin C contents modulate plant defense transcripts and regulate genes that control development through hormone signaling. *Plant Cell.* **15:** 939-951.

Pastori, G.M. and Trippi, V.S. (1992). Oxidative stress induces high rate of glutathione reductase synthesis in a drought resistant maize strain. *Plant Cell Physiol.* **33:** 957-961.

Pawlik-Skowronska, B. (2003). When adapted to high zinc concentrations the periphytic green alga *Stigeoclonium tenue* produces high amounts of novel phytochelatin-related peptides. *Aquat. Toxicol.* **62:** 155-163.

Payton, P., Webb, R., Kornyeyev, D., Allen, R. and Holaday, A.S. (2001). Protecting cotton photosynthesis during moderate chilling at high light intensity by increasing chloroplastic antioxidant enzyme activity. *J. Exp. Bot.* **52:** 2345-2354.

Pekker, I., Tel-Or, E. and Mittler, R. (2002). Reactive oxygen intermediates and glutathione regulate the expression of cytosolic ascorbate peroxidase during iron-mediated oxidative stress in bean. *Plant Mol. Biol.* **49:** 429-438.

Pell, E.J. and Dann, M.S. (1991). Multiple stress and plant senescence. In: Mooney, H.A., Winner, W.E. and Pell, E.J. (Eds.). Integrated Response of Plants to Stress. Academic Press, San Diego, CA, pp. 189-204.

Peltzer, D., Dreyer, E. and Polle, A. (2002). Differential temperature dependencies of antioxidative enzymes in two contrasting species: *Fagus sylvatica* and *Coleus blumei*. *Plant Physiol. Biochem.* **40:** 141-150.

Peuke, A.D. and Rennenberg, H. (2005). Phytoremediation: molecular biology, requirements for application, environmental protection, public attention and feasibility. *EMBO Rep.* **6:** 497-501.

Pfister-Sieber, M. and Braendle, R. (1994). Aspects of plant behavior under anoxia and post-anoxia. *Proc. R. Soc. Edinb.* **102B:** 313-324.

Pignocchi, C., Fletcher, J.M., Wilkinson, J.E., Barnes, J.D. and Foyer, C.H. (2003). The function of ascorbate oxidase in tobacco. *Plant Physiol.* **132:** 1631-1641.

Pignocchi, C. and Foyer, C.H. (2003). Apoplastic ascorbate metabolism and its role in the regulation of cell signalling. *Curr. Opin. Plant Biol.* **6:** 379-389.

Pineros, M.A. and Kochian, L.V. (2001). A patch-clamp study on the physiology of aluminum toxicity and aluminum tolerance in maize. Identification and characterization of Al3þ-induced anion channels. *Plant Physiol.* **125:** 292-305.

Piqueras, A., Olmos, E., Martýnez-Solano, J.R. and Hellýn, E. (1999). Cd-induced oxidative burst in tobacco BY2 cells: time course, subcellular location and antioxidant response. *Free Rad. Res.* **31** (Suppl.)**:** S33-S38.

Plochl, M., Lyons, T., Ollerenshaw, J. and Barnes, J. (2000). Simulating ozone detoxification in the leaf apoplast through the direct reaction with ascorbate. *Planta.* **210:** 454-467.

Pnueli, L., Liang, H., Rozenberg, M. and Mittler, R. (2003). Growth suppression, altered stomatal responses, and augment induction of heat shock proteins in cytosolic ascorbate peroxidase (Apx1)-deficient *Arabidopsis* plants. *Plant J.* **34:** 187-203.

Polle, A. (1998). Photochemical oxidants: uptake and detoxification mechanisms. In: De Kok, L.J. and Stulen, I. (Eds.). Responses of plant metabolism to air pollution and global change. Leiden, The Netherlands, Backhuys Publishers, pp. 95-116.

Powles, S.B. (1984). Photoinhibition of photosynthesis induced by visible light. *Annu. Rev. Plant Physiol.* **35:** 15-44.

Prasad, K.V.S.K., Saradhi, P.P. and Sharmila, P. (1999). Concerted action of antioxidant enzymes and curtailed growth under zinc toxicity in *Brassica juncea*. *Environ. Exp. Bot.* **42:** 1-10.

Prasad, T.K., Anderson, M. and Stewart, C.R. (1995). Localisation and characterization of peroxidases in the mitochondria of chilling-acclimated maize seedlings. *Plant Physiol.* **108:** 1597-1605.

Price, A.H., Taylor, A., Ripley, S.J., Griffiths, A., Trewavas, A.J. and Knight, M. (1994). Oxidative signals in tobacco increase cytosolic calcium. *Plant Cell.* **6:** 1301-1310.

Puppo, A., Groten, K., Bastian, F., Carzaniga, R., Soussi, M., Lucas, M.M., de Felipe, M.R., Harrison, J., Vanacker, H. and Foyer, C.H. (2005). Legume nodule senescence: roles for redox and hormone signalling in the orchestration of the natural aging process. *New Phytol.* **165:** 683-701.

Pyon, J.Y., Piao, R.Z., Roh, S.W., Shin, S.Y. and Kwak, S.S. (2004). Differential levels of antioxidants in paraquat-resistant and susceptible *Erigeron canadensis* biotypes in Korea. *Weed Biol. Manag.* **4:** 75-80.

Qadir, S., Qureshi, M.I., Javed, S. and Abdin, M.Z. (2004). Genotypic variation in phytoremediation potential of *Brassica juncea* cultivars exposed to Cd stress. *Plant Sci.* **167:** 1171-1181.

Quartacci, M.F., GlisÏicÂ, O., StevanovicÂ, B. and Navari-Izzo, F. (2002). Plasma membrane lipids in the resurrection plant *Ramonda serbica* following dehydration and rehydration. *J. Exp. Bot.* **53:** 2159-2166.

Qureshi, M.I., Abdin, M.Z., Qadir, S. and Iqbal, M. (2007). Lead-induced oxidative stress and metabolic alterations in *Cassia angustifolia* Vahl. *Biol. Plant.* **51:** 121-128.

Raeymaekers, T., Potters, G., Asard, H., Guisez, Y. and Horemans, N. (2003). Copper-mediated oxidative burst in *Nicotiana tabacum* L. cv. Bright Yellow 2 cell suspension cultures. *Protoplasma.* **221:** 93-100.

Rahman, M.A., Hasegawa, H., Rahman, M.M., Islam, M.N., Miah, M.A.M. and Tasmin, A. (2007). Effect of arsenic on photosynthesis, growth and yield of five widely cultivated rice (*Oryza sativa* L.) varieties in Bangladesh. *Chemosphere.* **67:** 1072-1079.

Ranieri, A., Castagna, A., Baldan, B., Sebastiani, L. and Soldatini, G.F. (2002). Increased H_2O_2 accumulation in sunflower plants as a consequence of iron deprivation, In: *Proceedings of the Eleventh International Symposium on Iron Nutrition and Interaction in Plants* (Book of Abstracts), Udine, Italy, 23-28 June, pp. 112.

Ranieri, A., Castagna, A., Scebba, F., Careri, M., Cagnoni, I.., Predieri, G., Pagliari, M. and Sanita' di Toppi, L. (2005). Oxidative stress and phytochelatin characterisation in bread wheat exposed to cadmium excess. *Plant Physiol. Biochem.* **43:** 45-54.

Rao, M.V. and Davis, K.R. (2001). The physiology of ozone induced cell death. *Planta.* **213:** 682-690.

Rao, M.V. (1992). Cellular detoxifying mechanisms determine the age dependent injury in tropical trees exposed to SO_2. *J. Plant Physiol.* **140:** 733-740.

Rao, M.V., Koch, J.R. and Davis, K.R. (2000). Ozone: a tool for probing programmed cell death in plants. *Plant Mol. Biol.* **44:** 345-358.

Rao, M.V., Paliyath, G. and Ormrod, D.P. (1996). Ultraviolet-B- and ozone-induced biochemical changes in antioxidant enzymes of *Arabidopsis thaliana*. *Plant Physiol.* **110:** 125-136.

Ratnayaka, H., Molin, W.T. and Sterling, T.M. (2003). Physiological and antioxidant responses of cotton and spurred anoda under interference and mild drought. *J. Exp. Bot.* **54:** 2293-2305.

Reddy, A.R., Chaitanya, K.V. and Vivekanandan, M. (2004). Drought-induced responses of photosynthesis and antioxidant metabolism in higher plants. *J. Plant Physiol.* **161:** 1189-1202.

Reddy, A.M., Kumar, S.G., Jyothsnakumari, G., Thimmanaik, S. and Sudhakar, C. (2005). Lead induced changes in antioxidant metabolism of horsegram (*Macrotyloma uniflorum* (Lam.) (Verdc.) and bangalgram (*Cicer arietinum* L.). *Chemosphere.* **60:** 97-104.

Rennenberg, H. and Brunold, C. (1994) Significance of glutathione metabolism in plants under stress. *Prog. Bot.* **55:** 144-156.

Renuga, G. and Paliwal, K. (1995). Detoxification of SO_2 derivatives in chloroplasts of *Hardwickia binata*. *J. Biosci.* **20:** 59-68.

Requejo, R. and Tena, M. (2005). Proteome analysis of maize roots reveals that oxidative stress is a main contributing factor to plant arsenic toxicity. *Phytochem.* **66:** 1519-1528.

Ribera, D., Labrot, F., Tisnerat, G. and Narbonne, J.F. (1996). Uranium in the environment: occurrence, transfer and biological effects. *Rev. Environ. Contam. Toxicol.* **146:** 53-89.

Richman, P.G. and Meister, A. (1975). Regulation of gamma-glutamyl-cysteine synthetase by nonallosteric feedback inhibition by glutathione. *J. Biol. Chem.* **25:** 1422-1426.

Riiegsegger, A. and Brunold, C. (1992). Effect of cadmium on γ-glutamylcysteine synthesis in maize seedlings. *Plant Physiol.* **99:** 428-433.

Riiegsegger, A., Schmutz, D. and Brunold, C. (1990). Regulation of glutathione synthesis by cadmium in *Pisum sativum* L. *Plant Physiol.* **93:** 1579-1584.

Rivero, R.M., Ruiz, J.M. and Romero, L. (2004). Oxidative metabolism in tomato plants subjected to heat stress. *J. Hort. Sci. Biotech.* **79:** 560-564.

Rodrýguez-Serrano, M., Romero-Puertas, M.C., Zabalza, A., Corpas, F.J., Gomez, M., del Rio, L.A. and Sandalio, L.M. (2006). Cadmium effect on oxidative metabolism of pea (*Pisum sativum* L.) roots. Imaging of reactive oxygen species and nitric oxide accumulation *in vivo*. *Plant Cell Environ.* **29:** 1532-1544.

Romero-Puertas, M.C., Corpas, F.J., Rodrýguez-Serrano, M., Gomez, M., del Rio, L.A. and Sandalio, L.M. (2007). Differential expression and regulation of antioxidative enzymes by cadmium in pea plants. *J. Plant Physiol.* **164:** 1346-1357.

Romero-Puertas, M.C., Corpas, F.J., Sandalio, L.M., Leterrier, M., Rodriguez-Serrano, M., del Rio, L.A. and Palma, J.M. (2006). Glutathione reductase from pea leaves: response to abiotic stress and characterization of the peroxisomal isozyme. *New Phytol.* **170:** 43-52.

Romero-Puertas, M.C., McCarthy, I., Sandalio, L.M., Palma, J.M., Corpas, F.J. and Gomez, M. (1999). Cadmium toxicity and oxidative metabolism of pea leaf peroxisomes. *Free Rad. Res.* **31:** 25-32.

Romero-Puertas, M.C., Rodrýguez-Serrano, M., Corpas, F.J., Gomez, M., del Rio, L.A. and Sandalio, L.M. (2004). Cadmium induced subcellular accumulation of O_2^- and H_2O_2 in pea leaves. *Plant Cell Environ.* **27:** 1122-1134.

Rout, G.R. and Das, P. (2003). Effect of metal toxicity on plant growth and metabolism: I. Zinc. *Agronomie.* **23:** 3-11.

Roxas, V.P., Lodhi, S.A., Garrett, D.K., Mahan, J.R. and Allen, R.D. (2000). Stress tolerance in transgenic tobacco seedlings that overexpress glutathione S-transferase/glutathione peroxidase. *Plant Cell Physiol.* **41:** 1229-1234.

Roxas, V.P., Smith, R.K., Allen, E.R. and Allen, R.D. (1997). Overexpression of glutathione S-transferase/glutathione peroxidase enhances the growth of transgenic tobacco seedlings during stress. *Nat. Biotechnol.* **15:** 988-991.

RoyChoudhury, A., Roy, C. and Sengupta, D.N. (2007). Transgenic tobacco plants overexpressing the heterologous *lea* gene *Rab16A* from rice during high salt and water deficit display enhanced tolerance to salinity stress. *Plant Cell Rep.* **26:** 1839-1859.

Roychoudhury, A., Basu, S., Sarkar, S.N. and Sengupta, D.N. (2008). Comparative physiological and molecular responses of a common aromatic indica rice cultivar to high salinity with non-aromatic indica rice cultivars. *Plant Cell Rep.* **27:** 1395-1410.

Roychoudhury, A., Basu, S. and Sengupta, D.N. (2009). Effects of exogenous abscisic acid on some physiological responses in a popular aromatic indica rice compared with those from two traditional non-aromatic indica rice cultivars. *Acta Physiol. Plant.* **31:** 915-926.

Ruiz, J.M. and Blumwald, E. (2002). Salinity-induced glutathione synthesis in *Brassica napus*. *Planta.* **214:** 965-969.

RuizSanchez, M.C., Domingo, R., Morales, D. and Torrecillas, A. (1996). Water relations of Fino lemon plants on two rootstocks under flooded conditions. *Plant Sci.* **120:** 119-125.

Ruley, A.T., Sharma, N.C. and Sahi, S.V. (2004). Antioxidant defense in a lead accumulating plant, *Sesbania drummondii*. *Plant Physiol. Biochem.* **42:** 899-906.

Sairam, P.K. and Saxena, D.C. (2000). Oxidative stress and antioxidants in wheat genotypes: possible mechanism of water stress tolerance. *J. Agron. Crop Sci.* **184:** 55-61.

Sairam, R.K., Shukla, D.S. and Saxena, D.C. (1997/98). Stress-induced injury and antioxidant enzymes in relation to drought tolerance in wheat genotypes. *Biol. Plant.* **40:** 357-364.

Sairam, R.K., Srivastava, G.C., Agarwal, S. and Meena, R.C. (2004). Differences in antioxidant activity in response to salinity stress in tolerant and susceptible wheat genotypes. *Biol. Plant.* **49:** 85-91.

Sairam, RK, Srivastava, G.C. and Saxena, D.C. (2000). Increased antioxidant activity under elevated temperature: a mechanism of heat stress tolerance in wheat genotypes. *Biol. Plant.* **43:** 245-251.

Salt, D.E., Blaylock, M., Kumar, N.P.B.A., Dushenkov, V., Ensley, D., Chet, I. and Raskin, I. (1995a). Phytoremediation: a novel strategy for the removal of toxic metals from the environment using plants. *Biotechnology.* **13:** 468-474.

Salt, D.E., Prince, R.C., Pickering, I.J. and Raskin, I. (1995b). Mechanisms of cadmium mobility and accumulation in Indian mustard. *Plant Physiol.* **109:** 1427-1433.

Sandalio, L., Dalurzo, H.C., Gomez, M., Romero-Puertas, M. and del Rio, L.A. (2001). Cadmium-induced changes in the growth and oxidative metabolism of pea plants. *J. Exp. Bot.* **52:** 2115-2126.

Sanmartin, M., Drogoudi, P.D., Pateraki, T.L.I., Barnes, J. and Kanellis, A.K. (2003). Over-expression of ascorbate oxidase in the apoplast of transgenic tobacco results in altered ascorbate and glutathione redox states and increased sensitivity to ozone. *Planta.* **216:** 918-928.

Sato, Y., Murakami, T., Funatsuki, H., Matsuba, S., Saruyama, H. and Tanida, M. (2001). Heat shock-mediated APX gene expression and protection against chilling injury in rice seedlings. *J. Exp. Bot.* **52:** 145-151.

Schickeler, H. and Caspi, H. (1999). Response of antioxidative enzymes to nickel and cadmium stress in hyperaccumulator plants of the genus *Alyssum*. *Physiol. Plant.* **105:** 39-44.

Schoner, S. and Krause, G.H. (1990). Protective systems against active oxygen species in spinach: response to cold acclimation in excess light. *Planta.* **180:** 383-389.

Schroeder, J.I., Allen, G.J., Hugouvieux, V., Kwak, J.M. and Waner, D. (2001a). Guard cell signal transduction. *Annu. Rev. Plant Physiol. Plant Mol. Biol.* **52:** 627-658.

Schroeder, J.I., Kwak, J.M. and Allen, G.J. (2001b). Guard cell abscisic acid signaling and engineering drought hardiness in plants. *Nature.* **410:** 327-330.

Schutzendubel, A. and Polle, A. (2002). Plant responses to abiotic stress: heavy metal-induced oxidative stress and protection by mycorrhization. *J. Exp. Bot.* **53:** 1351-1365.

Schutzendubel, A., Schwanz, P., Teichmann, T., Gross, K., Langenfeld-Heyser, R., Godbold, D.L. and Polle, A. (2001). Cadmium-induced changes in antioxidative systems, hydrogen peroxide content, and differentiation in Scots Pine roots. *Plant Physiol.* **127:** 887-898.

Selote, D.S. and Chopra, R.K. (2004). Drought-induced spikelet sterility is associated with an inefficient antioxidant defense in rice panicles. *Physiol. Plant.* **121:** 462-471.

Selote, D.S. and Chopra, R.K. (2006). Drought acclimation confers oxidative stress tolerance by inducing coordinated antioxidant defense at cellular and subcellular level in leaves of wheat seedlings. *Physiol. Plant.* **127:** 494-506.

Sen Gupta, A., Heinen, J.L., Holaday, A.S., Burke, J.J. and Allen, R.D. (1993b). Increased resistance to oxidative stress in transgenic plants that overexpress chloroplastic Cu/Zn superoxide dismutase. *Proc. Natl. Acad. Sci. USA.* **90:** 1629-1633.

Sen Gupta, A., Webb, R.P., Holaday, A.S. and Allen, R.D. (1993a). Overexpression of superoxide dismutase protects plants from oxidative stress. *Plant Physiol.* **103:** 1067-1073.

Sen Gupta, A.S., Alscher, R.G. and McCune, D. (1991). Response of photosynthesis and cellular antioxidants to ozone in Populus leaves. *Plant Physiol.* **96:** 650-655.

Setter, T.L., Kupkanchanakul, T., Waters, I. and Greenway, H. (1988). Evaluation of factors contributing to diurnal changes in floodwater in deepwater rice fields. *New Phytol.* **110:** 151-162.

Setter, T.L., Ramakrishnayya, G., Ram, P.C. and Singh, B.B. (1995). Environmental characteristics of floodwater in Eastern India: relevance to flooding tolerance of rice. *Indian J. Plant Physiol.* **38:** 34-40.

Sgherri, C.L.M., Loggini, B., Bochicchio, A. and Navari-Izzo, F. (1994). Antioxidant system in *Boea hygroscopica*: changes in response to desiccation and rehydration. *Phytochem.* **37:** 377-381.

Shalata, A., Mittova, V., Volokita, M., Guy, M. and Tal, M. (2001). Response of the cultivated tomato and its wild salt-tolerant relative *Lycopersicon pennellii* to salt-dependent oxidative stress: the root antioxidative system. *Physiol. Plant.* **112:** 487-494.

Shalata, A. and Neumann, P.M. (2001). Exogenous ascorbic acid (Vitamin C) increases resistance to salt stress and reduces lipid peroxidation. *J. Exp. Bot.* **52:** 2207-2211.

Shanker, A.K., Djanaguiraman, M., Sudhagar, R., Chandrashekar, C.N. and Pathmanabhan, G. (2004). Differential antioxidative response of ascorbate glutathione pathway enzymes and metabolites to chromium speciation stress in green gram (*Vigna radiata* (L.) R.Wilczek. cv CO4) roots. *Plant Sci.* **166:** 1035-1043.

Sharma, D.C. and Sharma, C.P. (1996). Chromium uptake and toxicity effects on growth and metabolic activities in wheat *Triticum aestivum* L. cv. UP 2003. *Indian J. Exp. Biol.* **34:** 689-691.

Sharma, P. and Dubey, R.S. (2005). Drought induces oxidative stress and enhances the activities of antioxidant enzymes in growing rice seedlings. *Plant Growth Regul.* **46:** 209-221.

Sharma, P. and Dubey, R.S. (2007). Involvement of oxidative stress and role of antioxidative defense system in growing rice seedlings exposed to toxic concentrations of aluminium. *Plant Cell Rep.* **26:** 2027-2038.

Shaw, B.P. (1995). Effect of mercury and cadmium on the activities of antioxidative enzymes in the seedlings of *Phaseolus aureus*. *Biol. Plant.* **37:** 587-596.

Sheinberg, O., Rubin, B., Rabinowich, H.D., Libal, Y. and Tel-Or, E. (2000). Acclimation of beans to oxidative stress by treatment with sublethal iron levels. *J. Plant Physiol.* **157:** 93-99.

Shen, W., Nada, K. and Tachibana, S. (1999). Effect of cold treatment on enzymic and non-enzymic antioxidant activities in leaves of chilling-tolerant and chilling-sensitive cucumber cultivars. *J. Japan Soc. Hortic. Sci.* **68:** 967-973.

Shi, G.X., Xu, Q.S. and Xie, K.B. (2003). Physiology and ultrastructure of *Azolla imbricate* as affected by Hg^{2+} and Cd^{2+} toxicity. *Acta Bot. Sin.* **45:** 437-444.

Shigenaga, T., Yamauchi, N., Funamoto, Y. and Shigyo, M. (2005). Effects of heat treatment on an ascorbate-glutathione cycle in stored broccoli (*Brassica oleracea* L.) florets. *Post Harvest Biol. Technol.* **38:** 152-159.

Shigeoka, S., Ishikawa, T., Tamoi, M., Miyagawa, Y., Takeda, T., Yabuta, Y. and Yoshimura, K. (2002). Regulation and function of ascorbate peroxidase isoenzymes. *J Exp. Bot.* **53:** 1305-1319.

Shri, M., Kumar, S., Chakrabarty, D., Trivedi, P.K., Mallick, S., Misra, P., Shukla, D., Mishra, S., Srivastava, S., Tripathi, R.D. and Tuli, R. (2009). Effect of arsenic on growth, oxidative stress, and antioxidant system in rice seedlings. *Ecotoxicol. Environ. Safety.* **72:** 1102-1110.

Simonovicova, M., Huttova, J., Mistrik, I., Siroka, B. and Tamas, L. (2004). Peroxidase mediated hydrogen peroxide production in barley roots grown under stress conditions. *Plant Growth Regul.* **44:** 267-275.

Simova-Stoilova, L., Demirevska, K., Petrova, T., Tsenov, N. and Feller, U. (2008). Antioxidative protection in wheat varieties under severe recoverable drought at seedling stage. *Plant Soil Environ.* **54:** 529-536.

Singh, D.V., Srivastava, G.C. and Abding, M.Z. (2001). Amelioration of negative effect of water stress in *Cassia angustifolia* by benzyladenine and/or ascorbic acid. *Biol. Plant.* **44:** 141-143.

Singh, H.P., Batish, D.R., Kohlo, R.K. and Arora, K. (2007a). Arsenic-induced root growth inhibition in mung bean (*Phaseolus aureus* Roxb.) is due to oxidative stress resulting from enhanced lipid peroxidation. *Plant Growth Regul.* **53:** 65-73.

Singh, N., Kumar, D. and Sahu, A.P. (2007b). Arsenic in the environment: effects on human health and possible prevention. *J. Environ. Biol.* **28:** 359-365.

Singh, S., Eapen, S. and D'Souza, S.F. (2006). Cadmium accumulation and its influence on lipid peroxidation and antioxidative system in an aquatic plant, *Bacopa monnieri* L. *Chemosphere.* **62:** 233-246.

Singla-Pareek, S.L., Reddy, M.K. and Sopory, S.K. (2003). Genetic engineering of the glyoxalase pathway in tobacco leads to enhanced salinity tolerance. *Proc. Natl. Acad. Sci. USA.* **100:** 14672-14677.

Skutnik, M. and Rychter, A.M. (2009). Differential response of antioxidant systems in leaves and roots of barley subjected to anoxia and post-anoxia. *J. Plant Physiol.* **166:** 926-937.

Slooten, L., Capau, C., Van Camp, W., Van Montagu, M., Sybesma, C. and Inzé, D. (1995). Factors affecting enhancement of oxidative stress tolerance in transgenic tobacco overexpressing manganese superoxide dismutase in the chloroplast. *Plant Physiol.* **107:** 737-750.

Smeets, K., Cuypers, A., Lambrechts, A., Semane, B., Hoet, P., Van Laer, A. and Vangronsveld, J. (2005). Induction of oxidative stress and antioxidative mechanisms in *Phaseolus vulgaris* after Cd application. *Plant Physiol. Biochem.* **43:** 437-444.

Smirnoff, N. (1993). The role of active oxygen in the response of plants to water deficit and desiccation. *New Phytol.* **125:** 27-58.

Smirnoff, N. (2005). Ascorbate, tocopherol and carotenoids: metabolism, pathway engineering and functions. In: Smirnoff, N. (Ed.). Antioxidants and reactive oxygen species in plants. Blackwell Publishing Ltd., Oxford, UK, pp. 53-86.

Smirnoff, N. and Colombe, S.V. (1988). Drought influences the activity of enzymes of the chloroplast hydrogen peroxide scavenging system. *J. Exp. Bot.* **39:** 1097-1108.

Smirnoff, N. and Wheeler, G. (2000). Ascorbic acid in plants: biosynthesis and function. *Crit. Rev. Biochem. Mol. Biol.* **35:** 291-314.

Sofo, A., Dichio, B., Xiloyannis, C. and Masia, A. (2005a). Antioxidant defences in olive trees during drought stress: changes in activity of some antioxidant enzymes. *Funct. Plant Biol.* **32:** 45-53.

Sofo, A., Tuzio, A.C., Dichio, B. and Xiloyannis, C. (2005b). Influence of water deficit and rewatering on the components of the ascorbate-glutathione cycle in four interspecific *Prunus* hybrids. *Plant Sci.* **169:** 403-412.

Sreenivasulu, N., Grimm, B., Wobus, U. and Weschke, W. (2000). Differential response of antioxidant compounds to salinity stress in salt-tolerant and salt-sensitive seedlings of foxtail millet (*Setaria italica*). *Physiol. Plant.* **109:** 435-442.

Srivalli, B., Sharma, G. and Chopra, R.K. (2003). Antioxidative defense system in an upland rice cultivar subjected to increasing intensity of water stress followed by recovery. *Physiol. Plant.* **119:** 503-512.

Srivastava, S., Mishra, S., Tripathi, R.D., Dwivedi, S., Trivedi, P.K. and Tandon, P.K. (2007). Phytochelatins and antioxidant systems respond differentially during arsenite and arsenate stress in *Hydrilla verticillata* (L.f.) Royle. *Environ. Sci. Technol.* **41:** 2930-2936.

Srivastava, S., Tripathi, R.D. and Dwivedi, U.N. (2004). Synthesis of phytochelatins and modulation of antioxidants in response to cadmium stress in *Cuscuta reflexa* - an angiospermic parasite. *J. Plant Physiol.* **161:** 665-674.

Stevens, R.G., Creissen, G.P. and Mullineaux, P.M. (1997). Cloning and characterization of a cytosolic glutathione reductase cDNA from pea (*Pisum sativum* L.) and its expression in response to stress. *Plant Mol. Biol.* **35:** 641-654.

Stohs, S.J. and Bagchi, D. (1995). Oxidative mechanisms in the toxicity of metal ions. *Free Radical Biol. Med.* **18:** 321-336.

Stoyanova, Z. and Doncheva, S. (2002). The effect of zinc supply and succinate treatment on plant growth and mineral uptake in pea plant. *Braz. J. Plant Physiol.* **14:** 111-116.

Streb, P., Michael-Knauf, A. and Feierabend, J. (1993). Preferential photoinactivation of catalase and photoinhibition of photosystem II are common early symptoms under various osmotic and chemical stress conditions. *Physiol. Plant.* **88:** 590-598.

Strohm, M., Eiblmeier, M. and Langebartels, C. (1999). Responses of transgenic poplar (*Populus tremula* × *P. alba*) over-expressing glutathione synthetase or glutathione reductase to acute ozone stress: visible injury and leaf gas exchange. *J. Exp. Bot.* **50:** 365-374.

Stroinski, A., Kubis, J. and Zieleziriska, M. (1999). Effect of cadmium on glutathione reductase in potato tubers. *Acta Physiol. Plant.* **21:** 201-207.

Sudhakar, C., Lakshmi, A. and Giridarakumar, S. (2001). Changes in the antioxidant enzyme efficacy in two high yielding genotypes of mulberry (*Morus alba* L.) under NaCl salinity. *Plant Sci.* **161:** 613-619.

Sumithra, K., Jutur, P.P., Dalton, C.B. and Reddy, A.R. (2006). Salinity-induced changes in two cultivars of *Vigna radiata*: responses of antioxidative and proline metabolism. *Plant Growth Regul.* **50:** 11-22.

Sun, Q., Ye, Z.H., Wang, X.R. and Wong, M.H. (2005). Increase of glutathione in mine population of *Sedum alfredii*: A Zn hyperaccumulator and Pb accumulator. *Phytochem.* **66:** 2549-2556.

Sundar, D., Perianayaguy, B. and Reddy, A.R. (2004). Localization of antioxidant enzymes in the cellular compartments of sorghum leaves. *Plant Growth Regul.* **44:** 157-163.

Syvertsen, J.P., Zablotowicz, R.M. and Smith, M.L. (1983). Soil-temperature and flooding effects on 2 species of citrus.1. Plant-growth and hydraulic conductivity. *Plant Soil.* **72:** 3-12.

Szal, B., Drozd, M. and Rychter, A.M. (2004). Factors affecting determination of superoxide anion generated by mitochondria from barley roots after anaerobiosis. *J. Plant Physiol.* **161:** 1339-1346.

Tambussi, E.A., Bartoli, C.G., Beltrano, J., Guiamet, J.J. and Araus, J.L. (2000). Oxidative damage to thylakoid proteins in water-stressed leaves of wheat (*Triticum aestivum*). *Physiol. Plant.* **108:** 398-404.

Tanaka, K., Otsubo, Ô. and Kondo, Í. (1982). Participation of hydrogen peroxide in the inactivation of Calvin cycle SH enzymes in SO_2 fumigated spinach leaves. *Plant Cell Physiol.* **23:** 1009-1018.

Tanaka, K., Suda, Y., Kondo, N. and Sugahara, K. (1985). O_3 tolerance and the ascorbate-dependent H_2O_2 decomposing system in chloroplasts. *Plant Cell Physiol.* **26:** 1425-1431.

Teisseire, H. and Guy, V. (2000). Copper-induced change in antioxidant enzymes activities in fronds of duckweed (*Lemna minor*). *Plant Sci.* **153:** 65-72.

Teixeira, F.K., Menezes-Benavente, L., Galvao, V.C., Margis, R. and Margis-Pinheiro, M. (2006). Rice ascorbate peroxidase gene family encodes functionally diverse isoforms localized in different subcellular compartments. *Planta.* **224:** 300-314.

Tewari, R.K., Kumar, P. and Sharma, P.N. (2008). Morphology and physiology of zinc stressed mulberry plants. *J. Plant Nutr. Soil Sci.* **171:** 286-294.

Tiryakioglu, M., Eker, S., Ozkutlu, F., Husted, S. and Cakmak, I. (2006). Antioxidant defense system and cadmium uptake in barley genotypes differing in cadmium tolerance. *J. Trace Elem. Med. Biol.* **20:** 181-189.

Torres-Franklin, M.L., Contour-Ansel, D., Zuily-Fodil, Y. and Pham-Thi, A-T. (2008). Molecular cloning of glutathione reductase cDNAs and analysis of GR gene expression in cowpea and common bean leaves during recovery from moderate drought stress. *J. Plant Physiol.* **165:** 514-521.

Treshow, M. and Anderson, F.K. (1989). Plant stress from air pollution, New York: John Wiley & Sons, Chichester.

Tripathi, R.D., Srivastava, S., Mishra, S., Singh, N., Tuli, R., Gupta, D. and Maathuis, F.J.M. (2007). Arsenic hazards: strategies for tolerance and remediation by plants. *Trends Biotechnol.* **25:** 158-165.

Tsai, Y.C., Hong, C.Y., Liu, L.F. and Kao, C.H. (2005). Expression of ascorbate peroxidaes and glutathione reductase in roots of rice seedlings in respons to NaCl and H_2O_2. *J. Plant Physiol.* **162:** 291-299.

Tsai, Y.C. and Kao, C.H. (2004). The involvement of hydrogen peroxide in abscisic acid-induced activities of ascorbate peroxidase and glutathione reductase in rice roots. *Plant Growth Regul.* **43:** 207-212.

Tu, S. and Ma, L.Q. (2004). Comparison of arsenic and phosphate uptake and distribution in arsenic hyperaccumulating and non-hyperaccumulating fern. *J. Plant Nutr.* **27:** 1227-1242.

Tukendorf, A. (1993). The role of glutathione in detoxification of cadmium and excess copper in spinach plants. *Acta Physiol. Plant.* **15:** 175-183.

Tukendorf, A. and Rauser, W.E. (1990). Changes in glutathione and phytochelatins in roots of maize seedlings exposed to cadmium. *Plant Sci.* **70:** 155-166.

Turcsanyi, E., Lyons, T., Plochl, M. and Barnes, J. (2000). Does ascorbate in the mesophyll cell walls form the first line of defence against ozone? Testing the concept using broad bean (*Vicia faba* L.). *J. Exp. Bot.* **51:** 901-910.

Ueki, S. and Citovsky, V. (2001). Inhibition of systemic onset of posttranscriptional gene silencing by non-toxic concentrations of cadmium. *Plant J.* **28:** 283-291.

Upadhyaya, H., Panda, S.K. and Dutta, B.K. (2008). Variation of physiological and antioxidative responses in tea cultivars subjected to elevated water stress followed by rehydration recovery. *Acta Physiol. Plant.* **30:** 457-468.

Urano, J., Nakagawa, T., Maki, Y., Masumura, T., Tanaka, K. and Murata, N. (2000). Molecular cloning and characterization of a rice dehydroascorbate reductase. *FEBS Lett.* **466:** 107-111.

Ushimaru, T., Maki, Y., Sano, S., Koshiba, K., Asada, K. and Tsuji, H. (1997). Induction of enzymes involved in the ascorbate-dependent antioxidative system, namely, ascorbate peroxidase, monodehydroascorbate reductase and dehydroascorbate reductase, after exposure to air of rice. *Plant Cell Physiol.* **38:** 541-549.

Ushimaru, T., Shibasaka, M. and Tsuji, H. (1992). Development of the O_2^- detoxification system during adaptation to air of submerged rice seedlings. *Plant Cell Physiol.* **33:** 1065-1071.

Ushimaru, T., Nakagawa, T., Fujioka, Y., Daicho, K., Naito, M., Yamauchi, Y., Nonaka, H., Amako, K., Yamawaki, K. and Murata, N. (2006). Transgenic *Arabidopsis* plants expressing the rice dehydroascorbate reductase gene are resistant to salt stress. *J. Plant Physiol.* **163:** 1179-1184.

Vacca, R.A., de Pinto, M.C., Valenti, D., Passarella, S., Marra, E. and Gara, L.D. (2004a). Production of reactive oxygen species, alteration of cytosolic ascorbate peroxidase, and impairment of mitochondrial metabolism, are early events in heat shock-induced programmed cell death in tobacco bright-yellow 2 cells. *Plant Physiol.* **134:** 1100-1112.

Vacca, R.A., de Pinto, M.C., Valenti, D., Passerella, S., Marra, E. and De Gara, L. (2004b). Reactive oxygen species production, impairment of glucose oxidation and cytosolic ascorbate peroxidase are early events in heat-shock induced programmed cell death in tobacco BY-2 cells. *Plant Physiol.* **134:** 1100-1112.

Vaidyanathan, H., Sivakumar, P., Chakrabarty, R. and Thomas, G. (2003). Scavenging of reactive oxygen species in NaCl-stressed rice (*Oryza sativa* L.) - differential response in salt-tolerant and sensitive varieties. *Plant Sci.* **165:** 1411-1418.

Van de Mortel, J.E., Villanueva, L.A., Schat, H., Kwekkeboom, J., Coughlan, S., Moerland, P.D., Van Themaat, E.V.L., Koornneef, M. and Aarts, M.G.M. (2006). Large expression differences in genes for iron and zinc homeostasis, stress response, and lignin biosynthesis distinguish roots of *Arabidopsis thaliana* and the related metal hyperaccumulator *Thlaspi caerulescens*. *Plant Physiol.* **142:** 1127-1147.

Van Rensburg, L. and Kruger, G.H.J. (1994). Evaluation of components of oxidative stress metabolism for use in selection of drought tolerant cultivars of *Nicotiana tobacum* L. *J. Plant Physiol.* **143:** 730-736.

Vandenhove, H., Cuypers, A., Van Hees, M., Koppen, G. and Wannijn, J. (2006). Oxidative stress reactions induced in beans (*Phaseolus vulgaris*) following exposure to uranium. *Plant Physiol. Biochem.* **44:** 795-805.

Vansuyt, G., Lopez, F., Inzé, D., Briat, J.F. and Fourcroy, P. (1997). Iron triggers a rapid induction of ascorbate peroxidase gene expression in *Brassica napus*. *FEBS Lett.* **410:** 195-200.

VanToai, T. and Bolles, C.S. (1991). Postanoxic injury in soybean (*Glycine max*) seedlings. *Plant Physiol.* **97:** 588-592.

Veljovic-Jovanovic, S.D., Pignocchi, C., Noctor, G. and Foyer, C.H. (2001). Low ascorbic acid in the *vtc*-1 mutant of *Arabidopsis* is associated with decreased growth and intracellular redistribution of the antioxidant system. *Plant Physiol.* **127:** 426-435.

Vogeli-Lange, R. and Wagner, G.J. (1990). Subcellular localization of cadmium-binding peptides in tobacco leaves. *Plant Physiol.* **92:** 1086-1093.

Volkov, R.A., Panchuk, I.I., Mullineaux, P.M. and Schoffl, F. (2006). Heat stress-induced H_2O_2 is required for effective expression of heat shock genes in *Arabidopsis*. *Plant Mol. Biol.* **61:** 733-746.

Vranova, E., Inze, D. and Van Breusegem, F. (2002). Signal transduction during oxidative stress. *J. Exp. Bot.* **53:** 1227-1236.

Walker, M.A. and McKersie, B.D. (1993). Role of the ascorbate-glutathione antioxidant system in chilling resistance of tomato. *J. Plant Physiol.* **141:** 234-239.

Wang, C., Zhang, S.H., Wang, P.F., Hou, J., Zhang, W.J., Li, W. and Lin, Z.P. (2009). The effect of excess Zn on mineral nutrition and antioxidative response in rapeseed seedlings. *Chemosphere.* **75:** 1468-1476.

Wang, K.H. and Jiang, Y.W. (2007). Antioxidant responses of creeping bentgrass roots to waterlogging. *Crop Sci.* **47:** 232-238.

Wang, R., Okamoto, M., Xing, X. and Crawford, N.M. (2003). Microarray analysis of the nitrate response in *Arabidopsis* roots and shoots reveals over one thousand rapidly responding genes and new linkages to glucose, trehalose-6-P, iron and sulfate metabolism. *Plant Physiol.* **132:** 556-567.

Wang, Y., Wisniewski, M., Meilan, R., Cui, M. and Fuchigami, L. (2006). Transgenic tomato (*Lycopersicon esculentum*) overexpressing *cAPX* exhibits enhanced tolerance to UV-B and heat stress. *J. Appl. Horticul.* **8:** 87-90.

Wang, Y., Wisniewski, M., Meilan, R., Cui, M., Webb, R. and Fuchigami, L. (2005). Overexpression of cytosolic ascorbate peroxidase in tomato confers tolerance to chilling and salt stress. *J. Am. Soc. Hortic. Sci.* **130:** 167-173.

Weckx, J.E.J. and Clijsters, H.M.M. (1996). Oxidative damage and defense mechanisms in primary leaves of *Phaseolus vulgaris* as a result of root assimilation of toxic amounts of copper. *Physiol. Plant.* **96:** 506-512.

Willekens, H., Camp, W.V., Montagu, M.V., Inzé, D., Langebartels, C. and Jr, H.S. (1994). Ozone, sulfur dioxide, and ultraviolet B have similar effects on mRNA accumulation of antioxidant genes in *Nicotiana plumbaginifolia* L. *Plant Physiol.* **106:** 1007-1014.

Willekens, H., Chamnongpol, S., Davey, M., Schraudner, M., Van Montagu, M., Inzé, D. and van Camp, W. (1997). Catalase is a sink for H_2O_2 and is indispensable for stress defence in C_3 plants. *EMBO J.* **16:** 4806-4816.

Williams, P.N., Villada, A., Deacon, C., Raab, A., Figuerola, J., Green, A.J., Feldmann, J. and Meharg, A.A. (2007). Greatly enhanced arsenic shoot assimilation in rice leads to elevated grain levels compared to wheat and barley. *Environ. Sci. Technol.* **41:** 6854-6859.

Wingate, V.M.P., Lawton, M.A. and Lamb, C.J. (1988). Glutathione causes a massive and selective induction of plant defense genes. *Plant Physiol.* **87:** 206-210.

Wise, R. and Naylor, A.W. (1987). Chilling-enhanced photooxidation. Evidence for the role of singlet oxygen and superoxide in the breakdown of pigments and endogenous antioxidants. *Plant Physiol.* **83:** 278-282.

Wisniewski, M., Fuchigami, L., Wang, Y., Srinivasan, C. and Norilli, J. (2002). Overexpression of a cytosolic ascorbate peroxidase gene in apple improves resistance to heat stress. XXXVI International Horticultural Congress & Exhibition. p. 147 (Abstr.).

Wohlgemuth, H., Mittelstrass, K., Kschieschan, S., Bender, J., Weigel, H-J., Overmyer, K., Kangasjaarvi, J., Sandermann, H. and Langebartels, C. (2002). Activation of an oxidative burst is a general feature of sensitive plants exposed to the air pollutant ozone. *Plant Cell Environ.* **25:** 717-726.

Wójcik, M., Skórzynska-Poli, E. and Tukiendorf, A. (2006). Organic acids accumulation and antioxidant enzyme activities in *Thlaspi caerulescens* under Zn and Cd stress. *Plant Growth Regul.* **48:** 145-155.

Wollenweber-Ratzer, B. and Crawford, R.M.M. (1994). Enzymatic defence against post-anoxic injury in higher plants. *Proc. R. Soc. Edinburgh.* **102B:** 381-390.

Wong, Y-Y., Ho, C-L., Nguyen, P.D., Teo, S-S., Harikrishna, J.A., Rahim, R.A. and Wong, M.C.V.L. (2007). Isolation of salinity tolerant genes from the mangrove plant, *Bruguiera cylindrica* by using suppression subtractive hybridization (SSH) and bacterial functional screening. *Aquatic Bot.* **86:** 117-122.

Wu, F-B., Chen, F., Wei, K. and Zhang, G-P. (2004). Effect of cadmium on free amino acid, glutathione and ascorbic acid concentrations in two barley genotypes (*Hordeum vulgare* L.) differing in cadmium tolerance. *Chemosphere.* **57:** 447-454.

Xiang, Ch. and Oliver, D.J. (1998). Glutathione metabolic genes coordinately respond to heavy metals and jasmonic acid in *Arabidopsis*. *Plant Cell.* **10:** 1539-1550.

Xiang, C., Werner, B.L., Christensen, E.M. and Oliver, D.J. (2001). The biological functions of glutathione revisited in *Arabidopsis* transgenic plants with altered glutathione levels. *Plant Physiol.* **126:** 564-574.

Yadav, S.K., Singla-Pareek, S.L., Reddy, M.K. and Sopory, S.K. (2005b). Transgenic tobacco plants overexpressing glyoxalase enzymes resist an increase in methylglyoxal and maintain higher reduced glutathione levels under salinity stress. *FEBS Lett.* **579:** 6265-6271.

Yamamoto, A., Bhuiyan, Md. N.H., Waditee, R., Tanaka, Y., Esaka, M., Oba, K., Jagendorf, A.T. and Takabe, T. (2005). Suppressed expression of the apoplastic ascorbate oxidase gene increases salt tolerance in tobacco and *Arabidopsis* plants. *J. Exp. Bot.* **56:** 1785-1796.

Yamamoto, Y., Kobayashi, Y., Devi, S.R., Rikiishi, S. and Matsumoto, H. (2003). Oxidative stress triggered by aluminum in plant roots. *Plant and Soil.* **255:** 239-243.

Yamauchi, N. and Kusabe, A. (2001). Involvement of ascorbate–glutathione cycle in senescence of stored broccoli (*Brassica oleracea* L.). *J. Jpn. Soc. Hort. Sci.* **70:** 704-708.

Yan, S-P., Zhang, Q-Y., Tang, Z-C., Su, W-A. and Sun, W-N. (2006). Comparative proteomic analysis provides new insights into chilling stress responses in rice. *Mol. Cell Proteomics.* **5:** 484-496.

Ye, B. and Gressel, J. (2000). Transient, oxidant-induced antioxidant transcript and enzyme levels correlate with greater oxidant-resistance in paraquat-resistant *Conyza bonariensis*. *Planta.* **211:** 50-61.

Yetisir, H., Caliskan, M.E., Soylu, S. and Sakar, M. (2006). Some physiological and growth responses of watermelon [*Citrullus lanatus* (Thunb.)] Matsum and Nakai grafted onto *Lagenaria siceraria* to flooding. *Environ. Exp. Bot.* **58:** 1-8.

Yordanova, R.Y., Christov, K.N. and Popova, L.P. (2004). Antioxidative enzymes in barley plants subjected to soil flooding. *Environ. Exp. Bot.* **51:** 93-101.

Yoshimura, K., Yabuta, Y., Ishikawa, T. and Shigeoka, S. (2000). Expression of spinach ascorbate peroxidase isoenzymes in response to oxidative stresses. *Plant Physiol.* **123:** 223-234.

Younis, M.E., Hasaneen, M.N.A., Ahmed, A.R. and El-Bialy, D.M.A. (2008). Plant growth, metabolism and adaptation in relation to stress conditions XXI. Reversal of harmful NaCl effects in lettuce plants by foliar application with urea. *Aust. J. Crop Sci.* **2:** 83-95.

Younis, M.E., Hasaneen, M.N.A. and Kazamel, A.M.S. (2009). Plant growth, metabolism and adaptation in relation to stress conditions. XXVII. Can ascorbic acid modify the adverse effects of NaCl and mannitol on amino acids, nucleic acids and protein patterns in *Vicia faba* seedlings? *Protoplasma.* **235:** 37-47.

Zabalza, A., Gaston, S., Sandalio, L.M., del Rio, L.A. and Royuela, M. (2007). Oxidative stress is not related to the mode of action of herbicides that inhibit acetolactate synthase. *Environ. Exp. Bot.* **59:** 150-159.

Zaharieva, T. and Abadia, A. (2003). Iron deficiency enhances the levels of ascorbate, glutathione and related enzymes in sugar beet roots. *Protoplasma.* **221:** 269-275.

Zaharieva, T.B., Gogorcena, Y. and Abadia, J. (2004). Dynamics of metabolic responses to iron deficiency in sugar beet roots. *Plant Sci.* **166:** 1045-1050.

Zaharieva, T., Yamashita, K. and Matsumoto, H. (1999). Iron deficiency induced changes in ascorbate content and enzyme activities related to ascorbate metabolism in cucumber roots. *Plant Cell Physiol.* **40:** 273-280.

Zayed, A., Lytle, C.M., Terry, N. and Qian, J.H. (1998). Chromium accumulation, translocation and chemical speciation in vegetable crops. *Planta.* **206:** 293-299.

Zeng, X.M., Shi, G.X., Xu, N., Xu, Q.S. and Zhang, X.L. (2003). Effects of mercury on active oxygen metabolism and chromosome in *Sagittaria trifolia. J. Plant Physiol. Mol. Biol.* **29:** 227-232.

Zenk, M.H. (1996). Heavy metal detoxification in higher plants. *Gene.* **179:** 21-30.

Zhang, F., Shi, W., Jin, Z. and Shen, Z. (2002). Response of antioxidative enzymes in cucumber chloroplasts to cadmium toxicity. *J. Plant Nutr.* **26:** 1779-1788.

Zhang, X., Zhang, L., Dong, F., Gao, J., Galbraith, D.W. and Song, C.P. (2001). Hydrogen peroxide is involved in abscisic acid induced stomatal closure in *Vicia faba. Plant Physiol.* **126:** 1438-1448.

Zhao, S. and Blumwald, E. (1998). Changes in oxidation reduction state and antioxidant enzymes in the roots of jack pine seedlings during cold acclimation. *Physiol. Plant.* **104:** 134-142.

Zhao, Z., Cai, Y., Zhu, Y. and Kneer, R. (2005). Cadmium-induced oxidative stress and protection by L-galactono-1, 4-lactone in winter wheat (*Triticum aestivum* L.). *J. Plant Nutr. Soil Sci.* **168:** 1-5.

Zheng, Y., Lyons, T., Ollerenshaw, J.H. and Barnes, J.D. (2000). Ascorbate in the leaf apoplast is a factor mediating ozone resistance in *Plantago major. Plant Physiol. Biochem.* **38:** 403-411.

6

Ascorbate-Glutathione Related Redox Regulation in Plant Stress Tolerance

Kun-Ming Chen[a] **and Hai-Jun Gong**[b]

[a]College of Life Sciences, Northwest A&F University, Yangling–712100, Shaanxi, China
[b]College of Horticulture, Northwest A&F University, Yangling–712100, Shaanxi, China
[*]***Corresponding author:*** *kunmingchen@163.com*

ABSTRACT

Environmental stresses, such as drought, salinity, extreme temperature, and heavy metal pollution, adversely affect the growth, metabolism, dry matter, and production of plants. As a common response of plants to a number of biotic and abiotic stresses, the oxidative burst of reactive oxygen species (ROS) can lead to oxidative stress and then unbalance redox homeostasis of plant cells. Therefore, it is very important for plants to develop physiological and molecular mechanisms to cope with varied environmental conditions. To this end, it is now well established that redox regulation is a central element in adjusting plant metabolism and development to the prevailing environmental conditions. As the major components of cellular redox network, ascorbate (ASC), glutathione (GSH) and their related redox couples such as NADPH/NADP and ferredoxin/thioredoxin systems in plants are reviewed here. The biosyntheses and metabolism of these substances as well as their redox cycling under varied stress conditions are presented. Emphasis is placed on redox regulation of ASC and GSH in the chloroplast and apoplast under varied environmental conditions.

Keywords: Ascorbate, glutathione, redox regulation, stress tolerance.

1. INTRODUCTION

The responses of plants to varied environmental stresses include the constant readjustment of physiology and metabolism throughout the life cycle. Plants have developed specific strategies to cope with ever-changing environmental conditions by monitoring their surroundings and adjusting their metabolic systems to maintain homeostasis. By triggering a cascade or network of events that starts with stress perception and ends with the expression of a set of target genes, plants can acclimate to a series of biotic and abiotic stresses Anjum et al. (2010). The key components of the stress-response relationship are stress stimulus, signals, transducers, transcript regulators, target genes, and stress responses, including morphological, biochemical, and physiological alterations (Pastori and Foyer 2002). Usually, factors that act at early stages are more critical because they often play common pathways and components in the stress-response relationship. This allows plants to adapt/acclimate to a range of different stresses after exposure to one specific stress. It has been found that the oxidative burst of reactive oxygen species (ROS) is a common response of plants to a number of biotic and abiotic stresses (Fleury et al., 2002; Hancock et al., 2002; Neill et al., 2002; Mothofer et al., 2004; Reddy et al., 2004; Kangasjarvi et al., 2005; Torres and Dang 2005). ROS, mainly hydrogen peroxide (H_2O_2) and superoxide anions (O_2^-), have been reported to mediate the induction of systematic acquired resistance, control, and regulation of biological processes such as growth, cell cycle, programmed cell death, hormone signalling, stress responses, and development (Mittler et al., 2004; Foyer and Noctor 2005a; Torres and Dang 2005). Therefore, it is central to cross-tolerance phenomena and has been widely considered a key component in the stress survival network of plants.

The reduction/oxidation (redox) state of cells is a consequence of the balance between the levels of oxidizing and reducing equivalents. A reducing intracellular environment is often associated with cell survival, however, redox unbalance is necessary since it represents a regulatory sensor for surviving since ROS also is a major component in the stress response and since the activities of many important redox modulating transcription factors are controlled by redox balance (Filomeni et al., 2002). The intracellular redox state is determined by the contribution of different redox couples. Each couple can exchange electrons in such a way that, by giving or accepting reducing equivalents, may represent good cofactors in redox enzymatic reactions. The relative amount of reduced and oxidized form of each system may determine changes in cellular redox state by shifting this value towards oxidizing or reducing conditions. To date, the three most important redox systems have been widely studied in plant cells: nicotinammide adenine dinucleotide phosphate (NADPH/NADP), ascorbate (ASC) (ASC/dehydroascorbate (DHA)) and glutathione (GSH) (GSH/oxidized glutathione (GSSG)) (May et al., 1998). It has been found that the changes in ASC and GSH pools including the increases in ASC and GSH level, ASC/DHA and GSH/GSSG ratios, and activities of the enzymes related to their biosynthesis and metabolism, are tightly related to the responses of plants to a wide range of environmental stresses (May et al., 1998; Noctor et al., 1998; Horemans et al., 2000a; Foyer and Noctor 2005a; Foyer et al., 2008). Levels and redox status of ASC and GSH also regulate expression of genes and activities of the redox sensitive transcription factors and enzymes (Sen 1999; Davey et al., 2000; Cnubben et al., 2001; Pastori and Foyer 2002). However, the coordinated regulatory mechanism of the three major redox couples during plant development and stress responses is still under investigation.

In this review, the cellular redox network which is tightly asscociated with ascorbate, glutathione, pyridine nucleotides and their redox status in plants are addressed. The biosyntheses and metabolism of these substances as well as their redox cycling under environmental stress conditions are summarized and particularly, the redox regulation of ASC and GSH in the chloroplast and apoplast under varied environmental conditions is emphasized.

2. REDOX NETWORK IN PLANTS

The redox network in plants consists of redox sensing and signalling associated with reactive oxygen in chloroplasts, cytoplasm, peroxisomes, mitochondria and nuclei (Foyer and Noctor 2003; Oelze et al., 2008). Chloroplasts and mitochondria are the powerhouses of photosynthetic cells. The redox cascades of the photosynthetic and respiratory electron transfer chains not only provide the driving forces for metabolism but also generate redox signals, which participate in and regulate every aspect of plant biology from gene expression and translation to enzyme chemistry.

The photosynthetic apparatus of higher plants and algae is composed of plastid- and nuclear-encoded components. Therefore, the expression of photosynthesis genes needs to be highly coordinated. In this process, the redox state of photosynthetic electron transport components and redox-active soluble molecules act as important regulating parameters (Pfannschmidt 2003). An increasing number of reports have shown that the redox state of plastoquinone (PQ), the cytochrome b_6f complex, other unidentified components of the electron transport chain and electron-flow-dependent soluble components such as thioredoxins, glutathione and ROS affect chloroplast gene expression (see the review by Pfannschmidt 2003). The two photosystems (PSI and PSII) of the chloroplast have slightly different absorption maxima, which can lead to an imbalance in the energy distribution from absorbed light between them. To optimize photosynthetic productivity particularly at limited light availability, it is necessary to balance and control light harvesting between the two photosyntems, the photosynthetic energy conversion and the consumption of reducing power and ATP in metabolic reactions. On a short-term basis, a regulatory mechanism for energy balance distribution between PSI and PSII called state transition was proposed (Allen 2005). When the light conditions favour the excitement of PSII (PQ pool reduced), part of the LHCII moves to PSI (State 2) to gather more light to PSI. Similarly, when light conditions favour PSI (PQ pool oxidized), the LHCII moves back to PSII (State 1) distributing the light energy evenly between the two photosystems (Wollman, 2001). Imbalance in energy distribution between the two photosystems is also counteracted by adjusting photosystem stoichiometry. State transition is a fast and thus short-term mechanism to rapidly improve energy distribution, particularly under low light conditions, whereas photoacclimation is a long-term process that involves modification of gene expression and degradation of proteins.

In the photoacclimation process, a redox regulation based on PQ pool redox changes and ROS production is crucial. Besides its direct effect on state transition by regulation of the LHCII kinase (STN7), the redox state of the plastoquinone pool (PQ/PQH2) also affects nuclear and chloroplast gene expression and thus contributes to photoacclimation and regulation of stress defence genes. It has been suggested that PQ pool has a control of chloroplast targeted genes on several occasions (Escoubas et al., 1995; Pfannschmidt et al., 2001). There are also several indications where redox status of some intersystem electron carrier is likely to control nuclear gene expression even though the exact molecule was unknown (Petracek et al., 1998; Pfannschmidt et al., 2001; Sherameti et al., 2002). The mechanisms by which these redox changes would transmit signals remain unclear. One possibility involves the protein phosphorylation cascades. Protein phosphorylation by kinases is a ubiquitous signalling mechanism in all organisms. The first identified thylakoid protein kinase was a TAK1 that is associated with the cyt b_6f complex (Snyders and Kohorn 1999). Recently, at least two thylakoid bound kinases were found in plants: STN7 kinase that phosphorylates the LHCII proteins (Bellafiore et al., 2005) and kinase that phosphorylates the PSII core proteins (Bonardi et al., 2005). Although it is possible that the phosphorylation of the target proteins of these thylakoid kinases might activate signalling cascade from chloroplasts to the nuclei, no downstream target proteins of the phosphorylated target proteins have been identified, leaving the key mechanisms of signal transduction unanswered. On the stromal side, the ferredoxin (fd)/thioredoxin (Trx) system is capable of light-dependent reduction and therefore activates several chloroplast enzymes, especially those functioning in the Calvin cycle. This has direct effect on the levels of triose phosphates generated by Calvin cycle and other derived metabolites, such as sugars, that can activate metabolite dependent signalling pathways in cytosol. It has even been suggested that chloroplasts and mitochondria could communicate by metabolites that are generated through enzyme activation by thioredoxins in chloroplasts and mitochondria (Balmer et al., 2006). In addition, recently it has been shown that chloroplast derived signals could affect the expression of mitochondria targeted alternative oxidase protein (Yoshida et al., 2007) and nuclear photosynthetic gene expression is synergistically modulated by rates of protein synthesis in chloroplasts and mitochondria (Pesaresi et al., 2006). Fd/Trx system also activates a malate-forming enzyme, NADP-malate dehydrogenase (NADP-MDH), which is generally used as an indicator of the redox state of the chloroplast stroma (Miginiac-Maslow and Lancelin 2002; Scheibe 2004). Since production of malate consumes NADPH, it can also be seen as a mechanism to balance the stromal redox state by exporting malate to the cytosol (Scheibe, 2004). Apparently, as discussed above, the network of redox signals orchestrates from energy-generating organelles to utilizing ones. The network of redox signals and regulation is shown in Fig. 1.

3. COMPONENTS OF REDOX SIGNALLING AND MAJOR PLAYERS

3.1 Components of Redox Signalling in Plants

The concept that redox signals are key regulators of plant metabolism, morphology, and development is widely accepted. Evidence obtained recently indicates that H_2O_2

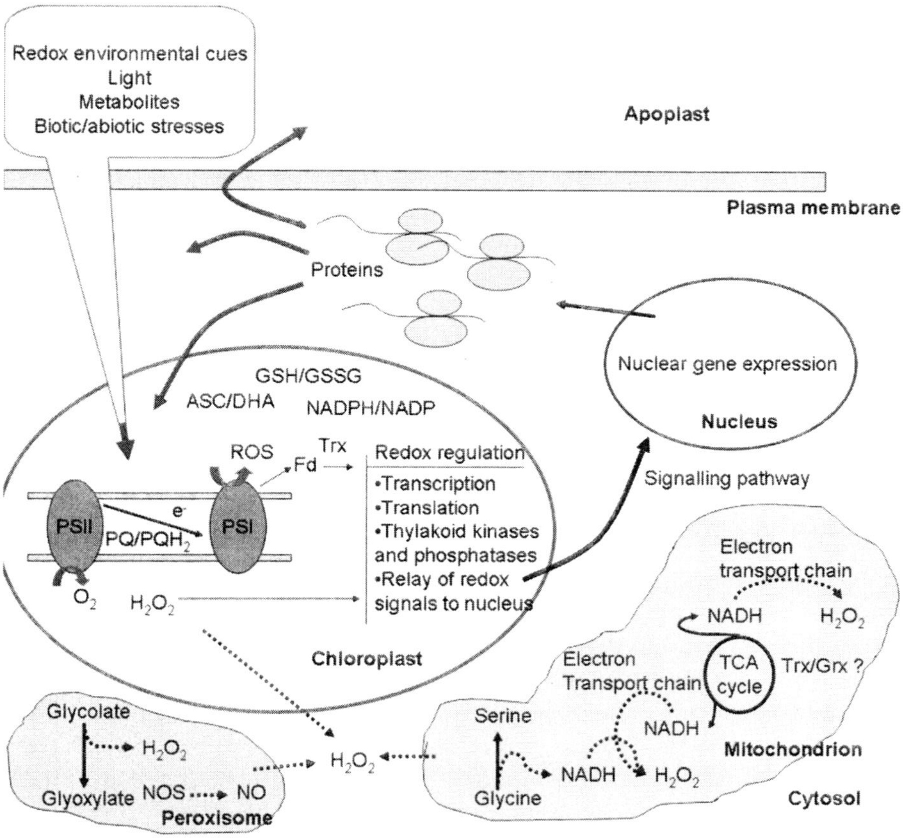

Fig. 1. Redox network in plant cells. 1) Environmental cues such as light, metabolites, and biotic/abiotic stresses can lead to unbalance of redox homeostasis during photosynthesis. The light-driven chemistry of photosynthesis consists of a series of redox steps involving structural components or functionally coupled pools of redox-active components, such as thioredoxin (Trx), peroxiredoxin (Prx), NADPH, ascorbate (ASC) and glutathione (GSH). The production of reative oxygen species (ROS) from Photosystem II and I (PSII and PSI) is crucial redox signal which together with its resulted changes of redox couples play important roles in redox regulation of chloroplast. The redox state of the plastoquinone pool (PQ/PQH_2) is a major regulator in this process. Besides its direct effect on state transition by regulation of the LHCII kinase (STN7), PQ/PQH_2 also affects photoacclimation and regulation of stress defence genes (see Oelze et al., 2008). 2) Peroxisomes are known to release signals that regulate nuclear gene expression. Signals arising from peroxisomes such as H_2O_2 and nitric oxide (NO) regulate photomorphogenesis, plant development peroxisomal biogenesis, light signalling and stress responses (Corpas et al., 2001, Hu et al., 2002). 3) Mitochondia have not traditionally been regarded as a major source of ROS in plant leaves since as compared to chloroplast and peroxisome, the formation rate of H_2O_2 in mitochondria is 30-100 times lower (see Foyer and Noctor, 2003). H_2O_2 is produced from electron transport chain where NADH from TCA cycle is involved. A Thioredoxin/Glutaredoxin (Trx/Grx) might exist in mitochondria but this still needs to be verified in plant cells.

acts as a signal transducing molecule playing a significant role in the signal transduction of CDPK (Xing et al., 2001) and other signals such as Ca^{2+} (Evans et al., 2005), NO (Delledonne et al., 2002), MAPK (Kovtun et al., 2000; Ikner and Shiozaki 2005), and hormones (Overmyer et al., 2003) in both optimal and stress conditions. The lifetime of H_2O_2 in plants is determined by the capacity of the two major antioxidant buffers of the plant cell, ascorbate (ASC) and glutatione (GSH), together with the antioxidant enzymes that use these antioxidants (Noctor and Foyer 1998). Most plant cells contain very large quantities of ASC (10-100 mM) and GSH (1-10 mM), and most intracellular compartments, hence, have the capacity to deal with very high fluxes of H_2O_2 production (Pastori and Foyer 2002). As key components of redox signalling, ASC and GSH participate in the specific compartment-based signalling by differential changes in their amounts and redox status (Noctor et al., 2000). The ratios of ASC/DHA and GSH/GSSG and their redox cycling can modulate expression of genes and activities of the redox sensitive transcription factors and enzymes (Sen 1999, Davey et al., 2000; Cnubben et al., 2001; Pastori and Foyer 2002). Besides ASC and GSH, other redox related components including ROS, NAD(P)H/NAD(P), oxidases, thioredoxins (Trxs), ferredoxins (Fd), glutaredoxins (Grxs), and peroxiredoxins (Prxs) are also involved in the redox regulation of plant development and stress responses (Figs. 2 and 3).

3.2 Major Redox Couples in Redox Regulation

3.2.1 NAD(P)H/NAD(P)

Pyridine nucleotides are key redox carriers in the soluble phase of all living cells, and both NAD and NADP play crucial roles in pro-oxidant and antioxidant metabolism (Noctor et al., 2006). They mediate hundreds of redox reactions, and thus impact on virtually every metabolic pathway in the cell. NAD and NADP participate in the regulation of both ROS-producing and ROS-consuming systems in plants so they have been considered the key players in redox signalling (May et al., 1998; Noctor et al., 2006). Chloroplast ROS production is influenced by NADPH/NADP ratios. Mitochondrial NAD(P) status is important in determining ROS formation by the respiratory electron transport chain, and NAD(P)H oxidases are key players in the generation of ROS at the plasmalemma. ROS processing partly depends on ASC and GSH, and redox flux through these pools is maintained by NAD(P)H (May et al., 1998). Thus, as cofactors in numerous energy-producing and biosynthetic reactions, NAD and NADP are crucial in redox regulation linked to stress and development.

In plant stress reactions, NADPH is particularly important as a branch point between ROS-promoting and ROS-processing metabolism (Noctor et al., 2006). In the stroma, the concentration of NADP available for reduction by ferredoxin-NADP reductase is considered important in determining electron flow to alternative acceptors such as oxygen (Scheibe et al., 2005). In the cytosol, NAD(P)H provides the reductant for production of ROS through NAD(P)H oxidase (Bedard et al., 2007). Growing evidence indicates that ROS interacts with nitric oxide (NO) in defence signalling and growth responses, and a specific NADPH-dependent nitric oxide synthase (NOS) is involved in the NO production (Zeier et al., 2004; Crawford and Guo 2005; Hu et al., 2005). The scavenging process of ROS is also tightly dependent on NAD(P)H. Reductive detoxification of ROS and organic peroxidase occurs through the ASC and GSH pools (Noctor et al., 2006). In addition, oxo-derivatives of lipid peroxides are detoxified by NAD(P)-dependent oxidation or NADPH-dependent reduction catalysed by aldehyde dehydrogenases (Sunkar *et al.*, 2003), aldo-keto reductases or alkenal reductases (Mano

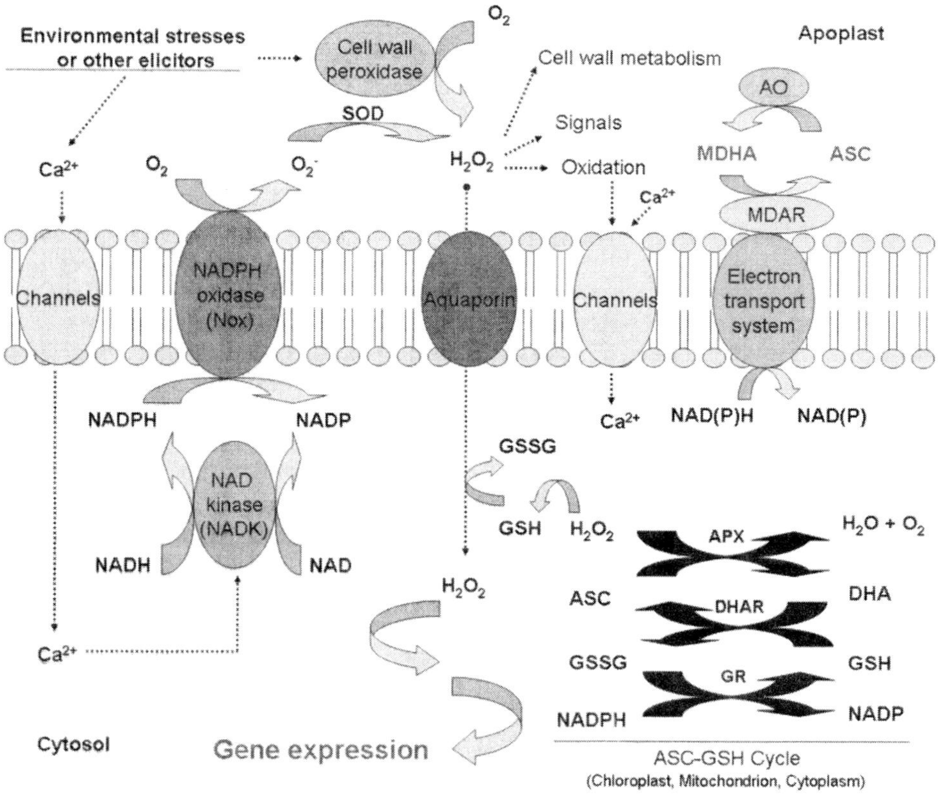

Fig. 2. Network of H_2O_2 signalling and major redox components in plant cells. 1) Upon elicitation, H_2O_2 is produced in the apoplast by several processes, including activation of NADPH oxidases, cell wall peroxidases, or other related enzymes. In the apoplast, H_2O_2 has several possible fates: used directly in cell wall metabolism, oxidized by ascorbate (ASC), or act directly as a local or systemic signal. 2) Most importantly, H_2O_2 can trigger at least three different routes of signal transduction simultaneously: a) Transport through the aquaporins allows rapid entry of H_2O_2 in the "sensing" cell, where it can directly trigger local signal transduction events, b) H_2O_2 modifies calcium flux into the cells and, hence, changes calcium-induced signalling pathways, c) For regeneration, the oxidized forms of ASC have to interact with the plasma membrane or return to the cytosol because the apoplast has little or no reducing power. 3) In the cytosol, H_2O_2 can also be generated in normal metabolism via the Mehler reaction in chloroplasts, electron transport in mitochondria and photorespiration in peroxisomes. 4) The influx of H_2O_2 can induce a series of responses in plant cells including gene express, specific enzyme activation and calcium release. 5) Environmental stresses or other elicitors stimulate the influx of the extracellular Ca^{2+} into cytoplasm which can regulate the activity of plasma membrane NADPH oxidase by activating the cytoplasmic NAD kinase. 6) Cellular H_2O_2 levels are determined by the rates of H_2O_2 production and metabolism via catalase (CAT) and the ubiquitous ascorbate-glutathione cycle (ASC-GSH cycle), which involves ascorbate peroxidase (APX), dehydroascorbate reductase (DHAR) and glutathione reductase (GR). 7) H_2O_2 also reacts with glutathione to convert it from its reduced state (GSH) to its oxidized state (GSSG). 8) The reduced and oxidized forms of ascorbate can themselves act as signal-transducing molecules, for example, monodehydroascorbate (MDHA) is involved in cell cycle regulation, whereas dehydroascorbate (DHA) regulates cell growth. Moreover, by interaction with the glutathione pool in the cell, DHA can trigger other signalling sequences because the reduced (GSH) and oxidized (GSSG) forms of glutathione also have effects on gene expression. 9) A plasma membrane associated monodehydroascorbate reductase (MDAR) could be involved in signal transduction perhaps via a plasma membrane electron transport chain. 10) Superoxide itself could trigger an independent series of signalling events, but most will be converted to H_2O_2 by the action of an apoplastic superoxide dismutase (SOD). Acorbate oxidase (AO) is a key regulator for determining the redox state of the ascorbate pool in the apoplast. (Neill et al., 2002; Pastori and Foyer, 2002).

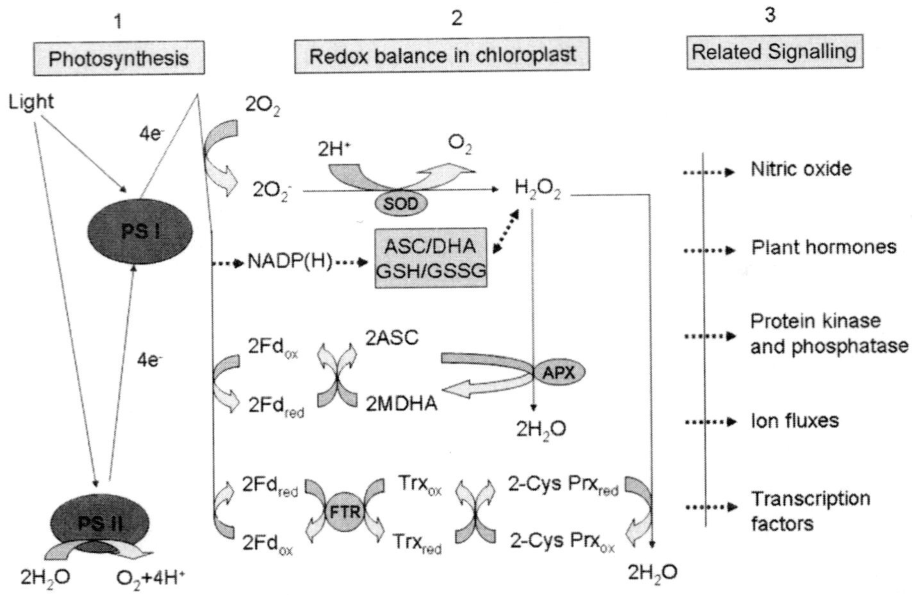

Fig. 3. Redox reaction and signalling in plants during photosynthesis. 1) ROS are produced from photosystem I (PSI) and photosystem II (PSII). 2) ROS are processed by superoxide dismutase (SOD) and reductive systems in which NADP(H), ascorbate (ASC), and glutathione (GSH) play a key part. Ascorbate peroxidase (APX) is commonly considered to exclusively detoxify H_2O_2. In this water-water cycle, monodehydroascorbate (MDHA) is reduced by monodehydroascorbate reductase (MDHAR) linked to ferredoxin (Fd) or, after disproportionation of two moleculars, by MDHAR to ASC and dehydroascorbate (DHA) via dehydroascorbate reductase (DHAR) linked to glutathione (GSH). Thioredoxin (Trx) or other reductants reduce oxidized peroxiredoxin (Prx) and need to be regenerated themselves, for example by ferredoxin-dependent thioredoxin reductase (FTR). 3) Interactions between ROS, ASC and GSH are important in acclimatory signaling mechanisms by which the plant perceives and responds to environmental changes. These mechanisms involve interplay with many other cell signalling components. (Scheibe et al., 2005; Dietz et al., 2006; Noctor, 2006).

et al., 2005). NADPH-dependent hydroxylation reactions catalysed by cytP450s are important in xenoboitic detoxification (Mezzari et al., 2005). Usually, these NADP(H) related enzymes act together with other enzymes such as GSTs (Baerson et al., 2005) and MDHAR (Noctor et al., 2006).

3.2.2 GSH/GSSG

Glutathione is the most abundant low molecular weight thiol inside mammalian cells. It is present as a reduced form (GSH) and two oxidased species: glutathione disulphide (GSSG) and glutathione mixed disulphide with protein thiol (GS-R) (Filomeni et al., 2002). Both these oxidized forms can be catalytically reduced back to GSH by the NADPH-dependent GR and Trx/Grx system, respectively (Filomeni et al., 2002). GSH is synthesized in two ATP-dependent reactions catalysed by γ-glutamyl cysteine synthetase (γ-GCS) and glutathione synthetase (GSHS) both with isoforms in the cytosol and the chloroplast with nearly equal activities. The synthesis is regulated in dependence on the cellular glutathione concentration and the GSH/GSSG ratio. Besides its function as the major reservoir of reduced non-protein thiol in plants, glutathione plays crucial roles in protection from cellular damage and in redox regulation (Filomeni et al., 2002). Glutathione cannot functionally be replaced by other metabolites, and at low concentrations it is essential for normal plant development (Foyer et al., 2008). Under non-stress conditions, the ratio of GSH/GSSG is generally more than 20 and total glutathione concentrations range between 1 and 5 mM (Foyer et al., 2001). The high GSH concentration is necessary for preventing protein thiol oxidation and for its function as substrate of GSH-dependent enzyme activities (Foyer and Noctor 2005a). Under stress conditions the photosynthetic electron transfer chain turns overreduced either due to a lack of metabolic electron acceptors or as a consequence of redox imbalances, and thus ROS are generated from photosystem I (PSI) and photosystem II (PSII) (Foyer and Noctor, 2003). High rates of ROS production shift the chloroplast redox environment to a more oxidized state by accumulation of GSSG. Now it has been considered that the dropping GSH/GSSG ratio is a stress signal triggering response reactions in the cell and the GSH/GSSG redox cycle is a key player in redox regulation (May et al., 1998; Filomeni et al., 2002; Foyer et al., 2008).

3.2.3 ASC/DHA

In addition to glutathione, plants accumulate high concentrations of ascorbate (ASC), particularly in photosynthetic tissues. It is tempting to categorize ascorbate primarily as a low molecular weight antioxidant. However, considerable data have shown the central importance of this metabolite in plant biology, with roles in hormone synthesis, gene expression, cell division, and growth (Foyer et al., 2008). The product of ASC oxidation is the short-lived radical MDHA, which can be reduced to ASC or undergo dismutation to produce DHA. In the apoplast, MDHA and DHA can also be produced by ascorbate oxidase (AO) (Pignocchi and Foyer 2003). Numerous routes exist for MDHA and DHA reduction. As well as reduction by ferredoxin and cytochromes b561, MDHA can be reduced by monodehydroascorbate reductase (MDHAR) that uses either NADH or NADPH (Noctor 2006). The most important enzyme for DHA reduction is dehydroascorbate reductase (DHAR). It has been found that the *Arabidopsis* genome consists of five annotated DHARs, including predicted chloroplastic and cytosolic isoforms (Conklin and Barth 2004). Two subclasses of glutathione S-transferase (GST) have also been shown to possess DHAR activity (Dixon et al., 2002). DHA can also be reduced by GSH (Polle 2001) and other GSH-dependent mechanisms such as Trxs and Grxs (Potters et al., 2004). In the apoplast, the reduction of MDHA and DHA is largely related to an ascorbate-reducible transmembrane *b*-type cytochrome *c* (PM cyt b561) and MDHAR (Foyer et al., 2008). As a major redox buffer, the ratio of ASC/DHA and its redox cycle could be crucial in redox signalling both in normal development and stress conditions in plants (May et al., 1998; Pignocchi and Foyer 2003; Noctor 2006).

3.2.4 Ferredoxin/thioredoxin system

In terms of plant cell redox regulation the ferredoxin/thioredoxin (Fd/Trx) system is a classical and the best studied example of thiol-disulphide dependent control of enzyme activities. Thioredoxins (Trxs) are abundant small disulphide oxidoreductases of approximately 12 kD with the conserved active site motif W-C-G(P)-C and a common 3D-structure. Trxs are found in almost all living organisms. More than 20 genes code for canonical Trxs in the *Arabidopsis* genome and an even larger set of genes encodes thioredoxin-like proteins and proteins with Trx domain. Based on the primary amino acid structure, this multigene family can be grouped into six different classes, namely, cytosolic and partly mitochondrial Trx-h, mitochondrial Trx-o, and plastidic Trx-f, -m, -x and –y. The chloroplast contains the largest number and highest amounts of Trxs, for instance, the *Arabidopsis* chloroplast contains 2 members of Trx-f, 4 Trx-m, 1 Trx-x and 2 Trx-y, respectively. Trx reduces regulatory disulphides of target proteins and thereby often changes the activation state of their targets (see the review by Oelze et al., 2008). The evidence suggests that individual organs express characteristic members of the thioredoxin *b* family. In plants, the disulphide formed between the two cysteines is reduced by ferredoxin and FTR in chloroplasts and in the other cell compartments by NADPH via a flavin enzyme, NTR. By reducing disulphide groups, Trxs, in turn, function as electron (hydrogen) donors for the regulation of enzymes, such as those of chloroplasts, or for the reduction of chemical substrates such as ribonucleotides or methionine sulphoxide (Buchanan and Balmer 2005). Oxidized target proteins can be reduced *in vitro* by Trxs, thioredoxin reductase and NADPH. Novel proteomic analysis showed that total 92 proteins in chloroplast were identified as putative Trx targets and these proteins could be assigned to ten functional categories. The large number of Trx targets underlines the importance of redox regulation via Trx in cellular and in particular photosynthetic metabolism (see the review of Oelze et al., 2008). The function of Trx proteins in the chloroplast are also dependent on peroxiredoxins (Prxs), which are a kind of thiol-based peroxide reductases with a broad substrate specificity (Dietz et al., 2006). It has been found that several Prx proteins exist in plants. The *Arabidopsis* genome encodes 10 Prx genes and the rice genome encodes at least 9 (Dietz et al., 2006). In the chloroplast, 2-Cys Prx/Trx and Mehler-peroxidase reaction would allow very high rates of H_2O_2 turnover and thus Prx/Trx system plays an important role for redox balance in this compartment (Figure 3; Dietz et al., 2006; Foyer et al., 2008).

4. ASC AND GSH RELATED REDOX REGULATION

4.1 Ascorbate

Ascorbate (ASC) is an abundant and ubiquitous small antioxidant in plants (Smirnoff 1996). It can directly react with most ROS such as singlet oxygen (O_2^1), O_2^-, H_2O_2 and hydroxy radical ($HO^•$), and thus plays an important role in the resistance of plants to oxidative stress. Besides its role in antioxidant defence, ASC also has a broad range of functions in plants: 1) participates in redox regulation and is an important redox buffer, 2) regulates gene transcription and translation, 3) modulates enzyme activities and acts as an enzyme cofactor, 4) takes part in cell wall synthesis, elongation and cross-linking processes, and 5) controls cell division and cell elongation. These functions that ASC has are tightly related to its redox status, biosynthesis, metabolism, regeneration and transport and associated enzymes activities (Smirnoff 1996; Horemans et al., 2000a, b).

4.1.1 Ascorbate biosynthesis and metabolism

ASC is present in most cell compartments. For example, ASC can accumulate in chloroplasts to very high concentrations and reach 12-30% of total ASC content of the leaves (Foyer et al., 1983; Foyer and Lelandais 1996). ASC also substantially exists in the apoplast, cytosol and vacuole (Rautenkranz et al., 1994; Vanacker et al., 1998a, b). It has been found that the ultimate step of ASC synthesis catalysed by L-galactono-γ-lactone dehydrogenase (GLDH) occurs in the mitochondria and isolated mitochondria produce large amounts of ASC in the presence of the precursor, L-galactono-γ-lactone (GL) (Bartoli et al., 2000). The ASC pool of intracellular compartments mainly maintains reduction except under stress conditions (de Pinto et al., 1999).

The biosynthesis pathway of ASC in plants was clarified in 1998 although it was ound to be present in plants quite early (Wheeler et al., 1998). It is a very complex process by which many intermediate precursors are needed. Despite the subcellular ocalization of different steps in the ASC biosynthesis remains to be elucidated, the nzyme GLDH which catalyses the final conversion of GL to ASC has been found to xist on the inner mitochondria membrane (Mutsuda et al., 1995; Siendones et al., 1999). GLDH, with a molecular mass of 56 kD, is highly specific for GL and requires cytochrome as a second substrate that is reduced as GL is converted to ASC (Ôba et al., 1995; Østergaard et al., 1997; Imai et al., 1998). GL is synthesized from L-Gal and then is apidly converted to ASC (Conklin et al., 1996). The pathway of ASC biosynthesis roposed by Wheeler et al. (1998) is very valuable for further study since many ntermediates for ASC biosynthesis can also be used for cell wall biosynthesis and protein lycosylation (Conklin et al., 1997).

ASC is generally involved as an electron donor transferring electrons in two equential steps. The first step results in the formation of ASC free radical MDHA and he second in the fully oxidized DHA. Both oxidation products are relatively unstable nd can be reduced rapidly by MDHAR and DHAR, respectively. Efficient cycling etween oxidized and reduced forms of ASC minimizes DHA degradation and decreases he requirement for high rates of *de novo* synthesis. ASC can feedbackly inhibit its own iosynthesis (Mozafar and Oertli 1993; Imai et al., 1999). The site of ASC degradation in lant cells is unknown but ASC has been shown to be involved in the synthesis of rtrate and oxalate (Loewus and Loewus 1987; De Gara and Tommasi 1999).

1.2 Transport of ascorbate in plant cells

SC is synthesized on the inner membrane of the mitochondria from where it is stributed to all the intracellular compartments and to the apoplast. Its concentration d rapid change of redox states between the oxidized and reduced forms are essential r those suborgans of plant cells (Horemans et al., 1997; Horemans et al., 2000a). Like e chloroplasts, the apoplast is rich in ASC. Together with apoplastic ascorbate oxidase O), ASC regulates the lignification of cell walls (Takahama and Onili 1992). The strict gulation of ASC concentrations in specific cellular compartments, according to quirements of physiology, development and metabolism, suggests that plant cells, e animal cells, have specific ASC and DHA transporters (Horemans et al., 2000b).

The transport of ASC in plants falls into two scales: transport between subcellular compartments and transport through the plasma membrane (Horemans et al., 2000a, b). Once ASC is synthesized in mitochondria, it will need to be transported to other compartments. The presence of ASC transport systems in plant cells makes ASC be transferred rapidly between the subcellular compartments according to the requirements of physiology, development and metabolism. At present, the ASC transport system has been found in chloroplast, thylakoid, tonoplast and plasma membrane, but no data are available in mitochondria or peroxisomes. Since the last ASC biosynthesis step occurs on the inner mitochondrial membrane, it is, therefore, conceivable that the transport of ASC from the intermembrane spaces of the mitochondria to the cytosol occurs by simple diffusion (Horemans et al., 2000a). ASC transport from the cytosol into the chloroplast stroma is a carrier-mediated process with a rather low but physiologically relevant affinity for ASC (Beck et al., 1983; Foyer and Lelandais 1996). Interestingly, the transport of ASC into intact chloroplasts shows cis-inhibition and trans-stimulation by DHA, suggesting that this transport can be driven by facilitated exchange diffusion (Bech et al., 1983; Foyer and Lelandais 1996). In contrast, transport of ASC through thylakoid membranes is not a carrier-mediated process but is a pH- and concentration gradient-dependent simple diffusion (Foyer and Lelandais 1996). In the thylakoid lumen, ASC has an important function in the synthesis of zeaxanthin and thus the changes in ASC availability via limitation of transport may be of regulatory influence for zeaxanthin formation (Foyer and Lelandais 1996). Transport of ASC also plays an essential role in the reactions regulating the transthylakoid proton gradient and electron transport in the chloroplast (Foyer and Noctor 1999).

Initial evidence for ASC transport through the plasma membrane was obtained in soybean by Mozafar and Oertli (1993). Then this phenomenon was continuously found in barley (Rautenkranz et al., 1994), pea (Foyer and Lelandais 1996) and tobacco BY-2 cell cultures (Horemans et al., 1998a), confirming the presence of ASC and DHA transporters in plants. Unlike the chloroplast membrane located ones, the plasma membrane-associated carrier shows high affinities for ASC and DHA (Horemans et al. 2000a). Although it has been sure that specific ASC and/or DHA transporters exist in the plasma membrane, the driven mechanism for these transporters is still not clarified. Now, it was suggested that at least three mechanisms for ASC and DHA transporter driving are present in the plasma membrane: facilitated exchange diffusion, proton electrochemical gradient, and ASC/DHA exchange mechanism (Horemans et al., 2000a). The apoplast of leaf cells can contain up to 10% of total foliar ASC pool and since n extracellular ASC biosynthesis exists, the transport of ASC from the cytoplasm to the apoplast is crucial for plant development and stress responses because ASC/DHA couple participates in many physiological processes such as cell wall lignification, cell division and redox regulation (Horemans et al., 2000a).

4.1.3 Ascorbate redox couple under varied environmental stresses

As the most important antioxidant in plant cells, ASC can protect chloroplasts from oxidative damage. ROS production from photosynthesis and photorespiration inevitable and, the chloroplast stromal and thylakoid-bound SOD and APX play an important role in the scavenging of O_2^- and H_2O_2 (Noctor and Foyer 1998). In chloroplasts, H_2O_2 can be reduced to H_2O and MDHA by APX with ASC present. The lifetime of MDHA is very short and it can rapidly be reduced to ASC or oxidized to DHA. In plant cells, MDHA is reduced to ASC either acting as eletron donor of ferredoxins located in the outside of thylakoid membranes or by ASC-GSH cycle (Noctor and Foyer 1998). In this ASC-GSH cycle, MDHA can also be reduced to ASC by monodehydroascorbate reductase (MDHAR). DHA, which is oxidized from MDHA, can use GSH as an electron donor to be reduced by dehydroascorbate reductase (DHAR). The resulted product GSSG can then be reduced by glutathione reductase (GR) returning to GSH. So, a GSH-dependent redox cycle for ASC is established in the cells. The cycle is also dependent on the NADPH, which comes from either photosynthetic electron transfer chain or the pentose phosphate pathway (PPP) (Noctor and Foyer 1998; Davey et al., 2000).

As mentioned above, the redox system of ascorbate consists of its reduced form (ASC), semi-reduced form (MDHA), oxidized form (DHA) and related enzymes in plant cells. Many enzymatic recycling processes function on the maintenance of reduction state of ASC in intracellular circumstance. However, the redox status of ASC is more determined by species variation and environmental conditions which the plants are in. The redox homeostasis and its maintenance of ASC pool as well as the changes of the related processes under varied environmental conditions are very important for plant resistance or tolerance. Salt stress led to decreases in the ASC content in wheat and the tolerant cultivar showed lesser decline in ASC content than the sensitive one while the APX activity was stimulated (Sairam and Srivastava 2002). In tomato roots, Shalata et al. (2001) found that salt stress decreased contents of ASC, ratio of ASC/DHA, and activities of APX and MDHAR in the sensitive cultivar whereas the effects were reverse in the tolerant cultivar, suggesting that the higher resistance of tomato to salt stress is, at least partially, from the relatively higher ASC content, ASC/DHA ratio and activities of related enzymes. Drought stress also led to the changes of ASC content and its metabolism. Li and van Staden (1998) found that the ASC content and the APX actvity were decreased by water deficit stress in maize. Schwanz *et al.* (1996) reported that the content of total ASC was decreased by drought in oak whereas the content of reduced ASC was increased. In spring wheat, we previously observed that a gradual field drought stress led to the increase of total ASC and its reduced form as well as the activity of APX (Chen et al., 2003a). In different reed ecotypes which grew in different environmental habitats, we found that, as compared to swamp reed ecotype, the two terrestrial reed ecotypes, namely, dune reed (DR) and heavy salt meadow reed (HSMR), possessed higher ASC/DHA ratio and activities of APX, DHAR and MDHAR, suggesting that upregulation of ASC metabolism and its redox cycling capacity contributes to the tolerance of reed plants to their long-term drought and saline prone habitats (Chen et al., 2007).

4.2 Glutathione

Glutathione (GSH) is an abundant and ubiquitous thiol. It is the major reservoir and transporter of non-protein reduced thiol in plants and plays an important role in the biosynthesis of nucleic acids and proteins as well as in modulation of enzyme activity (May et al., 1998). In addition, GSH is also a key player in redox signalling and regulation in plants (Foyer et al., 2008).

The level of GSH has been shown to correlate with the adaptation of plants to extreme temperature, the tolerance of plants to xenobiotics and to biotic and abiotic environmental stresses (May et al., 1998). Its notable stability derives from the γ-glutamyl linkage, while the strong nucleophilic nature of the central cystein makes it a powerful cellular reductant. GSH is one of the most efficient scavengers of peroxides arising as by-products of cellular metabolism or during oxidative stress. The efficiency with which the oxidized dithiol (GSSG) can be converted back to GSH during the reductive inactivation of peroxides contributes to the centrality of GSH in antioxidant defences. Alscher (1989) suggested that not only the elevation of GSH levels was correlated with environmental stress tolerance but also the increase of its synthesis appeared to be an intrinsic response of plants to stress. The ability of reduced GSH to participate in the redox regulation of cellular processes is thought to be largely dependent on its concentration and the ratio of GSH/GSSG. Thus, identification of the factors which act to modify the level, redox status and spatial distribution of GSH in plants will be crucial to understand the functions of GSH.

4.2.1 The biosynthesis and regulation of glutathione in plants

GSH is synthesized from glutamate, cysteine and glycine in two ATP-dependent reactions catalysed by γ-glutamylcysteine synthetase (γ-GCS) and glutathione synthetase (GSHS). These enzymes are found in the cytosol and chloroplasts of plant cell at approximately equal activities (Noctor and Foyer 1998). The potential to enhance GSH levels by genetic engineering of its own biosynthetic pathway is clearly an attractive proposition, so many studies have focused on the biosynthesis, gene expression, and regulation of these two enzymes. Through complementation of *E. coli* and *Saccharomyces cerevisiae*, mutants deficient in these enzymes, cDNAs encoding γ-GCS (May and Leaver 1994; Ullmann et al., 1997) and GSHS (Rawlins et al., 1995; Ullmann et al., 1996) have been isolated from *Arabidopsis thaliana*. ESTs encoding these enzymes have also been described from rice and cDNAs encoding γ-GCS from tomato (Kovari et al., 1997) and *Brassica juncea* (Schäfer et al., 1997) all display high sequence homology. Like animal γ-GCS, the enzyme in plants is susceptible to inhibition by GSH. This regulation may be an important homeostatic constraint that prevents excessive GSH accumulation, but it does not prevent the constitutive increases in tissue GSH contents obtained in transformed plants overexpressing γ-GCS (Notor and Foyer 1998; Foyer and Noctor 2001). The most important factors controlling GSH accumulation in plants are the activity of γ-GCS and the availability of cysteine (Foyer and Noctor 2001). As the first enzyme in the pathway, γ-GCS is clearly responsible for much of regulation of GSH concentration. The *Arabidopsis thaliana* mutant, *cad-2*, has a mutation in the gene encoding γ-GCS and has only one-third of the tissue GSH contents of the wild type. Increases in glutathione contents accompany increases in γ-GCS in tissues treated with xenobiotics and heavy metals such as cadmium. Most importantly, overexpression of an *E. coli* γ-GCS, but not GSHS, in poplar, mustard or tobacco substantially increases the glutathione pool (Notor and Foyer 1998; Foyer and Noctor 2001). Overexpression of γ-GCS (but not GSHS) led to constitutive increases in leaf glutathione of up to 400% (Notor and Foyer 1998). The leaf cysteine pool was not depleted in the transformed plants but was in fact slightly enhanced in response to increased γ-GCS activities. In these transformed plants, the pathways of sulphur assimilation and cysteine biosynthesis are up-regulated to compensate for the increased demand for cysteine, suggesting coordinated regulation of cysteine synthesis and glutathione synthesis (Foyer and Noctor 2001). However, the glutathione content of transformed plants overexpressing γ-GCS can be increased even further by the addition of cysteine, indicating that this substrate remains a key limiting factor.

4.2.2 The metabolism and transport of GSH in plants

There are three important enzymes related to GSH metabolism, namely, glutathione reductase (GR), glutathione peroxidase (GP), and glutathione S-transferases (GST). As an antioxidant and protector, GSH is oxidized to glutathione disulphide (GSSG), and then, in turn, GSSG is reduced efficiently back to GSH by the action of GR (Noctor et al., 1998). GR is a widely distributed member of the family of flavoprotein oxidoreductase and is found both in eukaryotes and prokaryotes. In plants, it plays a key role in the response to oxidative stress by maintaining intracellular glutathione pool primarily in the reduced state. Several isoforms of GR have been found in plants, and those isoforms encoded by different genes are crucial for the tolerance of plants to environmental stresses (Stevens et al., 1997; Noctor et al., 1998). GSH may also metabolize hydrogen peroxide by participating in the ascorbate-glutathione cycle or in the reaction catalysed by GP. GP is considered a key enzyme in the anti-ROS defence in animals, and in higher plants the expression of its encoding genes has been found to be tissue-specific (Leisinger et al., 2001). It catalyses the reduction of H_2O_2 by GSH. H_2O_2 is especially toxic in chloroplasts by inhibiting the Calvin-cycle enzymes possessing exposed sulfhydryl groups, such as glyceraldehydes-3-phosphate dehydrogenase (G3PD) and Fru-1,6-bisphosphatase (FBPase), and hence reduces the photosynthetic carbon dioxide assimilation (Takada et al., 1995). GP is encoded by a multiple gene family which can be grouped into two classes: selenium-dependent GPs and selenium-independent GPs (Cnubben et al., 2001). Selenium-dependent GPs consist of 5 members and can reduce H_2O_2 and organic hydroperoxide. The expression of genes encoding GPs can be upregulated by ROS and thus they play important roles in the stress defenses (Leisinger et al., 2001). GST is another GSH-dependent detoxifying enzyme in plants, and in both crops and weeds, it detoxifies electrophilic herbicides by catalysing their conjugation with GSH (Edwards et al., 2000). GST is also encoded by a multiple gene family and the expression of specific GSTs vary markedly during plant development and after exposure to pathogens, changes in environmental conditions and chemical treatments. Therefore, the enhancement of GST expression has become a marker for plant response to stress (Edwards et al., 2000).

Besides the three enzymes mentioned above, phytochelatin synthetase (PCS) also participates in the GSH metabolism (Edwards et al., 2000). Phytochelatin functions in detoxification of heavy metal pollution but the synthesis of phytochelatin (PC) needs a large GSH pool in plants (Edwards et al., 2000; Foyer et al., 2001).

The concentrations of glutathione in different cells and tissues are the product of an equilibrium between synthesis, degradation, use, and short- and long-distance transport. It has been shown that different cell types vary markedly in their GSH concentration and not all cell types have the same ability for GSH synthesis. Therefore, the transport of GSH both between compartments and organs occurs in plants (Foyer et al., 2001). As early as 1970s, it was shown that GSH could be transported both into and out of plant cells (Bergmann and Rennenberg 1978). As any peptide, the exchange of GSH between the cytosol, apoplast and phloem requires specific plasma membrane transport systems. Where GSH is transported against its concentration gradient, energy is required, and the transport is driven by ATP hydrolysis in the case of primary pumps or by secondary co-transport mechanisms that use the stored energy of transmembrane ion gradients (Foyer et al., 2001).

Multidrug resistance associated proteins (MRPs) belonging to the ATP binding cassette (ABC) protein family were first identified as ATP-driven pumps of drugs and GS-X in mammals, and were subsequently shown to export GSH across the plasma membrane (Cnubben et al., 2001). As a transporter of GSH, MRPs can pump GSSG out of cells and thus function in the stress defences by increasing the intracellular ratio of GSH/GSSG (Cnubben et al., 2001). GSH-specific uptake systems were not only found in yeast and animal cells, but also found in plant cells (Foyer et al., 2001). Early kinetic analysis of GSH transport into tobacco cells revealed that high affinity and low affinity systems in plants (Schneider et al., 1992). However, in the protoplasts of broad bean, GSH uptake followed single saturation kinetics and there were two systems for GSSG uptake (Jamaï et al., 1996). Further studies indicated that the plasma membrane of plant cells contains a specific transport system for GSH that imports GSSG and GS-X rather than GSH, which is distinct from known amino acid and peptide carries (Jamaï et al., 1996).

In plants, intracellular transport of GSH also exists (Foyer et al., 2001). A key parameter for intracellular GSH homeostasis is transport between organelles and the cytosol. Although chloroplasts synthesize GSH, GSH is also readily taken up by intact chloroplasts. The rate of GSH uptake is sufficient to play a significant role in determining the total pool on either side of the chloroplast envelope because uptake capacity is of the same order as the capacity for GSH synthesis. However, these capacities are much lower than those of other processes such as antioxidant recycling and hence might be predicted to play a relatively minor role in balancing GSH redox states between different compartments (Foyer et al., 2001). GSH transport system also exists in the tonoplast (Foyer et al., 2001).

4.2.3 Glutathione redox cycle under varied environmental stresses

As an important soluble antioxidant, GSH can directly reduce ROS in plants. However, since the reaction rate between GSH and H_2O_2 is very slow, the GSH-dependent reduction of ROS is not a major route of ROS detoxification in plants (Eshdat et al., 1997). However, GSH and its related physiological processes play an important role in the tolerance of plants to environmental stresses.

The function of GSH in plant stress responses is tightly related to its intracellular concentration, redox state (GSH/GSSG ratio), and the induction and activities of enzymes involved in its biosynthesis and metabolism (May et al., 1998; Noctor et al., 1998; Foyer et al., 2008). Many researches have shown that the increase of the activities of enzymes related to GSH biosynthesis and metabolism can elevate the resistance or tolerance of plants to environmental stresses (May et al., 1998). GP is induced in plants in response to stress for detoxifying lipid peroxides and H_2O_2 (Eshdat et al., 1997). The induction of GST is related to the detoxification of electrophilic herbicides (Cnubben et al., 2001). In plants, the major substrate for reductive detoxification of H_2O_2 is ascorbate, which must therefore be continuously regenerated from its oxidized forms. A major function of GSH in protection against oxidative stress is the re-reduction of ascorbate in the ASC-GSH cycle (Noctor and Foyer 1998). In this process, GSH acts as a recycled intermediate in the reduction of H_2O_2 using electrons derived, ultimately, form H_2O. Efficient recycling of GSH is ensured by GR activity. Under oxidative stress conditions, H_2O_2 can be reduced by GSH and simultaneously GSH is oxidized into GSSG, whereas under normal growth conditions, GSSG can be reduced by GR into GSH again when NADPH exists, therefore, a redox cycle is established (Lu et al., 2000). When the intracellular reduction ability is insufficient, the ratio of GSH/GSSG will be decreased, and this redox state changes of thiols, in turn, can activate a series of redox-sensitive transcript factors (Sen and Packer 1996). Therefore, the GSSG level has been considered a marker of intracellular oxidative stresses while GSH/GSSG ratio as an index of intracellular redox status. It has been found that the DNA-binding capacity of some transcript factors is dependent on the redox state of thiol groups of their cysteine residues (Kearns and Hall 1998). Owing to its regulatory ability in transcript factor's redox status, GSH has been proposed to play an important role in regulation of gene expression (Mulcahy and Gipp 1995). Recently, it is suggested that the enzymes of GSH synthesis and metabolism are induced together in response to stresses (Foyer et al., 2008). There is considerable overlap in the signal transduction cascades that induce the genes encoding the enzymes of GSH biosynthesis and metabolism. The evidence based on molecular genetics suggests that γ-GCS may fulfil a signalling function as well as a catalytic role (Ball et al., 2004).

In recent years, GSH levels have been used as a cue in the coordination of mechanisms for both the supply of cysteine and the maintenance of cellular NADPH pools. Co-ordination of these responses merely by lowering the GSH pool, provides plants with a simple mechanism to respond defensively to a wide range of stresses through a coordinated upregulation of the capacity for GSH synthesis and its redox cycling (May et al., 1998; Foyer et al., 2008). In physiological terms, the GSH functions mainly as a redox couple during plant development and stress responses. Under GSH pool deficit or oxidative stress conditions, for the removal of ROS, both the provision of NADPH and the assimilation of sulphate would be assured. This allows a rapid restoration of the GSH pool once the stress is alleviated or disappears. When GSH participates in detoxification of ROS with a redox couple, a sufficient supply of NADPH is essential. NADPH is supplied by the activity of the pentose phosphate pathway (PPP) in bacteria, yeast and mammals. Mutational analyses have shown that PPP pathway is essential for survival under oxidative stress (Storz et al., 1990; Pandolfi et al., 1995; Juhnke et al.,

1996; Slekar et al., 1996). In addition, genes encoding key enzymes of the PPP pathway are targets for transcription factors activated in response to oxidative stress (Storz et al., 1990) and an essential regulatory component of this event in yeast is the GSH level (Izawa et al., 1995). This implies that NADPH supply should be coordinated with the increases in the activity of the sulphate assimilation and GSH biosynthetic pathways during the responses of plants to stress to ensure the supply and functioning of GSH with which it functions intimately as a redox couple (May et al., 1998).

The sizes of GSH and ASC pools and their regulatory and functionary relationship during plant stress resistance have been paid much attention. In the roots of *Arabidopsis*, it has been found that a reduction of the meristematic activity caused by GSH depletion can be alleviated by exogenous supply of ASC (Sanchez-Fernandez et al., 1997). Similarly, an 80% depletion of the GSH pool in GSH pool serious deficit *Arabidopsis* plants did not lower the threshold of sensitivity of the plants to oxidative stress, implying that ASC/DHA may function effectively as a redox couple in the absence of a sufficient GSH pool in plants (May et al., 1998). However, in yeast, GSH is essential and GSH-deficient yeast cannot be rescued by exogenously supplied ASC (Grent et al., 1996). Recently, a GSH-independent pathway of ASC regeneration has been found in tobacco BY-2 cell (Potters et al., 2004). The phenomenon of GSH-independent ASC regeneration was also observed in animals (Guaiquil et al., 1997). However, more data have shown that GSH and ASC redox states and their related enzymes participate in the plant development and stress responses, and the functions of ASC is tightly dependent on the GSH supply and redox status (May et al., 1998; Pastori and Foyer 2002; Foyer et al., 2008). GSH can regenerate ASC from DHA both chemically and by acting as the reductant for several enzymes with DHAR activity as mentioned above. The relative redox potentials of the ASC and GSH couples strongly favour net electron flow from GSH to DHA, a reaction that can occur at significant rates even in the absence of DHAR, particularly at alkaline pH values (Polle 2001). Under conditions that cause high fluxes of ASC peroxidation, however, DHAR activity is necessary to ensure effective maintenance of the reduced ASC pool (Polle 2001). GSSG generated as a result of this reaction is subsequently reduced by GR. The significance of the degree of coupling of the ASC and GSH pools in different leaf compartments, particularly in relation to signal transduction, has been established previously (Polle 2001). Recently, we found that upregulations of GSH and ASC biosynthesis and their redox cycling capacities, as well as the elevating ratios of GSH/GSSG, ASC/DHA and the ratio of the reduced to oxidized forms of pyridine nucleotides, are essential for adaptation of the terrestrial reed ecotypes to their long-term drought or saline habitats by maintaining a more reduction state of redox status in the plant cells (Chen et al., 2003a; Chen et al., 2007). The high ratios of ASC/DHA and GSH/GSSG and their redox cycling capacity were found to be closely related with the high reduction redox status of cells in wheat under drought stress (Chen et al., 2003b; Chen et al., 2004). Therefore, we suggest that coordination of the three redox couples in the stress response to ensure a more reduction state of the cellular redox status might be an efficient strategy for the plant stress tolerance.

4.3. ASC and GSH Related Redox Regulation in Photosynthesis

Photosynthesis is a high rate redox metabolic pathway that generates many redox intermediates with extreme redox potentials. It functions optimally when the prevailing environmental conditions meet its species-specific requirements. However, environmental conditions can fluctuate on a timescale ranging from seconds to days or even years, and many of these changes decrease photosynthesis. As the central role of photosynthesis is energy production, photosynthetic organisms have developed many strategies to acclimate to a broad range of environmental conditions. In plants, the main generation sites of ROS are the reaction centres of photosystem I (PSI) and photosystem II (PSII) in chloroplast thylakoids. Under steady-state conditions, the intercellular level of ROS is tightly buffered and controlled by antioxidant systems (Noctor and Foyer 1998; Mittler 2004). However, under various environmental conditions, i.e., high light, cold, drought, and saline stresses, as well as pathogen attack, a limitation of linear electron transport due to over-reduction of stromal electron acceptors can be induced and thus the production of different kinds of ROS such as O_2^-, H_2O_2, and hydroxyl radicals ($HO^•$) by transfer of electrons to O_2 can be accelerated. In addition, singlet oxygen (O_2^1) can be formed by energy transfer from triplet P680. Redox regulation and over-reduction control in the photosynthesizing cells, therefore, are very important for photosynthetic performance. The states of defined redox reactions in the chloroplast provide the predominant information and thus directly and indirectly the decisive signals for the multilevel control of cell activity in the chloroplast. Moreover, many photosynthesis genes are affected by chloroplast redox signals (Pfannschmidt 2003). So, it has been proposed that redox-dependent signalling is the core of photosynthesis-dependent regulation and that the other involved control mechanisms either function in developmental control or act as ancillary measures modulating the redox signalling pathways (Oelze et al., 2008).

To date, as mentioned above, genetic and biochemical analyses have identified plastoquinone (PQ), ROS, ASC and the ferredoxin/thioredoxin (Fd/Trx) system as redox-active signalling components. Many components with redox potentials participate in the redox regulation in photosynthesis. In *Arabidopsis*, at least 31 Grxs (Rouhier et al., 2004), 10 Prxs (Dietz 2003) and about 45 Trxs and Trx-like proteins (Meyer et al., 2005) have been found. Fd/Trx system is an important mechanism in chloroplast redox regulations (Scheibe et al., 2005). One of the chloroplast enzymes, NADP-malate dehydrogenase (NADP-MDH), which can be reduced via the Fd/Trx system in the light, has been found serving as a redox valve in the balance of ATP/NADPH ratio in the chloroplast (Backhausen et al., 1998). As a result of the light-dependent primary reactions of photosynthesis, vectorial proton transfer acidifies the thylakoid lumen. The generated proton motive force is used to synthesize ATP via ATP synthase. Simultaneously, in linear electron transport, Fd-oxidoreductase transfers electrons from PSI to Fd-the main distributor of reduction energy to the various acceptors including Fd-dependent NADP reductase (FNR), Trx reductase (FTR), sulphite reductase (SR), nitrite reductase (NIR), cyclic electron transfer and oxygen reduction in the Mehler reaction (Oelze et al., 2008). In this process, reduced Fd plays the central distributor role in chloroplast redox metabolism. In terms of plant cell redox regulation, the Fd/Trx system is the classical and best studied example of thiol-disulphide dependent control of enzyme activities. It has been reported that many important enzymes including at least 11 Calvin cycle enzymes and up to 300 putative targets of Trxs are activation/deactivation enzymes by thiol/disulphide interchanges (see review of Oelze et al., 2008). Besides their regulatory role in metabolic processes by adjusting the redox status of targets, the Trx proteins also participate in antioxidant defence mechanisms as electron donors to Peroxiredoxins (Prxs) (Dietz et al., 2006).

Grxs are glutathione-dependent reductases which fall into two groups based on different active sites. The early discovered dithiol Grxs contain the characteristic Cys-Pro-Tyr-Cys active motif whereas the second group of Grxs, commonly named monothiol Grxs, lack the C-terminal active site thiol in its Cys-Gly-Phe-Ser active site but contain all structural and functional elements to bind and utilize GSH as substrate (see the review by Lillig et al., 2008). Grx are ideal candidates for the regulation of cellular processes associated with changes in the GSH-GSSG redox state and were first identified for their ability to deliver electrons to ribonucleotide reductase (RNR) (Holmgren 1976). Two distinct but functionally connected reaction mechanisms evolved for Grxs - the dithiol and the monothiol mechanisms, that both rely on the proteins inherent affinity for the GSH moiety. Grxs catalyse the reversible reduction of protein disulphides utilizing both cysteinyl residues in their Cys-Pro-Tye-Cys active site. In the first step, the N-terminal Cys residue performs a nucleophilic attack on the target disulphide. Next, the mixed disulfide intermediate formed between the two proteins is attacked by the second active site thiolate. The resulting disulphide in the active site is reduced by one molecule of GSH, leading to a mixed disulphide between GSH and the N-terminal active site Cys. This mixed disulphide is subsequently reduced by a second GSH molecule. The reduction of disulphides formed between GSH and proteins or small molecular weight components requires only the N-terminally located active site Cys residue. In this reaction, Grxs show a clear preference for the non-GSH molecule as living thiol. As described for the dithiol mechanism, this disulphide is reduced by the second molecule of GSH. The resulting GSSG is regenerated by glutathione reductase (GR) at the expense of NADPH (Lillig et al., 2008). In chloroplasts, two kinds of important proteins, chloroplastic f-type Trxs (Michelet et al., 2005) and chloroplastic glyceralaldehyde 3-phosphate dedydrogenase (Zaffagnini et al., 2007), were found to undergo reversible glutathionylation, which is tightly related to Grxs.

Peroxiredoxins (Prxs) are thiol-based peroxide reductases with a broad substrate specificity, ranging from H_2O_2 to alkyl hydroperoxides and peroxinitrite (Dietz et al., 2006). All Prxs have a similar basic protein structure with a thioredoxin fold. Their molecular masses range between 17 and 22 kDa and based on sequence similarities and catalytic mechanisms, four types of Prxs proteins are distinguished (Dietz 2003). The redox state of the photosynthetic electron transport carriers needs to be poised in order to avoid over-reduction and damage to the photosynthetic apparatus. It is reported that thylakoid bound and stroma APX take over the peroxidase function. However, Prxs can partly replace APX to establish an alternative pathway for the water-water cycle (Dietz et al., 2002). There are several Prxs proteins in the chloroplast including 2-Cys

PrxA, 2-Cys PrxsB, PrxQ, PrxII E, etc. It has been reported that the plastid 2-Cys Prxs proteins and PrxQ were estimated to amount to ~0.6% and ~0.3% of the chloroplast proteins, respectively, indicating important roles of these proteins in the chloroplast. Together with APX, these Prxs proteins mediate alternative water-water cycle and then contribute to the detoxification of Mehler-derived H_2O_2 in the chloroplast (Dietz et al., 2002).

The GSH-ASC redox pathway is a complex network of spontaneous, photochemical, and enzymatic reactions for detoxifying H_2O_2 in chloroplasts. The H_2O_2 produced from PSI and PSII in chloroplast thylakoids, at first, is reduced to water by ASC catalysed with APX. The resulted dehydroascorbate (DHA) and monodehydroascorbate (MDHA) are then reduced by DHA redutase (DHAR) and MDHA reductase (MDHAR) respectively to regenerate ASC. Glutathione reductase (GR) also participates in this process (Asada 1999). Although numerous studies have been performed on the GSH-ASC cycle, it is still a matter of intensive research. Valero et al., (2009) used a computer modelling of metabolic system to analyze the behaviour of the GSH-ASC redox cycle in chloroplasts and found that a sequential depletion of antioxidant power exists in chloroplasts in the order NADPH, GSH, ASC and their recovery in the reverse order. Decreasing levels of GR, APX, and SOD led to the irreversible photoinactivation of APX and the subsequent increase in H_2O_2 concentration, preceded by a maximum in DHAR activity (Valero et al., 2009). Figure 3 shows a simple scheme situating redox reactions during photosynthesis.

4.4. Redox Regulation in Apoplast

The apoplast is the extraprotoplastic matrix of plant cells and has important functions in plant biology. It is composed of all compartments from the external face of the plasmalemma to the cell wall. In recent years, the apoplast has been considered not only as a bridge from the environment to the symplast, but also as a fundamental reservoir of metabolic information concerning the biotic and abiotic environment that surrounds the cell. There are many enzymatic and non-enzymatic components in apoplastic spaces. The most important components which are related to stress response and redox signalling are SOD, ascorabte oxidase (AO), ASC, and ROS (mainly O_2^- and H_2O_2) (Pignocchi and Foyer 2003, Robert et al., 2009).

Many environmental and metabolic triggers, including pathogen attack, ozone, and physical and chemical assaults, alter the redox state of the apoplast by triggering an oxidative burst at the plasmalemma. The plasma membrane NADPH oxidase is a major ROS source in the apoplast (Ranieri et al., 2003). Although much attention has been focused on the roles of ROS as signals influencing plant responses to environmental stimuli and developmental cues (Mothofer et al., 2004; Kangasjarvi et al., 2005; Torres and Dang 2005), relatively little attention has been paid to the redox buffering capacity of the apoplast and how it is modulated. Despite the presence of flavonoids and polyamines in the cell wall that can act as antioxidants, the redox buffering capacity of the apoplast is weak (Horemans et al., 2000a; Pignocchi et al., 2003). The apoplast contains no NAD(P)H or very little GSH, so ASC is considered the major antioxidant (Foyer and

Noctor 2005b). It has been found that the apoplast of leaf cells contains only about 2% of total leaf GSH pool in oat leaves (Vanacker et al., 1999) and less than 1% in barley (Vanacker et al., 2000). In plant stress responses, such as the hypersensitive response and heat shock, GSH has been shown to stimulate the transcription of various genes, including those encoding cell wall hydroxyproline-rich glycoproteins, phenylalanineammonia lyase and chalcone synthase (Wingate et al., 1988). Moreover, a xenobiotic-inducible GST has been observed in the apoplast of soybean hypocotyls (Flury et al., 1996). In common with certain intracellular isoforms, this enzyme exhibits peroxidase activity that might contribute to GSSG production in the apoplast. Despite that most of the cellular ASC pool is localized within the cytoplasm, up to 5% is transported across the plasma membrane and into the apoplast, where it constitutes the major sink for gaseous oxidants such as ozone (Barnes et al., 2002). In addition to providing the major redox buffering capacity of the apoplast, ASC also participates in cell wall formation, controlling phenoxy radical-mediated cross-linking and other processes that affect cell growth (Smirnoff 2000).

The ASC pool in the apoplast is a key determinant of oxidant signal duration and strength as well as overall redox status of the apoplast (Foyer and Noctor 2005a). Low levels of ASC and antioxidant buffering would allow oxidative signals to persist and accumulate within the apoplast. The redox status of the extracellular ASC pool is regulated by AO (Pignocchi et al., 2003; Sanmartin et al., 2003). This enzyme is considered to be involved in the first step in the ASC degradation pathway in the apoplast (Green and Fry 2005) and has been found to influence cell enlargement via the modulation of redox control of the apoplast (Kato and Esaka 2000; Smirnoff 2000). AO is a cell wall-localized enzyme that uses oxygen to catalyse the oxidation of ASC to the unstable radical MDHA which disproportionately yield DHA and ASC, and thus contributes to the regulation of the ASC redox state. AO expression is modulated by complex transcriptional and translational controls (Esaka et al., 1992). The activity and expression of AO are closely correlated with cell expansion (Kato and Esaka 2000), with transcript levels increased by growth promoters and reduced by growth suppressors (Pignocchi et al., 2003). Moreover, AO has been shown to catalyse the oxidative decarboxylation of auxin, which may have impacts on root development (Kerk et al., 2000). These findings imply a role for AO in hormone mediated cell wall loosening, but a causal relationship remains to be established.

Together with AO, ASC transporters probably participate in a futile cycle that regulates the accumulation of both reduced ASC and its oxidized forms (MDHA and DHA) in the apoplastic environment for functioning in metabolism and growth. Such cycle has been suggested to be useful in photosynthetic energy dissipation (Nanasato et al., 2005). Since the redox buffering capacity of the apoplast is always low compared to that of the cytoplasm, a steep redox gradient will arise across the plasma membrane when the apoplast ASC pool is oxidized. It has been proposed that this gradient could influence gene expression through altered calcium release and modified channel activity (Pignocchi and Foyer 2003). One way that shifts in apoplast redox status could facilitate such effects is through interactions between receptor proteins containing oxidizable thiols that are situated in, or near, the membrane surface (Pignocchi and Foyer 2003). A further study more clearly showed that the redox state of the apoplast modulates plant growth and defence responses by regulating signal transduction cascades and gene expression patterns (Pignocchi et al., 2006). And, AO activity, which modulates the redox state of the apoplastic ASC pool, strongly influences the responses of plant cells to external and internal stimuli (Pignocchi et al., 2006).

Actually, although some studies reported that the enzymes involved in the ASC-GSH cycle such as MDHAR, APX, and GR are present in apoplast (Vanacker et al., 1998a; García-Limones et al., 2009), the presence of these enzymes in this compartment remains a debate because of their relatively low activities. But anyway, the results obtained already have shown that the oxidation of the apoplast, with inherent changes in redox state, is a primary result of the oxidative burst induced by environmental stresses. Oxidation of the apoplast influences the functions of all redox-sensitive proteins, particularly those with activities that are affected by thiol-disulphide exchange reactions (Pignocchi and Foyer 2003). The redox state of the ascorbate pool is a major determinant in the redox regulation of apoplast which is regulated by AO and the control of AO by light and other triggers such as gibberellic acid (GA), salicylic acid (SA) and auxin could modulate growth and defence processes in plants (Figure 2, Pignocchi and Foyer 2003; Pignocchi et al., 2006).

5. CONCLUDING REMARKS

ASC and GSH are central components of plant antioxidant defences acting together to limit ROS lifetime and accumulation. Together with NADPH/NAD and Fd/Trx redox couples, ASC/DHA and GSH/GSSG redox couples as well as their related processes such as ASC and GSH biosynthesis, metabolism, and their redox cycling are crucial for plant development and stress responses. Although we already know that ASC and GSH redox couples have a tight interaction in functions with each other, the details concerning this interaction is not clear yet because of the diverse functions and complicated regulatory mechanisms of each couple in plants, particular under stress conditions. The coordination of ASC and GSH in biosynthesis, metabolism, and redox cycling should be essential for plant life performance under varied stress conditions. ASC and GSH redox couples play key roles in redox regulation of photosynthesis in which the coordinated expression of photosynthesis genes is crucial for photosynthesis functions. Very recently, a computer simulation of the dynamic behaviour of the ASC-GSH cycle in chloroplasts has been done (Valero et al., 2009). Using a model system in which many terms related to ASC-GSH cycle were set up, the authors found a sequential depletion of antioxidant power in chloroplasts in the order NADPH, GSH, ASC and their recovery in the reverse order. Decreasing levels of GR, APX, and SOD led to the irreversible photoinactivation of APX and the subsequent increase in H_2O_2 concentration, preceded by a maximum in DHAR activity (Valero et al., 2009). This mathematical model, therefore, is a very powerful tool for improving the theoretical understanding of the kinetic behaviour of the ROS-detoxifying pathway in chloroplasts. The model, besides including new physiologically important aspects such as the balance between the photogeneration and consumption of NADPH and APX inactivation, can easily be expanded to include new steps such as regulatory devices for enzymes, dark-light interactions, the peroxiredoxin system, etc. It may also allow the guidance of future experimental work to evaluate those parameters involved in the ASC-GSH cycle that have not been evaluated yet, and thus offers a new strategy for studying plant defence mechanisms against oxidative stress.

REFERENCES

Allen, J.F. (2005). A redox switch hypothesis for the origin of two light reactions in photosynthesis. *FEBS Lett.* **579:** 963-968.

Alscher, R.G. (1989). Biosynthesis and antioxidant function of glutathione in plants. *Physiol. Plant.* **77:** 457-464.

Anjum, N.A., Umar, S. and Chan, M.-T. (2010). Eds. *Ascorbate-Glutathione Pathway and Stress Tolerance in Plants*, The Netherlands: Springer, ISBN 978-90-481-9403-2.

Asada, K. (1999). The water-water cycle in chloroplasts: Scavenging of active oxygens and dissipation of excess photons. *Annu Rev. Plant Physiol. Plant Mol. Biol.* **50:** 601-639.

Backhausen, J.E., Emmerlich, A., Holtgrefe, S., Horton, P., Nast, G., Roggers, J.J.M., Müller-Röber, B. and Scheibe, R. (1998). Transgenic potato plants with altered expression levels of chloroplasts NADP-malate dehydrogenase: interactions between photosynthetic electron transport and malate metabolism in leaves and in isolated intact chloroplasts. *Planta.* **207:** 105-114.

Baerson, S.R., Sánchez-Moreiras, A., Pedrol-Bonjoch, N., Schulz, M., Kagan, I.A., Agarwal, A.K., Reigosa, M.J. and Duke, S.O. (2005). Detoxification and transcriptome response in Arabidopsis seedlings exposed to the allelochemical benzoxazolin-2-(3H)-one. *J. Biol. Chem.* **280:** 21867-21881.

Ball, L., Accotto, G., Bechtold, U., Creissen, G., Funck, D., Jimenez, A., Kular, B., Leyland, N., Mejia-Carranza, J., Reynolds, H., Karpinski, S. and Mullineaux, P.M. (2004). Evidence for a direct link between glutathione biosynthesis and stress defense gene expression in *Arabidopsis*. *Plant Cell.* **16:** 2448-2462.

Balmer, Y., Vensel, W.H., Hurkman, W.J. and Buchanan, B.B. (2006). Thioredoxin target proteins in chloroplast thylakoid membranes. *Antioxid. Red. Sign.* **8:** 1829-1834.

Barnes, J.D., Zheng, Y. and Lyons, T.M. (2002). Plant resistance to ozone: the role of ascorbate. In Omasa, K., Saji, H., Youssefian, S. and Kondo, N. (Eds.). Air Pollution and Plant Biotechnology. Springer-Verlag, Tokyo, pp. 235-252.

Bartoli, C., Pastori, G.M. and Foyer, C.H. (2000). Ascorbate biosynthesis in mitochondria is linked to the electron transport chain between complexes III and IV. *Plant Physiol.* **123:** 335-343.

Beck, E., Burkert, A. and Hofmann, M. (1983). Uptake of L-ascorbate by intact spinach chloroplasts. *Plant Physiol.* **73:** 41-45.

Bedard, K., Lardy, B. and Krause, K.H. (2007). NOX family NADPH oxidases: Not just in mammals. *Biochimie.* **89:** 1107-1112.

Bellafiore, S., Barneche, F., Peltier, G. and Rochaix, J.D. (2005). State transitions and light adaptation require chloroplast thylakoid protein kinase STN7. *Nature.* **433:** 892-895.

Bergmann, L. and Rennenberg, H. (1978). Efflux and production of glutathione in suspension cultures of *Nicotiana tabacum*. *Z. Planzenphysiol.* **88:** 175-185.

Bonardi, V., Pesaresi, P., Becker, T., Schleiff, E., Wagner, R., Pfannschmidt, T., Jahns, P. and Leister, D. (2005). Photosystem II core phosphorylation and photosynthetic acclimation require two different protein kinases. *Nature.* **437:** 1179-1182.

Buchanan, B.B. and Balmer, Y. (2005). Redox regulation: A broadening horizon. *Annu. Rev. Plant Biol.* **56:** 187-220.

Chen, K.M., Gong, H.J., Chen, G.C., Wang, S.M. and Zhang, C.L. (2003a). The responses of antioxidant defense system at different developmental stages of spring wheat to gradual drought stress in field. *Acta Bot. Boreali-Occidentalia Sin.* **239:** 1497-1505.

Chen, K.M., Gong, H.J., Chen, G.C., Wang, S.M. and Zhang, C.L. (2003b). Upregulation of glutathione metabolism and changes in redox status involved in adaptation of reed (*Phragmites communis*) ecotypes to drought-prone and saline habitats. *J. Plant Physiol.* **160:** 293-301.

Chen, K.M., Gong, H.J., Chen, G.C., Wang, S.M. and Zhang, C.L. (2004). Gradual drought under field conditions influences the glutathione metabolism, redox balance and energy supply in spring wheat. *J. Plant Growth Regul.* **23:** 20-28.

Chen, K.M., Gong, H.J., Wang, S.M. and Zhang, C.L. (2007). Antioxidant defense system in *Phragmites communis* Trin. ecotypes. *Biol. Plant.* **51:** 754-758.

Cnubben, N.H.P., Rietjens, I.M.C.M., Wortelboer, H., van Zanden, J. and van Bladeren, P.J. (2001). The interplay of glutathione-related processes in antioxidant defense. *Environ. Toxicol. Pharmacol.* **10:** 141-152.

Conklin, P.L. and Barth, C. (2004). Ascorbic acid, a familiar small molecule intertwined in the response of plants to ozone, pathogens, and the onset of senescence. *Plant Cell Environ.* **27:** 959-970.

Conklin, P.L., Pallanca, J.E., Last, L. and Smirnoff, N. (1997). L-Ascorbic acid metabolism in the ascorbate-deficient *Arabidopsis* mutant *vtc1. Plant Physiol.* **115:** 1277-1285.

Conklin, P.L., Williams, E.H. and Last, R.L. (1996). Environmental stress sensitivity of an ascorbic acid-deficient Arabidopsis mutant. *Proc. Natl. Acad Sci. USA.* **93:** 9970-9974.

Corpas, F.J., Barroso, J.B. and del Rio, L.A. (2001). Peroxisomes as a source of reactive oxygen species and nitric oxide signal molecules in plant cells. *Trend Plant Sci.* **6:** 145-150.

Crawford, N.M. and Guo, F.Q. (2005). New insights in nitric oxide metabolism and regulatory functions. *Trends Plant Sci.* **10:** 195-200.

Davey, M.W., Montagu, M.V., Inzé, D., Sanmartin, M., Kanellis, A., Smirnoff, N., Benzie, I.J.J., Strain, J.J., Favell, D. and Fletcher, J. (2000). Plant L-ascorbic acid: chemistry, function, metabolism, bioavailability and effects of processing. *J. Sci Food Agric.* **80:** 825-860.

De Gara, L. and Tommasi, F. (1999). Ascorbate redox enzymes: a network of reactions involved in plant development. *Recent Dev. Phytochem.* **3:** 1-15.

de Pinto, M.C., Francis, D. and De Gara, L. (1999). The redox state of the ascorbate-dehydroascorbate pair as a specific sensor of cell division in tobacco BY-2 cells. *Protoplasma.* **209:** 90-97.

Delledonne, M., Murgia, I., Ederle, D., Sbicego, P.F., Biondani, A., Polverari, A. and Lamb, C. (2002). Reactive oxygen intermediates modulate nitric oxide signaling in the plant hypersensitive disease-resistance response. *Plant Physiol. Biochem.* **40:** 605-610.

Dietz, K.J. (2003). Plant peroxiredoxins. *Annu. Rev. Plant Biol.* **54:** 93-107.

Dietz, K.J., Horling, F., König, J. and Baier, M. (2002). The function of the chloroplast 2-cysteine peroxiredoxin in peroxide detoxification and its regulation. *J. Exp. Bot.* **53:** 1321-1329.

Dietz, K.J., Jacob, S., Oelze, M.L., Laxa, M., Tognetti, V., Marina, S., de Miranda, N., Baier, M. and Finkemeier, I. (2006). The function of peroxiredoxins in plant organelle redox metabolism. J. Exp. Bot. **57:** 1697-1709.

Dixon, D.P., Davis, B.G. and Edwards, R. (2002). Functional divergence in the glutathione transferase superfamily in plants. Identification of two classes with putative functions in redox homeostasis in Arabidopsis thaliana. *J. Biol. Chem.* **277:** 30859-30869.

Edwards, R., Dixon, D.P. and Walbot, V. (2000). Plant glutathione S-transferases: enzymes with multiple functions in sickness and in health. *Trends Plant Sci.* **5:** 193-198.

Esaka, M., Fujisawa, K., Goto, M. and Kisu, Y. (1992). Regulation of ascorbate oxidase expression in pumpkin by auxin and copper. *Plant Physiol.* **100:** 231-237.

Escoubas, J., Lomas, M., LaRoche, J. and Falkowski, P.G. (1995). Light intensity regulation of *cab* gene transcription is signaled by the redox state of the plastoquinone pool. *Proc. Natl. Acad Sci. USA.* **92:** 10237-10241.

Eshdat, Y., Holland, D., Faltin, Z. and Ben-Haygim, G.. (1997). Plant glutathione peroxidases. *Physiol. Plant.* **100:** 234-240.

Evans, N.H., McAinsh, M.R., Hetherington, A.M. and Knight, M.R. (2005). ROS perception in Arabidopsis thaliana: the ozone-induced calcium response. *Plant J.* **41:** 615-626.

Filomeni, G., Rotilio, G. and Ciriolo, M.R. (2002). Cell signalling and the glutathione redox system. *Biochem. Pharmacol.* **64:** 1057-1064.

Fleury, C., Mignotte, B. and Vayssiere, J. (2002). Mitochondrial reactive oxygen species in cell death signaling. *Biochimie.* **84:** 131-141.

Flury, T., Wagner, E. and Kreuz, K. (1996). An inducible glutathione S-transferase in soybean hypocotyl is localized in the apoplast. *Plant Physiol.* **112:** 1185-1190.

Foyer, C.H. and Lelandais, M. (1996). A comparison of the relative rates of transport of ascorbate and glucose across the thylakoid, chloroplast and plasmalemma membranes of pea leaf mesophyll cells. *J. Plant Physiol.* **148:** 391-398.

Foyer, C.H. and Noctor, G. (1999). Oxygen processing in photosynthesis: a molecular approach. *New Phytol.* **146:** 359-392.

Foyer, C.H. and Noctor, G. (2003). Redox sensing and signalling associated with reactive oxygen in chloroplasts, peroxisomes and mitochondria. *Physiol. Plant.* **119:** 355-364.

Foyer, C.H. and Noctor, G. (2005a). Oxidant and antioxidant signalling in plants: a re-evaluation of the concept of oxidative stress in a physiological context. *Plant Cell Environ.* **28:** 1056-1071.

Foyer, C.H. and Noctor, G. (2005b). Redox homeostasis and antioxidant signaling: a metabolic interface between stress perception and physiological responses. *Plant Cell.* **17:** 1866-1875.

Foyer, C.H., Rowell, J. and Walker, D.A. (1983). Measurements of the ascorbate content of spinach leaf protoplasts and chloroplasts during illumination. *Planta.* **157:** 239-244.

Foyer, C.H., Theodoulou, .FL. and Delrot, S. (2001). The functions of inter- and intracellular glutathione transport systems in plants. *Trends Plant Sci.* **6:** 1360-1385.

Foyer, C.H., Trebst, A. and Noctor, G. (2008). Signaling and integration of defense functions of tocopherol, ascorbate and glutathione. In: Demming-Adams, B., Adams III, W.W. and Mattoo, A.K. (Eds.). Photoprotection, photoinhibition, gene regulation, and environment. Springer Science + Business Media B.V., pp. 241-268.

García-Limones, C., Dorado1, G., Navas-Cortés, J.A., Jiménez-Díaz, R.M. and Tena, M. (2009). Changes in the redox status of chickpea roots in response to infection by *Fusarium oxysporum f.sp. ciceris*: apoplastic antioxidant enzyme activities and expression of oxidative stress-related genes. *Plant Biol.* **11:** 194-203.

Grant, C.M., Maclver, F.H. and Dawes, I.W. (1996). Glutathione is an essential metabolite required for resistance to oxidative stress in the yeast *Saccharomyces cerevisiae. Curr. Genetics.* **29:** 511-515.

Green, M.A. and Fry, S.C. (2005). Vitamin C degradation in plant cells via enzymatic hydrolysis of 4-O-oxalyl-L-threonate. *Nature.* **433:** 83-87.

Guaiquil, V.H., Farber, C.M., Golde, D.W. and Vera, J.C. (1997). Efficient transport and accumulation of vitamin C in HL-60 cells depleted of glutathione. *J. Biol. Chem.* **272:** 9915-9921.

Hancock, J.T., Desikan, R., Clarke, A., Hurst, R.D. and Neill, S.J. (2002). Cell signaling following plant/pathogen interactions involves the generation of reactive oxygen and reactive nitrogen species. *Plant Physiol. Biochem.* **40:** 611-617.

Holmgren, A. (1976). Hydrogen donor system for *Escherichia coli* ribonucleoside-diphosphate reductase dependent upon glutathione. *Proc. Natl. Acad Sci. USA.* **73:** 2275-2279.

Horemans, N., Asard, H. and Caubergs, R.J. (1997). The ascorbate carrier of higher plant plasma membranes preferentially translocates the fully oxidized (dehydroascorbate) molecule. *Plant Physiol.* **114:** 1247-1253.

Horemans, N., Asard, H. and Caubergs, R.J. (1998b). Carrier mediated uptake of dehydroascorbate into higher plant plasma membrane vesicles shows trans-stimulation. *FEBS Lett.* **421:** 41-44.

Horemans, N., Foyer, C.H. and Asard, H. (2000b). Transport and action of ascorbate at the plant plasma membrane. *Trends Plant Sci.* **5:** 263-267.

Horemans, N., Foyer, C.H., Potters, G. and Asard, H. (2000a). Ascorbate function and associated transport systems in plants. *Plant Physiol. Biochem.* **38:** 531-540.

Horemans, N., Potters, G., Asard, H. and Caubergs, R.J. (1998a). Transport of ascorbate into protoplasts of Nicotiana tabacum Bright Yellow 2 cell line. *Protoplasma.* **205:** 114-121.

Hu, J., Aguirre, M., Peto, C., Alonso, J., Ecker, J. and Chory, J. (2002). A role for peroxisomes in photomorphogenesis and development of *Arabidopsis. Science.* **297:** 405-409.

Hu, X., Neill, S.J., Tang, Z. and Cai, W. (2005). Nitric oxide mediates gravitropic bending in soybean roots. *Plant Physiol.* **137:** 663-670.

Ikner, A., Shiozaki, K. (2005). Yeast signaling pathways in the oxidative stress response. *Mutat. Res.* **569:** 13-27.

Imai, T., Karita, S., Shiratori, G., Hattori, M., Nunome, T., Ōba, K. and Hirai, M. (1998). L-Galactono-¦Ã-lactone dehydrogenase from sweet potato: purification and cDNA sequence analysis. *Plant Cell Physiol.* **39:** 1350-1358.

Imai, T., Kingston-Smith, A.H. and Foyer, C.H. (1999). Inhibition of endogenous ascorbate synthesis in potato leaves supplied with exogenous ascorbate. *Free Radical Res.* **31:** 171-179.

Izava, S., Inoue, Y. and Kimura, A. (1995). Oxidative stress response in yeast: effect of glutathione on adaptation to hydrogen peroxide stress in *Saccharomyces cerevisiae. FEBS Lett.* **368:** 73-76.

Jamaï, A., Tommasini, R., Martinoia, E. and Delrot, S. (1996). Characterization of glutathione uptake in broad bean leaf protoplasts. *Plant Physiol.* **111:** 1145-1152.

Juhnke, H., Krems, B., Kötter, P. and Entian, K-D. (1996). Mutants that show increased sensitivity to hydrogen peroxide reveal an important role for the pentose phosphate pathway in protection of yeast against oxidative stress. *Mol. Gen. Genet.* **252:** 456-464.

Kangasjarvi, J., Jaspers, P. and Kollist, H. (2005). Signalling and cell death in ozone-exposed plants. *Plant Cell Environ.* **28:** 1021-1036.

Kato, N. and Esaka, M. (2000). Expansion of transgenic tobacco protoplasts expressing pumpkin ascorbate oxidase is more rapid that that of wild type protoplasts. *Planta.* **210:** 1018-1022.

Kearns, P.R. and Hall, A.G. (1998). Glutathione and the response of malignant cells to chemotherapy. *Drug Discovery Today.* **3:** 113-121.

Kerk, N.M., Jiang, K. and Feldman, L.J. (2000). Auxin metabolism in the root apical meristem. *Plant Physiol.* **122:** 925-932.

Kovari, I.A., Cobbett, C. and Goldsbrough, P.B. (1997). Expression of tomato gamma-Glu-Cys synthetase in the *Arabidopsis cad2* mutant restores cadmium tolerance. *Plant Physiol.* **114 (Suppl.):** 126.

Kovtun, Y., Chiu, W.L., Tena, G. and Sheen, J. (2000). Functional analysis of oxidative stress-activated mitogen-activated protein kinase cascade in plants. *Proc. Natl. Acad Sci. USA.* **97:** 2940-2945.

Leisinger, U., R¨1fenacht, K., Fischer, B., Pesaro, M., Spengler, A., Zehnder, A.J.B. and Eggen, R.I.L. (2001). The glutathione peroxidase homologous gene from *Chlamydomonas reinhardtii* is transcriptionally up-regulated by singlet oxygen. *Plant Mol. Biol.* **46:** 395-408.

Li, L. and van Staden, J. (1998). Effects of plant growth regulators on the antioxidant system in callus of two maize cultivars subjected to water stress. *Plant Growth Regul.* **24:** 55-66.

Lillig, C.H., Berndt, C. and Holmgren, A. (2008). Glutaredoxin systems. *Biochim. Biophys. Acta.* **1780:** 1304-1317.

Loewus, F.A. and Loewus, M.W. (1987). Biosynthesis and metabolism of ascorbic acid in plants. *Crit Rev. Plant Sci.* **5:** 101-119.

Lu, S.C. (2000). Regulation of glutathione synthesis. *Curr. Top Cell Regul.* **36:** 95-116.

Mano, J., Belles-Boix, E., Babiychuk, E., Inzé, D., Torii, Y., Hiraoka, E., Takimoto, K., Slooten, L., Asada, K. and Kushnir, S. (2005). Protection against photooxidative injury of tobacco leaves by 2-alkenal reductase. Detoxification of lipid peroxide-derived reactive carbonyls. *Plant Physiol.* **139:** 1773-1783.

May, M.J. and Leaver, C.J. (1994). *Arabidopsis thaliana* γ-glutamylcysteine synthetase is structurally unrelated to mammalian, yeast and *E. coil* homologs. *Proc. Natl. Acad Sci. USA.* **91:** 10059-10063.

May, M.J., Vernoux, T., Leaver, C., Montagu, M.V. and Inzé, D. (1998). Glutathione homeostasis in plant: Implications for environmental sensing and plant development. *J. Exp. Bot.* **49:** 649-667.

Meyer, Y., Reichheld, J.P. and Vignols, F. (2005). Thioredoxins in Arabidopsis and other plants. *Photosynthesis Res.* **86:** 419-433.

Mezzari, M., Walters, K., Jelinkova, M., Shih, M.C., Just, C.L. and Schnoor, J.L. (2005). Gene expression and microscopic analysis of Arabidopsis exposed to chloroacetanilide herbicides and explosive compounds. A phytoremediation approach. *Plant Physiol.* **138:** 858-869.

Michelet, L., Zaffagnini, M., Marchand, C., Collin, V., Decottignies, P., Tsan, P., Lancelin, J.M., Trost, P., Miginiac-Maslow, M., Noctor, G. and Lemaire, S.D. (2005). Glutathionylation of chloroplast thioredoxin f is a redox signaling mechanism in plants. *Proc. Natl. Acad Sci. USA.* **102:** 16478-16483.

Miginiac-Maslow, M. and Lancelin, J.M. (2002). Intrasteric inhibition in redox signalling: light activation of NADP-malate dehydrogenase. *Photosynthesis Res.* **72:** 1-12.

Mittler, R., Vanderauwera, S., Gollery, M. and Van Breusegem, F. (2004). Reactive oxygen gene network of plants. *Trends Plant Sci.* **9:** 490-498.

Mothofer, A., Schulze, B. and Boland, W. (2004). Biotic and heavy metal stress response in plants: evidence for common signals. *FEBS Lett.* **566:** 1-5.

Mozafar, A. and Oertli, J.J. (1993). Vitamin C, (ascorbic acid): Uptake and metabolism by soybean. *J. Plant Physiol.* **141:** 316-321.

Mulcahy, R.I. and Gipp, J.J. (1995). Identification of a putative antioxidant responsive element in the 5′-flanking region of the human γ-glutamylcysteine synthetase heavy subunit gene. *Biochem. Biophys. Res. Comm.* **209:** 227-233.

Mutsuda, M., Ishikawa, T., Takeda, T. and Shigeoka, S. (1995). Subcellular localization and properties of L-galactono-g-lactone dehydrogenase in spinach leaves. *Biosci. Biotech. Biochem.* **59:** 1983-1984.

Nagalakshmi, N. and Prasad, M.N.V. (2001). Responses of glutathione cycle enzymes and glutathione metabolism to copper stress in *Scenedesmus bijugatus*. *Plant Sci.* **160:** 291-299.

Nanasato, Y., Akashi, K. and Yokota, A. (2005). Co-expression of cytochrome b561 and ascorbate oxidase in leaves of wild watermelon under drought and high light conditions. *Plant Cell Physiol.* **46:** 1515-1524.

Neill, S., Desikan, R. and Hancock, J. (2002). Hydrogen peroxide signalling. *Curr. Opin. Plant Biol.* **5:** 388-395.

Noctor, G. (2006). Metabolic signalling in defence and stress: the central roles of soluble redox couples. *Plant Cell Environ.* **29:** 409-425.

Noctor, G., Arisi, A-C. M., Jouanin, L., Kunert, K-J., Rennenberg, H. and Foyer, C.H. (1998). Glutathione: biosynthesis, metabolism and relationship to stress tolerance explored in transformed plants. *J. Exp. Bot.* **49:** 623-647.

Noctor, G. and Foyer, C.H. (1998). Ascorbate and glutathione: keeping active oxygen under control. *Annu. Rev. Plant Physiol. Plant Mol. Biol.* **49:** 249-279.

Noctor, G., Queval, G. and Gakière, B. (2006). NAD(P) synthesis and pyridine nucleotide cycling in plants and their potential importance in stress conditions. *J. Exp. Bot.* **57:** 1603-1620.

Ôba, K., Ishikawa, S., Nishikawa, M., Mizuno, H. and Yamamoto, T. (1995). Purification and properties of L-galactono-γ-lactone dehydrogenase, a key enzyme for ascorbic acid biosynthesis, from sweet potato roots. *J. Biochem.* **117:** 120-124.

Oelze, M-L., Kandlbinder, A. and Dietz, K-J. (2008). Redox regulation and overreduction control in the photosynthesizing cell: Complexity in redox regulatory networks. *Biochim. Biophys. Acta.* **1780:** 1261-1272.

Ostergaard, J., Persiau, G., Davey, M.W., Bauw, G. and Van Montagu, M. (1997). Isolation of a cDNA coding for L-galactono-γ-lactone dehydrogenase, an enzyme involved in the biosynthesis of ascorbic acid in plants. *J. Biol. Chem.* **272:** 30009-30016.

Overmyer, K., Brosché, M. and Kangasjärvi, J. (2003). Reactive oxygen species and hormonal control of cell death. *Trends Plant Sci.* **8:** 335-342.

Pandolfi, P.P., Sonati, F., Rivi, R., Mason, P., Grosveld, F. and Luzzatto, L. (1995). Targeted disruption of the housekeeping gene encoding glucose 6-phosphate dehydrogenase (G6PD): G6PD is dispensable for pentose synthesis but essential for defense against oxidative stress. *EMBO J.* **14:** 5209-5215.

Pastori, G.M. and Foyer, C.H. (2002). Common components, networks, and pathways of cross-tolerance to stress. The central role of "redox" and abscisic acid-mediated controls. *Plant Physiol.* **129:** 460-468.

Pesaresi, P., Masiero, S., Eubel, H., Braun, H.P., Bhushan, S., Glaser, E., Salamini, F. and Leister, D. (2006). Nuclear photosynthetic gene expression is synergistically modulated by rates of protein synthesis in chloroplasts and mitochondria. *Plant Cell.* **18:** 970-991.

Petracek, M.E., Dickey, L.F., Nguyen, T.T., Gatz, C., Sowinski, D.A., Allen, G.C. and Thompson, W.F. (1998). Ferredoxin-1 mRNA is destabilized by changes in photosynthetic electron transport. *Proc. Natl. Acad Sci. USA.* **95:** 9009-9013.

Pfannschmidt, T. (2003). Chloroplast redox signals: how photosynthesis controls its own genes. *Trends Plant Sci.* **8:** 33-41.

Pfannschmidt, T., Schutze, K., Brost, M. and Oelmuller, R. (2001). A novel mechanism of nuclear photosynthesis gene regulation by redox signals from the chloroplast during photosystem stoichiometry adjustment. *J. Biol. Chem.* **276:** 36125-36130.

Pignocchi, C., Fletcher, J.M., Wilkinson, J.E., Barnes, J.D. and Foyer, C.H. (2003). The function of ascorbate oxidase in tobacco. *Plant Physiol.* **132:** 1631-1641.

Pignocchi, C. and Foyer, C.H. (2003). Apoplastic ascorbate metabolism and its role in the regulation of cell signalling. *Curr. Opin. Plant Biol.* **6:** 379-359.

Pignocchi, C., Kiddle, G., Hernández, I., Foster, S.J., Asensi, A., Taybi, T., Barnes, J. and Foyer, C.H. (2006). Ascorbate oxidase-dependent changes in the redox state of the apoplast modulate gene transcript accumulation leading to modified hormone signaling and orchestration of defense processes in tobacco. *Plant Physiol.* **141:** 423-435.

Polle, A. (2001). Dissecting the superoxide dismutase-ascorbate-glutathione-pathway in chloroplasts by metabolic modeling. Computer simulations as a step towards flux analysis. *Plant Physiol.* **126:** 445-462.

Potters, G., Horemans, N., Bellone, S., Caubergs, J., Trost, P., Guisez, Y. and Asard, H. (2004). Dehydroascorbate influences the plant cell cycle through a glutathione-independent reduction mechanism. *Plant Physiol.* **134:** 1479-1487.

Ranieri, A., Castagna, A., Pacini, J., Baldan, B., Mensuali Sodi, A. and Soldatini, G.F. (2003). Early production and scavenging of hydrogen peroxide in the apoplast of sunflower plants exposed to ozone. *J. Exp. Bot.* **54:** 2529-2540.

Rautenkranz, A.A.F., Li, L., M?chler, F., M?rtinoia, E. and Oertli, J.J. (1994). Transport of ascorbic and dehydroascorbic acid cross protoplast and vacuole membranes isolated from Barley (*Hordeum vulgare* L. cv Gerbel) leaves. *Plant Physiol.* **106:** 187-193.

Rawlins, M.R., Leaver, C.J. and May, M.J. (1995). Characterization of a cDNA encoding *Arabidopsis* glutathione synthetase. *FEBS Lett.* **376:** 81-86.

Reddy, A.R., Chaitanya, K.V. and Vivekanandan, M. (2004). Drought-induced responses of photosynthesis and antioxidant metabolism in higher plants. *J. Plant Physiol.* **161:** 1189-1202.

Robert, G., Melchiorre, M., Racca, R., Trippi, V. and Lascano, H.R. (2009). Apoplastic superoxide level in wheat protoplast under photooxidative stress is regulated by chloroplast redox signals: Effects on the antioxidant system. *Plant Sci.* **177:** 168-174.

Rouhier, N., Gelhaye, E. and Jacquot, J.P. (2004). Plant glutaredoxins: still mysterious reducing systems. *Cell Mol. Life Sci.* **61:** 1266-1277.

Sairam, P.K. and Srivastava, G.C. (2002). Changes in antioxidant activity in sub-cellular fractions of tolerant and susceptible wheat genotypes in response to long term salt stress. *Plant Sci.* **162:** 897-904.

Sanchez-Fernandez, R., Fricker, M., Corben, L.B., White, N.S., Sheard, N., Leaver, C.J., Van Montagu, M., Inzé, D. and May MJ (1997). Cell proliferation and hair tip growth in the *Arabidopsis* root are under mechanistically different forms of redox control. *Proc. Natl. Acad Sci. USA.* **94:** 2745-2750.

Sanmartin, M., Drogoudi, P.D., Lyons, T., Pateraki, I., Barnes, J. and Kanellis, A.K. (2003). Over-expression of ascorbate oxidase in the apoplast of transgenic tobacco results in altered ascorbate and glutathione redox states and increased sensitivity to ozone. *Planta.* **216:** 918-928.

Schäfer, H.J., Greiner, S., Rausch, T. and Haag-Kerwer, A. (1997). In seedlings of the heavy metal accumulator *Brassica juncea* Cu^{2+} differentially affects transcript amounts for γ-glutamylcysteine synthetase (ã-ECS) and metallothionein (MT2). *FEBS Lett.* **404:** 216-220.

Scheibe, R. (2004). Malate valves to balance cellular energy supply. *Physiol. Plant.* **120:** 21-26.

Scheibe, R., Backhausen, J.E., Emmerlich, V. and Holtgrefe, S. (2005). Strategies to maintain redox homeostasis during photosynthesis under changing conditions. *J. Exp. Bot.* **56:** 1481-1489.

Schneider, A., Schatten,T. and Rennenberg, H. (1992). Reduced glutathione (GSH) transport in cultured tobacco cells. *Plant Physiol. Biochem.* **30:** 29-38.

Schwanz, P., Picon, C., Vivin, P., Dreyer, E., Guehl, J-M. and Polle, A. (1996). Responses of antioxidative systems to drought tress in penduncculate oak and maritime pine as modulated by elevated CO_2. *Plant Physiol.* **110:** 393-402.

Sen, C.K. (1999). Glutathione homeostasis in response to exercise training and nutritional supplements. *Mol. Cell Biochem.* **196:** 31-42.

Sen, C.K. and Packer, L. (1996). Antioxidant and redox regulation of gene transcription. *FASEB J.* **10:** 709-720.

Shalata, A., Mittova, V., Volokita, M., Guy, M. and Tal, M. (2001). Response of the cultivated tomato and its wild salt-tolerant relative *Lycopersicon pennellii* to salt-dependent oxidative stress: The root antioxidative system. *Physiol. Plant.* **112:** 487-494.

Sherameti, I., Sopory, S.K., Trebicka, A., Pfannschmidt, T. and Oelmüller, R. (2002). Photosynthetic electron transport determines nitrate reductase gene expression and activity in higher plants. *J. Biol. Chem.* **277:** 46594-46600.

Siendones, E., Gonz¨¢lez-Reyes, J.A., Santos-Ocana, C., Navas, P. and Cordoba, F. (1999). Biosynthesis of ascorbic acid in kidney beans. $_L$-Galactono-ã-lactone dehydrogenase is an intrinsic protein located at the mitochondrial inner membrane. *Plant Physiol.* **120:** 907-912.

Slekar, K.H., Kosman, D.J. and Culotta, V.C. (1996). The yeast copper/zinc superoxide dismutase and the pentose phosphate pathway play overlapping roles in oxidative stress protection. *J. Biol. Chem.* **271:** 28831-28836.

Smirnoff, N. (1996). The function and metabolism of ascorbic acid in plants. *Ann. Bot.* **78:** 661-669.

Smirnoff, N. (2000). Ascorbic acid: metabolism and functions of a multifaceted molecule. *Curr. Opin. Plant Biol.* **3:** 229-235.

Snyders, S. and Kohorn, B.D. (1999). TAKs, thylakoid membrane protein kinases associated with energy transduction. *J. Biol. Chem.* **274:** 9137-9140.

Stevens, R.G., Creissen, G.P. and Mullineaux, P.M. (1997). Cloning and characterisation of a cytosolic glutathione reductase cDNA from pea (*Pisum sativum* L.) and its expression in response to stress. *Plant Mol. Biol.* **35:** 641-654.

Storz, G., Tartaglia, L.A. and Ames, B.N. (1990). Transcriptional regulator of oxidative stress-inducible genes: direct activation by oxidation. *Science.* **248:** 189-194.

Sudhakar, C., Lakshmi, A. and Giridarakumar, L.S. (2001). Changes in the antioxidant enzyme efficacy in two high yielding genotypes of mulberry (*Morus alba* L.) under NaCl salinity. *Plant Sci.* **161:** 613-619.

Sunkar, R., Bartels, D. and Kirch, H.H. (2003). Overexpression of a stress-inducible aldehyde dehydrogenase gene from *Arabidopsis thaliana* in transgenic plants improves stress tolerance. *Plant J.* **35:** 452-464.

Takahama, U. and Oniki, T. (1992). Regulation of peroxidase-oxidation of phenolics in the apoplast of spinach leaves by ascorbate. *Plant Cell Physiol.* **33:** 379-387.

Takeda, T., Yokota, A. and Shigeoka, S. (1995). Resistance of photosynthesis to hydrogen peroxide in algae. *Plant Cell Physiol.* **36:** 1089-1095.

Tommasi, F., Paciolla, C., de Pinto, M.C. and De Gara, L. (2001). A comparative study of glutathione and ascorbate metabolism during germination of *Pinus pinea* L. seeds. *J. Exp. Bot.* **52:** 1647-1654.

Torres, M.A. and Dang, J.L. (2005). Functions of the respiratory burst oxidase in biotic interactions, abiotic stress and development. *Curr. Opin. Plant Biol.* **8:** 397-403.

Ullmann, P., Gondet, L. and Bach, T.J. (1997). Molecular analysis of glutathione biosynthesis enzymes in higher plants. In: Hatzios, K.K., Ed. *Regulation of enzymatic systems detoxifying xenobiotics in plants*. NATO ASI series. Kluwer Academic Publishers.

Ullmann, P., Gondet, L., Potier, S. and Bach, T.J. (1996). Cloning of the *Arabidopsis thaliana* glutathione synthetase (*GSH2*) by functional complementation of a yeast *gsh2* mutant. *Euro. J. Biochem.* **236:** 662-669.

Valero, E., González-Sánchez, M.I., Maciá, H. and Garcia-Carmona, F. (2009). Computer simulation of the dynamic behavior of the glutathione-ascorbate redox cycle in chloroplasts. *Plant Physiol.* **149:** 1958-1969.

Vanacker, H., Carver, T.L.M. and Foyer, C.H. (2000). Early H_2O_2 accumulation in mesophyll cells leads to induction of glutathione during the hyper-sensitive response in the barley-powdery mildew interaction. *Plant Physiol.* **123:** 1289-1300.

Vanacker, H., Carver, T.L.W. and Foyer, C.H. (1998a). Pathogen-induced changes in the antioxidant status of the apoplast in barley leaves. *Plant Physiol.* **117:** 1103-1114.

Vanacker, H., Foyer, C.H. and Carver, T.L.W. (1999). Changes in apoplastic antioxidants induced by powdery mildew attack in oat genotypes with race non-specific resistance. *Planta.* **208:** 444-452.

Vanacker, H., Harbinson, J., Ruisch, J., Carver, T.L.W. and Foyer, C.H. (1998b). Antioxidant defences of the apolast. *Protoplasma.* **205:** 129-140.

Wheeler, G.L., Jones, M.A. and Smirnoff, N. (1998). The biosynthetic pathway of vitamin C in higher plants. *Nature.* **393:** 365-369.

Wingate, V.P.M., Lawton, M.A. and Lamb, C.J. (1988). Glutathione causes a massive and selective induction of plant defense genes. *Plant Physiol.* **85:** 6738-6742.

Wollman, F.A. (2001). State transitions reveal the dynamics and flexibility of the photosynthetic apparatus. *EMBO J.* **20:** 3623-3630.

Xing, T., Wang, X.J., Malik, K. and Miki, B.L. (2001). Ectopic expression of an Arabidopsis calmodulin-like domain protein kinase-enhanced NADPH oxidase activity and oxidative burst in tomato protoplasts. *Mol. Plant Microbe In.* **14:** 1261-1264.

Yoshida, K., Terashima, I. and Noguchi, K. (2007). Up-regulation of mitochondrial alternative oxidase concomitant with chloroplast overreduction by excess light. *Plant Cell Physiol.* **48:** 606-614.

Zaffagnini, M., Michelet, L., Marchand, C., Sparla, F., Decottignies, P., Marechal, P.L., Miginiac-Maslow, M., Noctor, G., Trost, P. and Lemaire, S.D. (2007). The thioredoxin-independent isoform of chloroplastic glyceraldehyde-3-phosphate dehydrogenase is selectively regulated by glutathionylation. *FEBS J.* **274:** 212-226.

Zeier, J., Delledonne, M., Mishina, T., Severi, E., Sonoda, M. and Lamb, C. (2004). Genetic elucidation of nitric oxide signaling in incompatible plant–pathogen interactions. *Plant Physiol.* **136:** 2875-2886.

7

Detoxification of Heavy Metal Induced Oxidative Stress in Plants – Enzymatic and Non-Enzymatic Mechanisms—A Mini Review

S. SREEDEVI[1] AND P.N. KRISHNAN[2*]

[1]Department of Botany, Sree Narayana College, Kollam, Kerala-691001, India
[2]Plant Biotechnology and Bioinformatics Division, Tropical Botanic Garden and Research Institute, Kerala-695562, India
[*]**Corresponding author:** peringattulli @yahoo.co.uk

ABSTRACT

Oxidative stress in plants is buffered by the activation of antioxidant defences which include, antioxidant enzymes such as superoxide dismutase (SOD), catalase (CAT), peroxidase (POD), ascorbate peroxidase (AP), monodehydroascorbate reductase (MDHAR), dehydroascorbate reductase (DHAR) and glutathione reductase (GR). The non-enzymatic defences involve antioxidants such as ascorbate, glutathione and phenolic compounds along with lipid-soluble molecules such as carotenoids and tocopherols. This review will provide a detailed account of the current knowledge of the role of enzymatic and non-enzymatic mechanisms in detoxification of heavy metal induced oxidative stress.

Keywords: Detoxification, heavy metals, oxidative stress.

1. INTRODUCTION

Metals are omnipresent in the environment occurring naturally, in various concentrations in parent rock, soil, water, air and all biological matter. Anthropogenic sources include smelting of metallic ores, industrial fabrication, commercial application of metals, agrochemical pesticides as well as burning of fossil fuels (Galloway et al., 1982; Angelone and Bini 1992).

Metals occur in various chemical forms exhibiting different physical and chemical behaviour, in terms of chemical interactions, mobility, bioavailability and also their potential toxicity. The solubility and mobility of metals is affected by adsorption, desorption and complexation processes which in turn are dependent on the soil type. Soil contamination by heavy metals is today a serious ecological problem. Industrially contaminated soils consist of a mixture of heavy metals (cadmium, zinc and lead), where cadmium is the first metal of concern. Metal ions turn toxic as soon as their concentration exceeds a metal-specific threshold which varies strongly among plant species and ecotypes, and also with metal properties. The biochemical impact of metal ions on the cells is as diverse as their chemical nature.

Heavy metals (HMs) are chemical elements with a density higher than 5 g cm^{-3}. Based on their solubility under physiological conditions, 17 heavy metals may be available for living cells and of importance for organisms and ecosystems (Weast 1984). They are non-degradable and are often accumulated by plant parts, biologically magnified through trophic levels causing deleterious biological effects. Exposure of plants even to minute concentrations may lead to the alteration of many cellular processes and structures (Hall 2000). Among the pollution-producing metals, cadmium (Cd), lead (Pb), arsenic (As) and mercury (Hg) are regarded as non-essential elements, with no known physiological functions. They are extremely toxic to plants and animals, have a long half-life and are persistent in the environment (Salt et al., 1995).

Ever since Brenchely (1914) investigated the poisonous effect of an excess supply of certain HMs on barley, numerous studies have been made on the toxic effects resulting from an excess supply of heavy metals. In small quantities certain HMs like iron (Fe), copper (Cu), manganese (Mn), and zinc (Zn) are nutritionally essential for plants, animals and man, being important cofactors of many enzymes, constituent of cells, etc. But others such as Cd, Pb and Hg have no known biological function. A major factor governing the toxicity of a metal in soil is its bioavailability (Dong et al., 2007) which is largely governed by (i) soil type—its physical and chemical characteristics such as soil pH and redox potential, (ii) metal speciation and (iii) nature of microorganisms. In addition rainfall, evaporation, and plant transpiration can change trace element concentrations in soil solution (Fischerov et al., 2006).

Excess of heavy metals in cells cause molecular damage to plants either directly or indirectly through the formation of reactive oxygen species (ROS) such as hydrogen peroxide (H_2O_2), hydroxyl radical ($^{\bullet}OH$) and superoxide radicals ($O_2^{\bullet -}$) (Scandalios et al., 1997; Noctor and Foyer 1998; Lin and Kao 2000). The partial reduction of molecular oxygen occurs in one electron step, by addition of one, two and four electrons to O_2 which leads to the successive formation of intermediates—$O_2^{\bullet -}$, $^{\bullet}OH$ and H_2O_2 which are highly reactive compared to molecular oxygen.

Excess concentration of redox active heavy metals such as iron (Fe) and copper (Cu) result in oxidative injury (De Vos et al., 1992; Gallego et al., 1996; Weckx and Clijsters 1996; Raeymaekers et al., 2003). The principal mechanism of Cu toxicity involves the

Haber–Weiss reaction, characterized by a heavy metal-catalysed production of hydroxyl radicals from hydrogen peroxide (Hammond-Kosack and Jones 1996; Briat and Lebrun 1999). This type of reaction has not been described for either Cd or Hg in plants (Schützendübel and Polle 2002), despite the evidence of oxidative stress induction in different plants after exposure to Cd (Lozano-Rodríguez et al., 1997; Dixit et al., 2001; Sreedevi et al., 2008) and to Hg (Cho and Park 2000). Unlike Cu and Fe, Cd is a redox-inactive metal that is incapable of producing reactive oxygen species (ROS) directly, but can indirectly promote ROS generation by disrupting physiological processes (Sanità di Toppi and Gabbrielli 1999). Several studies have demonstrated that Cd treatment increases lipid peroxidation levels in common crops, indicating that oxidative stress is induced by Cd (Somashekaraiah et al., 1992; Gallego et al., 1996; Choui et al., 1997; Sandalio et al., 2001).

Reactive oxygen species (ROS) are produced by all aerobically respiring cells. As ROS are produced in significant quantities in subcellular compartments or organelles, each organelle has potential targets for oxidative stress as well as mechanisms for eliminating the oxyradicals. ROS are potentially destructive chemical species, cause oxidative damage to biomolecules such as proteins, membrane lipids and nucleic acids (Elstner 1987; Gille and Singler 1995; Halliwell and Gutteridge 2007) and disrupt the cellular metabolism (Asada 1994, Gille and Singler 1995). It is therefore essential for aerobic organisms to modulate ROS levels and activities in order to protect themselves against toxicity. Because of their sessile nature, plants have developed sophisticated metabolic responses to cope and survive, rather than avoiding stressful conditions as the mobile organisms (Madlung and Comai 2004). As a response to the toxic action of metals various protective mechanisms have been elaborated in plants (Kleizaite et al., 2004). The control of oxidant levels is achieved through a defence system composed of non-enzymatic metabolites and enzymatic scavengers (Noctor and Foyer 1998; Asada 1999).

2. ANTIOXIDANT DEFENCE MACHINERY

Oxidative stress in plants is buffered by the activation of antioxidant defences, including antioxidant enzymes such as superoxide dismutase (SOD, E.C. 1.15.1.1), catalase (CAT, E.C. 1.11.1.6), and peroxidases (POD, E.C. 1.11.1.7). The enzymes of Halliwell-Asada pathway or ascorbate-glutathione cycle such as ascorbate peroxidase (APX, EC 1.11.1.11), monodehydroascorbate reductase (MDAR, EC 1.6.5.4), dehydroascorbate reductase (DHAR, EC 1.8.5.1) and glutathione reductase (GR, EC 1.6.4.2) also play a significant role in scavenging H_2O_2 mainly in chloroplasts and in maintaining the redox status of the cell (Foyer et al., 1997). The defences may also involve water-soluble antioxidants such as ascorbate, glutathione and phenolic compounds and lipid-soluble molecules such as tocopherols (Foyer et al., 1994; Rice-Evans et al., 1997; Hodges et al., 1997; Hodges and Forney 2000; Pastori et al., 2000).

2.1 Superoxide Dismutase

Superoxide dismutase is a crucial component of the plant antioxidative defence system

(Cakmak and Horst 1991; Scandalios 1993; Boscolo et al., 2003) and is present in all aerobic organisms and in most subcellular compartments that generate oxidative stress (Bowler et al., 1992). The SOD family of enzymes represents a primary line of defence against the superoxide radical and derived ROS in all organisms. Three distinct types of SOD have been detected in plants, which can be classified according to their metal cofactors—Mn, Fe and Cu/Zn. Mn-SOD is located in the mitochondria and peroxisomes, Fe-SOD, which is not found in all plants, is associated with chloroplasts, and the abundant Cu/Zn- SODs are located in the cytosol, chloroplasts and peroxisomes (Bowler et al., 1992; Del Río et al., 1998). All three isozymes are nuclear encoded and SOD genes have been shown to be sensitive to environmental stresses.

The increase in SOD activity on exposure to Cd is a response to the accumulation of ROS particularly H_2O_2 and O_2^- induced by Cd. The growth of pea plants for a long period of time with 50 mM $CdCl_2$ produced reductions in the activities of Mn-SOD, Fe-SOD, and CuZn-SOD of 60%, 80%, and 90%, respectively (Sandalio et al., 2001). The decline in SOD activity may be due to replacement of Zn by Cd in CuZn- SOD. Cd-dependent reduction of SOD activity has been reported in bean (Cardinaels et al., 1984), pea (Sandalio et al., 2001) and wheat (Milone et al., 2003), although the opposite effect was observed in *Alyssum* plants (Schickler and Caspi, 1999), radish roots (Vitoria et al., 2001), sunflower (Laspina et al., 2005) and coffee (Gomes-Junior et al., 2006). These discrepancies are due to differences in the metal concentration and also in the period of treatment used in each case, in addition to the plant tissue studied. Thus, in garlic (*Allium sativum*) plants, SOD increased under short- term Cd treatment but decreased after long-term exposure (Zang et al., 2005).

2.2 Catalase

Catalase is the most universal oxidoreductase, which scavenges H_2O_2 to O_2 and H_2O and it is one of the key enzymes involved in removal of toxic peroxides. Conflicting reports exist on the activities of catalase on plant tissues. Cakmak and Horst (1991) and Luna et al. (1994) reported decrease in its activity in roots and leaves respectively. This may be an adaptive process as roots emerge out in an environment of stress. Meng et al. (2007) indicated that in *Allium sativum* exposed to Cu^{2+} the CAT activity response differed between roots and leaves, and that the reaction was dependent on the level and duration of stress. Reduced CAT activity has been regarded as a general response to many stresses and was suggested to be due to inhibition of enzyme synthesis or to a change in assembly of enzyme subunits, indicating that the ability of these antioxidant enzymes to eliminate ROS is limited (Hertwig, 1992).

Decline in CAT activity has been reported in bean, pea, rice, and cabbage (Chaoui et al., 1997; Sandalio et al., 2001; Shah et al., 2001; Pandey and Sharma 2002) under Cd stress, and this was correlated with an increase in H_2O_2. Increased CAT activity under Cd stress has been reported in *Phaseolus aureus* (Shaw 1995), *Pisum sativum* (Dalurzo et al., 1997), *Lemna minor* (Mohan and Hosetti 1997), barley (Patra and Panda 1998) and sunflower (Gallego et al., 1996, 1999). CAT activity in leaves and roots increased with Cd treatment in *Agropyron repens* (Brej 1998) and in radish (Vitüria et al., 2001). A decrease in CAT activity due to heavy metals has been attributed to inhibition of the CAT synthesis and other oxidase proteins in *Oryza sativa* (Das et al., 1978).

2.3 Peroxidase

Peroxidases (POD) are heme containing proteins that utilize H_2O_2 in the oxidation of various organic and inorganic substrates (Asada 1992). Peroxidases utilizing guaiacol as electron donor *in vitro* are guaiacol peroxidases and participate in developmental processes, lignification, ethylene biosynthesis, defence, wound healing, etc. (Asada 1992). The second group of peroxidases, scavenges H_2O_2 in the cell and utilizes glutathione, Cyt c, pyridine nucleotide and ascorbate as electron donors *in vitro*. Guaiacol peroxidases are glycoproteins, located in cytosol, vacuole, cell wall and in extracellular space, while the other group is non-glycosylated and localized in chloroplasts and cytosol (Asada 1992).

The role of POD as a stress enzyme in plants has been widely accepted. POD activity has been suggested as a potential biomarker for sub-lethal metal toxicity in plant species (Agawal and Pandey 2004). Induction, POD activity has been documented under toxic concentrations of Al and Cd (Cakmak and Horst 1991; Radotic et al., 2000; Guo et al., 2004). An increase in POD activity is a common response to oxidative and abiotic stresses. In plants, POD protects cells against harmful concentrations of hydroperoxides and helps in a variety of cellular functions (Beauchamp and Fridovich 1971).

Enhancement of POD activity in garlic reflects its great capacity to adapt to lower Cu stress by rapidly engaging an antioxidative defence system (Meng et al., 2007). However, the decline in POD activity in roots after prolonged Cu exposure indicates that the scavenging function of POD was impaired with prolonged severe Cu stress. That POD activity increased much more in roots than in leaves can be explained by the probability that the glutathione/ascorbate cycle was operating at a high rate in order to detoxify the ROS formed in the roots. In roots, the increase in POD activity correlated with the Cd-sensitivity of the pea genotypes (Meng et al., 2007), supporting previous suggestions to use POD activity as a biomarker for metal toxicity in plants (Radotic et al., 2000). A similar increase in POD was also observed after the application of Ni to the leaves of *Silene italica* (Gabbrielli et al., 1987).

2.4 Ascorbate Peroxidase and Glutathione Reductase

Ascorbate peroxidase (APX) and glutathione reductase (GR) are indispensable components of ascorbate-glutathione pathway, required to scavenge H_2O_2 produced mainly in chloroplasts and other cell organelles and to maintain the redox state of the cell (Asada 1992). APX utilizes the reducing power of ascorbic acid to eliminate potentially harmful H_2O_2. Enhancement in the activity of APX in response to Pb stress was reported by Verma and Dubey (2003). APX along with catalase and SODs are considered the key enzymes within the antioxidative defence mechanism, which directly determine the cellular levels of $O_2^{\bullet -}$ and H_2O_2.

Being a major water-soluble antioxidant in plant cells, glutathione (GSH) directly reduces most active oxygen species, while GR uses NADPH to reduce oxidized glutathione (GSSG) to reduced glutathione GSH (Noctor and Foyer 1998). Various free radicals and oxidants are able to oxidize GSH to GSSG. GSH is involved in the redox regulation of the cell cycle and has often been considered to play an important role in defence of plants and other organisms against oxidative stress (Grant et al., 1996; Sanchez-Fernandez et al., 1997). GR contains a highly conserved disulphide bridge between Cys^{76} and Cys^{81} (Creissen et al., 1992; Lee et al., 1998), which may undergo cleavage by heavy metals. Increased GR activity in Pb treated rice seedlings suggests possible involvement of GR in regenerating GSH from GSSG to increase GSH/GSSG ratio and the total glutathione pool (Noctor and Foyer 1998). The oxidative stress induced by Pb in growing rice seedlings (Verma and Dubey 2003) appears to be an indirect effect of Pb toxicity leading to production of ROS with a simultaneous increase in tissue levels of SOD, POD and GR. Induction in GR activity has been reported in the leaves of Cd and Zn stressed *Phaseolus* plants (Chaoui et al., 1997). Heavy metal-induced loss in glutathione reductase has been observed in pea exposed to Zn, Cu and Fe (Bielawski and Joy 1986), in sunflower exposed to Fe, Cu and Cd (Gallego et al., 1996), in *Lemna minor* exposed to Cu (Teisseire and Guy 2000).

Oxidative stress is characterized by the synthesis of hydrogen peroxide, which is detoxified by CAT activity in the peroxisomes and by APX in the cytosol, mitochondria and chloroplasts (Foyer et al., 1997; Asada 1999). The activity of APX produces dehydroascorbate, which is normally reduced to ascorbate by DHAR with the concomitant production of oxidized glutathione. Subsequently, oxidized glutathione is reduced to glutathione by GR, which consumes NADPH (Asada 1999).

The cycle coupling APX, DHAR and GR in plants is called the Haliwell–Asada cycle and its function is to detoxify hydrogen peroxide through the consumption of NADPH. On the other hand, antioxidant molecules such as glutathione, ascorbate and soluble phenolic compounds are able to directly scavenge reactive oxygen radicals and, in the case of ascorbate and glutathione, they are also substrates for the antioxidant enzymes APX and GP, respectively. Ascorbate and glutathione might play a key role in buffering oxidative stress in most eukaryotic systems (Noctor and Foyer 1998; Smirnoff et al., 2001).

2.5 Glutathione

Glutathione plays a central role in protecting plants from environmental stresses, including oxidative stress, xenobiotics, and some heavy metals (May et al., 1998; Noctor and Foyer 1998; Xiang and Oliver 1998; Foyer and Noctor 2003; Freeman et al., 2005). GSH, the most abundant low-molecular-weight thiol compound in plants (Bergmann and Rennenberg 1993; Hell 1997), is synthesized through a two-step ATP-dependent enzymatic pathway (Noctor et al., 2002). GSH acts as an antioxidant, quenching the (ROS) generated in response to stress, before ROS cause damage to cells (Navari-Izzo et al., 1997).

Glutathione has been shown to act as a regulator of gene expression (Wingate et al., 1988; Baier and Dietz 1997), is the precursor of phytochelatins, which bind supra-optimal concentrations of heavy metals (Grill et al., 1987, 1989), and is a substrate for the GSH S-transferases, which catalyse the conjugation of GSH with potentially dangerous xenobiotics (Marrs 1996). Glutathione may also be involved in the redox regulation of the cell cycle (Russo et al., 1995; Shaul et al., 1996; Sanchez-Fernàndez et al., 1997). Owing to its redox-active thiol group, GSH has often been considered to play an important role in defence of plants and other organisms against oxidative stress (Alscher 1989; Grant et al., 1996; Anjum et al., 2008, 2011). In all cells where GSH is found, the reduced tripeptide form exists interchangeably with the oxidized form (glutathione disulphide: GSSG). The reduced form of glutathione may act as an important redox buffer, preventing enzyme inactivation by protecting potentially susceptible protein thiol groups (Foyer and Halliwell 1976; Halliwell and Foyer 1978; Alscher 1989).

In most cases, exposure to heavy metals initially resulted in a severe depletion of GSH (Cd: *Rauvolfia serpentina:* Grill et al., 1987; carrot: Sanita di Toppi et al., 1999; tobacco: Vögeli-Lange and Wagner 1996; Cu: *Silene cucubalus*: de Vos et al., 1992; Cu or Cd: *Arabidopsis*: Xiang and Oliver 1998; Ni and Zn: pigeon pea: Rao and Sresty 2000; Fe, Cu or Cd: sunflower leaves: Gallego et al., 1996). This is a common response to Cd caused by an increased consumption of glutathione for phytochelatin production (Zenk 1996).

Glutathione serves an additional function in plant responses to heavy metal stress as a precursor of phytochetatins (PCs) (Rauser 1995; Cobbett and Goldsbrough 2002). PCs are synthesized from GSH by the enzyme phytochelatin synthase (Cobbett 2000) and comprise a family of intracellular heavy metal binding peptides with the same basic structure [(γ-L-Glutamyl-L-Cysteinyl)2–11-glycine] (Cobbett et al., 1998; Nocito et al., 2006). The formation of PCs is induced following the exposure of plants to a number of metal(loid)s such as Cd and As (Clemens et al., 1999; Schmoger et al., 2000; Hartley-Whitaker et al., 2001; Heiss et al., 2003). In support of the involvement of PCs in Cd tolerance, *Arabidopsis cad1* mutants were shown to be Cd-sensitive because of their lack in GSH and PCs (Howden et al., 1995). The overexpression of phytochelatin synthase in *Arabidopsis* increased As tolerance, but led to Cd hypersensitivity (Lee et al., 2003), perhaps due to GSH depletion.

The significance of phytochelatins for protection from heavy metals has frequently been reviewed (Rauser 1995; Zenk 1996). Phytochelatins sequester heavy metals. For Cd, the formation of Cd–thiolate (Cd–S) complexes in phytochelatins has been shown (Strasdeit et al., 1991). The chelated metals are transported to the tonoplast, taken up by active transport systems, and deposited in the vacuole (Tommasini et al., 1998; Rea 1999). This mechanism contributes to the protection from heavy metal toxicity in several plant species and in some fungi as well (Ishikawa et al., 1997). The reduction in GSH may be due to the active channelling of this compound to PC synthesis. Tsuji et al. (2002) reported that the ROS scavenging activity of phytochelatins was higher than GSH. Enhanced synthesis of phytochelatins could play a dual role in metal tolerance by directly sequestering heavy metals (Cobbett 2000) and functioning as ROS scavengers.

Glutathione is probably the most important cellular antioxidant. Fahey and Sundquist (1991) found strong evidence for an evolutionary link between glutathione and aerobic eukaryotic metabolism. The findings indicate that glutathione evolved as a molecule that protects cells against oxygen toxicity. Cells that are deprived of glutathione typically suffer severe oxidative damage associated with mitochondrial degeneration. The ability to synthesize glutathione appears to be crucial for protection from Cd, as shown by the increased tolerance of plants with elevated levels of GSH as well as a decreased tolerance in plants with diminished levels of GSH (Howden et al., 1995; Zhu et al., 1999a, b; Anjum et al., 2008, 2011). However, the threshold required to enhance protection seems to be plant-specific (Arisi et al., 2000).

Since glutathione is also an important component for the redox balance of the cell, as it is involved in the regulation of the cell cycle, the detoxification of oxidants, and acts as a transport form of reduced sulphur (Bergmann and Rennenberg 1993; May et al., 1998; Noctor and Foyer 1998; Vernoux et al., 2000), it may be suspected that a short-term lack of GSH may favour the accumulation of ROS and disturb developmental processes.

2.6 Ascorbate

Ascorbate and glutathione are major antioxidants and redox buffers in plant cells and also play key functions in growth, development, and stress responses (Smirnoff 2000; Pastori et al., 2003; Anjum et al., 2008, 2011). Ascorbate is an important antioxidant, which reacts not only with H_2O_2 but also with O^{2-}, OH and lipid hydroperoxidases (Igamberdiev and Hill 2004; Shao et al., 2006). The ascorbate/ GSH cycle reduces H_2O_2 to water (Foyer and Halliwell 1976; Alscher 1989). The maintenance of the cellular redox homeostasis is the result of a balance between the synthesis, consumption, and recycling of the most abundant antioxidants GSH and ASC. Ascorbate is also important in maintaining α-tocopherol in the reduced state and therefore links GSH to the dominant free radical scavenger in membranes (Hess 1994). The maintenance of ascorbate in its reduced form is achieved by monodehydroascorbate radical reductase (MDAR) and NAD (P) H or ferredoxin as reductant or by the operation of the ascorbate–glutathione pathway.

In the ascorbate-glutathione cycle, two molecules of ASC are utilized by APX to reduce H_2O_2 to water with generation of monodehydroascorbate. Monodehydroascorbate is a radical with a short lifetime and can disproportionate into dehydroascorbate and ascorbic acid. The electron donor is usually NADPH and catalysed by monodehydroascorbate reductase or ferredoxin in water-water cycle in the chloroplasts (Gapper and Dolan 2006).

Ascorbate is water-soluble and also has an additional role in protecting or regenerating oxidized carotenoids or tocopherols (Shao et al., 2005a; Shao et al., 2006). It can regenerate tocopherol from tocopheroxyl radicals, thus providing membrane protection (Shao et al., 2005b; Li and Jin 2007). Ascorbate also acts as a co-factor of violaxanthin de-epoxidase, thus sustaining dissipation of excess excitation energy (Li and Jin 2007; Pourcel 2007). It influences many enzyme activities, and minimizes the damage caused by oxidative process through synergic function with other antioxidants (Pourcel 2007; Foyer and Noctor 2003; Foyer and Noctor 2005a,b). Along with glutathione it is involved in the neutralization of secondary products of ROS reactions (Hare et al., 1998; Foyer and Noctor 2003).

2.7 α–Tocopherols

Oxidative stress has been reported to activate the expression of genes responsible for the synthesis of tocopherols in higher plants (Shao and Chu 2005; Wu et al., 2007). The most important function of tocopherols is their involvement in various mechanisms to protect polyunsaturated fatty acids (PUFA) from oxidation (Ledford and Niyogi 2005). α-tocopherols scavenge and quench various ROS and lipid oxidation products, stabilize membranes, and modulate signal transduction (Shao and Chu 2005; Noctor 2006). α-tocopherol, scavenges lipid peroxy radicals through the concerted action of other antioxidants (Hare et al., 1998; Kiffin et al., 2006), protect lipids and other membrane components by physically quenching and chemically reacting with O_2 in chloroplasts (Ivanov 2003; Igamberdiev and Hill 2004). Alpha tocopherols form the major lipid soluble antioxidant present in the PUFA-enriched membranes of chloroplasts and are proposed to be an essential component of the plastid antioxidant network (Shao et al., 2008).

3. CONCLUSION

The ability of plants to increase antioxidative protection to combat negative consequences of heavy metal stress appears to be limited since many studies showed that exposure to elevated concentrations of redox reactive metals resulted in decreased and not in increased activities of antioxidative enzymes. Autoxidation and Fenton reaction may cause the oxidative loss of defence enzymes. For example, catalase activity is directly inhibited by $O_2^{\bullet-}$ (Kono and Fridovich 1982). Cu-Zn-superoxide dismutase is fragmented by $HO^{\bullet-}$ radicals (Casano et al., 1997). If uptake of excess Fe or Cu preferentially drives the formation of $HO^{\bullet-}$ radicals, protection mediated by antioxidative enzymes is unlikely (Polle 1997).

In an environment of molecular oxygen, all living cells are confronted with the reactivity and toxicity of active and partially reduced forms of oxygen which can lead to the complete destruction of cells (Mittler et al., 2004). An unavoidable consequence of membrane-linked electron transport in different plant cell compartments, especially peroxisomes, mitochondria, and chloroplasts is the spilling of electrons onto molecular oxygen with the resultant generation of highly toxic ROS. This has led to the selection of antioxidative defence mechanisms, which place all aerobic cells in a state of pro-oxidant/antioxidant balance that prevents cellular damage. The control of the levels of endogenous ROS has opened the evolutionary possibility of using ROS as signalling and effector molecules in growth, development, stress responses, and biotic interactions (Mittler et al., 2004). Reactive oxygen species form central components of stress response and the levels of ROS determine the nature of the plant response. At low levels of ROS induction of defence genes and initiation of adaptive physiological response occurs, while at high levels of ROS cell death is initiated. Reactive oxygen species have multiple functions and fulfilment of the functions depends on a tight control of cellular ROS levels. The innumerable sources of ROS and the intricate and complex system of oxidant scavengers provide flexibility to these functions. Sub-lethal levels of ROS help the plant to acclimatize to stress conditions but the biochemical and molecular mechanisms of acclimatization is still not clearly known and the signalling pathways also remain elusive. The multiple functions exerted by ROS have been reviewed by a number of authors (Mittler 2002; Overmyer et al., 2003; Apel and Hirt 2004; Mittler et al., 2004; Foyer and Noctor 2006). In contrast to earlier views, it is becoming increasingly evident that even during stress, ROS production is not necessarily a symptom of cellular dysfunction but might represent a necessary signal in adjusting the cellular machinery to the altered conditions, and the intricacies of this signalling network are only beginning to be elucidated (Jaspers and Kangasjärvi 2009).

The term oxidative stress implies a harmful process, but it is probably in many cases quite the opposite. As suggested by Foyer and Noctor (2003), (2005a&b) enhanced oxidation is an essential component of the repertoire of signals that higher plants use to make appropriate adjustments of gene expression and cell structure in response to environmental and developmental cues to maintain homeostasis for stress tolerance and survival in a fluctuating environment.

ACKNOWLEDGEMENTS

The authors acknowledge The Director, TBGRI for the facilities and encouragement extended for this work.

REFERENCES

Agawal, S. and Pandey, V. (2004). Antioxidant enzyme responses to NaCl stress in *Cassia angustifolia*. *Biol. Plant.* **48:** 555-560.

Alscher, R. G. (1989). Biosynthesis and antioxidant function of glutathione in plants. *Physiol. Plant.* **77:** 457-64.

Angelone, M. and Bini, C. (1992). Trace elements in soils and plants of Western Europe. In *Biogeochemistry of trace metals.* pp. 19-60 Ed. D.C. Adriano (Boca Raton, FL Lewis, Publ).

Anjum, N.A., Umar, S., Iqbal, M. and Khan, N.A. (2011). Cadmium causes oxidative stress in moongbean [*Vigna radiata* (L.) Wilczek] by affecting antioxidant enzyme systems and ascorbate-glutathione cycle metabolism. *Russ. J. Plant Physiol.* **58:** 92-99.

Anjum, N.A., Umar, S., Ahmad, A., Iqbal, M. and Khan, N.A. (2008). Sulphur protects mustard (*Brasasica campestris* L.) from cadmium toxicity by improving leaf ascorbate and glutathione. *Plant Growth Regul.* **54:** 271-279.

Apel, K. and Hirt, H. (2004). Reactive oxygen species: Metabolism, oxidative stress, and signal transduction. *Annu. Rev. Plant Biol.* **55:** 373-399.

Arisi, A. C. M., Mocquot, B., Lagriffoul, A., Mench, M., Foyer, C. H. and Jouanin, L. (2000). Responses to cadmium in leaves of transformed poplars overexpressing γ-glutamylcysteine synthetase. *Physiol. Plant.* **109(2):** 143-149.

Asada, K. (1992). Ascorbate peroxidase - a hydrogen peroxide-scavenging enzyme in plants. *Physiol. Plant.* **85:** 235-241.

Asada, K. (1994). Production and action of active oxygen species in photosynthetic tissues In *Causes of Photo-Oxidative Stress and Amelioration of Defense Systems in Plants.* pp. 77-104 Eds. Foyer, C.H. and Mullineaux, P. (CRC Press , Boca Raton, FL).

Asada, K. (1999). The water-water cycle in chloroplasts: scavenging of active oxygens and dissipation of excess photons. *Annu. Rev. Plant Physiol. Plant Mol. Biol.* **50:** 601-639.

Baier, M. and Dietz, K.J. (1997). The plant 2-cys peroxiredoxin BAS1 synthesis is a nuclear encoded chloroplast protein: its expression, regulation, phylogenetic origin and implications for its specific physiological function in plants. *The Plant J.* **12:** 179-90.

Beauchamp, C. and Fridovich, I. (1971). Superoxide dismutase: improved assays and an assay applicable to acrylamide gels. *Anal. Biochem.* **44:** 276-287.

Briat, J.F. and Lebrun, M. (1999). Plant responses to metal toxicity. *Comptes Rendus Academie des Sciences de Paris.* **322:** 43-54.

Brenchley, W. E. (1914). Inorganic plant poisons and stimulants. (University Press Cambridge, England).

Bergmann, L. and Rennenberg, H. (1993). Glutathione metabolism in plants. In: *Sulfur Nutrition and Assimilation in Higher Plants*. pp. 102-123 Eds. DeKok, L.J., Stulen, I., Rennenberg, H., Brunold, C. and Rauser, W. (The Hague, The Netherlands, SPB Academic).

Bielawski, W. and Joy K.W. (1986). Reduced and oxidized glutathione and glutathione reductase activity in tissues of *Pisum sativum*. *Planta*. **169:** 267-72.

Brej, T. (1998). Heavy Metal Tolerance in *Agropyron repens* (L.) P. Bauv Populations from Legnica Copper Smelter Area, Lower Silesia. *Acta Societatis Poloniae*. **67:** 325-333

Boscolo, P.R.S., Menossi, M. and Jorge, R.A. (2003). Aluminum-induced oxidative stress in maize. *Phytochemistry*. **62:** 181-189.

Bowler, C., Van Montagu, M. and Inzé, D. (1992). Superoxide dismutase and stress tolerance. *Annu. Rev. Plant Physiol. Plant Mol. Biol.* **43:** 83-116

Cakmak, I. and Horst, W. J. (1991). Effect of aluminium on lipid peroxidation, superoxide dismutase, catalase and peroxidase activities in root tip of soybean (*Glycine max*) *Physiol. Plant.* **83:** 463-468.

Cardinaels, C., Put, C., Van Asche, F. and Clijsters, H. (1984). The superoxide dismutase as a biochemical indicator, discriminating between zinc and cadmium toxicity. *Biol. Plant.* **40:** 131-135.

Casano, L.M., Gomez, L.D., Lascano, H.R., Gonzales, C.A. and Trippi, V.S. (1997). Inactivation and degradation of CuZn–SOD by active oxygen species in wheat chloroplasts exposed to photo-oxidative stress. *Plant Cell Physiol.* **38:** 433-440.

Chaoui, S., Mazhoudi, M.H., Ghorbal, E. and El-Ferjani (1997). Cadmium and zinc induction of lipid peroxidation and effects on antioxidant enzyme activities in bean (*Phaseolus vulgaris* L). *Plant Sci.* **127:** 139-147.

Clemens, S., Kim, E.J., Neumann, D. and Schroeder, J.I. (1999). Tolerance to toxic metals by a gene family of phytochelatin synthases from plants and yeast. *J. EMBO*. **18:** 3325-3333.

Cho, U.H. and Park, J.O. (2000). Mercury-induced oxidative stress in tomato seedlings *Plant Sci.* **156:** 1-9.

Cobbett, C..S. (2000). Phytochelatin biosynthesis and function in heavy metal detoxification. *Curr. Opinion Plant Biol.* **3:** 211-216.

Cobbett, C.S. and Goldsbrough, P. (2002). Phytochelatins and metallothioneins: Roles in heavy metal detoxification and homeostasis. *Annu. Rev. Plant Biol.* **53:** 159-182.

Cobbett, C.S., May, M.J., Howden, R. and Rolls B. (1998). The glutathione-deficient, cadmium sensitive mutant, cad 2–1, of *Arabidopsis thaliana* is deficient in gamma-glutamylcysteine synthethase. *The Plant J.* **16:** 73-78.

Creissen, G., Edwards, E.A., Enard, C., Wellburn, A. and Mullineaux, P. (1992). Molecular characterisation of glutathione reductase cDNAs from pea (*Pisum sativum* L.) *The Plant J.* **2:** 129.

Dalurzo, H.C., Sandalio, L.M., Gomez, M. and Del Rio, L.A. (1997). Cadmium Infiltration of Detached Pea Leaves: Effect on Its Activated Oxygen Metabolism. *Phyton-Annales Rei Botaniae.* **37:** 59-64.

Das, P.K., Kar, D. and Mishra, D. (1978). Nickel nutrition of plants. 1. Effects of nickel on some oxidase activities during rice (*Oryza sativa* L.) seed germination. *Z Pflanzenphysiol.* **90:** 225-233.

Del Río, L.A., Pastori, G.M., Sandalio, J.M., and Hernandez, J.A. (1998). The activated oxygen role of peroxisomes in senescence. *Plant Physiol.* **116:** 1195-1200.

De Vos, C.H.R., Vonk, M.J., Vooijs, R. and Schat, H. (1992). Glutathione depletion due to copper-induced phytochelatin synthesis causes oxidative stress in *Silene cucubalus*. *Plant Physiol.* **98:** 853-858.

Dong, J. Mao, W.H., Zhang, G.P., Wu, F.B. and Cai, Y. (2007). Root excretion and plant tolerance to cadmium toxicity – a review, *Plant Soil Environ.* **53(5):** 193-200.

Dixit, V., Pandey, V. and Shyam, R. (2001). Differential antioxidative responses to cadmium in roots and leaves of pea (*Pisum sativum* L. cv. Azad) *J. Expt. Bot.* **52:** 1101-1109.

Eltsner, E.F. (1987). Metabolism of activated oxygen species. In *The Biochemistry of Plants.* pp. 253-315 Ed. Davis, D.D. (Academic Press, San Diego).

Fahcy, R.C. and Sundquist, A.R. (1991). Evolution of glutathione metabolism. *Adv. Enzymol.* **64:** 1-53.

Fischerová, Z., Tlustoš, P., Száková, J. and Šichorová, K. (2006). A comparison of phytoremediation capability of selected plant species for given trace elements. *Environ. Pollut.* **144:** 93-100.

Foyer, C.H. and Halliwell, B. (1976). The presence of glutathione and glutathione reductase in chloroplasts: a proposed role in ascorbic acid metabolism. *Planta.* **133:** 21-25.

Foyer, C.H., Lelandais, M. and Kunert, K.J. (1994). Photooxidative stress in plants. *Physiol. Plant.* **92:** 696-717.

Foyer, C.H., Lopez-Delgado, H., Dat, J-F. and Scott, I.M. (1997). Hydrogen peroxide and glutathione-associated mechanisms of acclimatory stress tolerance and signaling. *Physiol. Plant.* **100:** 241-254.

Foyer, C. H. and Noctor, G. (2003). Redox sensing and signaling associated with reactive oxygen in chloroplasts, peroxisomes and mitochondria. *Physiol. Plant.* **119(3):** 355-364.

Foyer, C.H. and Noctor, G. (2005a) Redox homeostasis and antioxidant signaling: A metabolic interface between stress perception and physiological responses. *Plant Cell.* **17:** 1866-1875.

Foyer, C.H. and Noctor, G. (2005b) Oxidant and antioxidant signaling in plants: a re-evaluation of the concept of oxidative stress in a physiological context. *Plant Cell and Environ.* **28:** 1056-1071.

Freeman, J.L., Garcia, D., Kim, D., Hopf, A. and Salt, D.E. (2005). Constitutively elevated salicylic acid signals glutathione-mediated nickel tolerance in *Thlaspi* nickel hyperaccumulators. *Plant Physiol.* **137(3):** 1082-1091.

Gabbrielli, R., Pandolfini, T. and Vergnano, O. (1987). Peroxidase involvement in tolerance mechanisms. *G. Bot. Ital.* **21:** 200-201.

Gallego, S.M., Benavides, M.P. and Tomaro, M.L. (1996). Effect of heavy metal ion excess on sunflower leaves: evidence for involvement of oxidative stress. *Plant Sci.* **121:** 151-159.

Gallego, S.M., Benavides, M.P. and Tomaro, M.L. (1999). Effect of Cadmium Ions on Antioxidant Defense System in Sunflower Cotyledons. *Biol. Plant.* **42:** 49-55.

Galloway, J.N., Thornton, J.D., Norton, S.A., Volcho, H.C. and McLean, R.A. (1982). Trace metals in atmospheric deposition : a review and assessment. *Atmosp. Environ.* **16:** 1677.

Gapper, C. and Dolan, L. (2006). Control of plant development by reactive oxygen species. *Plant Physiol.* **141:** 341-345.

Gille, G. and Singler, K. (1995). Oxidative stress in living cells. *Folia Microbiol.* **2:** 131-152.

Gomes-Junior, R.A., Moldes, C.A., Delite, F.S., Pompeu, G.B., Gratao, P.L., Mazzafera, P., Lea, P.J. and Azevedo, R.A. (2006). Antioxidant metabolism of coffee cell suspension cultures in response to cadmium. *Chemosphere.* **65:** 1330-1337.

Grant, C.M., MacIvr, F.H. and Dawes, I.W. (1996). Glutathione is an essential metabolite required for resistance to oxidative stress in the yeast *Saccharomyces cerevisiae. Curr. Gen.* **29:** 511-515.

Grill, E., Löffler, S., Winnacker, E.L. and Zenk, M.H. (1989). Phytochelatins, the heavy-metal binding peptides of plants are synthesized from glutathione by a specific c-glutamylcysteine dipeptidyl transpeptidase (phytochelatin synthase).c-glutamylcysteine synthetase. *Proc. Natl. Acad. Sci .USA.* **86:** 6838-6842.

Grill, E., Winnacker, E.L. and Zenk, M.H. (1987). Phytochelatins, a class of heavy-metal-binding peptides from plants, are functionally analogous to metallothioneins. *Proc. Natl. Acad. Sci. USA.* **84(2):** 439-443.

Guo, T.R., Zhang, G.P., Zhou, M.X., Wu, F.B. and Chen, J.X. (2004). Effect of aluminum and cadmium toxicity on growth and antioxidant enzyme activities of two barley genotypes with different Al tolerance. *Plant Soil.* **258(1):** 241-248.

Hall, J.L. (2000). Cellular mechanisms for heavy metal detoxification and tolerance. *J Exp. Bot.* **53:** 1-11.

Halliwell, B. and Foyer, C.H. (1978). Properties and physiological function of a glutathione reductase purified from spinach leaves by affinity chromatography. *Planta.* **139:** 9-17.

Halliwell, B. and Gutteridge, J. M. C. (2007). Free radicals in biology and medicine 4th edn. (Oxford, UK, Oxford University Press).

Hammond-Kosack, K.E. and Jones, J.D.G. (1996). Resistance gene-dependent plant defense responses. *Plant Cell.* **8:** 1773-1791.

Hare, P.D., Cress, W.A., Van Staden, J. (1998). Dissecting the roles of osmolyte accumulation during stress. *Plant Cell and Environ.* **21:** 535-553.

Hartley-Whitaker, J., Ainsworth, G., Vooijs, R., Ten Bookum, W., Schat, H. and Meharg, A. A. (2001). Phytochelatins are involved in differential arsenate tolerance in *Holcus lanatus. Plant Physiol.* **126(1):** 299-306.

Heiss, S., Wachter, A., Bogs, J., Cobbett, C. and Rausch, T. (2003). Phytochelatin synthase (PCS) protein is induced in *Brassica juncea* leaves after prolonged Cd exposure. *J. Exp. Bot.* **54:** 1833-1839

Hell, R. (1997). Molecular physiology of plant sulfur metabolism. *Planta.* **202:** 138-148.

Hertwig, B., Streb, P. and Feierabend J. (1992). Light dependence of catalase synthesis and degradation in leaves and the influence of interfering stress conditions. *Plant Physiol.* **100:** 1547-1553.

Hess, J.L. (1994). Free radical scavenging In: *Plant Responses to the Gaseous Environment.* pp. 165-180. Eds. Alscher, R.G. and Wellburn, A.R. (Chapman and Hall, London).

Hodges, D.M., Andrews, C.J., Johnson, D.A. and Hamilton, R.I. (1997). Antioxidant enzyme responses to chilling stress in differentially sensitive inbred maize lines. *J. Exp. Bot.* **48:** 1105-1113.

Hodges, D.M. and Forney, C.F. (2000). The effect of ethylene, depressed oxygen and elevated carbon dioxide on antioxidant profiles of senescing spinach leaves. *J. Exp. Bot.* **51:** 645-655.

Howden, R., Goldsbrough, P.B., Andersen, C.R. and Cobbett, C.S. (1995). Cadmium-sensitive, *cad1* mutants of *Arabidopsis thaliana* are phytochelatin deficient. *Plant Physiol.* **107:** 1059-1066.

Igamberdiev, A.U. and Hill, R.D. (2004). Nitrate, NO and haemoglobin in plant adaptation to hypoxia: an alternative to classic fermentation pathways. *J. Exp. Bot.* **55:** 2473-2482.

Ishikawa, T., Li, Z.S., Lu, Y.P. and Rea, P.A. (1997). The GS-X pump in plant, yeast, and animal cells: structure, function, and gene expression. *Bio. Science Reports.* **17:** 189-207.

Ivanov, B. (2003). Participation of photosynthetic electron transport in production and scavenging of reactive oxygen species. *Antioxidants & Redox Sig.* **5(1):** 43-53.

Jaspers, P. and Kangasjärvi, J. (2009). Reactive oxygen species in abiotic stress signaling – review. *Physiol. Plant.* **138(4):** 405-413.

Kiffin, R., Bandyopadhyay, U. and Cuervo, A.M. (2006). Oxidative stress and autophagy *Antioxidants & Redox Sig.* **8(1-2):** 152-162.

Kleizaite, V., Cesniene, T. and Rancelis, V. (2004). The use of cobalt-induced chlorophyll morphoses for studying Co^{2+} interactions with cysteine and SOD. *Plant Sci.* **167:** 1249-1256.

Kono, Y. and Fridovich, I. (1982). Superoxide radical inhibits catalase. *J. Biol. Chem.* **257:** 5751-5754.

Laspina, N.V., Groppa, M.D., Tomaro, M.L. and Benavides, M.P. (2005). Nitric oxide protects sunflower leaves against Cd-induced oxidative stress. *Plant Sci.* **169:** 323-330

Lee, H., Jo, J. and Son, D. (1998). Molecular cloning and characterization of the gene encoding glutathione reductase in *Brassica campestris. Biochimica et Biophysica Acta.* **1395:** 309-314.

Lee, S., Moon, J.S., Ko, T., Petros, D., Goldsbrough, P.B. and Korban, S.S. (2003). Overexpression of *Arabidopsis* phytochelatin synthase paradoxically leads to hypersensitivity to cadmium stress. *Plant Physiol.* **131(2):** 656-663.

Ledford, H.K. and Niyogi, K.K. (2005). Singlet oxygen and photo-oxidative stress management in plants and algae. *Plant Cell Environ.* **28:** 1037-1045

Li, J.M. and Jin, H. (2007). Regulation of brassinosteroid signaling. *Trends in Plant Sci.* **12(1):** 37-41.

Lin, C.C. and Kao, C.H. (2000). Effect of NaCl stress on H_2O_2 metabolism in rice leaves. *Plant Growth Regul.* **30:**151-155.

Lozano-Rodriguez, E., Hernández, L.E., Bonay, P. and Carpena-Ruiz, R.O. (1997). Distribution of Cd in shoot and root tissues of maize and pea plants: physiological disturbances. *J. Exp. Bot.* **306:** 123-128.

Luna, C.M., Gonzalez, C.A. and Trippi, V.S. (1994). Oxidative damage caused by excess copper in oat leaves. *Plant Cell Physiol.* **35:** 11-15.

.Madlung, A. and Comai, L. (2004). The effect of stress on genome regulation and structure. *Ann. Bot.* **94:** 481-495.

Marrs, K. (1996). The functions and regulation of glutathione S-transferases in plants. *Annu. Rev. Plant Physiol. Plant Mol. Biol.* **47:** 127-158.

May, M.J., Vernoux, T., Leaver, C.J., Van Montagu, M. and Inzé, D. (1998**).** Glutathione homeostasis in plants: Implications for environmental sensing and plant development. *J. Exp. Bot.* **49:** 649-667.

Meng, Q., Zou, J., Zou, J. jiang, W., and Liu, D. (2007). Effect of Cu^{2+} concentration on growth, antioxidant enzyme activity and Malondialdehyde content in garlic (*Allium sativum* L.). *Acta Biologica Cracoviensia Series Botanica.* **49(1):** 95-101.

Milone, M.T., Sgherri, C., Clijsters, H. and Navarri-Izzo, F. (2003). Antioxidant responses of wheat treated with realistic concentration of cadmium. *Environ. Exp. Bot.* **50:** 265-276.

Mittler, R., Vanderauwera, S., Gollery, M. and Van Breusegem, F. (2004). Reactive oxygen gene network of plants. *Trends Plant Sci.* **9:** 490-498.

Mittler, R. (2002). Oxidative stress, antioxidants and stress tolerance. *Trends Plant Sci.* **7:** 405-410.

Mohan, B.S. and Hosetti, B.B. (1997). Potential Phytotoxicity of Leaf and Cadmium to *Lemna minor* Grown in Sewage Stabilization Ponds. *Environ. Pollut.* **98:** 233-238.

Navari-Izzo, F., Meneguzzo, S., Loggini, B., Vazzana, C. and Sgherri, C. (1997). The role of the glutathione system during dehydration of *Boea hygroscopica. Plant Physiol.* **99:** 23-30.

Nocito, F.F., Lancilli, C., Crema, B., Fourcroy, P., Davidian, J-C. and Sacchi, G.A. (2006). Heavy metal stress and sulfate uptake in maize roots. *Plant Physiol.* **141:** 1138-1148.

Noctor, G. (2006). Metabolic signaling in defence and stress: the central roles of soluble redox couples. *Plant Cell and Environ.* **29:** 409-425.

Noctor, G. and Foyer, C.H. (1998). Ascorbate and glutathione: keeping active oxygen under control. *Annu Rev Plant Physiol. Plant Mol. Biol.* **49:** 249-279.

Noctor, G., Gomez, L., Vanacker, H. and Foyer, C.H. (2002). Interactions between biosynthesis, 437 compartmentation and transport in the control of glutathione homeostasis and signaling. *J, Exp. Bot.* **53:** 1283-1304.

Overmyer, K., Brosche, M. and Kangasjärvi, J. (2003). Reactive oxygen species and hormonal control of cell death. *Trends Plant Sci.* **8:** 335-342.

Pandey, N. and Sharma, C.P. (2002). Effect of heavy metals Co^{2+}, Ni^{2+} and Cd^{2+} on growth and metabolism of cabbage. *Plant Sci.* **163:** 753-758.

Pastori, G.M., Foyer, C.H. and Mullineaux, P. (2000). Low temperature induced changes in the distribution of H_2O_2 and antioxidants between the bundle sheath and mesophyll cells of maize leaves. *J. Exp. Bot.* **51:** 107-113.

Pastori, G., Kiddle, G., Antoniw, J., Bernard, S., Veljovic-Jovanovic, S., Verrier, P., Noctor, G. and Foyer, C.H. (2003). Leaf vitamin C contents modulate plant defense transcripts and regulate genes that control development through hormone signaling. *Plant Cell.* **15:** 939-951.

Patra, J. and Panda, B.B. (1998). A comparison of biochemical responses to oxidative and metal stress in seedlings of barley (*Hordeum vulgare* L). *Environ. Pollut.* **101:** 99-105

Polle, A. (1997). Defense against photo-oxidative damage in plants. In Oxidative stress and the molecular biology of antioxidant defenses. pp. 623-666 Ed. Scandalios, J.G. (Cold Spring Harbor Laboratory Press).

Pourcel, L., Routaboul, J.M. and Cheynier, V. (2007). Flavonoid oxidation in plants: from biochemical properties to physiological functions. *Trends in Plant Sci.* **12(1):** 29-36.

Raeymaekers, T., Potters, G., Asard, H., Guisez, Y. and Horemans, N. (2003). Copper-mediated oxidative burst in *Nicotiana tabacum* L. cv. Bright Yellow 2 cell suspension cultures. *Protoplasma.* **221:** 93-100.

Radotic, K., Ducic, T. and Mutavedzic, D. (2000). Changes in peroxidase activity and isoenzymes in spruce needles after exposure to different concentrations of cadmium. *Environ. Exp. Bot.* **44:** 105-113

Rao, K.V.M. and Sresty, T.V.S. (2000). Antioxidative parameters in the seedlings of pigeon pea (*Cajanus cajan*) in response to Zn and Ni stresses. *Plant Sci.* **157:** 113-128.

Rauser, W.E. (1995). Phytochelatins and related peptides: Structure, biosynthesis, and function. *Plant Physiol.* **109:** 1141-1149.

Rea, P. (1999). MRP subfamily ABC transporters from plants and yeast. *J. Exp. Bot.* **50:** 895-913.

Rice-Evans, C.A., Miller, N.J. and Paganga, G. (1997). Antioxidant properties of phenolic compounds. *Trends Plant Sci.* **2:** 152.

Russo, T., Zambrano, N., Esposito, F., Ammendola, R., Cimino, F., Fiscella, M., Jackman, J., O'Connor, M., Anderson, C.W. and Apella, E. (1995). A p53-independent pathway for activation of WAF1/C1P1 expression following oxidative stress. *Biological Chemistry.* **270:** 29 386-391.

Salt, D.E., Prince, R.C., Pickering, I.J. and Raskin, I. (1995). Mechanisms of cadmium mobility and accumulation in Indian mustard. *Plant Physiol.* **109:** 1427-1433.

Sanchez-Fernandez, R., Fricker, M., Corben, L.B., White, N.S., Sheard, N., Leave, C.J., Van Montagu, M., Inze, D. and May, M.J. (1997). Cell proliferation and hair tip growth in the *Arabidopsis* root are under mechanistically different forms of redox control. *Proc. Natl. Acad. Sci. USA.* **94(6):** 2745-2750.

Sandalio, L.M., Dalurzo, H.C., Gomez, M., Romero-Puertas, M.C. and Del Rio, L.A. (2001). Cadmium-induced changes in the growth and oxidative metabolism of pea plants. *J. Exp. Bot.* **52:** 2115-2126.

Sanita di Toppi, L. and Gabbrielli, R. (1999). Response to cadmium in higher plants. *Environ. Exp. Bot.* **41:** 105-130.

Scandalios, J.G. (1993). Oxygen stress and superoxide dismutases. *Plant Physiol.* **101:** 7-12.

Scandalios, J.G., Guan, L., and Polidoros, A.N. (1997). Catalases in plants: gene structure, properties, regulation, and expression. In Oxidative stress and the molecular biology of antioxidant defenses pp. 343-406 Ed. Scandalios, J.G. (Cold Spring Harbor Laboratory Press New York).

Schmöger, M.E.V., Oven, M. and Grill, E. (2000). Detoxification of arsenic by phytochelatins in plants. *Plant Physiol.* **122(3):** 793-802.

Schickler, H. and Caspi, H. (1999). Response of antioxidative enzymes to nickel and cadmium stress in hyperaccumulator plants of the genus *Alyssum. Physiol. Plant.* **105:** 39-44.

Schützendübel, A. and Polle, A. (2002). Plant responses to abiotic stress: heavy metal - induced oxidative stress and protection by mycorrhization. *J. Exp. Bot.* **53:** 1351-1365.

Shah, K., Kumar, R.G., Verma, S. and Dubey, R.S. (2001). Effect of cadmium on lipid peroxidation, superoxide anion generation and activities of antioxidant enzymes in growing rice seedlings. *Plant Sci.* **161:** 1135-1144.

Shao, H.B. and Chu, L.Y. (2005). Plant molecular biology in China: Opportunities and Challenges. *Plant Mol. Biol. Rep.* **23:** 345-358.

Shao, H.B., Chu, L.Y., Zhao, C.X., Guo, Q.J., Liu, X.A., and Ribaut. J.M. (2006). Plant gene regulatory net work system under abiotic stress. *Acta Biol. Sezeged.* **50(1-2):** 1-9.

Shao, H.B., Liang, Z.S., Shao, M.A., Sun, Q. and Hu, Z.M. (2005a). Investigation on dynamic changes of photosynthetic characteristics of 10 wheat (*Triticum aestivum L.*) genotypes during two vegetative-growth stages at water deficits. *Biointerfaces.* **43(3-4):** 221-227.

Shao, H.B., Liang, Z.S. and Shao, M.A. (2005b). Dynamic changes of anti-oxidative enzymes of 10 wheat genotypes at soil water deficits. *Biointerfaces.* **42(3-4):** 187-195.

Shao, H.B., Chu, L.Y. Lu, Z-H. and Kang, C-M. (2008**).** Primary antioxidant free radical scavenging and redox signaling pathways in higher plant cells. *Int. J. Biol. Sci.* **4(1):** 8-14.

Shaul, O., Mironov, V., Burssens, S., Van Montagu, M.V., Inzé, D. (1996). Two *Arabidopsis* cyclin promoters mediate distinctive transcriptional oscillation in synchronised tobacco 3Y-2 cells. *Proc. Natl. Acad. Sci. USA.* **93:** 4868-4872.

Shaw, B.P. (1995). Effect of mercury and cadmium on the activities of antioxidative enzymes in the seedling of *Phaseolus aureus. Biol. Plant.* **37:** 587-596.

Smirnoff, N. (2000). Ascorbic acid: metabolism and functions of multi-facetted molecule. *Curr. Opin. Plant Biol.* **3:** 229.

Smirnoff, N., Conklin, P.L. and Loewus, F.A. (2001). Biosynthesis of ascorbic acid in plants: a renaissance. *Annu. Rev. Plant Physiol. and Plant Mol. Biol.* **52:** 437-467.

Somashekaraiah, B.V., Padmaja, K. and Prasad, A.R.K. (1992). Phytotoxicity of cadmium ions on germinating seedlings of mung bean (*Phaseolus vulgaris*): involvement of lipid peroxides in chlorophyll degradation. *Physiol. Plant.* **85:** 85-89.

Sreedevi, S., Krishnan, P.N., and Pushpangadan, P. (2008). Cadmium induced oxidative damage and antioxidant responses in roots of black gram (*Vigna mungo* (L.) Hepper. *Indian J. Plant Physiol.* **13(1):** 1-7

Strasdeit, H., Duhme, A.K., Kneer, R., Zenk, M.H., Hermes, C. and Nolting, H.F. (1991). Evidence for discrete $Cd(SCys)_4$ units in cadmium phytochelatin complexes from EXAFS spectroscopy. *J. Chem. Soc. Series of Chemical Communications* 1129-1130.

Teisseire, H. and Guy, V. (2000). Copper-induced change in antioxidant enzymes activities in fronds of duckweed (*Lemna minor*). *Plant Sci.* **153:** 65-72.

Tommasini, R., Vogt, E., Fromenteau, M., Hoertensteiner, S., Matile, P., Amrhein, N. and Martinoia, E. (1998). An ABC-transporter of *Arabidopsis thaliana* has both glutathione-conjugate and chlorophyll catabolite transport activity. *The Plant J.* **13:** 773-780.

Tsuji, N., Hirayanagi, N., Okada, M., Miyasaka, H., Hirata, K., Zenk, M.H. and Miyamoto, K. (2002). Enhancement of tolerance to heavy metals and oxidative stress in *Dunaliella tertiolecta* by Zn induced phytochelatin synthesis. *Biochem. Biophys. Res. Commun.* **293:** 653-659.

Verma, S. and Dubey, R.S. (2003). Lead toxicity induces lipid peroxidation and alters the activities of antioxidant enzymes in growing rice plants. *Plant Sci.* **164:** 645-655.

Vernoux, T., Wilson, R.C., Seeley, K.A., Reichheld, J.P., Muroy, S., Brown, S., Maughan, S. C., Cobbett, C.S., Van Montagu, M., Inzé, D., May, M.J. and Sung, Z.R. (2000). The *Root Meristemless1/Cadmium Sensitive2* gene defines a glutathione-dependent pathway involved in initiation and maintenance of cell division during postembryonic root development. *The Plant Cell.* **12:** 97-109.

Vitória, A.P., Lea, P.J. and Azevedo, R.A. (2001). Antioxidant Enzymes Responses to Cadmium in Radish Tissues. *Phytochemistry.* **57:** 701-710.

Vögeli-Lange, R. and Wagner, G.W. (1996). Relationship between cadmium, glutathione and cadmium-binding peptides (phytochelatins) in leaves of intact tobacco seedlings. *Plant Sci.* **114:** 11-18.

Weast, R.C. (1984). CRC Handbook of chemistry and physics 64th ed. Boca Raton, Florida, USA.

Weckx, J.E.J. and Clijster, H.M.M. (1996). Oxidative damage and defense mechanisms in primary leaves of *Phaseolus vulgaris* as a result of root assimilation of toxic amounts of copper. *Physiol. Plant.* **96:** 506-512.

Wingate, V.P.M., Lawton, M.A. and Lamb, C.J. (1988). Glutathione causes a massive and selective induction of plant defense. *Plant Physiol.* **31:** 205-210.

Wu, G., Wei, Z.K. and Shao, H.B. (2007). The mutual responses of higher plants to environment: physiological and microbiological aspects. *Biointerfaces.* **59:** 113-119.

Xiang, C. and Oliver, D.J. (1998). Glutathione metabolic genes coordinately respond to heavy metals and jasmonic acid in *Arabidopsis. Plant Cell.* **10:** 1539-1550.

Zang, H., Jiang, Y., He, Z., Ma, M. (2005). Cadmium accumulation and oxidative burst in garlic (*Allium sativum*). *J. Plant Physiol.* **162:** 977-984.

Zenk, M.H. (1996). Heavy metal detoxification in higher plants: a review. *Gene.* **179:** 21-30.

Zhu, Y.L., Pilon-Smits, E., Jouanin, L. and Terry, N. (1999a). Overexpression of glutathione synthetase in Indian mustard enhances cadmium accumulation and tolerance. *Plant Physiol.* **119:** 73-79.

Zhu, Y.L., Pilon-Smits, E., Tarun, A.S., Weber, S., Jouanin, L. and Terry, N. (1999b). Cadmium tolerance and accumulation in Indian mustard is enhanced by overexpressing gamma-glutamylcysteine synthetase. *Plant Physiol.* **121:** 1169-1177.

8

Significance of Glutathione in Higher Plants for Combating Abiotic Stress-Effects

SHAO HONG-BO[1, 2, 3*], LIU ZENG-HUI[2], CHU LI-YE[2], YAN KUN[1], ZHANG LI-HUA[1]

[1] The CAS/Shandong Provincial Key Laboratory of Coastal Environmental Process, Yantai Institute of Coastal Zone Research, Chinese Academy of Sciences (CAS), Yantai–264 003, China
[2]Institute for Life Sciences, Qingdao University of Science & Technology (QUST), Qingdao–266 042, China
[3]Present address: Yantai Institute of Coastal Zone Research, Chinese Academy of Sciences (CAS), Yantai–264 003, China
*Corresponding author: E-mail: shaohongbochu@126.com

ABSTRACT

Glutathione is one of the most important proteins, which plays many functions in plant metabolism. It can play role in signal transduction and help plant resist the harm from environment stress. In current paper, we review the synthesis of glutathione, its functions and the change of content under drought, metal ion and low-temperature stress. Meanwhile, we also concern the mechanism how glutathione performed its functions.

Keywords: Abiotic stress, Ascorbate peroxidase, glutathione, higher plants.

1. INTRODUCTION

Glutathione is distributed widely in animals and plants, especially in yeast and wheat embryo and animal liver. It was found in 1929 by Hopkins. It is a kind of tripeptide making up of glutamic acid, cysteine and glycine (Dixon et al., 2005). It was classed into reduced glutathione (GSH) (Fig. 1) and oxidized glutathione (GSSG) (Fig. 2).

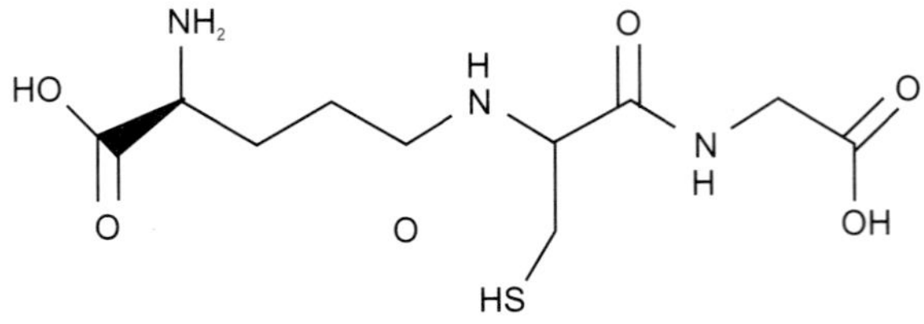

Fig. 1. The structure of GSH.

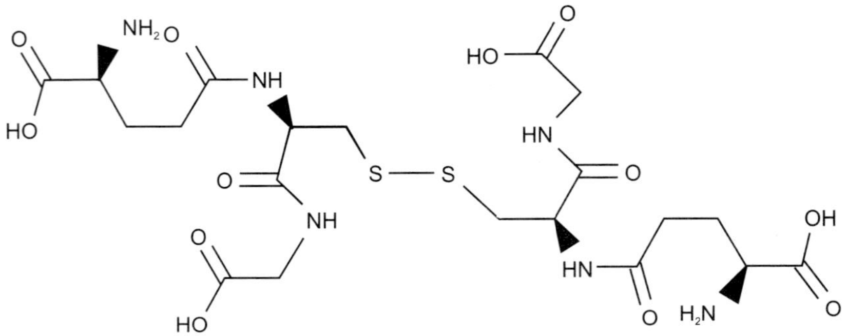

Fig. 2. The structure of GSSH.

Its chemical essence is thioalcohol. The peptide bond of N-terminal was made up of r-carboxyl of glutamic and amino of cysteine, which can help glutathione keep stable in cells (Xu and Song 2001). There is a hydrosulphonyl in its structure, which was oxidized easily. This special structure is very significant for organisms. Because, it has the ability to eliminate free radicals in cells. And this is the reason why glutathione was studied by people so widely. The content of GSH is very rich in plants, which is an important active substance. It mainly takes part in the synthesis of disulphide, thioether, sulpholipid and eliminates the free radicals in body (Zetterstrm et al., 2006).

2. SYNTHESIS OF GLUTATHIONE AND ITS REGULATION IN CELLS

The synthesis of GSH has two steps (Fig. 3) (Peng et al., 2004): Stage I, Synthesis of glutamylcysteine from L-Glu and L-Cys catalysed by γ-glutamylcysteine synthetase (γ-ECS); Stage II, a Gly would be added to the glutamylcysteine under the γ-glutathione synthetase (γ-GS) catalyse. ATP was required in this reaction. Glutamylcysteine synthetase is the main rate-limiting factor in this synthesis course (May et al., 1998; Noctor et al., 2002).

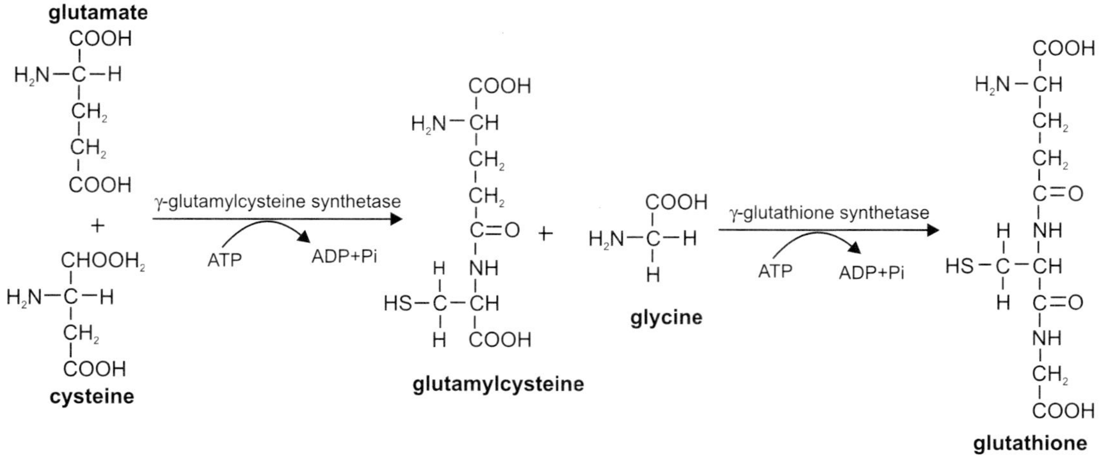

Fig. 3. The synthesis of glutathione.

Under different abiotic stress, the content of glutathione pool would expand in plants. In many plants, the synthesis of glutathione was limited by γ-ECS. In Arabidopsis, gshl has the duty to control the synthesis of γ-ECS. In Arabidopsis gshl mutants, the content of glutathione is one-third of wild type (Du and Zhang 2008). In addition, the opposite restrain of γ-ECS can lead to the decrease of glutathione in Arabidopsis leaf. In poplar and tobacco, the overexpression of γ-ECS and its congener would increase the glutathione tremendously in their leaf. But γ-GS could not produce this phenomenon (Noctor et al., 2002). The synthesis of GSH could feedback inhibition of the γ-ECS. It means that γ-ECS is the critical factor in the regulation of GSH content in plants (Noctor et al., 2002, Peng et al., 2004).

3. TRANSPORT AND DEGRADATION

The rate of glutathione synthesis is different in different plant histiocytes. So the distribution of glutathione in cells is not average. There are three main places to store GSH in plant cells. 90 per cent of GSH was stored in cytoplasm, 10% in mitochondria and the rest in endoplasmic reticulum (Hu et al., 2004). In order to play functions, GSH must transport effectively. The first GSH transport protein was separated in yeast (Bourbouloux et al., 2000). The cloning and identification of genes of GSH transport protein in plants proved that there is a special GSH transmembrane transport system in plant cells (Bogs et al., 2003; Mai et al., 2005). Through related study, the Km value of GSH transport protein does yeast, tomato cells and wheat chloroplast is similar. The transport protein does not only transport GSH but also keeps the GSH content stable and the oxidation-reduction reaction balanced in cells (Noctor et al., 2002; Bogs et al., 2003; Mai et al., 2005).

The degradation of GSH is not clear. But there are some pathways about its degradation (Edwards et al., 2000; Chen et al., 2004). First, Gly was removed from GSH or the GSH combined into GS-X under the vacuole carboxypeptidase (Cpase) catalysis.

Second, removing the Glu to other dipetid under the catalysis of 7-glutamyltranspeptidase (7-GTase) produces the substrate of 7-glutamylcycle. Third, the dipetid (Cys-Gly) produced after 7-GTase catalysis was decomposed to free amino acid by dipeptidase (DPase).

4. FUNCTION OF GLUTATHIONE

Glutathione has many roles in plants (Fig. 4). It is a stored form of reduced sulphur, while it can regulate inter-organ sulphur allocation (Herschbach and Rennenberg 1991, 1994). It could be the precursor of phytochelatins, which bind to some heavy metals (Grill et al., 1987, 1989). It is also a substrate for the GSH S-transferases, which can catalyse the joining of GSH with dangerous substances such as herbicides (Marrs 1996). In this course, glutathione performs its detoxification role. In some situations, it may take part in the stress signal transduction in plants (Ullmann et al., 1996).

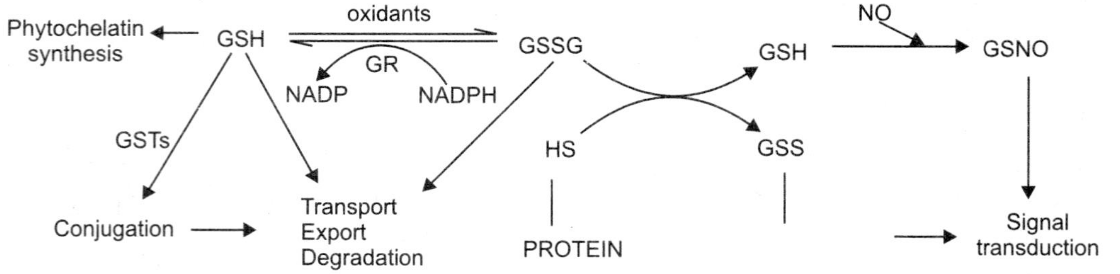

Fig. 4. The course of glutathione with others factors.

Glutathione joins many metabolisms in plants. It plays an important role for the synthesis of proteins and nucleic acids. It is a regulator of gene expression. It may also take part in the redox regulation of cell cycle (Gyuris et al., 1993; Russo et al., 1995; Shaul et al., 1996; Sanchez-Fernandez et al., 1997). Meanwhile, it is very significant for the plant resistance to different stresses from environment (Foyer and Halliwell 1976; Alscher 1989; Grant et al., 1996; Shao et al., 2008). The tolerance is close relation to the content of GSH (reduced glutathione) in plants. GSH can be oxidized by free radicals to GSSG (oxidized glutathione). This reaction can protect protein and lipids from oxidation with the interrelated enzymes under stresses. In addition, it can play buffering roles in redox reaction. Under certain conditions, GSSG can be reduced to GSH. The ratio of GSH/GSSG can indicate the ability of plant stress resistance (Foyer and Halliwell 1976; Halliwell and Foyer 1978; May et al., 1998; Alscher 1989). These reactions are as follows:

$$2GSH + H_2O_2 \longrightarrow GSSH + 2H_2O$$

$$GSSG + NADPH + H^+ \xrightarrow{GR} 2GSH + NADP^+$$

There is an ascorbate-glutathione cycle in plants (Fig. 5) (Potters et al., 2004). The cycle mainly exists in plant chloroplast. Its aim is to eliminate H_2O_2 production in

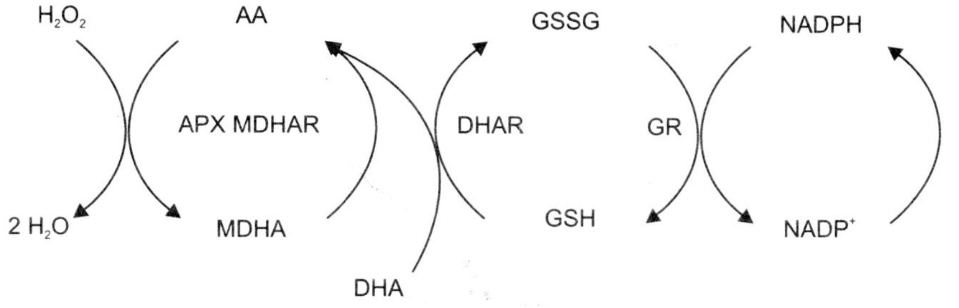

Fig. 5. Ascorbate-glutathione cycle.
APX: Ascorbate Peroxidase, MDHA: Monodehydroascorbate reductase
DHA: Dehydroascorbate reductase, GR: Glutathione reductase

photoreaction. These enzymes in this cycle have high activity and many roles in plants. So the ascorbate-glutathione cycle is the main metabolic pathway of H_2O_2 in photosynthetic organelles (Chew et al., 2003). This cycle exists in plastosome well.

5. GLUTATHIONE UNDER STRESS

Because glutathione has many multiple roles in plant metabolism and defences against both biotic and abiotic stress, it is an attractive target for engineering stress tolerance in plants. Glutathione accumulation is observed frequently under stress condition, which would compensate the decrease of other antioxidants (Smith 1985; Smith et al., 1984; Chamnongpol et al., 1996; Willekens et al., 1997).

Stresses can induce the increase of free radical content, which would bring to oxidation stress for plant cells. So the form of plant anti-stress ability has close relation with the activity of anti-oxidation systems. Under stress, plant photorespiration would be strengthened with the decreasing of photosynthesis rate. Though plants have corresponding protect systems, the content of active oxygen would be increasing still (Ball et al., 2004; Shao et al., 2008). In addition, some anti-oxidation enzymes (SOD, CAT, GR) and anti-oxidation substance (GSH, ASA) would be produced to eliminate active oxygen and free radicals when plant encounters environment stress (Qian et al., 2006). The accumulation of ROS in plants would activate the defensive genes expression (Foyer and Noctor 2005). Transgenic research has proved that the defence ability of a plant is related with the level of the anti-oxidation system. For example, the decrease of GSH content would cause the decline of plant defence ability. This fact means that the content and form of glutathione could affect the plant stress response (Ball et al., 2004).

5.1 Glutathione Under Drought Stress

Drought is a main stress for plant growth, which can result in water deficit in plant issues. It can lead to an imbalance in the redox poise of plant cells and cause osmotic stress. The reason is that drought stress could cause the accumulation of free radicals and azotic acid and the membrane lipid peroxidation (Noctor and Foyer 1998; Smirnoff 1998). The plant metabolism would be in disorder. In order to reduce the harm from drought stress for cells, the cell would produce many kinds of proteins to eliminate the free radicals. Glutathione is one of these proteins in plants. Glutathione is a cellular protectant in plants (Noctor and Foyer 1998; Polle 2001).

The ascorbate-glutathione cycle is critical for eliminating active oxygen and free radicals and keeping the glutathione steady state, which was made up of three interdependent oxidation-reduction reactions (Fig. 5) (Potters et al., 2004). The change of the level of glutathione can stimulate an upregulation machine, which can coordinate the ascorbic acid, glutathione and NADPH change. In this course, glutathione plays a sensor role (May et al., 1998; Jaleel et al., 2009). It has been found that GSH may act to ensure maintenance of the cellular NADPH pool under stress. This means that plants have adopted to ensure a supply of NADPH for stress defences (May et al., 1998; Shao et al., 2008; Jaleel et al., 2009). After investigating the activity of key enzymes of GSH biosynthesis, degradation and regeneration is very important for investigating the GSH metabolism under stress. Chen et al. (2008) found that the activity of γ-GS, GSHS and GR was increased significantly in spring wheat (*Triticum aestivum L.*). It means that a higher rate of GSH biosynthesis is involved in the response of wheat to drought stress. In contrast, the content of GSH, NADPH and $NADP^+$ decreased. But the ratios of NADPH/$NADP^+$, GSH/GSSH and ATP/ADP have markedly increased. This indicates that a more reduced state of glutathione pools was involved in the response of wheat to drought. And the increased ratio of ATP/ADP indicates that upregulation of energy supply is another important event for wheat cells under drought stress. In addition, Wingate and Dhindsa found that GSH is the signal molecule in plant drought stress. It may be the second and third (Foyer et al., 1995; Creissen et al., 1999; Shao et al., 2008).

5.2 Glutathione Under Metal Ion Stress

With the emission of industrial waste, domestic sewage and the wide use of pesticides and fertilizers in agriculture, more and more soil was polluted by metal ions, especially heavy metal ions, which made the content of ion in soil increased sharply. It affects the plant growth seriously and brings serious harm to plant growth. There is a kind of protein in plant cells, whose function is chelating these metal ions, named phytochelatin (PC). GSH can be catalysed to phytochelatin by plant chelating peptide synthetases (Rauser 1990). Plant phytochelatin has strong affinity with heavy metal ions. It can chelate heavy metal ions, like Cu and Cd, to produce atoxic-chemical compounds. This reaction would decrease the free metal ion density in cells and protect plant cells from harm. These chelating molecules were transported to vacuole and out of cells finally (Rauser 1990). This system is an important mechanism for plant resisting the metal ion harm. The content of phytochelatin in some Arabidopsis mutants (for example, lack of cad2 and RML1 for GSH synthesizing) is lower than normal value, which would be sensitive to Cu and Cd ions. With the Cu and Cd increase in environment, γ-EC would be increased gradually in wild or γ-EC sensitive plant cells (Vernoux et al., 2000). It is proved that the overexpression of γ-EC in Cd sensitive Arabidopsis mutants can strength the tolerance to Cd stress (Cobbett et al., 1998). Cu or Cd stress could induce the expression of γ-EC genes and GSH synthetase genes (Xiang and Oliver 1998). At the same time, the γ-EC gene expression was regulated by the ratio of GSH/GSSG. And the concentration of cysteine can limit the synthesis of GSH (Xiang et al., 2001; Noctor et al., 2002). In summary, Cd stress can induce the transcription of γ-ECS. In addition, the low content of GSH can reduce the feedback inhibition of γ-ECS and induce the synthesis of cysteine. Report of Anjum et al., (2008) has also demonstrated a positive correlation of sulphur nutrition-induced glutathione content with photosynthesis and plant dry weight in *B. campestris* under Cd stress.

5.3 Glutathione Under Cold Stress

Low-temperature stress can bring to harm for many ordinary plants. But some cold-tolerance plants can resist this stress with their protecting systems. There are many evidences proving that GSH takes part in the low-temperature protection of plants. The study about polar and alpine plant indicates that GSH is increasing in plants in winter. This means GSH may be involved into the cold stress-resisting as the protective agents (Anderson et al., 1992; Wildi and Lutz 1996).

Under the low temperature stress, the content of GSH and the ratio of GSH/GSSG in tomato and wheat are rising. It proved that high ratio of GSH/GSSG can help improve plant cold-tolerance (Kingston-Smith et al., 1999). Kocsy et al. (2000) proved that the change of the level of GSH is the signal for inducing the cold-tolerance gene expression. The sharp changing of ratio GSH/GSSG can lead to the increasing of GR in maize roots. The changing of GSH synthesis could affect this signal-transduction pathway too (Foyer et al., 1997). Under low-temperature stress, the activity of GR is increasing in European spruce, tomato and wheat (Kocsy et al., 1997). The reason about raising the activity of GR to strength the cold tolerance of plants may be as follows: first, the deoxidization of GSSG is produced in the ascorbate-glutathione cycle (Foyer and Halliwell 1976; Foyer et al., 1994); second, GR plays an important role in the synthesis of cysteine and is catalysed to GSSG by ascorbate peroxidase (APX) (Suter et al., 2000).

The ascorbate-glutathione cycle is the main defence system of plants for cold stress when the content of GSH and GR in cells is very low. But the system would be effective when the content is very high. Overexpression of γ-ECS in cells and chloroplast can improve the level of GSH in cells, but it does not enhance the cold tolerance (Arisi et al., 1997; Noctor et al., 1998). It indicates that increasing the content of GSH cannot reinforce plant's cold tolerance. With the total of GSH increasing and the overexpression of GR in poplar, its cold tolerance would be improved (Foyer et al., 1995). It must be pointed out that overexpression of γ-EC and overfull GSH in plants may bring harm for plants. Creissen et al., (1995) found that overexpression of γ-EC and overfull GSH in tobacco could lead it to cyanosis and death.

6. SUMMARY

From the above we can find that GSH plays a very important role in plant anti-stress. GSH mainly takes part in the plant response to environment stress, the development of plants and stress signal transduction. The distribution of GSH between subcellular organelles and the change of GSH redox state are significant for stress signal transduction.

In addition, the ascorbate-glutathione cycle is one of the important pathways for performing glutathione functions. The ratio of GSH/GSSG is an index of plant stress tolerance. Though some genes were cloned by researchers, the study about the GSH inducing anti-stress gene expression is scanty.

ACKNOWLEDGEMENTS

This work is jointly supported by One Hundred-Talent Plan of Chinese Academy of Sciences (CAS), the CAS/SAFEA International Partnership Programme for Creative Research Teams, the Important Direction Project of CAS (KZCX2-YW-JC203) and CAS YOUNG SCIENTISTS FELLOWSHIP (2009Y2B211).

REFERENCES

Alscher, R.G. (1989). Biosynthesis and antioxidant function of glutathione in plants. *Physiol Plant.* **77:** 457-64.

Anderson, J.V., Chevone, B.I. and Hess, J.L. (1992). Reactive oxygen species and antioxidants: relationships in green cells. *Physiol. Plant.* **100:** 224-233.

Anjum, N.A., Umar, S., Ahmad, A., Iqbal, M. and Khan, N.A. (2008). Sulphur protects mustard (*Brassica campestris* L.) from cadmium toxicity by improving leaf ascorbate and glutathione. *Plant Growth Regul.* **54:** 271-279.

Arisi, A.C., Noctor, G., Foyer, C.H. and Jouanin, L. (1997). Modification of thiol contents in poplars (*Populus tremula* XP. alba) overexpressing enzymes involved in glutathione synthesis. *Planta.* **203:** 362-372.

Ball, L., Accotto, G.P., Bechtold, U., Creissen, G., Funck, D., Jimenez, A., Kular, B., Leyland, N., Jaime, M.C., Reynolds, H., Karpinski, S. and Mullineaux, P.M. (2004). Evidence for a Direct Link between Glutathione Biosynthesis and Stress Defense Gene Expression in Arabidopsis. *Plant Cell.* **16:** 2448-2462.

Bogs, J., Bourbouloux, A., Cagnac, O., Wachter, A., Rausch, T. and Delrot, S. (2003). Functional characterization and expression analysis of a glutathione transporter, BjGT1, from *Brassica juncea*: evidence for regulation by heavy metal exposure. *Plant Cell Environ.* **26:** 1703-1711.

Bourbouloux, A., Shahi, P. and Chakladar, A. (2000). Hgtl, a high affinity glutathione transporter from the yeast *Saccharomyces cerevisae. J. Biol. Chem.* **275:** 13259-13265.

Chamnongpol, S., Willekens, H., Langebartels, C., Van Montagu, M., Inze, D. and Van Camp, W. (1996). Transgenic tobacco with a reduced catalase activity develops necrotic lesions and induces pathogenesis-related expression under high light. *Plant J.* **10:** 491-503.

Chen, K.M., Gong, H.J. and Wang, S.M. (2004a). Glutathione metabolism and environmental stresses in plants. *Acta Bot Boreali-Occid Sin.* **24:** 1119-1130.

Chen, K.M., Gong, H.J., Chen, G.C., Wang, S.M., Zhang, C.L. (2004b). Gradual Drought Under Field Conditions In?uences the Glutathione Metabolism, Redox Balance and Energy Supply in Spring Wheat. *J Plant Growth Reggul.* **23:** 20-28.

Chew, O., Whelan, J. and Millar, A.H. (2003). Molecular Definition of the Molecular Definition of the Ascorbate-Glutathione Cycle in Arabidopsis Mitochondria Reveals Dual Targeting of Antioxidant Defenses in Plants. *J. Biol. Chem.* **278:** 46869-46877.

Cobbett, C.S., May, M.J., Howden, R. and Rolls, B. (1998). The glutathione-deficient, cadmium-sensitive mutant, cad2-1, of *Arabidopsis thaliana* is deficient in gamma-glutamylcysteine synthetase. *Plant J.* **1:** 73-78.

Creissen, G., Firmin, J., Fryer, M., Kular, B., Leyland, N., Reynolds, H., Pastori, G., Wellburn, F., Baker, N., Wellburn, A. and Mullineaux, P. (1999). Elevated glutathione biosynthetic capacity in the chloroplasts of transgenic tobacco plants paradoxically causes increased oxidative stress. *Plant J.* **11:** 1277-1292.

Dhindsa, R.S. (1991). Drought Stress, Enzymes of Glutathione Metabolism, Oxidation Injury, and Protein Synthesis in *Tortula ruralis. Plant Physiol.* **95:** 648-651.

Dixon, D.P., Skipsey, M., Grundy, N.M. and Edwards, R. (2005). Stress-Induced Protein S-Glutathionylation in Arabidopsis. *Plant Physiol.* **138:** 2233-2244.

Du, L. and Zhang, Q. (2008). Glutathione and Oxidative Stress Tolerance of Plants. *Shandong Sci.* **21:** 2008.

Edwards, R., Dixon, D.P. and Walbot, V. (2000). Plant glutathione S-transferases: enzymes with multiple functions in sickness and in health. *Trends Plant Sci.* **5:** 193-198.

Foyer, C.H. and Halliwell, B. (1976). The presence of glutathione and glutathione reductase in chloroplasts: a proposed role in ascorbic acid metabolism. *Planta.* **133:** 21-25.

Foyer, C.H. and Noctor, G. (2005). Redox Homeostasis and Antioxidant Signaling: A Metabolic Inter face between Stress Perception and Physiological Responses. *Plant Cell.* **17:** 1866-1875.

Foyer, C.H., Descourvierse, P. and Kunert, K.J. (1994). Protection against oxygen radicals : an important defence mechanism studied in transgenic plants. *Plant Cell Environ.* **17:** 507-523

Foyer, C.H., Lopez-Delgado, H., Dat, J.F. and Scott, I.M. (1997). Hydrogen peroxide and glutathione-associated mechanisms of acclimatory stress tolerance and signalling. *Physiol Plant.* **100:** 241-254.

Foyer, C.H., Souriau, N., Perret, S., Lelandais, M., Kunert, K.J., Pruvost, C. and Jouanin, L. (1995). Overexpression of glutathione reductase but not glutathione synthetase leads to increases in antioxidant capacity and resistance to photoinhibition in poplar trees. *Plant Physiol.* **109:** 1047-1057.

Grant, C.M., MacIver, F.H. and Dawes, I.W. (1996). Glutathione is an essential metabolite required for resistance to oxidative stress in the yeast *Saccharomyces cerevisiae. Curr. Genet.* **29:** 511-515.

Grill, E., Winnacker, E.L. and Zenk, M.H. (1987). Phytochelatins, a class of heavy metal binding peptides from plants, are functionally analogous to metallothioneins. *PNAS.* **84:** 439-443.

Grill, E., Löffler, S., Winnacker, E.L. and Zenk, M.H. (1989). Phytochelatins, the heavy-metal-binding peptides of plants, are synthesized from glutathione by a specific γ-glutamylcysteine dipeptidyl transpeptidase (phytochelatin synthase). *PNAS.* **86:** 6838-6842.

Gyuris, J., Golemis, E., Chertkov, H. and Brent, R. (1993). Cdi1, a human G1 and S phase protein phosphatase that associates with Cdk2. *Cell.* **75:** 791-803.

Ha, S.B., Smith, A.P., Howden, R., Dietrich, W.M., Bugg, S., Connell, M.J.O., Goldsbrough, P.B. and Cobbett, C.S. (1999). Phytochelatin synthase genes from Arabidopsis and the yeast Schizosaccharomyces pombe. *Plant Cell.* **11:** 1153-1164.

Halliwell, B. and Foyer, C.H. (1978). Properties and physiologica function of a glutathione reductase puri?ed from spinach leaves by a?nity chromatography. *Planta.* **139:** 9-17.

Herschbach, C. and Rennenberg, H. (1991). Influence of Glutathione (GSH) on Sulphate Influx, Xylem Loading and Exudation in Excised Tobacco Roots. *J. Exp. Bot.* **42:** 1021-1029.

Herschbach, C. and Rennenberg, H. (1994). Influence of glutathione (GSH) on net uptake of sulphate and sulphate transport in tobacco plants. *J. Exp. Bot.* **45:** 1069-1076.

Hu, W.Q., Wang, T. and Meng, Q.L. (2004). Research advance of antioxidative bioactive peptides. *China Oils Fats.* **29:** 42-45.

Jaleel, C.A., Riadh, K., Gopi, R., Manivannan, P., Ines, J., Al-Juburi, A.J., Zhao, C.X., Shao, H.B. and Panneerselvam. R. (2009). Antioxidant defense responses: physiological plasticity in higher plants under abiotic constraints. *Acta Physiol. Plant.* **31:** 427-436.

Kingston-Smith, A.H., Harbinson, J. and Foyer, C.H. (1999). Acclimation of glutathione synthetase by buthionine sulfoximine (S-n_buty1 homocysteine sulfoximine). *J. Biol. Chem.* **102:** 42-51.

Koscy, G., Owttrim, G., Brander, K. and Brunold, C. (1997). Effect of chilling on the diurnal rhythm of enzymes involved in protection against oxidative stress in a chilling tolerant and a chilling sensitive maize genotype. *Physiol. Plant.* **99:** 249-254.

Kocsy, G., VonBallmoos, P., Suter, M., Rüegsegger, A., Galli, U., Szalai, G., Galiba, G. and Brunold, C. (2000). Inhibition of glutathione synthesis reduces chilling tolerance in maize. *Planta.* **211:** 528-536

Mai, W.J., Wang, Y., Liang, C.Y. and Zhang, M.Y. (2005). Glutathione and its stress tolerance in plants. *Guihaia.* **25:** 570-575.

Marrs, K. (1996). The functions and regulation of glutathione S-transferases in plants. *Ann. Rev. Plant Physiol. Plant Mol. Biol.* **47:** 127-158.

May, M., Vernoux, T., Leaver, C., Montagu, M.V. and Inze, D. (1998b). Glutathione Homeostasis in Plants: Implications for Environmental Sensing and Plant Development. *J. Exp. Bot.* **49:** 649-667.

Noctor, G. and Foyer, C.H. (1998a). Ascorbate and glutathione: Keeping active oxygen under control. *Ann. Rev. Plant Physiol. Plant Mol. Biol.* **49:** 249-279.

Noctor, G., Arisi, A.C., Jouanin, L., Kunert, K., Rennenberg, H. and Foyer, G.H. (1998b). Glutathione: biosynthesis, metabolism and relationship to stress tolerance explored in transformed plants. *J. Expe. Bot.* **49:** 623-647.

Noctor, G., Gomez, L., Vanacker, H. and Foyer, C.H. (2002). Interactions between biosynthesis, compartmentation and transport in the control of glutathione homeostasis and signaling. *J. Exp. Bot.* **53:** 1283-1304.

Peng, L.T., Wang, J., Li, L., An, L.S. and Zhang, J.L. (2004). Structure and Expression Analysis of the γ-glutamylcysteine Synthetase Gene in Rice. *J. Plant Physiol. Mol. Biol.* **30:** 533-540.

Polle, A. (2001). Dissecting the superoxide dismutase-ascorbate-glutathione-pathway in chloroplasts by metabolic modelling. Computer simulations as a step towards ?ux analysis. *Plant Physiol.* **126:** 445-462.

Potters, G., Horemans, N., Bellone, S., Caubergs, R.J., Trost, P., Guisez, Y. and Asard, H. (2004). Dehydroascorbate influences the plant cell cycle through a glutathione-independent reduction mechanism. *Plant Physiol.* **134:** 1479-1487.

Qian, Q.Q., Zai, W.S., Zhu, Z.J. and Yu, J.Q. (2006). Effects of Exogenous Silicon on Active Oxygen Scavenging Systems in Chloroplasts of Cucumber (*Cucumis sativus* L.) Seedlings under Salt Stress. *J. Plant Physiol. Mol. Biol.* **32:** 107-112.

Rauser, W.E. (1990). Phytochelatin. *Annu. Rev. Biochem.* **59:** 61-86.

Russo, T., Zambrano, N., Esposito, F., Ammendola, R., Cimino, F., Fiscella, M., Jackman, J., O'Connor, P.M., Anderson, C.W. and Appella, E. (1995). A p53-independent Pathway for Activation of WAF1/CIP1 Expression Following Oxidative Stress. *J. Biol. Chem.* **270:** 386-391.

Sanchez-Fernandez, R., Fricker, M., Corben, L.B., White, N.S., Sheard, N., Leaver, C.J., Van Montagu, M., Inze, D. and May, M.J. (1997). Cell proliferation and hair tip growth in the Arabidopsis root are under mechanistically di?erent forms of redox control. *PNAS.* **94:** 2745–2750.

Shao, H.B., Chu, L.Y., Lu, Z.H. and Kang, C.M. (2008a). Primary antioxidant free radical scavenging and redox signaling pathways in higher plant cells. *Int. J. Biol. Sci.* **4:** 8-14.

Shao, H.B., Chu, L.Y., Shao, M.A., Jaleel, C.A. and Mi, H.M. (2008b). Higher plant antioxidants and redox signaling under environmental stresses. *Comptes Rendus. Biol.* **331:** 433-441

Shaul, O., Mironov, V., Burssens, S., Van Montagu, M.V. and Inze, D. (1996). Two Arabidopsis cyclin promoters mediate distinctive transcriptional oscillation in synchronized tobacco BY-2 cells. *PNAS.* **93:** 4868-4872.

Smirnoff, N. (1998). Plant resistance to environmental stress. *Curr. Opin. Biotechnol.* **9:** 214-219.

Smith, I.K. (1985). Stimulation of glutathione synthesis in photorespiring plants by catalase inhibitors. *Plant Physiol.* **79:** 1044-1047.

Smith, I.K., Kendall, A.C., Keys, A.J., Turner, J.C. and Lea, P.J. (1984). Increased levels of glutathione in a catalase-de?cient mutant of barley (*Hordeum vulgare L.*). *Plant Sci. Let.* **37:** 29-33.

Suter, M., Von Ballmoos, P., Kopriva, S., Camp, R.O., Schaller, J., Kuhlemeier, C., Sch¨¹rmann, P. and Brunold, C. (2000). Adenosine 5´-Phosphosulfate Sulfotransferase and Adenosine 5´-Phosphosulfate Reductase are Identical Enzymes. *J. Biol. Chem.* **275:** 930-936.

Ullmann, P., Gonder, L. and Potier, S. (1996). Cloneing of the Arabidopsis thaliana glutathione synthetase (GSH_2) by functional complementation of a yeast gsh2 mutant. *Eur. J. Biochem.* **236:** 662-669.

Vernoux, T., Wilson, R.C., Seeley, K.A., Reichheld, J.P., Muroy, S., Brown, S., Maughan, S.C., Cobbett, C.S., Montagu, M.V., Inze, D., May, M.J. and Sung, Z.R. (2000). The root meristemless1/cadmium sensitive 2 gene defines a glutathione-dependent pathway involved in initiation and maintenance of cell division during postembryonic root development. *Plant Cell.* **12:** 97-110.

Willekens, H., Chamnongpol, S., Davey, M., Schraudner, M., Langebartels, C., Van Montagu, M., Inze, D. and Van Camp, W. (1997). Catalase is a sink for H_2O_2 and is indispensable for stress defence in C_3 plants. *EMBO J.* **16:** 4806-4816.

Wildi, B. and Lutz, C. (1996). Antioxidant composition of selected high alpine plant species from different altitudes. *Plant Cell Environ.* **2:** 138-146.

Wingate, V.P.M., Lawton, M.A. and Lamb, C.J. (1988). Glutathione Causes a Massive and Selective Induction of Plant Defense Genes. *Plant Physiol.* **87:** 206-210.

Xiang, C.B. and Oliver, D.J. (1998). Glutathione Metabolic Genes Coordinately Respond to Heavy Metals and Jasmonic Acid in Arabidopsis. *Plant Cell.* **10:** 1539-1550.

Xiang, C.B., Werner, B.L., Chritensen, E.M. and Oliver, D.J. (2001). The biological functions of glutathione revisited in Arabidopsis transgenic plants with altered glutathione levels. *Plant Physiol.* **126:** 564-574

Xu, D.Y. and Song, Y.G. (2001). Glutathione regulate the signal transduction by the ROS. *Heath Toxicology.* **15:** 52-54.

Zetterstrm, R., Eijkman, C. and Hopkins, S.F.G. (2006). The dawn of vitamins and other essential nutritional growth factors. *Acta Paediatrica.* **95:** 1331-1333.

9

Heavy Metals and Oxidative Stress in Plants

Dmitrij I. Bashmakov[1], **Mikhail E. Stepanov**[1], **Naser A. Anjum**[2], **Alexander S. Lukatkin**[1*]

[1]*Department of Botany and Plant Physiology, Mordovian N.P.Ogariov State University, Bolhevistskaja Str., 68. Saransk–430 005, Russia*

[2]*Academia Sinica, Agricultural Biotechnology Research Center, Taiwan; Division of Plant Physiology and Plant Biochemistry, Department of Botany, Aligarh Muslim University, Aligarh–202 002, India;*

Present address: Center for Environmental and Marine Studies (CESAM) & Department of Chemistry, University of Aveiro, 3810-193 Aveiro, Portugal

[*]***Corresponding author:*** *aslukatkin@yandex.ru*

ABSTRACT

Heavy metal (HM) stress is one of the important abiotic stress factors affecting flora and fauna on the globe. The concentration of HMs is gradually increasing through rapid increase in industrialization. In addition, to a number of morphological and physiological changes HMs cause oxidative stress in plants and lead to activate the generation of reactive oxygen species (ROS). ROS has already been shown to severely affect plant metabolism to a great extent and disturb normal vital cellular functions. However, plants possess an efficient antioxidant defence system to counteract the ROS-induced anomalies. The present chapter reviews the available information on important HMs, their occurrence and toxicity in plant system in detail. In addition, the chapter also throws light on the responses of various components of antioxidant system in plants growing under HMs stress condition.

Keywords: Heavy metals, oxidative stress, reactive oxygen species, antioxidants, antioxidative enzymes.

1. INTRODUCTION

The contamination of environment with heavy metals has started to sharply increase since 1900 (Nriagu 1979) and be of a serious problem for biosphere and human health on the globe (Ensley 2000). In fact, HMs are elements with atomic mass above 50, possessing properties of metals and having density over than 5 $g\cdot cm^{-3}$. They are characterized by variable valence, low solubility of their oxyhydroxides, high ability to form complexes.

It is considered that HMs are the most toxic among chemical elements (Prasad 2006). They invoke infringements in plants at cell, tissue, organism, population levels (Doncheva 1998; Briat and Lebrun 1999; Seryogin and Ivanov 2001). Unlike organic pollutants, HMs cannot be removed from the environment by natural chemical or biological transformation and cannot be degraded in the biosphere (Prasad and Freitas 2006).

Gravity of HMs accruing to biosphere consists that: (1) HM enter into an organic matter cycles for a long time, variate geochemical constants, enhance or liquidate natural geochemical anomalies and create new technogenic anomalies, (2) speed of entered anthropogenic HM resumes natural migration rates many times, (3) HM technogenic accumulation in soils descends non-uniformly. With some errors, it is possible to consider that HM intake to the environment is proportional to the population density. Therefore, technogenic dissipate of HM in the environment exceeds their connatural concentrations in soils in hundreds or thousands times within the most densely populated territories.

Plants constantly sustain unfavourable environmental conditions. Action of unfavourable environmental factors leads to occurrence in plant cells of a complex of the reactions united by the term oxidative stress (Scandalios 1990). In these conditions, there is an uncontrollable enhancement of reactive oxygen species (ROS) generation. Oxidative processes occur in cellular compartments at normal conditions always; where, antioxidative system keeps the amount of ROS at stationary level. The last one includes low-molecular weight antioxidants (such as, amino acids, carotenoids, polyamines, glutathione, ascorbate, tocopherols, etc.) and antioxidative enzymes (mono- and dehydroasc orbate reductases, guaiacol peroxidase, glutathione peroxidase, glutathione reductase, ascorbate peroxidase, catalase and superoxide dismutase) (Pich and Scholz 1993; Herbick et al., 1996) and glutathione reductase (GR) (Schickler and Caspi 1999). The interrelationship of prooxidants and antioxidants vary according to environmental stress factors. It is supposed that shift of this interrelationship starts reactions of plant cell damage. ROS are capable to initiate reactions of proteins degradation and lipid peroxidation of membrane lipids leading to breaking of cellular metabolism and finally the cell death. However, various components of antioxidant defence system protect plant cells against oxidative stress (Caldwell 2001). In addition, efficiency of components of plant antioxidant system has been shown to enhance detoxification of ROS-induced oxidative damage through use of plant mineral nutrients, growth regulators and various biotechnological approcahes (Anjum et al., 2008a-c; Lukatkin et al., 2007; Anjum et al., 2010).

1.1 Occurrence of HMs in Soil-Plant System

HMs enter into soils naturally (with volcanic ash, from ore bodies occurring close to the surface), and also by various anthropogenic ways (from open pits and ore mines, from the metallurgical factories, from the thermal power station) (Prasad 2006). HMs may be available to plants through washout by rainfall (Linzon et al., 1976; Beckett et al., 1978) and phytovolatilization (Briat and Lebrun 1999) as Paribok et al. (1983) have shown that HM may be present in volatile condensates (complexes with terpenes) of coniferous trees. A large amount of HMs-input to plant-soil system is through various anthropogenic activities. Anthropogenic ways of input of HMs to biosphere exceeds tenfold over the natural ways of HMs-input (Prasad and Freitas 2006). HM technogenic spreading during mineral fuel combustion has started long before these elements have begun to use. HM accumulation descends in humus horizons of soils. Basic soil pools of HM are fine grained mineral fractions, humus, and Fe and Mn oxides, partly. HM absorbed by roots from the deepest soil horizons, is passed to humus with tree waste. A migratory ability of HM ions in soil are varied and depend on concrete physiographic conditions (sorption barriers, pH changes, etc.) (Prasad 2006).

The main edaphic factors of sorption and accumulation of HMs by plants are the mechanical composition and organic matter of soils, water and carbonate content in soil, pH of soil solutes and metal physiological activity (Barton et al., 2000; Seryogin and Ivanov 2001; Prasad 2006). HM content in plants is caused by physiological features of plants. An organism's physiological state is its ancestral feature, and plants in any conditions take up HMs that is necessary or confused for its necessity for them, even from poor soil (Schickler and Caspi 1999). Ions of HMs absorbed by roots can be loaded into the xylem exudates and moved to overground plant parts with transpiration stream. Various amino acids and some low-molecular weight substances (such as, nicotineamide) are main potential HM chelators (Cataldo et al., 1988; Kramer et al., 1996). It is observed that histidine participates in long-range HM transport (Kerkeb and Kramer 2003). In addition, the translocation of HMs from soil to plant system is governed by various factors including plant roots which mainly regulate HM sorbtion (Coughtrey and Martin 1979; Patra and Sharma 2000). This process is connected to cell wall negatively or positively charged sites (Briat and Lebrun 1999). In this regard, rhizosphere plays a key role in the sorption of HM-ions; whereas endodermis actively controls their transport to stellar region. Casparian strips cells carry out the barrier function in roots, they restrict water, ions and low-molecular mobility, and limit their transfer in vascular tissues (Parker et al., 2001; Woitke et al., 2001). The sequestration of HMs is basically done in vacuole. The cellular walls separate HMs-ions and restrict their inclusion to endocellular metabolic sites and prevent their binding with SH-, N- and O-containing groups (Turner and Marshall 1971). The tonoplast functions as an effective selective barrier for metals diffusion (Salt and Rauser 1995). Compartmentation to vacuoles is most typical at HM excess (Frey et al., 2000). Nonspecific, HM contact with organic acids—malonic, succinic (Parker et al., 2001; Doncheva et al., 2001), histidine and cysteine (Doncheva et al., 1998), create resistant complexes with carbohydrates and other organic compounds and play a significant role as a 'metal's ions transmitting agent' from cytoplasm to vacuoles (Pelosi et al., 1976). The chelation is HM immobilisation in the strong solvable complexes. Thus, metals concentration in the most active ionic form drops considerably. Most HM toxicity in the chelates form is lower than of its salts (Pelosi et al., 1976). It is known that at excess HM in medium the plant cells actively synthesize metal-binding cytoplasmic proteins (metallotioneins, MTs and phytochelatins, PCs) (Rauser 1999; Cobbett 2000; Clemens 2001). However, it is shown, that PCs are not involved in resistance to some HMs (as nickel). But histidine (nickel chelator) is involved in nickel tolerance and detoxication in symplast (Kramer et al., 1996; Kerkeb and Kramer 2003). Except for histidine, low molecular weight organic acids participate in HM chelation. For example, citric acid has greater HM chelation capacity (Sagner et al., 1998).

1.1.1 Important HMs and Their Effects on Various Physiological Functions in Plants

1.1.1.a Cobalt (Co)

It is concentrated in generative organs and accumulated in anther dust, accelerating pollen-grain germination (Smirnoff 1989). Cobalt enhances protein biosynthesis, controls growth processes, removes auxin inhibiting action on a cellular reproduction, influences quantity of chloroplasts and inhibits ethylene biosynthesis (Hevitt 1983). Co is a part of vitamin B_{12}, raises immune resistance to some illnesses (at the expense of methylation), Co-amide coenzymes participate in DNA synthesis and in cellular division (Smirnoff 1989). Co stimulates reproduction of nodule bacteria, participates in leghemoglobin synthesis, activates dehydrogenases, hydrogenase, nitrate reductase, the maintenance of a chlorophyll and the general hematin increases (Polevoi, 1989).

1.1.1.b Nickel (Ni)

Nickel is a chemical analogue of Co. It has non-specific effect on some metal-enzymatic complexes. It activates arginase, oxaloacetate decarboxylase, nitrate reductase, a number of peptidases, stabilises ribosomes structure (Voinar 1962). It influences nutrients acquisition and transport in plants (Saprykin 1984). Lack of nickel results in metabolic destruction of a nitrogen metabolism through destructive acetylurea and amino acids metabolism, and ornithine cycle intermediate joints. It leads to accumulation of xanthine, allantoic acid, ureidoglycolates and citrulline in leaves, but total concentration of acetylurea, urea and urease activity reduces. Amino acids metabolism destruction results in accumulation of glycine, valine, isoleucine, thyrosinum, indole amino-propionic acid, arginine and total of free amino acids, and to decrease in concentrations of histidine and glutamic acid. Deficiency of nickel also breaks a citric acid cycle, and breath. Destruction of a metabolism of Carboneum descends through accumulation lactic and oxalic acids. Results specify, that "mouse ear" is key morphological sign in leaves which is possibly bound to toxic accumulation oxalic and lactic acids in quickly growing tips and along the edges of leaves (Bai et al., 2006). High Ni maintenance in cells of endodermis and pericycle was the cause of cellular divisions inhibition in a pericycle and as result was the cause of root branching stops (Seregin and Kojevnikova 2006). Excess of nickel in root apex reduces meristem mitotic activity, that conducts to absence of root meristem wholeness (Robertson 1985; L'Huillier et al., 1996). In chickpea seeds sugar content increases at augmentation of Ni dose. The protein also was depressed at augmentation of a metal's doses, except for initial Ni dose (0,25 mM) (Tandon and Srivastava 2004).

1.1.1.c Copper (Cu)

Copper promotes chlorophyll formation, is a part plastocyanine, is the activator of the ferments connected to redox reactions in plant cells, influences nitrogenous metabolism, being a part nitrite reductase, hyponitrite reductase and nitrogen oxide reductases, influences leghemoglobin synthesis, enhances process of nitrogen molecules binding, reduces chlorophyll destruction in darkness (Polevoi, 1989). Copper-bearing enzymes inactivate auxin (polyphenol oxidase), oxidised thyrosinum to melanin (tyrosinase) (Smirnoff, 1989). Cu raises drought, frost and heat resistance, promotes manganese, zinc, boron uptake to plant (Alekseeva-Popova, Il'inskaya 1983). The copper lack invokes a growth inhibition, turgor loss, a flowering detention, albinism (Smirnoff 1989). An excess of Cu^{2+} disturbs membranes permeability (Babalakova and Traykova 1997). Excess of Cu^{2+} interferes with cells transferring to mitosis, inhibites the photosystem II, reduces thylacoides quantity, inactivates ribulose-1,5-diphosphate carboxylase, reduces chlorophyll concentration in leaves and invokes a reduction of thylakoides membranes (Doncheva 1998; Bernal et al., 2004; Lidon 1999; Pätsikkä et al., 2002). Peach plants treatment with Cu invokes apoptosis and supersensitivity reaction in peach leaves (Goto et al., 2005). In roots of *Panax ginseng* at 50 μM of Cu^{2+} the maintenance of malonic dialdehyde, activity of lipoxygenase, generation of superoxide anion and H_2O_2 maintenance increased, resulting in ascorbate and glutathione oxidation, specifying in oxidative stress, activity of catalase and guaiacol peroxidase was inhibited. Changes in peroxide oxidation intensity, the maintenance of glutathione and activity of antioxidative ferments testify that Cu toxicity can be involved in oxidative damage (Babar et al., 2006; Devi and Prasad, 2005). It is shown, that Cu^{2+} at 60 mg/kg of soil increase polyphenol oxidase activity in roots of *Prunus armeniaca,* but reduce its activity in other organs (Dongmei et al., 2004). Activity of superoxide dismutase, catalase and polyphenol oxidase and peroxidases raised with growth of Cu^{2+} ions concentration (from 5 to 40 μM) (Dong-Hang et al., 2007). Excess of copper in a growth medium does not invoke serious oxidative stress, crash of antioxidative protection. It is shown, that in the presence of ions Cu^{2+} the MDA maintenance in leaves decreased, activity of SOD, CAT and peroxidases activity slightly increased, but the system of protective ferments remained balanced (Pätsikkä et al., 2002).

1.1.1.d Cadmium (Cd)

Cadmium is chemical analogue of Zn, it is capable to substitute it in the enzymatic system participating in carbohydrates development and breaking up (Zimakov, 1981). Cd is a nonessential element. However, low temperatures and metabolism inhibitors reduce cadmium transport through a plasma membrane testifying that absorption of Cd^{2+} is bound to a plant metabolism. The gene responsible for encoding of the vacuolar trensporter of Cd-phytochelatin complexes was sequenced (Ortiz et al., 1992). Cd toxicity mainly leads to chlorosis and severe plant growth reductions. Cadmium negatively influences practically all physiological processes of a plant: growth, water balance, photosynthesis. It alters a metastructure of chloroplasts, reduces plastid pigments concentration, inhibits chlorophyll biosynthesis, blocks transport of electrons in PS II, and also in PS I at concentrations more than 50-75 mM (Franco et al., 1999; Kuznetsova et al., 2008). Cadmium negatively influences on photosynthesis dark reactions (especially on binding of CO_2), alter some key enzymes of Calvin cycle (Vassilev et al., 1997; Vassilev and Manolov 1999). After the 48-h of treatment the ultrastructural organisation of chloroplasts in mesophyll cells of *Lemna minor* as a result of descending destructive changes considerably variated. Under the same conditions of experiment chloroplasts of *Salvinia natans* showed a high level of structural stability, irrespective of Cd concentration and treatment exposition (Stoyanova and Baldjiev 2002). Cd influence on plants water exchange. So, Cd treatment increased accumulation of osmoprotector proline in sheet and root tissues (Panda 2001). Cd variate the hormonal status of plants. So, Cd^{2+} inhibites indoleacetic and gibberellic acids synthesis in roots and stimulated zeatin and abscisic acid synthesis not only in roots, but also in above-ground parts. Cd^{2+} at low concentrations stimulated indoleacetic and gibberellic acids synthesis in above-ground parts (Yun-xrand et al., 2006). Cd^{2+} variates activity of antioxidative enzymes: catalase, ascorbate peroxidase, guaiacol peroxidase, superoxide dismutase and others (Fang-ming et al., 2007; Chen et al., 2007; Fornazier et al., 2002; Cordova et al., 2003; Xue-zhi et al., 2004; Israr et al., 2006). Cd leads to ROS and lipids peroxidation increase (Yan, Hong et al., 1998; Chen et al., 2000). The usual response to cadmium is the induction of specific metabolites synthesis: synthesis: amino acids (proline), organic acids, phytochelatines and metallotioneines (Maier et al., 2003; Knecht et al., 1994; Lee, Leustek, 1999; Zhang et al., 2000).

1.1.1.e Chromium (Cr)

Cr^{4+} stimulates the maintenance of a chlorophyll and carotenoides while Cr^{3+} is more toxic (Panda and Patra 1998). Cr^{4+} and Cr^{3+} inactivate nitrate reductase (Panda, Patra, 2000). With augmentation of CrO_3 concentration the maintenance of proline, the solvable protein and malonic dialdehyde increased, and activities of SOD, peroxidases and catalases gradually raised (Xiqin and Qianhua, 2005). Cr^{6+} reduces anabolic abilities of leaves cells, invoking symptoms of their "starvation" (Xu, Shi et al., 2000). Ions of Cr^{6+} inhibited growth of a root and damaged DNA, in wheat seedlings cells (Shu-mei, Xiao-ping, 2005). $Cr_2O_7^{2-}$ reduces concentration of inorganic phosphorus, a biomass of plants the chlorophyll maintenance and activity of catalase and peroxidase, simultaneously raising activity of acid phosphatase and ribonuclease (Sharma and Sharma 1996). Low ion densities Cr^{4+} and Cr^{3+} increase activity of peroxidase (Panda and Patra 2000). Plants of *Convolvulus arvensis* treated with Cr (VI) at concentrations from 0 to 40 mg/l showed higher ascorbate peroxidase specific activity in roots, than in shoots. Cathalase activity in shoots decreased almost right after the treatment beginnings (Montes-Holguin et al., 2006). Now it is proved that the strontium exchange is interconnected with a calcium exchange (Kretovitch, 1986).

1.1.1.f Lead (Pb)

In small amounts Pb raises starch maintenance, accelerates plants germination (Saprykin 1984; Smirnoff 1989; Yagodin 1989). In concentration from 0,1 to 2,4 mM Pb is capable to inhibit photosystem II activity at an oxidation stage. Pb influences plant water absorption, inhibits respiration, chlorophyll biosynthesis and mineral nutrition (Obuhov 1987; Kozarenko 1987; Singh et al., 1996). At Pb^{2+} treatment the concentration of chlorophyll and root activity dropped (Li-hong et al., 2006). Pb inhibits absorption of cations (To^{+}, $Cà^{2+}$, Mg^{2+}, Mn^{2+}, Zn^{2+}, Cu^{2+}, Fe^{3+}), and anions (NO_3^-), breaks potassium and calcium transport through plasma membranes (Seryogin and Ivanov 2001; Rucinska and Gwozdz 2005). As a result of Pb and Ca antagonism, Ca^{2+} ion concentration in cells decreases, and calmodulin does not activate Ca-ATPase. Through the calmodulin-mediated reactions lead interferes with formation of a spindle of division and normal mitosis, induces formation of two-nuclear cells (Jiang and Liu 2000). High concentrations of lead and manganese strongly complicate growth of mycorrhizal fungy hyphas, however, this effect is eliminated at root exudate addition to the medium (Malcová and Gryndler 2003). High concentration of Pb induced formation of adventive roots. Maintenance of abscisic acid in roots increased with Pb concentration in leaves (Su-chun et al., 2006). Lead increases the ABA and zeatine maintenance. In germinating seeds in the presence of Pb ions maintenance of gibberellic acid decreases (Atici et al., 2005). Quantity of peroxidase, superoxide dismutases and catalases isoenzymes and amount of expressed peroxidases gradually dropped with increasing of Pb concentration. Activity of superoxide dismutase increased with the increasing of Pb concentration from 0 to 500 mg/l, reaching a maximum on 7th day, then was quickly depressed (below the control). At Pb concentration of 900 mg/l SOD activity dropped in due course. Quantity of isoenzymes and expressed SOD have been negatively bound to concentration of Pb (Suchun et al., 2006). In cells activity of SOD gradually increased at low concentrations (< 2,5 mM/l) of Pb^{2+} ions. At high concentrations (> 10 mm/l) of Pb^{2+} ions activity of SOD increased in the beginning, and after 4 days was depressed, with increasing of ROS induced by lead and denaturation of detoxifying enzymes (Hui-zhong et al., 2006).

1.1.1.g Mercury (Hg)

Low concentrations of Hg^{2+} ions (0,1 mg/l) during short time can stimulate seeds germination and seedlings growth (Lan-fang et al., 2004). At concentration of 0,1 mkg/l mercury oppresses photosynthetic activity, growth and cell division. Hg inhibits transmembrane transport of water, breaking water balance (Wan, Zwiazek 1999). Principal causes of mercury toxicity are change of cellular membranes permeability, reactions with sulfhydric bunches, phosphatic bunches of ADP and ATP and replacement of essential cations (Kositsin and Alekseeva-Popova 1983). As the majority of proteins contain sulfhydryl groups or disulphide bridges, practically Hg can disturb any function in which proteins are involved. Mercury ions can distort a protein macromolecule or bridge two chains together. Hg can be bound with DNA and RNA (Patra and Sharma 2000; Lan-Fang et al., 2004; Rudakova et al., 1987). Mercury inhibits transport of electrons and chlorophyll fluorescence in a photosystem II, a dark repair of cytochrome C_{554}, inhibits ferredoxin and a ferredoxin-NADP-reductase, is capable to substitute Mg in a chlorophyll molecule, inhibits catalase at water photolysis (Lenti et al., 2002; Patra and Sharma 2000). At Hg treatment the mitotic index of root apical meristem drops and sharply the number of chromosomes aberations because of disorganisation of microtubules (the least efficiency Hg concentration is 10 pM) (Patra and Sharma 2000). Ions of Hg^{2+} damage membranes of chloroplasts and thylakoids: the nuclear membrane breaks up and nucleoluses disappear, also mitochondrial membranes are damaged and gradually degrade (Yi-qun et al., 2004). Increasing Hg^{2+}concentrations inactivate a superoxide dismutase and other antioxidative enzymes that results in ROS accumulation in plant tissues (Hao et al., 2000). Hg^{2+} ions activate antioxidative enzymes. The augmentation of Hg^{2+} ions concentration essentially increased peroxidase activity. However, activity of peroxidase dropped at 100 mg/l of Hg^{2+} ions (Yi-xian, Li-ping, 2005, Lan-fang et al., 2004). At simultaneous mercurating and other metals the last cannot compete to Hg for binding sites on enzymes because of higher affinity of mercury to organic ligands and stability of chelates (Patra and Sharma 2000).

2. OXIDATIVE STRESS IN PLANTS

2.1 General Aspects

Oxidative stress occurs when there is a serious imbalance in any cell compartment between the production of reactive oxygen species (ROS) and antioxidant defence, leading to dramatic physiological challenges (Foyer and Noctor 2000). During the reduction of O_2 to H_2O, there may transfer of one, two or three electron(s) to O_2 from the ROS superoxide ($O_2^{\bullet-}$), hydrogen peroxide (H_2O_2), and the hydroxyl radical ($^{\bullet}OH$), respectively. $O_2^{\bullet-}$ and H_2O_2 have a relatively low reactivity in comparison with other ROS. Acceptance of excess energy by O_2 can additionally lead to the formation of singlet oxygen ($^1O_2$), a highly reactive molecule compared with O_2.

2.2 Sites of ROS Production

The major sites of ROS synthesis are the reaction centres of PSI and PSII in chloroplast thylakoids (Edreva 2005; Asada 2006). Superoxide radical is formed through univalent photoreduction of molecular oxygen in PSI, either directly or via ferredoxin. Spontaneous or enzymatic disproportionation of $O_2^{\bullet-}$ leads to H_2O_2 production. In PSII, energy transfer from the triplet state of excited chlorophyll to oxygen of the ground state results in production of $^1O_2$ (Edreva 2005; Foyer et al., 2006). Besides photosynthesis, mitochondrial respiration is also believed to be an important source of ROS production in plants. In complexes I and III of the respiratory chain single electrons can be transferred to molecular oxygen, leading to the generation of $O_2^{\bullet-}$ (Foyer et al., 2006). Considerable amounts of H_2O_2 are formed through P-oxidation of fatty acids and the action of glycolate oxidase in the photorespiratory pathway in peroxisomes (Mittler 2002; Foyer and Noctor 2003). Reactive oxygen species are considered to play a dual role in plant cells (Dat et al., 2000). Most of them show high reactivity towards various macromolecules and therefore excess accumulation of ROS in the tissues may result in the oxidative damage to the cell structure. Aside from their destructive nature, ROS as ubiquitous messengers of stress responses are likely to play a signalling role in various adaptive processes (Apel and Hirt 2004; Pitzschke et al., 2006), into appropriate cellular responses with the help of some redox-sensitive protein (Bhattacharjee 2005; Vranova et al., 2002). Moreover, plants may use ROS as second messengers in signal transduction cascades in processes as diverse as mitosis, tropisms and cell death, and their accumulation is crucial to plant development as well as defence (Foyer and Noctor 2005) and immune response (Kovalchuk et al., 2003) (Tables 1-2).

Table 1: Reactive oxygen species and their sources in plants. (After Mittler 2002)

Compound	Shorthand notation(s)	Structural representation(s)	Sources
Molecular oxygen (triplet ground state)	O_2, $^3\Delta$	$\ddot{O}=\ddot{O}$ $1s^2 2s^2 (\sigma_s)^2 (\sigma_s^*)^2 (\sigma_x)^2 (\pi_y)^2 (\pi_z)^2 (\pi_y^*)^1 (\pi_z^*)^1$	Most common form of dioxygen gas
Singlet oxygen (first excited singlet state)	$^1O_2$, $^1\Delta$	:O=O: $1s^2 2s^2 (\sigma_s)^2 (\sigma_s^*)^2 (\sigma_x)^2 (\pi_y)^2 (\pi_z)^2 (\pi_y^*)^2$	UV irradiation, photoinhibition, photosystem II e^- transfer reactions (chloroplasts)
Superoxide anion	$O_2^{\bullet-}$	[:O ∓ O:]⁻	Mitochondrial e^- transfer reactions, Mehler reaction in chloroplasts (reduction of O_2 by iron-sulphur centre F_x of Photosystem I), glyoxysomal photorespiration, peroxisome activity, plasma membrane, oxidation of paraquat, nitrogen fixation, defence against pathogens, reaction of O_3 and OH^- in apoplastic space
Hydrogen peroxide	H_2O_2	H–:O=O:–H	Photorespiration, P-oxidation, proton-induced decomposition of $O_2^{\bullet-}$, defence against pathogens
Hydroxyl radical	$^{\bullet}OH$	: O·H	Decomposition of O_3 in presence of protons in apoplastic space, defence against pathogens
Perhydroxyl radical	O_2H^-	:O ∓ O–H	Reaction of O_3 and OH^- in apoplastic space
Ozone	O_3	:O=O⁺–:O:⁻	Electrical discharge or UV radiation in stratosphere, reactions involving combustion products of fossil fuels and UV radiation in troposphere

Table 2: Production, scavenging and avoidance of ROS in plants (After Mittler et al., 2004).

Mechanism	Localization	Primary ROS
Production		
Photosynthesis ET and PSI or II	Chl	$O_2^{\bullet-}$
Respiration ET	Mit	$O_2^{\bullet-}$
Glycolate oxidase	Per	H_2O_2
Excited chlorophyll	Chl	$^1O_2$
NADPH oxidase	PM	$O_2^{\bullet-}$
Fatty acid β-oxidation	Per	H_2O_2
Oxalate oxidase	Apo	H_2O_2
Xanthine oxidase	Per	$O_2^{\bullet-}$
Peroxidases, Mn^{2+} and NADH	CW	H_2O_2, $O_2^{\bullet-}$
Amine oxidase	Apo	H_2O_2
Scavenging		
Superoxide dismutase	Chl, Cyt, Mit, Per, Apo	$O_2^{\bullet-}$
Ascorbate peroxidase	Chi, Cyt, Mit, Per, Apo	H_2O_2
Catalase	Per	H_2O_2
Glutathione peroxidase	Cyt,	H_2O_2, ROOH
Peroxidases	CW, Cyt, Vac	H_2O_2
Thioredoxin peroxidase	Chl, Cyt, Mit	H_2O_2
Ascorbic acid	Chi, Cyt, Mit, Per, Apo	H_2O_2, $O_2^{\bullet-}$
Glutathione	Chi, Cyt, Mit, Per, Apo	H_2O_2
α-Tocopherol	Membranes	ROOH, $^1O_2$
Caretenoids	Chl	$^1O_2$
Avoidance		
Anatomical adaptations	Leaf structure epidermis	$O_2^{\bullet-}$, H_2O_2, $^1O_2$
C_4 or CAM metabolism	Chl, Cyt, Vac	$O_2^{\bullet-}$, H_2O_2
Chl movement	Cyt	$O_2^{\bullet-}$, H_2O_2
Suppression of photosynthesis	Chl	$O_2^{\bullet-}$, H_2O_2
PS and antenna modulations	Chl	$O_2^{\bullet-}$, $^1O_2$
Alternative oxidases	Chl, Mit	$O_2^{\bullet-}$

Abbreviations: Apo, apoplast, Chl, chloroplast, CW, cell wall, Cyt, cytosol, ET, electron transport, Mit, mitochondria, $^1O_2$, singlet oxygen, Per, peroxisome, PM, plasma membrane, PS, photosystem, Vac, vacuole

2.3 ROS and Their Effects in Plant Cells

Plants exposed to biotic and abiotic stresses generate more ROS than their capacity to scavenge them which lead to the activation of both local programmed cell death and including the activation of various components of plant antioxidant defence system (Pandhair and Sekhon 2006).

Results of numerous studies provide evidence for HM-induced oxidative stress in plant tissues. Enhanced production of oxidized macromolecules such as lipid peroxides has been reported for many plants subjected to HM stress (Hao et al., 2006; Gajewska and Sklodowska 2007a; Prasad et al., 2005). It is generally considered that virtually all the biochemical effects of heavy metals may ultimately lead to damage of cells and tissues (Shaw et al., 2004). Lipid peroxidation is one of general factors which tap severity of oxidative stress occurring in cells. Lipid peroxides are formed as a result of the oxidation of polyunsaturated fatty acids by $^1O_2$, $^{\bullet}OH$ and HO_2, the protonated form of $O_2^{\bullet-}$ (Mano et al., 2002; Temple et al., 2005). In addition to the direct reaction of ROS with fatty acids lipid peroxides may be also produced enzymatically by lipoxygenase (LOX, EC 1.13.11.12). Lipid peroxidation under heavy metal stress might result from the increase in LOX activity (Gallego et al., 1996; Weckx and Clijsters 1997). Recently, induction of this enzyme activity accompanied by the accumulation of lipid peroxidation products has been found in the roots and leaves of HM-treated plants (Saeidi-Sar et al., 2007). Peroxidation of membrane lipids contributes to the breakdown of the functional and structural integrity of plasma membranes. This process is considered to be responsible for the injury of plasmalemma and organelle membranes. The results obtained by Molas (1997) implied the involvement of lipid peroxidation in damage of chloroplast thylakoid membranes in the leaves of HM-exposed plants. This author found that Ni was associated with chloroplast membranes, which could facilitate the toxic effect of this metal on the studied structures. Direct reaction of ROS with unsaturated lipids has been suggested to result in alterations in lipid composition, particularly the decrease in unsaturated fatty acid levels (Ouariti et al., 1997). Apart from peroxidation of unsaturated fatty acids ROS are also able to stimulate de-estrification of phospholipids, which may also result in modification of the membrane structure (McKersie et al., 1988). Excess HM concentrations were shown to induce changes in the lipid composition of plasma membranes and to enhance membrane Mg^{2+}-ATPase activity in rice shoots (Ros et al., 1990).

Malondialdehyde (MDA) content is often used as an indicator of oxidative degradation of cell membranes. Increase in MDA content in response to HM treatment was found (Sinha and Pandey 2003; Saeidi-Sar et al., 2007; Hao et al., 2006; Gajewska and Sklodowska 2007b; Gomes-Junior et al., 2006). It was reported that the enhancement of lipid peroxidation found in the shoots and roots of HM-exposed maize occurred before these organs revealed decrease in growth (Baccouch et al., 2001). According to these authors lipid peroxidation could be one of the primary phytotoxic effects of HM on membrane integrity resulting in suppression of growth. Although lipid peroxidation is considered to be an important factor of growth restriction, enhancement of lipid peroxides content is not essential for HM-induced reduction in growth. Membrane damage arising from increased lipid peroxidation is probably responsible for HM-evoked alteration of plant water status. HM was reported to decrease tissue water content in wheat (Pandolfmi et al., 1992; Gajewska et al., 2006a) and pea (Gabbrielli et al., 1999). Apart from induction of lipid peroxidation HM are also implicated in oxidative DNA damage, which is considered to be responsible for mutagenic effects of this metal (Gajewska 2009).

2.4 Antioxidative System in Plants

To maintain the crucial balance between generation and removal of ROS plants possess an efficient antioxidant system thus to continue undisturbed functioning of cells (Schitzendubel and Polle 2002; Kovalchuk et al., 2003; Gratao et al., 2005; Pitzschke et al., 2006). The battery of plant antioxidant system comprises enzymatic antoxidants such as superoxide dismutase (SOD), ascorbate peroxidase (APX), glutathione reductase (GR), peroxidase (POX), catalase (CAT), and the low molecular weight non-enzymatic antioxidants like ascorbic acid (AsA) and glutathione (GSH) (Chen et al., 2000; Schützendübel et al., 2001).

2.4.1 Enzymatic antioxidants

The major ROS-scavenging mechanisms of plants include enzymes such as superoxide dismutase (SOD, EC 1.15.1.1), catalase (CAT, 1.11.1.6), ascorbate peroxidase (APX,), glutathione reductase (GR) and glutathione peroxidase (GPX, EC 1.11.1.9). Cell oxidative stress levels are determined by the amounts of $O_2^{\bullet-}$, H_2O_2 and $^{\bullet}OH$ radicals (Foyer and Noctor 2003). H_2O_2 can be directly metabolized by peroxidases, particularly those in the apolplast (Zoller et al., 2003), and by CAT in the peroxisome (del Rio et al., 2002; Igamberdiev and Lea 2002). Though, the expression for antioxidant enzymes is altered under stress conditions, their upregulation has a key role in combating the abiotic stress-induced oxidative stress. However, the level of upregulation is subjected to type, and magnitude of the stress.

2.4.1.a Superoxide dismutase (SOD, EC 1.15.1.1)

Superoxide dismutases, located in various cell compartments (Cakmak and Horst, 1991) especially in chloroplasts, are responsible for the dismutation of $O_2^{\bullet-}$ into H_2O_2 and O_2, forming the first line defence against ROS, and influencing the concentration of $O_2^{\bullet-}$ and H_2O_2, the two Haber-Weiss reaction substrates. The abiotic stress-induced production of superoxides might take place at many locations where an electron transport chain proceeds, and hence their activation has been detected in different cellular compartments (Elstner 1991). Presence of SODs is crucial for the removal of superoxides produced in the compartments in time (Takahashi and Asada 1983).

Superoxide dismutase (SOD) has been considered to play the most significant role in active oxygen species scavenging because its action prevents the accumulation of $O_2^{\bullet-}$ radical, which could lead to generation of toxic $HO^{\bullet}$ (Shaw et al., 2004). It is also the most widely studied enzyme in the context of environmental stresses (Breusegem et al., 1999; Thomas et al., 1999; Scandalios 1997). SOD isozymes are compartmentalized in higher plants and play a major role in combating oxygen radical mediated toxicity (Gratao et al., 2009). Three distinct types of SOD isoenzymes have been detected in plants, which are classified according to their metal cofactor, Mn, Fe and Cu/Zn (Alscher et al., 2002). Mn-SODs are located in the mitochondria and peroxisomes and have been reported in pea, watermelon, cucumber, tomato and other plant species (del Rio et al., 2002, 2003). Fe-SOD has been shown to be associated with the chloroplasts (Alscher et al., 2002), while the Cu/Zn-SODs are located in the cytosol, chloroplasts and peroxisomes (del Rio et al., 2002). Recently, a distinct SOD class that contains Ni (Ni-SOD) was discovered in cyanobacteria and *Streptomyces* (Barondeau et al., 2004). The enzyme is also considered to be involved actively in heavy metal tolerance in plants (Cakmak and Horst 1991; Chongpraditium et al., 1992; Przymusinski et al., 1995). In response to heavy metals the activity of SOD is increased more in the metal-tolerant plants than in the sensitive ones (Hendry et al., 1992). Exposure of pea plants to Ni resulted in the decrease in Cu/Zn-SOD as well as total SOD activity in both leaves and roots (Gajewska and Sklodowska 2005). This speaks further in favour of an important role of the enzyme in heavy metal tolerance. The SOD activity responses to metal stress vary considerably depending on plant species, tissue and exposure time. In shoots and leaves of wheat plants, SOD activity decreased signicantly in response to HM treatment (Gajewska et al., 2006; Gajewska and Sklodowska 2007). However, SOD activity increases were observed in coffee (*Coffea arabica*) cell suspension subjected to HM stress (Gomes et al., 2006). Schickler and Caspi (1999) have investigated the role of the antioxidative enzyme system in relation to Ni and Cd stress. In both species analyzed, *Alyssum argenteum* and *Alyssum maritimum,* SOD activity was elevated at high Cd concentrations. However, in the presence of Ni, *A. maritimum* exhibited an elevated SOD activity, while *A. argenteum* exhibited a different enzyme response pattern, with a significant reduction in SOD activity at the highest Ni concentration used. Also higher SOD activity in Zn and Ni treatments were observed when compared to the controls (Rao and Sresty 2000).

2.4.1.b Catalase (CAT, EC 1.11.1.6)

Catalase is a haem-containing enzyme that catalyses the dismutation of hydrogen peroxide into water and oxygen. The enzyme is found in all aerobic eukaryotes and is important in the removal of hydrogen peroxide generated in peroxisomes (microbodies) by oxidases involved in β-oxidation of fatty acids, the glyoxylate cycle (photorespiration) and purine catabolism. Catalase was one of the first enzymes to be isolated in a highly purified state. All forms of the enzyme are tetramers in excess of 220,000 molecular weight.

Like SOD, CAT activity varies considerably in response to HM stress. A decline in the activity of catalase due to Cd has been reported for pepper by Leon et al. (2002). In contrast, CAT activity increased considerably in Agropyron repens (Brej 1998) and sugarcane (Fornazier et al., 2002a, b), but remained unaltered in soybean (Ferreira et al., 2002) when exposed to cadmium. Reports are also available showing enhancement in the activity of catalase, probably the main H_2O_2 scavenging enzyme, in response to heavy metal treatment (Shaw et al., 2004), suggesting possible involvement of the enzyme in heavy metal tolerance (Gratao et al., 2009). CAT activity descreased significantly in response to HM treatment in shoots and leaves of wheat (Gajewska et al., 2006; Gajewska and Sklodowska 2007), whereas in leaves and roots of pea plants, CAT activity was shown to remain unaltered under HM stress (Gajewska and Sklodowska 2005). The effects of combined exposure to enhanced ultraviolet-B irradiance and Ni in soybean seedlings exhibited a CAT activity decline, even more than in the control, while the individual treatments resulted in increased CAT activity (Prasad et al., 2005). Moreover, *Scenedesmus armatus* grown in the presence of different heavy metals (Cd, Mn and Ni) exhibited increased CAT activity, as observed in calli derived from *L. leucocephalla* growing on contaminated soil with Cr, Ni and Zn (Rout et al., 1999). The perimedullar tuber tissue of potato cultivar resistant to Cd showed a higher constitutive level of catalase and also significantly greater increase in the activity of the enzyme in response to exposure to Cd when compared to the Cd-sensitive cultivar (Stroinski, Kozlowska 1997). The observation is in contrast to that observed for peroxidase and ascorbate peroxidase and suggests that the enzyme could be of greater importance than peroxidase and ascorbate peroxidase in reducing the metal-induced increase in H_2O_2 level, leading to tolerance of the plant (Cd-tolerant cultivar) to the metal (Cd).

2.4.1.c Ascorbate peroxidase (APX, EC 1.11.1.11)

Ascorbate peroxidase is a heme protein, and its primary function is the rapid removal of H_2O_2 at the site of generation (Asada 1992). Ascorbate peroxidase (APX) catalyses the conversion of H_2O_2 to H_2O, indispensable to protect chloroplasts and other cell constituents from damage by H_2O_2 and ^{-}OH, using ascorbate as the specific electron donor in this enzymatic reaction (Teixeira et al., 2006; Gratao et al., 2009). APX can be inhibited by thiol-modifying reagents, and the Cys residue participates in the binding of AsA to the heme (Raven 2003). The APX family consists of at least five different isoforms (Asada 1992), including thylakoid, soluble stromal, cytosolic, peroxisomal and apoplastic isoenzymes (Jimenez et al., 1997). The various isoforms respond differentially to metabolic and environmental signals (Kubo et al., 2000; Yoshimura et al., 2002). Within the chloroplasts, a stromal (sAPX) and a thylakoid-bound (tAPX) forms function to remove superoxide anions produced as a result of univalent reduction of molecular oxygen by the photosynthetic electron transport chain and to participate in the regeneration of membrane antioxidants such as alpha tocopherols and xanthophylls. The sAPX is thought to function primarily in the stromal AGC (Nakano and Asada 1981).

Significant enhancement in the activity of peroxidase in response to heavy metals has also been reported (Subhadra et al., 1991; Hendry et al., 1992; Mazhoudi et al., 1997; Shaw and Rout 1998; Teisseire and Guy 2000). Hendry et al. (1992) observed enhancement in the activity of peroxidase in the Cd-sensitive, but not in the Cd-tolerant, plants of *Holcus lanatus* in response to Cd. The activity of ascorbate peroxidase has also been reported to increase in plants in response to heavy metals (Shaw and Rout 1998; Schicker and Caspi 1999; Teisseire and Guy 2000). Working on *Phaseolus aureus,* Shaw and Rout (1998) observed metal-specific (Hg and Cd) differences in the response of the enzyme by the older seedlings when compared to the younger ones, with Hg inducing the activity of the enzyme while Cd had not. This was accompanied by death of the Cd-treated older seedlings after exposure for more than 36 h, suggesting that the enhanced synthesis of ascorbate peroxidase in response to Hg could be protecting the plant against the oxidative stress induced by the metal, this was not possible in the case of Cd treatment Like peroxidase, ascorbate peroxidase is also induced more in the metal-sensitive plant (*Alssum maritinum*) than in the tolerant plant (*A. argentums,* a nickel hyperaccumulator) (Schicker and Caspi 1999). The response of *AtPX* to Ni exposure indicated that APX activity increases as part of the defence against Ni stress, an alteration which has been shown to be often dose dependent and variable over time. In wheat *(Triticum durum* Desf.) cultivar 'Adamello' (drought sensitive (DS) APX was increased in response to Ni-induced stress, which was not observed in the other cultivar 'Ofanto' (drought tolerant, (DT)) (Pandolfini et al., 1996). The activity of APX has also been shown to increase following HM treatment of leaves and cell suspension cultures of wheat, coffee tea, pea and maize (Gajewska and Sklodowska 2007; Gomes et al., 2006; Ghanati and Ishka 2006; Gajewska and Sklodowska 2005; Baccouch et al., 2001). However, the negative effect, decrease of APX activity, has also been reported in pea roots in response to Ni exposure (Gajewska and Sklodowska 2005).

2.4.1.d Monodehydroascorbate reductase (MDHAR, EC 1.6.5.4) and dehydroascorbate reductase (DHAR, EC 1.6.5.4)

One of the key enzymes in ascorbate-glutathione pathway is MDHAR, a FAD enzyme that catalyses the reduction of the monodehydroascorbate radical, found in the mitochondria, peroxisome (Jimenez et al., 1997), cytosol, chloroplast, in vascular plants (Lunde et al., 2006). Expression of PpMdharl and PpMdhar3 was increased up to 5-fold under salt stress, osmotic stress or upon exposure to abscisic acid (Lunde et al., 2006). DHAR links ascorbate to glutathione, the other major soluble antioxidant in plant cells (Smirnoff 1996; Gratao et al., 2009). In contrast to other antioxidant enzymes, the role of MDHAR and DHAR in response to toxic metals is still unclear probably due to limited analysis of these enzymes in response to stressful conditions, compared to other enzymatic systems. Therefore, only very limited information is available in response to HM. MDHAR activity in *Zea mays* cv. LG 23/01 plants exhibited an enhanced activity as a response to the stress induced by Ni, while other enzymes of the ascorbate-glutathione cycle, like DHAR and GR, remained around the control values (Baccouch et al., 2001).

2.4.1.e Glutathione reductase (GR, EC 1.6.4.2)

Glutathione reductase is a flavoprotein that catalyses the NADPH-dependent reduction of oxidized GSSG to the reduced form GSH. The enzyme protein although synthesised in the cytoplasm, can be targeted to both the chloroplast and mitochondria (Mullineaux and Creissen 1997). In higher plants, GR is involved in defence against oxidative stress, whereas GSH plays an important role within the cell system, which includes participation in the ascorbate-glutathione cycle (AGC), maintenance of the sulfhydryl (-SH) group and a substrate for glutathione-S-transferases (Noctor et al., 2002). The majority of the studies determining the response of GR to Cd exposure have shown that GR activity increases as part of the defence against the Cd-exposure, an alteration that has often been dose-dependent and variable over time (Fornazier et al., 2002a, b). GR activity increased in the presence of Cd in *Phaseolus vulgaris* (Chaoui et al., 1997), *Piper nigrum* (Leon et al., 2002) and *Glycine max* (Ferreira et al., 2002). More enhanced GR activity was noticed by Dixit et al., (2001) in *Pisum sativum* exposed to 40 μM than at 4 μM Cd. However, a decrease in GR activity following the application of Cd has also been reported for a few plant species such as sunflower (Gallego et al., 1996a, b). GR activity in response to HM stress has also been shown to be often dose dependent and variable over time and this increased activity help in maintaining glutathione in the reduced form, prior to incorporation into PCs, and/or the activation of the ascorbate-glutathione cycle operating in order to detoxify the ROS induced by the exposure of the Ni (Gomes et al., 2006). Increase in GR activity in response to HM has also been reported for wheat (Pandolfini et al., 1996) *Alyssum* species (Schickler and Caspi 1999) and *Cajanus cajan* (Rao and Sresty 2000). However, maize plants GR activity was not affected by Ni exposure (Baccouch et al., 2001) and in *Crotalaria juncea,* GR activity was reduced (Cardoso et al., 2005).

2.4.1.f Glutathione peroxidase (GPX, EC 1.11.1.9)

GPX can metabolize lower rates of H_2O_2 generation in plants (Noctor et al., 2002). Seven related *Arabidopsis* proteins named AtGPXl-AtGPX7 were identified in the cytosol, chloroplast, mitochondria and endoplasmic reticulum (Millar et al., 2003). The abundance of some GPXs has shown their involvement in response to both biotic and abiotic stress conditions (Navrot et al., 2006; Chen et al., 2004; Kang et al., 2004). There are few reports considering the effects of HM on GPX activity. For instance, accumulation of Ni in the roots of maize plants exposed to Ni has been observed, but GPX activity in this tissue was unaltered (Baccouch et al., 2001). However, an increased GPX activity was demonstrated in wheat leaves on the 3rd, 6th and 9th days after Ni treatment (Gajewska and Sklodowska 2007). A cDNA (OsGPX1) encoding an *Oryza sativa* GPX has been considered a stress-inducible gene of rice GPX family that protects cells against both metabolic and environmental oxidative stresses (Kang et al., 2004). The addition of Cd also led to an increase in GPX activity in cultivars of pepper plants (Leon et al., 2002), while the GPX activity decreased in roots and remained almost unaltered in leaves of pea plants (Dixit et al., 2001).

2.4.1.g Guaiacol peroxidase (GOPX, EC)

Phytophenolics can act as antioxidants by donating electrons to guaiacol-type peroxidases (GPOX, EC 1.11.1.7) for the detoxification of H_2O_2 produced under stress conditions in plant systems, as a result of such enzymatic as well as non-enzymatic antioxidant reactions (Sakihama et al., 2002; Asada 1999; Gratao et al., 2009). The response of GPOX to HM stress varies considerably, depending on plant species and therefore the role of the enzyme in plant metabolism is not clear. HM treatment increases enzyme activity in seedlings cell suspension cultures of soybean, wheat and coffee (Saeidi-Sar et al., 2007; Gajewska and Sklodowska 2007; Gomes et al., 2006). But in maize GPOX activity remained unaffected by HM stress (Baccouch et al., 2001). Increases in GPOX activity were detected in *Arabidopsis thaliana* (Cho and Seo, 2004) and *Ceratophyllum demersum* (Arvind and Prasad 2003) following the addition of Cd. In contrast, pea plants exposed to Cd exhibited a decrease in GPOX and other antioxidant enzymes (Sndalio et al., 2001). While, Radotic et al. (2000) noted increase in GPOX activity in spruce needles subjected to Cd treatment initially; while a decrease in GPOX activity was found in subsequent Cd-treatments.

2.4.1.h Glutathione-S-Transferase (GST, EC2.5.1.18)

Glutathione-S-transferases (GST, EC 2.5.1.18) are major phase II detoxification enzymes found mainly in cytosol, with numerous roles in cellular defence, sulphur metabolism and in the use of GSH to reduce peroxides (Droog 1997; Noctor et al., 2002). Plant GSTs are a heterogeneous superfamily of multifunctional proteins, grouped into six classes (Moons 2005). Five homologous τ-class GSTs have been identified in tomato as proteins, involved in the oxidative stress response (Kilili et al., 2004). GSTs play roles in normal cellular metabolism as well as in the detoxification of a wide variety of compounds, implicating in numerous stress responses, including those arising from pathogen attack, oxidative stress and toxic metal (Marrs 1996; Gratao et al., 2009). In rice roots, Cd, Zn, Co, Ni and the strong oxidant H_2O_2 rapidly induced *osgstu4* and *osgstu3* suggesting their involvement in stress response regulations (Moons 2003). It has been demonstrated that changes in GST activity have a direct relation to the effect of stress on plant metabolism. For instance, in horseradish hairy roots cultures, the GST activity increased after Ni and diaminonitrotoluenes (DANTs) exposure (Nepovim et al., 2004). Significant GST activity increases in response to Ni-induced stress have also been observed in wheat shoots (Gajewska et al., 2006) and leaves and roots of pea plants, but more markedly in roots (Gajewska and Sklodowska 2005).

2.4.1.i Aldehyde dehydrogenases

Aldehyde dehydrogenases (ALDHs) also play a role in the detoxification of aldehydes generated in plants exposed to HM-stress. *Arabidopsis thaliana* ALDH3I1 gene was transcriptionally activated by abiotic stress, and overexpression of the ALDH3I1 gene confered stress tolerance in transgenic plants (Kotchoni et al., 2006). The protective antioxidant role of NADP-dependent isocitrate dehydrogenase (NADP-ICDH) in plants, which catalyses the production of NADPH, may be equally important against certain environmental stresses, including HM stress, in plants (Leterrier et al., 2007).

2.4.2 Non-enzymatic antioxidants

2.4.2.a Ascorbic acid (AsA)

Ascorbate is one of the most powerful antioxidants (Noctor and Foyer 1998), reacts directly with hydroxyl radicals, superoxide and singlet oxygen and reduces H_2O_2 to water via ascorbate peroxide reaction (Noctor and Foyer 1998). AsA is a key primary antioxidant that reacts with $^{\bullet}OH$ radicals, $O_2^{\bullet -}$, and $^1O_2$ (Chen and Gallie 2004). AsA is the major, probably the only, antioxidant buffer in the apoplast and is an essential metabolite involved with vital cell functions. The AsA content of plant tissues is modulated by a large number of factors internal or external to plants. Amongst the external factors, highest sensitivities are to heavy metal and water availability (Hancock and Viola 2005). Cadmium toxicity reduces the AsA content in plants (Arvind and Prasad 2005; Ozturk et al., 2003), and also does the drought stress (Gong et al., 2005). AsA together with GSH affects plant tolerance to reactive oxygen species (ROS) by participation in the detoxification of ROS in plant cells (Noctor and Foyer 1998). On reacting with ROS, AsA itself is oxidized through sequential electron transfer from AsA, to a partially oxidized monodehydroascorbate (MDHA) radical and the fully oxidized molecular dehydoascorbate (DHA) (Hancock and Viola 2005). The role of AsA in the antioxidative defence system provides a reason for its use as a stress marker. For instance, the Ni-stressed soybean seedlings exposed to AsA or gibberellin GA(3), especially to GA(3) plus AsA, exhibited an improved growth when compared to Ni-treated plants (Saeidi-Sar et al., 2007). GA(3) decreased Ni uptake by roots, while AsA considerably reduced root-to-shoot translocation of Ni. The interaction of AsA and GA(3) prevented the decrease in chlorophyll content and lipid peroxidation as well as increased the activities of the antioxidant enzymes (Saeidi-Sar et al., 2007).

2.4.2.b Glutathione (GSH)

Glutathione, glutamyl cysteinyl glycine (GSH) is a major low molecular weight thiol (-SH) compound in most plants (Kunert and Foyer 1993). It is a major non-protein thiol in plant cells and also represents a storage form of the reduced S. A regulatory role has been suggested for glutathione (GSH) in influencing the expression of many genes important in both abiotic and biotic stresses (Mullineaux and Rausch 2005), such as its involvement in the signalling pathway leading to plant cell death (PCD), probably contributing to guaranteeing the cellular redox conditions required for successful PCD (de Pinto et al., 2006). Glutathione acts as disulphide reductant to protect thiol (-SH) groups on enzymes, regenerate ascorbate and react with singlet oxygen and hydroxyl radicals (Noctor et al., 2002). In some plants, such as legumes, homoglutathione (having alanine in place of glycine, i.e., glutamyl cysteinyl alanine) may partly or wholly replace glutathione (Klapheck 1988). GSH acts as a protein disulphide reductant, which detoxifies herbicides by conjugation, either spontaneously or by the activity of one of a number of glutathione-S-transferases, and regulates gene expression in response to environmental stress and pathogen attack (Kunert and Foyer, 1993). It also participates in the regeneration of ascorbate from dehydro-ascorbate via the enzyme dehydro-ascorbate reductase (E.C. 7.8.5.1). In such reactions GSH is oxidized to glutathione disulphide (GSSG). GSH is regenerated by glutathione reductase (GR) in a NADPH-dependent reaction where NADPH is used as an electron source. De Vos et al. (1992) observed that 20 μM Cu reduces the GSH pool in the roots of Cu-sensitive *Silene cucubalus,* and this is accompanied by enhanced MDA accumulation, indicating oxidative damage of membrane. However, even at 30 μM concentration, Cd caused less decrease in the GSH pool than copper and no lipid peroxidation despite inducing nearly 13-fold increase in PC levels, compared to only 6-fold increase induced by Cu. In the Cu-tolerant plants, the decrease of GSH level was significantly less than in the Cu-sensitive plants. Because GSH is an important antioxidant, it may be concluded that tolerance to copper in Cu-tolerant *S. cucubalus* could be because of its more efficient antioxidative system than that of Cu-sensitive plants. Gallego et al. (1999) also correlated oxidative damage in sunflower cotyledons induced by Cd with decrease in the GSH pool. In suspension-cultured tobacco (*Nicotiana tabacum* cv. Bright Yellow-2) cells, the induction of PC biosynthesis in response to Cd was not affected by the coexistence of Ni, showing that Cd and Ni formed complexes with different compounds in cells (Nakazawa et al., 2001). In transgenic *Arabidopsis,* GSH concentrations are strongly correlated with increased resistance to both the growth inhibitory and oxidative stress induced effects of Ni, involved in conferring tolerance to Ni-induced oxidative stress in *Thlaspi* Ni hyperaccumulators (Freeman et al., 2004). In *Arabidopsis,* elevation of free salicylic acid (SA) levels enhances the specific activity of serine acetyltransferases, leading to elevated GSH and increased HM resistance (Freeman et al., 2005).

2.4.2.c Amines and Polyamines

Accumulation of high millimolar concentrations of N-containing metabolites is a common metabolic adaptation preferentially synthesized under toxic metal stress such as Cd, Cu, Ni, and Zn (Sharma and Dietz 2006; Yevdokimova 1994; Gratao et al., 2009). These solutes protect proteins and membranes against damage by high concentrations of inorganic ions and the metabolic machinery against oxidative damage (Rathinasabapathi 2000). Important non-enzymatic antioxidant is a plant derived chelator of metals, nicotianamine, produced by the trimerization of S-adenosylmethionine catalyzed by nicotianamine synthase (Kim et al., 2005). It was observed that nicotianamine plays a critical role in metal detoxification, through the promoter activities of four nicotianamine synthase genes in *Arabidopsis,* which increased in response to excess Ni, suggesting that nicotianamine plays a role in the detoxification of Ni in plants (Kim et al., 2005).

Several studies have indicated the involvement of polyamines in signalling and antioxidative processes (Sharma and Dietz 2006). In many plant species under various forms of stress the concentration of proline increases up to 80% suggesting an important role as an osmolyte and a reservoir of carbon and nitrogen and to protect plants against free radical-induced damage (Matysik et al., 2002). In HM-treated plants shoots, accumulation of proline was observed (Gajewska et al., 2006; Gajewska and Sklodowska 2005; Prasad et al., 2005; Singh et al., 2004). Such as proline, glycine betaine (GB), an amphoteric quaternary amine, is an organic osmolyte that accumulates in a variety of plant species in response to environmental stresses such as drought, salinity, extreme temperatures, UV radiation and toxic metals (Sakamoto and Murata 2002). Plant species vary in their capacity to synthesize GB and some plants, such as spinach and barley, accumulate relatively high levels of GB in their chloroplasts while others, such as *Arabidopsis* and tobacco, do not synthesize this compound (Sakamoto and Murata 2002).

2.4.2.d Phenolics

Phenolics in plant cells can form an antioxidant system equivalent to that of AsA. In contrast to their antioxidant activity, phytophenolics also have the potential to act as pro-oxidants under certain conditions (Sakihama et al., 2002). Plants produce a range of ligands for Cd, copper (Cu), Ni and zinc (Zn). In the xylem sap moving from roots to leaves, citrate and histidine are the principal ligands for Cu, Ni, and Zn (Rauser 1999; Gratao et al., 2009).

3. CONCLUSION

Although the influence of excess HM concentrations on plants has been extensively investigated by many researchers, our knowledge of the mechanisms of HM phytotoxicity is still incomplete. Numerous data indicate that HM toxicity to plants cannot be attributed to any single factor, but results from the combined effects of this element on metabolism and structure of plant cells. They include disturbances in mineral nutrition, physiological processes such as photosynthesis and respiration, alterations in plant water content and photoassimilate transport, destruction of various cellular organelles, reduction in mitotic activity of cells as well as induction of mutagenesis (Gajewska 2009).

Available data provide evidence that HM-induced oxidative stress is one of the most important components of the complex mechanism of toxicity for plants. Oxidative damage triggered by HM excess seems to be related to diverse toxic effects in plants. Peroxidation of membrane lipids has been suggested to be responsible for perturbations in tissue water status in plants subjected to HM stress. In addition, increased production of lipid peroxides in response to HM treatment contributes to destruction of many cellular organelles, mainly due to alterations in the structure and function of their membranes. HM mutagenic effects may be related to oxidative injury of nucleic acids. Occurrence of oxidative stress in plants exposed to HM seems to be closely associated with other mechanisms of phytotoxicity. HM-evoked disturbances in mineral nutrition leading to deficiency of essential microelements such as Fe, Mn, Cu and Zn may account for the inhibition of some metalloenzyme activities involved in scavenging of ROS. This may result in accumulation of reactive oxygen intermediates and consequently in oxidative stress. Perturbations in photosynthetic electron transport caused by HM are believed, at least partly, to contribute to increased production of ROS in plant cells under conditions of heavy metal stress.

In conclusion, it may be expected that the information available in the literature aims to improve our understanding of the basic events of the phytotoxicity caused by HM contamination, the induction of oxidative stress in plants, and the integration of multiple systems at the level of gene expression. Many questions related to ROS mode of regulation, its protective roles and its modulation of signalling networks such as biological processes that control growth, development, stress response, adaptation and programmed cell death, still remain unanswered (Mittler et al., 2004; Gechev et al., 2006). Furthermore, future studies should also include political and social actions reducing the release of pollutants in to the environment. This must include better information about the construction of genetically altered plants in a responsible way, which may help to overcome contaminated environmental conditions (Gratao et al., 2009). Oxidative stress response in plants is still poorly understood in comparison with the correspondent phenomenon in bacteria, yeast and mammals.

ACKNOWLEDGEMENTS

The work was supported by the Russian Federal Education's Agency under the Analytical Departmental Special-Purpose Programme "Development of Scientist Potential of Higher Education", Project 2.1.1/624. Naser A. Anjum is grateful to the Portuguese Foundation for Science and Technology (FCT) for financial assistance (SFRH/BPD/64690/2009).

REFERENCES

Alekseeva-Popova, N.V. and Il'inskaya, N.L. (1983). Reaction of some species and populations on the high maintenance of copper in the environment. In: *Plants in extreme conditions of mineral nutrition*. pp. 42-54 Eds. Shkol'nik, M.Ya. and Alekseeva-Popova, N.V. (Nauka: Leningrad) (In Russ.).

Alscher, R.G., Erturk, N. and Heath, L.S. (2002). Role of superoxide dismutases (SODs) in controlling oxidative stress in plants. *J. Experimental Botany*. **53**: 1331-1341.

Anjum, N.A., Umar, S., Iqbal, M. and Khan, N.A. (2011). Cadmium causes oxidative stress in moongbean [*Vigna radiata* (L.) Wilczek] by affecting antioxidant enzyme systems and ascorbate-glutathione cycle metabolism. *Russ. J. Plant Physiol.* **58**: 92-99.

Anjum, N.A., Umar, S., Singh, S., Nazar, R. and Khan, N.A. (2008a). Sulfur assimilation and cadmium tolerance in plants. In: Sulfur Assimilation and Abiotic Stress in Plants (Eds. Khan, N.A. et al.) Springer-Verlag Berlin Heidelberg, pp. 271-302.

Anjum, N.A., Umar, S., Ahmad, A. and Iqbal, M. (2008b). Responses of components of antioxidant system in moongbean [*Vigna radiata* (L.) Wilczek] genotypes to cadmium stress. *Commun. Soil Sci. Plant Anal.* **39**: 2469-2483.

Anjum, N.A., Umar, S., Ahmad, A., Iqbal, M. and Khan, N.A. (2008c). Sulphur protects mustard (*Brassica campestris* L.) from cadmium toxicity by improving leaf ascorbate and glutathione. *Plant Growth Regulation.* **54:** 271-279.

Apel, K. and Hirt, H. (2004). Reactive oxygen species: metabolism, oxidative stress, and signal transduction. *Ann. Rev. Plant Bio.* **55**: 373-399.

Arvind, P. and Prasad, M.N.V. (2003). Zinc alleviates cadmium-induced oxidative stress in *Ceratophyllum demersum* L.: a free floating freshwater macrophyte. *Plant Physiol. Biochem.* **41:** 391-397.

Arvind, P. and Prasad, M.N.V. (2005). Cadmium and zinc interactions in a hydroponic system using *Ceratophylum demersum* L.: adaptive ecophysiology, biochemistry and molecular toxicology. *Braz. J. Plant Physiol.* **17:** 3-20.

Asada, K. (1999). The water cycle in chloroplast: scavenging of active oxygens and dissipation of excess photons. *Ann. Rev. Plant Physio. Plant Mole. Bio.* **50:** 601-639.

Asada, K. (2006). Production and scavenging of reactive oxygen species in chloroplasts and their functions. *Plant Physiology.* **141:** 391-396.

Asada, K., (1992). Ascorbate peroxidase—a hydrogen peroxide-scavenging enzyme in plants. *Physiologia Plantarum.* **85:** 235-241.

Atici, Ö., Ağar, G. and Battal, P. (2005). Changes in phytohormone contents in chickpea seeds germinating under lead or zinc stress. *J. Biologia Plantarum.* **49:** 215-222.

Babalakova, N.K. (1997). Proton pump ion sensitivity change after uptake of copper and calcium by roots of barley plants. *Comptes Rendus de I'Academie Bulgare des Sciences.* **50:** 107-110.

Babar, A.M., Eun-Joo, H. and Kee-Yoeup, P. (2006). Copper-induced changes in the growth, oxidative metabolism, and saponin production in suspension culture roots of *Panax ginseng* in bioreactors. *Plant Cell Repts.* **25:** 1122-1132.

Baccouch, S., Chaoui, A., El Ferjani, E. (2001). Nickel toxicity induces oxidative damage in *Zea mays* roots. *J. Plant Nutrition.* **24:** 1085-1097.

Bai, Ch., Reilly., Ch. C. and Wood, B.W. (2006). Nickel deficiency disrupts metabolism of ureides, amino acids, and organic acids of young pecan foliage. *Plant Physiol.* **140:** 433-443.

Barondeau, D.P., Kassmann, C.J., Bruns, C.K., Tainer, J.A. and Getzoff, E.D. (2004). Nickel superoxide dismutase structure and mechanism. *Biochemistry.* **43:** 8038-8047.

Barton, L.L., Johnson, G.V., CNan, A.G. and Wagener, B.M. (2000). Inhibition of ferric chelate reductase in Alfalfa roots by cobalt, nickel, chromium, and copper. *J. Plant Nutr.* **23:** 1833-1845.

Beckett, P.H.T. and Davis, R.D. (1978). Additivity detoxic effect of Cu, Ni and Zn in young barley. *New Phytol.* **81:** 155-173.

Bernal, M., Roncel, M., Ortega, J.M., Picorel, R. and Yruela, I. (2004). Copper effect on cytochrome b559 of photosystem II under photoinhibitory conditions. *Physiol. Plant.* **120:** 686-694.

Bhattacharjee, S. (2005). Reactive oxygen species and oxidative burst: roles in stress, senescence and signal transduction in plants. *Curr. Sci.* **89:** 1113-1121.

Brej, T. (1998). Heavy metal tolerance in *Agropyron repens* (L.) p. Bauv. Populations from the Legnica copper smelter area, Lower Silesia. *Acta Societatis Botanicorum Poloniae.* **67:** 325-333.

Breusegem, F.V., Slooten, L., Stassart, J.-M., Moens, T., Botterman, J., Montagu, M.V., and Inze, D. (1999). Overproduction of *Arabidopsis thaliana* FeSOD confers oxidative stress tolerance to transgenic maize. *Plant Cell Physiol.* **40:** 515.

Briat, J.F. and Lebrun, M. (1999). Plant Responses to Metal Toxicity. *Plant Biol. Pathol.* **322:** 43-54.

Cakmak, I. and Horst, W.J. (1991). Effect of aluminum on lipid peroxidation, superoxide dismutase, catalase, and peroxidase activities in root tips of soybean *(Glycine max). Physiol. Plant.* **83:** 463.

Caldwell, Ch. (2001). Effect of elevated copper on the ultraviolet light-absorbing compounds of cucumber cotyledon and leaf tissues. *J. Plant Nutr.* **24:** 283-295.

Cataldo, D.A., McFadden, K.M., Garland, T.R. and Wildung, R.E. (1988). Organic constituants and complexation of nickel (II), iron (III), cadmium (II) and plutonium (IV) in soybean xylem exudates. *Plant Physiol.* **86:** 734-739.

Chaoui, A., Mazhoudi, S., Ghorbal, M.H. and El Ferjani, E. (1997). Cadmium and zinc induction of lipid peroxidation and effects on antioxidant enzyme activities in bean (*Phaseolus vulgaris* L). *Plant Sci.* **127:** 139-147.

Chen, Z. and Gallie, D.R. (2004). The ascorbic acid redox state controls guard cell signaling and stomatal movement. *The Plant Cell.* **16:** 1143-1162.

Chen, L.M., Lin, C.C. and Kao, C.H. (2000). Copper toxicity in rice seedlings: Changes in antioxidative enzyme activities, H_2O_2 level, and cell wall peroxidase activity in roots. *Bot. Bull. Acad. Sin.* **41:** 99-103.

Chen, S.R., Vaghchhipawala, Z., Li, W., Asard, H., Dickman, M.B. (2004). Tomato phospholipid hydroperoxide glutathione peroxidase inhibits cell death induced by Bax and oxidative stresses in yeast and plants. *Plant Physiology.* **135:** 1630-1641.

Chen, J., Zhu, C., Lin, D., Sun, Z.-X. and Can, J. (2007). The effects of Cd on lipid peroxidation, hydrogen peroxide content and antioxidant enzyme activities in Cd-sensitive mutant rice seedlings. *Plant Sci.* **87:** 49-57.

Cho, U. and Seo, N. (2004). Oxidative stress in *Arabidopsis thaliana* exposed to cadmium is due to hydrogen peroxide accumulation. *Plant Sci.* **168:** 113-120.

Chongpraditium, P., Mori, S., and Chino, M. (1992). Excess copper induces a cytosolic Cu, Zn-superoxide dismutase in soybean root. *Plant Cell Physiol.* **33:** 239.

Clemens, S. (2001). Molecular mechanisms of plant metal tolerance and homeostasis. *Planta.* **212:** 475-486.

Cobbett, Ch. S. (2000). Phytochelatins and their roles in heavy metal detoxification. *Plant Physiol.* **123:** 825-832.

Cordova, R.E.V., Valgas, C., Souza-Sierra, M. M., Correa, A.X.R. and Radetski, C.M. (2003). Biomass growth, micronucleus induction, and antioxidant stress enzyme responses in *Vicia faba* exposed to cadmium in solution. *Environ. Toxicol. Chem.* **22:** 645-649.

Coughtrey, P.J. and Martin, M.H. (1979). Cadmium, lead and zinc interactions and tolerance in two populations of *Holcus lanatus* L. grown in solution culture. *Environ. Exp. Bot.* **19:** 285-290.

Dat, J., Vandenabeele, S., Vranova, E., Van Montagu, M., Inze, D. and Van Breusegem, F. (2000). Dual action of the reactive oxygen species during plant stress responses. *Cell Mol. Life Sci.* **7:** 779-795.

de Pinto, M.C., Paradiso, A., Leonetti, P., De Gara, L. (2006). Hydrogen peroxide, nitric oxide and cytosolic ascorbate peroxidase at the crossroad between defence and cell death. *Plant J.* **48:** 784-795.

De Vos, C.H.R., Vonk, M.J., Vooijs, R. and Schat, H. (1992). Glutathione depletion due to copper-induced phytochelatin synthesis causes oxidative stress in *Silene cucubalus. Plant Physiol.* **98:** 853.

del Rio, L.A., Corpas, F.J., Sandalio, L.M., Palma, J.M., Gomez, M. and Barroso, J.B. (2002). Reactive oxygen species, antioxidant systems and nitric oxide in peroxisomes. *J. Experime. Bot.* **53:** 1255-1272.

del Rio, L.A., Sandalio, L.M., Altomare, D.A., Zilinskas, B.A. (2003). Mitochondrial and peroxisomal manganese superoxide dismutase: differential expression during leaf senescence. *J. Experim. Bot.* **54:** 923-933.

Devi, S.R. and Prasad, M.N.V. (2005). Antioxidative activity of *Brassica juncea* plants, the high concentrations of copper treated to action. *Russ. J. Plant physiol.* **52:** 233-237.

Dixit, V., Pandey, V. and Shyam, R. (2001). Differential oxidative responses to cadmium in roots and leaves of pea (*Pisum sativum* L cv. Azad). *J. Exp. Bot.* **52:** 1101-1109.

Doncheva, S. (1998). Copper-induced alterations in structure and proliferation of maize root meristem Cells. *J. Plant Physiol.* **153**: 482-487.

Doncheva, S., Stoyanova, Z. and Velikova, V. (2001). Influence of succinate on zinc toxicity of pea plants. *J. Plant Nutr.* **24:** 789-804.

Doncheva, Sn., Stoyanova, Z. and Gradinarski, L. (1998). Effect of succinate on zinc toxicity of pea plants. *Bulg. J. Plant Physiol.* pp. 269 (Special Issue) Abstr. Hth Congress of the Federation of European Societies of Plant Physiologists, Varna, Sept. 1998.

Dong-Hang, Q., Guo-jun, Z., Dong, Y., Teng-fei, P., Yong-yu, L. and Li-qin, Z. (2007). Influence of stress of the high copper maintenance on permeability of cytoplasmic membrane and activity of ferments in leaves of citrus. *Nonqye humijiny kexue xuebuo=J. Agro-Environ. Sci.* **26:** 1008-1013.

Dongmei, K., Hailong, S. and Yantao, Y. (2004). Influence of Ñu and Zn on polyphenol oxidase and superoxide dismutase activity in *Prunus armeriaca. Dongbei linye daxue xuebao = J. North-East Forest. Univ.* **32:** 72-73.

Droog, F. (1997). Plant glutathione S-transferases, a tale of theta and tau. *Journal of Plant Growth Regul.* **16:** 95-107.

Edreva, A. (2005). Generation and scavenging of reactive oxygen species in chloroplasts: a submolecular approach. *Agri. Ecosys.Environ.* **106:** 119-133.

Elstner, E.F. (1991). Mechanisms of oxygen activation in different compartments of plant cells. In: Active oxygen/oxidative stress and plant metabolism. Pell, E.J. and Steffen, K.L. (Eds.). *Am. Soc. Plant Physiol.* Rockville MD. pp. 13-25.

Ensley, B.D. (2000). Rational for use of phytoremediation. In *Phytoremediation of Toxic Metals – Using Plants to Clean Up the Environment.* pp. 3-12 Eds. Raskin, I. and Ensley, B.D. (John Wiley & Sons, Inc.: New York).

Fang-ming, Y., Rong-liang, Q., Peng-jie, Í., Xuen, Z. and Yu-dou, W. (2007). Influence of various levels of cadmium on activity of antioxidative enzymes in *Brassica chineiisis* leaves. *J. Agro-Environ. Sci.* **26:** 950-954.

Fornazier, R.Ê, Ferreira, R.R., Pereira, G.J.G., Molina, S.M.G., Smith, J., Lea, P.J. and Azevedo, R.A. (2002). Cadmium stress in sugar cane callus cultures: Effect of antioxidant enzymes. *Plant Cell, Tissue and Organ Cult.* **71:** 125-131.

Foyer, C.H. and Noctor, G. (2000). Oxygen processing in photosynthesis: regulation and signaling. *New Phytol.* **146:** 359-388.

Foyer, C.H. and Noctor, G. (2003). Redox sensing and signaling associated with reactive oxygen in chloroplasts, peroxisomes and mitochondria. *Physiol. Plant.* **119:** 355-364.

Foyer, C.H. and Noctor, G. (2005). Oxidant and antioxidant signalling in plants: a re-evaluation of the concept of oxidative stress in a physiological context. *Plant Cell and Environ.* **28:** 1056-1071.

Foyer, C.H., Trebst, A. and Noctor, G. (2006). Signaling and integration of defence functions of tocopherol, ascorbate and glutathione. In *Photoprotection, Photoinhibition, Gene Regulation, and Environment*. pp. 241-268. Eds. Demmig-Adams, B., Adams III, W.W. and Mattoo, A.K. (Springer: Dordrecht).

Franco, E., Alessandrelli, S., Masojidek, J., Margonelli, A. and Giardi, M.T. (1999). Modulation of D1 protein turnover under cadmium and heat stresses monitored by [^{35}S]metionine incorporation. *Plant Sci.* **144:** 53-61.

Freeman, J.L., Garcia, D., Kim, D.G., Hopf, A. and Salt, D.E. (2005). Constitutively elevated salicylic acid signals glutathione-mediated nickel tolerance in *Thlaspi* nickel hyperaccumulators. *Plant Physiol.* **137:** 1082-1091.

Freeman, J.L., Persans, M.W., Nieman, K., Albrecht, C, Peer, W., Pickering, I.J. and Salt, D.E. (2004). Increased glutathione biosynthesis plays a role in nickel tolerance in *Thlaspi* nickel hyperaccumulators. *Plant Cell.* **16:** 2176-2191.

Frey, B., Keller, C, Zierold, K. and Schulin, R. (2000). Distribution of Zn in functionally different leaf epidermal cells of the hyperaccumulator *Thiaspi caerulescens*. *Plant Cell Environ.* **23:** 675-687.

Gabbrielli, R., Pandolfini, T., Espen, L. and Palandri, M.R. (1999). Growth, peroxidase activity and cytological modifications in *Pisum sativum* seedlings exposed to Ni^{2+} toxicity. *J. Plant Physiol.* **155:** 639-645.

Gajewska, E. (2009). Nickel toxicity in Plants: oxidative stress. In *Nickel in relation to rlants*. pp. 89-116. Eds, Barket, A., Hayat, S. and Ahmad, A. (Narosa Publishing House Pvt. Ltd.).

Gajewska, E. and Sklodowska, M. (2005). Antioxidative responses and proline level in leaves and roots of pea plants subjected to nickel stress. *Acta Physiologiae Plantarum.* **27:** 329-339.

Gajewska, E. and Sklodowska, M. (2007). Effect of nickel on ROS content and antioxidative enzyme activities in wheat leaves. *Biometals.* **20:** 27-36.

Gajewska, E., Sklodowska, M., Slaba, M. and Mazur, J. (2006a). Effect of nickel on antioxidative enzyme activities, proline and chlorophyll contents in wheat shoots. *Biol. Plant.* **50:** 653-659.

Gajewska, E., Slaba, M., Andrzejewska, R. and Sklodowska, M. (2006b) Nickel-induced inhibition of wheat root growth is related to H_2O_2 production but not to lipid peroxidation. *Plant Growth Regul.* **49:** 95-103.

Gallego, S.M., Benavides, M.P. and Tomaro, M.L. (1999). Effect of cadmium ions on antioxidant defense system in sunflower cotyledons, *Biol. Plant.* **42:** 49.

Gallego, S.M., Benavides, M.P., Tomaro, M.L. (1996). Effect of heavy metal ion excess on sunflower leaves: evidence for involvement of oxidative stress. *Plant Sci.* **121:** 151-159.

Gechev, T.S., Van Breusegem, F., Stone, J.M., Denev, I., Laloi, C. (2006). Reactive oxygen species as signals that modulate plant stress responses and programmed cell death. *Bioessays.* **28:** 1091-1101.

Ghanati, F. and Ishka, M.R. (2006). Improvement of antioxidant system and decrease of lignin by nickel treatment in tea plant. *J. Plant Nutr.* **29:** 1649-1661.

Gomes, R.A., Moldes, C.A., Delite, F.S., Gratao, P.L., Mazzafera, P., Lea, P.J. and Azevedo, R.A. (2006). Nickel elicits a fast antioxidant response in *Coffea arabica* cells. *Plant Physio. Biochem.* **44:** 420-429.

Gomes-Junior, R.A., Moldes, C.A., Delite, F.S., Gratao, P.L., Mazzafera, P., Lea, P.J. and Azevedo, R.A. (2006). Nickel elicits a fast antioxidant response in *Coffea arabica* cells. *Plant Physiol. Biochem.* **44:** 420-429.

Gong, H., Zhu, X., Chen, K., Wang, S. and Zhang, C. (2005) Silicon alleviates oxidative damage of wheat plants in pots under drought. *Plant Sci.* **169:** 313.321.

Goto, P., Kobayaahi, M., Shoji, K., Shimada, H. and Yoshihara, T. (2005). Analysis of the mechanisms of lesion formation caused by copper treatment in peach: Thesis [46 Annual Meeting of the Japanese Society of Plant Physiologists (JSPP), Niigata, March 24-26, 2005]. *Plant and Cell Physiol.* **46:** 58.

Gratao, P.L., Monteiro, C.C. and Azvedo, R.A. (2009). Antioxidant stress responses of plants to nickel. pp. 117-140 Eds. Barket, A., Hayat, S. and Ahmad, A. (Narosa Publishing House Pvt. Ltd.).

Hancock, R.D. and Viola, R. (2005). Biosynthesis and catabolism of L-ascorbic acid in plants. *Crit. Rev. Plant Sci.* **24:** 167-188.

Hao, F., Wang, X. and Chen, J. (2006). Involvement of plasma-membrane NADPH oxidase in nickel-induced oxidative stress in roots of wheat seedlings. *Plant Sci.* **170:** 151-158.

Hao, H., Shi, G. and Xu, Q. (2000). The Toxic effect of Hg^{2+} on *Hydrocharis dubia* BL. Backer *J. Nanjing Normal University (Natural Science).* **23:** 87-90.

Hendry, G.A.F., Baker, A.J.M. and Ewart, C.F. (1992). Cadmium tolerance and toxicity, oxygen radical processes and molecular damage in cadmium-tolerant and cadmium-sensitive clones of *Holcus lanatus* L., *Acta Bot. Need.* **41:** 271.

Herbick, A., Giritch, A., Hortsmann, C., Becker, R., Balzer, H.J., Blumlein, H. and Stephan, U.W. (1996). Iron and copper nutrition-dependent changes in protein expression in a tomato wild type and the nicotianamine-free mutant chloronerva. *Plant Physiol.* **Ill:** 533-540.

Hevitt, B.J. (1983). Metals and Micronutrients: Uptake and Utilization bay Plants. pp. 277-323. (Acad. Press: London).

Hui-zhong W., Xin-quan Z. and Cur-ping H. (2006). Influence of lead on activity of a superoxide dismutase in roots of *Agrostis* seedlings. *J. Agro-Environ. Sci.* **25:** 644-647.

Igamberdiev, A.U. and Lea, P.J. (2002). The role of peroxisomes in the integration of metabolism and evolutionary diversity of photosynthetic organism. *Phytochem.* **60:** 651-674.

Israr, M., Sahi, S.V. and Jain, J. (2006). Cadmium accumulation and antioxidative responses in the *Sesbania drummondii* callus. *Arch. Environ. Contam. Toxicol.* **50:** 121-127.

Jiang, W. and Liu, D. (2000). Effects of Pb^{2+} on Root Growth, Cell Division, and Nucleolus of *Zea mays* L. *Bull. Environ. Contam. Toxicol.* **65(6):** pp. 786-793.

Jimenez, A., Hernandez, J.A., Del Rio, L.A., Sevilla, F. (1997). Ascorbate-glutathione cycle in mitochondria and peroxisomes of pea leaves: Changes induced by leaf senescence. *Phyton-Ann. Rei. Botanicae.* **37:** 101-107.

Kang, S.G., Jeong, H.K. and Suth, H.S. (2004). Characterization of the new member of the glutathione peroxidase gene family in *Oryza sativa. Molecules and Cells.* **17:** 23-28.

Kilili, K.G., Atanassova, N., Vardanyan, A., Clatot, N., Al-Sabarna, K., Kanellopolos, P.N., Makris, A.M. and Kampranis, S.C. (2004). Differential roles of Tau class glutathione S-transferases in oxidative stress. *J. Biolo. Chem.* **279:** 24540-24551.

Kim, S., Takahashi, M., Higuchi, K., Tsunoda, K., Nakanishi, H., Yoshimura, E., Mori, S. and Nishizawa, N.K. (2005). Increased nicotianamine biosynthesis confers enhanced tolerance of high levels of metals, in particular nickel, to plants. *Plant and Cell Physiol.* **46:** 1809-1818.

Klapheck, S. (1988). Homoglutathione: Isolation, quantification and occurrence in legumes. *Physiol. Plant.* **74:** 727-732.

Knecht, J.A., Dillen, M., Koevoets, P.L.M., Schat, H., Verklij, J.A.C. and Ernst, W.H.O. (1994). Phytochelatins in Cadmium-Sensitive and Cadmium-Tolerant *Silene vulgaris. Plant Physiol.* **104:** 255-261.

Kositsin, A.V. and Alekseeva-Popova, N.V. (1983). Heavy metals effects on plants and mechanisms of metal-tolerance. In *Plants in extreme conditions of mineral nutrition.* pp. 5-22. Eds. Shkol'nik, M.Ya. and Alekseeva-Popova, N.V. (Nauka: Leningrad) (In Russ.).

Kotchoni, S.O., Kuhns, C, Ditzer, A., Kirch, H.H., Bartels, D. (2006). Over-expression of different aldehyde dehydrogenase genes in *Arabidopsis thaliana* confers tolerance to abiotic stress and protects plants against lipid peroxidation and oxidative stress. *Plant Cell Environ.* **29:** 1033-1048.

Kovalchuk, I., Filkowski,. J, Smith, K. and Kovalchuk, O. (2003). Reactive oxygen species stimulate homologous recombination in plants. *Plant Cell Environ.* **26:** 1531-1539.

Kozarenko, A.E. (1987). Lead in plants. In *Lead in the environment.* pp. 71-76. (Nauka: Moscow) (In Russ.).

Kramer, U., Cotter-Howells, J.D., Baker, A.J.M. and Smith, J.A.C. (1996). Free histidine as a metal chelator in plants that accumulate nickel. *Nature.* **379:** 635-638.

Kretovitch, V.L. (1986). Biochemistry of plants. pp. 502 (Visshaja shkola: Moscow) (In Russ.).

Kukkola, E., Rautio, P. and Huttunen, S. (2000). Stress indications in copper- and nickel-exposed Scots pine seedlings. *Environ. Exp. Bot.* **43:** 197-210.

Kunert, K.J. and Foyer, C.H. (1993). In: Sulphur Nutrition and Assimilation in Higher Plants: Regulatory, Agricultural and Environmental Aspects (Eds. De Kok, L.J. et al.). SPB Academic Publishers, The Hague, pp. 139-151.

Kuznetsova, T. Yu., Vetchinnikova, L.V., Titov, A.F., Il'Inova, M.K. (2008) Effect of cadmium on fatty acid composition of lipids in the shoots of Karelian birch cultured *in vitro. Russ. J. Plant Physiol.* **55:** 657-662.

L'Huillier, L., d'Auzac, J., Durand, M., Michaud-Ferribre, N. (1996) Nickel effects on two maize (*Zea mays*) cultivars growth, structure, Ni concentration, and localization. *Can. J. Bot.* **74:** 1547-1554.

Lan-fang, D., Zong-gen, S., Da, Y., Wei-hong, L., Yan, D. and Yin-fen, Z. (2004) Toxic influence of mercury on peas seeds. *Xibei zhiwu xuebao = Acta bot. boreali-occident. sin.* **24:** 2266-2271.

Lee, S. and Leustel, T. (1999). The affect of cadmium on sulfate assimilation enzymes in *Brassica juncea. Plant Sci.* **141:** 201-207.

Lenti, K., Fodor, F. and Boddi, B. (2002). Mercury inhibits the activity of the NADPH: Protochlorophyllide oxidoreductase (FOR). *Photosynthetica.* **40:** 145-151.

Leon, A.M., Palma, J.M., Corpas, F.J., Gomez, M. et al. (2002). Antioxidant enzymes in cultivars of pepper plants with different sensitivity to candium. *Plant Physiol. Biochem.* **40:** 813-820.

Leonard, S.S., Harris, G.K., Shi, X. (2004). Metal-induced oxidative stress and signal transduction. *Free Rad. Biol. Med.* **37:** 1921-1942.

Leterrier, M., Del Rio, L.A., Corpas, F.J. (2007). Cytosolic NADP-isocitrate dehydrogenase of pea plants: genomic clone characterization and functional analysis under abiotic stress conditions. *Free Rad. Res.* **41:** 191-199.

Lidon, F.C. (1999). An overview of the effects of excess copper on the photosynthesis of rice plants. *Agronomia Lusitana.* **47:** 69-88.

Li-hong, Z., Xue-mei, L., Qiang, C. and Xing-yuan, H. (2006). Lead influence on antioxidative ferments and root activity at maize cv. *Jilin nonyye daxue xuebao=J. Jilin Agr. Univ.* **28:** 119-122.

Lukatkin, A.S., Gracheva, N.V., Grishenkova, N.N., Dukhovskis, P.V., Brazaitite, A.A. (2007) Cytokinin-like growth regulators mitigate toxic action of zinc and nickel ions on maize seedlings. *Russ. J. Plant Physiol.* **54:** 381-387.

Lunde, C., Baumann, U., Shirley, N.J., Drew, D.P. and Fincher, G.B. (2006). Gene structure and expression pattern analysis of three monodehydroascorbate reductase (MDHAR) genes in *Physcomitrella* patens: Implications for the evolution of the MDHAR family in plants. *Plant Mole. Bio.* **60:** 259-275.

Maier, E.A., Matthews, R.D., McDowell, J.A., Walden, R.R. and Ahner, B.A. (2003). Environmental cadmium levels increase phytochelatin and glutathione in lettuce grown in a chelator-buffered nutrient solution. *J. Environ. Qual.* **32:** 1356-1364.

Malcová, R. and Gryndler, M. (2003). Amelioration of Pb and Mn Toxicity to Arbuscular Mycorrhizal Fungus *Glomus intraradices* by Maize Root Exudates. *J. Biologia Plantarum.* **47:** 297-299.

Mano, J., Torii, Y., Hayashi, S., Takimoto, K., Matusi, K., Nakamura, K., Inze, D., Babiychuk, E., Kushnir, S. and Asada, K. (2002). The NADPH:quinone oxidoreductase P1-ξ-crystallin in *Arabidopsis* catalyzes the α,β-hydrogenation of 2-alkenals: detoxication of the lipid peroxide-derived reactive aldehydes. *Plant Cell Physiol.* **43:** 1445-1455.

Marrs, K.A. (1996). The functions and regulation of glutathione S-transferases in plants. *Ann. Rev. Plant Physio. Plant Mole. Bio.* **47:** 127-158.

Matysik, J., Alia, Bhalu, B. and Mohanty, P. (2002). Molecular mechanisms of quenching of reactive oxygen species by proline under stress in plants. *Curr. Sci.* **82:** 525-532.

Mazhoudi, S., Chaoui, A., Ghorbal, M.H. and Ferjani, E.E. (1997). Response of antioxidant enzymes to excess copper in tomato *(Lycopersicon esculentum* Mill.). *Plant Sci.* **127:** 129.

McKersie, B.D., Senaratna, T., Walker, M.A., Kendall, E.J. and Hetherington, P.R. (1988). Deterioration of membranes during aging in plants: evidence for free radical mediation. In *Senescence and Aging in Plants.* pp. 441-464 Eds. Nooden, L.D. and Leopold, A.C. (Academic Press: New York).

Millar, A.H., Mittova, V., Kiddle, G, Heazlewood, J.L., Bartoli, C.G., Theodoulou, F.L. and Foyer, C.H. (2003). Control of ascorbate synthesis by respiration and its implication for stress responses. *Plant Physiol.* **133:** 443-447.

Mittler , R. (2002). Oxidative stress, antioxidants and stress tolerance. *Trends Plant Sci.* **7:** 405-410.

Mittler, R., Vanderauwera, S., Gollery, M. and Van Breusegem, F. (2004). Reactive oxygen gene network of plants. *Trends Plant Sci.* **9:** 490-498.

Molas, J. (1997). Changes in morphological and anatomical structure of cabbage (*Brassica oleracea* L.) outer leaves and in ultrastructure of their chloroplasts caused by an *in vitro* excess of nickel. *Photosynthetica.* **34:** 513-522.

Montes-Holguin, M.O., Peralta-Videa, J.R, Meitzner, G., Martinez-Martinez, A., De La Rosa, G., Castillo-Michel, H.A., Gradea-Torresdey, J.L. (2006). Biochemical and spectroscopic studies of the response of *Convolvulus arvensis* L. to chromium(III) and chromium(VI) stress. *Environ. Toxicol. Chem.* **25:** 220-226.

Moons, A. (2003). Osgstu3 and osgtu4, encoding tau class glutathione S-transferases, are heavy metal-and hypoxic stress-induced and differentially salt stress-responsive in rice roots. *Febs. Lett.* **553:** 427-432.

Moons, A. (2005). Regulatory and functional interactions of plant growth regulators and plant glutathione S-transferases (GSTS). *Plant Hormones.* **72:** 155-202.

Mullineaux, P.M. and Creissen, G.P. (1997). Glutathione reductase: regulation and role in oxidative stress. In: *Oxidative stress and the molecular biology of antioxidant denfenses.* pp. 667-713. Ed. Scandalios, J.C. (Cold Spring Harbor Laboratory Press: Cold Spring Harbor, NY).

Mullineaux, P.M. and Rausch, T. (2005). Glutathione, photosynthesis and the redox regulation of stress-responsive gene expression. *Photosynthesis Res.* **86:** 459-474.

Nakano, Y. and Asada, K. (1981). Hydrogen peroxide is scavenged by ascorbate-specific peroxidase in spinach chloroplasts. *Plant Cell Physiol.* **22:** 867-880.

Nakazawa, R., Ozawa, T., Naito, T., Kameda, Y. and Takenaga, H. (2001). Interactions between cadmium and nickel in phytochelatin biosynthesis and the detoxification of the two metals in suspension-cultured tobacco cells. *Biologia Plantarum.* **44:** 627-630.

Navrot, N., Collin, V., Gualberto, J., Gelhaye, E., Hirasawa, M., Rey, P., Rnaff, D.B., Issakidis, E., Jacquot, J.P., Rouhier, N. (2006). Plant glutathione peroxidases are functional peroxiredoxins distributed in several subcellular compartments ans regulated during biotic and abiotic stresses. *Plant Physiol.* **142:** 1364-1379.

Nepovím, A., Hubálek, M., Podlipná, R., Zeman, S., Vanek T. (2004). *In vitro* degradation of 2,4,6-trinitrotoluene by plant tissue culture of *Solanum aviculare* and *Rheum palmatum. Eng. Life Sci.* **24:** 46-49.

Noctor, G. (2006). Metabolic signalling in defence and stress: the central roles of soluble redox couples. *Plant Cell and Environ.* **29:** 409-425.

Noctor, G. and Foyer, C.H. (1998). Ascorbate and glutathione: Keeping active oxygen under control. *Annu. Rev. Plant Physiol. Plant Mol. Biol.* **49**: 249-279.

Noctor, G., Dutilleul, C., De Paepe, R. and Foyer, C.H. (2003). Use of mitochondrial electron transport mutants to evaluate the effects of redox state on photosynthesis, stress tolerance and the integration of carbon/nitrogen metabolism. *J. Experi. Bot.* **55:** 49-57.

Noctor, G., Gomez, L., Vanacker, H. and Foyer, C.H. (2002). Interactions between biosynthesis, compartmentation and transport in the control of glutathione homeostasis and signalling. *J. Experi. Bot.* **53:** 1283-1304.

Nriagu, J. O. (1979). Global inventory of natural and anthropogenic emissions of trace metals to the atmosphere. *Nature.* **279:** 409-411.

Obuhov, A.I. (1987) Lead availability to plants. In *Lead in the environment.* pp. 109-116. (Nauka: Moscow) (In Russ.).

Ortiz, D.F., Kreppel, L., Speiser, D.M., Scheel, G., McDonald, G., Ow, D.W. (1992). Heavy metal tolerance in the fission yeast requires in ATP-binding cassette-type vacuolar membrane transporter. *EMBO J.* **11:** 3491-3499.

Ouariti, O., Boussama, N., Zarrouk, M., Cherif, A., Ghorbal, M.H. (1997). Cadmium and Copper-Induced Changes in Tomato Membrane Lipids. *Phytochem.* **45:** 1343-1350.

Ozturk, L., Eker, S., Ozkutlu, F. and Cakmak, I. (2003). Effect of cadmium on growth and concentrations of cadmium, ascorbic acid and sulfhydril groups in durum wheat cultivars. *Turk. J. Agric. For.* **27:** 161-168.

Panda, S.K. (2001). Proline accumulation and lipid peroxidation in wheat seedlings exposed to cadmium toxicity. *Proc. Nat. Acad. Sci., India. B.* **71:** 255-258.

Panda, S.K. and Patra, H.K. (1998). Alteration of Nitrate Reductase Activity by Chromium Ions in Excised Wheat Leaves. *Indian J. Agric. Biochem.* **11:** 56-57.

Panda, S.K. and Patra, H.K. (2000). Nitrate and ammonium ions effect on the chromium toxicity in developing wheat seedlings. *Proc. Nat. Acad. Sci. India.* **70:** 75-80.

Pandolfini, T., Gabbrielli, R. and Ciscato, M. (1996). Nickel toxicity in two durum wheat cultivars differing in drought sensitivity. *J. Plant Nutri.* **19:** 1611-1627.

Pandolfini, T., Gabbrielli, R. and Comparing, C. (1992). Nickel toxicity and peroxidase activity in seedlings of *Triticum aestivum* L. *Plant Cell Environ.* **15:** 719-725.

Paribok, T.A., Leina, G.D., Sazykina, N.A. and Temp, G.A. (1983). Accumulation of metals in herbaceous and woody plants in conditions of city. In *Plants in extreme conditions of mineral nutrition.* pp. 100-116. Eds. Shkolnik, M.Ya. and Alekseeva, N.V. (Nauka: Leningrad) (in Russ.).

Parker, D.R., Pedler, J.F., Ahnsnrom, Z.A.S. and Resketo, M. (2001). Re-evaluating the free-ion activity model of trace metal toxicity toward higher plants: experimental evidence with copper and zinc. *Environ. Toxicol. Chem.* **20:** 899-906.

Patra, M. and Sharma, A. (2000). Mercury Toxicity in Plants. *The Bot. Rev.* **66:** 379-422.

Pätsikkä E., Kairavuo M., Šeršen F., Aro E.-M. and Tyystjärvi E. (2002). Excess Copper Predisposes Photosystem II to Photoinhibition *in vivo* by Outcompeting Iron and Causing Decrease in Leaf Chlorophyll. *Plant Physiol.* **129:** 1359-1367.

Pauly, N., Pucciariello, C, Mandon, K., Innocenti, G., Jamet, A., Baudouin, E., Herouart, D., Frendo, P. and Puppo, A. (2006). Reactive oxygen and nitrogen species and glutathione: key players in the legume—*Rhizobium symbiosis. J. Experi. Bot.* **57:** 1769-1776.

Pelosi, P., Fiorentini, R. and Galoppini, C. (1976). On the nature of nickel compounds in *Alyssum bertolonii* Desv. *Agr. Biol. Chem.* **40:** 1641-1642.

Pich, A. and Scholz, G. (1993). The relationship between the activity of various iron-containing and iron free enzymes and the presence of nicotianamine in tomato seedlings. *Physiol. Plant.* **88:** 172-178.

Pitzschke, A., Fornazi, C. and Hirt H. (2006). Reactive oxygen species signalling in plants. *Antioxidants & Redox Signalling*. **8:** 1757-1764.

Prasad, M.N.V. and Freitas, H. (2006). Metal-tolerant plants: biodiversity prospecting for phytoremediation technology. In *Trace elements in the environment: biogeochemistry, biotechnology, and bioremediation*. pp. 483-506. Ed. Prasad, M.N.V., Sajwan, K.S. and Naidu, R. (CRC Press, Taylor & Fransis Group: Boca Raton, London, New York).

Prasad, M.N.V. (2006). Stabilisation, remediation, and integratid management of metal-contaminated ecosystems by grasses (Poaceae). In *Trace Elements in the Environment: Biogeochemistry, Biotechnology, and Bioremediation*. pp. 405-424. Eds. Prasad, M.N.V. Sajwan, K.S. and Naidu, R. (CRC Press, Taylor & Fransis Group: Boca Raton, London, New York).

Prasad, S.M., Dwivedi, R. and Zeeshan, M. (2005). Growth, photosynthetic electron transport, and antioxidant responses of Young soybean seedlings to simultaneous exposure of nickel and UV-B stress. *Photosynthetica.* **43:** 177-185.

Przymusinski, R., Rucinska, R. and Gwozdz, E. (1995). The stress-stimulated 16-kDa polypeptide from lupine roots has properties of cytosolic Cu:Zn-superoxide dismutase, *Environ. Exp. Bot.* **35:** 485.

Radotic, K., Ducic, T. and Mutavdzic, D. (2000). Changes in perxidase activity and isoenzymes in spruce needles after exposure to different concentrations of cadmium. *Env. Exp. Bot.* **44:** 105-113.

Rao, K.V.M. and Sresty, T.V.S. (2000). Antioxidative parameters in the seedlings of pigeonpea *(Cajanus cajan* (L.) Millspaugh) in response to Zn and Ni stresses. *Plant Sci.* **157**: 113-128.

Rathinasabapathi, B. (2000). Metabolic engineering for stress tolerance: installing osmoprotectant synthesis pathways. *Ann. Bot.* **86:** 709-716.

Rauser, W.E. (1999). Structure and function of metal chelators produced by plants—the case for organic acids, amino acids, phytinm and metallothioneins. *Cell Biochem. & Biophy.* **31:** 19-48.

Raven, E.L. (2003). Understanding functional diversity and substrate specificity in haem peroxidases: what can we learn from ascorbate peroxidase? *Natural Product Reports.* **20:** 367-381.

Robertson, A.I. (1985). The poisoning of roots of *Zea mays* by nickel ions, and the protection afforded by magnesium and calcium. *New Phytol.* **100:** 173-189.

Ros, R., Cooke, D.T., Burden, R.S. and James, C.S. (1990). Effects of the herbicide MCPA, and the heavy metals, cadmium and nickel on the lipid composition, Mg^{2+}-ATPase activity and fluidity of plasma membranes from rice, *Oryza saliva* (cv. Bahia) shoots. *J. Exp. Bot.* **41:** 457-462.

Rout, G.R., Samantaray, S. and Das, P. (1999). Chromium, nickel and zinc tolerance in *Leucaena leucocephalla* (K8). *Silvae Genetica.* **48:** 151-157.

Rucinska, R. and Gwozdz, E. A. (2005). Influence of lead on membrane permeability and lipoxygenase activity in lupine roots. *J. Biologia Plantarum.* **49:** 617-619.

Rudakova, E.V., Karakis, K.D. and Sidorkina, T.N. (1987) Trace elements: uptake, transport and physiological role in plants. pp. 180. Ed. Ostrovskaya, L.K. (Naukova dumka: Kiev) (In Russ.).

Saeidi-Sar, S., Khavari-Nejad, R.A., Fahimi, H., Ghorbanli, M. and Majd, A. (2007). Interactive effects of gibberellin A(3) and ascorbic acid on lipid peroxidation and antioxidant enzyme activities in *Glycine Max* seedlings under nickel stress. *Russ. J. Plant Physiol.* **54:** 74-79.

Sagner, S., Kneer, R., Wanner, G., Cosson, J.P., Deus-Neumann, B. and Zenk, M.H. (1998). Hyperaccumulation, complexation and distribution of nickel in *Sebertia acumnata. Phytochem.* **47:** 339-347.

Sakamoto, A. and Murata, N. (2002). The role of glucine betaine in the protection of plants from stress: clues from trangenic plants. *Plant Cell Environ.* **25:** 163-171.

Sakihama, Y., Cohen, M.F., Grace, S.C. and Yamasaki, H. (2002). Plant phenolic antioxidant and prooxidant activities: phenolics-induced oxidative damage mediated by metals in plants. *Toxico.* **177:** 67-80.

Salt, D.E., Rauser, W.E. (1995). MgATP-dependent transport of phytochelatins across the tonoplast of oat roots. *Plant Physiol.* **107:** 1293-1301.

Saprykin, F.Ya. (1984). Soils geochemistry and nature protection. pp. 231. (Nedra: Leningrad) (In Russ).

Scandalios, J.G. (1990). Response of plant antioxidant defense genes to environmental stress. *Adv. Genet.* **28:** 1-41.

Scandalios, J.G. (1997). Molecular genetics of superoxide dismutases in plants. In *Oxidative stress and the molecular biology of antioxidant defenses.* pp. 527. Ed. Scaldalios, J.G. (Cold Spring Harbor Laboratory Press, Cold Spring Harbor: New York).

Schickler, H. and Caspi H. (1999). Response of antioxidative enzymes to nickel and cadmium stress in hyperaccumulator plants of the genus *Alyssum. Physiologia Plantarum.* **105:** 39-44.

Schtitzendubel, A. and Polle, A. (2002). Plant responses to abiotic stresses: heavy metal-induced oxidative stress and protection by mycorrhization ion. *J. Experi. Bot.* **53:** 1351-1365.

Schützendübel, A., Schwanz, P., Teichmann, T., Gross, K., Langenfeld-Heyser, R. et al. (2001). Cadmium-induced changes in antioxidative systems, hydrogen peroxide content, and differentiation in scots pine roots. *Plant Physiol.* **127:** 887-898.

Seregin, I.V. and Kozhevnikova, A.D. (2006). Physiological role of nickel and its toxic effects on higher plants. *Russ. J. Plant Physiol.* **53:** 257-277.

Seryogin, I.V. and Ivanov, V.B. (2001). Physiological aspects of cadmium and lead toxic effect on higher plants. *Russ. J. Plant Physiol.* **48:** 606-630.

Sharma, D.C. and Sharma, C.P. (1996). Chromium uptake and toxicity effects on growth and metabolic activities in wheat, *Triticum aestivum* L. cv. UP 2003. *Indian J. Experi. Bio.* **34:** 689-691.

Sharma, S.S. and Dietz, K.J. (2006). The significance of amino acids and amino acid-derived molecules in plant responses and adaptation to heavy metal stess. *J. Experi. Bot.* **57:** 711-726.

Shaw, B.P and Rout, N.P. (1998). Age-dependent responses of *Phaseolus aureus* Roxb. to inorganic salts of mercury and cadmium. *Acta Physiol. Plant.* **20:** 85.

Shaw, B.P., Sahu, S.K. and Mishra, R.K. (2004). Heavy metal induced oxidative damage in terrestrial plants. In *Heavy Metal Stress in Plants—from Biomolecules to Ecosystems.* Chap. 4. Ed. Prasad, M.N.V. (Springer-Verlag: Heidelberg).

Shu-mei, H. and Xiao-ping, S. (2005). Inhibition of root growth and DNA damage, invoked by Cr^{6+}-stress at wheat seedlings. *Acta bot. boreali-occident. sin.* **25:** 541-545.

Singh, R.P. Dabas, S. and Choudhary, A. (1996). Recovery of Pb^{2+} caused inhibition of chlorophyll biosynthesis in leaves of *Vigna radiata* (L.) Wilczek by inorganic salts. *Indian J. Exp. Bio.* **34:** 1129-1132.

Singh, S., Kayastha, A.M., Asthana, R.K. and Singh, S.P. (2004). Response of garden pea to nickel toxicity. *J. Plant Nutri.* **27:** 1543-1560.

Sinha, S. and Pandey, K. (2003). Nickel induced toxic effects and bioaccumulation in the submerged plant, *Hydrilla verticillata* (l.f.) Royle under repeated metal exposure. *Bull. Environ. Contam. Toxicol.* **71:** 1175-1183.

Smirnoff, N. (1996). The function and metabolism of ascorbic acid in plants. *Anna. Bot.* **78:** 661-669.

Smirnoff, P.M. (1989). Mineral fertilizers. In *Agrochemistry*. pp. 367-427 (Agropromizdat: Moscow) (In Russ.).

Stoyanova, D. and Baldjiev, G. (2002). Ultrastructural responses of the chloroplasts of *Lemna minor* L. and *Salvinia natans* (L.) ALL. to cadmium. *Phytol. Balcan.* **8:** 369-377.

Stroinski, A. and Kozlowska, M. (1997). Cadmium-induced oxidative stress in potato tuber. *Acta Soc. Bot. Pol.* **66:** 189.

Subhadra, A.V., Nanda, A.K., Behera, P.K. and Panda, B.B. (1991). Acceleration of catalase and peroxidase activities in *Lemna minor* L. and *Allium cepa* L. in response to low levels of aquatic mercury. *Environ. Pollut.* **69:** 169.

Su-chun, L., Lang-tao, X., Bo-Han, L., Ke-qin, W. and Mao-feng, Í. (2006). Association between lead stress and growth and endogenic phytohormones at cucumber seedlings. *Nongye huanjing kexue xuebao. J. Agro-Environ. Sci.* **25:** 592-596.

Suchun, L., Langtao, X., Bohan, L., Xudong, L., Fengchun, K., Wenkui, Z. and Jianhua, T. (2006). Lead stress influence on activity and isoenzymes of antioxidative ferments in cucumber seedlings. *Yingyɔng shengtai xuebao=Chin. J. Appl. Ecol.* **17:** 300-304.

Takahashi, M.A. and Asada, K. (1983). Superoxide anion permeability of phospholipids membranes and chloroplast thylakoids. *Arch. Biochem. Biophys.* **225:** 558-566.

Tandon, P. K. and Srivastava, Mini. (2004). Effect of cadmium and nickel on metabolism during early stages of growth in gram (*Cicer arietinum* L.) seeds. *Indian J. Agr. Biochem.* **17:** 31-34.

Teisseire, H. and Guy, V. (2000). Copper-induced changes in antioxidant enzymes activities in fronds of duckweed *(Lemna minor). Plant Sci.* **153:** 65.

Teixeira, F.K., Menezes-Benavente, L., Galvao, V.C., Margis, R. and Margis-Pinheiro, M. (2006). Rice ascorbate perooxidase gene family encondes functionally diverse isoforms localized in different subcellular compartments. *Planta.* **224:** 300-314.

Temple, M.D., Perrone, G.G. and Dawes, I.W. (2005). Complex cellular responses to reactive oxygen species. *Trends Cell Biol.* **15:** 319-326.

Thomas, D.J., Thomas, J.B., Prier, S.D., Nasso, N.E. and Herbert, S.K. (1999). Iron superoxide dismutase protects against chilling damage in the cyanobacterium *Synechococcus* species PCC7942. *Plant Physiol.* **120:** 275.

Turner, R.G. and Marshall, C. (1971). The accumulation of Zn by root homogenates of zinc-tolerant and non-tolerant clones of *Agrostis tenuis* Sibth. *New Phytol.* **70:** 539-545.

Vassilev, A. and Manolov, P. (1999). Chlorophyll fluorescence of barley (*H. vulgare* L.) seedlings grown in excess of Cd. *Bulg. J. Plant Physiol.* **25:** 67-76.

Vassilev, A., Yordanov, I. and Tsonev, Ò. (1997). Effects of Cd^{2+} on the Physiological State and Photosynthetic Activity of Young Barley Plants. *Photosynthetica.* **34:** 293-302.

Voinar, A.I. (1962). Microelement in wildlife. pp. 91. (Vysshaya shkola: Moskow) (in Russ.).

Vranova, E., Inze, D. and Van Breusegem, F. (2002). Signal transduction during oxidative stress. *J. Exp. Bot.* **53:** 1227-1236.

Wan, X. and Zwiazek, J.J. (1999). Mercuric chloride effects on root water transport in aspen seedlings. *Plant Physiol.* **121:** 939-946.

Weckx, J.E.J. and Clijsters, H.M.M. (1997). Zn phytotoxicity induces oxidative stress in primary leaves of *Phaseolus vulgaris. Plant Physiol Biochem.* **35:** 405-410.

Woitke, M., Degenhardt, B., Gade, B. and Gimmler, H. (2001). Content and internal distribution of heavy metals in roots of plants grown at alkaline pH on slag from municipal solid waste incineration. *J. Appl. Bot.* **75:** 59-66.

Xiqin, Z. and Qianhua, J. (2005). Responses of maize seed and seedlings to chromium stress and their relation to chromium accumulation. *Shengtaixue zazhi. Chin. J. Ecology.* **24:** 1048-1052.

Xu, G., Shi, G., Liu, L., Ding, X., Cheng, G. and Wu, G. (2000). Study on the toxic effects of chromium-VI on brasenia schreberi Gmel. leaves. *J. Nanjing Norm. Univ. (Nat. Sci.).* **23:** 67-71.

Xue-zhi, F., Zhu-jun, Z. and Guang-wen, S. (2004). Influence of various cadmium concentrations on *Brassica campestris* ssp. chinensis growth and antioxidative system. *J. Agra-Environ. Sci.* **23:** 877-880.

Yagodin, B.A. (1989). Application of fertilizings and environmental control. In *Agricultural Chemistry,* pp. 626-649. Ed. Yagodin, B.A. (Agropromizdat: Moskow) (in Russ.).

Yan, Ch., Hong, Y., Fu, Sh., Fang, Ch., Lei, J., Shen, Q. (1998). Effect of Cd, Pb stress on the activated oxygen scavenging system in tobacco leaves. *Chinese J. Geochemistry.* **17:** 372-378.

Yevdokimova, G.A. (1994). Accumulation of nitrates in plants on soils with increased heavy-metal content. *Eurasian Soil Sci.* **26:** 116-122.

Yi-qun, W., Jin-gui, Z., Wen-lie, C. and Lian-yun, C. (2004). Ultrastructural studying of rice mesophyll cell, damaged by contamination with Hg^{2+} and Cd^{2+}. *J. Fujian Agr. Forest. Univ. Natur. Sci.* **33:** 409-413.

Yi-xian, Z. and Li-ping, Z. (2005). Effects of antioxidative enzymes activity in *Hordeum vulgare* seedlings at Cd^{2+}, Đb^{2+}, Hg^{2+}, Ni^{2+}stresses. *Nongye huanjing kexue xuebao = J. Agro-Environ. Sci.* **24:** 217-221.

Yoshimura, K., Yabuta, Y., Ishikawa, T. and Shigeoka, S. (2002). Identification of *cis* element for tissue,specific alternative splicing of chloroplast ascorbate peroxidase. *J. Bio. Chem.* **277:** 40623-40632.

Yun-xrand, H., Bo-han, L., Lang-tao, X., Su-chun, L. and Zhi-kun, W. (2006). Influence of Cd^{2+} on growth and the maintenance of phytohormones in *Glycine max* seedlings. *Huanjing kexue= Environ. Sci.* **27:** 1398-1401.

Zhang, F., Li, X., Wang, Ch., Shen, Zh. (2000). Effect of cadmium on autoxidation rate of tissue and inducing accumulation of free proline in seedlings of mung bean. *J. Plant Nutri.* **23:** 357-368.

Zimakov, I.E. (1981). Transformation of sparingly soluble mercury species in soils and their availability to plants. In *Migrations of pollutants in soils and related environments: Proceedings of III all-USSR session.* pp. 97-126. Eds. Bobovnikova, C.I. and Malahov, S.G. (Gidrometeoizdat: Leningrad) (In Russ.).

Zoller, T., Skroppa, T., Johnsen, O. and Pole, A. (2003). Apoplastic peroxidases in needles of Norway spruce (*Picea abies*) progenies from different crossing environments. *Forstwissenschaftliches Centralbatt.* **122:** 153-159.

10

Ascorbate and Glutathione: A Defence System Against Heavy Metal Stress in Plants

ZHONGQIU ZHAO

School of Land Science and Technology, China University of Geosciences, No. 29 Xueyuan Rd, Haidian District, Beijing 100083, China; Key Laboratory of Land Consolidation and Rehabilitation Ministry of Land and Resources, Beijing 100035, China

Corresponding author: *zhongqiuzhao@163.com*

ABSTRACT

Many studies have been published which conclusively show that during adaptation and response to a variety of stress conditions, there is an upregulation in the plant's antioxidant defence systems, including an increase in the cellular concentrations of both ascorbate (AsA) and glutathione (GSH). Indeed activation of both AsA and GSH biosynthesis appears to be an essential component of plant stress tolerance. The current review discusses some untouched aspects of the oxidative stress in plants and the role of AsA and GSH in detail.

Keywords: Ascorbate-glutathione, defence system, heavy metals.

1. NEED FOR ANTIOXIDANT SYSTEMS IN PLANTS

In either photosynthetic or non-photosynthetic organisms, the aerobic metabolism continuously generates reactive oxygen species (ROS) in mitochondria, peroxisomes, cell wall, and so on. They include, primarily, superoxide ions ($O_2^{\bullet -}$), hydrogen peroxides (H_2O_2) and hydroxyl radicals ($^{\bullet}OH$) which can at high concentration cause oxidative damage to macromolecules such as proteins, lipids, and nucleic acids (Navrot et al.,

2007). Plants, with photosynthetic capacities, have to deal with a second source of oxygen-related molecules generated at the level of the photosynthetic electron chain. The generation of these ROS, together with singlet oxygen ($^1O_2$) is greatly enhanced in reactions occurring mostly at the level of photosystem I and photosystem II in chloroplast (Foyer and Noctor 2003). On the other hand, ROS has also been thought to serve as signals to activate the defence system in plants (Mittler 2002). Rapid activation of the protective mechanisms is of great importance to alleviate the hazardous effects imposed by stress (Rouhier et al., 2008).

Efficient antioxidant defence system to control the level of oxidative stress in the cell is very important for the organism, especially for plants. The overwhelming majority of plant species are essentially immobile and have to deal with various abiotic environmental adversity such as drought, chilling, high temperature, heavy metal pollution, etc. Most of these stresses are accompanied by an oxidative burst that needs to be controlled. Membrane lipid peroxidation and/or protein oxidation are the simplest indicator of assessing the extent of oxidative stress in the tissue (Noctor and Foyer 1998; Mittler 2002).

Production and removal of ROS are strictly controlled by an array of non-enzymatic and enzymatic antioxidant system in plants. It is described in detail in the Section 8.3.

2. HEAVY METALS AND OXIDATIVE STRESS IN PLANTS

Heavy metals (HMs) are defined as a group of elements having a density higher than 5 g/cm^3. 53 of 90 naturally occurring elements are heavy metals (Alloway 1995). Based on their solubility under physiological conditions, 17 heavy metals may be available for living cells of importance for organism. Among these heavy metals, Fe, Mn and Mo are important as micronutrients. Zn, Cu, Ni, Cr, Co, V and W are toxic elements with high or low importance as trace elements. Pb, As, Hg, Cd, Ag, Sb and U have no known biological function as nutrients and seem to be more or less toxic to organism (Nies, 1999; Schützendübel and Polle 2002).

In most terrestrial ecosystems, there are two main sources of heavy metals: natural sources and anthropogenic sources. Natural sources are underlying parent material, volcanoes and continental dusts, etc. Anthropogenic sources include all the anthropogenic activities such as mining, smelting, phosphate fertilizing, sewage irrigation, etc., that lead to heavy metals emission and accumulation in ecosystems (Zhao et al., 2005a). In recent years, heavy metals contamination in soils has been widely spread and becoming more and more serious over the world. For example, Cd concentrations are about 0.1-0.5 mg kg^{-1} in natural or uncontaminated soils, in some heavily contaminated soils, however, Cd concentrations have reached up to 150 mg kg^{-1}.

The availability of heavy metals to plants depends on the solubility and mobility of metals. The solubility and mobility of metals is affected by adsorption, desorption and complexation process of metals in the soils which depends on soil properties such as soil pH, organic materials, Eh, etc (Sign and Kristen 1998). There are two parallel pathways for metals uptake and transport through the root cortex toward the shoot:

one pathway of active transport from cell to cell in the symplast (selective transport across membranes) and another pathway of passive transport by diffusion and convection through the apoplast, namely cell walls and intercellular spaces (Steudle and Peterson 1998; Stephan 2002).

High concentrations of heavy metals in plants could be very toxic for plants resulting in varied effects on plant physiology (Schützendübel et al., 2001). Among other things, oxidative stress was found to be directly related to high concentrations of heavy metals (Schützendübel and Polle 2002; Zhao et al., 2005b; Stobrawa and Lorenc-Plucińska 2008). To understand the action mode of heavy metals leading to toxicity in organisms, their chemical properties and their relations with the physiological redox of organism cells must be taken into account. The physiological redox of aerobic cells ranges between –420 mV and +800 mV. So heavy metals of biological significance can be divided into two groups: redox active and inactive metals. Heavy metals that have lower redox potentials compared with biological molecules cannot participate in the biological redox reaction (Schützendübel and Polle 2002).

There is ample evidence that exposure of plants to excess concentration of heavy metals can result in oxidative stress (Shaw, 1995; Schützendübel et al., 2001; Olmos et al., 2003; Zhao et al., 2005b; Liu et al., 2007; Stobrawa and Lorenc-Plucińska 2008)). Uncontrolled and excessive production of free radicals can cause peroxidation of lipids, inactivation of enzymes, and damaged DNA and other constituents of cells. Malondialdehyde (MDA), a product of lipid peroxidation, has been seen to be greatly accumulated after heavy metal exposure (Zhao et al., 2005b; Tripathi et al., 2006; Singh et al., 2006; Anjum et al., 2008). All these changes accelerate senescence in the affected plants or even kill them (Shah et al., 2001). Taking Cu, redox active metal, and Cd, redox inactive metal as example, to elucidate the process heavy metals inducing oxidative stress.

Cu is an essential micronutrient for plants that is a component of several electron transport enzymes and is involved in catalysing the redox reactions in mitochondria and chloroplasts (Marschner 1995). However, Cu also induces toxicity at tissue concentrations slightly above its optimal levels. Many studies reported copper toxicity in plants (Mediouni et al., 2006). Cu phytotoxicity could be a consequence of its interference with a number of metabolic processes associated with normal development like nitrogenous and lipid metabolisms and photosynthesis. In fact, presence of high concentrations of Cu in the *Chlamydomonas reinhardtii* growth medium induces a NO^{3-} uptake inhibition, more pronounced with the supply of Cu^{2+}. Ouariti et al. (1997) reported that excess copper led to an efficient reduction of lipid content as well as to a strong change in fatty acid composition, compared to excess cadmium. Those changes, particularly in the leaves, would mainly concern the photosynthetic apparatus, reducing, therefore, the redox power and the metabolic energy. On the other hand, photosynthetic apparatus alteration may be the consequence of the disturbance of the chloroplast structure via a degradation of polar lipids. Indeed, heavy metals are known to involve a reak down of cells membrane lipid due to the increased accumulation of ROS mediated-oxidative stress (Rellán-Á lvarez et al., 2006; Stobrawa and Lorenc-Plucińska 2008).

Cadmium (Cd) is a widely spread pollutant with no known biological function. It can reach high levels in agricultural soils and is easily assimilated by plants. As a non-redox-metal, Cd is not able to generate directly ROS through Haber–Weiss reactions. There is ample evidence that the toxicity of Cd to plants is due to Cd inducing oxidative stress in plant cells and consequently ROS especially hydrogen peroxide (H_2O_2) accumulation. The first and common response of plants to Cd stress is decrease of glutathione (GSH) and accumulation of H_2O_2, and subsequently series of changes in antioxidant system (Zhao and Xi 2007). Overproduction of ROS and occurrence of oxidative stress in plants could be the indirect consequence of Cd toxicity (Stobrawa and Lorenc-Plucińska 2008). The mechanisms include interacting with the antioxidant system, disrupting the electron transport chain or disturbing the metabolism of essential elements (Dong et al., 2006). One of the most deleterious effects induced by Cd is lipid peroxidation which can directly cause biomembrane deterioration. MDA, one of the decomposition products of polyunsaturated fatty acids of membrane, is regarded as a reliable indicator of oxidative stress (Demiral and Türkan 2005). A hypothetical model of Cd action on the cellular redox control was suggested by Schützendübel and Polle (2002) to illustrate the mechanism of Cd-induced oxidative stress in plant cells. (Fig 1). Cd induces the synthesis of PCs when entered plant cells and bind with the thiol in PCs, resulting in a rapid depletion of GSH and activity decrease of GR and APX, and consequently accumulation of H_2O_2. Activities of many antioxidant enzymes are regulated by H_2O_2 acting as a signal molecule. H_2O_2 also induces various secondary metabolites as a signal molecule of programmed cell death (PCD), resulting in rigidification and lignification of cell wall, and in programmed cell death finally.

Recently, Chamseddine et al. (2009) had taken a comparative research on cadmium and copper induction of oxidative stress and antioxidative response in tomato (*Solanum lycopersicon*) leaves. The findings add new kinetic and comparative information on the

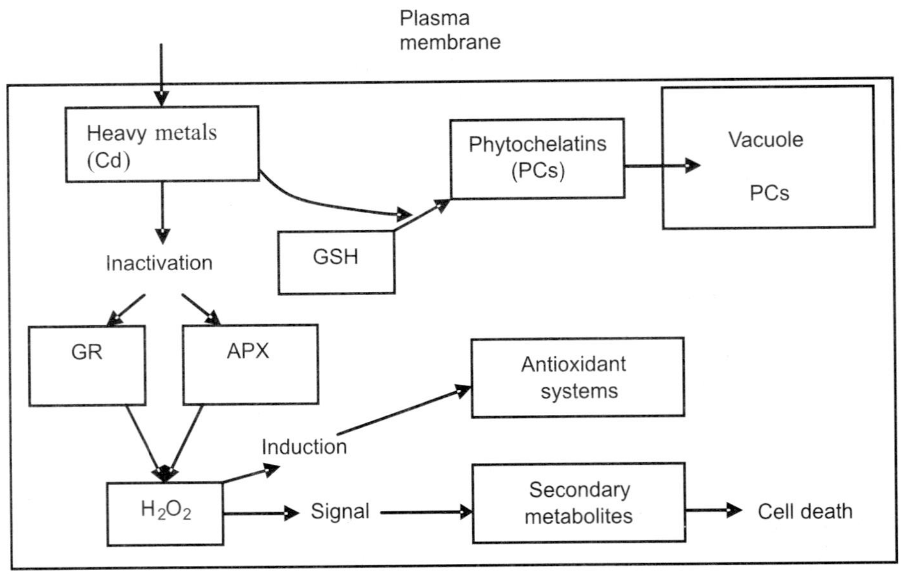

Fig. 1. Hypothetical model of Cd action on the cellular redox systems.

metabolism response of plants under heavy metal stress. The results indicated that the early oxidative stress induced was more pronounced with cadmium than with copper, however, after prolonged heavy metal exposure, oxidative stress was mainly enhanced under copper exposure. Tomato leaves seems to be able to respond to heavy metal by the stimulation of the antioxidative enzyme activity that can be implicated in oxidative stress alleviation. In addition, copper induces more cadmium than the activation of the antioxidative response. This can be related to the ability of copper to provoke a more important oxidative stress situation in tomato leaves and therefore suggest that antioxidative response is related to the intensity of oxidative stress induced rather than to the plant sensitivity to cadmium and copper. Difference in cadmium and copper plant response can be explained by a divergence in some other heavy metal defence mechanisms. A long-time exposure to cadmium and copper resulted in a decline in the antioxidative activity probably due to an excess both metals-induced ROS formation that can alter enzyme function.

3. ASCORBATE AND GLUTATHIONE: A DEFENSE SYSTEM AGAINST HEAVY METAL STRESS IN PLANTS

Aerobic organisms respond to oxidative stress either by non-enzymatic or enzymatic ROS scavenging systems. Non-enzymatic defence involves ascorbic acid (AsA), glutathione (GSH), α-tocopherol, β-carotene and other compounds capable of quenching ROS. Enzymatic ROS scavenging system includes the major $O_2^{\bullet -}$-scavenger-superoxide dismutase (SOD, EC 1.15.1.1), which dismutate $O_2^{\bullet -}$ to H_2O_2. The ascorbate peroxidase (APX, EC 1.11.1.11), catalase (CAT, EC 1.11.1.6) and peroxidase (POX, 1.11.1.7), subsequently scavenge H_2O_2 (Mittler 2002; Razinger et al., 2007).

All plants can synthesize AsA. In plants, AsA can accumulate to millimolar concentrations in both photosynthetic and non-photosynthetic tissues. AsA is one of the major primary as well as the most abundant low molecular weight antioxidants present in plant tissues, reacting directly with $O_2^{\bullet -}$, H_2O_2 and $^{\bullet}OH$. In addition to its importance in photo-protection and the regulation of photosynthesis, Ascorbate plays an important role in preserving the activities of enzymes that contain prosthetic transition metal ions. It is also a powerful secondary antioxidant, reducing the oxidized form of α-tocopherol, an important antioxidant in nonaqueous phases (Noctor and Foyer 1998). The recent identification of the pathway of plant AsA biosynthesis as well as the cloning of various enzymes of the pathway has refocused attention on this important micronutrient, particularly as a potential route to improve plant abiotic stress resistance (Davey et al., 2004). Interestingly, plants appear to possess alternative biosynthetic routes (Smirnoff et al., 2001), although the simultaneous operation of both pathways has to date only been demonstrated in *Arabidopsis* (Davey et al., 1999). The existence of these alternative routes has recently gained some genetic support with the cloning of a D-galacturonate dehydrogenase (Agius et al., 2003), but the physiological relevance of these conversions still has to be demonstrated *in vivo* (Davey et al., 2004).

GSH represents the major store of reduced sulphur in plant cells and performs many functions ranging from the scavenging of reactive oxygen species (ROS), to the detoxification of heavy metals, which can be divided into two categories: sulphur metabolism and defence. GSH is the predominant non-protein thiol regulating sulphur uptake at root level, and it is used by GSH S-transferases in detoxification of xenobiotics. GSH is also a precursor of phytochelatins which are crucial in controlling cellular heavy metal levels (Noctor and Foyer, 1998). In both plants and animals, GSH is a very important antioxidant and redox buffer. Plant GSH biosynthesis involves two reactions, catalysed by the enzymes g-glutamyl-cysteine synthase and GSH synthetase, localized in the chloroplast and the cytosol.

Many studies have been published which conclusively show that during adaptation and response to a variety of stress conditions, there is an upregulation in the plant's antioxidant defence systems, including an increase in the cellular concentrations of both AsA and GSH. Indeed activation of both AsA and GSH biosynthesis appears to be an essential component of plant stress tolerance. Further, AsA- or GSH-deficient plants have been shown to be hypersensitive to environmental stress conditions (Cobbett et al., 1998). Finally, stress conditions can also alter the redox status of the AsA and GSH pools, leading to an accumulation of the oxidized forms (DHA and glutathione disulphide or GSSG, respectively), and this is considered to be a marker for the degree of oxidative stress being experienced by the cell (Davey et al., 2004).

The AsA and GSH pools are intimately linked *in vivo* through the AsA-GSH (Halliwell–Asada) cycle (Fig. 2) (Noctor and Foyer 1998), whereby GSH is used as a

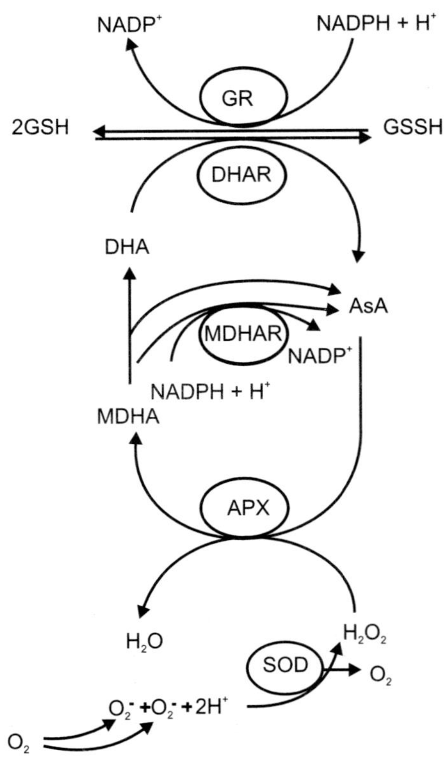

Fig. 2. The glutathione-ascorbate cycle.

source of reducing power for the enzymatic regeneration of oxidized AsA (dehydroascorbate or DHA) (Foyer and Halliwell 1976). Among the antioxidant mechanisms, the AsA-GSH cycle has been shown to play a crucial role in removing ROS and maintaining the cellular redox status. In plant cells, components of AsA-GSH cycle, which comprises AsA, GSH, APX, mono dehydroascorbate reductase (MDHAR, EC 1.6.5.4), dehydroascorbate reductase (DHAR, EC 1.8.5.1), and glutathione reductase (GR, EC 1.6.4.2), represent more effective H_2O_2 detoxification mechanism. In this cycle, H_2O_2 is reduced to water by APX in the presence of reduced AsA and the resulting dehydroascorbate (DHA) can be converted to AsA using reduced GSH as an electron donor leading to the production of oxidized glutathione (GSSG). GSH is then regenerated from the NADPH-dependent reduction of GSSG catalyzed by GR (Liu et al., 2007).

The AsA-GSH cycle has been reported in various subcellular compartments including chloroplasts, cytosol, mitochondria, peroxisome and plasma membrane, implying its predominant role in the detoxification of ROS in these compartments (Noctor and Foyer 1998). Reduced GSH and AsA, which are found at high concentrations in chloroplasts and other cellular compartments, are crucial for plant defense against oxidative stress.

REFERENCES

Agius, F., Gonzalez-Lamothe, R., Caballero, J.L., Munoz-Blanco, J., Botella, M.A. and Valpuesta V. (2003). Engineering increased vitamin C levels in plants by overexpression of a D-galacturonic acid reductase. *Nature Biotechnology.* **21:** 177-181.

Alloway, B.J. (1995). Heavy metals in soils. 2nd edition. USA: Blackie Academic Professional.

Anjum, N.A., Umar, S., Ahmad, A., Iqbal, M. and Khan, N.A. (2008). Sulphur protects mustard (*Brassica campestris* L.) from cadmium toxicity by improving leaf ascorbate and glutathione. *Plant Growth Regul.* **54:** 271-279.

Chamseddine, M., Wided, B.A., Guy, H., Marie-Edith, C., Fatma, J. (2009). Cadmium and copper induction of oxidative stress and antioxidative response in tomato (*Solanum lycopersicon*) leaves. *Plant Growth Regul.* **57:** 89-99.

Cobbett, C.S., May, M.J., Howden, R. and Rolls, B. (1998). The glutathione-deficient, cadmium-sensitive mutant, cad2-1, of *Arabidopsis thaliana* is deficient in gamma-glutamylcysteine synthetase. *Plant J.* **16:** 73-78.

Davey, M.W., Franck, C., Keulemans, J. (2004). Distribution, developmental and stress responses of antioxidant metabolism in *Malus* [J]. *Plant, Cell and Environment.* **27:** 1309-1320.

Davey MW, Gilot C, Persiau G, Østergaard J, Han Y, Bauw GC, Van Montagu MC (1999) Ascorbate biosynthesis in Arabidopsis cell suspension culture. *Plant Physiol.* **121:** 535–543. doi:10.1104/pp. 121.2.535.

Demiral, T. and Tu¨rkan, I. (2005). Comparative lipid peroxidation, antioxidant defense systems and proline content in roots of two rice cultivars differing in salt tolerance. *Environ. Exp. Bot.* **53:** 247-257.

Dong, J., Wu, F. and Zhang, G. (2006). Influence of cadmium on antioxidant capacity and four microelement concentrations in tomato seedlings (*Lycopersicon esculentum*). *Chemosphere.* **64:** 1659-1666.

Foyer, C.H. and Halliwell, B. (1976). Presence of glutathione and glutathione reductase in chloroplasts: a proposed role in ascorbic acid metabolism. *Planta.* **133:** 21-25.

Foyer, C.H., Noctor, G. (2003). Redox sensing and signalling associated with reactive oxygen in chloroplasts, peroxisomes and mitochondria. *Physiol. Plant.* **119:** 355-364.

Liu, Y.G., Wang, X., Zeng, G.M., Qu, D., Gu, J.J., Zhou, M., Chai, L.Y. (2007). Cadmium-induced oxidative stress and response of the ascorbate-glutathione cycle in *Bechmeria nivea* (L.) Gaud. *Chemosphere.* **69:** 99-107.

Marschner, H.. (1995). Mineral nutrition of higher plants. Academic Press, London, pp. 337-347.

Mediouni, C., Benzarti, O., Tray, B., Ghorbel, M.H. and Jemal, F. (2006). Cadmium and copper toxicity for tomato seedlings. *Agron. Sustain Dev.* **26:** 227-232.

Mittler, R.. (2002). Oxidative stress, antioxidants and stress tolerance. *Trends Plant Sci.* **7:** 405-410.

Murzaeva, S.V. (2004). Effect of heavy metals on wheat seedlings: activation of antioxidant enzymes. *Appl. Biochem. Microbiol.* **40:** 98-103.

Navrot, N., Rouhier, N., Gelhaye, E. and Jacquot, J.P. (2007). Reactive oxygen species generation and antioxidant systems in plant mitochondria. *Physiol Plant.* **129:** 185-195.

Nies, D.H. (1999). Microbial heavy-metal resistance. *Applied Microbiology and Biotechnology.* **51(6):** 730-750.

Noctor, G. and Foyer, C.H. (1998). Ascorbate and glutathione: keeping active oxygen under control. *Annu. Rev. Plant Physiol. Plant Mol. Biol.* **49:** 249-279.

Olmos, E.O., Martínez-Solano, J.R., Piqueras, A. and Hellín, E. (2003). Early steps in the oxidative burst induced by cadmium in cultured tobacco cells (BY-2 line). *J. Exp. Bot.* **54:** 291-301.

Ouariti, O., Boussama, N., Zarrouk, M., Cherif, A. and Ghorbal, M.H. (1997). Cadmium- and copper-induced changes in tomato membrane lipids. Phytochem. **45:** 1343-1350.

RellánÁ lvarez, R., Ortega-Villasante, C., A´ lvarez-Fernández, A., Del Campo, F.F., Herna´ndez, L.E. (2006). Stress Responses of Zea mays to Cadmium and Mercury. *Plant Soil.* **279:** 41-50.

Rouhier, N., San, K.C., Gelhaye, E., Corbier, C., Favier, F., Didierjean, C. and Jean-Pierre, J. (2008). Redox based antioxidant systems in plants: Biochemical and structural analyses. *Biochimica Biophysica Acta.* **1780:** 1249-1260.

Schützendübel, A., Polle, A. (2002). Plant responses to abiotic stresses: heavy metal-induced oxidative stress and protection by micorrhization. *J. Exp. Bot.* **53:** 1351-1365.

Schützendübel, A., Schwanz, P., Teichmann, T., Gross, K., Langenfeld-Heyser, R., Godbold, D.L. and Polle, A. (2001). Cadmium-induced changes in antioxidative systems, hydrogen peroxide content, and differentiation in Scots Pine roots. *Plant Physiol.* **127:** 887-98.

Shah, K., Kumar, R.G., Verma, S., Dubey, R.S. (2001). Effect of cadmium on lipid peroxidation, superoxide anion generation and activities of antioxidant enzymes in growing rice seedlings. *Plant Sci.*

Shaw, B.P. (1995). Effects of mercury and cadmium on the activities of antioxidant enzymes in seedlings of *Phaseolus aureus. Biologia Plantarum.* **37:** 587-596.

Sign, B.R. and Kristen, M. (1998). Cadmium uptake by barley as affected by Cd sources and PH levels[J]. *Geoderma.* **84:** 185-194.

Singh, S., Eapen, S.D. and Souza, S.F. (2006). Cadmium accumulation and its influence on lipid peroxidation and antioxidative system in an aquatic plant, *Bacopa monnieri* L. *Chemosphere.* 233-246.

Smirnoff, N., Conklin, P.L., Loewus, F.A. (2001). Biosynthesis of ascorbic acid in plants: a Renaissance. *Ann. Rev. Plant Physiol. and Plant Mole. Bio.* **52:** 437-467.

Stephan, U.W. (2002). Intra- and intercellular iron trafficking and subcellular compartmentation within roots. *Plant Soil.* **241:** 19-25.

Steudle, E. and Peterson, C.A. (1998). How does water get through roots? *J. Exp. Bot.* **49(322):** 775-788.

Stobrawa, K. and Lorenc-Pluciñska, G. (2008). Thresholds of heavy-metal toxicity in cuttings of European black poplar (*Populus nigra* L.) determined according to antioxidant status of fine roots and morphometrical disorders. *Science of the Total Environment.* **390:** 86-96.

Tripathi, B.N., Mehta, S.K., Amar, A. and Gaur, J.P. (2006). Oxidative stress in *Scenedesmus* sp. during short- and long-term exposure to Cu^{2+} and Zn^{2+}. *Chemosphere.* **62:** 538-544.

Zhao, Z.Q., Zhu, Y.G. and Cai, Y.L. (2005a). Transport and transformation of cadmium in soil-plant systems and the influence factors. *Ecology and Environment.* **14(2):** 282-286 (in Chinese).

Zhao, Z.Q., Zhu, Y.G., Kneer, R., Smith, S.E. and Smith, F.A. (2005b). Effect of Zn on Cd Toxicity-induced Oxidative Stress in Winter Wheat Seedlings. *J. Plant Nutrition.* **28(11):** 1947-1959.

Zhao, Z. Q. and Xi, M. Z. (2007). Advances in research on mechanisms of Cd-induced oxidative stress in plants. *J. Agro-Environment Sci.* **26:** 47-51.

11

Role of Ascorbate Peroxidase and Glutathione in Higher Plants to Abiotic-Induced Oxidative Stress

WEIFENG XU AND WEIMING SHI*

State Key Laboratory of Soil and Sustainable Agriculture, Institute of Soil Science, Chinese Academy of Sciences, Nanjing–210008, China

***Corresponding author:** wmshi@issas.ac.cn*

ABSTRACT

APX (ascorbate peroxidase) and GSH (glutathione) plays the important roles in abiotic-induced oxidative stress in higher plants. APX has a high substrate specificity for ascorbate and is the primary H_2O_2-scavenging enzyme in higher plants. APXs are ubiquitous in the plant cell, being localized in chloroplasts, mitochondria, cytosol and peroxisomes. Glutathione, a tripeptide with the sequence γ-Glu-Cys-Gly, exists in either a reduced (GSH) or oxidized (GSSG) state. Under heavy metal stress, GSH functions as a heavy metal ligand and an antioxidant. Also, GSH acts as a cellular signal, mediating the control of enzyme and/or regulatory protein activities. So, our results show that: (1) peroxisomal type APX plays the important roles in the Arabidopsis tolerance to heat, salt, zinc and cadmium stress; (2) GSH plays an important role in the cadmium tolerance of cattail (*Typha angustifolia* L.). Thus, these results make the substantial contribution to the field of the roles APX and GSH in higher plants to abiotic-induced oxidative stress.

Keywords: Ascorbate peroxidase, glutathione, higher plants, oxidative stress.

1. PEROXISOMAL ASCORBATE PEROXIDASE (pAPX) IN HIGHER PLANTS

The abiotic-induced oxidative stresses, particularly due to heat, salt and heavy metal, were very serious environmental problems with potentially harmful consequences for agriculture and human health. As a consequence, the exposure of higher plants to these abiotic stresses can enhance the intracellular level of reactive oxygen species (ROS) including partially reduced forms of oxygen (H_2O_2, $OH^{\bullet}$, O_2^-), as well as oxygen in a higher excited state ($^1O_2$). So, higher plants also have developed the antioxidant defence systems to cope with ROS. The systems are a cadre of compounds including ascorbate, glutathione, polyamines and carotenoids, as well as the antioxidant enzymes such as superoxide dismutase (SOD), catalase (CAT), glutathione reductase (GR) and ascorbate peroxidase (APX). In higher plants, ascorbate peroxidase (APX) has the high substrate specificity for ascorbate and is the primary H_2O_2-scavenging enzyme in higher plants (Zhang et al., 1997; Smirnoff 2005). APX are ubiquitous in the higher plants, being localized in the chloroplasts, mitochondria, cytosol, and in the peroxisomes (Shi et al., 2001). Peroxisomes are single-membrane-bound organelles that in higher plants participate in major tissue-specific functions such as ureide metabolism, fatty acid beta oxidation and photorespiration. Reactive oxygen byproducts of these processes are removed in part by a peroxisomal type APX (pAPX). pAPX as a single-pass peroxisomal membrane protein with its N-terminal active site domain facing the cytosol, where it is thought to scavenge H_2O_2 that escapes the peroxisome or that is produced in the cytosol. Thus, pAPX may play an important role in the tolerance of abiotic-induced oxidative stresses in higher plants.

2. ROLE OF pAPX IN HIGHER PLANTS TO HEAT-INDUCED OXIDATIVE STRESS

Heat stress is one of the main factors constraining the growth and productivity of plants. Under heat stress, excessive ROS are formed which cause oxidative damage to cell constituents. A major hydrogen peroxide detoxifying system in plant chloroplast and cytosol is the so-called ascorbate glutathione cycle, in which ascorbate peroxidases (APXs) are the key enzymes (Asada, 1992). APXs, especially cytosolic APX, are thought to play an essential role in protecting plants from oxidative stress. However, under heat stress, more H_2O_2 is generated in the microbody matrix and may readily diffuse into the cytosol. Thus, pAPX (peroxisomal type ascorbate peroxidase) that is considered to be bound to the outside of the membrane may play an important role in decomposing H_2O_2 and protecting cells from oxidative damage.

In our study (Shi et al., 2001), a full-length cDNA clone (HvAPX1) encoding a pAPX from barley (*Hordeum vulgare* cv. *Haruna-nijyo*) leave was isolated by DDRT-PCR (differential display reverse transcription-PCR). The deduced amino acid sequence of the HvAPX1 gene had the 72.1% homology to the *Arabidopsis thaliana* peroxisomal APX gene, APX3. Southern blot analysis indicated that a single-copy gene in the barley genome encoded HvAPX1. Northern blot analysis showed that the HvAPX1 transcript increased remarkably in response to heat stress. The HvAPX1 gene was introduced into *A. thaliana* under control of the 35S RNA promoter of the cauliflower mosaic virus. The transgenic plants were significantly more tolerant to heat stress as compared to the wild-type

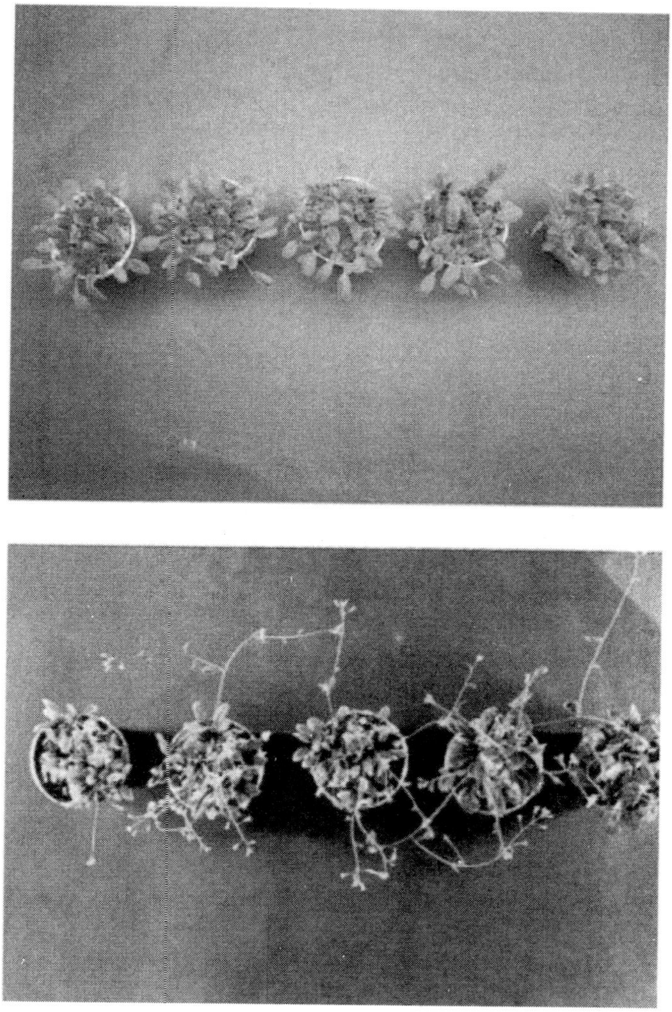

WT pAPX3 pAPX2 pAPX1 pAPX5

Fig. 1. Thermotolerance of wild-type and transgenic *HvAPX1 A. thaliana.* Transgenic Arabidopsis overexpressing HvAPX1 shows more tolerance to heat stress than wild-type (lower), and no obvious growth ference at normal conditions (upper). (Upper) Twenty-day-old Arabidopsis grown at 23°C; (lower) 3-week-old plants were exposed to 35°C for 5 days and transferred back to a 23°C chamber for an extra 5 days.

ig. 1). These results suggest that pAPX plays an important role in the higher plants to at stress.

Enhancing higher plant thermotolerane by genetic engineering method has been ed through different strategies such as overexpressing a heat shock protein (Lee and hoffl 1996) or a heat shock transcriptional factor (Prandle et al., 1998) or by synthesizing zyme producing osmoprotectant, glycinebetaine (Alia et al., 1998). Also, overexpression the pAPX gene from Arabidopsis in tobacco to provide additional protection against oxidative stress has been reported (Wang et al., 1999). However, enhanced heat-induce oxidative stress tolerance was observed only in the transgenic plants in the presence c a catalase inhibitor, aminotriazole, indicating the importance of pAPX to scavengin H_2O_2 overproduced in peroxisome. However, thermotolerance was not tested in thes researches. In this study, we showed directly that overexpression of the pAPX gene HvAPX1, is also quite helpful for enhancing plant thermotolerance. However, the degre of thermotolerance is not positively proportional to the HvAPX1 transcript level in th transgenic plants. Too high transcript levels give weaker plant thermotolerance. Mor detailed analyses on this point should be carried out.

3. ROLE OF pAPX IN HIGHER PLANTS TO SALT-INDUCED OXIDATIVE STRESS

Soil salinity, particularly due to NaCl, is a major abiotic stress in agricultur worldwide. Although Na^+ is a major cation present in the soil, it is not considere an essential mineral for most plants. Moreover, excessive Na^+ in the soil are harmfu to most plants (Xu and Shi 2007). The high concentration Na^+ disrupts the balanc of mineral nutrients in saline soils, thereby creating ionic stress, osmotic stress, an secondary stress such as nutritional disorders and oxidative stress. Some studie have revealed some of the mechanisms evolved by plants to adjust their physiolog and metabolisms to salt stress. Thus, one of them is the antioxidant defence syste (Zhu, 2001; Xu et al., 2008). In the recent years, in higher plants, the role of APX i the tolerance of salt stress has been studied. For example, overexpression of ascorba peroxidase in tobacco chloroplasts enhances the tolerance of salt stress (Badawi al., 2004). The results of Tsugane et al. (1999) also showed that when grown phot autotrophically, the *pst1* (a recessive Arabidopsis mutant) seedling were more tolera to salt stress than were wild-type seedlings and the enzyme activity of APX w enhanced significantly in *pst1* plants.

Although some studies have indicated that APX plays an important role in sa tolerance in higher plants, little work has been done on the role of pAPX in modulati plants adapted to salt tolerance. It is very clear that H_2O_2 are produced at high leve in peroxisomes (Lisenbee et al., 2003). A large proportion of the formed H_2O_2 peroxisome are catalysed by both APX and catalase through the reduction of H_2O_2 water so as to avoid damage to plants by the excessive H_2O_2. However, under oxidati stress, catalase activity is sharply decreased and then the predominant roles of H_2(detoxifying in peroxisome are supposed to rely on pAPX (Shi et al., 2001). Thus, w propose that pAPX may also play an important role in scavenging H_2O_2 under sa stress. In order to test this hypothesis and to better clarify the contribution of pAPX the regulation of H_2O_2 levels and in the protection against salt-induced oxidative stres we used the Arabidopsis and rice lines overexpressing *HvAPX1*, and tested them und salt stress conditions.

In our study, we found that the transgenic *Arabidopsis thaliana* plant (pAPX3) w more tolerant to salt stress than the wild type. Irrespective of salt stress, there were r significant difference in Na^+, K^+, Ca^{2+} and Mg^{2+} contents and the ratio of K^+ to N between pAPX3 and the wild type. Clearly, the mechanism of salt tolerance in transger

Arabidopsis thaliana (pAPX3) is not due to the maintenance and reestablishment of cellular ion homeostasis. However, the degree of H_2O_2 and MDA (measured as the levels of lipid peroxidation) accumulation under salt stress was more in the wild type than in transgenic line (pAPX3). The mechanism of salt tolerance in transgenic *Arabidopsis thaliana* (pAPX3) can thus be explained by reduction of oxidative stress injury. Under all conditions tested, activities of superoxide, glutathione reductase and catalase were not significantly different between pAPX3 and the wild type. In contrast, the activity of APX was significantly higher in the transgenic plant than in wild type under salt stress. These results suggest that in higher plants, peroxisomal ascorbate peroxidase (pAPX) gene plays an important role in salt tolerance and is a candidate gene for developing salt-tolerant crop plants. Therefore, we overexpressed the *HvAPX1* gene in rice plants by genetic method. Our unpublished results show that compared with the wild-type rice plants, rice plants overexpressing *HvAPX1* is more tolerant to salt stress (Fig. 2).

Fig. 2. Visual symptoms of 30-day-old wild-type (WT) and transgenic rice lines (APX3 and APX7) under the 100 μM NaCl.15-day-old rice plants were grown for 15 days under the 100 μM NaCl.

In conclusion, this study demonstrated that the pAPX gene, *HvAPX1*, is useful for enhancing the salt tolerance of plants. However, comprehensive analyses including micro-arrays for the pattern of gene expression, and proteomic studies to provide information on the accurate location of pAPX in peroxisomes during salt-induced oxidative stress will help to answer the many questions that remain. To improve salt stress tolerance, future work should focus on understanding many factors in great details, such as the isozymes, strength and quality of overexpression, and intracellular targeting and the nature of salt stress per se.

4. ROLE OF pAPX IN HIGHER PLANTS TO HEAVY-METAL-INDUCED OXIDATIVE STRESS

Heavy metal pollution is a very serious environmental problem with potentially harmfu consequences for agriculture and human health (Salt et al., 1998). The modern agricultura practices and the industrial activities have polluted soils with cadmium (Cd) or Zin (Zn) in areas that are favourable for crop production. Excess Cd or Zn causes severa symptoms of toxicity in plants, including growth retardation, inhibition c photosynthesis, altered stomatal movement, enzyme inhibition, and the generation c lipid peroxidation products. Therefore, plants must adapt to contaminated environment containing excess Cd or Zn. Thus, different mechanisms have been proposed to explai the tolerance of plants to Cd or Zn toxicity (Xu et al., 2008b). Firstly, regulation of Cd c Zn uptake at the root, as well as long distance and intracellular partitioning are necessar to control overloading. Secondly, environmental fluctuations experienced by plants ca lead to an increase in the internal concentrations of Cd or Zn. Mechanisms maintainin intracellular Cd or Zn in non-toxic forms are, therefore, of primary importance. Variou subcellular compartments participate in this control, by harbouring various molecule bearing affinity for Cd or Zn, and involved in their storage and detoxification. Thirdly Cd or Zn overload can lead to intracellular concentrations beyond the storage an detoxification capacities of plant cells. As a consequence, the exposure of plant cells t excessive Cd or Zn can enhance the intracellular level of ROS. So, plants also hav developed the antioxidant defence systems to cope with ROS. The systems are a cadr of compounds including ascorbate, glutathione, polyamines and carotenoids, as well a the antioxidant enzymes such as superoxide dismutase (SOD), catalase (CAT), glutathion reductase (GR) and ascorbate peroxidase (APX).

Recently, some researches have showed that in higher plants, APX may play a rol in excessive Cd or Zn stress (Chaoui et al., 1997; Xu et al., 2008b). Moreover, som works using transgenic plants have shown that APX plays an important role in oxidativ stress metabolism in higher plants (Allen et al., 1997; Yabuta et al., 2002; Lee et al 2007). However, little work has been done on the role of pAPX in modulating plants t adapt to excessive Cd or Zn stress. Here, the aim of the study is to determine whethe pAPX can reduce the damage of oxidative stress induced by heavy metal such a excessive Cd or Zn by using transgenic Arabidopsis plants.

In this study, in order to determine the role of pAPX in protection against excessive zinc-induced oxidative stress, transgenic *Arabidopsis* plants constitutively overexpressin a peroxisomal ascorbate peroxidase gene (HvAPX1) from barley were analyzed. Ou results showed that transgenic plants were more tolerant to zinc stress than wild-typ plants (Fig. 3). Under zinc stress, the concentration of hydrogen peroxide an malondialdehyde accumulation were higher in wild-type plants than in transgenic plant Therefore, the mechanism of zinc tolerance in transgenic plants may be due to reduce oxidative stress damage. Under zinc stress, the activities of APX were significantly highe in transgenic plants than in wild-type plants. We also found that the zinc accumulatio in the shoots were much higher in transgenic plants than in wild-type plants under zin stress. In addition, we found that compared with wild-type plants, transgenic plant

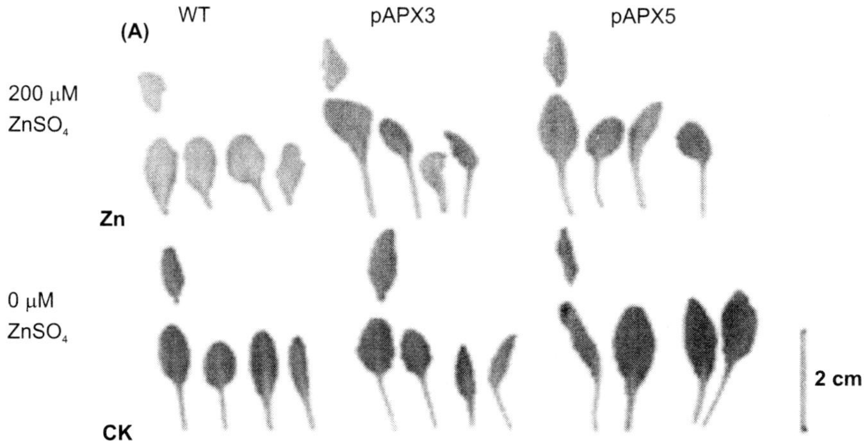

Fig. 3. Phenotypes of leaves of 15-day-old seedlings of wild-type and HvAPX1-overexpressing Arabidopsis treated with 0 or 200 μM $ZnSO_4$ for 10 d.

were more tolerant to excessive cadmium stress and accumulated more cadmium in shoots. Also, we overexpressed the HvAPX1 in rice plants (Duan et al., 2006). Our results indicated that with increasing *HvAPX1* transcript in transgenic rice, elongation rate of main root, total biomass, chlorophyll content and activity of the APX were significantly higher in transgenic rice plants than in wild type rice. The transgenic rice was significantly more tolerant to cadmium stress than the wild type rice and can accumulate more cadmium. Taken together, these results suggest that HvAPX1 plays an important role in zinc and cadmium tolerance and might be a candidate gene for developing high-biomass tolerant plants for phytoremediation of zinc- and cadmium-polluted environments.

5. GLUTATHIONE AND ITS ROLES IN HIGHER PLANTS TO CADMIUM-INDUCED OXIDATIVE STRESS

Glutathione, a tripeptide with the sequence γ-Glu-Cys-Gly, exists in either a reduced (GSH) or oxidized (GSSG) state in higher plants. Its protective role is threefold (Foyer et al., 2001). Firstly, GSH is a key antioxidant. Secondly, conjugation of glutathione to electrophilic molecules by glutathione S-transferases play a protective role in detoxification. Thirdly, GSH can chelate heavy metal, facilitating their sequestration in the vacuole. GSH biosynthesis and accumulation are enhanced in response to stress; for example, in catalase-deficient metants, dramatic increases in leaf glutathione are observed in high light when H_2O_2 is generated via photorespiration. Although glutathione can be synthesised in many plant organs, green leaves are the major sites of synthesis. The glutathione content in different cells and tissues are the product of equilibrium between synthesis, degradation, use, and short- and long-distance transport. GSH is the major transported form of reduced sulphur in plants and is exchanged between phloem and xylem. For instance, photosynthetic cells produce and export GSH which is supplied to other organs through phloem. Also, GSH is transported both into and out of plant cells by plasma membrane GSH transport systems. What is more, GSH is not merely a passive transport metabolite. Rather, it is a signal sensing and mediating the response systems of plants to stress conditions (Smirnoff 2005).

Under Cd stress, GSH functions as a Cd ligand and an antioxidant in higher plants. In addition, GSH acts as a cellular signal, mediating the control of enzyme and/or regulatory protein activities. Also, plants synthesize phytochelatins (PC_n, γ-(Glu-Cys)$_n$-Gly, n = 2–11) from GSH (PC_1) under Cd stress (Rausch et al., 2007; Chekmeneva et al., 2008; Rouhier et al., 2008). However, compared with GSH, the ability of PC_n synthesis is insufficient to cope with the Cd toxicity in higher plants (Sharma and Dietz 2009). Our unpublished results also showed that in cattail plants (*Typha angstifolia* L.), compared with the normal growth condition, the GSH content in cattail plants were increased under the Cd stress (Table 1).

Table 1: The effect of Cd on the biochemical characteristics of cattail (*Typha angustifolia* L.) plants.

Cd^{2+} (mg L^{-1})	MDA content (nmol g^{-1} FW)	Proline content (µg g^{-1} FW)	ASA content (µg g^{-1} FW)	GSH content (µg g^{-1} FW)
0	10.54±0.74[a]	42.09±2.14[a]	105.84±9.81[a]	569.77±13.34[a]
5	16.29±1.82[b]	43.24±1.89[a]	99.18±9.62[ab]	816.98±33.11[b]
50	30.31±2.90[c]	45.39±6.41[a]	87.82±11.79[b]	650.73±13.67[c]

In a word, glutathione has many diverse and important functions in higher plants respond to abiotic stress. In particular, glutathione is required for root development. Although much work has focused on the regulation role of GSH, it is only recently that molecular genetic works have been on the direction. In different cell and tissue of higher plant, GSH have the different regulation roles. Thus, this will inform future attempts to manipulate glutathione in higher plant.

6. CONCLUSION AND PERSPECTIVES

In conclusion, it is very clear that ascorbate peroxidase and glutathione play the indispensable roles in higher plants to abiotic-induced (heat, salt, heavy metal) oxidative stress and are two candidate genes for developing the tolerant plants for remediation of abiotic-stressed environments. However, to improve abiotic-stressed tolerance, future work should focus on understanding many factors in great details, such as the isozymes, strength and quality of ascorbate peroxidase and glutathione and their intracellular targets and these processes regulation.

ACKNOWLEDGEMENTS

These studies were supported financially by grants from the National Basic Research Programme of China (2008ZX07101-006-06), the Ministry of Science and Technology (2006BAD17B03) and the Chinese Academy of Sciences Innovation Program (ISSASIP0713). Also, we sincerely thank the DAAD (German Academic Exchange Service) scholarship to Dr. Weifeng Xu for working in the lab of Prof. Dr. S. Schubert (Justus-Liebig University, Giessen, Germany).

REFERENCES

Alia, Hayashi, H., Sakamoto, A. and Murata, N. (1998). Enhancement of the tolerance of Arabidopsis to high temperature by genetic engineering of the synthesis of glycinebetaine. *Plant J.* **16:** 155-161.

Allen, R.D., Webb, R.P., and Schake, S.A. (1997). Use of transgenic plants to study antioxidant defenses. *Free Radic. Biol. Med.* **23:** 473-479.

Asada, K. (1992). Ascorbate peroxidase: a hydrogen peroxide-scavenging enzyme in plants. *Physiol. Plant.* **85:** 235-241.

Badawi, G.H., Kawano, N., Yamauchi, Y., Shimada, E., Sasaki, R., Kubo, A. and Tanaka, K. (2004). Overexpression of ascorbate peroxidase in tobacco chloroplasts enhances the tolerance of salt stress and water deficit. *Physiologia Plantarum.* **121:** 231-238.

Chaoui, A., Mazhoudi, S., Ghorbal, M.H., and Ferjani, E.E. (1997). Cadmium and zinc induction of lipid peroxidation and effects on antioxidant enzyme activities in bean (*Phaseolus vulgaris* L.). *Plant Sci.* **127:** 139-147.

Chekmeneva, E., Prohens, R., Diaz-Cruz, J.M., Arino, C., Esteban, M. (2008). Thermodynamics of Cd^{2+} and Zn^{+} binding by the phytochelatin $(\beta\text{-GLu-Cys})_4$-Gly and its precursor glutathione. *Anal. Biochem.* **375:** 82-89.

Duan, S.R., Shi, W.M., Wang, J.R. (2006). Overexpressing peroxisomal type ascorbate peroxidase gene in rice enhanced tolerance to cadmium stress. *Acta Pedologica Sinica.* **43:** 111-116 (in Chinese).

Foyer, C.H., Theodoulou, F.L., Delrot, S. (2001). The functions of inter- and intracellular glutathione transport systems in plants. *Trends in Plant Sciences.* **6:** 486-492.

Lee, J.H., Schoffl, F. (1996). A Hsp70 antisense gene affects the expression of HSP70/HSC70, the regulation of HSF, and the acquisition of thermotolerance in transgenic *Arabidopsis thaliana. Mol. Gen. Genet.* **252:** 11-19.

Lee, Y.P., Kim, S.H., Bang, J.W., Lee, H.S., Kwak, S.S., and Kwon, S.Y. (2007). Enhanced tolerance to oxidative stress in transgenic tobacco plants expressing three antioxidant enzymes in chloroplasts. *Plant Cell Rep.* **26:** 591-598.

Lisenbee, C.S., Heinze M. and Trelease, R.N. (2003). Peroxisomal ascorbate peroxidase resides within a subdomain of rough endoplasmic reticulum in wild-type arabidopsis cells. *Plant Physiol.* **132:** 870-882.

Prandle, R., Hinderhofer, K., Eggers-Schumacher, G. and Schoffl, F. (1998). HSF3, a new heat shock factor from Arabidopsis thaliana, derepresses the heat shock response and confers thermotolerance when overexpressed in transgenic plants. *Mol. Gen. Genet.* **258:** 269-278.

Rausch, T., Gromes, R., Liedschulte, V., Muller, I., Bogs, J., Galovic, V. and Wachter, A. (2007). Novel insight into the regulation of GSH biosynthesis in higher plants. *Plant Biol.* **9:** 565-572.

Rouhier, N., Lemaire, S.D., Jacquot, J.P. (2008). The role of glutathione in photosynthetic organisms: emerging functions for glutaredoxins and glutathionylation. *Annu. Rev. Plant Biol.* **59:** 143-166.

Salt, D.E., Smith, R.D., Raskin, I. (1998). Phytoremediation. *Annu. Rev. Plant Physiol. Plant Mol. Biol.* **49:** 643-668.

Sharma, S.S., Dietz, K.J. (2009). The relationship between metal toxicity and cellular redox imbalance. *Trends Plant Sci.* **14:** 43-50.

Shi, W.M., Maramoto, Y., Ueda, A., and Takabe, T. (2001). Cloning of peroxisomal ascorbate peroxidase gene from barley and enhanced thermotolerance by overexpressing in *Arabidopsis thaliana. Gene.* **273:** 23-27.

Smirnoff, N. (2005). Antioxidants and Reactive Oxygen Species in Plants. Blackwell, Oxford.

Tsugane, K., Kobayashi, K., Niwa, Y., Ohba, Y., Wada K. and Kobayashi, H. (1999). A recessive Arabidopsis mutant that grows photoautotrophically under salt stress shows enhanced active oxygen detoxification. *Plant Cell.* **11:** 1195-1206.

Wang, J., Zhang, H., Randy, D.A. (1999). Overexpression of an Arabidopsis peroxisomal ascorbate peroxidase gene in tobacco increases protection against oxidative stress. *Plant Cell Physiol.* **40:** 725-732.

Xu, W.F., Shi, W.M. (2007). Mechanisms of salt tolerance in transgenic *Arabidopsis thaliana* constitutively overexpressing the tomato 14-3-3 protein TFT7. *Plant Soil.* **301:** 17-28.

Xu, W.F., Shi, W.M., Liu, F., Ueda, A., Takabe, T. (2008). Mechanisms of salt tolerance in the transgenic *Arabidopsis thaliana* carrying a peroxisomal ascorbate peroxidase gene from barley (*Hordeum vulgare* L.). *Pedosphere.* **18:** 486-495.

Xu, W.F., Shi, W.M., Liu, F., Ueda, A., Takabe, T. (2008b). Enhanced zinc and cadmium tolerance and accumulation in transgenic *Arabidopsis* plants constitutively overexpressing a barley gene (HvAPX1) that encodes a peroxisomal ascorbate peroxidase. *Bot.* **86:** 567-575.

Yabuta, Y., Motoki, T., Yoshimura, K., Takeda, T., Ishikawa, T., and Shigeoka, S. (2002). Thylakoid membrane-bound ascorbate peroxidase is a limiting factor of antioxidative systems under photo-oxidative stress. *Plant J.* **32:** 915-925.

Zhang, H., Wang, J., Nickel, U., Allen, R.D., and Goodman, H.M. (1997). Cloning and expression of an Arabidopsis gene encoding a putative peroxisomal ascorbate peroxidase. *Plant Mol. Biol.* **34:** 967-971.

Zhu, J.K. (2001). Plant salt tolerance. *Trends in Plant Sci.* **6:** 66-71.

12

Defence Strategies of Plants Against UV-B- Induced Oxidative Stress

S.B. Agrawal*, Suruchi Singh, Ruchika Tripathi and Madhoolika Agrawal

Lab of Air Pollution and Global Climate Change, Ecology Research Circle, Department of Botany, Banaras Hindu University, Varanasi–221 005, India

***Corresponding author:** *sbagrawal56@gmail.com*

ABSTRACT

Destruction of stratospheric ozone layer leads to increased influx of biologically effective ultraviolet-B (UV-B) radiation (280-315 nm) to the Earth's surface. Plants perceive and respond to UV-B by altering their metabolism in order to optimize their performance and defend themselves against the stressful condition. The perception of low levels of UV-B probably occurs via a UV-B photoreceptor followed by different signalling pathways. These pathways include second messengers such as calcium, kinases and the catalytic formation of reactive oxygen species (ROS). UV-B induced ROS formation has been suggested to be the primary events in signal transduction in plant cells. High levels of UV-B cause cellular damage and oxidative stress. UV-B is known to upregulate the synthesis of protective pigments, antioxidative enzymes, PR-genes and senescence induced genes. This chapter deals with UV-B perception, epidermal screening, cellular changes, ROS formation and its effects, ROS metabolism and redox couples.

Keywords: UV-B, photoreceptor, ROS, antioxidative enzymes, protective pigments.

1. INTRODUCTION

Due to stratospheric ozone depletion, induced by anthropogenic emission of halocarbons, increasing level of ultraviolet-B (UV-B) radiation (280-315 nm) is reaching the earth's surface (McKenzie et al., 2007). This has led to a major concern about the impact of enhanced UV-B radiation on agricultural crop plants (Caldwell et al., 2007). UV-B radiation can cause direct and indirect effects including degradation and conformational changes in DNA, proteins and cell membranes, alterations in transpiration and photosynthesis and changes in growth, development and morphology and ultimately the yield (Singh et al., 2009). UV-B radiations induce oxidative stress (Panagopoulos et al., 1990) although the mechanism of reactive oxygen species (ROS) formation is not well known (Rao et al., 1996). Since plants do not have locomotion and are forced to live with harmful UV-B radiation throughout their evolutionary history (Jansen et al., 1998) they have acquired different strategies to counteract the effects of solar UV-B radiation. The ability of higher plants to scavenge toxic ROS seems to be an important determinant for their tolerance against environment stress (Tang et al., 2010). To prevent injuries, plants have developed different mechanisms by which they can convert ROS into less-toxic products. These defence mechanisms are based on the presence of various metabolic compounds and enzymes in plants (Pagano et al., 2000). Earlier studies have already shown that high levels of antioxidant enzymes can lead to increased stress tolerance in plants including transgenic ones (Zhang et al., 2008; Agrawal et al., 2009).

1.1 UV-B Perception

A wide range of biologically active molecules like nucleic acid, lipids, phenolic compounds and aromatic amino acids are there which strongly absorb the light of UV-B spectrum. Previously it was speculated that UV-B may be perceived by a distinguished class of photoreceptors but till date no specific molecule has ever been identified and the concept of UV-B receptor is under vague. Various mutant studies have been conducted to identify the molecule behind the UV-B perception. According to earlier hypothesis phytochromes and cryptochromes served as receptors of UV-B. Mutants which are devoid of these photoreceptors show influence of UV-B on hypocotyl elongation. But this hypothesis was discarded because the response was mainly specific to some parameters (Frohnmeyer and Staiger 2003). Inferences from some important studies suggest that UV-B receptor consists of a protein with a bound pterin or flavin as chromophores (Frohnmeyer and Staiger 2003). Recently an important factor ULI3 has been identified as a component of the UV-B signalling pathway which encodes for an unknown protein containing putative haem and diacylglycerol binding sites (Lariguet and Dunand 2005). But the presence of ULI3 was mainly reported in the members belonging to the family Brassicaceae. However, other reports suggest that activation of receptors for UV-B defence signals co-opts other stress signalling pathway. Yalamanchili and Stratmann (2002) have given evidence of some membrane bound receptors in plants. Important cell surface receptor, SR 160 (a peptide systemin) becomes phosphorylated on the intracellular kinase domain in response to UV-B resulting in activation of several defence signalling steps (Stratmann 2003).

1.2 Epidermal Screening

Plants have developed various defence strategies to cope with UV-B stress. Increased production of UV-B absorbing compounds like flavonoids, higher amount of reflective waxes and thicker and smaller leaves are some important adaptive response of plants against UV-B (Qi et al., 2003). Flavonoids constitute a family of aromatic compound mainly derived from phenylalanine and malonyl coenzyme mediated pathways. These compounds include six major subgroups as chalcone, flavones, flavonols, flavondiols, anthocyanin and condensed tannins and beside these a seventh one which is not ubiquitous in nature, i.e., aurones (Shirley 2001). All these molecules play an important protective role by reducing epidermal transmittance of UV-B radiation because of their strong absorbance in this wavelength region and by accumulation of their high concentrations in the epidermal layer (Qi et al., 2003). General concept is that leaves absorb over 90% of incident UV-B and reflect generally below 10% while there is negligible transmission through leaves. Qi et al. (2002) have studied leaves of four tree species and reported that little or no UV-B is transmitted to palisade and mesophyll tissue layer and it is the upper epidermal layer which is responsible for absorbing 90% UV-B. Along with flavonoids, epicuticular waxes, various polyamines and alkaloids are responsible for UV-B screening. UV-B is generally reflected through cuticle or absorbed by soluble flavonoids localized in waxy layer or cutin having polymerization of ferulic acid (Lingakumar and Kulandaivelu 2006). Kakani et al. (2003) have studied UV-B induced anatomical changes in cotton leaves and reported quantitative and qualitative changes in wax content along with increased thickness of epidermal layer. However, in two pine species, Laakso et al. (2000) have also noticed increased thickness in epidermal layer and significant reduction in mesophyll cells. Epidermal cells were also filled with increased amount of tannin in the lumen and cell wall bound phenolics in outer wall (Fig. 1). Among all the flavonoids kaempferol glycosides is the most abundant compound in adaxial epidermis, however, in abaxial layer mainly hydroxyl cinnamic acid derivatives are found. Quercetin glycosides are the main compound abundant in both epidermal layer (Kalbin et al., 2001). Increase in flavonoids and sinapic esters are mainly accompanied with UV-B triggered transcription of key enzymes of phenylpropanoid pathway. Phenylalanine ammonia lyase (PAL), cinnamate-4-hydroxylase (C4H), 4-coumarate: CoA ligase (4CL), acetyl-COA carboxylase (ACC), chalcone synthase (CHS), and chalcone flavanone isomerase (CHI) are enzymes of flavonoid synthesis pathway and among these CHS is the main enzyme that catalyses the deciding step and is mainly triggered by UV-B (Lingakumar and Kulandaivelu 2006) (Fig. 1). CHS have two UV-B responding units—cis and trans which provide binding sites for transcription factors. Upregulation and downregulation of CHS gene mainly depends on response of plant to UV-B (Jansen et al., 1998). Ryan et al. (2002) have reported upregulation of genes encoding key enzymes of phenylpropanoid pathway including PAL, C4H, CHS, CHI and F3′H in case of petunia plant (Fig. 1). Transcripts of these enzymes show accumulation in vacuolar region of epidermal cell. Similarly, a specific subgroup of R2R3-MYB genes like C1 gene of maize, AN2 in petunia and PA1/PA2 in *Arabidopsis* have been reported which regulate anthocyanin biosynthesis in plants (Guo et al., 2008). Figure 1 shows the synthesis of flavonoids and anthocyanin in epidermal layer of plants.

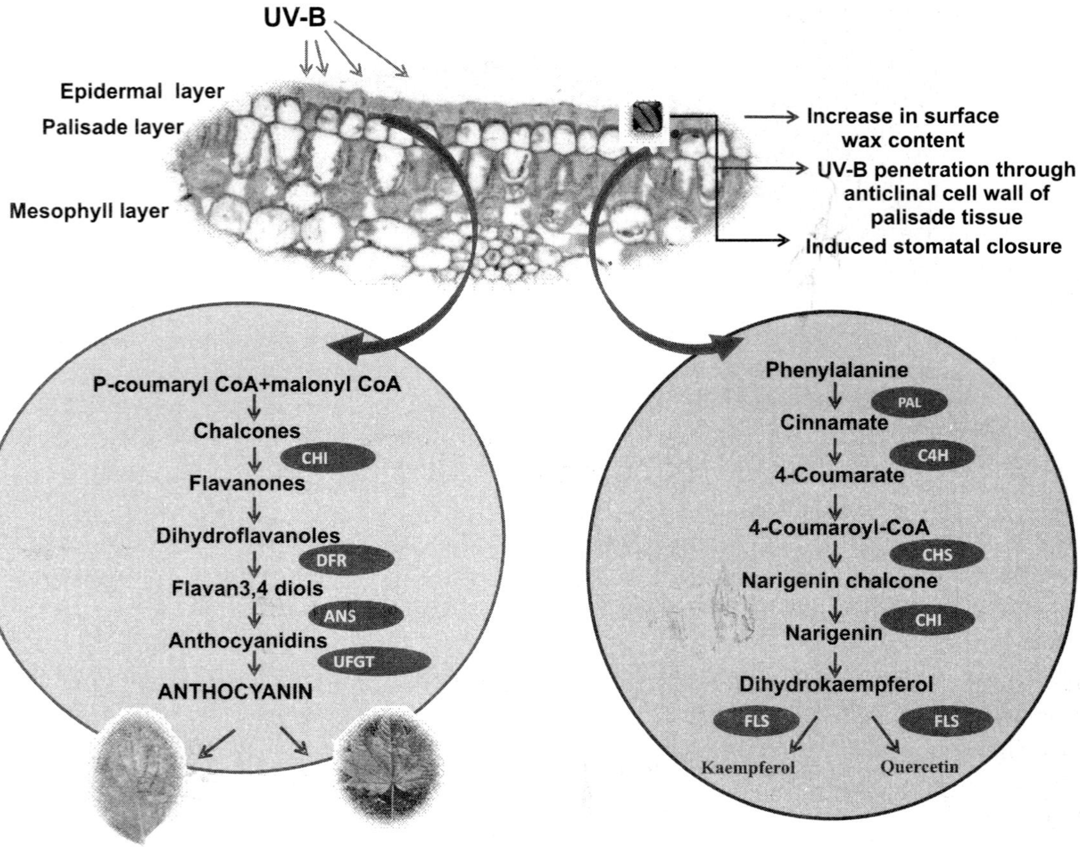

Fig. 1. UV-B induced synthesis of flavonoids and anthocyanin in epidermal layer and alterations in leaf tissue. PAL, phenylalanine ammonia lyase; C4H, cinnamate-4-hydroxylase; CHS, chalcone synthase; CHI, chalcone flavanone isomerase; FLS, flavonol synthase; DFR, dihydroflavanol-4-reductase; ANS, anthocyanidin synthase; UFGT, UDP flavonoid glucosyl transferase.

After screening through the protoplast of epidermis containing phenolic compounds, relatively little transmission of UV-B occurs and then it passes through the underlying mesophyll and palisade cells by passage through the anticlinal cell wall (Jordan 2002) (Fig. 1).

1.3 Cellular Changes

Besides inducing various protective mechanisms, UV-B also imparts several changes at ultra-structural level in plants (Fig. 2). Membranes of various organelles show different structural and functional alterations. UV-B induces damage of proteins and pigment protein complexes of plasmalemma and photosynthetic apparatus leading to altered transport and ion permeability across the membrane (Murphy and Wilson 1982; Chow et al., 1992) (Fig. 2). An important consequence of UV-B damage is loss of stomatal conductance due to UV-B induced loss of K^+ ions from guard cells (Nagesh and Bjorn

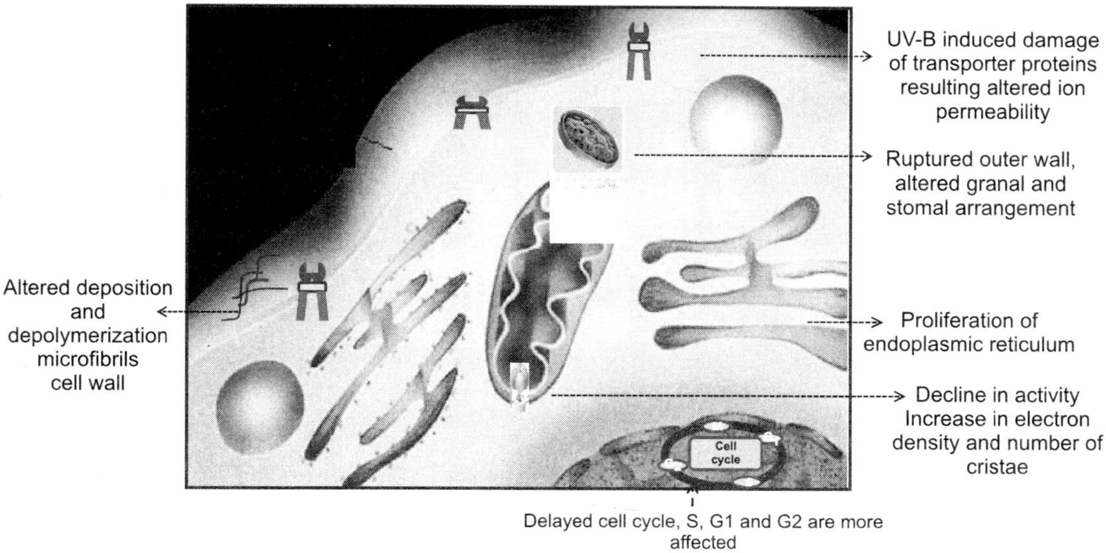

Fig. 2. Alterations induced by UV-B in various cell organelles of plants.

1986). Among all the cell organelles, chloroplast appears to be the most sensitive to UV-B (Bornman et al., 1983), however in *Zea mays*, mitochondria showed more alterations. Chloroplasts generally show swelling and its outer wall becomes ruptured. The orderly arrangement of thylakoids in grana and stroma becomes changed and appear slightly dilated (Hollosy 2002). PSI mediated cyclic photophosphorylation, water oxidizing complex (D1 and D2), ATPase enzyme are the most prone target sites of UV-B. Alterations were also reported in endoplasmic reticulum and dictyosomes. Since the enzyme of flavonoid synthesis, chalcone synthase is often associated with endoplasmic reticulum (ER) so the ER shows proliferation in its network (Hollosy 2002). RuBisco the key enzyme of carbon dioxide assimilation is one of the most important UV-B prone parts of cell. UV-B may induce modification of peptide chain, degradation of protein, photo modification and diminished transcription of gene are some important causes that impart a sharp decline in RuBisco enzyme (Jordan et al., 1992). Staxen and Bornman (1994) have reported altered deposition of cell wall microfibrils and its depolymerization after UV-B exposure. Staxen (1993) have also seen delayed progression through the cell cycle with the phases S, G1 and G2 being most affected (Fig. 2).

2. ROS FORMATION AND EFFECTS

Exposure to UV-B leads to the generation of ROS (Agrawal et al., 2009). In general, oxidative stress results from the disruption of cellular homeostasis of ROS production from the excitation of O_2 to form $^1O_2$ and the transfer of 1, 2 or 3 e^-(s) to O_2 to form superoxide anion ($^{\bullet}O_2^-$), hydrogen peroxide (H_2O_2) and the hydroxyl radical ($^{\bullet}OH^{\bullet}$), respectively (Halliwell and Gutteridge 1999). Ascorbate radicals (Hideg et al., 1997), long-lived chlorophyll radicals (Kumagai et al., 1999) as well as oxygen-, carbon- and nitrogen-centred free radicals (He et al., 2005) were detected with electron paramagnetic resonance (EPR) spectroscopy in a variety of plants upon UV-B exposure. UV-B also induces NADPH-oxidase activity in *Arabidopsis*, leading to peroxide formation (Rao and Omrod 1995). Peroxide scavenging by anionic peroxidases results in the formation of phenoxy radicals that can spontaneously polymerize, leading to lignification. The main sources of ROS in plants are electron transport chain (ETC) in chloroplast and mitochondria, some peroxidases and oxidases (NADPH oxidase, xanthine oxidase, lipoxygenase, glycolate oxidase, amine oxidase, etc.), photosensitizers such as chlorophyll molecules (Dat et al., 2000) and peroxisomes (Foyer and Noctor 2003).

When the photon intensity is within the range of the utilizing capacity of chloroplasts for CO_2 fixation, the generation of ROS is very low and these molecules are effectively scavenged. However, when either PAR intensity is high or the utilizing capacity is lowered due to environmental stress, excess photons promote the production of ROS and cause the photoinhibition of photosynthesis (Fig. 3). Under such condition, PSII is preferentially damaged (Powles 1984; Barber and Anderson 1992). Environmental stress conditions reduce $NADP^+$ generation by the Calvin cycle, consequently the photosynthetic electron transport chain is overreduced, forming $^{\bullet}O_2^-$ and $^1O_2$ in the chloroplast (Krause 1994). Rubisco activity as well as content is found to get reduced by prolonged exposure to UV-B radiation (Strid et al., 1990; Jordan et al., 1992). Decline in

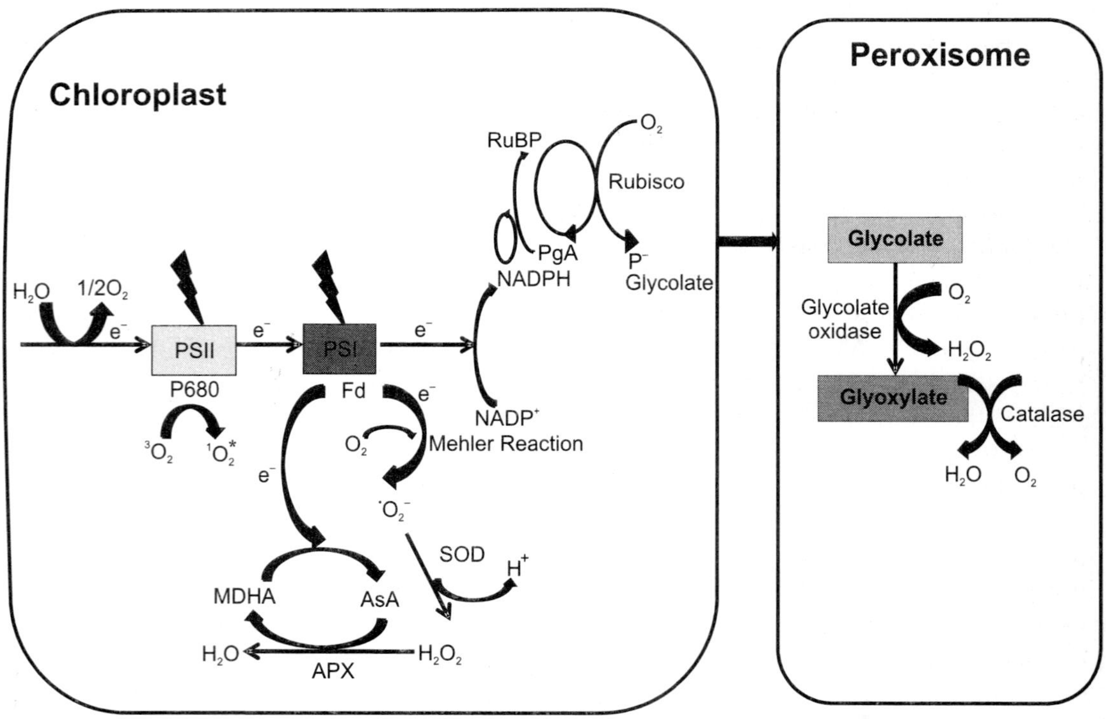

Fig. 3. ROS formation and scavenging in chloroplast and peroxisome (drawn with changes from Apel and Hert, 2004). PSII, photosystem II; PSI, photosystem I; AsA, ascorbic acid; Fd, ferredoxin; SOD, superoxide dismutase; APX, ascorbate peroxidase; MDHA, monodehydroascorbate.

the rate of CO_2 assimilation induced by exposure to elevated UV-B radiation in oilseed rape has been shown due to the loss of Rubisco (Allen et al., 1997), which may also be associated with loss of activities of other enzymes of Calvin cycle (Baker et al., 1997). Wilson et al. (1995) reported in oilseed rape, pea, tomato and tobacco that the appearance of 66 kDa protein during UV-B exposure is the photomodified form of the large subunit of Rubisco. The ETC in chloroplast operates in an O_2 rich environment, so that leakage of electrons from overloaded ETC will lead to ROS production. The e-flow from excited PS centres is directed to NADPH, this then enters the Calvin cycle and reduces the final electron acceptor, CO_2. In conditions of overloading of ETC, a part of the electron flow is diverted from ferredoxin to O_2, reducing it to superoxide free radical (Fig. 3).

Dai et al. (1997) observed that excessive accumulation of ROS in leaves following UV-B treatment strongly inhibited photosynthetic electron activity. The outlet of e^-(s) from ferredoxin to O_2 is called the Mehler reaction (Fig. 3). The rate of H_2O_2 production in the Mehler reaction is sufficiently high to cause an accumulation of 10 μM H_2O_2 within 0.5 s which leads up to 50% inhibition of CO_2 fixation when the scavenging enzymatic system of chloroplast does not function. The acceptor side of ETC in PSII also provides sites (QA, QB) of electron leakage to O_2 producing $O_2^{\bullet-}$ (Dat et al., 2000). Once superoxide anions are produced, they follow different pathways. On the internal lumen membrane surface, $O_2^{\bullet-}$ may be protonated to H_2O_2 which initiates lipid peroxidation. On the external membrane surface, $O_2^{\bullet-}$ is enzymatically or spontaneously dismutated to H_2O_2 and O_2. In dismutation, however, the scavenging of one ROS type, $O_2^{\bullet-}$ yields another ROS type, H_2O_2 (Foyer et al., 1994). At the level of Fe-S centres where Fe^{2+} is available, H_2O_2 is transformed through the Fenton reaction into $^{\bullet}OH$. Highly energetic photons of the UV-B range are absorbed by the chromophore groups of many biologically important molecules such as chlorophyll, phycobiliproteins and quinones. These molecules can act as photosensitizers for the production of ROS (Caldwell et al., 1998; Jordan 1996). Under normal growth condition, the energy of the excited chlorophylls or phycobilins is utilized efficiently for photosynthesis. However, the inhibition of photosynthesis or electron transport chain under excess of UV-B may lead to photosensitization process as well as the formation of ROS. Normally, the excited singlet status of the chlorophyll serves to transfer energy or electrons. To emit energy, chlorophyll uses fluorescence or conversion to the triplet state which leads to formation of singlet oxygen (Arora et al., 2002).

The formation of singlet oxygen *via* photosensitization is known to play a crucial role in damaging the D1 protein (Hideg et al., 1994). Under UV-B stress, the inhibition of the electron transport chain due to the degradation of the D1 protein of PSII might promote the energy transfer from triplet chlorophyll to oxygen to form singlet oxygen (Jordan 1996). The imbalance between the light phase and the Calvin cycle probably due to the decreased activity of ribulose-1,5-bisphosphate carboxylase/oxygenase (Rubisco) by UV radiation promotes the formation of superoxide radicals at the level of ferredoxin in photosystem I (PSI) (Bischof et al., 2000) (Fig. 3). Another important source of ROS formation especially of H_2O_2 is the photorespiration in the peroxisomes (Fig. 3). During CO_2 fixation, Rubisco uses CO_2 to carboxylate RuBP. This enzyme can also use molecular O_2 to oxygenate RuBP. During this reaction, glycolate is formed and transported from chloroplast to peroxisomes. The glycolate is then oxidized and H_2O_2 is formed as a byproduct. Oxygen reduction absorbs significant levels of the photosynthetic electron flux not only through its role in photorespiration but also by its direct reduction by PSI (Asada 1999). H_2O_2 is generated as a result of direct UV-B photochemical reaction in the plants (Larson 1988), and also due to increase in SOD activity under UV-B radiation (Balakumar et al., 1997) (Fig. 3).

Evidence exist for the involvement of ROS in some morphological changes and gene expression responses initiated by UV-B (Jenkins 2009). Cotyledon curling in *Brassica napus* was inhibited by ascorbate whereas stimulated by UV-B and H_2O_2 (Gerhardt et al., 2005). A number of reports have suggested that ROS are involved in the photodamage of PSII (Prásil et al., 1992; Aro et al., 1993; Andersson and Barber 1996). ROS such as $O_2^{\bullet-}$ (Chen et al., 1992), H_2O_2 (Ananyev et al., 1992; Miyao et al., 1995) and $^{\bullet}OH$ (Miyao et al., 1995) have shown to induce the cleavage of the D1 protein directly *in vitro*. Besides such ROS, $^1O_2$, which is generated by transfer of excitation energy in the reaction centre of PSII, is often considered to trigger degradation of the D1 protein (Vass et al., 1992; Hideg et al., 1994; Telfer et al., 1994; Okada et al., 1996). High ROS levels initiate signalling responses that include enzyme activation, gene expression, programmed cell death and cellular damage (A.-H.-Mackerness et al., 2001). The involvement of ROS in regulation of photosynthetic genes by UV-B radiation requires a relationship between ROS levels within a tissue and the extent of downregulation in green tissue (A.-H.-Mackerness et al., 1998). UV-B irradiation results in the oxidative damage of structures (membrane, lipid, protein) or functions (redox component, enzyme) causing smaller bioproduction and under extreme condition even the irreversible damage (Hideg et al., 2008). Changes in the activity of antioxidant enzymes strongly suggest the involvement of ROS. Oxidative stress occurs when there is serious imbalance in any cell compartment between the production of ROS and antioxidant defence, leading to significant physiological challenges (Ahmed et al., 2008; Foyer and Noctor 2000; Tuteja et al., 2009). $^{\bullet}OH$ reacts rapidly with all types of cellular components at any place, while $^{\bullet}O_2^-$ reacts primarily with Fe-S centers and $^1O_2$ is particularly reactive with conjugated double bonds as found in polyunsaturated fatty acid (PUFA).

Excess ROS cause damage to proteins, lipids, carbohydrates, DNA and ultimately results in cell death (Mittler et al., 2004; Foyer and Noctor 2005a). Amino acid residues vary in their susceptibility to oxidative modification, the most vulnerable including Cys, Tyr, Trp and His (Dröge, 2002). The maintenance of sulphur-containing amino acid is particularly important. Methionine is readily oxidized to the sulphoxide (MetSO). The S-stereoisomer of MetSO is reduced back to Met by type A protein MetSO reductase (PMSR), enzymes that use thioredoxins (TRX) as reductant (Romero et al., 2004). Various mechanisms may act to regenerate protein Cys, including protein disulphide reductases such TRXs and glutaredoxins (Lemaire 2004).

Polyunsaturated Fatty Acid, that is, linolenic acid and levulonic acid are the major fatty acids in the plant membrane, galactolipids (thylakoid membrane), and phospholipids (all other membranes). Relatively higher content of PUFA is found in thylakoid membranes which facilitates photosynthesis (He and Hader 2002). It has been suggested that exposure to UV-B results in the generation of ROS within the chloroplasts, as thylakoid membranes are very rapidly perturbed upon exposure to UV-B radiation (Chow et al., 1992). The PUFA are particularly susceptible to attack by $_1O^2$ and $^{\bullet}OH$, giving rise to complex mixtures of lipid hydroperoxides (Mueller 2004). $\cdot OH$ radicals may attack PUFA to form carbon centred (lipid) radicals by hydrogen abstraction. In the presence of oxygen, these are converted into lipid peroxyl radicals, which are capable of hydrogen abstraction and thus of propagating oxidative stress in the membrane. Polyunsaturated lipids may be converted into carbon-centred lipids upon hydrogen abstraction. The unpaired electron is usually formed on bis-allyl methyl groups. Two carbon-centred radical groups may form cross links and stabilize as conjugated double bonds. These carbon-centred radical groups decrease membranes fluidity and alter hydrophobicity. Carbon-centred radicals may also combine with oxygen, forming ROS capable of hydrogen abstraction and thus propagating oxidative damage in the membrane. PUFA peroxidation decreases the fluidity of the membrane, increases the leakiness and causes secondary damage to the membranes. The lipid peroxidation which is indicative of damage to cellular membranes, interferes with the function of membranes (Panagopoulos et al., 1990; Kramer et al., 1991), and ultimately results in bleaching of chlorophyll (Elstner 1982). Lipid hydroperoxides, formed as a result of lipid peroxidation, can affect membrane properties by increasing the membrane hydrophilicity of the internal side of the bilayer (Frenkel 1991). This phenomenon is very important for the membrane favouring the generation of tocopherol by ascorbate.

ROS generated under UV-B treatment, especially $HO^{\bullet}$ and $^1O_2$ react with sugar and purines (Apel and Hert 2004). DNA can be modified by ROS in many ways, $HO^{\bullet}$ is the most reactive, $^1O_2$ primarily attacks guanine, while H_2O_2 and $O_2^{\bullet-}$ do not react at all (Wiseman and Halliwell 1996). The $HO^{\bullet}$ radical is an electrophilic, highly reactive radical. OH ion can abstract hydrogen from the deoxyribose sugar units in DNA which, if not repaired, leads to cleavage of the sugar-phosphate backbone of the DNA (Breen and Murphy 1995). DNA bases suffer photooxidative damage and in this respect guanine is particularly sensitive (Kanvah and Schuster 2006). A direct photon hit in a base may expel an electron, and the electron hole can mitigate along the DNA chain unless it encounters a guanine residue, via interaction with water, resulting in 2,6-diamino-4-hydroxy-5-formamidopyrimidine. Deprotonation instead of hydration may also occur, and gives another product. Another possibility is that a photoexcitation enters a triplet excited state and reacts with oxygen, resulting in generation of $^1O_2$. The $^1O_2$ then attacks a DNA base, preferably guanine. In addition to direct DNA oxidation, ROS can also indirectly modify DNA. A common type of damage involves conjugation of PUFA breakdown product MDA with guanine (Jeong et al., 2005). UV-B induced DNA damage and oxidative stress as potential causes of growth reduction. Furthermore, there is evidence that ROS are involved in signalling mechanisms that activate some UV-induced defence genes (Green and Fluhr 1995). The oxidation of sugars with $^{\bullet}OH$ often releases formic acid as the main breakdown product (Isbell et al., 1973).

ROS may also damage proteins yielding various carbon-centred (alkoxyl or $CO_2^{\bullet-}$ anion) free radicals and bound amino acid hydroperoxides (Davies et al., 1995). By reacting with neighbouring amino acids, these later may spread oxidative damage along the protein. This promotes protein fragmentation or may also produce free radicals

(Hawkins and Davies 2001). Phenolic compounds may also act as antioxidants. Similarly to ascorbate, they react with ROS while being converted into a free radical. Radicals formed from antioxidants are less harmful than ROS and are usually recycled, reduced back to their original protective form by other antioxidants or enzymes. Their recycle may be delayed when ROS production is intense, or when enzymes are damaged. In this way, the accumulation of antioxidant radicals, such as ascorbate (Heber et al., 1996; Hideg et al., 1997) or phenolic radical (Sakihama et al., 2000) may serve as stress marker.

3. ROS METABOLISM

Plants are well equipped with both enzymatic and non-enzymatic antioxidants. Antioxidant defence system consists of low molecular weight antioxidants such as ascorbate, glutathione (GSH), α-tocopherol and carotenoids (Alscher and Hess 1993), as well as of protective enzymes as superoxide dismutase (SOD), catalase (CAT), peroxidase (POD), monodehydroascorbate (MDHAR), dehydroascorbate reductase (DHAR) and glutathione reductase (GR). In the enzymatic ROS-scavenging pathways, SOD (EC 1.15.1.1) converts $O_2^{\bullet -}$ to H_2O_2 and APX (EC 1.11.1.11) and GR (EC 1.6.4.2.) in the ascorbate-glutathione (AA-G) cycle are responsible for H_2O_2 removal (Asada 1999). It is well known that APX utilizes AsA to reduce H_2O_2 by the oxidation of AsA to monodehydroascorbate (MDHA) radical, which can be reduced by photosynthetic electron (e^-) flow through ferredoxin or reduced to AsA by NAD(P)H dependent MDHAR (EC 1.6.5.4) and MDHA would be spontaneously disproportionated to AsA and dehydroascorbate (DHA) (Mittler 2002). DHA will then be reduced to generate AsA by dehydroascorbate reductase (DHAR, EC 1.8.5.1) using glutathione as an e^- donor, the glutathione disulphide (GSSG) formed after glutathione oxidation is reduced to glutathione (GSH) by GR utilizing reducing equivalents from NAD(P)H (Fig. 4). Catalase (CAT, EC 1.11.1.6) (Willekens et al., 1997) and peroxidase (POD, EC 1.11.1.7) (Asada and Takahashi 1987) are also involved in H_2O_2 removal. ROS play a role in mediating UV-damage (Strid et al., 1994; Landry et al., 1995). Oxidative stress occurs when there is serious imbalance in any cell compartments between the production of ROS and antioxidant defence, leading to significant physiological challenges (Ahmad et al., 2008; Foyer and Noctor 2000; Tuteja et al., 2009). The transcriptional and translational regulations for the antioxidative gene are closely associated with the change of *in vivo* ROS generation (Foyer et al., 1994; Niyogi 1999). SOD is a metallo-enzyme containing either Cu and Zn or Mn which catalyses the dismutation reaction of superoxide anion ($O_2^{\bullet -}$) into H_2O_2 and O_2. In plants, three different SODs are present: in the cytosol (Cu/Zn SOD), in mitochondria (Mn-SOD) and in chloroplasts (Cu/Zn SOD).

These SODs can be easily differentiated on the basis of mRNA transcripts, as well as activity levels with *in situ* staining technique on gel (Rao et al., 1996). H_2O_2 is generated as a result of direct UV-B photochemical reaction in plants (Larson 1988) and also due to the increase in SOD activity under UV-B irradiation (Balakumar et al., 1997). Xu et al. (2008) and Yanqun et al. (2003), however, reported decrease in SOD activity in UV-B sensitive *Glycine max*. Rao et al. (1996) showed induction of cytosolic as well chloroplastic

Cu, Zn-SOD, due to preferential expression of Cu/Zn-SOD-3, -4 and -5 isoforms, while UV-B exposure did not significantly affect Mn-SOD. Strid et al. (1994) reported a decrease in mRNA transcript of chloroplastic SOD of *Pisum sativum* grown under supplemental UV-B. Selvakumar (2008) reported linear increase in SOD activity in *Crotalaria juncea* with corresponding increase in UV-B dose. Increase in SOD activity was found associated with concurrent decline in CAT activity suggesting accumulation of H_2O_2 in the plant cells exposed to UV-B (Selvakumar 2008).

SOD activity was increased by UV-B radiation in pea and wheat (Alexieva et al., 2001), *Arabidopsis* (Rao and Ormrod 1995), *Lemna gibba* (Babu et al., 2003), rice (Dai et al., 1997), *Acorus* (Kumari et al., 2010) and *Abelmoschus* (Kumari et al., 2009) but not affected in Barley (Mazza et al., 1999) and soybean (Malanga et al., 1999), and was decreased in sunflower cotyledon (Costa et al., 2002). Also, SOD expression was not affected by UV-B radiation in *Nicotiana plumbaginifolia* L. (Willekens et al., 1994), but was decreased in *Pisum sativum* (Strid 1993). In a field study, supplemental UV-B increased SOD activity in wheat and mung bean (Agrawal and Rathore 2007), and caused differential responses among soybean cultivars (Yanqun et al., 2003).

APX isoenzymes are distributed in at least four distinct cellular compartments: stromal APX (sAPX), and thylakoid-membrane bound (tAPX) in chloroplasts, microbody (including glyoxysomes and peroxisome), membrane- bound APX (mAPX) and cytosolic APX (cAPX) (Chen and Asada 1989) (Fig. 3). A fifth APX isoenzyme (mit APX) occurs in mitochondria as a membrane-bound form (Leonardis et al., 2000) (Fig. 3). The isoforms are encoded by distinct genes and differ in size, specificity for their electron donors and sensitivity to inactivation (Chen and Asada 1989). In *Arabidopsis*, a total of five genes have been identified coding for various isoforms in the chloroplast (APX4 and APX5), cytosol (APX1 and APX2) and microbodies (APX3) (Santos et al., 1996). This probably represents a full gene set but other isoforms may arise by some form of post-transcriptional or post-translational processes. A higher APX activity is correlated with physiological processes such as the removal of H_2O_2, biosynthesis of lignin and general plant development (Asada 1992). Though elevated activities of APX result in the depletion of AsA, it has to be regenerated for maintaining the redox status of the cells.

To accomplish the cellular regeneration, UV-B irradiation increased APX activity in *Arabidopsis thaliana* (Rao et al., 1996). However, Yannarelli et al. (2006) and Costa et al. (2002) reported no significant effect on APX activity of cotyledons of sunflower exposed with UV-B in growth chamber, and Dai et al. (1997) in rice leaves grown under greenhouse conditions. Yao and Liu (2006) also reported decline in APX activity in *Acer Mono Maxim* leaves at a high UV-B dose under field conditions.

Catalases are tetrameric heme-containing enzymes and dismutate H_2O_2 to H_2O and O_2. This eliminates H_2O_2 accumulation and prevents reaction between H_2O_2 and $O_2^{\bullet -}$ which generate the extremely potent oxidant $^{\bullet}OH$. In most plants, CAT occurs as multiple isoforms (Chandlee et al., 1983; Havir and McHale 1990; Willekens et al., 1994). In maize, CAT exists as three biochemically distinct isozymes (CAT-1, CAT-2 and CAT-3) encoded by three unlinked nuclear genes, Cat1, Cat2 and Cat3 (Scandalios et al., 1980). Catalases are present in peroxisomes, glyoxysomes and related organelles where H_2O_2-generating enzymes such as glycolate oxidase are found. In developing maize leaves, the catalase genes exhibit different responses to light. The expression of Cat1 gene is light independent throughout development (Scandalios et al., 1997). In contrast, the Cat2 gene is positively regulated by light (Skadsen and Scandalios 1987) whereas the light regulation of the Cat3 gene is quiet complex. Induction of Cat1 and Cat2 transcripts as well as of the enzyme activities by UV-B radiation are well known (Willekens et al., 1994; Alexieva et al., 2001). Willekens *et al.* (1994) reported that under UV-B treatment Cat1 is highly expressed in photosynthesizing cells of *Nicotiana plumbaginifolia* where it can scavenge the H_2O_2 that is produced during photorespiration. In maize, the expression and accumulation of Cat2 and Cat3 catalase genes is induced by UV-B suggesting that both the genes may act together to scavenge ROS generated by UV-B to protect the plants from oxidative damage (Boldt and Scandalios 1997). It is known that the limited protective action of catalase is attributed to its poor affinity for its substrate, its sensitivity to light induced inactivation (Foyer et al., 1994) and inhibition of its activity by high $O_2^{\bullet-}$ or H_2O_2 concentrations (Lardinois 1995). When the production of ROS exceeds the capacity of antioxidant metabolism to remove them, oxidative damage to cellular macromolecules and structure occurs, which if unchecked leads to cell death. CAT activity mostly increases in leaves of relatively resistant plants such as *Lycopersicum esculentum, Picea asperata, Acer Mono Maxim, Triticum aestivum,* under natural field conditions (Balakumar et al., 1997; Yao and Liu 2006; Ambasht and Agrawal 2003; Agrawal and Rathore 2007). In sensitive leguminous plants, however, CAT activity declines when UV-B irradiation increases under field conditions (Agrawal and Mishra 2009).

The different affinities of APX (μM range) and CAT (mM range) for H_2O_2 suggest that they belong to two different classes of H_2O_2-scavenging enzymes, where APX might be responsible for the removal of excess of ROS during stress. Peroxidases (POD) are monomeric hemoprotein that catalyse the oxidation of a range of substrates by H_2O_2. POD usually occurs as multiple molecular forms. Anionic PODs are believed to utilize phenolic compounds to initiate the chain reaction that leads to lignifications. UV-B is known to alter the distribution of peroxidase (POD) isoforms (Murali et al., 1988; Yannarelli et al., 2006). Balakrishnan et al. (2005) reported increased POD activity under elevated UV-B radiation and the increasing trend reached a maximum (64.5%) on the fourth day in treated seedlings. Kumari et al. (2009) reported decline in POD activity in *Abelmoschus esculentus* under field conditions with exposure of 1.8 kJ m^{-2} d^{-1} UV-B dose.

Generally, plants with suppressed APX production induce SOD, CAT and GR to compensate for the loss of APX, whereas plants with suppressed CAT production induce APX, GPX and mitochondrial AOX (Willekens et al., 1997). Pyro A expression increases following exposure to acute UV-B dose, and this has been associated with the singlet oxygen scavenging properties of pyridoxine (Brosche et al., 2002). Pyro A expression is downregulated in plants acclimated to low chronic UV-B (Hectors et al., 2007). Yannarelli et al. (2006) observed one CAT isoform (Cat1) and four APX isoforms (APX1, APX2, APX3 and APX4) in sunflower cotyledons. APX isoform activities responded in a different way, while APX4 activity significantly decreased whereas APX2 significantly increased under 15 and 30 kJ m^{-2} UV-B radiation. Two isoforms of POD (POD1 and POD2) were observed, while the activity of POD2 increased under 15 kJ m^{-2} and decreased with 30 kJ m^{-2} UV-B radiation. A higher APX activity is correlated with physiological processes such as the removal of H_2O_2, biosynthesis of lignin and general plant development (Asada 1992). Though elevated activities of APX result in the depletion of AsA, it has to be regenerated for maintaining the redox status of the cells. Dai et al. (1997) demonstrated that rice leaves under supplemental UV-B radiation showed enhanced CAT and SOD activities with no change in the activity of APX. In sunflower cotyledons, the induction of CAT and GPX activities indicated that H_2O_2 participates actively on UV-B plant response (Yannarelli et al., 2006).

Ascorbate and glutathione act in the aqueous phase, whereas the lipophilic antioxidants α-tocopherols and carotenoids are active in the membrane environment. Tocopherols are lipid soluble plant antioxidants and are precursors of Vit. E. In plants, tocopherols are involved in the reduction of polyunsaturated fatty acid radicals that are formed in UV-B stressed plants (DeLong and Steffen 1998). Acute exposure of UV-B leads to decrease in α-tocopherol levels in plants (DeLong and Steffen 1998; Galatro et al., 2001; Carletti et al., 2003; Jain et al., 2003), possibly reflecting reactions with lipid radicals. Tocopherol is restricted to the chloroplast where it is synthesized from homogentisic acid. Chloroplasts contain all of the tocopherol biosynthetic enzymes. The genes encoding tocopherol-synthesizing proteins are located in the nucleus and their respective proteins are imported into the chloroplast. Tocopherol is localized largely to the thylakoid membranes (Soll et al., 1985). Tocopherol is slowly oxidized to radical forms in air, but reacts rapidly with singlet oxygen to form a hydroperoxychromanone (Neely et al., 1988). This intermediate can then be hydrolyzed to tocopherylquinone, shown to occur in the chloroplast (Kruk and Strzalka 1995), in which the chromane ring is irreversibly opened. However, the hydroperoxychromanone intermediate can be re-reduced to tocopherol by ascorbate (Neely et al., 1988) (Fig. 4).

Ascorbate also functions in the systems that reduce the probability of ROS generation in the chloroplast, particularly the xanthophylls cycle, in which ascorbate provides the reducing power necessary to convert violaxanthin to zeaxanthin in the de-epoxidation sequence (Foyer and Harbinson 1999). Chloroplasts contain high concentrations of all components associated with H_2O_2 detoxification. The thylakoid contain superoxide dismutase and APX, and isoforms of these enzymes also exist in the stroma together with glutathione, ascorbate, MDHAR, DHAR and GR (Foyer and Noctor 2000). Chloroplasts contain other proteins that may also function in reduction/detoxification of H_2O_2. Such enzymes include peroxiredoxins (PRX, Dietz et al., 2008), glutathione-peroxidases (GRX, Eshdat et al., 1997). Along with certain glutathione-S-transferases, some GPXs genes are strongly induced by oxidative stress (Willekens et al., 1997). GSH induces pathogenesis-related (PR) gene expression in a similar manner to salicylic acid (SR, Mou et al., 2003). AsA has to be regenerated by cells. This is accomplished by the ascorbic acid-glutathione (AA-G) cycle or the Halliwell-Asada pathway (Halliwell and Gutteridge 1999). The AA-G cycle is an efficient way for plants to dispose of H_2O_2 in certain cellular compartments where no CAT is present. To accomplish the cellular regeneration of AsA from MDHA produced by the activity of APX, three enzymes function in a concerted manner. They are MDHAR, DHAR and GR (Hossain and Asada 1984).

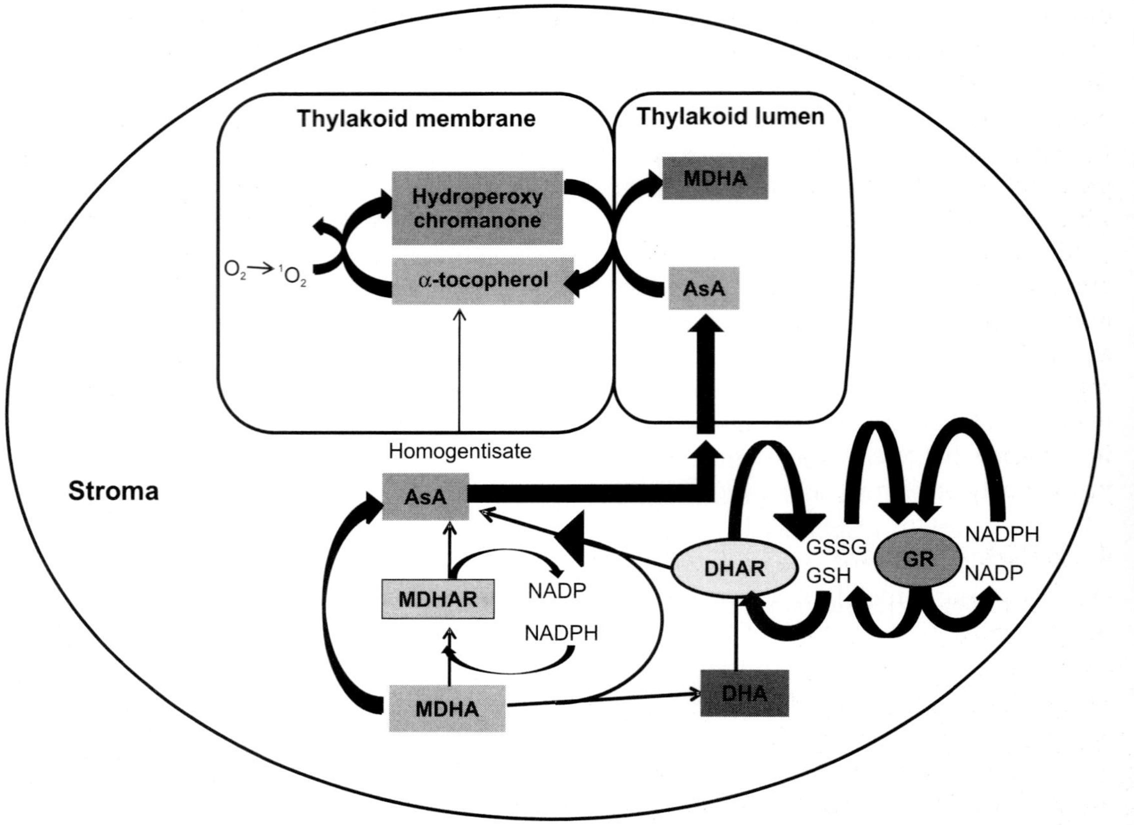

Fig. 4. Halliwell-Asada pathway and low molecular weight antioxidant synthesis in different compartments of chloroplast (redrawn with changes after Foyer et al., 2008). MDHA, monodehydroascorbate; AsA, ascorbic acid; MDHAR, monodehydroascorbate reductase; DHA, dehydroascorbate; DHAR, dehydroascorbate reductase; GR, glutathione reductase.

In AA-G cycle, the GR activity which favours the production of GSH seems to outweigh the activity of DHAR which leads to the formation of GSSG in the UV-B treated plants (Selvakumar 2008). Apparent lack of pace between GR and DHAR might influence the regeneration of AsA. There are also other means of recycling AsA through MDHAR activity and *de novo* synthesis, so that equimolar levels of AsA and glutathione are sustained. Equimolar concentration of AsA and glutathione are essential to sustain a high redox state in the AA-G cycle. The ability of a plant to maintain a high redox state of AsA and glutathione has been attributed to coordination between SOD that generates H_2O_2 and GR-APX that metabolizes H_2O_2 (Foyer et al., 1994). High redox state of the AA-G cycle is believed to provide resistance against oxidative stress in plants (Smirnoff 1996; Singh et al., 2006).

A major role of glutathione is as a thiol/disulphide buffer, i.e., to allow appropriate condition for protein function. Glutathione plays an important role in peroxide detoxification through several enzymes systems. First, GSH can regenerate ascorbate, either directly or through the action of various enzymes able to catalyse DHA reduction, including glutaredoxins (GRXs) and glutathione-S-transferases (GSTs) (Dixon et al., 2002). Secondly, both GRX and GSTs can also reduce hydroperoixdes to the corresponding alcohol and water (Collinson et al., 2002). Third, certain peroxiredoxins are regenerated by a glutathione reductase/glutaredoxins system (Rouhier et al., 2002). Fourth, specific glutathione peroxidases (GPXs) exist in plants (Eshdat et al., 1997). Along with certain glutathione-S-transferases, some GPXs genes strongly induced by oxidative stress (Willekens et al., 1997). GSH induces pathogenesis-related (PR) gene expression in a similar manner to salicylic acid (Mou et al., 2003). Activation of GSH synthesis and accumulation of glutathione is general feature of enhanced oxidation of cytosol. Recent evidence suggests that the enzymes of GSH synthesis and metabolism are induced together in response to stress (Mittova et al., 2003). This implies that there is considerable overlap in the signal transduction cascades that induce GSH synthesis and those involved in defence functions that use GSH, such as GSTs and GPX, some of which show particularly strong response to ROS (Levine et al., 1994).

4. REDOX COUPLES

Central to the efficient removal of ROS and the maintenance of cellular homeostasis is the AA-G cycle, which comprises three independent redox couples: AsA/DHA, GSH/GSSG and NADPH/NADP (May et al., 1998). Central to efficient removal of ROS and maintenance of cellular homeostasis is the ascorbic acid—glutathione cycle, which comprises three independent redox couples: AsA/DHA, GSH/GSSG and NADPH/NADH (May et al., 1998) (Fig. 5). Efficient flux through plant electron transport cascades requires the simultaneous presence of both oxidized and reduced forms of e^- carriers. This requirement, known as redox poising, involves a continuous flux of e^- to molecular oxygen from multiple sites in the photosynthetic and respiratory e^- transport chains (Foyer and Noctor 2005b). Redox homeostasis is governed by the presence of large pools of low molecular weight antioxidants like ascorbate, glutathione and tocopherols that absorb and buffer reductants and oxidants. The ability of the ascorbate, glutathione and tocopherol pools to act as redox buffers in plant cells is one of their most important attributes.

The cellular redox state is envisaged as the sum of reducing and oxidizing redox active molecules, as well as redox capacities in the cell. This redox state maintains general cell homeostasis, and in particular, the ability of the cell to deal with redox events (such as oxidative stress) (Foyer and Noctor 2005b). The oxidation of AsA yields monodehydroascorbate (MDHA) and dehydroascorbate (DHA), which can be re-reduced to AsA through the AA-G cycle using electrons from GSH and eventually NADPH (Noctor and Foyer 1998). DHA reduction has also been linked to other redox components, such as the thioredoxin and glutaredoxin systems. Exposure to acute UV-B stress caused net oxidation of AsA and the accumulation of monodehydroascorbate (Hideg et al., 1997). Under stress-free condition, the glutathione pool is almost completely reduced, but detoxification of ROS by AA-G cycle leads to transient glutathione oxidation, impacting the cellular redox environment through the glutathione redox state (Foyer

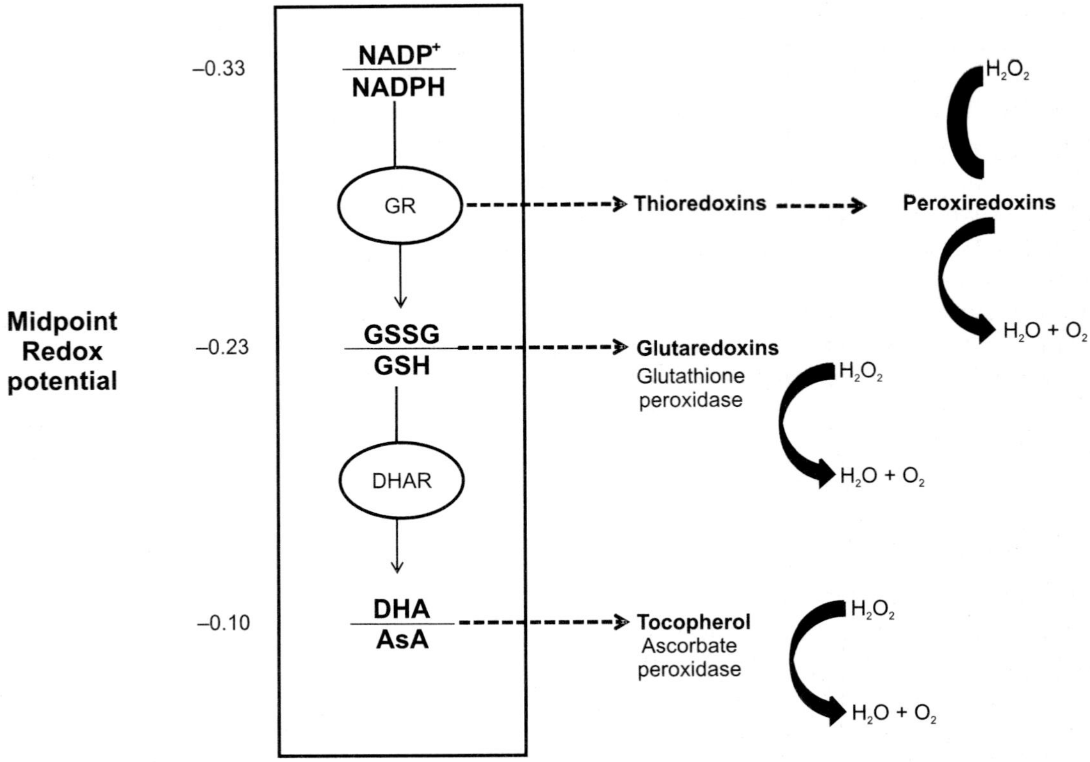

Fig. 5. Soluble reductant couples arranged according to their midpoint potentials and processes with which they interact (redrawn with changes after Foyer et al., 2008). GR, glutathione reductase; DHAR, dehydroascorbate reductase; DHA, dehydroascorbate; AsA, ascorbic acid.

and Noctor 2005b). Acclimation of plants under a relatively high, acute UV-B resulted in an increase in total glutathione as well as simultaneous increase in the oxidation state of GSH pool (Kalbin et al., 1997). The buffering capacity of the glutathione system is lost above a redox potential of –160 mV. At that point, GSH will lose the capacity to protect macromolecules from ROS (Apel and Hert 2004; Oracz et al., 2009).

5. CONCLUSION

Low level of UV-B is perceived by a distinguished class of photoreceptors. UV-B induces several changes at ultrastructural level in plants. It damages pigment-protein complexes, ion channels and transporters of plasmalemma, alters endoplasmic reticulum, chloroplast, dictyosomes and mitochondria. UV-B also affect S, G_1 and G_2 phases of cell cycle thereby delays its progression. The damaging effects can be minimized by epidermal screening such as deposition of wax, thickening of leaf, accumulation of flavonoids. Flavonoids play an important protective role by reducing epidermal transmittance of UV-B radiation. UV-B upregulates various key enzymes of phenylpropanoid pathway. Enzymatic and non-enzymatic antioxidants of plant defence system play a very important role in counteracting the ROS induced by UV-B. ROS are capable of altering the redox potential of various redox couples thereby affecting cell homeostasis.

REFERENCES

A.-H.-Mackerness, S., John, C.F., Jordan, B. and Thomas, B. (2001). Early signaling components in ultraviolet-B responses: distinct roles for different reactive oxygen species and nitric oxide. *FEBS Lett.* **489:** 237-242.

A.-H.-Mackerness, S., Surplus, S.L., Jordan, B.R. and Thomas, B. (1998). Effects of supplementary ultraviolet-B radiation on photosynthetic transcripts at different stages of leaves development and light levels in pea (*Pisum sativum* L.): role of active oxygen species and antioxidant enzymes. *Photochem. Photobiol.* **68:** 88-96.

Agrawal, S.B. and Mishra, S. (2009). Effects of supplemental ultraviolet-B and cadmium on growth, antioxidants and yield of *Pisum sativum* L. *Ecotoxicol. Environ. Saf.* **72:** 610-618.

Agrawal, S.B. and Rathore, D. (2007). Changes in oxidative stress defense system in wheat (*Triticum aestivum* L.) and mung bean (*Vigna radiata* L.) cultivars grown with and without mineral nutrients and irradiated by supplemental ultraviolet-B. *Environ. Exp. Bot.* **59(1):** 21-33.

Agrawal, S.B., Singh, S. and Agrawal, M. (2009). Ultraviolet-B induced changes in gene expression and antioxidants in plants. *Adv. Bot. Res.* **52:** 47-86.

Ahmad, P., Jhon, R., Sarwat, M. and Umar, S. (2008). Responses of proline, lipid peroxidation and antioxidative enzymes in two varieties of *Pisum sativum* L. under salt stress. *Int. J. Plant Prod.* **2:** 353-366.

Alexieva, V., Sergiev, I., Mapelli, S. and Karanov, E. (2001). The effect of drought and ultraviolet radiation on growth and stress markers in pea and wheat. *Plant Cell Environ.* **24:** 1337-1344.

Allen, D.J., McKee, I.F., Farage, P.K. and Baker, N.R. (1997). Analysis of the limitation to CO_2 assimilation on exposure of leaves of two *Brassica napus* cultivars to UV-B. *Plant Cell Environ.* **20:** 633-664.

Alscher, R.G. and Hess, J.L. (1993). Antioxidants in higher plants. CRC Press, Boca Raton.

Ambasht, N.K. and Agrawal, M. (2003). Interactive effects of ozone and ultraviolet-B radiation on physiological and biochemical characteristics of soybean plants. *J. Plant Bio.* **30:** 37-45.

Ananyev, G.M., Wydrzynski, T., Renger, G. and Klimov, V.V. (1992). Transient peroxide formation by the Mn-containing, redoxactive donor side of photosystem II upon inhibition of O2 evolution with lauroylcholine chloride. *Biochimica et Biophysica Acta.* **1100:** 303-311.

Anderson, J.M., Park, Y.I. and Chow, W.S. (1997). Photoinactivation and photoprotection of photosystem II in nature. *Physiol. Plant.* **100:** 214-223.

Andersson, B. and Barber, J. (1996). Mechanisms of photodamage and protein degradation during photoinhibition of photosystem II. In: Baker, N.R. (Ed.). Photosynthesis and the Environment. Kluwer Academic Publishers, Dordrecht, The Netherlands, pp. 101-121.

Apel, K., and Hert, H. (2004). Reactive oxygen species: metabolism, oxidative stress and signal transduction. *Ann. Rev. Plant Bio.* **55:** 373-399.

Aro, E.M., Virgin, I. and Andersson, B. (1993). Photoinhibition of photosystem II. Inactivation, protein damage and turnover. *Biochim. Biophy. Acta.* **1143:** 113-134.

Arora, A., Sairam, R.K. and Srivastava, G.C. (2002). Oxidative stress and antioxidative system in plants. *Curr. Sci.* **82:** 1227-1238.

Asada, K. (1992). Ascorbate peroxidase—a hydrogen peroxide-scavenging enzyme in plants. *Physiol. Plant.* **85:** 235-241.

Asada, K. and Takahashi, M. (1987). Production and scavenging of active oxygen in photosynthesis. In: Kyle, D.J., Osmond, C.B. and Amtzen, C.J. (Eds.). Photoinhibition, Elsevier, Amsterdam, The Netherlands, pp. 227-228.

Babu, T.S., Akhtar, T.A., Lampi, M.A., Tripuranthakam, S., Dixon, D.G. and Greenberg, B.M. (2003). Similar Stress Responses are Elicited by Copper and Ultraviolet Radiation in the Aquatic Plant *Lemna gibba*: Implication of Reactive Oxygen Species as Common Signals. *Plant Cell Physio*. **44(12):** 1320-1329.

Baker, N.R., Nogués, S. and Allen, D.J. (1997). Photosynthesis and Photoinhibition. In: Lumsden, P.J. (Ed.). Plants and UV-B, Responses to Environmental Change, Cambridge University Press, Cambridge, UK, pp. 95-112.

Balakrishnan, V., Venkatesan, K., Ravindran, K.C. and Kulandaivelu, G. (2005). Protective mechanism in UV-B treated *Crotalaria juncea* L. seedlings. *Plant Protec. Sci*. **41:** 115-120.

Balakumar, T., Gayathri, B. and Anbudurai, P.R. (1997). Oxidative stress injury in tomato plants induced by supplemental UV-B radiation. *Biol. Plant*. **39:** 215-221.

Barber, J. and Andersson, B. (1992). Too much of a good thing: Light can be bad for photosynthesis. *Trend Biochem. Sci*. **17:** 61-66.

Bischof, K., Hanelt, D. and Wiencke, C. (2000). Effects of ultraviolet radiation on photosynthesis and related enzyme reactions of marine macroalgae. *Planta*. **211:** 555-562.

Boldt, R. and Scandalios, J.G. (1997). Influence of UV-light on the expression of Cat2 and Cat3 catalase genes in maize. *Free Rad. Bio. Medi*. **23:** 505-514.

Bornman, J.F., Evert, R.F. and Mierzwa, R.J. (1983). The effect of UV-B and UV-C radiation on sugar beet leaves. *Protoplasma*. **117:** 7-16.

Caldwell, M.M., Bjorn, L.O., Bornman, J.F., Flint, S.D., Kulandaivelu, G., Teramura, A.H. and Tevini, M. (1998). Effects of solar ultraviolet radiation on terrestrial ecosystems. *J. Photochem. Photobio. B. Bio*. **46:** 40-52.

Caldwell, M.M., Bornmann, J.F., Ballaré, C.L., Flint, S.D. and Kulandaivelu, G. (2007). Terrestrial ecosystems, increased solar ultraviolet radiation, and interactions with other climate change factors. *Photochem. Photobio. Sci*. **6:** 252-266.

Carletti, P., Masi, A., Wonisch, A., Grill, D., Tausz, M. and Ferretti, M. (2003). Changes in antioxidant and pigment pool dimensions in UV-B irradiated maize seedlings. *Environ. Experi. Bot*. **50:** 149-157.

Chandlee, J.M., Tsaftaris, A.S. and Scandalios, J.G. (1983). Purification and partial characterization of three genetically defined catalases of maize. *Plant Sci. Lett*. **29:** 117-131.

Chen, G.X. and Asada, K. (1989). Ascorbate peroxidase in tea leaves: occurrence of two isozymes and the differences in their enzymatic and molecular properties. *Plant Cell Physiol*. **30:** 987-998.

Chen, G.X., Kazimir, J. and Cheniae, G.M. (1992). Photoinhibition of hydroxylamine-extracted photosystem II membranes: studies of the mechanism. *Biochem*. **31:** 11072-11083.

Chow W.S., Strid, A. and Anderson J.M. (1992). Short-term treatment of pea plants with supplementary ultraviolet-B radiation: Recovery time-courses of some photosynthetic functions and components. In: Murata, N. (Ed.). Research in Photosynthesis, Vol. 4, Kluwer Academic Publishers, Dordrecht, The Netherlands, pp. 361-364.

Chow, W.S., Strid, A. and Anderson, J.M. (1992). Short-term treatment of peas with supplementary ultraviolet-B radiation: recovery time-courses of some photosynthetic functions and components. In: Murata, N. (Ed.). Research in photosynthesis, Vol. 4, Kluwer Academic Publishers, Dordrecht. pp. 361-364.

Collinson, E.J., Wheeler, G.L., Garrido, E.O., Avery, A.M., Avery, S.V. and Grant, C.M. (2002). The yeast glutaredoxins are active as glutathione peroxidases. *J. Biolo. Chem*. **277:** 16712-16717.

Costa, H., Gallego, S.M. and Tomaro, S.L. (2002). Effect of UV-B radiation on antioxidant defense system in sunflower cotyledons. *Plant Sci.* **162:** 939-945.

Dai, Q., Yan, B., Huang, S., Liu, X., Peng, S., Miranda, M.L.L., Chavez, A.Q., Vergara, B.S. and Olszyk, D.H. (1997). Response of oxidative stress defence systems in rice (*Oryza sativa* L.) leaves with supplemental UV-B radiation. *Physiol. Plant.* **101:** 301-308.

Dat, J., Vandenabeele, S., Vranova, E., Van Montage, M, Inze, D. and Breusegem Van, F. (2000). Dual action of the active oxygen species during plant stress. *Cell. Mole. Life Sci.* **57:** 779-795.

Davies, M.J., Fu, S. and Dean, R.T. (1995). Protein hydroperoxides can give rise to reactive free radicals. *Biochem. J.* **305:** 643-649.

DeLong, J.M. and Steffen, K.L. (1998). Lipid peroxidation and α-tocopherol content in α-tocopherol–supplemented membranes during UV-B exposure. *Environ. & Experi. Bot.* **39:** 177-185.

Dixon, D.P., Davis, B.G. and Edwards, R. (2002). Functional divergence in the glutathione transferase superfamily in plants. Identification of two classes with putative functions in redox homeostasis in *Arabidopsis thaliana. J. Biolo. Chem.* **277:** 30859-30869.

Dröge, W. (2002). Free radicals in the physiological control of cell function. *Physiological. Rev.* **82:** 47-95.

Elstner, E.F. (1982). Oxygen activation and oxygen toxicity. *Ann. Rev. Plant Physiol.* **33:** 73-96.

Eshdat, Y., Holland, D., Faltin, Z. and Ben-Hayyim, G. (1997). Plant glutathione peroxidases. *Physiol. Plant.* **100:** 234-240.

Foyer, C.H. and Harbinson, J. (1999). Relationships between antioxidant metabolism and carotenoids in the regulation of photosynthesis. In: Frank, H.A., Young, A.J., Britton, G. and Cogdell, R.J. (Eds.). The Photochemistry of Carotenoids, Kluwer Academic Publishers, Dordrecht, pp. 305-325.

Foyer, C.H., Lelandais, M. and Kunert, K.J. (1994). Photooxidative stress in plants. *Physiol. Plant.* **92:** 696-717.

Foyer, C.H. and Noctor G. (2005b). Oxidant and antioxidant signalling in plants: a re-evaluation of the concept of oxidative stress in a physiological context. *Plant Cell Environ.* **28:** 1056-1071.

Foyer, C.H. and Noctor, G. (2005a). Redox homeostasis and antioxidant signaling: a metabolic interface between stress perception and physiological responses. *Plant Cell.* **17:** 1866-1875.

Foyer, C.H. and Noctor G. (2000). Oxygen processing in photosynthesis: regulation and signalling. *New Phytol.* **146:** 359-388.

Foyer, C.H. and Noctor, G. (2003). Redox sensing and signalling associated with reactive oxygen in chloroplasts, peroxisomes and mitochondria. *Physiol. Plant.* **119:** 355-364.

Foyer, C.H., Descourvieres, P. and Kunert, K.J. (1994). Protection against oxygen radicals: an important defence mechanism studied in transgenic plants. *Plant Cell Environ.* **17:** 507-523.

Foyer, C.H., Trebst, A. and Noctor, G. (2008). Signalling and integration of defense functions of tocopherol, ascorbate and glutathione. In: Demmig-Adams, B., Adams III, W.W. and Mattoo, A.K. (Eds.). Photoprotection, photoinhibition, gene regulation and environment, Springer. pp. 241-268.

Frankel, E.N. (1991). Recent advances in lipid oxidation. *J. Sci. Food Agri.* **54:** 495-551.

Frohnmeyer, H. and Staiger, G. (2003). Ultraviolet-B radiation-mediated responses in plants: balancing damage and protection. *Plant Physiol.* **133:** 1420-1428.

Galatro, A., Simontacchi, M. and Puntarulo, S. (2001). Free radical generation and antioxidant content in chloroplasts from soybean leaves exposed to ultraviolet-B. *Physiol. Plant.* **113:** 564-570.

Gerhardt, K.E., Wilson, I. and Greenberg, B.M. (1999). Tryptophan photolysis leads to a UVB-induced 66 kDa photoproduct of ribulose-1, 5-biphosphate carboxylase/oxygenase (Rubisco) *in vitro* and *in vivo*. *Photochem. & Photobio.* **70:** 49-56.

Green, R. and Fluhr, R. (1995). UV-B-induced PR-1 accumulation is mediated by active oxygen species. *Plant Cell.* **7:** 203-212.

Guo, J., Han, W. and Wang, M.H. (2008). Ultraviolet and environmental stresses involved in the induction and regulation of anthocyanin biosynthesis: A review. *African J. Biotechnology* . **7(25):** 4966-4972.

Halliwell, B. and Gutteridge, J.M.C. (1999). Free radicals in biology and medicine, 3rd ed. Oxford University Press, Oxford.

Havir, E.A. and McHale, N.A. (1990). Purification and characterization of an isozyme of catalase with enhanced-peroxidatic activity from leaves of *Nicotiana sylvestris. Arch. Biochem. & Biophy.* **283:** 491-495.

Hawkins, C.L. and Davies, M.J. (2001). Generation and propagation of radical reactions on proteins. *Biochim. Biophys. Acta.* **1504**: 196-219.

He, J.M., Xu, H., She, X.P., Song, X.G. and Zhao, W.M. (2005). The role and the interrelationship of hydrogen peroxide and nitric oxide in the UV-B-induced stomatal closure in broad bean. *Funct. Plant Bio.* **32:** 237-247.

He, Y.Y. and Häder, D.P. (2002). Reactive oxygen species and UV-B: effect on cyanobacteria. *Photochem. Photobio. Sci.* **1:** 729-736.

Heber, U., Miyake, C., Mano, J., Ohno, C. and Asada, K. (1996). Monodehydroascorbate radical as a sensitive probe of oxidative stress in intact leaves as detected by electron paramagnetic resonance spectrometry, *Plant Cell Physio.* **37:** 1066-1072.

Hectors, K., Prinsen, E., De Coen, W., Jansen, M.A.K. and Guisez, Y. (2007). *Arabidopsis thaliana* plants acclimated to low dose rates of ultraviolet-B radiation show specific changes in morphology and gene expression in the absence of stress symptoms. *New Phytol.* **175:** 255-270.

Hideg, E., Mano, J., Ohno, C. and Asada, K. (1997). Increased levels of monodehydroascorbate radical in UV-B irradiated broad bean leaves. *Plant Cell Physio.* **38:** 684-690.

Hideg, E., Spetea, C. and Vass, I. (1994). Singlet oxygen and free radical production during acceptor and donor side induced photoinhibition studies with spin trapping EPR spectroscopy. *Biochem. Biophys. Advances.* **1186:** 143-152.

Hideg, E., Török, K., Šnyrychová, I., Sándor, G., Szegedi, E. and Horváth, G.V. (2008). Response of ferritin overexpressing tobacco plants to oxidative stress. In: Allen, J.F., Gantt, E., Golbeck, J.H. and Osmond, B. (Eds.). Photosynthesis, energy from the sun, Springer, Netherlands, pp. 1469-1472.

Hollosy, F. (2002). Effects of Ultraviolet radiation on plant cells. *Micron.* **33:** 179-197.

Hossain, M.A. and Asada, K. (1984). Inactivation of ascorbate peroxidase in spinach chloroplasts on dark addition of hydrogen peroxide: its protection by ascorbate. *Plant Cell Physiol.* **25:** 1285-1295.

Isbell, H.S., Frush, H.L. and Martin, E.T. (1973). Reactions of carbohydrates with hydroperoxides. 1. Oxidation of aldoses with soiudm peroxide. *Carbohydrate Res.* **26:** 287-295.

Jain, K., Kataria, S. and Guruprasad, K.N. (2003). Changes in antioxidant defenses of cucumber cotyledons in response to UV-B and to the free radicals generating compound AAPH. *Plant Sci.* **165:** 551-557.

Jansen, M.A.K, Gaba V. and Greenberg B.M. (1998). Higher plants and UV-B radiation: balancing damage, repair and acclimation. *Trends in Plant Sci.* **3:** 131-135.

Jenkins, G.I. (2009). *Signal transduction in responses to UV-B radiation. Ann. Rev. Plant Bio.* **60(1):** 407-431.

Jeong, Y.C., Nakamura, J., Upton, P.B. and Swenberg, J.A. (2005). Pyrimido [1,2-α]-purin-10(3H)-one, M1G, is less prone to artifact than base oxidation. *Nucleic Acids Res.* **33:** 6426-6434.

Jordan, B.R. (1996). The effects of ultraviolet-B radiation on plants: a molecular perspective. *Adv. Botanical Res.* **22:** 97-162.

Jordan, B.R. (2002). Molecular response of plant cells to UV-B stress. *Funct. Plant Bio.* **29:** 909-916.

Jordan, B.R., He, J., Chow, W.S. and Anderson, J.M. (1992). Changes in mRNA levels and polypeptide subunits of Ribulose-1,5-bisphosphate carboxylase in response to supplementary ultraviolet-B radiation. *Plant Cell & Environ.* **15:** 91-98.

Kakani, V.G., Reddy, K.R., Zhao, D. and Mohammed, A.R. (2003). Effects of Ultraviolet-B radiation on cotton (*Gossypium hirsutum* L.) morphology and anatomy. *Ann. Bot.* **91:** 817-826.

Kalbin, G., Hidema, J., Brosche, M., Kumagai, T., Bornman, J.F. and Strid, A. (2001). UV-B-induced DNA damage and expression of defence genes under UV-B stress: tissue-specific molecular marker analysis in leaves. *Plant Cell & Environ.* **24:** 983-990.

Kalbin, G., Ohlsson, A.B., Berglund, T., Rydström, J. and Strid, Å. (1997). Ultraviolet-B-radiation induced changes in nicotinamide and glutathione metabolism and gene expression in plants. *European J. Biochem.* **249:** 465-472.

Kanvah, S. and Schuster, G.B. (2002). Long-range oxidative damage to DNA: protection of guanines by a nonspecifically bound disulfide. *J. Am. Chem. Soc.* **124:** 11286-11287.

Kramer, G.F., Norman, H.A., Krizek, D.T. and Mirecki, R.M. (1991). Influence of UV-B radiation on polyamines, lipid peroxidation and membrane lipids in cucumber. *Phytochem.* **30:** 2101-2108.

Krause, G.H. (1994). The role of oxygen in photoinhibition of photosynthesis. In: Foyer, C.H. and Mullineaux, P.M. (Eds.). Causes of photooxidative stress and amelioration of defence systems in plants, CRC Press, Boca Raton, FL, pp. 43-76.

Kruk, J. and Strzalka, K. (1995). Occurrence and function of alphatocopherol quinone in plants. *J. Plant Physiol.* **145:** 405-409.

Kumagai, J., Katoh, H., Miyazaki, T., Hidema, J. and Kumagai, T. (1999). Differences in the sensitivity to UVB radiation of two cultivars of rice (*Oryza sativa* L.) based on observation of long-lived radicals. *J. Rad. Res.* **40:** 303-310.

Kumari, R., Singh, S. and Agrawal, S.B. (2009). Combined effects of Psoralens and ultraviolet-B on growth, pigmentation and biochemical parameters of *Abelmoschus esculentus* L. *Ecotoxicol. & Environ. Saf.* **72:** 1129-1136.

Kumari, R., Singh, S. and Agrawal, S.B. (2010). Responses of ultraviolet-B induced antioxidant defense system in a medicinal plant *Acorus calamus*. *J. Environ. Bio.* **30:** (in press).

Laakso, K. and Huttunen, S. (2000). Effects of the ultraviolet-B radiation (UV-B) on conifers: a review. *Environ. Pollu.* **99:** 319-328.

Laakso, K., Sullivan, J.H. and Huttunen S. (2000). The effects of UV-B radiation on epidermal anatomy in loblolly pine (*Pinus taeda* L.) and Scots pine (*Pinus sylvestris* L.). *Plant Cell & Environ.* **23:** 461-472.

Landry, L.G., Chapple, C.C.S. and Last, R.L. (1995). *Arabidopsis* mutants lacking phenolic sunscreens exhibit enhanced ultraviolet-B injury and oxidative damage. *Plant Physiol.* **109:** 1159-1166.

Lardinois, O.M. (1995). Reactions of bovine liver catalase with superoxide radicals and hydrogen-peroxide. *Free Rad. Res.* **22:** 251-274.

Lariguet, P. and Dunand, C. (2005). Plant Photoreceptors: Phylogenetic Overview. *J. Mole. Evolu.* **61:** 559-569.

Larson, R.A. (1988). The antioxidants of higher plants. *Phytochem.* **27:** 969-978.

Lemaire, S.D. (2004). The glutaredoxin family in oxygenic photosynthetic organisms. *Photosyn. Res.* **79:** 305-318.

Levine, A., Tenhaken, R., Dixon, R. and Lamb, C. (1994). H_2O_2 from the oxidative burst orchestrates the plant hypersensitive disease resistance response. *Cell.* **79:** 583-593.

Malanga, G., Calmanovici, G. and Puntarulo, S. (1997). Oxidative damage to chloroplasts from *Chlorella vulgaris* exposed to ultraviolet-B radiation. *Physiol. Plant.* **101:** 455-467.

May, M.J., Vernoux, T., Leaver, C., van Montagu, M. and Inzé, D. (1998). Glutathione homeostasis in plants: implications for environmental sensing and plant development. *J. Experi. Bot.* **49:** 649-667.

Mazza, C.A., Battista, D., Zima, A.M., Szwarcberg, B.M., Giodano, C.V., Acevedo, A., Scopel, A.L. and Ballare, C.L. (1999). The effects of solar ultraviolet-B radiation on the growth and yield of barley are accompanied by increased DNA damage and antioxidant responses. *Plant Cell Environ.* **22:** 61-70.

McKenzie, R.L., Aucamp, P.J., Bais, A.F., Björn, L.O. and Ilyas, M. (2007). Changes in biologically-active ultraviolet radiation reaching the Earth's surface. *Photochem. Photobio. Sci.* **6:** 218-231.

Mittler, R. (2002). Oxidative stress, antioxidant and stress tolerance. *Trends Plant Sci.* **7:** 405-410.

Mittler, R., Vanderauwera, S., Gollery, M. and Van Breusegem, F. (2004). Reactive oxygen gene network of plants. *Trends Plant Sci.* **9:** 490-498.

Mittova, V., Kiddle, G., Theodoulou, F.L., Gomez, L., Volokita, M., Tal, M., Foyer, C.H. and Guy, M. (2003). Co-ordinate induction of glutathione biosynthesis and glutathione-metabolising enzymes is correlated with salt tolerance in tomato. *FEBS Lett.* **554:** 417-421.

Miyao, M., Ikeuchi, M., Yamamoto, N. and Ono, T. (1995). Specific degradation of the D1 protein of photosystem II by treatment with hydrogen peroxide in darkness: implication for the mechanism of degradation of the D1 protein under illumination. *Biochem.* **34:** 10019-10026.

Mou, Z., Fan, W.H. and Dong, X.N. (2003). Inducers of plant systemic acquired resistance regulate NPR1 function through redox changes. *Cell.* **113:** 935-944.

Mueller, M.J. (2004). Archetype signals in plants: The phytoprostanes. *Curr. Opin. Plant Bio.* **7:** 441-448.

Murali, N.S., Teramura, A.H. and Randall, S.K. (1988). Response differences between two soybean cultivars with contrasting UV-B radiation sensitivities. *Photochem. & Photobio.* **48:** 653-657.

Murphy, T.M. and Wilson, C. (1982). The UV stimulated K^+ efflux from rose cells. *Plant Physiol.* **70:** 709-713.

Nagesh, L. and Bjorn, L.O. (1986). Stomatal closure by ultraviolet radiation. *Physiol. Plant.* **66:** 360-364.

Neely, W.C., Martin, J.M. and Barker, S.A. (1988). Products and relative reaction rates of the oxidation of tocopherols with singlet molecular oxygen. *Photochem. Photobiol.* **48:** 423-428.

Niyogi, K.K. (1999). Photoprotection revisited: Genetic and molecular approaches. *Ann. Rev. Plant Physiol. Plant Mole. Bio.* **50:** 333-359.

Noctor G. and Foyer C.H. (1998). Ascorbate and glutathione: keeping active oxygen under control. *Ann. Rev. Plant Physiol. Plant Mole. Bio.* **49:** 249-279.

Okada, K., Ikeuchi, M., Yamamoto, N., Ono, T. and Miyao, M. (1996). Selective and specific cleavage of the D1 and D2 proteins of photosystem II by exposure to singlet oxygen: factors responsible for the susceptibility to cleavage of the proteins. *Biochim. Biophys. Acta.* **1274:** 73-79.

Oracz, K., El-Maarouf-Bouteau, H. Kranner, I., Bogatek, R., Corbineau, F. and Bailly, C. (2009). The mechanisms involged in seed dormancy alleviation by hydrogen cyanide unravel the role of reactive oxygen species as key factors of cellular signalling during germination. *Plant Physiol.* **150:** 494-505.

Pagano, E.A., Chueca,A. and Lopéz-Gorgé, J. (2000). Expression of thioredoxins f and m, and of their targets fructose-1,6-bisphosphatase and NADP-malatedehydrogenase, in pea plants grown under normal and light/temperature stress conditions. *J. Exp. Bot.* **51:** 1299-1307.

Panagopoulos, I., Bornman, J.F. and Bjorn, L.O. (1990). Effects of ultraviolet radiation and visible light on growth, fluorescence induction, ultraweak luminescence and peroxidase activity in sugar beet plants. *J. Photochem. Photobio. B: Bio.* **8:** 73-87.

Powles, S.B. (1984). Photoinhibition induced by visible light. *Ann. Rev. Plant Physiol.* **135:** 15-44.

Prásil, O., Adir, N. and Ohad, I. (1992). Dynamics of photosystem II: mechanism of photoinhibition and recovery processes. In: Barber, J. (Ed.). Topics in Photosynthesis, Vol. 11, The Photosystems: Structure, Function and Molecular Biology. Elsevier Science Publishers, Amsterdam, pp. 295-348.

Qi, Y., Bai, S. and Heisler, G. (2003). Changes in ultraviolet-B and visible optical properties and absorbing pigment concentrations in pecan leaves during a growing season. *Agri. For. Metro.* **120:** 229-240.

Qi, Y.D., Bai, S., Vogelmann, T.C., Heisler, G.M. and Qin, J. (2002). Methodology for comprehensive evaluation of UV-B tolerance in trees. In: Slusser, J.R., Herman, J.R. and Gao, W. (Eds.). Ultraviolet Ground- and Space-based Measurements, Models, and Effects, Proceedings of SPIE 4482. International Society of Optical Engineering, Bellingham, WA, pp. 367-380.

Rao, M.V. and Ormrod, D.P. (1995). Impact of UV-B and ozone on oxygen free radical scavenging system in *Arabidopsis thaliana* genotypes differing in flavonoid biosynthesis. *Photochem. Photobio.* **62:** 719-726.

Rao, M.V., Paliyath, G. and Ormrod, D.P. (1996). Ultraviolet-B and ozone-induced biochemical changes in antioxidant enzymes of *Arabidopsis thaliana. Plant Physiol.* **110:** 125-136.

Romero, H.M., Berlett, B.S., Jensen, P.J., Pell, E.V. and Tien, M. (2004). Investigations into the role of the plastidial peptide methionine sulfoxide reductase in response to oxidative stress in Arabidopsis. *Plant Physiol.* **136:** 3784-3994.

Rouhier, N., Gelhaye, E. and Jacquot, J.P. (2002). Glutaredoxindependent peroxiredoxin from poplar. Protein-protein interaction and catalytic mechanism. *J. Biolo. Chem.* **277:** 13609-13614.

Ryan, K.G. Swinny, E.E., Markham, K.R. and Winefield, C. (2002). Flavonoid gene expression and UV photoprotection in transgenic and mutant Petunia leaves. *Phytochem.* **59:** 23-32.

Sakihama, Y., Mano, J., Sano, S., Asada, K. and Yamasaki, H. (2000). Reduction of phenoxyl radicals mediated by monodehydroascorbate reductase. *Biochem. Biophy. Res. Commun.* **279:** 949-954.

Santos, M., Gousseau, H., Lister, C., Foyer, C., Creissen, G. and Mullineaux, P. (1996). Cytosolic ascorbate peroxidase from *Arabidopsis thaliana* L. is encoded by a small multigene family. *Planta.* **198:** 64-69.

Scandalios, J.G., Guan, L. and Polidoros, A.N. (1997). Catalases in plants: gene structure, properties, regulation, and expression. In: Scandalios, J.G. (Ed.). Oxidative Stress and the Molecular Biology of Antioxi dant Defenses, Cold Spring Harbor Laboratory Press, New York, pp. 343-406.

Scandalios, J.G., Tong, W.F. and Roupakias, D.G. (1980). *Cat3,* a third gene locus coding for the tissue specific catalase in maize: Genetics, intracellular location, and some biochemical properties. *Mole. Gene. Genomics.* **179:** 33-41.

Selvakumar, V. (2008). Ultraviolet-B radiation (280-315 nm) invoked antioxidant defence systems in *Vigna unguiculata* (L.) Walp and *Crotalaria juncea* L. *Photosynthetica.* **46(1):** 98-106.

Singh, N., Ma, L.Q., Srivastava, M. and Rathinasabapathi, B. (2006). Metabolic adaptations to arsenic induced oxidative stress in *Pteris vittata* and *Pteris ensiformis* L. *Plant Sci.* **170:** 274-282.

Singh, S., Kumari, R., Agrawal, M. and Agrawal, S.B. (2009). Modification of growth and yield responses of *Amaranthus tricolor* L. to sUV-B under varying mineral nutrient supply. *Scientia Horticul.* **120:** 173-180.

Skadsen, R. W. and Scandalios, J. G. (1987). Translational control of photoinduced expression of the *Cat2* catalase gene during leaf development in maize. *Proc. Nat. Acad. Sci. USA.* **84:** 2785-2789.

Smirnoff, N. (1996). The function and metabolism of ascorbic acid in plants. *Ann. Bot.* **78:** 661-667.

Soll, J., Schultz, G., Joyard, J., Douce, R. and Block, M.A. (1985). Localization and synthesis of prenylquinones in isolated outer and inner envelope membranes from spinach chloroplasts. *Arch. Biochem. Biophys.* **238:** 290-299.

Staxen, I. and Bornman, J.F. (1994). A morphological and cytological study of *Petunia hybrida* exposed to UV-B radiation. *Physiol. Plant.* **91:** 735-740.

Staxen, I., Bergounioux, C. and Bornman, J.F. (1993). Effect of ultraviolet radiation on cell division and microtubule organization in *Petunia hybrid* protoplast. *Protoplasma.* **173:** 70-76.

Stratman, J. (2003). Ultraviolet-B radiation co-opts defense signaling pathways. *Trends Plant Sci.* **8:** 526-533.

Strid, A. (1993) Alteration in expression of defence genes in *Pisum sativum* after exposure to supplementary ultraviolet-B radiation. *Plant Cell Physiol.* **34**: 949-953.

Strid, A., Chow, W.S. and Anderson, J.M. (1990). Effects of supplementary ultraviolet-B radiation on photosynthesis in *Pisum sativum. Biochim. Biophys. Acta.* **1020:** 260-268.

Strid, A., Chow, W.S. and Anderson, J.M. (1994). UV-B damage and protection at the molecular level in plants. *Photosyn. Res.* **39:** 475-489.

Tang, K., Zhan, J.C., Yang, H.R. and Huamg, W.D. (2010). Changes of resveratol and antioxidant enzymes during UV-induced plant defense response in peanut seedlings. *J. Plant Physiol.* **167:** 95-102.

Telfer, A., Bishop, S.M., Phillips, D. and Barber, J. (1994). The isolated photosynthetic reaction center of PS II as a sensitiser for the formation of singlet oxygen, detection and quantum yield determination using a chemical trapping technique. *J. Biolo. Chem.* **269:** 13244-13253.

Tuteja, N., Ahmad, P., Panda, B.B. and Tuteja, R. (2009). Genotoxic stress in plants: shedding light on DNA damage, repair and DNA repair helicases. *Mutat. Res. Rev. Mutat. Res.* **681:** 134-149.

Vass, I., Styring, S., Hundal, T., Koivuniemi, A., Aro, E.M. and Andersson, B. (1992). The reversible and irreversible intermediates during photoinhibition of photosystem II- stable reduced QA species promote chlorophyll triplet formation. *Proc. Nat. Acad. Sci. USA.* **89:** 1408-1412.

Willekens, H., Camp, W.V., Van Montagu, M., Inze, D., Langebartels, C.H. and Sandermann, H. (1994). Ozone, sulphur dioxide and ultraviolet-B have similar effects on mRNA accumulation of antioxidant genes in *Nicotiana plumbaginifolia* L. *Plant Physiol.* **106:** 1007-1014.

Willekens, H., Chamnogopol, S., Davey, M., Schraudner, M. and Langebartels, C. (1997). Catalase is a sink for H_2O_2 and is indispensable for stress defense in C3 plants. *EMBO J.* **16:** 4806-4816.

Wilson, M.I., Ghosh, S., Gerhardt, K.E., Holland, N., Babu, T.S., Edelman, M., Dambroff, E.B. and Greenburg, B.M. (1995). *In vivo* photomodification of ribulose-1,5-bisphosphate carboxylase oxygenase holoenzyme by ultraviolet-B radiation-formation of a 66-kilodalton variant of the large subunit. *Plant Physiol.* **109:** 221-229.

Wiseman, H. and Halliwell, B. (1996). Damage to DNA by reactive oxygen and nitrogen species: role in inflammatory disease and progression to cancer. *Biochem. J.* **313:** 17-29.

Xu, C., Natarajan, S. and Sullivan, J.H. (2008). Impact of solar ultraviolet-B radiation on the antioxidant defense system in soybean lines differing in flavonoid contents. *Environ. Experi. Bot.* **63:** 39-48.

Yalamanchili, R.D. and Stratmann, J. (2002) Ultraviolet-B activates components of the system in signaling pathway in *Lycopersicon peruvianum* suspension-cultured cells. *J. Biolo. Chem.* **277:** 8424-28430.

Yannarelli, G.G., Gallego, S.M. and Tomaro, M.L. (2006). Effect of UV-B radiation on the activity and isoforms of enzymes with peroxidase activity in sunflower cotyledons. *Environ. Experi. Bot.* **56:** 174-181.

Yanqun, Z., Yuan, L., Haryan, C.J. and Ianjun, C. (2003). Intraspecific differences in physiological response to 20 soybean cultivars to enhanced ultraviolet-B radiation under field conditions. *Environ. Exp. Bot.* **50:** 87-97.

Yao, X. and Liu, Q. (2006). Changes in morphologal, photosynthetic and physiological responses of Mono Maple seedlings to enhanced UV-B and to nitrogen addition. *Plant Growth Regul.* **50:** 165-177.

Zhang, Y.M., Yang, J.F., Lu, S.Y., Cai, J.L. and Guo, Z.F. (2008). Overexpressing SgNCED1 in tobacco increases ABA level, antioxidant enzyme activities, and stress tolerance. *J. Plant Growth Regul.* **27:** 151-158.

13

Role of Ascorbate-Glutathione During Salt Stress-Induced Oxidative Stress Tolerance in Plants

SUDESH KUMAR YADAV*

Biotechnology Division, Institute of Himalayan Bioresource Technology, CSIR, Palampur–176 061 (HP), India

***Corresponding author:** skyt@rediffmail.com*

ABSTRACT

Environmental stresses including salt stress considerably limit plant productivity. Plant survival under salt stress requires the activation of proper defence mechanisms to avoid the impairment of metabolic functions. Salt stress induces the overproduction of reactive oxygen species (ROS) in the cell. Excess ROS accumulation produces oxidative stress to plant cell and causes cell damage. Hence, it was hypothesized that increased tolerance to salt stress could be achieved through enhanced tolerance to oxidative stress. In plants, the ascorbate (ASC)-glutathione (GSH) cycle plays a pivotal role in controlling ROS levels and cellular redox homeostasis. Most importantly, they provide essential information on cellular redox state, and regulate gene expression associated with salt stress responses to optimize defence and survival. The present review emphasizes the roles of ASH and GSH in plant metabolism and salt stress tolerance. Further, it presents the integrated regulation and role of several genes in providing salt stress induced oxidative stress tolerance to plants.

Keywords: Ascorbate-glutathione, environmental stress, oxidative stress, salt stress, tolerance.

1. INTRODUCTION

Approximately, 400 million hectares of land throughout the world are salt or salinity stress affected. This is over 3% of the world's total land area. The reason for such conditions are partially natural and partially because of land clearing or irrigation. Of the 1500 million ha of land farmed by dryland agriculture, 32 million (2%) are affected by secondary salinity to varying degrees. Of the current 230 million ha of irrigated land, 45 million ha are salt-affected (20%) (FAO, 2005).

Environmental stresses including salt stress induce harmful effect through the induction of oxidative stress to plants. Oxidative stress is measured in terms of excess reactive oxygen species (ROS) accumulation. ROS include singlet oxygen, superoxide radical and hydrogen peroxide (H_2O_2). Overproduction of such reactive species cause disturbance in the metabolic processes of plants and lead to the early senescence or death. However, plants have enzymatic and non-enzymatic antioxidant mechanisms to scavenge the ROS and protect cells from oxidative stresses. The enzymatic antioxidant defence system includes several enzymes, such as superoxide dismutase (SOD), catalase (CAT), ascorbate peroxidase (APX), glutathione peroxidase, glutathione-*S*-transferase (GST) and glutathione reductase (GR). Among the non-enzyme antioxidants, ascorbate (ASH) and glutathione (GSH) are the most important (Noctor and Foyer, 1998).

The ASH–GSH constitutes a cycle and is operative in various subcellular compartments including chloroplasts, cytosol, mitochondria, peroxisome and plasma membrane. This cycle plays a crucial role in the detoxification of ROS (Jimenez et al., 1997; Berczi and Moller 1998; Noctor and Foyer 1998). Additionally, they help in maintaining the cellular redox status (Drazkiewicz et al., 2003). Operation of this cycle facilitates the conversion of a most stable ROS as H_2O_2 to water. APX catalyses this reduction reaction and during the process reduced ascorbate (ASH) is oxidized to dehydroascorbate (DHA). DHA is then reduced to ASH that leads to oxidation of reduced glutathione (GSH) to oxidized glutathione (GSSG). To complete the cycle, GR is involved in the reduction of GSSG to GSH using NADPH as a reductant. Hence, both low molecular weight antioxidants ASH and GSH are important in ROS detoxification through ASH-GSH cycle. Additional, both can react directly with ROS. Therefore, both are very important in maintaining the various metabolic processes at a required pace (Kuzniak and Maria, 2001; Chaparzadeh et al., 2004).

In addition to the role of ASH and GSH in ROS quenching, they are very crucial molecules to maintain the cellular redox balance by regulating the ratios of DHA/ASH and GSSG/GSH. The alteration in these ratios under any environmental response may function as signals. Such signals are important for the regulation of antioxidant mechanisms inside a cell (Mittler, 2002). Therefore, maintenance of such ratios plays a detrimental role in plants survival during stresses including salt stress (Srivastava et al., 2004).

2. ASCORBATE (ASH)

Ascorbate (ASH) is highly abundant in plants. It accumulates in between 2-25 mM concentrations inside a cell (Davey et al., 2000). Though ASH is known for its antioxidant properties, but it also acts as a cofactor for various enzymes. ASH also contributes to the regulation of cell division and expansion (Smirnoff and Wheeler 2000). Its role has been justified in plant growth (Alhagdow et al., 2007; Dowdle et al., 2007). The ASH role during environmental stress is through its oxidized form dehydroascorbic acid (DHA). DHA acts as signalling agent (Fotopoulos et al., 2008) and then ASH participates in the interaction with the environment such as oxidizing stress (Fotopoulos et al., 2006).

This documents that the benefits of increasing ASH production in plants are profound. For increasing the ASH *in vivo*, understanding of its biosynthesis, metabolism, and functions is very important. These aspects have been resolved recently to a significant extent (Hancock and Viola 2005; Ishikawa et al., 2006). ASH is synthesized from mannose-6-phosphate via GDP-mannose and GDP-L-galactose. Upto this step, pathway is common for the synthesis of ASH as well as certain cell wall polysaccharides containing mannose, L-galactose, and L-fucose (Hancock and Viola 2005). From this point onward ASH is synthesized through the following pathway. The L-galactose is released from GDP-L-galactose by the action of two enzymes GDP-L-galactose phosphorylase and L-galactose-1-phosphate phosphatase (GPP). L-galactose is then oxidized by L-galactose dehydrogenase to form L-galactono-1, 4-lactone. Finally, ASH is produced through the oxidation of L-galactono-1,4-lactone by L-galactono-1,4-lactone dehydrogenase. Last enzyme is located on the inner mitochondrial membrane. Unlikely, all the other enzymes are cytosolic in location. Various genes of ASH synthesizing pathway have been cloned (Wheeler et al., 1998; Ishikawa et al., 2006). Studies have also demonstrated the importance of ASH in plant abiotic stress resistance.

Alternate pathways for ASH biosynthesis have been demonstrated (Agius et al., 2003; Ishikawa et al., 2006). The existence of these alternative routes has gained genetic support (Jain and Nessler 2000) with the cloning of a D-galacturonate dehydrogenase (Agius et al., 2003) and myo-inositol oxygenase (MIOX) genes (Lorence et al., 2004). However, *in vivo* physiological relevance of these routes is still to know.

ASH is synthesized in the mitochondria and is transported to the other cell components through a proton electrochemical gradient or through facilitated diffusion. Most of the plant tissues possess ASH and are observed to be higher in photosynthetic cells and meristems. Around 40% of the total ascorbate is present in the chloroplast and stroma (Foyer and Noctor 2005). ASH has effects on many physiological processes such as regulation of growth, development and metabolism of plants.

During environmental stresses, a fundamental role of ASH in plants is to protect metabolic processes against oxidative stress. ASH essentially acts as a reductant and reacts non-enzymatically to scavenge many types of free radicals such as superoxide, hydrogen peroxide, and singlet oxygen. It can react indirectly by regenerating tocopherol or in the synthesis of zeaxanthin in the xanthophylls cycle. Therefore, ASH regulates many enzyme activities of several metabolic pathways and minimizes the damage caused by oxidative process through synergic function with other antioxidants (Foyer and Noctor 2003, 2005; Pourcel et al., 2007).

3. GLUTATHIONE (GSH)

Glutathione (GSH) is the major low molecular weight thiol in all living organisms. GSH is a tripeptide (α-glutamyl cysteinylglycine) and in plants, it has been detected virtually in all cell compartments such as cytosol, chloroplasts, endoplasmic reticulum, vacuoles and mitochondria (Millar et al., 2003). GSH is present in millimolar concentrations in leaves and is the major source of non-protein thiols in most plant cells (Foyer and Halliwell 1976; Kunert and Foyer 1993). The chemical reactivity of the thiol group of GSH makes it particularly suitable to serve a broad range of biochemical functions in all organisms including plants. The nucleophilic nature of GSH due to its thiol group facilitate in the formation of mercaptide bonds with metals and for reacting with selected electrophiles. Along with this reactivity, the relative stability and high water solubility of GSH makes it an ideal antioxidant to protect plants against stresses including salt stress and oxidative stress (Millar et al., 2003; Foyer and Noctor 2005).

Glutathione exists in two different forms in the living cells, i.e., reduced glutathione (GSH) and oxidized glutathione (GSSG). Glutathione is maintained almost exclusively in the reduced form in plants. The relative proportion of reduced glutathione (GSH) is normally greater than 0.9 under natural growth conditions (Noctor et al., 1998). The physiological functions of glutathione have been mainly attributed to its reduced form in plants (Alscher 1989). Therefore, maintenance of high proportion of reduced glutathione (GSH) in plants is an essential requirement (Noctor et al., 1998; May et al., 1998; Maughan and Foyer, 2006). Glutathione acts as an major antioxidant in plant cells and during its reaction GSH is oxidized to GSSG. However, plants have evolved the mechanisms to maintain a high proportion of the GSH. GSH is synthesized in both the chloroplast and the cytosol of plant leaves. The GSSG is reduced to GSH by glutathione reductase (GR) enzyme with the accompanying oxidation of NADPH (Noctor and Foyer 1998; Asada, 1999).

GSH plays a number of important physiological functions in plants (May et al., 1998; Noctor et al., 1998; Maughan and Foyer 2006). Some major functions of GSH are as follows: it is an important sulphur sink of a plant cell, a transporter of reduced sulphur and a regulator of sulphur assimilation (Hell 1997). GSH in conjunction with glutathione-S-transferases catalyses the detoxification of xenobiotics (Marrs 1996). Additionally, GSH is also involved in the regulation of plant growth and development through the regulation of gene expression, cell signalling, cell cycle and cell death (May et al., 1998; Noctor et al., 1998; Gomez et al., 2004; Maughan and Foyer 2006).

Glutathione also plays an important role in the protection of plants against oxidative stress (Alscher 1989; Noctor et al., 1998). Glutathione takes part in the control of hydrogen peroxide levels (Foyer and Noctor 2005; Shao et al., 2005a). The change in the GSH/GSSG ratio during the degradation of hydrogen peroxide is important in certain redox signalling pathways (Millar et al., 2003). The GSH/GSSG ratio is an indicator of the cellular redox balance and involved in ROS perception (Millar et al., 2003; Foyer and Noctor 2005). In addition, reduced glutathione (GSH) acts as an antioxidant and is directly involved in the reduction of most active oxygen radicals generated due to environmental stresses (Foyer and Noctor 2005; Rio et al., 2006).

4. OXIDATIVE STRESS AND REDOX SIGNALLING

Higher plants are aerobic organisms and require oxygen for the survival (Shcolnick and Keren 2006). Inside the cell, reduction of oxygen to water also resulted in the production of reactive oxygen species (ROS). ROS could be superoxide radical, hydrogen peroxide, and hydroxyl radical (Fath et al., 2002; Cvetkovska et al., 2005). ROS are generated by most of cellular compartments in higher plants. Several environmental stresses including salt stress limit CO_2 fixation and reduce the $NADP^+$ regeneration by the Calvin cycle. As a consequent, the photosynthetic electron transport chain is overreduced that induces production of superoxide radicals and singlet oxygen in the chloroplasts (Wu and Tang, 2004; Bechtold et al., 2005; Shao et al., 2005a). To circumvent, this overreduction of the electron transport chain under such adverse conditions, higher plants evolved a photorespiratory pathway to regenerate $NADP^+$ (Ledford and Niyogi 2005). The photorespiratory pathway produces H_2O_2 in the peroxisomes. The H_2O_2 is also generated from the catabolism of lipids as a by-product of β–oxidation of fatty acids (Foyer and Noctor, 2005; Wu et al., 2007).

Overproduced ROS are sensed, transduced and translated as signals into appropriate cellular responses. Commonly influenced processes are those which require redox-sensitive proteins or those undergo reversible oxidation/reduction. The processes that may switch on and off depending upon the cellular redox state are also considered in this category (Shao et al., 2005b; Foyer and Noctor 2003). The highly cytotoxic and reactive nature of ROS requires a tight control under cellular environment. In this regard, plants possess very efficient enzymatic and non-enzymatic antioxidant defence systems. These systems provide protection to plant cells from oxidative damage by scavenging ROS (Foyer and Noctor 2003; 2005; Shao et al., 2007). Various components of such defence processes are localized in different compartments of a plant cell. Further, they are differentially regulated at enzymatic activity and gene expression level. Importantly, non-enzymatic scavengers such as ASH and GSH render the antioxidant defence systems a very versatile and flexible unit that provides temporal and spatial control on ROS accumulation (Millar et al., 2003; Foyer and Noctor 2003; 2005; Shao et al., 2007).

ROS transduces its signal through oxidizing either directly or indirectly the redox-sensitive proteins. In the indirect mechanism, ubiquitous redox-sensitive molecules such as GSH or thioredoxins are involved (Foyer and Noctor 2005; Shao et al., 2005a). The signal is translated further by redox-sensitive signalling proteins to downstream signalling components such as kinases, phosphatases, and transcription factors (Foyer and Noctor, 2005; Li et al., 2007). So far studies have documented the occurrence of two molecular mechanisms of redox-sensitive regulation of protein function (Delledome et al., 2003; Link 2003; Rouhier et al., 2003; Foyer and Noctor 2005). In one of the mechanism hetero-trimeric G-proteins are involved in the ROS signal transmission (Pfannschmidt et al., 2003). While in the second, MAP kinases and protein Tyr phosphatases are involved that transmit ROS signal through protein phosphorylation (Kiffin et al., 2006). However, in plants, the best-characterized redox signal transduction system is the stromal ferredoxin-thioredoxin system (Yabuta et al., 2004).

Plant hormones are also involved in the ROS signal transduction. ROS such as H_2O_2 induces accumulation of stress hormones, i.e., salicylic acid (SA) and ethylene (Yabuta et al., 2004; Kiffin et al., 2006). Involvement of ABA, jasmonates (JA) and ethylene hormones are also reported (Cvetkovska et al., 2005). Generally, plant hormones are considered to be located downstream of the ROS signal in signal transduction pathway. But ROS are also known to act as secondary messengers in many hormone signalling pathways (Coupe et al., 2006). The feedback or feedforward interactions are also assumed between different plant hormones and ROS (Terman and Brunk, 2006).

5. ROLE OF VARIOUS GENES IN SALT STRESS-INDUCED OXIDATIVE STRESS TOLERANCE IN PLANTS

Soil salinity is one of the most important environmental factors limiting plant growth and development. The first physiological change in plants due to salt stress is known to limit the internal CO_2 concentrations of leaves because of stomatal closure. This is responsible for inhibition of CO_2 reduction by Calvin cycle. This in turn leads to over-reduction of Photosystem I. This overreduction encourages electron leakage to oxygen and leads to the formation of superoxide radical (Hsu and Kao, 2003). Secondly, limiting the internal CO_2 concentration induces the oxygenase activity of Rubisco. This in turn increases the rate of photorespiration. High rate of photorespiration enhances H_2O_2 production in the peroxisome (Hernandez et al., 2000; Van Breusegem et al., 2001). Hence, the primary effect of salt stress is ROS production at higher levels. These ROS can attack cellular macromolecules, generate lesions in DNA, cause membrane damage, and affect protein synthesis and stability (Wang et al., 2005). Therefore, it is important for plants to have effective ROS scavenging mechanisms.

Plants have an array of enzymes and metabolites that work individually as well as in concert to scavenge ROS in different cellular compartments. Superoxide radical in the chloroplast is dismutated by superoxide dismutase (SOD) into a relatively less toxic H_2O_2. H_2O_2 is detoxified by a variety of peroxidases into water by using various reducing substrates. These peroxidases include ascorbate peroxidase (APX), glutathione peroxidase (GPX) and phenol peroxidase (POD) (Hernandez et al., 2001). Ascorbate is preferably used as an electron donor by APX. However, APX can oxidize artificial substrates such as guaiacol and pyrogallol at considerable rates during the reduction of H_2O_2 (Asada 1999; Chen and Asada 1989). On the contrary, H_2O_2 produced as a consequence of photorespiration in the peroxisome is decomposed by catalase (CAT) (Dat et al., 2000). Although both POD and CAT decompose H_2O_2, POD has much higher affinity to H_2O_2 than CAT. This is suggesting that POD recognizes even lower concentration of H_2O_2 (Mittler 2002). Hence, POD is involved in fine regulation of H_2O_2 level whereas CAT is involved in its bulk removal under stress conditions (Mittler 2002; Noctor and Foyer 1998; Willekens et al., 1997).

Superoxide dismutases (SODs) are metalloproteins that catalyse the dismutation of superoxide radicals to hydrogen peroxide and oxygen. Therefore, SODs are a critical defence mechanism of plants during stresses (Rubio et al., 2002). SODs are classified according to their subcellular localization and their metal cofactor requirement. The predominant forms of SOD are the cytosolic Cu/ZnSOD, mitochondrial MnSOD and the chloroplastic Cu/Zn and FeSODs (Bowler et al., 1994). Among these, MnSODs are ubiquitous metalloenzymes involved in cell defence against both endogenous and exogenous ROS (Frealle et al., 2005). Several genes encoding SOD enzymes have been cloned and characterized from plant species such as radish (Kwon and An 2003), wheat (Baek and Skinner 2006), mangrove (Prashanth et al., 2008), red pepper (Kwon and An 2003), rice (Feng et al., 2006) and cassava (Shin et al., 2005). The genes individually as well in combination have been used for transformation into several plant species. Such transformants showed tolerance to various stresses including salt stress. A MnSOD gene transformed into rice and provides resistance to oxidative stress (Wang et al., 2005). Similarly, a tomato SOD gene transformed into sugar beet also provided tolerance to oxidative stress (Tertivanidis et al., 2004). While simultaneous overexpression of both Cu/ZnSOD and APX in transgenic tall fescue plants resulted in significantly lower amounts ROS and reduced ion leakage compared to control plants, under stresses (Lee et al., 2007). The maize Cu/ZnSOD and CAT genes have also been introduced into Chinese cabbage for stress tolerance (Tseng et al., 2007). A salt tolerance test of these transgenic plants showed that the simultaneous expression of these genes sufficiently increased stress tolerance.

Plants overexpressing foreign genes for antioxidant enzymes provided tolerance to stresses (Allen et al., 1997; Payton et al., 1997; Chen et al., 2003). However, transgenic tobacco plants expressing both Cu/ZnSOD and APX in chloroplasts showed a very high degree of oxidative stress tolerance (Kwon et al., 2002). Transgenic plants expressing human DHAR in their chloroplasts showed tolerance to salt stress induced oxidative stress. This has been documented due to increase in ASH levels in the transgenic plants (Kwon et al., 2003).

Ascorbate peroxidase (APX) is also an important enzyme to play the most essential role in scavenging ROS. Several isoenzymes of APX are known that are distributed in distinct cellular compartments such as the cytosol, mitochondria and peroxisomes. Increased activity of different APX isoforms in response to environmental stresses including salt stress has been reported in different plant species. This has indicated the probable functional specialization of the respective isoenzymes in eliminating H_2O_2 from cells (Sharma and Dubey 2005; Teixeira et al., 2006; Tsai et al., 2005). Among the various isoforms, cytosolic APX have been studied extensively. They are activated by different types of stresses and seem to have a more general stress protective function (Davletova et al., 2005). In agricultural importance crop rice, the APX gene family has eight members. These genes encode two cytosolic and other subcellular isoforms. These various APX genes are differentially upregulated during salt stress. During salt stress, the accumulation of cytosolic OsAPx2 transcript has been found to be more remarkable than that of OsAPx1 in rice (Teixeira et al., 2006). However, recognizing the individual importance of these isoforms is very difficult and is yet to decide.

Influence of stress in plants is primarily seen on photosynthetic organs. Recently, analysis of transgenic tobacco overexpressing GST/GPXD together have shown substantial improvement in seed germination and seedling growth under stressful conditions (Roxas et al., 1997). The growth of such transgenic seedlings was observed to be higher under salt stress conditions that inhibited the growth of control seedlings. Elucidating the mechanism of tolerance in these transgenics indicated that overexpression of GST/GPX in seedlings provides enhanced capacity to scavenge ROS. This protective effect appears to prevent cellular damage and allow seedlings to retain high levels of metabolic activity and growth. The acclimation of GSH biosynthesis and GSH-utilizing enzymes to salt stress have also been studied in tomato. The increase in GSH content and GSH:GSSG ratio during salt stress is found to be a associated with salt-induced upregulation of GSH synthesizing enzyme. Salt treatment induced glutathione peroxidase and glutathione-S-transferase but not glutathione reductase activities in tomato. A coordinated upregulation of synthesis and metabolism of GSH has been suggested for the salt tolerance in tomato (Mittova et al., 2003).

Indeed, many studies have demonstrated that GR plays an important role in protecting plants against oxidative stress (Kocsy et al., 2000; Mittova et al., 2003). Stress-tolerant plants tend to have high activities of GR. Specifically, enhanced chloroplastic GR activity in transgenic plants resulted in increased protection against oxidative stress (Foyer et al., 1995). While decrease in GR activity in tobacco plants led to an increased sensitivity to oxidative stress (Aono et al., 1995). Hence, GR might be providing tolerance to salt stress induced oxidative stress by maintaining higher levels of GSH that involved in ROS quenching and redox homeostasis maintenance.

Phytoene synthase (PSY) is known to be a rate-limiting enzyme of carotenoid biosynthesis. It catalyses the conversion of two molecules of geranylgeranyl pyrophosphate into phytoene (Cunningham and Gantt, 1998). Overexpression of PSY gene from *Salicornia europaea* alters response to reactive oxygen species under salt stress in transgenic Arabidopsis. The transgenics displayed higher activities of SOD and POD. Transgenics also showed lower contents of H_2O_2 and MDA than the control plant. Additionally, transgenic lines showed higher tolerance to salt stress because of increased photosynthesis efficiency and antioxidative capacity. This study has documented the integrated regulation of carotenoid biosynthesis and antioxidative system during salt stress induced oxidative stress tolerance in plants (Han et al., 2008).

Tolerance to salt stress in plants is also conferred by the overexpression of glyoxalase pathway enzymes. Salt stress induces higher level accumulation of methylglyoxal (MG), a potent cytotoxin and primary substrate for glyoxalase pathway, in various plant species (Yadav et al., 2005a). The transgenic plants overexpressing glyoxalase pathway genes showed enhanced basal activity of various GSH-related antioxidative enzymes. These plants suffered minimal salt stress induced oxidative damage measured in terms of the lipid peroxidation. The GSH content was high in these transgenic plants and also maintained a higher GSH:GSSG ratio under salinity. Hence, glyoxalase transgenics resist an increase in MG together with maintaining higher GSH levels provide an interesting example of these two pathways integration for salt stress tolerance (Yadav et al., 2005b).

Helicases are ubiquitous enzymes that catalyse the unwinding of energetically stable duplex DNA (DNA helicases) or duplex RNA secondary structures (RNA helicases). Most helicases are members of DEAD-box protein superfamily. They are essential for basic cellular processes such as DNA replication, repair, recombination, transcription, ribosome biogenesis and translation initiation. Additionally, helicases play an important role in regulating plant growth and development under stress conditions by regulating stress-induced pathways. Helicases are also involved in salt stress tolerance (Vashisht et al., 2005; Vashisht and Tuteja 2006). However, a functional involvement of helicases in the antioxidative response of plants has been described recently. For this, a helicase from *Medicago sativa* (*MH1*) was ectopically expressed in *Arabidopsis*. The transgenic Arabidopsis showed improved seed germination and plant growth under salt and oxidative stress. The capacity for osmotic adjustment, superoxide dismutase (SOD) and ascorbate peroxidase (APX) activities and proline content were also elevated in the transgenic *Arabidopsis* plants. Salt stress tolerance in such transgenics was found to be due to enhanced capacities for ROS scavenging (Luo et al., 2009).

6. CONCLUDING REMARKS

Agricultural productivity is greatly suffering due to salt stress induced oxidative stress all over the world. In view of this, there is a need to step forward for improving the plants through genetic modification for oxidative stress tolerance. Towards this effort, studies have documented the crucial role of two important antioxidants GSH and ASH. These two molecules involved directly in the quenching of ROS generated during salt stress. Secondly, they are involved in the mechanism of antioxidative enzymes for ROS detoxification. Additionally, several enzymes and enzymatic pathways are emerging that involves directly or indirectly these molecules in their mechanism of salt stress tolerance. Through these various studies it emerges that individual component has little effect towards salt stress tolerance. An integrated approach involving more than one enzyme or pathway and metabolite is required to substantially improve the salt stress induced oxidative stress tolerance in plants.

REFERENCES

Agius, F., Gonzalez-Lamothe, R., Caballero, J.L., Munoz-Blanco, J., Botella, M.A. and Valpuesta, V. (2003). Engineering increased vitamin C levels in plants by overexpressing of a D-galacturonic acid reductase. *Nat. Biotechnol.* **21:** 177-181.

Alhagdow, M., Mounet, F., Gilbert, L., Nunes-Nesi, A. Garcia, V., Just, D., Petit, J., Beauvoit, B., Fernie, A.R., Rothan, C. and Baldet P. (2007). Silencing of the mitochondrial ascorbate synthesizing enzyme L-galactono-1,4-lactone dehydrogenase affects plant and fruit development in tomato. *Plant Physiol.* **145:** 1408-1422.

Allen, R.D., Webb, R.P. and Schake, S.A. (1997). Use of transgenic plants to study antioxidant defenses. *Free Radic. Biol. Med.* **23:** 473-479.

Alscher, R.G. (1989). Biosynthesis and antioxidant function of glutathione in plants. *Physiol. Plant.* **77:** 457-464.

Aono, M., Saji, H., Fujiyama, K., Sugita, M., Kondo, N. and Tanaka, K. (1995). Decrease in activity of glutathione reductase enhances paraquat sensitivity in transgenic *Nicotiana tabacum. Plant Physiol.* **107:** 645-648.

Asada, K. (1999). The water-water cycle in chloroplasts: scavenging active oxygen species and dissipation of excess photons. *Annu. Rev. Plant Physiol. Mol. Biol.* **50:** 601-639.

Baek, K.H. and Skinner, D.Z. (2006). Differential expression of manganese superoxide dismutase sequence variants in near isogenic lines of wheat during cold acclimation. *Plant Cell Rep.* **25:** 223-230.

Bechtold, U., Karpinski, S. and Mullineaux, P.M. (2005). The influence of the light environment and photosynthesis on oxidative signaling responses in plant-biotrophic pathogen interactions. *Plant Cell Environ.* **28:** 1046-1055.

Berczi, A. and Moller, I.M. (1998). NADH–monodehydroascorbate oxidoreductase is one of the redox enzymes in spinach leaf plasma membranes. *Plant Physiol.* **116:** 1029-1036.

Bowler, C., Van Camp, W., Van Montagu, M. and Inze, D. (1994). Superoxide dismutase in plants. *CRC Crit. Rev. Plant Sci.* **13:** 199-218.

Chaparzadeh, N., D'Amico, M.L., Khavari-Nejad, R.A., Izzo, R. and Izzo, F.N. (2004). Antioxidative responses of *Calendula offýcinalis* under salinity conditions. *Plant Physiol. Biochem.* **42:** 695-701.

Chen, G.X. and Asada, K. (1989) Ascorbate peroxidase in tea leaves: occurrence of two isoforms and the differences in their enzymatic and molecular properties. *Plant Cell Physiol.* **30:** 987-998.

Chen, Z., Young, T.E., Ling, J., Chang, S.C. and Gallie, D.R. (2003). Increasing vitamin C content of plants through enhanced ascorbate recycling. *Proc. Natl. Acad. Sci. USA.* **100:** 3525-3530.

Coupe, S.A., Palmer, B.G. and Lake, J.A. (2006). Systemic signaling of environmental cues in *Arabidopsis* leaves. *J. Exp. Bot.* **57:** 329-341.

Cunningham, J.F.X. and Gantt, E. (1998). Genes and enzymes of carotenoid biosynthesis in plants. *Annu. Rev. Plant Physiol.* **49:** 557-583.

Cvetkovska, M., Rampitsch, C., Bykova, N. and Xing, T. (2005). Genomic analysis of MAP kinase cascades in *Arabidopsis* defense responses. *Plant Mol. Biol. Rep.* **23:** 331-343.

Dat, J., Vandenabeele, S., Vranova, E., Van Montagu, M., Inzé, D. and Van Breusegem, F. (2000) Dual action of the active oxygen species during plant stress responses. *Cell Mol. Life Sci.* **57:** 779-795.

Davey, M.D., Van Montagu, M., Inze, D., Sanmartin, M., Kanellis, A.K., Smirnoff, N., Benzie, I.J.J., Strain, J.J., Favell, D. and Fletcher, J. (2000). Plant L-ascorbic acid: chemistry, function, metabolism, bioavailability and effects of processing. *J. Sci. Food Agr.* **80:** 825-860.

Davletova, S., Rizhsky, L., Liang, H., Shengqiang, Z., Oliver, D.J., Coutu, J., Shulaev, V., Schlauch, K. and Mittler, R. (2005). Cytosolic ascorbate peroxidase 1 is a central component of the reactive oxygen gene network of Arabidopsis. *Plant Cell.* **17:** 268-281.

Delledome, M., Polverari, A. and Murgia, I. (2003). The functions of nitric oxide-mediated signaling and changes in gene expression during the hypersensitive response. *Antioxid. Red. Sig.* **5:** 33-41.

del Rio, L.A., Sandalio, L.M., Corpas, F.J., Palma, J.M. and Barrosa, J.B. (2006). Reactive oxygen species and reactive nitrogen species in peroxisomes. Production, scavenging, and role in cell signaling. *Plant Physiol.* **141:** 330-335.

Dowdle, J., Ishikawa, T., Gatzek, S., Rolinski, S. and Smirnoff, N. (2007). Two genes in *Arabidopsis thaliana* encoding GDP-L-galactose phosphorylase are required for ascorbate biosynthesis and seedling viability. *Plant J.* **52:** 673-689.

Drazkiewicz, M., Polit, E.S., Krupa, Z. (2003). Response of the ascorbate–glutathione cycle to excess copper in *Arabidopsis thaliana* (L.). *Plant Sci.* **164:** 195-202.

F.A.O. (2005). *Global Network on Integrated Soil Management for Sustainable Use of Salt-affected Soils.* Rome, Italy: FAO Land and Plant Nutrition Management Service. http://www.fao.org/ag/agl/agll/spush

Fath, A., Bethke, P., Beligni, V. and Jones, R. (2002). Active oxygen and cell death in cereal aleurone cells. *J. Exp. Bot.* **53:** 1273-1282.

Feng, W., Hongbin, W., Bing, L. and Jinfa, W. (2006). Cloning and characterization of a novel splicing isoform of the iron-superoxide dismutase gene in rice (*Oryza sativa* L). *Plant Cell Rep.* **24:** 734-742.

Fotopoulos, V., De Tullio, M.C., Barnes, J. and Kanellis, A.K. (2008). Altered stomatal dynamics in ascorbate oxidase over-expressing tobacco plants suggest a role for dehydroascorbate signalling. *J. Exp. Bot.* **59:** 729-737.

Fotopoulos, V., Sanmartin, M. and Kanellis, A.K. (2006). Effect of ascorbate oxidase over-expression on ascorbate recycling gene expression in response to agents imposing oxidative stress. *J. Exp. Bot.* **57:** 3933-3943.

Foyer, C.H. and Halliwell, B. (1976). The presence of glutathione and glutathione reductase in chloroplast: a proposed role in ascorbic acid metabolism. *Planta.* **133:** 21-25.

Foyer, C.H. and Noctor, G. (2003). Redox sensing and signaling associated with reactive oxygen in chloroplasts, peroxisomes and mitochondria. *Physiol. Plant.* **119:** 355-364.

Foyer, C.H. and Noctor, G. (2005). Redox homeostasis and antioxidant signaling: A metabolic interface between stress perception and physiological responses. *Plant Cell.* **17:** 1866-1875.

Foyer, C.H., Souriau, N., Perret, S., Lelandais, M., Kunert, K.J., Pruvost, C. and Jouanin, L. (1995). Overexpression of glutathione reductase but not glutathione synthetase leads to increases in antioxidant capacity and resistance to photoinhibition in poplar trees. *Plant Physiol.* **109:** 1047-1057.

Frealle, E., Noel, C., Viscogliosi, E., Camus, D., Dei-Cas, E. and Delhaes, L. (2005). Manganese superoxide dismutase in pathogenic fungi: an issue with pathophysiological and phylogenetic involvements. *FEMS Immunol. Med. Microbiol.* **45:** 411-422.

Gomez, L.D., Noctor, G., Knight, M.R. and Foyer, C.H. (2004). Regulation of calcium signalling and gene expression by glutathione. *J. Exp. Bot.* **55:** 1851-1859.

Han, H., Li, Y. and Zhou, S. (2008). Overexpression of phytoene synthase gene from *Salicornia europaea* alters response to reactive oxygen species under salt stress in transgenic Arabidopsis. *Biotechnol Lett.* **30:** 1501-1507.

Hancock, R.D. and Viola, R. (2005). Improving the nutritional value of crops through enhancement of L-ascorbic acid (vitamin C) content: rationale and biotechnological opportunities. *J. Agr. Food Chem.* **53:** 5248-5257.

Hell, R. (1997). Molecular physiology of plant sulfur metabolism. *Planta.* **202:** 138-148.

Hernandez, J.A., Ferrer, M.A., Jiménez, A., Barceló, A.R. and Sevilla, F. (2001). Antioxidant systems and production in the apoplast of pea leaves. Its relation with salt-induced necrotic lesions in minor veins. *Plant Physiol.* **127:** 817-831.

Hernandez, J.A., Jiménez, A., Mullineaux, P. and Sevilla, F. (2000). Tolerance of pea (*Pisum sativum* L.) to long-term salt stress is associated with induction of antioxidant defences. *Plant Cell Environ.* **23:** 853-862.

Hsu, S.Y. and Kao, C.H. (2003). Differential effect of sorbitol and polyethylene glycol on antioxidant enzymes in rice leaves. *Plant Growth Regul.* **39:** 83-90.

Ishikawa, T., Dowdle, J. and Smirnoff, N. (2006). Progress in manipulating ascorbic acid biosynthesis and accumulation in plants. *Physiol. Plant.* **126:** 343-355.

Jain, A.K. and Nessler, C.L. (2000). Metabolic engineering of an alternative pathway for ascorbic acid biosynthesis in plants. *Mol. Breed.* **6:** 73-78.

Jimenez, A., Herna'ndez, J.A., del Rýo, L.A., Ros Barcelo, A. and Sevilla, F. (1997). Evidence for the presence of the ascorbate-glutathione cycle in mitochondria and peroxisomes of pea (*Pisum sativum* L.) leaves. *Plant Physiol.* **114:** 275-284.

Kiffin, R., Bandyopadhyay, U. and Cuervo, A.M. (2006). Oxidative stress and autophagy. *Antioxid. Red. Sig.* **8:** 152-162.

Kocsy, G., von Ballmoos, P., Suter, M., Ruegsegger, A., Galli, U., Szalai, G., Galiba, G. and Brunold, C. (2000). Inhibition of glutathione synthesis reduces chilling tolerance in maize. *Planta.* **211:** 528-536.

Kunert, K.J. and Foyer, C.H. (1993). Thiol/disulphide exchange in plants. In: De Kok, L.J. (Ed.). Sulfur nutrition and assimilation in higher plants. SPB Academic Publishing, The Hague, The Netherlands, pp. 139-151.

Kuzniak, E. and Maria, S. (2001). Ascorbate, glutathione and related enzymes in chloroplasts of tomato leaves infected by *Botrytis cinerea. Plant Sci.* **160:** 723-731.

Kwon, S.I. and An, C.S. (2003). Cloning and expression of mitochondrial MnSOD from the small radish (*Raphanus sativus* L). *Mol. Cell.* **16:** 194-200.

Kwon, S.Y., Choi, S.M., Ahn, Y.O., Lee, H.S., Lee, H.B., Park, Y.M. and Kwak, S.S. (2003). Enhanced stress-tolerance of transgenic tobacco plants expressing a human DHAR gene. *J. Plant Physiol.* **160:** 347-353.

Kwon, S.Y., Jeong, Y.J., Lee, H.S., Kim, J.S., Cho, K.Y., Allen, R.D. and Kwak, S.S. (2002). Enhanced tolerances of transgenic tobacco plants expressing both superoxide dismutase and ascorbate peroxidase in chloroplasts against methyl viologen-mediated oxidative stress. *Plant Cell Environ.* **25:** 873-882.

Ledford, H.K. and Niyogi, K.K. (2005). Singlet oxygen and photo-oxidative stress management in plants and algae. *Plant Cell Environ.* **28:** 1037-1045.

Lee, S.H., Ahsan, N., Lee, K.W., Kim, D.H., Lee, D.G., Kwak, S.S., Kwon, S.Y., Kim, T.H. and Lee, B.H. (2007). Simultaneous overexpression of both Cu/Zn superoxide dismutase and ascorbate peroxidase in transgenic tall fescue plants confers increased tolerance to a wide range of abiotic stresses. *J. Plant Physiol.* **164:** 1626-1638.

Li, J.M. and Jin, H. (2007). Regulation of brassinosteroid signaling. *Trends Plant Sci.* **12:** 37-41.

Link, G. (2003). Redox regulation of chloroplast transcription. *Antioxid. Red. Sig.* **5:** 79-87.

Lorence, A., Chevone, B.I., Mendes, P. and Nessler, G.L. (2004). Myoinositol oxygenase offers a possible entry point into plant ascorbate biosynthesis. *Plant Physiol.* **134:** 1200-1205.

Luo, Y., Liu, Y.B., Dong, Y.X., Gao, X.Q. and Zhang, X.S. (2009). Expression of a putative alfalfa helicase increases tolerance to abiotic stress in Arabidopsis by enhancing the capacities for ROS scavenging and osmotic adjustment. *J. Plant Physiol.* **166:** 385-394.

Marrs, K. (1996). The functions and regulation of gluthathione-S-transferases in plants. *Annu. Rev. Plant Physiol. Plant Mol. Biol.* **48:** 127-158.

Maughan, S. and Foyer, C.H. (2006). Engineering and genetic approaches to modulating the glutathione network in plants. *Physiol. Plant.* **126:** 382-397.

May, M.J., Vernoux, T., Leaver, C., Montagu, M.V. and Inze, D. (1998). Glutathione homeostasis in plants: implications for environmental sensing and plant development. *J. Exp. Bot.* **49:** 649-667.

Millar, A.H., Mittova, V. and Kiddle, G. (2003). Control of ascorbate synthesis by respiration and its implications for stress responses. *Plant Physiol.* **133:** 443-447.

Mittler, R. (2002). Oxidative stress, antioxidants and salt tolerance. *Trends Plant Sci.* **7:** 405-410.

Mittova, V., Theodouloub, F.L., Kiddleb, G., Gomezb, L., Volokitaa, M., Talc, M., Foyer, C.H. and Guya, M. (2003). Coordinate induction of glutathione biosynthesis and glutathione metabolizing enzymes is correlated with salt tolerance in tomato. *FEBS Lett.* **554:** 417-421.

Noctor, G., Arisi, A.M., Jouanin, L., Kunert, K.J., Rennenberg, H. and Foyer, C.H. (1998). Glutathione: biosynthesis, metabolism and relationship to stress tolerance explored in transformed plants. *J. Exp. Bot.* **49:** 623-647.

Noctor, G. and Foyer, C.H. (1998). Ascorbate and glutathione: keeping active oxygen under control. *Annu. Rev. Plant Physiol. Plant Mol. Biol.* **49:** 249-279.

Payton, P., Allen, R.D., Trolinder, N. and Holaday, A.S. (1997). Over-expression of chloroplast-targeted Mn superoxide dismutase in cotton (*Gossypium hirsutum* L., cv. Coker 312) does not alter the reduction of photosynthesis after short exposures to low temperature and high light intensity. *Photosynth. Res.* **52:** 233-244.

Pfannschmidt, T., Schutze, K., Fey, V., Sherameti, I. and Oelmüller, R. (2003). Chloroplast redox control of nuclear gene expression-A new class of plastid signals in interorganellar communication. *Antioxid. Red. Sig.* **5:** 95-101.

Pourcel, L., Routaboul, J.M. and Cheynier, V. (2007). Flavonoid oxidation in plants: from biochemical properties to physiological functions. *Trends Plant Sci.* **12:** 29-36.

Prashanth, S.R., Sadhasivam, V. and Parida, A. (2008). Over expression of cytosolic copper/zinc superoxide dismutase from a mangrove plant *Avicennia marina* in indica Rice var Pusa Basmati-1 confers abiotic stress tolerance. *Transgenic Res.* **17:** 281-291.

Rouhier, N., Vlamis-Gardicas, A. and Lilling, C.H. (2003). Characerization of redox properties of poplar glutaredoxin. *Antioxid. Red. Sig.* **5:** 15-22.

Roxas, V.P., Smith, R.K., Allen, E.R. and Allen, R.D. (1997). Overexpression of glutathione S-transferase/glutathione peroxidase enhances the growth of transgenic tobacco seedlings during stress. *Nat. Biotechnol.* **15:** 988-991.

Rubio, M.C., Gonzalez, E.M., Minchin, F.R., Webb, K.J., Arrese-Igor, C., Ramos, J. and Becana, M. (2002). Effects of water stress on antioxidant enzymes of leaves and nodules of transgenic alfalfa overexpressing superoxide dismutases. *Physiol. Plant.* **115:** 531-540.

Shao, H.B., Guo, Q.J., Chu, L.Y., Zhao, X.N., Su, Z.L., Hu, Y.C. and Cheng, J.F. (2007). Understanding molecular mechanism of higher plant plasticity under abiotic stress. *Biointerfaces.* **54:** 37-45.

Shao, H.B., Liang, Z.S. and Shao, M.A. (2005a) Adaptation of higher plants to stresses and stress signal transduction. *Acta Ecol. Sin.* **25:** 1871-1882.

Shao, H.B., Liang, Z.S., Shao, M.A., Sun, Q. and Hu, Z.M. (2005b). Investigation on dynamic changes of photosynthetic characteristics of 10 wheat (*Triticum aestivum* L.) genotypes during two vegetative-growth stages at water deficits. *Biointerfaces.* **43:** 221-227.

Sharma, P. and Dubey, R.S. (2005). Drought induces oxidative stress and enhances the activities of antioxidant enzymes in growing rice seedlings. *Plant Growth Regul.* **46:** 209-211.

Shcolnick, S. and Keren, N. (2006). Metal homeostasis in cyanobacteria and chloroplasts to. Balancing benefits and risks to the photosynthetic apparatus. *Plant Physiol.* **141:** 805-810.

Shin, S.Y., Lee, H.S., Kwon, S.Y., Kwon, S.T. and Kwak, S.S. (2005). Molecular characterization of a cDNA encoding copper/zinc superoxide dismutase from cultured cells of *Manihot esculenta. Plant Physiol. Biochem.* **43:** 55-60.

Smirnoff, N. and Wheeler, G.L. (2000). Ascorbic acid in plants: biosynthesis and function. *Crit. Rev. Biochem. Mol. Biol.* **35:** 291-314.

Srivastava, S., Tripathi, R.D. and Dwivedi, U.N. (2004). Synthesis of phytochelatins and modulation of antioxidants in response to cadmium stress in *Cuscuta reflexa* – An angiospermic parasite. *J. Plant Physiol.* **161:** 665-674.

Teixeira, F.K., Menezes-Benavente, L., Galvao, V.C., Margis, R. and Margis-Pinheiro, M. (2006). Rice ascorbate peroxidase gene family encodes functionally diverse isoforms localized in different subcellular compartments. *Planta.* **224:** 300-314.

Terman, A. and Brunk Uft, T. (2006). Oxidative stress, accumulation of biological 'Garbage', and aging. *Antioxid. Red. Sig.* **8:** 197-204.

Tertivanidis, K., Goudoula, C., Vasilikiotis, C., Hassiotou, E., Perl-Treves, R., Tsaftaris, A. (2004). Superoxide dismutase transgenes in sugarbeets confer resistance to oxidative agents, the fungus *C. beticola. Transgenic Res.* **13:** 225-233.

Tsai, Y.C., Hong, C.Y., Liu, L.F. and Kao, C.H. (2005). Expression of ascorbate peroxidase and glutathione reductase in roots of rice seedlings in response to NaCl and H_2O_2. *J. Plant Physiol.* **162:** 291-299.

Tseng, M.J., Liu, C.W. and Yiu, J.C. (2007). Enhanced tolerance to sulfur dioxide and salt stress of transgenic Chinese cabbage plants expressing both superoxide dismutase and catalase in chloroplasts. *Plant Physiol. Biochem.* **45:** 822-833.

Van Breusegem, F., Vranova, E., Dat, J.F. and Inzé, D. (2001). The role of active oxygen species in plant signal transduction. *Plant Sci.* **161:** 405-414.

Vashisht, A.A., Pradhan, A., Tuteja, R. and Tuteja, N. (2005). Cold- and salinity stress-induced bipolar pea DNA helicase 47 is involved in protein synthesis and stimulated by phosphorylation with protein kinase C. *Plant J.* **44:** 76-87.

Vashisht, A.A. and Tuteja, N. (2006). Stress responsive DEAD-box helicases: a new pathway to engineer plant stress tolerance. *J. Photochem. Photobiol.* B. **84:** 150-160.

Wang, F.Z., Wang, Q.B., Kwon, S.Y., Kwak, S.S. and Su, W.A. (2005). Enhanced drought tolerance of transgenic rice plants expressing a pea manganese superoxide dismutase. *J. Plant Physiol.* **162:** 465-472.

Wheeler, G.L., Jones, M.A. and Smirnoff, N. (1998). The biosynthetic pathway of vitamin C in higher plants. *Nature.* **393:** 365-369.

Willekens, H., Chamnongpol, S., Davey, M., Schraudner, M., Langebartels, C., Van Montagu, M., Inzé, D. and Van Camp, W. (1997). Catalase is a sink for H_2O_2 and is indispensable for stress defense in C3 plants. *EMBO J.* **10:** 1723-1732.

Wu, G., Wei, Z.K. and Shao, H.B. (2007). The mutual responses of higher plants to environment: physiological and microbiological aspects. *Biointerfaces.* **59:** 113-119.

Wu, Y.S. and Tang, K.X. (2004). MAP Kinase cascades responding to environmental stress in plants. *Acta Bot. Sin.* **46:** 127-136.

Yabuta, Y., Maruta, T. and Yoshimura, K. (2004). Two distinct redox signaling pathways for cytosolic APX induction under photooxdidative stress. *Plant Cell Physiol.* **45:** 1586-1594.

Yadav, S.K., Singla-Pareek, S.L., Ray, M., Reddy, M.K. and Sopory, S.K. (2005a) Methylglyoxal levels in plants under salinity stress are dependent on glyoxalase I and glutathione. *Biochem. Biophys. Res. Commun.* **337:** 61-67.

Yadav, S.K., Singla-Pareek, S.L., Reddy, M.K. and Sopory, S.K. (2005b). Transgenic tobacco plants overexpressing glyoxalase enzymes resist an increase in methylglyoxal and maintain higher reduced glutathione levels under salinity stress. *FEBS Lett.* **579:** 6265-6271.

14

Secondary Metabolites as Non-Enzymatic Plant Protectors from Oxidative Stress

Ana C. Ramos-Valdivia[1*], **Ariana A. Huerta-Heredia**[1], **Gabriela Trejo-Tapia**[3] and **Carlos M. Cerda-García Rojas**[2]

[1]*Departamento de Biotecnología y Bioingeniería.*
[2]*Departamento de Química. Centro de Investigación y de Estudios Avanzados del Instituto Politécnico Nacional (CINVESTAV-IPN). Av. Instituto Politécnico Nacional 2508. Col. San Pedro Zacatenco, 07360 México, D.F., Mexico*
[3]*Departamento de Biotecnología, Centro de Desarrollo de Productos Bióticos, Instituto Politécnico Nacional (CEPROBI-IPN), P.O. Box 24 62730, Yautepec, Morelos, Mexico*
[*]***Corresponding author:*** *aramos@cinvestav.mx*

ABSTRACT

Plants in their environment are exposed to biotic, mechanical, or abiotic stress, which can change their physiology, morphology and development. It is proposed that stress can stimulate oxidative burst in which there is a transient increase in reactive oxygen species (ROS) such as hydrogen peroxide (H_2O_2). There are two detoxification mechanisms to regulate the concentration of ROS. The enzymatic one is formed by superoxide dismutase (SOD), catalase (CAT), peroxidase (POD), glutathione peroxidase (GPx), glutathione reductase (GR) and ascorbate peroxidase. The non-enzymatic mechanism is composed of molecules such as pigments, glutathione, ascorbic acid and several secondary metabolites. In this chapter, we review the information supporting the secondary metabolites, in particular alkaloids, flavonoids and saponins, through their diversity of functions, can be involved in the non-enzymatic plant protection from oxidative stress.

Keywords: Secondary metabolites, oxidative stress, defence, ROS, non-enzymatic mechanisms.

1. INTRODUCTION

Secondary metabolites in plants are those compounds which are not essential to growth and life of the organisms, but play a crucial role in the interaction of the species with their environment, ensuring their survival in their ecosystem (Verpoorte 2000). These compounds have restricted occurrence according to the taxonomic groups and are biosynthesized from one or more primary metabolites (Fig. 1) by a wider variety of pathways in which they can be defined as intermediates or final products (Wink 1999). Therefore, it has been regarded at the functional level of plant metabolism that secondary metabolism is dispensable for growth and development but indispensable for the survival of the species (Fig. 2) (Hartmann 1996, 2007). The ability of plants for synthesizing secondary metabolites has been selected through the course of evolution in most taxonomic lineages where such compounds address specific needs. Thus, floral scent volatiles and pigments have evolved to attract insect pollinators improving fertilization; toxic chemicals have also evolved to protect species against pathogens and herbivores or to suppress the growth of neighbouring plants. Chemicals found in fruits, which give the colour, aroma, and flavour, prevent spoilage and act as signals of help to disperse

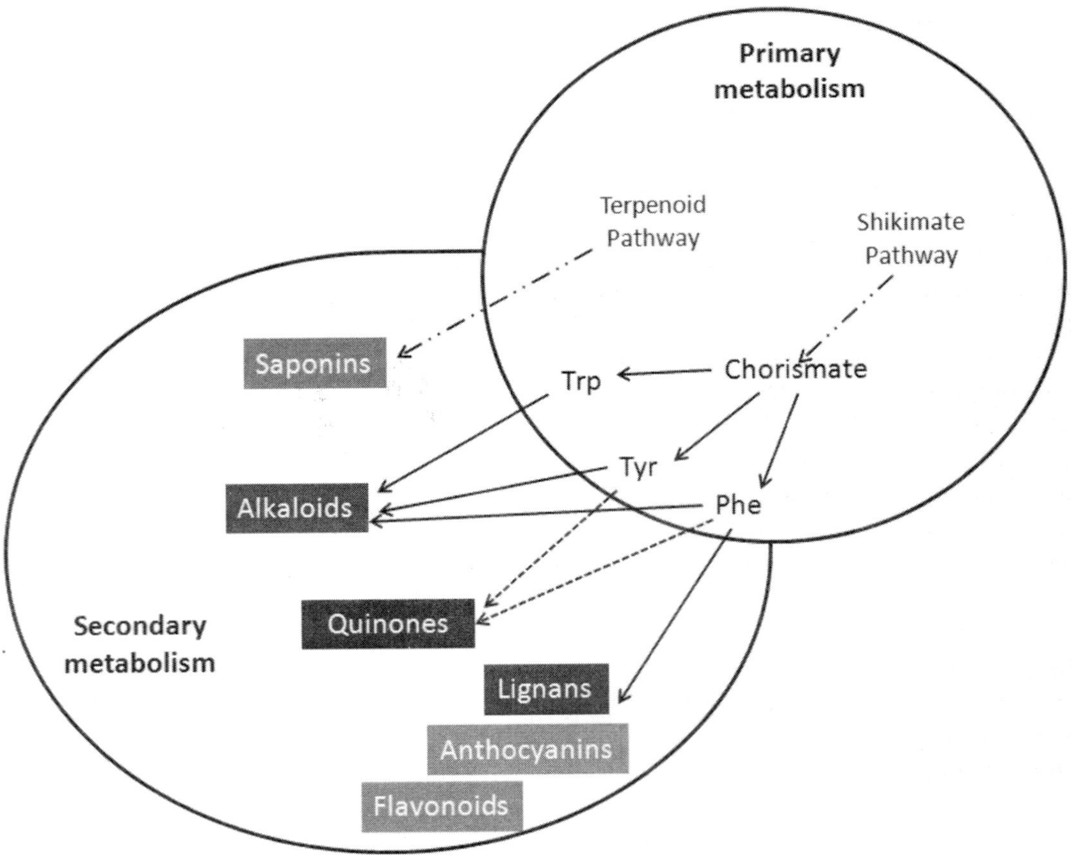

Fig. 1. Schematic view of the primary and secondary metabolism in plants.

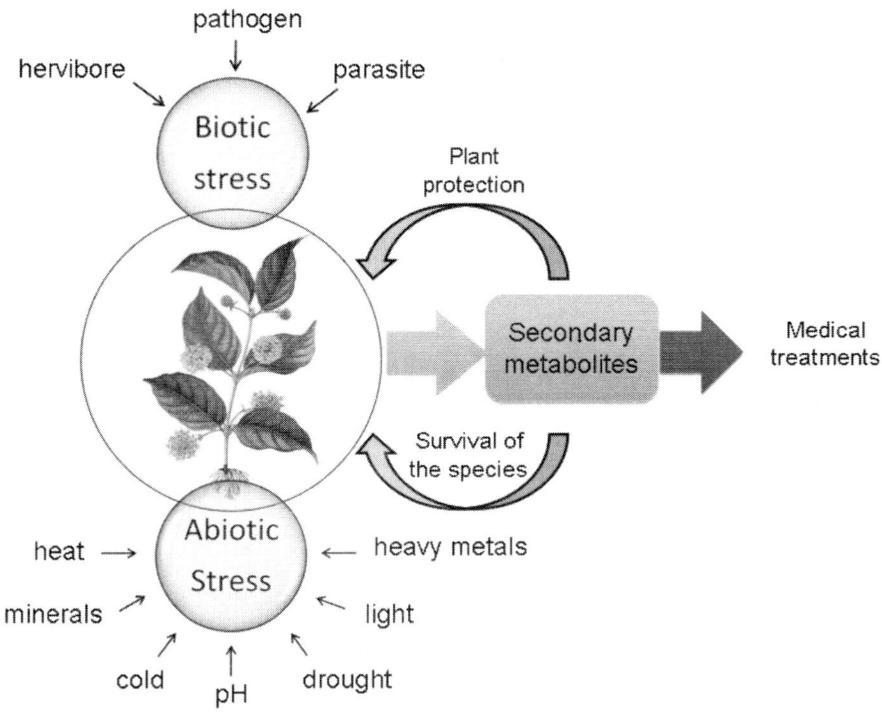

Fig. 2. Biotic and abiotic stress influence production of secondary metabolites.

the seeds. These requirements for secondary metabolites to have highly diverse biological activities have led plants to accumulate an immense diversity of compounds (Wink 1999).

Most of the secondary metabolite structural diversity is generated by differentially modifying common backbone structures, with the derived compounds having potentially divergent biological activities. Therefore, the catalogue in vascular plants is at least several tens thousand of secondary metabolites (Table 1) including many nitrogen-free substances (terpenes, polyketides, phenolics, saponins and polyacetylenes) and nitrogen-containing compounds (alkaloids, amines, cyanogenic glycosides, non-protein amino acids, glucosinolates, alkamides and peptides) (Wink 2008). In plants, secondary metabolites form complex mixtures, which differ from organ to organ and are accumulated and stored in high concentrations at regularly levels of 1–3% dry weight. In general, hydrophilic compounds are stored in the vacuole, whereas lipophilic substances are deposited in resin ducts, laticifers, trichomes, oil cells, or in the cuticle. In many cases the sites of synthesis are still unknown, but in some instances the sites of storage, long-distance transport by xylem, phloem or via the apoplast, have been already discovered (Wink 2008, 2010).

In the present review, the emphasis of discussion is placed on understanding the role of secondary metabolites as plant protectors against oxidative stress caused by biotic and/or abiotic factors.

Table 1: Approximate number of known secondary metabolites from higher plants classified by chemical family.

Secondary metabolites	Number[a]
Aliphatics	5 200
Polyketides	2 442
Carbohydrates	3 210
Oxygenated heterocycles	1 348
Simple aromatics	4 527
Benzofuranoids	387
Benzopyranoids	2 694
Flavonoids	8 128
Tannins	750
Lignans	1 565
Polycyclic aromatics	2 448
Terpenoids	27 463
Amino acids and peptides	3 921
Alkaloids	15 765
Total	85 058

[a]Taken from Verpoorte, 2000.

2. SECONDARY METABOLITES: FUNCTIONS AND CHALLENGES

Plants exhibit a wide array of defence strategies towards pathogen attack. The resistance against pathogens is performed by both preexisting constitutive (phythoanticipins) and induced defence systems (Bednarek and Osbourn 2009). Inducible defence responses are triggered following recognition of a range of chemical factors termed 'elicitors'. Originally, the term elicitor was used for molecules capable of inducing the production of phytoalexins, i.e., antimicrobial secondary compounds but it is now commonly used for compounds stimulating any type of plant defence (Nürnberger 1999). They may be of biotic or abiotic origin and are usually capable to induce various modes of plant defence including the production of reactive oxygen species, the hypersensitive response, and the production of phytoalexins (Mittler 2002). Secondary metabolites also play an important role in the adaptation of plants to the changing environment overcoming stress constraints, e.g., by polymerization and condensation of phenols and quinones, accumulation of polyamines and formation of phenylamides in tobacco and beans subjected to water stress and heat shocks (Edreva et al., 2008). Most of orders, families, and species evolved singular pathways and accomplished protection mechanisms through different biochemical adjustments. In addition to aiding plant defence and survival, the chemical diversity found in secondary metabolites has led to the development of numerous medical treatments and assisted in significant nutritional advancements (Caporale 1995). In spite of their importance, the studies about the secondary metabolites function in plants have still several major challenges as: (a) the extreme complexity of secondary metabolism demand novel strategies and tools to be developed; (b) only a fraction of the total number of different secondary metabolites present in plants are currently known and structurally characterized and, for most of the known compounds, commercially available standards are generally lacking. This limits an easy and accurate identification of individual components (Lehmann et al., 2009); (c) the diversity, specificity, and variability of secondary metabolites within and among plant families or species have been major impediments in the elucidation of many secondary pathways; (d) regulatory properties in secondary metabolism, such as cell-type-specific localization and transient expression, can in some cases complicate the biosynthetic potential achieved by plant tissue cultures; (e) many of the metabolic secondary pathways are present in more than one compartment (Lunn 2007); and (f) many plant species have complex genome interactions.

3. OXIDATIVE AND ANTIOXIDATIVE RESPONSES

Plants have developed complex regulatory mechanisms during their adaptation to various environmental stresses. One of the consequences of biotic or abiotic stress is a disturbance in the cellular redox balance with an increase in the cellular concentration of reactive oxygen species (ROS) (Apel and Hirt 2004) which include superoxide radical, hydrogen peroxide (H_2O_2) and singlet oxygen, ubiquitous molecules produced as a consequence of normal cellular metabolism (Fujita et al., 2006; Kotchoni and Gachomo 2006). Under normal conditions, ROS are rapidly metabolized. However, when plants are subjected to environmental stresses such as cold, high light intensity, ozone (O_3), drought, high salt concentration, pathogens, and UV irradiation, an excessive amount of ROS are generated by the affected species which require the activation of additional defences (Mittler et al., 2004). Unless these ROS are efficiently metabolized, they rapidly oxidize and damage lipids in cellular membranes, proteins, and other cellular components leading consequently to cellular dysfunction and ultimately to cell death, or appearance of necrotic lesions (Foyer and Noctor 2005).

For these reasons plants have built up protective enzymatic and non-enzymatic mechanisms to scavenge the reactive oxidative species (ROS) and minimize their harmful effects. The control of oxidant levels is then achieved by inducing antioxidative defence mechanisms. These defence systems are composed of metabolites such as ascorbate (ASC), glutathione (GSH), and tocopherol, and enzymatic scavengers of ROS such as superoxide dismutase (SOD), catalases (CAT) and peroxidases (POD). The majority of enzymes of the ascorbate-glutathione cycle, glutathione peroxidase (GPx), glutathione reductase (GR) and ascorbate peroxidase (APX) have been found in chloroplasts, cytosol, mitochondria, and peroxisomes (Noctor et al., 1998). Catalase and the ascorbate-glutathione cycle are important in H_2O_2 scavenging. The maintenance of ascorbate in its reduced form is achieved by dehydroascorbate reductase (DHAR) or by the operation of the ASC-GSH pathway. Ascorbate plays a crucial role acting as direct scavenger and as an electron donor of APX. Glutathione reductase also plays a key role in the antioxidant defence processes by reducing oxidized glutathione (GSSG) to GSH which consumes NADPH, thus allowing a high GSH/GSSG ratio to be maintained (Dat et al., 2000). Hydrogen peroxide is produced in response to such a variety of stimuli and mediates crosstalk between signalling pathways (Apel and Hirt 2004; Chesseman 2007). However, the total mechanism of H_2O_2 as a signal elicitor or regulator molecule is still under study (Neill et al., 2002; Hung et al., 2005). Some authors propose that reactive oxygen species, and H_2O_2 in particular, are not specific signal molecules during the plant development because they are not sensed through a direct receptor–ligand interaction. Recently, other authors have been regarded H_2O_2 as a signalling molecule contributing to the cross-tolerance towards another types of stress and as a regulator of the expression of some genes in cells. These include genes encoding antioxidants, cell rescue/defence proteins, and signalling proteins (Hancock et al., 2006) including kinases, phosphatases and transcription factors (Neill et al., 2002; Hung et al., 2005; Quan et al., 2008).

Under normal conditions, ROS are routinely generated at low levels in non-stressed plant cell chloroplasts and mitochondria, and also by cytoplasm, membrane bound or exocellular enzymes involved in redox reactions (Wojtaszek 1997). Consequently, the majority of cells have acquired the protective mechanisms to maintain the lowest possible level of ROS inside the cell. In some cases, however, especially under stress conditions such as pathogen attacks, these protective mechanisms are overdriven by a rapid transient production of huge amount of ROS, named 'oxidative burst' (Lamb and Dixon 1997). It is necessary to mention that in contrast to this negative term, implying a state to be avoided, it has been proposed to make use of the term 'oxidative signalling', which is more related with the mechanisms by which plant cells sense the environment and make appropriate adjustments to gene expression, metabolism, and physiology (Foyer and Noctor 2005).

The purpose of ROS production in plants includes: (i) direct killing of pathogens by generation of extremely reactive chemical species like the hydroxyl radical; (ii) involvement in cell wall stiffening by catalysing intra- and inter-molecular cross-links between structural components of cell walls and lignin polymerization; (iii) promotion of the programmed cell death; and (iv) generation of signal molecules to activate further plant defences as phytoalexins (Scandalios 2002).

The production of ROS has been established as one of the earliest signalling events involved in the response of plant to biotic (Fujita et al., 2006) and abiotic stress (Mittler 2006) such as metals (Leopoldini et al., 2006), wounding (Orozco-Cardenas et al., 2001) and mechanical stress (Gould et al., 2002). Plants use ROS as second messengers in signal transduction cascades in diverse processes as mitosis, tropisms and cell death being crucial in plant development and defence. Direct ROS signal transduction will ensue only if ROS escape from destruction by antioxidants or are otherwise consumed in a ROS cascade. Moreover, there is awesome evidence that many antioxidants play a key role in plant adaptation to both abiotic and biotic stresses (Lamb and Dixon 1997).

It has been proposed that an antioxidant is a molecule that (i) can donate electrons or hydrogen atoms; (ii) yields an antioxidant-derived radical that is efficiently quenched by other electron or hydrogen sources to prevent cellular damage; and (iii) its properties are spatially and temporally correlated with oxidative stress events (Hernandez et al., 2009).

Plants typically produce a diverse group of antioxidants as a protective mechanism against oxidative compounds which are generated in response to various stresses while a significant part of them are secondary metabolites. The plant defence response is also associated with the production of salicylic acid, methyl salicylate, jasmonic acid, methyl jasmonate and other small molecules produced as a result of stress that can also serve as signalling molecules activating systemic defence and acclimation responses (Apel and Hirt 2004). Higher plant cells can be considered a series of interconnecting compartments with diverse antioxidant buffering capacities determined in part by the nature of their secondary metabolites, their synthesis, transport and/or degradation. Antioxidants are not passive in this crosstalk, but rather function as key signalling components that constitute a dynamic metabolic interface between higher plant cell stress perception and physiological responses. This will require a better understanding of the degradation and transport of antioxidants in higher plants, and also elucidation of the molecular events responsible for stress perception and stress-related signal transduction. Hence, plant secondary metabolites are easily accumulated under stress conditions and are involved in versatile functions at different levels of the plant organism, scavenging of free radicals and the protection against excessive oxidation caused by UV irradiation, chemical oxidants, or pathogen attacks or other kinds of stress. The involved substances have characteristic features in its chemical structure which determine the mode of action as antioxidants. In spite of the wide variety of these compounds, there are some representative groups that have been studied. In this review, they have been classified in three major groups (a) phenolic compounds mostly hydroxylated flavonoids; (b) indole and oxindole alkaloids; and (c) saponins.

4. FLAVONOIDS

Flavonoids are widely distributed plant secondary metabolites resulting from addition of malonyl CoA to the phenylpropanoid molecule coumaroyl CoA. These polyphenolic compounds are characterized by two aromatic cycles (A and B) linked by a heterocyclic ring (C) (Fig. 3). They are classified, according to the oxidation degree of the C-ring, as flavonones, flavonols, flavan-3-ols, flavan-4-ols, flavan-3,4-diols, and anthocyanins (Fig. 3). Modifications of their aromatic cycles, including hydroxylations, methylations, glycosylations, acylations or prenylations originate a high diversity within these molecules (Winkel-Shirley 2002). Self-condensation of flavan-3-ols leads to the formation of proanthocyanidins (PAs), also called condensed tannins, while condensation of flavan-4-ols gives phlobaphenes. Both compounds can be oxidized to afford polymeric materials, namely brown derivatives. Some flavonoids with high *in vitro* antioxidant capacity are hydrophobic, supporting a biological function associated with membranes as flavan-3-ols which have been shown to adsorb to microsomal membranes.

One of the most important roles of flavonoids is related to their interactions with the environment, including defence, allelopathy, modulating the levels of ROS and protecting the plant against ultraviolet radiations. In this respect, it has been found that glycosylated flavonoids are present in chloroplasts (Schwinn and Davies 2004). Flavonoids also influence the transport of the plant hormones named auxins, provide flower colouring to attract pollinators, and in many species are required for pollen viability (Buer et al., 2010). It has been reported that flavonoid-deficient mutants are susceptible to UV irradiation, but some specimens can compensate this deficiency by over-production of sinapate esters (Sheahan 1996). In the same way, the pollen of

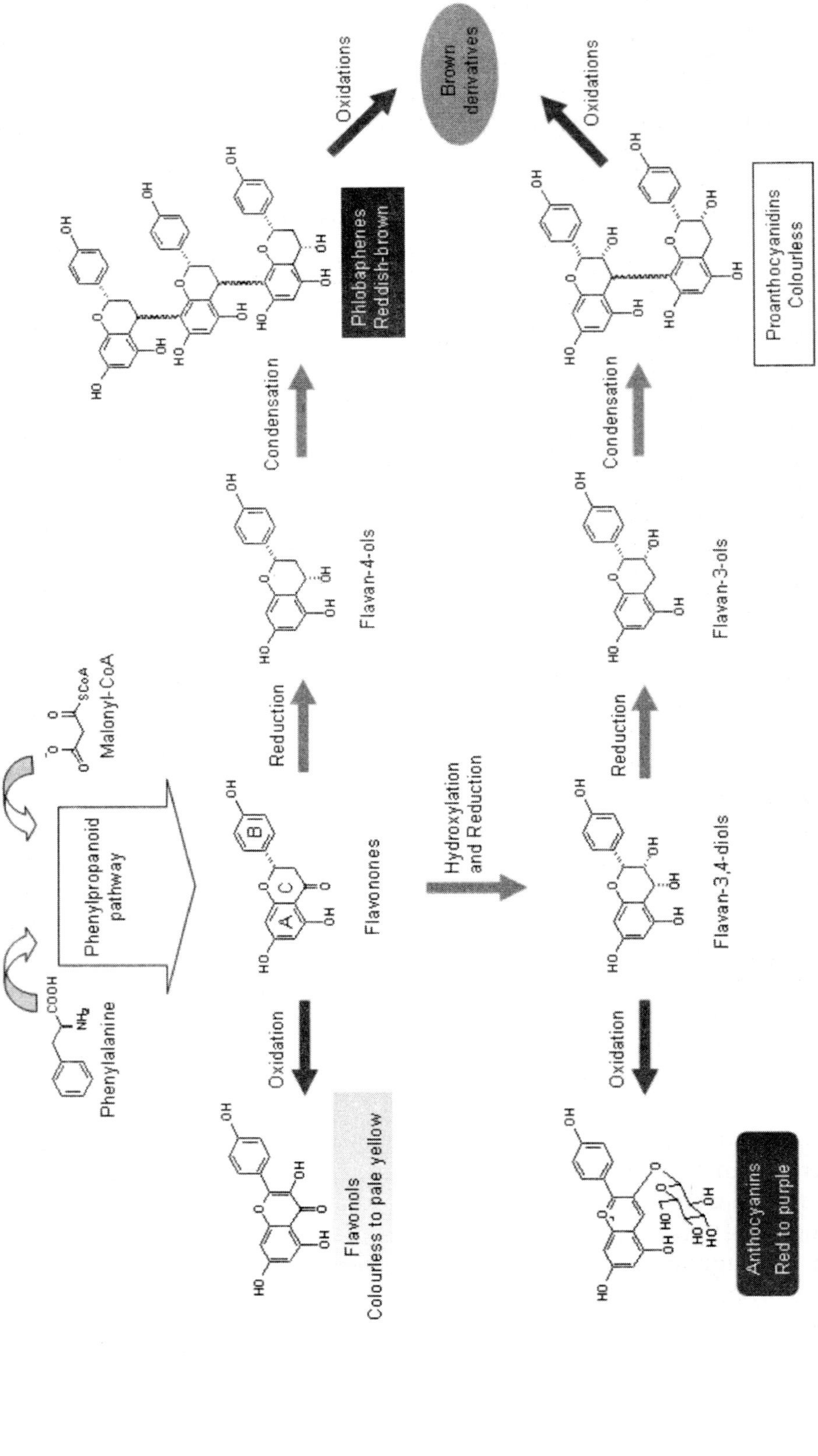

Fig. 3. Scheme for the most relevant pathways of flavonoid biosynthesis.

Arabidopsis thaliana mutants lacking flavonoids is not sterile as occurs in other plant species and this role is probably compensated by other molecules (Burbulis et al., 1996).

Biological properties of flavonoids are associated to their potential cytotoxicity and as scavengers of ROS, preventing their formation by chelating metals and quinones which can alkylate proteins. Flavonoids are the most common secondary metabolites in plants and are primarily localized in vacuoles. Their *in vivo* scavenging capacity has been questioned because the largest part of ROS effective scavenging mechanisms take place in the chloroplast, mitochondria, cytosol, peroxisomes and the cell wall (Bienert et al., 2006). Once inside the vacuoles, flavonoids are physically separated from the main sources of ROS and, hence, their relevance as antioxidants is minor in intact cells, despite their outstanding antioxidant capacity *in vitro*. A vacuolar scavenging system could be important under conditions of elevated stress to increase detoxification capacity. Hence, vacuolar flavonoids can only exert their potential antioxidative function when the physical barrier, created by the tonoplast, is disrupted and they come in contact with cytosolic oxidizing agents. Therefore, the antioxidative function of vacuolar flavonoids (e.g. anthocyanins and proanthocyanidins) should be considered in a specific context, such as after mechanical injury, piercing or feeding by herbivores. Takahama and Egashira (1991) could demonstrate that vacuoles of mesophyll cells of *Vicia faba* have the capacity to detoxify H_2O_2 and peroxidases involved in the scavenging process. Several peroxidases are localized in vacuoles of different plants (Morales and Ros Barceló 1997) and may be localized at the inner surface of the tonoplast (Thomas and Jen 1980). Thus, the permeating H_2O_2 entering the vacuoles could be converted by peroxidases using flavonoids as electron donors even if the concentration of them is up to 100 times higher than the K_m values for vacuolar peroxidases (Takahama and Egashira 1991). Interestingly, flavonoid synthesis is additionally induced in a wide range of stresses (UV light, wounding, pathogen infection, chilling, ozone, nutrient deprivation, etc.) where H_2O_2 levels increase significantly.

It has been reported that red anthocyanin-rich parts of *Pseudowintera colorata* leaves have a high H_2O_2 scavenging capacity after mechanical injury. In contrast, green leaves only slowly decrease H_2O_2 levels (Gould et al., 2002). Other flavonoid contribution to a powerful antioxidant/antiradical performance is by chelation of transition metals (Fe^{2+}) such as quercetin which interferes with the generation of ROS via the Fenton reaction (Leopoldini et al., 2006). While there has been a major focus on the antioxidant properties, there is an emerging view that flavonoids and their *in vivo* metabolites do not act as conventional hydrogen-donating antioxidants but may exert modulatory actions in cells through actions at protein kinase and lipid kinase signalling pathways (Williams et al., 2004).

Evidence is accumulating to support that flavonoid oxidation as browning reaction has a protection function during plant development and growth. Flavonoid oxidation also plays a role in defending the plant against various biotic and abiotic stresses. The complex process of flavonoid oxidation involves different mechanisms, thus further research on the regulation of flavonoid oxidase biosynthesis and activity will be important to elucidate the precise functions and regulatory factors that induce or repress flavonoid oxidation in plants (Pourcel et al., 2007).

Accumulation of anthocyanins in leaves of cotton suffering Na/K imbalance, shift from mono- to orthodihydroxy substitution in the B-ring of anthocyanin aglycones, with this conferring a higher ROS-scavenging capacity in relation to drought tolerance (Edreva et al., 2008). The presence of *o*-diphenol groups allows formation of the corresponding oxidized *o*-quinoid moieties. On the other hand, *in vitro* covalent interactions between proteins and oxidation products of caffeoylquinic acid were recently reported (Prigent et al., 2007).

A related flavonoid compound, *trans*-resveratrol, a stilbene phythoalexin, plays a role in the induction of protective mechanisms involved in tolerance to UV irradiation. It has been proposed that the increase in endogenous resveratrol along with changes in antioxidant enzymes may be involved in the defence response induced by UV-C exposure in peanut seedlings (Tang et al., 2010). These are several illustrative examples for the protective roles which have been studied so far in relation with the antioxidant properties of flavonoids and derived substances, although in many cases further studies are needed to elucidate the detailed mechanisms involved in the oxido-reduction processes.

5. INDOLE AND MONOTERPENOID INDOLE ALKALOIDS

Alkaloids are low molecular weight nitrogen-containing substances with characteristic toxicity and pharmacological activity accumulated in about 20% of plant species. The alkaloids are mostly derived from the amino acids, Phe, Tyr, Trp, Lys and Orn, being the monoterpenoid indole alkaloids (MIA) a large class of complex compounds derived from Trp and terpenoid precursors (Verpoorte 2000). Although over 12,000 different alkaloids have been described, little is known about their biosynthesis or about the factors that regulate its production. Intensive efforts have been done towards elucidating the basic biochemistry and molecular biology of some MIA pathways generally those from *Catharanthus roseus*. More than 100 *C. roseus* alkaloids that have been identified share many biosynthetic steps. The early stages of alkaloid biosynthesis in *C. roseus* involve the formation of secologanin, derived from the terpenoid biosynthesis and its condensation with tryptamine to produce the central intermediate strictosidine, as a common precursor. After that, several specific MIA end-products are formed by activity of strictosidine-β-D-glucosidase (SGD) (De Luca and St Pierre 2000).

As occurs in other secondary metabolites, in MIA biosynthesis, the diversification is originated from the elaboration of a few central intermediates that could be modified through oxidation, followed by protection of hydroxyl groups, by acylation, glucosylation, methylation, or among others, prenylation (Chou and Kutchan 1998). Recent studies have begun to illustrate that alkaloid biosynthesis is a highly ordered process with respect to plant development, controlling the expression of pathways within organs, within specific cells, or within organelles inside those cells and in response to stress factors (De Luca and St Pierre 2000).

The alkaloid biosynthesis involves multiple oxidations catalysed in a stereo- and regiospecific manner, suggesting that only specific oxidases are involved in biosynthesis *in vivo*. In more recent years, it has been found that some oxidations in alkaloid biosynthesis are catalysed by microsomal cytochrome P-450-dependent enzymes, peroxidases, 2-oxoglutarate dependent dioxygenases and flavoproteins. These have been described as highly substrate specific enzymes dedicated to alkaloid biosynthesis. However, *in vitro* studies have shown the capacity of plant peroxidases to accept alkaloids as substrates as well as a number of vacuolar metabolites, such as phenols and flavonoids (Sottomayor and Ros Barceló 2004; Takahama 2004).

Different studies have shown that exogenous addition of H_2O_2 to plants leads to a rapid production of endogenous H_2O_2 and disruption of the homeostatic redox status (Allan and Fluhr 1997). Therefore, in *C. roseus* different concentrations of H_2O_2 were used to describe the effects on vinblastine (VBL) production, one of the most important antitumour alkaloids, and the effect on the redox state with respect to POD activity and the ascorbate-glutathione cycle (Tang et al., 2009). In contrast to these findings, some authors propose that is still a question whether H_2O_2 is a signalling compound for indole alkaloid biosynthesis in cell suspension cultures (Zhao et al., 2001; Pauw et al., 2004). Calcium ions influx was identified as a prerequisite for fungal elicitor-induced oxidative stress, jasmonate biosynthesis, and indole alkaloid production (Zhao et al., 2001; Pauw et al., 2004; Lee-Parsons and Ertürk 2005). The implications of oxidative stress via nitric oxide burst were also observed in the production of the alkaloid catharanthine by elicitor-induced *C. roseus* cell cultures (Xu et al., 2005). In the same way other alkaloids, such like nicotine and berberine increase with oxidative stress induced by biotic and abiotic factors (Chou and Kutchan 1998; De Luca and St Pierre 2000).

In *C. roseus* cultured cells, accumulation of serpentine (oxidation product of ajmalicine) inside the vacuole is striking related with conformity between the time course of its content and that of the activity of basic peroxidases. Ajmalicine was converted efficiently into serpentine by basic peroxidases extracted from vacuoles and by intact isolated vacuoles (Blom et al., 1991). The results are consistent with the hypothesis that ajmalicine is accumulated by an ion-trap mechanism and that ajmalicine is converted into serpentine inside vacuoles. Transformation of ajmalicine into the charged serpentine creates a trap which retains the alkaloids more efficiently in the vacuole (Blom et al., 1991). Recently, it has been found that besides ion trapping and diffusion through membranes of MIA, a selective transport by ATP-binding cassette (ABC) transporters is also involved (Roytrakul and Verpoorte 2007). In the last years, Jaleel et al. (2006, 2007) have reported that, in *Catharanthus* plants, production of MIA is associated to the antioxidant metabolism and is induced under different oxidative stress conditions as salt stress, drought stress effects and application of systemic fungicides such as triazole compounds.

Studies of the metabolic adjustments of plants in response to oxidative stress are significant and can be substantially enriched with the newly developed tools. Recently, analyzing the transcriptome, metabolome and key metabolic fluxes on primary and secondary metabolism, the response of *Arabidopsis thaliana* cells and roots to menadione treatment were investigated (Baxter et al., 2007; Lehmann et al., 2009). An emergency survival strategy was uncovered in which the anabolic primary metabolism was largely downregulated in favour of the catabolic and antioxidant metabolism. After 2 and 6 h of treatment, the response of the *A. thaliana* roots was different from that of the cell cultures. Metabolite levels did not remain depressed, but instead recovered, that is the

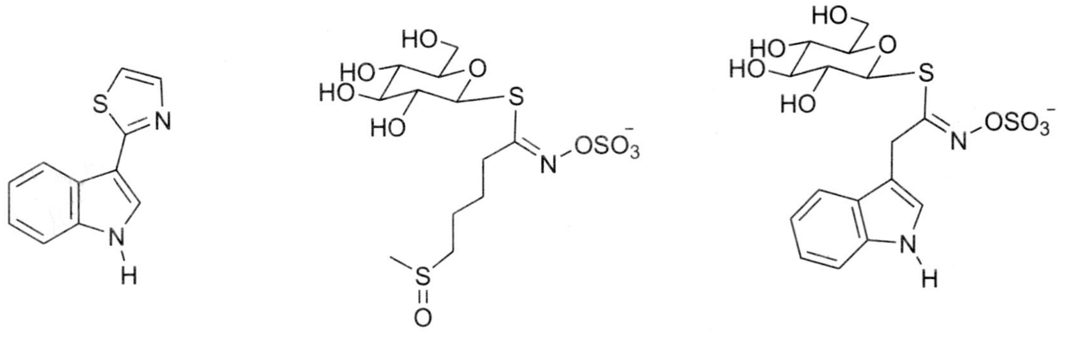

Fig. 4. Structures of camalexin, an aliphatic glucosinolate, and an indolylglucosinolate.

case of pyruvate, some amino acids, and aliphatic glucosinolates (Fig. 4). These last secondary metabolites showed a steady increase above control levels. However, after oxidative inhibition, no major changes in fluxes of central carbon metabolism were observed and metabolic transcripts changed largely and independently of the corresponding metabolites, suggesting that root tissues can recover metabolic activity for glycolysis and for the oxidative pentose phosphate pathway.

Another *Arabidopsis* secondary metabolite involved in plant/pathogen interactions is camalexin (Fig. 4), an indole phytoalexin with antioxidant capacity. This phytoalexin is induced under pathogen attacks and by abiotic elicitors that generate reactive oxygen species (Tsuji et al., 1992; Reuber et al., 1998; Roetschi et al., 2001). Camalexin is originated from tryptophan and its biosynthesis involves the cytochrome P450 enzymes CYP79B2 and CYP71B15 (PAD3) (Glawischnig 2007). In *Arabidopsis* many defence compounds are induced by jasmonate or salicylate, but these signalling compounds are not essential for camalexin induction as the ROS formation. Camalexin induction via amino acid starvation is likely to occur by damaged chloroplasts which cause photooxidative stress (Zhao et al., 1998). In general, phytoalexin formation is a non-systemic process while a local induction of phytoalexin biosynthetic genes is a concept that is found for several defence chemicals. The status of the antioxidant system in a range of ascorbate deficient *A. thaliana* vtc mutants showed that under the mild growth conditions used, other major antioxidants are not altered. Moreover, this mutant showed an elevated cell wall peroxidase activity, increased transcript level of PR proteins, and high accumulation of camalexin (Kliebenstein 2004).

6. GLUCOINDOLE AND OXINDOLE ALKALOIDS

Plant cell cultures in bioreactors often suffer from oxidative stress due to improper culture conditions such as oxygen supply, mechanical damages, and nutrition imbalance even without elicitation. Such oxidative stress can induce multiple effects on plant cell growth and secondary metabolism (Zhao et al., 2001; Zhao et al., 2005). Reactive oxygen species generated during oxidative stress may be the main factors for these effects (Trejo-Tapia

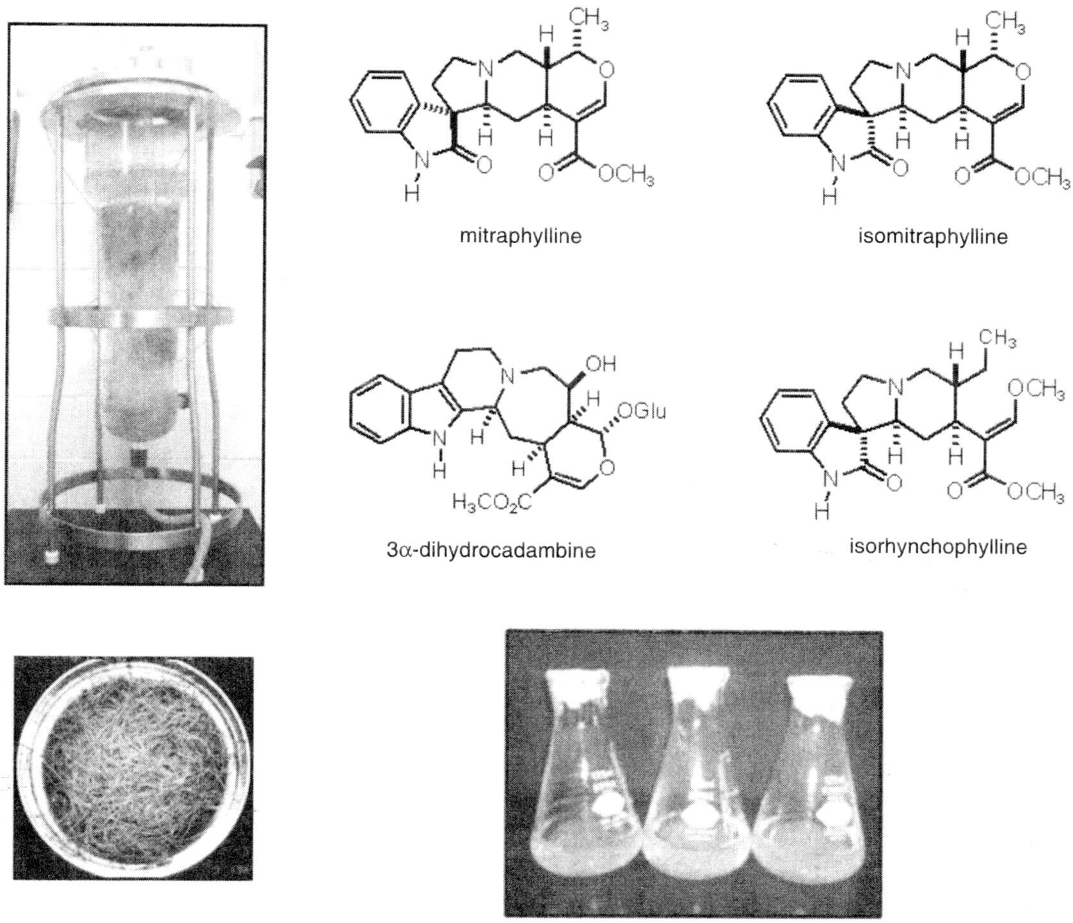

Fig. 5. *Uncaria tomentosa* root cultures (airlift bioreactor and flasks) and structures of their major alkaloids.

et al., 2007). Monoterpenoid oxindole alkaloids (MOA) (Fig. 5) are more oxidized than MIA. MOA possessed immunomodulatory, cytotoxic, anti-AIDS and antileukemic activities (Laus 2004, Garcia Prado et al., 2007). They are obtained from the powdered bark of at least 8 years old native plants of *Uncaria tomentosa* which is an indigenous plant from the Amazon rainforest known as "cat's claw" and affected by the deterioration of natural resources. As an alternative to favour *U. tomentosa* sustainability, these highly oxidized alkaloids have been produced in cell (Luna-Palencia et al., 2005; Trejo-Tapia et al., 2005) and tissue *in vitro* cultures (Luna-Palencia et al., 2006). The antioxidant activity of the MOA mitraphylline and pteropodine has been studied (Garcia Prado et al., 2007; Paniagua-Pérez et al., 2009).

Although the biosynthetic pathway of MOA in *U. tomentosa* still remains to be investigated, it has been proposed that in *Mitragyna* species, these alkaloids are formed from monoterpenoid indole alkaloids (MIA) through oxidation of the indole moiety

(Shellard et al., 1969). On the other hand it has been reported that the indole moieties can be oxidized to oxindole by several classes of enzymes like cytochrome P450 (Gillam et al., 2000), chloroperoxidases (van de Velde et al., 2001) and peroxidases requiring H_2O_2 (Folkes et al., 2002; Takemoto et al., 2004). Even though the biochemical (Luna-Palencia et al., 2006) or molecular mechanisms in *U. tomentosa* are still not fully elucidated, all the reports together suggested that an oxidative condition may stimulate MOA production. Therefore, Trejo-Tapia et al. (2007) demonstrated that in cell suspension cultures of *U. tomentosa* growing in stirred tank bioreactors, MOA production was stimulated by hydrodynamic stress via oxidative burst. In such cultures and in *U. tomentosa* roots cultures (Huerta-Heredia et al., 2009), it was proposed that shear stress and dissolved oxygen initiate an oxidative burst where the formed H_2O_2, by NAD(P)H oxidase and/or peroxidases, stimulates MOA production acting either as a signal molecule or as a donor to oxidize MOA precursors (Fig. 6).

In *U. tomentosa* differentiated tissues as root cultures, MOA production is accompanied by accumulation of 3α-dihydrocadambine (DHC) (Fig. 5) a glucoindole alkaloid with hypotensive (Endo et al., 1983), lipid lowering, and high antioxidant activities (Kumar et al., 2008). The structure of this compound was verified by 1D and

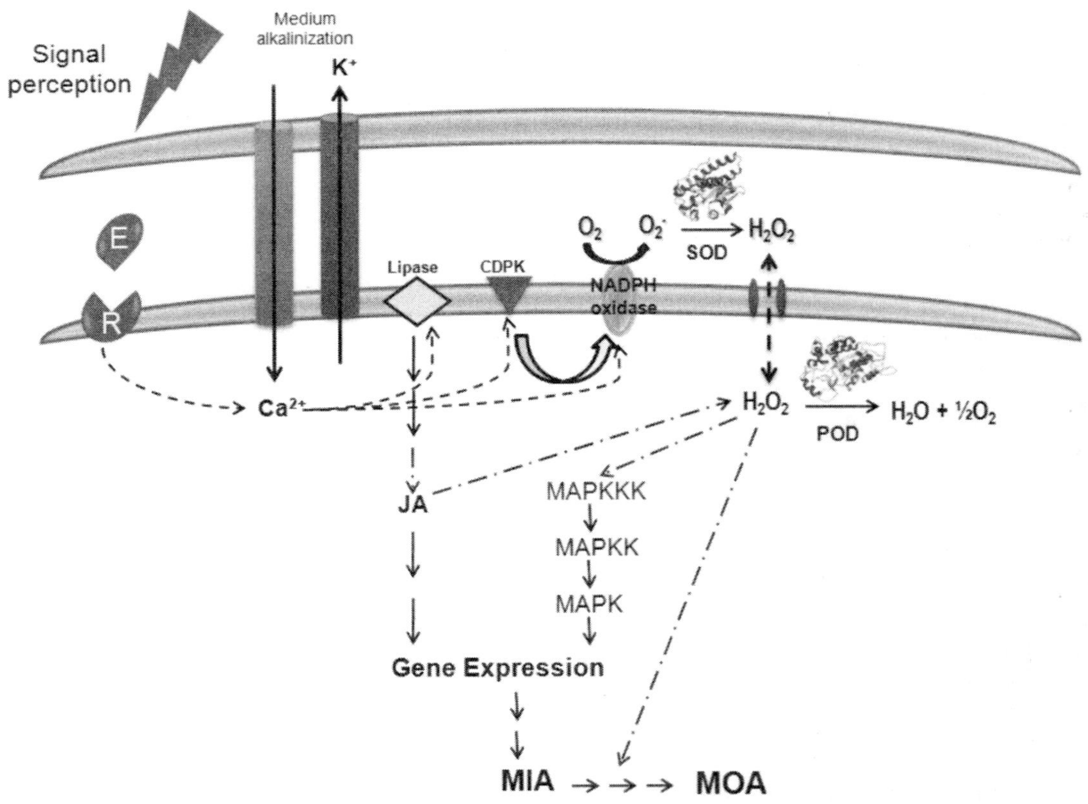

Fig. 6. An hypothetical model for the oxidative burst induced by biotic and abiotic stress that activates indole and oxindole alkaloid pathway.

2D NMR spectroscopy and mass spectrometry (Luna-Palencia et al., 2006). Huerta-Heredia et al. (2009) studied the effect of oxidative stress on MOA accumulation in cell and root cultures of *U. tomentosa* by adding H_2O_2 as an elicitor. To evaluate if MOA and DHC were stimulated by exogenous H_2O_2, this molecule was added in several amounts (200, 400, 800, and 1000 μM) to linearly growing 13-day-old root cultures in shaken flasks. Root viability remained 80% after H_2O_2 addition and it was found that H_2O_2 had a differential effect on DHC and MOA production. At 200 mM H_2O_2, MOA were enhanced by 56% with respect to the control (370.8 ± 15.3 μg/g) and DHC was increased by 265%; while addition of 400, 800 and 1000 mM H_2O_2, reduced by 30–40% DHC accumulation with respect to the control (428 ± 2.9 μg/g) (Fig. 7) without change in MOA. The DHC content reduction correlates with a high H_2O_2 concentration inside the

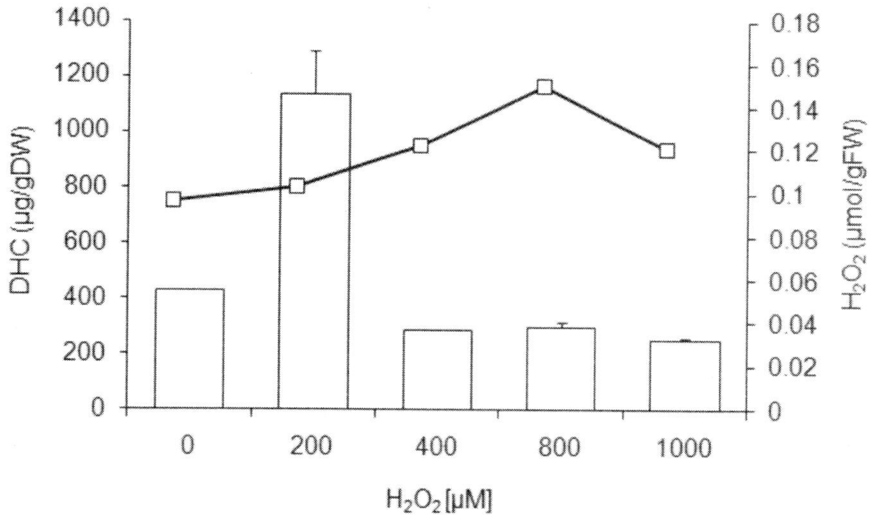

Fig. 7. Changes in 3α-dihydrocadambine (DHC) and in intracellular H_2O_2 concentrations after addition of H_2O_2 to root cultures of *Uncaria tomentosa* in shaken flasks. Bars indicate DHC amounts and the curve with (□)-symbols represents intracellular H_2O_2 concentrations. Error bars indicate standard deviation from the mean (n = 3).

roots (Fig. 7), which may oxidize this monoterpenoid glucoindole alkaloid. (Huerta-Heredia et al., 2009). As occurred in cell cultures (Trejo-Tapia et al., 2007), *U. tomentosa* root cultures in the airlift reactor produced extracellular H_2O_2 with a characteristic biphasic profile after changing aeration. Maximum MOA was 9.06 mg/L at day 60 while at this time roots reached *ca.* 1 mg/L of DHC. Intracellular H_2O_2 in root cultures growing in the bioreactor was 0.87 mmol/g DW compared to 0.26 mmol/g DW of shaken flasks cultures (Huerta-Heredia et al., 2009). It was concluded that *U. tomentosa* roots growing in the airlift bioreactor were exposed to an oxidative stress and their antioxidant system was highly active allowing them to produce oxindole alkaloids and liberated them to the culture medium while higher H_2O_2 inside the roots could be oxidating DHC (Huerta-Heredia et al., 2009).

7. OTHER INDOLIC COMPOUNDS

Many isolated monoterpene glucoindole alkaloids from *Rubiaceae* species have a high antioxidant activity *in vitro* as exemplified by some bioactive agents from Brazilian rainforest plants (Cardoso 2004). In this study, two new glucoalkaloids dehydro-strictosidine and 3,4-dehydrostrictosidinic acid were isolated from *Chimarrhis turbinate* along with seven known glucoalkaloids, cordifoline, strictosidinic acid, strictosidine, 5*R*-carboxystrictosidine, turbinatine, desoxycordifoline, and harman-3-carboxylic acid. Alkaloid cordifoline showed strong free radical scavenging activity against 1,1-diphenyl-2-picrylhydrazyl (DPPH) as well as relevant antioxidant activity evidenced by redox properties. Some oxidation experiments using indole derivatives, which are by far less complex substances than monoterpenoid indole alkaloids, have been carried out. Indeed, 5-fluoroindole-3-acetic is oxidized by horseradish peroxidase in the presence of H_2O_2 (Folkes et al., 2002), while indole derivatives are oxidized by adding H_2O_2 to suspension cultures of *Camellia sinensis, Nicotiana tabacum, Catharanthus roseus* and *Daucus carota* (Takemoto et al., 2004).

The β-carboline alkaloids found in medicinal plants and in a variety of foods, beverages and cigarette smoke have a range of action in various biological systems. *In vitro* and *in vivo* studies have demonstrated that the aromatic alkaloids harmane, harmine, harmol and the dihydro-β-carbolines, harmaline and harmalol can act as scavengers of reactive oxygen species with antioxidative properties. It has been shown that those alkaloids have a significant protective effect against H_2O_2 and paraquat oxidative agents in yeast cells, and that their ability to scavenge hydroxyl radicals contributes to their antimutagenic and antigenotoxic effects (Moura et al., 2007).

Alkaloids represent one of the most active natural product groups against a wide range of organisms. The main role of these substances is generally linked to plant defence from predators besides the important ecological factors associated to them. However, the close relationship between alkaloids and the oxido-reduction processes in plants containing them strongly suggests that these compounds play a fundamental role in protecting plants when they are subjected to oxidative stress.

8. SAPONINS

Saponins constitute an important group of secondary metabolites, widely distributed in higher plants, characterized by a skeleton derived from the 30-carbon precursor oxidosqualene to which glycosyl residues are attached (Vincken et al., 2007). Saponins are distinguished from other glycosides by their surfactant properties and for giving stable, soap-like foams in aqueous solutions (Osbourn 2003, Sparg et al., 2004). Based on the nature of their aglycone skeleton, saponins are classified into two groups. The first group consists of the steroidal saponins, which are almost exclusively present in monocotyledonous angiosperms highlighting the members of the Liliaceae, Dioscoraceae and Agavaceae families (Sparg et al., 2004) and in a few dicotyledonous angiosperms, such as Fabaceae and Scrophulariaceae (Gershenzon and Kreis 1999). The second group consists of the triterpenoid saponins, which are most common and occur mainly in the dicotyledonous angiosperms (Sparg et al., 2004). Cereals and grasses appear to be generally deficient in these secondary metabolites with the exception of oats (*Avena* spp.) (Osbourn 2003). Another class of compounds considered saponins are the glycosteroidalkaloids (Haralampidis et al., 2002). These compounds have the same biosynthetic precursor as saponins, and contain a steroid type skeleton linked to monosaccharide moieties buy they contain a nitrogen atom as an intrinsic and characteristic part of their aglycone structure so many authors classified them as a separate group and will not be considered in this review (Vincken et al., 2007).

Steroidal saponins consist of a steroidal aglycone, a C_{27} spirostane skeleton containing six rings (Fig. 8). In some cases, the hydroxyl group at the 26-position is engaged in a glycosidic linkage, and so the aglycone structure remains pentacyclic (furostane skeleton) (Fig. 8). Triterpenoid saponins consist of a C_{30} skeleton comprising a pentacyclic structure (Fig. 8) (Sparg et al., 2004).

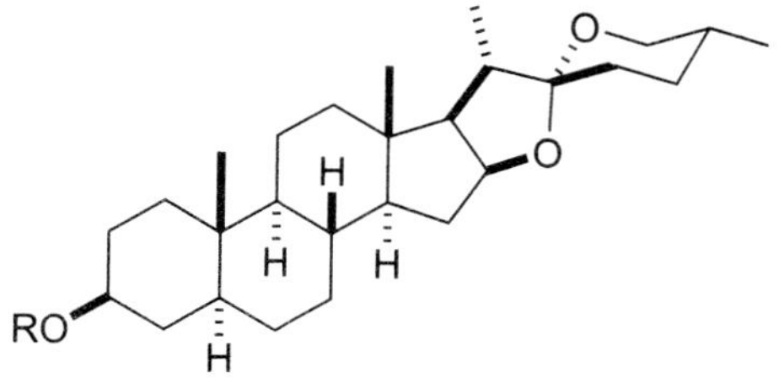

1

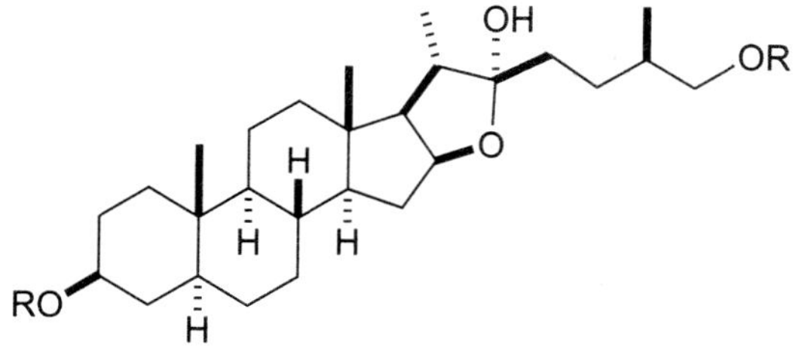

2

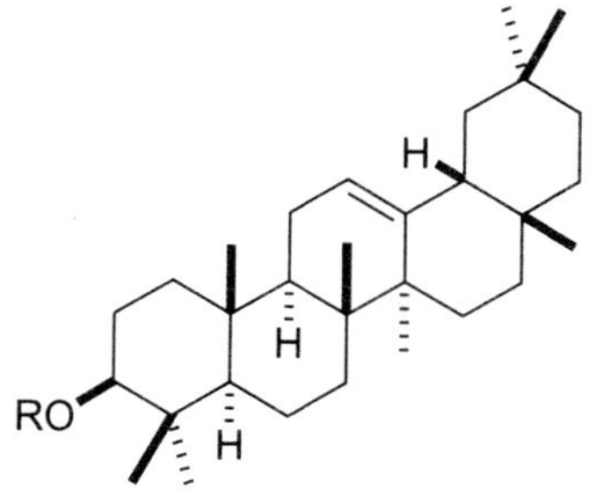

3

Fig. 8. Steroidal spirostane (**1**) steroidal furostane (**2**) and triterpenoid saponins (**3**) R = sugar moieties.

The enzymes, genes and biochemical pathways involved in the synthesis of these complex molecules are largely uncharacterized for any plant species. Saponin biosynthesis proceeds via the isoprenoid pathway in which three isopentene units (generally assumed to be provided by the mevalonate pathway) are first linked in a head-to-tail manner to each other, resulting in the 15-carbon atom molecule, called farnesyl pyrophosphate (FPP). Two FPPs are subsequently linked in a tail-to-tail manner to give a compound of 30 carbon atoms, called squalene. Subsequently, squalene is oxidized by squalene epoxidase to 2,3-oxidosqualene, which is the common starting point for cyclization reactions in triterpenoid biosynthesis. The latter compound is then transformed by a family of 2,3-oxidosqualene cyclases into tetra- or pentacyclic triterpenes, the immediate precursors of triterpene and steroidal sapogenins (saponins aglycones). For the synthesis of C_{30} triterpene sapogenins, squalene 2,3-epoxide is cyclized to one of several penta- or tetracyclic compounds such as α-amyrin, β-amyrin, lupeol, isomultiflorenol or dammaradienol (Flores-Sánchez et al., 2002; Kalinowska et al., 2005). Further diversity is given by modification of the cyclic products such as oxidation, rearrangements, glycosylation at C3 which is particularly important for their biological activity, esterification or hydroxylation (Jenner et al., 2005; Kuljanabhagavad and Wink 2009).

It is well documented that in plants, these molecules are involved in their chemical defence; its natural role is likely to be in protection against attack by pests and diseases (Bednarek and Osbourn 2009). They also may play a specific role in the interactions between various plant species as the case of allelopathy (Oleszek 1993; Hiradate et al., 1999; Bednarek and Osbourn 2009). Saponins are toxic to fungi (Osbourn et al., 2003; Chapagain et al., 2007), hervibores, insects (Adel et al., 2000; Shinoda et al., 2002), nematodes and mollusks (Joshi et al., 2008; Vasil'eva et al., 2009).

In recent years, the interest in saponins has grown rapidly because of their beneficial effects on human health. They present pharmacological properties, such as anti-HIV (Kashiwada et al., 1998), anti-diabetic (Wang et al., 2004; Xi et al., 2008), anti-inflammatory (Park et al., 2009), hepatoprotective (Wang et al., 2004) and hypolipidemic activities (Ji and Gong 2007). Besides, they are antioxidant agents (Zhang et al., 2008), anticancer (Lemeshko et al., 2006; Sun et al., 2010) and cytotoxic (Wang et al., 2007). Based on the increasing evidence that oxidative stress is implicated in the pathogenesis of many diseases, research has been focused on the protective effects of saponins against oxidative stress in mammalian cells. Little attention has been paid to the possible protective effect of saponins on plants themselves.

As stated in previous lines, the role of secondary metabolites in plants and the factors involved in their biosynthesis have been addressed by using several strategies, including plant cell culture and elicitation. Here we present some of the studies concerning triterpene and saponin biosynthesis by *in vitro* cultures. Special attention is paid to those investigations suggesting a link between oxidative stress and saponins biosynthesis.

Ginsenosides biosynthesis in *Panax ginseng* cultures has been observed after their exposure to biotic and abiotic stress factors. In many cases, oxidative stress has been involved. In *P. ginseng* cell cultures, saponin accumulation was stimulated as part of the cell response to the osmotic stress achieved by adding sorbitol to culture medium. In this case, sorbitol induced the production of ROS indicating an oxidative stress condition (Wu et al., 2005). In root cultures, saponin synthesis was stimulated at a copper concentration between 5 and 25 μM (Ali et al., 2006). At those concentrations, Cu^{2+} application is accompanied by a substantial increase in oxidative stress indicated by protein oxidation and reduced root growth (Ali et al., 2006). These reports agree with the observation that ginsenosides concentration increased when 25 and 50 mM H_2O_2 were applied (Ali et al., 2005). The results of this last work indicated that plant roots could grow and protect themselves from oxygen stress by coordinated induction of various antioxidant enzymes and saponins content. In adventitious roots of *Panax ginseng* it was examined the temporal effect of nitric oxide (NO) elicitation on root growth, saponin accumulation and antioxidant defence responses. It resulted that NO is involved in root growth and saponin production and this was related to oxidative stress and NADPH oxidase activity (Tewari et al., 2007). Concerning mechanical stress, in cell cultures of *Panax ginseng* low-energy ultrasound induced early events of plant defense responses such as ROS production and stimulated saponins biosynthesis (Wu and Lin 2002).

Metabolite profiling of *Medicago truncatula* cell cultures indicated that methyl jasmonate (MeJa) induced accumulation of the triterpene β-amyrin, a precursor of triterpene saponins, and LC-MS analysis confirmed the accumulation of triterpene saponins in MeJa-elicited samples (Broeckling et al., 2005). The link between oxidative conditions and triterpene biosynthesis agrees with the research performed using plant cell cultures of *Uncaria tomentosa* as demonstated by the increase in the levels of the triterpenes ursolic acid and oleanolic acid after osmotic stress (by increasing sucrose from 20 to 50 g/L) (Feria-Romero et al., 2005) or elicitation with pectin (Flores-Sánchez et al., 2002). Both, osmotic stress and pectin elicitation (Wu et al., 2005; Shohael et al., 2006) may be accompanied by oxidative stress.

There are examples of the role of saponins in protecting plants against biotic (fungal attack) and abiotic stress (salinity). The isolation of plant mutants defective in saponin biosynthesis and metabolic profiling has represented strategies for evaluating the importance of these compounds in plant defence (Shulaev et al., 2008). Oat varieties lacking the saponins avenacins were susceptible to the infection by the fungi *Gaeumannomyces graminis* var. *tritici*. (Osbourn et al., 1994). Similarly, saponin-deficient (avenacin A) mutants of *Avena strigosa* obtained with the chemical mutagen sodium azide were compromised in their resistance to a variety of fungal pathogens such as *Gaeumannomyces graminis* var. *tritici, Fusarium culmorum* and *Fusarium avenaceum* (Papadopoulou et al., 1999). Genetic evidence indicated the existence of a direct link between the gene encoding for oxidosqualene cyclase AsbAS1 enzyme, triterpenoid saponin biosynthesis and disease resistance (Haralampidis et al., 2001). Overall, these results support the role of these triterpenoid saponins as determinant of resistance to fungal attack.

In relation to salinity stress, by comparing the metabolic profile of salt-sensitive and salt-tolerant soybean varieties, saponins from group B were identified as key secondary metabolites correlated with salt tolerance and adaptation of plants (Wu et al., 2008).

Another example is that discussed by Vasil'eva et al. (2009), about the role of steroidal saponins such as furostanol glycosides, as adaptogenes. Furostanol glycosides, bisdesmoside glycosides of diosgenin, extracted from *Dioscorea deltoidea* cell culture, caused non-specific defence reactions of tomato plants *Lycopersicon esculentum* against the gall nematode *Meloidogyne incognita*. These defences reactions consisted among others in the change in isoprene content (phytosterines, tomatin, and carotenoids) and in the rate of oxidative processes in the leaves and roots of intact and treated plants resulting in the formation of systemic acquired resistance.

There is increasing evidence of the antioxidant properties of triterpenoid saponins isolated from plants, both using *in vitro* and *in vivo* models. Saponins have shown to be able to scavenge free-radicals (Xin et al., 2005; Dini et al., 2009; Nzowa et al., 2010), reduce membrane lipid peroxidation (Liang et al., 2010), and maintain redox status in cells or tissues (Dini et al., 2009). Even they can enhance the protection given by antioxidants such as gallic acid (Yoshiki et al., 2001) or vitamin E (Amzal et al., 2008). It has also been proposed that saponins may act as chelators of transition metals (Cu^{2+}) which are crucial catalyst of oxygen radical formation in biological systems. Thus, their removal or their intracellular sequestration may result in diminished cellular sensitivity to oxidant damage (Amzal et al., 2008).

Yoshiki et al. (1998) analyzed the relationships between chemical structure and biological activities of triterpenoid saponins from soybean (*Glycine max*). Based on the chemical structure of soyasaponin bg [possessing a DDMP (2,3-dihydro-2,5-dihydroxy-6-methyl-4*H*-pyran-4-one) moiety] they proposed that saponins are oxygen radical controllers in this species.

The extensive role of saponins in the chemical defence of plants against abiotic and biotic stress factors has been demonstrated and it has been proven that these molecules present antioxidant properties. Still, the possible role of saponins in plant protection from oxidative stress has not been reviewed until now. There is increasing evidence that plants and animals may share common components that lead to similar cellular events (e.g. apoptosis). Therefore, it seems reasonable that saponins may be involved in plant protection against oxidative stress using similar mechanisms as those demonstrated recently for mammalian cells. In the following lines, the protective role of saponins on mammalian cells against oxidative stress is discussed.

Triterpene saponins have been found to inhibit nitric oxide production in a nitric oxide (NO)-production bioassay (Tapondjou et al., 2003) and in mammalian cell culture (Tundis et al., 2009). Changkil saponins, isolated from the roots of *Platycodon grandiflorum* A.D.C. (Campanulaceae), provided good protection from intracellular ROS and DNA damage to liver cells and were able to quench 1,1-diphenyl-2-picrylhydrazyl free radicals and the superoxide radical (Lee et al., 2004). *In vivo* studies showed that pretreatment of mice liver with saponins prior to the administration of *tert*-butylhydroperoxide or carbon tetrachloride (for inducing hepatoxicity) significantly reduced oxidative stress may be by protecting mice liver against depletion of hepatic glutathione levels (Khanal et al., 2009) and reducing lipid peroxidation (Lee et al., 2004). These results support the anti-oxidative role of saponins, and demonstrate that these molecules can scavenge oxygen free radicals and protect mammalian cells from oxidative stress (Lee et al., 2004, 2008).

In PC12 cells, pretreatment with ginsenoside Rg1 (saponin from *Panax ginseng*) markedly reduced the generation of dopamine-induced reactive oxygen species and the release of mitochondrial cytochrome *c* into the cytosol, and subsequently inhibited the activation of caspase-3. In addition, Rg1 pretreatment also reduced inducible nitric oxide (NO) synthase protein level and NO production (Chen et al., 2003).

Other studies report that some gisenosides (Rb3, Rb1, Rc and Rg2) play an antioxidative role to protect erythrocytes from free-radicals induced haemolysis (Liu et al., 2002; Amzal et al., 2008). Moreover, this effect was potentiated by vitamin E which was explained by the authors based on the fact that saponins increase cell membrane permeability and thus may facilitate penetration of vitamin E into red blood cell membranes (RBC) which may explain the synergistic effect. Despite these observations, it should be noticed that saponins present pro-oxidative effects and might induce haemolysis of RBC; this activity has been linked to the structure of the aglycone, on the sugar moiety of the saponin and its concentration (Liu et al., 2002; Chwalek et al., 2006; Amzal et al., 2008).

Diosgenin (aglycone of steroidal saponins) has been proposed to modulate oxidative stress and prevent H_2O_2-induced apoptosis. In hyperlipidemic rats, diosgenin treatment attenuated oxygen free radicals by increasing superoxide dismutase (SOD), glutathione peroxidase (GSH-PX) and nitric oxide synthase (NOS) activities. Besides, the treatment with diosgenin significantly attenuated H_2O_2-induced apoptosis in human vein endothelium cells, intracellular ROS, GSH depletion, DNA fragment formation, and restored nitric oxide (NO) (Gong et al., 2010).

REFERENCES

Adel, M.M., Sehnal, F. and Jurzysta M. (2000). Effects of alfalfa saponins on the moth *Spodoptera littoralis*. *J. Chem. Ecol.* **26:** 1065-1078.

Ali, M.B., Hahn, E.J. and Paek, K.Y. (2006). Copper-induced changes in the growth, oxidative metabolism, and saponin production in suspension culture roots of *Panax ginseng* in bioreactors. *Plant Cell Rep.* **25:** 1122-1132.

Ali, M.B., Thanh, N.T., Yu, K.-W., Hahn, E.-J., Paek, K.-Y. and Lee, H.L. (2005). Induction in the antioxidative systems and lipid peroxidation in suspension culture roots of *Panax ginseng* induced by oxygen in bioreactors. *Plant Sci.* **169:** 833-841.

Allan, A.C. and Fluhr, R. (1997). Two distinct sources of elicited reactive oxygen species in Tobacco epidermal cells. *Plant Cell.* **9:** 1559-1572.

Amzal, H., Alaoui, K., Tok, S., Errachidi, A., Charof, R., Cherrah, Y. and Benjouad, A. (2008). Protective effect of saponins from *Argania spinosa* against free radical-induced oxidative haemolysis. *Fitoterapia.* **79:** 337-344.

Apel, K. and Hirt, H. (2004). Reactive oxygen species: metabolism, oxidative stress, and signal transduction. *Annu. Rev. Plant Biol.* **55:** 373-399.

Baxter, C.J., Redestig, H., Schauer, N., Repsilber, D., Patil, K.R., Nielsen, J., Selbig, J., Liu, J., Fernie, A.R. and Sweetlove, L.J. (2007). The metabolic response of heterotrophic *Arabidopsis* cells to oxidative stress. *Plant Physiol.* **143:** 312-325.

Bednarek, P. and Osbourn, A. (2009). Plant-microbe interactions: Chemical diversity in plant defence. *Sci.* **324:** 746-748.

Bienert, G.P., Schjoerring, J.K., and Jahn, T.P. (2006). Membrane transport of hydrogen peroxide. *Biochim. Biophys. Acta Biomembr.* **1758:** 994-1003.

Blom, T.J.M., Sierra, M., van Vliet, T.B., Franke-van Dijk, M.E.I., de Koning, P., van Iren, F., Verpoorte, R. and Libbenga, K.R. (1991). Uptake and accumulation of ajmalicine into isolated vacuoles of cultured cells of *Catharanthus roseus* (L.) G. Don and its conversion into serpentine. *Planta.* **183:** 170-177.

Broeckling, C.D., Huhman, D.V., Farag, M.A., Smith, J.T., May, G.D., Mendes, P., Dixon, R.A. and Sumner, L.W. (2005). Metabolic profiling of *Medicago truncatula* cell cultures reveals the effects of biotic and abiotic elicitors on metabolism. *J. Exp. Bot.* **56:** 323-336.

Buer, C.S., Nijat, I. and Djordjevic, M.A. (2010). Flavonoids: New Roles for Old Molecules. *J. Integr. Plant Biol.* **52:** 98-111.

Burbulis, I.E., Iacobucci, M. and Shirley, B.W. (1996). A null mutation in the first enzyme of flavonoid biosynthesis does not affect male fertility in *Arabidopsis. Plant Cell.* **8:** 1013-1025.

Caporale, L.H. (1995). Chemical ecology: a view from the pharmaceutical industry. *Proc. Natl. Acad. Sci. USA.* **92:** 75-82.

Cardoso, C.L., Castro-Gamboa, I., Silvia, D.H., Furlan, M., Epifanio, R.de A., Pinto, A. da C., Moraes de Rezende, C., Lima, J.A. and Bolzani V. da S. (2004). Indole glucoalkaloids from *Chimarrhis turbinata* and their evaluation as antioxidant agents and acetylcholinesterase inhibitors. *J. Nat. Prod.* **67:** 1882-1885.

Chapagain, B.P., Wiesman, Z., and Tsror, L. (2007). In vitro study of the antifungal activity of saponin-rich extracts against prevalent phytopathogenic fungi. *Ind. Crop. Prod.* **26:** 109-115.

Cheeseman, J. (2007). Hydrogen peroxide and plant stress: A challenging relationship. *Plant Stress.* **1:** 4-15.

Chen, X.-C., Zhu, Y.-G., Zhu, L.-A., Huang, C., Chen, Y., Chen, L.-M., Fang, F., Zhou, Y.-C. and Zhao, C.-H. (2003). Ginsenoside Rg1 attenuates dopamine-induced apoptosis in PC12 cells by suppressing oxidative stress. *Eur. J. Pharmacol.* **473:** 1-7.

Chou, W.M. and Kutchan, T.M. (1998). Enzymatic oxidations in the biosynthesis of complex alkaloids. *Plant J.* **15:** 289-300.

Chwalek, M., Lalun, N., Bobichon, H., Plé, K. and Voutquenne-Nazabadioko, L. (2006). Structure-activity relationships of some hederagenin diglycosides: Haemolysis, cytotoxicity and apoptosis induction. *Biochim. Biophys. Acta Gen. Subj.* **1760:** 1418-1427.

Dat J., Vandenabeele, S., Vranová, E., van Montagu, M., Inzé, D., and van Breusegem, F. (2000). Dual action of the active oxygen species during plant stress responses. *Cell Mol. Life Sci.* **57:** 779-795.

De Luca, V., and St Pierre, B. (2000). The cell and developmental biology of alkaloid biosynthesis. *Trends Plant Sci.* **5:** 168-173.

Dini, I., Tenore, G.C. and Dini, A. (2009). Saponins in *Ipomoea batatas* tubers: Isolation, characterization, quantification and antioxidant properties. *Food Chem.* **113:** 411-419.

Edreva, A., Velikova, V., Tsonev, T., Dagnon, S., Gürel, A., L. Akta°, L. and Gesheva, E. (2008). Stress-protective role of secondary metabolites: diversity of functions and mechanisms. *Gen. Appl. Plant Physiol.* **34:** 67-78.

Endo, K., Oshima, Y., Kikuchi, H., Koshihara, Y. and Hikino, H. (1983). Hypotensive principles of *Uncaria* hooks. *Planta Med.* **49:** 188-190.

Feria-Romero I, Lazo, E., Ponce-Noyola, T., Cerda-García-Rojas, C.M. and Ramos-Valdivia, A.C. (2005). Induced accumulation of oleanolic acid and ursolic acid in cell suspension cultures of *Uncaria tomentosa. Biotechnol. Lett.* **27:** 839-843.

Flores-Sanchez, I.J., Ortega-Lopez, J., Montes-Horcasitas, M.C. and Ramos-Valdivia, A.C. (2002). Biosynthesis of sterols and triterpenes in cell suspension cultures of *Uncaria tomentosa. Plant Cell Physiol.* **43:** 1502-1509.

Folkes, L.K., Greco, O., Dachs, G.U., Straford, M.R.L. and Wardman, P. (2002). 5-Fluoroindole-3-acetic acid: A prodrug activated by a peroxidase with potential for use in targeted cancer therapy. *Biochem. Pharmacol.* **63:** 265-272.

Foyer, C.H. and Noctor, G. (2005). Oxidant and antioxidant signalling in plants: a re-evaluation of the concept of oxidative stress in a physiological context. *Plant Cell Environ.* **28:** 1056-1071.

Fujita, M., Fujita, Y., Noutoshi, Y., Takahashi, F., Narusaka, Y., Yamaguchi-Shinozaki, K. and Shinozaki, K. (2006). Crosstalk between abiotic and biotic stress response: a current view from the points of convergence in the stress signaling networks. *Curr. Opin. Plant Biol.* **9:** 436-442.

Garcia Prado E., Garcia Gimenez, M. D., De la Puerta Vázquez, R., Espartero Sánchez, J. L. and Saenz Rodríguez, M.T. (2007). Antiproliferative effects of mitraphylline, a pentacyclic oxindole alkaloid of *Uncaria tomentosa* on human glioma and neuroblastoma cell lines. *Phytomedicine.* **14:** 280-284.

Gershenzon, J. and Kreis, W. (1999). Biochemistry of terpenoids: monoterpenes, sesquiterpenes, diterpenes, sterols, cardiac glycosides and steroid saponins. In: Wink, M. Ed. Biochemistry of plant secondary metabolism. *Annu. Rev.*, volume 2. Sheffield Academic Press. CRC Press. pp. 222-299.

Gillam, E.M.J., Notley, L.M., Cai, H., De Voss, J.J. and Guengerich, F.P. (2000). Oxidation of indole by cytochrome P450 enzymes. *Biochem.* **39:** 13817-13824.

Glawischnig, E.(2007). Camalexin. *Phytochem.* **68:** 401-406.

Gong, G., Qin, Y., Huang, W., Zhou, S., Wu, X., Yang, X., Zhao, Y. and Li, D. (2010). Protective effects of diosgenin in the hyperlipidemic rat model and in human vascular endothelial cells against hydrogen peroxide-induced apoptosis. *Chem. Biol. Interact.* **184:** 366-375.

Gould, K.S., McKelvie, J. and Markham, K.R. (2002). Do anthocyanins function as antioxidants in leaves? Imaging of H_2O_2 in red and green leaves after mechanical injury. *Plant Cell Environ.* **25:** 1261-1269.

Hancock, J., Desikan, R., Harrison, J., Bright, J., Hooley, R. and Neill, S. (2006). Doing the unexpected: proteins involved in hydrogen peroxide perception. *J. Exp. Bot.* **57:** 1711-1718.

Haralampidis, K., Bryan, G., Qi, X., Papadopoulou, K., Bakht, S., Melton, R. and Osbourn, A. (2001). A new class of oxidosqualene cyclases directs synthesis of antimicrobial phytoprotectants in monocots. *Proc. Natl. Acad. Sci.* **98:** 13431-13436.

Haralampidis, K., Trojanowska, M. and Osbourn, A. (2002). Biosynthesis of Triterpenoid Saponins in Plants. In: Scheper, Th. Editor. *Adv. Biochem. Eng. Biotechnol.* Vol. **75:** Springer-Verlag Berlin Heidelberg, pp. 31-49.

Hartmann, T. (1996). Diversity and variability of plant secondary metabolism: a mechanistic view. *Entomol. Exp. Appl.* **80:** 177-188.

Hartmann, T. (2007). From waste products to ecochemicals: fifty years research of plant secondary metabolism. *Phytochem.* **68:** 2831-2846.

Hernandez I., Alegre, L., Van Breusegem, F. and Munné-Bosch, S. (2009). How relevant are flavonoids as antioxidants in plants? *Trends Plant Sci.* **14:** 125-132.

Hiradate, S., Yada, H., Ishii, T., Nakajima, N., Ohnishi-Kameyama, M., Sugie, H., Zungsontiporn, S. and Fujii, Y. (1999). Three plant growth inhibiting saponins from *Duranta repens. Phytochem.* **52:** 1223-1228.

Huerta-Heredia, A.A., Marín-López, R., Ponce-Noyola, T., Cerda-García-Rojas, C.M., Trejo-Tapia, G. and Ramos-Valdivia, A.C. (2009). Oxidative stress induces alkaloid production in *Uncaria tomentosa* root and cell cultures in bioreactors. *Eng. Life Sci.* **9:** 211-218.

Hung, S.H., Yu, C.W. and Lin, C.H. (2005). Hydrogen peroxide functions as a stress signal in plants. *Bot. Bull. Acad. Sin.* **46:** 1-10.

Jaleel, C.A., Gopi, R., Alagulakshmanan, G.M. and Panneerselvam R. (2006). Triadimefon induced changes in the antioxidant metabolism and ajmalicine production in *Catharanthus roseus* (L.) G. Don. *Plant Sci.* **171:** 271-276.

Jaleel, C.A., Gopi, R., Manivannan, P. and Panneerselvam R. (2007). Responses of antioxidant defense system of *Catharanthus roseus* (L.) G. Don. to paclobutrazol treatment under salinity. *Acta Physiol. Plant.* **29:** 205-209.

Jenner, H., Townsend, B. and Osbourn, A. (2005). Unravelling triterpene glycoside synthesis in plants: phytochemistry and functional genomics join forces. *Planta.* **220:** 503-506.

Ji, W. and Gong, B.Q. (2007). Hypolipidemic effects and mechanisms of *Panax notoginseng* on lipid profile in hyperlipidemic rats. *J. Ethnopharmacol.* **113:** 318-324.

Joshi, R.C., San Martín, R., Saez-Navarrete, C., Alarcon, J., Sainz, J., Antolin, M.M., Martin, A.R. and Sebastian, L.S. (2008). Efficacy of quinoa (*Chenopodium quinoa*) saponins against golden apple snail (*Pomacea canaliculata*) in the Philippines under laboratory conditions. *Crop Protect.* **27:** 553-557.

Kalinowska, M., Zimowski, J., P[1]czkowski, C. and Wojciechowski, Z.A. (2005). The formation of sugar chains in triterpenoid saponins and glycoalkaloids. *Phytochem. Rev.* **4:** 237-257.

Kashiwada, Y., Wang, H.K., Nagao, T., Kitanaka, S., Yasuda, I., Fujioka, T., Yamagishi, T., Cosentino, L.M., Kozuka, M., Okabe, H., Ikeshiro, Y., Hu, C.Q., Yeh, E. and Lee, K.H. (1998). Anti-AIDS agents, 30, Anti-HIV activity of oleanolic acid, pomolic acid and structurally related triterpenoids. *J. Nat. Prod.* **61:** 1090-1095.

Khanal, T., Choi, J.H., Hwang, Y.P., Chung, Y.C. and Jeong, H.G. (2009). Saponins isolated from the root of *Platycodon grandiflorum* protect against acute ethanol-induced hepatotoxicity in mice. *Food Chem. Toxicol.* **47:** 530-535.

Kliebenstein, D.J. (2004). Secondary metabolites and plant/environment interactions: a view through *Arabidopsis thaliana* tinged glasses. *Plant Cell Environ.* **27:** 675-684.

Kotchoni, S.O. and Gachomo, E.W. (2006). The reactive oxygen species network pathways: an essential prerequisite for perception of pathogen attack and the acquired disease resistance in plants. *J. Biosci.* **31:** 389-404.

Kuljanabhagavad, T. and Wink, M. (2009). Biological activities and chemistry of saponins from *Chenopodium quinoa* Willd. *Phytochem. Rev.* **8:** 473-490.

Kumar, V., Singh, S., Khanna, A.K., Khan, M.M., Chander, R., Mahdi, F., Saxena, J.K., Singh, R. and Singh, R.K. (2008). Hypolipidemic activity of *Anthocephalus indicus* (kadam) in hyperlipidemic rats. *Med. Chem. Res.* **17:** 152-158.

Lamb, C. and Dixon, R.A. (1997). The oxidative burst in plant disease resistance. Annu. Rev. Plant Physiol. *Plant Mol. Biol.* **48:** 251-275.

Laus, G. (2004). Advances in chemistry and bioactivity of the genus *Uncaria., Phytother. Res.* **18:** 259-274.

Lee, K.J., Choi, C.Y., Chung, Y.C., Kim, Y.S., Ryu, S.Y., Roh, S.H. and Jeong, H.G. (2004). Protective effect of saponins derived from roots of *Platycodon grandiflorum* on tert-butyl hydroperoxide-induced oxidative hepatotoxicity. *Toxicol. Lett.* **147:** 271-282.

Lee, K.J., Choi, J.H., Kim, H.G., Han, E.H., Hwang, Y.P., Lee, Y.C., Chung, Y.C. and Jeong, H.G. (2008). Protective effect of saponins derived from the roots of *Platycodon grandiflorum* against carbon tetrachloride induced hepatotoxicity in mice. *Food Chem. Toxicol.* **46:** 1778-1785.

Lee-Parsons, C.W.T. and Ertürk, S. (2005). Ajmalicine production in methyl jasmonate-induced *Catharanthus roseus* cell cultures depends on Ca^{2+} level. *Plant Cell Rep.* **24:** 677-682.

Lehmann, M., Schwarzländer, M., Obata, T., Sirikantaramas, S., Burow, M., Olsen, C.E., Tohge T, Fricker, M.D., Møller, B.L., Fernie A.R., Sweetlove, L.J. and Laxa, M. (2009). The metabolic response of *Arabidopsis* roots to oxidative stress is distinct from that of heterotrophic cells in culture and highlights a complex relationship between the levels of transcripts, metabolites and flux. *Mol. Plant.* **2:** 390-406.

Lemeshko, V.V., Haridas, V., Quijano Pérez, J.C. and Gutterman, J.U. (2006). Avicins, natural anticancer saponins, permeabilize mitochondrial membranes. *Arch. Biochem. Biophys.* **454:** 114-122.

Leopoldini, M., Russo, N., Chiodo, S. and Toscano, M. (2006). Iron chelation by the powerful antioxidant flavonoid quercetin. *J. Agric. Food Chem.* **54:** 6343-6351.

Liang, Q., Yu, X., Qu, S., Xu, H. and Sui, D. (2010). *Acanthopanax* senticosides B ameliorates oxidative damage induced by hydrogen peroxide in cultured neonatal rat cardiomyocytes. *Eur. J. Pharmacol.* **627:** 209-215.

Liu, Z.Q., Luo, X.Y., Sun, Y.X., Chen, Y.P. and Wang, Z.C. (2002). Can ginsenosides protect human erythrocytes against free-radical-induced hemolysis? *Biochim. Biophys. Acta Gen. Subj.* **1572:** 58-66.

Luna-Palencia, G., Cerda-García-Rojas, C.M., Rodríguez-Monroy, M. and Ramos-Valdivia A.C. (2005). Influence of auxins and sucrose in monoterpenoid oxindole alkaloid production by *Uncaria tomentosa* cell suspension cultures. *Biotechnol. Prog.* **21:** 198-204.

Luna-Palencia, G., Cerda-García-Rojas, C.M., Romo-Cruz, I., Albarrán, J.G. and. Ramos-Valdivia, A.C. (2006). Advances in the biosynthesis of oxindole alkaloids in *Uncaria tomentosa* cultures. *Inter. Symp. Chem., Pharmacol. and Biosynthesis of Alkaloids*, Belek, Antalya, Turkey, April 25-29.

Lunn, J.E. (2007). Compartmentation in plant metabolism. *J. Exp. Bot.* **58:** 35-47.

Mittler, R. (2002). Oxidative stress, antioxidants and stress tolerance. *Trends Plant Sci.* **7:** 405-410.

Mittler, R. (2006). Abiotic stress, the field environment and stress combination. *Trends Plant Sci.* **11:** 15-19.

Mittler, R. Vanderauwera, S., Gollery, M. and Van Breusegem, F. (2004). Reactive oxygen gene network of plants. *Trends Plant Sci.* **9:** 490-498.

Morales, M. and Ros Barceló, A. (1997). A basic peroxidase isoenzyme from vacuoles and cell walls of *Vitis vinifera. Phytochem.* **45:** 229-232.

Moura, D.J., Richter, M.F., Boeira, J.M., Pêgas Henriques, J.A. and Saffi, J. (2007). Antioxidant properties of β-carboline alkaloids are related to their antimutagenic and antigenotoxic activities. *Mutagenesis* **22:** 293-302.

Neill, S.J., Desikan, R. and Hancock, J. (2002). Hydrogen peroxide signalling. *Curr. Opin. Plant Biol.* **5:** 388-395.

Neill, S.J., Desikan, R., Clarke, A., Hurst, R.D. and Hancock, J.T. (2002). Hydrogen peroxide and nitric oxide as signalling molecules in plants, *J. Exp. Bot.* **53:** 1237-1247.

Noctor, G., Arisi, A.C.M., Jouanin, L., Kunert, K.J., Rennenberg, H. and Foyer, C.H. (1998). Glutathione: biosynthesis, metabolism and relationship to stress tolerance explored in transgenic plants. *J. Exp. Bot.* **49:** 623-647.

Nürnberger, T. (1999).Signal perception in plant pathogen defence. *Cell Mol. Life Sci.* **55:** 167-182.

Nzowa, L.K., Barboni, L., Teponno, R.B., Ricciutelli, M., Lupidi, G., Quassinti, L., Bramucci, M. and Tapondjou, L.A. (2010). Rheediinosides A and B, two antiproliferative and antioxidant triterpene saponins from *Entada rheedii. Phytochem.* **71:** 254-261.

Oleszek, W. (1993). Allelopathic potentials of alfalfa (*Medicago sativa*) saponins: Their relation to antifungal and hemolytic activities. *J. Chem. Ecol.* **19(6):** 1063-1074.

Orozco Cárdenas, M.L., Narváez-Vásquez, J. and Ryan, C.A. (2001). Hydrogen peroxide acts as second messenger for the induction of defense genes in tomato plants in response to wounding, systemin and methyl jasmonate. *Plant Cell.* **13:** 179-191.

Osbourn, A.E. (2003). Saponins in cereals. *Phytochem.* **62:** 1-4.

Osbourn, A.E., Clarke, B.R., Lunness, P., Scott, P.R. and Daniels, M.J. (1994). An oat species lacking avenacin is susceptible to infection by *Gaeumannomyces graminis* var. *tritici. Physiol. Mol. Plant Pathol.* **45:** 457-467.

Osbourn, A.E., Qi, X., Townsend, B. and Qin, B. (2003). Dissecting plant secondary metabolism - constitutive chemical defences in cereals. *New Phytologist.* **159:** 101-108.

Paniagua-Pérez, R., Madrigal-Bujaidar, E., Molina-Jasso, D., Reyes-Cadena, S., Álvarez-González, I., Sánchez-Chapuland, L. and Pérez-Gallaga, J. (2009). Antigenotoxic, antioxidant and lymphocyte induction. Effects produced by pteropodine. *Basic Clin. Pharmacol. Toxicol.* **104:** 222-227.

Papadopoulou, K., Melton, R., Leggett, M., Daniels, M. and Osbourn, A. (1999). Compromised disease resistance in saponin-deficient plants. *Proc. Natl. Acad. Sci.* **96:** 12923-12928.

Park, J.-S., Park, E.-M., Kim, D.-H., Jung, K., Jung, J.-S., Lee, E.-J., Hyun, J.-W., Kang, J.L. and Kim, H.-S. (2009). Anti-inflammatory mechanism of ginseng saponins in activated microglia. *J. Neuroimmunol.* **209:** 40-49.

Pauw, B., van Dujin, B., Kijne, W.J. and Memelink, J. (2004). Activation of oxidative burst by yeast elicitor in *Catharanthus roseus* cells occurs independently of the activation of genes involved in alkaloid biosynthesis. *Plant Mol. Biol.* **55:** 797-805.

Pourcel, L., Routaboul, J., Cheynier, V., Lepiniec L. and Debeaujo, I. (2007). Flavonoid oxidation in plants: from biochemical properties to physiological functions. *Trends Plant Sci.* **12:** 29-36.

Prigent, S. V., Voragen, A.G., Visser, A.J.W.G., van Koningsveld, G.A., and Gruppen, H. (2007). Covalent interactions between proteins and oxidation products of caffeoylquinic acid (chlorogenic acid). *J. Sci. Food Agric.* **87:** 2502-2510.

Quan, L-H., Zhang, B., Shi, W-W. and Li H-Y. (2008). Hydrogen peroxide in plants: a versatile molecule of the reactive oxygen species network. *J. Integr. Plant Biol.* **50:** 2-18.

Reuber, T.L., Plotnikova, J.M., Dewdney, J., Rogers, E.E., Wood, W. and Ausubel F.M. (1998). Correlation of defence gene induction defects with powdery mildew susceptibility in *Arabidopsis* enhanced disease susceptibility mutants. *Plant J.* **16:** 473-487.

Roetschi, A., Si-Ammour, A., Belbahri, L., Mauch, F. and Mauch-Mani, B. (2001). Characterization of an *Arabidopsis*-Phytophthora pathosystem: resistance requires a functional PAD2 gene and is independent of salicylic acid, ethylene and jasmonic acid signalling. *Plant J.* **28:** 293-305.

Roytrakul, S. and Verpoorte R. (2007). Role of vacuolar transporter proteins in plant secondary metabolism: *Catharanthus roseus* cell culture. *Phytochem. Rev.* **6:** 383-396.

Scandalios, J.G. (2002). The rise of ROS. *Trends Biochem. Sci.* **27:** 483-86.

Schwinn, K.E. and Davies, K.M. (2004). Flavonoids. In Davies KM Editor. Plant Pigments and their Manipulation. *Annu. Plant Rev.* 14, Blackwell Publishing, Oxford, UK. pp. 92-149.

Sheahan, J.J. (1996). Sinapate esters provide greater UV-B attenuation than flavonoids in *Arabidopsis thaliana* (Brassicaceae). *Am. J. Bot.* **83:** 679-686.

Shellard, E.J., Phillipson, J.D. and Gupta, D. (1969). The *Mitragyna* species of Asia. Part XV. The alkaloid from bark of *Mitragyna parviflora* (Roxb.) Kort and a possible biogenetic route for oxindole alkaloids. *Planta Med.* **17:** 146-216.

Shinoda, T., Nagao, T., Nakayama, M., Serizawa, H., Koshioka, M., Okabe, H. and Kawai, A. (2002). Identification of a triterpenoid saponin from a crucifer, *Barbarea vulgaris*, as a feeding deterrent to the diamondback moth, *Plutella xylostella. J. Chem. Ecol.* **28:** 587-599.

Shohael, A.M., Chakrabarty, D., Ali, M.B., Yu, K.W., Hahn, E.J., Lee, H.L. and Paek, K.Y. (2006). Enhancement of eleutherosides production in embryogenic cultures of *Eleutherococcus sessiliflorus* in response to sucrose-induced osmotic stress. *Process Biochem.* **41:** 512-518.

Shulaev, V., Cortes, D., Miller, G. and Mittler, R. (2008). Metabolomics for plant stress response. *Physiol. Plant.* **132:** 199-208.

Sottomayor, M. and Ros Barceló, A. (2004). Plant peroxidases and phytochemistry. *Phytochem. Rev.* **3:** 1-256.

Sparg, S.G., Light, M.E. and van Staden, J. (2004). Biological activities and distribution of plant saponins. *J. Ethnopharmacology.* **94:** 219-243.

Sun, S., Wang, C-Z., Tong, R., Li, X-L., Fishbein, A., Wang, Q., He, T-C., Du, W. and Yuan, C-S. (2010). Effects of steaming the root of *Panax notoginseng* on chemical composition and anticancer activities. *Food Chem.* **118:** 307-314.

Takahama, U. (2004). Oxidation of vacuolar and apoplastic phenolic substrate by peroxidase: Physiological significance of the oxidation reactions. *Phytochem. Rev.* **3:** 207-219.

Takahama, U. and Egashira, T. (1991). Peroxidase in vacuoles of *Vicia faba* leaves. *Phytochem.* **30:** 73-77.

Takemoto, M., Iwakiri, Y., Suzuki, Y. and Tanaka, K. (2004). A mild procedure for the oxidative cleavage of substituted indoles catalyzed by plant cell cultures. *Tetrahedron Lett.* **45:** 8061-8064.

Tang, K., Zhan, J.C., Yang, H.R. and Huang, W.D. (2010). Changes of resveratrol and antioxidant enzymes during UV-induced plant defence response in peanut seedlings. *J. Plant Physiol.* **167:** 95-102.

Tang, Z., Yang, L., Zu, Y. and Guo, X. (2009). Variations of vinblastine accumulation and redox state affected by exogenous H_2O_2 in *Catharanthus roseus* (L.) G. Don. *Plant Growth Regul.* **57:** 15-20.

Tapondjou, L.A., Lontsi, D., Sondengam, B.L., Shaheen, F., Choudhary, M.I., Atta-ur-Rahman van Heerden, F.R., Park, H-J. and Lee, K-T. (2003). Saponins from *Cussonia bancoensis* and their inhibitory effects on nitric oxide production. *J. Nat. Prod.* **66:** 1266-1269.

Tewari, R., Lee, S., Hahn, E. and Paek, K. (2007). Temporal changes in the growth, saponin content and antioxidant defense in the adventitious roots of *Panax ginseng* subjected to nitric oxide elicitation. *Plant Biotechnol. Rep.* **1:** 227-235.

Thomas, R.L. and Jen, J.J. (1980). The cytochemical localisation of peroxidase in tomato fruit cells. *J. Food Biochem.* **4:** 247-259.

Trejo-Tapia, G., Cerda-García-Rojas, C.M., Rodríguez-Monroy M. and Ramos-Valdivia, A.C. (2005). Monoterpenoid oxindole alkaloid production by *Uncaria tomentosa* (Willd) D.C. cell suspension cultures in a stirred tank biorreactor. *Biotechnol. Prog.* **21:** 786-792.

Trejo-Tapia, G., Sepúlveda-Jiménez, G., Trejo-Espino, J.L., Cerda-García-Rojas, C.M., de la Torre, M., Rodríguez-Monroy, M. and Ramos-Valdivia, A.C. (2007). Hydrodynamic stress induces monoterpenoid oxindole alkaloid accumulation by *Uncaria tomentosa* (Willd) D.C. Cell suspension cultures via oxidative burst. *Biotechnol. Bioeng.* **98:** 230-238.

Tsuji, J., Jackson, E.P., Gage, D.A., Hammerschmidt, R. and Somerville, S.C. (1992). Phytoalexin accumulation in *Arabidopsis thaliana* during the hypersensitive reaction to *Pseudomonas syringae* pv *syringae*. *Plant Physiol.* **98:** 1304-1309.

Tundis, R., Bonesi, M., Deguin, B., Loizzo, M.R., Menichini, F., Conforti, F., Tillequin, F. and Menichini, F. (2009). Cytotoxic activity and inhibitory effect on nitric oxide production of triterpene saponins from the roots of *Physospermum verticillatum* (Waldst & Kit) (Apiaceae). *Bioorg. Med. Chem.* **17:** 4542-4547.

van de Velde, F., Bakker, M., van Rantwijk, F. and Sheldon, R.A. (2001). Chloroperoxidase catalyzed enantioselective oxidations in hydrophobic organic media. *Biotechnol. Bioeng.* **72:** 523-529.

Vasil'eva, I.S., Udalova, Z.V., Zinov'eva, S.V. and Paseshnichenko, V.A. (2009). Steroid furostanol glycosides: A new class of natural adaptogenes. *Appl. Biochem. Microbiol.* **45:** 463-472.

Verpoorte, R. (2000). Secondary metabolism. In Verpoorte, R. and Alfermann, A.W. Ed. Metabolic engineering of plant secondary metabolism. Dordrecht, Kluwer Academic Publisher. pp. 1-29

Vincken, J.-P., Heng, L., de Groot, A. and Gruppen, H. (2007). Saponins, classification and occurrence in the plant kingdom. *Phytochem.* **68:** 275-297.

Wang, B.J., Liu, C.T., Tseng, C.Y., Wu, C.P. and Yu, Z.R. (2004). Hepatoprotective and antioxidant effects of *Bupleurum kaoi* Liu (Chao et Chuang) extract and its fractions fractionated using supercritical CO_2 on CCl_4-induced liver damage. *Food Chem. Toxicol.* **42:** 609-617.

Wang, Y., Che, C.M., Chiu, J.F. and He, Q.Y. (2007). Dioscin (saponin)-induced generation of reactive oxygen species through mitochondria dysfunction: A proteomic-based study. *J. Proteome Res.* **6:** 4703-4710.

Williams, R.J., Spencer, J.P. and Rice-Evans, C. (2004). Flavonoids: antioxidants or signaling molecules? *Free Radic. Biol. Med.* **36:** 838-849.

Wink M. (1999). Introduction: Biochemistry, role and biotechnology of secondary metabolism. In Wink, M. Ed. Biochemistry of plant secondary metabolism. *Annu. Plant Rev.* **Vol 2:** Sheffield Academic Press and CRC Press, pp. 1-17.

Wink, M. (2008). Plant secondary metabolism: diversity, function and its evolution. *Nat. Prod. Commun.* **3:** 1205-1216.

Wink, M. (2010). Introduction: Ecological function of secondary metabolites. In Wink, M. Ed. *Annu. Plant Rev.* Vol. 39, 2nd edition: Functions and Biotechnology of Plant Secondary Metabolites. Blackwell Publishing, Oxford, UK. pp. 1-20.

Winkel-Shirley, B. (2002). Biosynthesis of flavonoids and effects of stress. *Curr. Opin. Plant Biol.* **5:** 218-223.

Wojtaszek, P. (1997). Oxidative burst: an early plant response to pathogen infection. *Biochem. J.* **322:** 681-692.

Wu, J. and L. Lin (2002). Elicitor-like effects of low-energy ultrasound on plant (*Panax ginseng*) cells: induction of plant defense responses and secondary metabolite production. *Appl. Microbiol. Biotechnol.* **59:** 51-57.

Wu, J.Y., Wong, K., Ho, K.P. and Zhou, L.G. (2005). Enhancement of saponin production in *Panax ginseng* cell culture by osmotic stress and nutrient feeding. *Enzym. Microb.Tech.* **36:** 133-138.

Wu, W., Zhang, Q., Zhu, Y., Lam, H.M., Cai, Z. and Guo, D. (2008). Comparative metabolic profiling reveals secondary metabolites correlated with soybean salt tolerance. *J. Agr. Food Chem.* **56:** 11132-11138.

Xi, M.M., Hai, C.X., Tang, H.F., Chen, M.S., Fang, K.Q. and Liang, X. (2008). Antioxidant and antiglycation properties of total saponins extracted from traditional Chinese medicine used to treat diabetes mellitus. *Phytother. Res.* **22:** 228-237.

Xin, X., Zhong, J., Wei, D. and Liu, J. (2005). Protection effect of 20(S)-ginsenoside Rg2 extracted from cultured *Panax notoginseng* cells on hydrogen peroxide-induced cytotoxity of human umbilical cord vein endothelial cells *in vitro. Process Biochem.* **40:** 3202-3205.

Xu, M., Dong, J. and Zhu, M. (2005). Effect of nitric oxide on catharanthine production and growth of *Catharanthus roseus* suspension cells. *Biotechnol. Bioeng.* **89:** 367-371.

Yoshiki, Y., Kahara, T., Okubo, K., Sakabe, T. and Yamasaki, T. (2001). Superoxide- and 1,1-diphenyl-2-picrylhydrazyl radical-scavenging activities of soyasaponin β g related to gallic acid. *Biosci. Biotechnol. Biochem.* **65:** 2162-2165.

Yoshiki, Y., Kudou, S. and Okubo, K. (1998). Relationship between chemical structures and biological activities of triterpenoid saponins from soybean. *Biosci. Biotechnol. Biochem.* **62:** 2291-2299.

Zhang, Q.H., Wu, C.F., Duan, L. and Yang, J.Y. (2008). Protective effects of total saponins from stem and leaf of *Panax ginseng* against cyclophosphamide-induced genotoxicity and apoptosis in mouse bone marrow cells and peripheral lymphocyte cells. *Food Chem. Toxicol.* **46:** 293-302.

Zhao, J., Davis, L. C. and Verpoorte, R. (2005). Elicitor signal transduction leading to production of plant secondary metabolites. *Biotechnol. Adv.* **23:** 283-333.

Zhao, J., Hu, Q., Guo, Y. Q. and Zhu, W.H. (2001). Elicitor induced indole alkaloid biosynthesis in *Catharanthus roseus* cell cultures in related to Ca^{2+} influx and the oxidative burst. *Plant Sci.* **161:** 423-431.

Zhao, J., Williams, C.C. and Last, R.L. (1998). Induction of Arabidopsis tryptophan pathway enzymes and camalexin by amino acid starvation, oxidative stress, and an abiotic elicitor. *Plant Cell.* **10:** 359-370.

15

Role of Ascorbate in Growth and Development of Cells During the Formation of Annual Rings in Coniferous Trees

G. F. ANTONOVA*

VN Sukachev Institute of Forestry, Siberian Branch of Russian Academy of Sciences, Akademgorodok, Krasnoyarsk– 660 036 Russia

***Corresponding author:** *institute_forest@ksc.krasn.ru*

ABSTRACT

Ascorbic acid (AsA) is actively utilized as an electron donor in metabolism of cells during their differentiation. It controls cell division, takes part in primary wall growth, regulating the formation of diferulic bridges between wall polysaccharides and interfering due to this in growth, regulates peroxide-dependent polymerisation of phenolic compounds, controlling an amount of p-cinnamic alcohols to be accessible to peroxides, what in turn controls the rate of lignin biosynthesis. The basic feature of coniferous wood is the presence of two types of xylem, early and late, which differed in the tracheid diameter and the cell wall thickness. The development of two types of wood in annual ring occurs by one programme: cell production by cambium, growth of primary cell walls (expansion growth) and maturation, including secondary cell wall and lignification. Physiological and biochemical events in the cells, leading to morphological differences, resulted from the changes in cell metabolism under influence of environmental factors. The main season of these factors is water stress, in particular, low water potential in tissues. Low water potential (internal water stress) in developing xylem can arise because of the lack of rainfalls (external stress), high temperature and evapotranspiration or even large excess of water in soil (physiological drought). The changes along the chain of metabolic reactions evoke the development of cells produced further by the cambium into the late tracheids. The cambium activity also slides down, so that the number of late tracheids can be diminished. Taking into account physiological functions of AsA in plants we have studied the content of AsA and its oxidized form, dehydroascorbic acid (DHA), important pair in ascorbate-glutathione pathway, at the differentiation stages of xylem and phloem cells during development of early and latewood in annual rings in larch (*Larix sibirica* Ldb.) and pine (*Pinus sylevestris* L.) trees. The cell layers of forming wood in larch and pine trees were collected in the periods of the formation of earlywood as well as latewood. There were great differences in the contents of AsA and DHA and in their ratio in dependence on the wood type developing in the trunks of pine and larch during growth period as well as on differentiation stages of xylem and phloem cells. The changes in the AsA content during xylem cell differentiation were accompanied by changes in the content of uronic acids. The AsA to DHA ratio changed in the course of cell growth and of early and late xylem lignification. It reduced from the start of lignification to early mature xylem cells and, in contrast, increased to late mature tracheids, indicating the differences in the level of redox processes in these tissues. The changes in the content of AsA interfered in the quantity of early and late tracheids in annual wood rings, in radial diameters of these cells and in the rate of lignification in two types of wood.

Keywords: Ascorbate, annual ring, cambium, cell division, coniferous tree, growth and development, wood.

1. INTRODUCTION

Ascorbic acid (AsA) is an important component of plant redox reactions and actively utilized as an electron donor in the metabolism of cells during their differentiation. AsA synthesis occurs in the cytosol and its initial precursors are successively hexoses and uronic acids (Isherwood and Mapson 1962; Baig et al., 1970; De Gara et al., 1994; Loewus 1999; Smirnoff 2001). Plants form ascorbate from GDP-D-mannose via GDP-L-galactose, L-galactose, and L-galactono-1,4-lactone (Wheeler et al., 1998; Smirnoff et al., 2001). The final oxidation of L-galactono-1,4-lactone to ascorbate is catalysed by a mitochondrial L-galactono-1,4-lactone dehydrogenase located on the inner membrane and using cytochrome c as electron acceptor (Siendones et al., 1999). This enzyme is the last in the AsA biosynthetic pathway (DeGara et al., 2000). From the cytosol, AsA is transported into the apoplast along the concentration gradient (Arrigoni 1994). In the apoplast, AsA, being an antioxidant, is successively oxidized to monodehydroascorbic and dehydroascorbic acid (DHA), thereafter, DHA is transported back into the cytosol, where AsA regeneration is catalysed by dehydroascorbate reductase (Rautenkranz et al., 1994). The two enzymes involved in the ascorbate recycling, ascorbate free radical (AFR) reductase and DHA reductase (Tommasi et al., 2001). In the apoplast, AsA could be directly reduced from monoDHA by corresponding reductase (Hossain et al., 1984), which is activated under drought conditions (Mukheljee and Choudhuri 1983). In the symplast AsA is regenerated in DHA via the ascorbate-glutathione cycle (Foyer and Halliwell 1976; Noctor and Foyer 1998). In higher plants DHA reductase is one of enzymes of the AsA-glutathione cycle where glutathione reductase is considered the key enzyme (Pinto et al., 1985; Asada 2006; Chalapathi Rao and Reddy 2006). Romero-Puertas et al. (2006) indicated that glutathione reductase is the rate-limiting enzyme in the ascorbate-glutathione cycle. In the same time glutathione-independent pathway for

DHA reduction has been found to exist *in vivo* and to be present together with a glutathione-driven DHA reduction (Potters et al., 2004). Both compounds were proposed to act independently in the growth control. AsA transfer to the apoplast against the concentration gradient depends on its concentration within the apoplast (Arrigoni 1994) and is controlled in particular by ascorbate oxidase (Ros-Barceló et al., 2006). Cell wall-localized ascorbate oxidase regulates the redox status of AsA in the apoplast oxidizing with reduction of O_2 to water (Pignocchi and Foyer 2003).

Along with ascorbate oxidase, ascorbate peroxidase I and ascorbate peroxidase III are involved in AsA oxidation in the apoplast and their functioning differs in dependence on substrate available and cell developmental stage (Córdoba-Pedregosa et al., 2003; Ros-Barceló et al., 2006). Ascorbate oxidase and ascorbate peroxidase were present in both apoplast and symplast of etiolated pea shoots and changed during cell differentiation in different ways (Pinto and De Gara 2004). Two peroxidases differing in their activity, the set of isoforms, and the degree of their association with the cell walls were found in the developing xylem layer of Scots pine (Sudachkova et al., 1980). The activity of these peroxidases and their isoform composition changes in dependence on tissue development and on the season. Acid, neutral, and alkaline peroxidases were shown to specifically catalyse oxidation of different monolignols (Marjamaa et al., 2006).

The ascorbate–dehydroascorbate cycle, including glutathione reductase along with ascorbate peroxidases, is involved in detoxification of drought-induced hydrogen peroxide (Carrol et al., 1988; Van Rensburg and Krüger 1995; Kärkönen and Fry 2006) as well as active oxygen (Noctor and Foyer 1998; De Gara et al., 2000). Being an antioxidant, AsA fulfils diverse functions in plant development (Smirnoff 1996; Horemans et al., 2000; Smirnoff et al., 2001). During cell division, it controls the synthesis of hydroxyproline-containing proteins required for successful progress of the cell cycle phases (Liso et al., 1988). It is high level AsA that provides for active proliferation of cells. Intracellular AsA was shown to play a key role in the reactivation of the apical meristem of white spruce somatic embryos (Stassolla and Yeung 2006). The high AsA content favoured pine hypocotyl growth and with a decline of its content along the axis and with age of that cell growth was retarded (Sánchez et al., 1997; Zarra et al., 1999). It is supposed that this growth retardation of hypocotyls cells occurred because of diferulic bridge formation between cell wall polysaccharides under the effect of peroxidases. Further AsA content has been found to present in the apoplastic fluid and it also decreased with the hypocotyl age and along the hypocotyl axis of 10-day-old seedlings (Pedreira et al., 2004). Authors showed a very significant correlation between the hydrogen peroxide apoplastic level and the growth rate as well as between the ascorbate/hydrogen peroxide molar ratio and the growth rate of hypocotyls and suggested that the redox state is the main factor controlling the cell wall stiffening mechanism and thus the growth of pine hypocotyls. An increased AsA content and a definite AA/DHA ratio were detected also in the active growth zone of onion root and these indices changed in the apoplast and symplast along the root axis (Córdoba-Pedregosa et al., 2003). It is accepted that AsA key role in the apoplast is related to the control of events dependent on hydrogen peroxide (Takahama 1993). The analysis of the kinetic characteristics of the different peroxidases, present in the apoplast in pea shoots, also showed that the presence of AsA and ascorbate peroxidase in the cell wall might play a critical role in regulating the wall stiffening process during cell differentiation by interfering with the activity of secretory peroxidases (Pinto and De Gara 2004). Until AsA is available for hydrogen peroxide, processes of cross-linking and polymerization in the cell wall are retarded (Takahama 1993; Zarra et al., 1999). According to this principle, AsA regulates also the amount of cinnamic acids accessible for peroxidase, which, in turn, controls the rate of lignin synthesis (Takahama 1993, 1994).

Two types of xylem, early and late, having both the tracheid diameter and the cell wall thickness, is a basic feature of coniferous wood. Although the development of two wood types in annual ring occurs by one programme: cell production by cambium, growth of primary cell walls (expansion growth) and maturation, including secondary cell wall and lignification, their morphological parameters as well as the number of cells in the annual rings depend on the conditions in which cells form and develop (Antonova and Stasova 1992, 1993, 1997). Physiological and biochemical events in the cells, leading to morphological differences, resulted from the changes in cell metabolism under influence of environmental factors. One of decisive factors affecting cell development transition from early to late type is an internal water stress (Zahner 1963; Zahner et al., 1964). The changes along the chain of metabolic reactions due to low water potential evoke the cells produced by the cambium to be further developed as late tracheids. The cambium activity also slides down, so that the number of late tracheids can be diminished. Internal water stress can arise because of external drought due to the absence of precipitations and exhaustion of water stores in soil and increased solar radiation (Antonova 1999) or strengthening of transpiration at high air temperature (Gregg et al., 1988) what reduces the internal water stores in trunk tissues or because of flooding of root system ("physiological drought"). Physiological drought provoked by oxygen deficiency in soil decreases a water absorption (Lyr et al., 1967) and disturbs some physiological and biochemical processes, in particular influencing activity of some reductases in pine seedling (Balmaeva 1986).

Taking into account physiological functions of AsA in plants the content of AsA and its oxidized form DHA (important pair in ascorbate-glutathione pathway) in the cells at various developmental stages, isolated from the trunks of the intact 20-25-year old larch (*Larix sibirica* Ldb.) (Antonova et al., 2005) and pine (*Pinus sylevestris* L.) (Antonova et al., 2009) trees, were studied.

2. METHODOLOGY AND SAMPLINGS

The cell layers were sampled during of early- and latewood development of annual ring. The distribution of differentiation stages of both cell types in the course of annual ring formation is rather complex: the stages of growth and development of early and late tracheids overlap with each other in time (Antonova 1999; Antonova and Stasova 2002). Because of that the sampling schedule was designed to obtain either early- or late tracheids at all developmental stages, what corresponded under the Middle Siberian climatic conditions to early June, late June, and late July-early August. Early in June, under the favourable environmental conditions, the cambium is most active, the cell growth zone (G) contains the greatest number of cells and tracheid growth rate is usually most intense (Antonova and Stasova 1992, 1993, 1997). Late June is particularly interesting as the period of transition from early to late xylem formation and as the period when the developing xylem cells represent the whole range of xylem cell differentiation – from cambium towards mature tracheids. In late July, the cambial activity in larch trunks is usually completed, and only the last layers of latewood tracheids are found in G zone. In early August in pine trunks, the termination of cambial activity is also observed. In the time of sampling G zone in pine tree trunk was absent but the cells in maturation zone adjusting to cambium zone were present as well as mature xylem tracheids.

The cell layers were sampled by scraping tissues off layer by layer from trunk segments with the anatomical and histochemical control. Metachromatic staining of the cross cut sections with cresyl violet permitted to visualize the differentiation stage of the cells (Antonova and Shebeko 1981) and the staining with phloroglucinol-hydrochloric acid was employed to test the beginning of lignification. In the course of sampling one situation should be taken into account. After removing of the phloem from stem segments, it is necessary to control constantly the disposition of cambium because it may be deposited either on xylem or phloem side what depends on growth period and on what site the cells of cambium zone are divided.

Early in June from both larch and pine tree trunks the cell layers of the conducting phloem (cPh), adjoining to the Cam, Cam per se (Cam) and two parts of growth expansion zone (G) were sampled. One of them adjoining to Cam corresponded to the first (G1) and other to the second (G2) stages of growth with the increased and reduced rate of tracheid growth (Antonova and Shebeko 1985; Antonova 1999). During sampling from pine stem the cells at the beginning stage of secondary thickening before lignification (D1) have been additionally obtained. Late in June all the cells of developing annual ring, including mature ones, were early tracheids. On the xylem side, the cells of the Cam, G, and maturation zone were obtained. The maturation zone was divided on three parts—the beginning of the secondary cell wall thickening without lignification (D1), two successive layers of the cells at the stage of lignification (D2a and D2b). Mature xylem cells (mX) as well as the cells on phloem side were also isolated from larch stem. From pine stem the cells of conductive phloem (cPh) and mature phloem (mPh) were sampled successively on the phloem side adjacent to cambium zone. They differed in the presence or absence contents within the sieve elements and the thickness of the cell walls. Late in July the xylem cells of Cam, G and four cell layers, namely, D1, three successive layers of lignified cells, D2a, D2b, D2c and mature xylem (mX) cells on xylem side and additionally the cells of cPh and the cells of cambium zone adjoining to Ph have been isolated from larch trunk. In early August the xylem cells of Cam, D1, D2a, D2b, and D2c as well as cPh and mPh were sampled from pine tree trunk.

The content of AsA, DHA and uronic acids, as a possible precursor of AsA in plants, was assessed in the composition of metabolites extracted by 80% ethanol from received samples (Antonova et al., 2005, 2009). The contents of the acids were calculated both per dry weight and per cell. The calculation per cell reflects more natural (real) changes in the pool of compounds during morphogenesis of tracheids compared with the calculation per dry weight because the accumulation of substances within cell walls decreases respectively the level of the acids. Tissue maceration was used to assess the number of cells in 1 g dry weight (Antonova et al., 2005, 2009).

The general peculiarities in the changes of the AsA and DHA during cell xylem differentiation were remarked in both pine and larch. The AsA and DHA contents depended on the stage and step of tracheid differentiation as well as on the forming wood type what is conditioned by the time in the season and by weather conditions. At the same time the species of tree can also interfere in the content of the acids. The reason of such effect may be in differences of morphological structure of trunk wood and some chemical peculiarities what results in water store in trunk and consequently in internal water potential. For example, larch and spruce trees are different in the time of water flow from soil in trunk after depletion of accessible moisture store in that due to transpiration (Schulze et al., 1985).

3. RESULTS AND DISCUSSION

Besides the changes in the AsA-DHA system depend on the mode of data calculation. The commonly accepted mode is the calculation of per dry weight basis or fresh weight. To calculate per fresh weight one should take into account that content moisture changes during plant cell development. The calculation per dry weight can be suitable when almost all cells are in the same developmental stage or have similar nature. In coniferous the tracheids occupy almost 90% of all xylem cells. To compare different developmental stages and their separate steps one should remember the accumulation of cell wall substances during growth expansion zone and especially in secondary thickening zone as far as that increases the cell weight and artificially enhances the weight of material taken for analysis what in turn results in the apparent decrease in the component content. In phloem there are at least two type of cells in phloem–sieve and parenchyma cells, the number of which vary and may be the equal in annual increment. Because of that it is rather difficult to present the data by the content of any compound per either sieve or parenchyma cell and in this case the calculation "per cell" means a result per "medium" cell. Sometimes they may be present per dry weight.

In the course of early xylem cell development in larch in late June the rather high level of ÀsÀ was observed in the functional phloem (cPh). Late in June the cells of cPh contained less AsA than of Cam (0,11 and 0,16 mg/g correspondingly). In late July the cells of cambium zone managed to divide on phloem and xylem cells. The former contained AsA also less than the latter (0,02 and 0,03 mg/g respectively). At the same time the content of DHA in the phloem cells of cambium zone was equal to that of AsA while DHA was absent in xylem cells in Cam. This means that there are differences in functioning of ascorbate-glutathione cycle through cambium zone, what is evidently resulted from the activity of processes either xylem or phloem side.

In pine stem in early and late June just as in early August the cPh cells contained AsA and DHA more than mature phloem (mPh) cells but less than Cam cells (per dry weight). The ratio of AsA/DHA changed with different manners from mPh towards Cam in late June and in early August. During early tracheid development this ratio reduced from mPh (0, 44) to cPh (0, 27), increasing then to 0, 64 in Cam. In the course of latewood formation the ratio decreased successively 4.6, 1.02 to 0.5 in the same direction. It seems evident from the low value of the AsA/DHA ratio, oxidative processes are more active in cPh as compared to mPh and they are especially active in late June. This means in turn that metabolic processes in conductive cells of phloem are more active in comparison with mature cells of that. AsA has been shown to be loaded into phloem of source leaves and transported to sink tissues (Franceschi and Tarlyn 2002). Authors supposed AsA transporters reside in the phloem and the phloem translocation of AsA may be one factor regulating sink development because AsA is critical to cell division and growth. At the same time active AsA biosynthesis has been shown to occur in the phloem and according to assumption of authors must be implicated in whole plant AsA distribution. (Hancock et al., 2003). But it is difficult to suppose that AsA synthesis is only going on phloem of coniferous as well as its transport along all differentiation stages of xylem cells.

The high values of the AsA/DHA ratio in Cam late in June indicates the predominance of AsA synthesis over its oxidation what ensures certain level of AsA in dividing cells of pine. The high AsA level is known to be required for active cell proliferation (Liso et al., 1988). In our case, Cam was very active in late June while its activity dropped to early August. In this time it can be in a dormant state. Low level of DHA in mPh cells may be connected with its destruction to oxalate (Green and Fry 2005, Debolt et al., 2007) which crystals are accumulated in mature phloem of some coniferous.

3.1 Cambium Zone

Through the season the highest content of ÀsÀ and DHA in Cam of larch trunk was observed in early June (Fig. 1) (Antonova et al., 2005). In this time the highest cambial activity is usually remarked and the ÀsÀ content was as high as 0.6 mg/g dry wt, four times more than late in June when the cambial activity declined, and the cambial initials started to produce the late xylem cells. Similar profile of DHA was observed. Late in July,

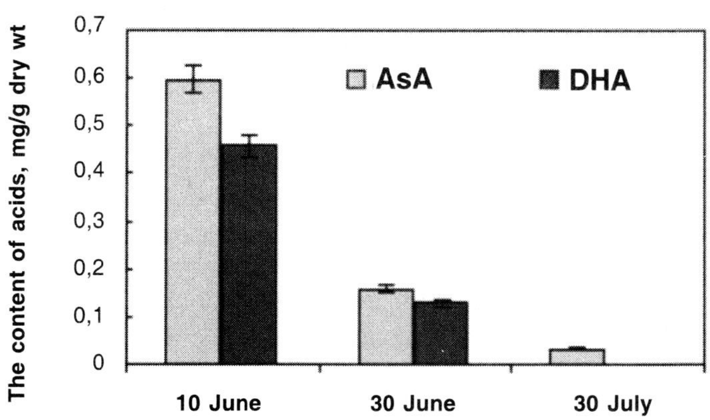

Fig. 1. The content of ascorbic (AsA) and dehydroascorbic (DHA) acids in cambium cells in the course of annual ring formation in Siberian larch tree trunks.

when the cambial activity was terminated, the ÀsÀ content dropped 4.8-fold as compared to the previous stage, and DHA in cambial xylem derivatives completely disappeared.

The considerable decrease in the ÀsÀ content in the Cam cells late in June was related to the transition from early to late wood production by the cambium cells. Such changes in cambial activity and the subsequent differentiation of cambium derivatives as latewood cells resulted from water deficit in the developing tree tissues caused either by drought (Zahner 1963; Zahner et al., 1964) or by enhanced transpiration under the conditions of high ambient temperatures and solar radiation (Gregg et al., 1988; Antonova 1999) or flooding of root system (Lyr et al., 1967). Water deficit can decrease the ÀsÀ content in developing tissues of tree via lowered AsA synthesis (Smirnoff and Colombe 1988) or an enhancement of ascorbate peroxidase and ascorbate oxidase activities by drought (Mukheljee and Choudhuri 1983). The latter enzyme did not change with the age of pine hypocotyls and along the axis of that (Sánchez et al., 1997), but ascorbate oxidase has been found to be in both apoplast and symplast in pea shoots (Pinto and De Gara 2004). It has been demonstrated that ascorbate oxydase regulates the apoplastic AsA pool, reducing its level, and when apoplastic oxydase is active the rate of AsA oxidation exceeds the capability of DHA transport in cytosol (Pignocchi and Foyer 2003).

Late in July, when the late xylem is developed and cambial activity is completed, the absence of DHA in Cam could result from the rapid reduction of monoDHA by monoDHA reductase to ÀsÀ in the apoplast (Hossain et al., 1984) where peroxidase oxidizes ÀsÀ to monoDHA and activates monoDHA reductase. Thus, the second stage of ÀsÀ oxidation to DHA can be absent or almost absent. Both the activities of enzymes, monoDHA reductase and peroxidase, have been shown to be promoted by drought (Smirnoff and Colombe 1988).

Thus, the contents of ÀsÀ and DHA in Cam of trunk declined through the season indicating the diminishing redox potential of the tissue. Such decline in the ÀsÀ content in Cam was in line with slowing cell proliferation what was resulted in lowering of the rate of cambial initial division. For example, in larch stem (60-year-old trees) the rate of division into xylem side in late June was 2.6 cells per day while in the end of July 1.2 cells only (Antonova 1999). This evidence supports the observations that high ÀsÀ content favours all phases of cell division cycle (Liso et al., 1988). In pine the decrease of AsA in Cam from late June to early August was also observed (Antonova et al., 2009).

Diminishng of AsA amount in cambium zone of trees through the season is probably connected with reducing of water potential in growing tissue. Negative influence of a drought on AsA content can be the result of decrease both of synthesis and reduction of AsA as well as increase of ascorbate oxidase activity. The changes in ascorbate content in Cam cells correlated with the activity of cell division process and corresponded to the observations that dividing cells require large amounts of ascorbic acid to be divided.

3.2 Growth Expansion Zone

In larch, the ÀsÀ and DHA contents (per dry weight) in growth zone changed during the season (Fig. 2). Both AsA and DHA decreased from early to late June. Further, in late July, the content of AsA dropped almost threefold while that of DHA increased in

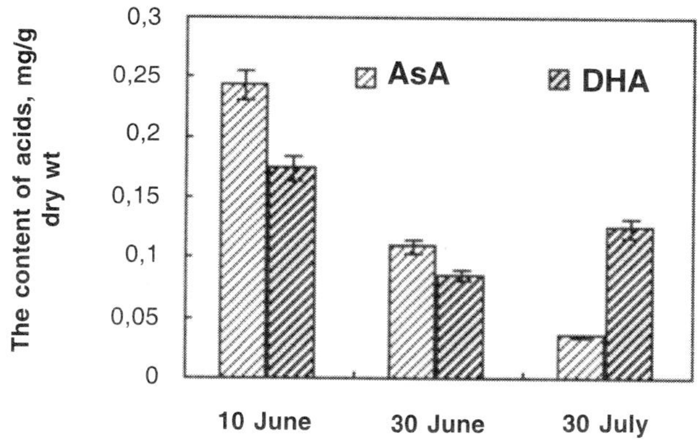

Fig. 2. The contents of ascorbic (AsA) and dehydroascorbic (DHA) acids in the radial growth zone in the course of annual ring formation in Siberian larch tree trunks.

1, 5 times. The ratio of AsA/DHA in G during the season diminished 1, 4/1, 3/0, 23 indicating the changes in ascorbate-glutathione cycle and strengthening of oxidation reactions as compared with reduction/synthesis of AsA. In the end of July, when the late xylem developed, the oxidation processes already dominated in G. As a result, the amount of AsA decreased and that of DHA increased by comparison with the end of June. A high AsA level in G in the course of early xylem formation corresponded to wider tracheids, whereas a low AsA level in the zone of late tracheid development was associated with a lower tracheid diameter. The average diameter of the early and late tracheids in the annual rings of the trees was 61.3 ± 1.1 and 28.9 ± 1.1 mm, respectively.

Within G zone itself, the changes in the ÀsÀ content depended on the method of calculation. Per dry weight it dropped down initially and did not considerably change later (Fig. 3a), whereas per cell, the ÀsÀ and DHA contents did not notably changed in G as compared with Cam at the beginning and considerably declined later (Fig. 3b). The DHA content decreased twofold in the course of tracheid growth. The changes in AsA in the course of tracheid growth were in line with the slow-down of cell growth rate (Fig. 3c). The radial size of tracheid was earlier shown to depend on cell enlargement rate (Antonova and Shebeko 1985). Cosgrove (1986, 1997) emphasized that the cell growth rate is closely related to the ability of cell walls to expand. Such property depends on numerous physical and biochemical events in the cell wall (Antonova 1999) including the formation of the diferuloyl bridges between the cell wall polysaccharides (Fry 1986, 1998, 2004).

In pine the changes in the AsA and DHA in the direction from cPh toward expanding cells also showed clear dependence of these acids contents on the growth stage and cell nature (Antonova et al., 2009). Both AsA and DHA contents decreased in Cam by comparison with cPh, increased sharply at the first growth stage and declined at the second (Fig. 4). But the ratio AsA/DHA rises from cPh to Cam and to the first step of growth and then is reduced at the second step (Fig. 5). As within larch growth zone, a synthesis of AsA and, possibly, its reduction seems to prevail during the G1 stage, whereas oxidative processes dominated in G2 stage. This is in agreement with the

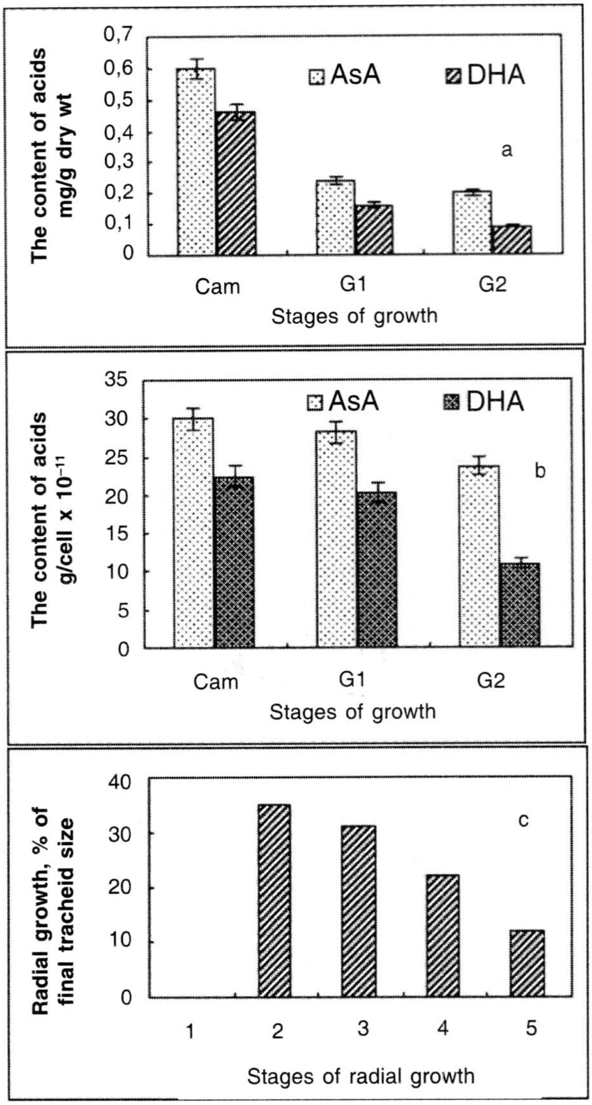

Fig. 3. The content of ascorbic (AsA) and dehydroascorbic (DHA) acids in cambium zone (Cam) and two parts of radial growth zone (G1 and G2) in the course of early xylem formation (June10): (a) per dry weight, (b) per cell, (c) the increment of tracheid diameter after cambium zone (1) at successive growth stages (2-5).

reduced growth rate at this stage and the approval that a prerequisite for growth retardation could be lowering of cell wall extensibility due to the formation of cross-links between cell wall hemicelluloses (Fry 1982, 1998, 2004).

This process involves hydrogen peroxide and is catalysed by peroxidase (Sánchez et al., 1996; Zarra et al., 1997; Pedreira et al., 2004). As a result, cell wall extensibility is diminished, and cell growth is terminated (Sánchez et al., 1996; Zarra et al., 1999; Cosgrove 1999; Fry 2004). The increase in peroxidase activity and hydrogen peroxide

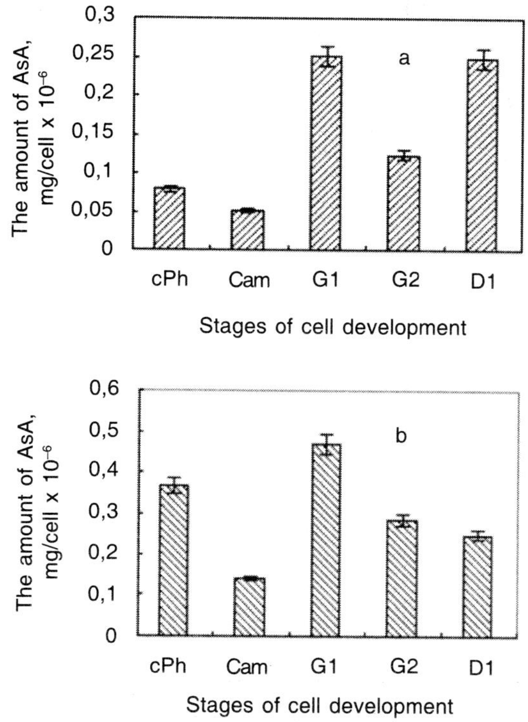

Fig. 4. The content of ascorbic (a) and dehydroascorbic (b) acids in the cells of conducting phloem (cPh), cambium (Cam), two steps of cell growth (G1 and G2) and the first stage of secondary thickening zone before lignification (D1) during annual wood ring formation in Scots pine trunk in early June.

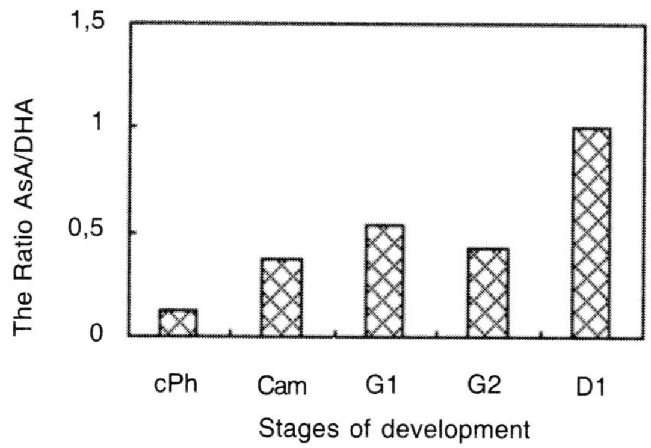

Fig. 5. The AsA / DHA ratios during annual ring formation in Scots pine trunk in early June. Designations as in Fig. 4.

content in line with growth retardation was observed along pine hypocotyls and correlated with the age of that (Sánchez et al., 1996; Zarra et al., 1999; Pedreira et al., 2004). Authors suggested the balance of AsA and hydrogen peroxide could control cell wall extensibility (Sánchez et al., 1996; Zarra et al., 1999). Because of lower content of AsA, peroxide is consumed less in AsA oxidation and can be used more for oxidation of ferulate what lead to diferuloyl bridges formation and stoppage of cell growth. As showed later studying the hydrogen peroxide content in apoplastic fluid increased while the AsA content in that decreased with the hypocotyl age and along the hypocotyl axis of 10-day-old seedlings of pine (Pedreira et al., 2004). A very significant correlation between the hydrogen peroxide apoplastic level and the growth rate as well as between the ascorbate/hydrogen peroxide molar ratio and the growth rate of hypocotyls have been found suggesting that the redox state is the main factor controlling the cell wall stiffening mechanism and thus growth in pine hypocotyls. Our data on pine and larch trunk suppose these observations.

There are important peculiarities in the changes in AsA, DHA and their ratio along growth zone cells in comparison with the beginning of mature zone during wood formation in pine trunk. In the course of early xylem formation the AsA/DHA ratio increased within the cells at the beginning of secondary wall thickening before lignification (D1) as compared with the G2 stage (Fig. 5) what resulted from rising of AsA and lowering of DHA (Fig. 3). Apparently AsA accumulation occurred not only due to its synthesis but also due to its reduction from DHA or even more from monoDHA just in apoplast. In any case this points to the changes in metabolism of cells at this differentiation stage and in ascorbate-glutathione cycle.

3.3 Zone of Secondary Wall Thickening

In larch stem when the secondary walls of early and late tracheids are thickened and xylem lignified, the AsA content in the xylem stays high up to lignification calculated both per dry weight and per cell. In the course of early xylem formation the content of ÀsÀ is high at the beginning of the secondary cell wall material deposition but that of DHA is higher (Fig. 6). Such combination in D1 presumes the ongoing oxidation processes, that is, the active ÀsÀ utilization as antioxidant oxidized into DHA. After the initial lignification (D2a), ÀsÀ was not be examined (probably, because of track

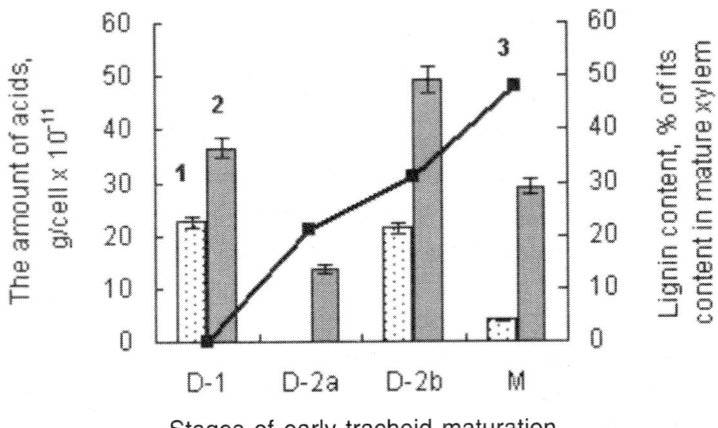

Fig. 6. The contents of ascorbic (1) and dehydroascorbic (2) acids, and lignin deposition (3) at the successive stages of early tracheid maturation: D1-before lignification, D2a, D2b - during lignification, M-mature xylem.

amount) and the DHA content was low, while the subsequent xylem maturation step (D2b) was accompanied by the accumulation of ÀsÀ and especially DHA.

The analysis of the late xylem tracheid maturation demonstrated that the ÀsÀ content, both per dry weight and per cell (Fig. 7a), reached also the maximum before the start of lignification, i.e., at the early step of the secondary thickening. During subsequent lignification, it dropped down more than tenfold and did not change later until tracheids matured (the M stage), when the ÀsÀ content per cell slightly increased. The DHA content in the differentiating late tracheids (D1) was almost six times lower than in the early tracheids at the beginning of the secondary cell wall formation. With lignification commenced (D2a), the amount of DHA in the cells increased dramatically while the AsA content declined presuming the ongoing oxidation processes. Later, the DHA content in the maturing cells gradually decreased, whereas the AsA level stayed low.

The difference in the oxidation processes could affect the commencement of lignification. In Figs. 6 and 7 the changes in the ÀsÀ and DHA contents are presented against the time-course of lignin accumulation in the maturing tracheids of the early and latewood thus indicating the lignification rate in the early and late xylem, respectively. The temporal patterns of lignification are different in the early and late xylem. The intensity of lignification in early xylem gradually enhances by the end of

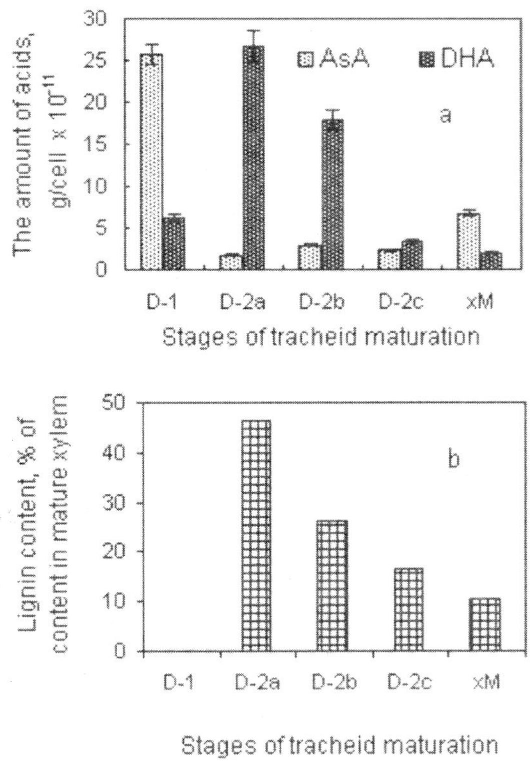

Fig. 7. The contents of ascorbic and dehydroascorbic acids (a) and lignin deposition (b) at successive stages of late tracheid maturation: D1 - before lignification, D2a D2b and D2c–during lignification, xM - mature xylem.

tracheid maturation (Fig. 6) while that in latewood is the highest at the outset of lignification (D2a) and declines as the tracheids mature (Fig. 7b). This observation supports the idea that polymerization of lignin precursors is delayed when enough AsA is present in the tissue (Takahama 1993, 1994).

The lignification involves the oxidation of p-hydroxycinnamoyl alcohols by peroxidase and laccase (Bao et al., 1993, McDougall et al., 1994). The former reaction employs hydrogen peroxide as a donor, whereas the latter does not require this toxic compound and may to be involved in lignification already at its earliest steps (Sterjiades et al., 1993). Freudenberg was first to indicate that both enzymes can cooperate in tissue lignification (Freudenberg 1959). Peroxidase may promote the late steps of xylogenesis, when hydrogen peroxide is generated, or in response to stress (Sterjiades et al., 1993). As far as the early and late xylems develop at different moisture contents in plant tissues, we presume that the activities of two enzymes are different in the course of early and late xylem lignification.

As suggested the dehydrogenization of p-hydroxycinnamic acids to corresponding alcohols can occur due to both hydrogen peroxide and oxygen, and the latter may also control the rate of lignification (Savidge and Udagama-Randeniya 1992). Both enzymes and oxidizing agents, influenced in initiation of lignin deposition, seems to be dissimilar in the early and late xylem, i.e., the lignification of the early wood involves primarily oxygen, which may be produced at the first stage AsA oxidation in apoplast or other pathway, while hydrogen peroxide may be the major agent in the latewood lignification. It is proposed that ascorbate can act as a pro-oxidant since several steps in the pathway potentially generate H_2O_2 (Green and Fry 2005). Hydrogen peroxide can be generated in the course of phenolic/phenolic radical reaction catalysed by H_2O_2-dependent peroxidase (Takahama 1993; Otter and Polle 1994).

Due to the oxidation either by oxygen radical into monoDHA and hydrogen peroxide (Hossain et al., 1984; Forti and Elli 1995) or by hydrogen peroxide with participation peroxidase AsA can interfere in lignification by restricting the oxidation of phenoxyl radicals. It is suggested that in addition to AsA, monoDHA radicals may also reduce phenoxyl radicals (Takahama and Oniki 1997). The reduction of oxidized compounds by the corresponding reductases (Hossain et al., 1984) would restore the AsA pool. There are the data that the activities of these reductases are enhanced under drought conditions (Smirnoff and Colombe 1988; Smirnoff 1996). It is interesting to see that water-stress has been found to enhance also the activity of such special enzymes as guaiacol peroxidase, coniferyl alcohol peroxidase, syringaldazine peroxidase, involved in lignification in white clover leaves (Lee et al., 2007). It must expect an increased AsA/DHA ratio under water stress, and we recorded such increase in the late xylem developed at the Dl stage preceding lignification. One should remind that the late xylem is developed under low internal water potential in trees.

When, at this stage of late tracheid development, the AsA pool is restored by DHA reduction, the AsA transport from the cytosol into apoplast is probably impeded. In such case, AsA is inaccessible for oxidants and cannot interfere with the process of lignification. The rate of AsA transport across the plasmalemma and tonoplast has been found to be lower than DHA (Rautenkranz et al., 1994). In addition, the AsA pool produced by DHA reduction is supposed to be probably not completely equivalent to the AsA pool produced via L-galacto-γ-lactone (Paciolla et al., 2001) and in this case might be less accessible to the AsA-consuming metabolic reactions. Such suggestion is in agreement with the low DHA level observed at the D1 stage of late xylem formation.

These arguments permit to understand why high AsA content in D1 cells of the late xylem did not hinder its subsequent lignification at the D2a stage (Fig. 7 b). The greatest lignin deposition at the beginning of lignification in the late xylem presumes that lignin precursors, phenoxyl radicals, could be accumulated at the previous stage of cell differentiation.

Other correlations are observed in the course of early xylem differentiation (Fig. 6). High contents of AsA and DHA, that is, the actively functioning system of AsA synthesis-oxidation-reduction, at the D1 stage must retard the formation of lignin precursors and, as a result, decrease the extent of lignification of the early xylem at the next stage (D2a), where AsA is absent and the presence of DHA is apparently determined by retarded DHA reduction into AsA and by diminished AsA synthesis itself. At the further step of xylem maturation (D2b), with a considerable amount of AsA present, the DHA content increased fourfold. Probably, such change was due to a local shift in cell metabolism evoked by rain, which preceded the sampling. The notable accumulation of AsA and especially DHA at this phase of xylem maturation (Fig. 6) was correlated with the promoted tissue lignification at this and further stages of tracheid maturation. The progressive lignin deposition presumes that AsA was localized already in the symplast. In this case, the DHA content was a measure of AsA utilization, and the temporal changes in the DHA content at the D2a and D2b stages were correlated with the time-course of lignin deposition. In the mature cells of early xylem, the low AsA content against the high DHA content indicated an inhibition of DHA reduction. In the course of late xylem lignification, starting from the D2a stage, the AsA level was already low and did not considerably change in maturing cells, whereas the level of DHA gradually decreased in parallel with declining lignification (Fig. 7). It follows that the DHA content reflects the rate of lignification.

Taking into account different compartmentalizations of AsA oxidation and reduction, one would presume that the accumulation of p-hydroxycinnamoyl acids, as precursors of corresponding alcohols, before lignification was also different in the early and late xylem. Studying the content of phenolic acids at the same stages of differentiation cells we found that there are considerable differences in the contents and composition of both free and bound phenolic acids at these stages (unpublished data).

In pine, as showed above, the forming earlywood cells in early June contained rather high AsA and less DHA at the beginning of secondary thickening zone before lignification (D1). In late June the AsA content per cell after steady increasing from Cam and G sharply reduced before lignification (D1) with sharp increasing at the beginning of lignification, and subsequent declining during cell maturation and lignification (Antonova et al., 2009). The content of DHA also increased from Cam to G but stronger than AsA did (1.8 vs. 1.2 times) and then sharp also diminishing in D1 with following rising and again dropping at next stages. In spite of low content of AsA and DHA in D1 stage in this sample in late June the ratio of AsA/DHA in D1 decreases only slightly

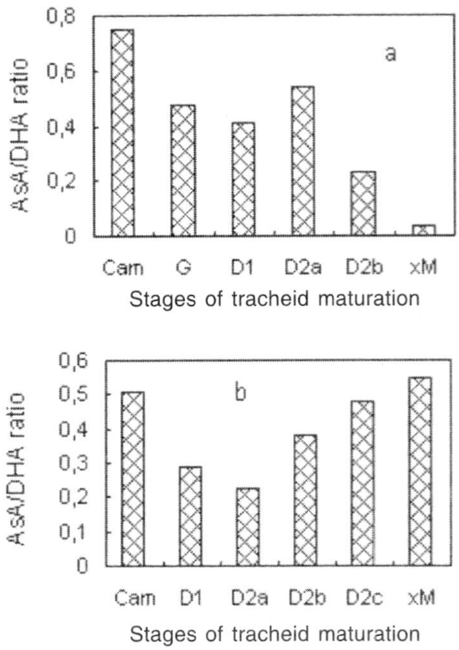

Fig. 8. The changes in AsA/DHA ratio in the cells of Cam, G, D1, D2a, D2b, D2c and xM during formation of early (a) and late (b) xylem in Scots pine trunk. Designations as in Figs. 4 and 7.

by comparison with growth zone and increases a few at the beginning of lignification (D2a) dropping successively during maturation (Fig. 8a). In early August, as was aforementioned, the growth of late tracheids was terminated and G zone was absent. By the moment of sampling, only the cells at the stage of maturation were present, starting from the stage of secondary cell wall compound deposition. With the start of secondary thickening the amount of both acids increased as compared with Cam, attained the highest values in D2a, and declined with tracheid maturation. The changes in the content of AsA and DHA occurred differently what affected AsA/DHA ratio (Fig. 8b). Before lignification of late xylem the AsA/DHA ratio is less as compared with earlywood cells (0.41 vs. 0.29). This indicates a decrease in the AsA content before lignification in the late xylem, which could affect the process of lignin deposition. So, in early xylem the ratio decreased from the start of lignification to xM, whereas in the late xylem, in contrast, it increased in this direction. This indicates that, in the early and late xylem, the redox state and the conditions of monolignol polymerization during lignification of two types of wood differ. Our data shows the dynamics of lignin deposition in early and late xylem are really different and diametrically opposed to AsA/DHA ratio. In the earlywood pine, as during in larch, the intensity of lignification occurs gradually, increasing from the beginning of the process to the end of tracheid maturation. In contrast, in the latewood, lignin deposition is most efficient in the beginning of the process and reduced gradually towards the xM. This suggests the essential changes in the principal metabolic reactions in lignifying xylem cells due to a variation in internal water potential in coniferous trees. This can be connected to the system of AsA synthesis-oxidation-reduction, an activity and nature of enzymes and the state of membranes what determines transport of AsA between symplast and apoplast.

The content and composition of bound and free phenolic acids in developing cells of early and late xylem in pine also showed the differences along differentiation stages. The differences in the content of free phenolic acid fraction were especially noticeable before lignin deposition while that of bound acids, both ethers and esters, were observed at lignification stages. In the course of cell maturation the content of ethers and esters in early and late xylem changed differently thus, indicating their distinctive use in lignification (Antonova et al., in press). Changes in the free phenolic acid composition can interfere in the state of membranes. So, ferulic and especially benzoic acids has been found to depolarize the membranes what influences their permeability and favours to metabolite transport (Glass and Dunlop, 1974). According to our data the pool of benzoic acid increased twofold during the transition to secondary thickening in pine early tracheids what in turn might intensify AsA transport (Antonova et al., in press).

3.4 Uronic Acids

During annual ring wood formation in pine trunk the amount of uronic acids, which are considered possible precursors of AsA (Isherwood and Mapson 1962; De Gara et al., 1994; Oba et al., 1995; Smirnoff 1996), depended on the differentiation stages and type of forming wood (Antonova et al., 2009). In growth zone in early June the level of uronic acids dropped at the transition from Cam to the first growth (G1) step and then increased in the G2 step and further in D1 (Fig. 9). Dropping of uronic acids with the beginning of

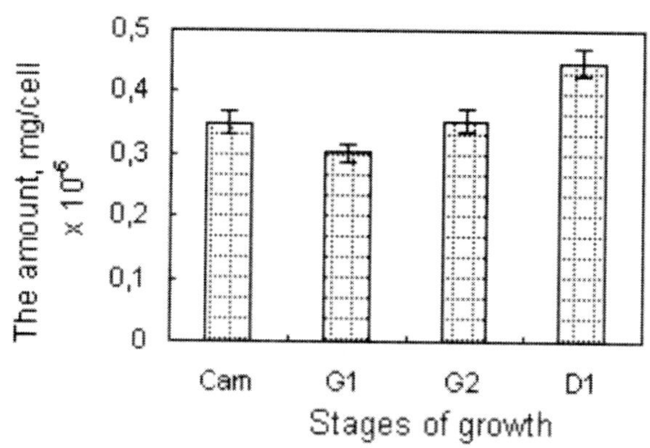

Fig. 9. The content of Uronic acids in the cells of Cam, G1, G2 and D1 stages during annual wood ring formation in Scots pine trunk in early June. Designations as in Fig. 4.

cell growth, i.e., in the course of the increase in primary cell walls surface, is mainly related to their usage for the synthesis of middle lamella pectin substances, the principal of which is galacturonic acid. But it seems to be impossible to exclude their usage also for AsA synthesis because of lower level of uronic acids at the first growth step and high amount of AsA in contrast. The increase of the amount of uronic acids in the cells at the beginning of secondary cell wall thickening before lignification (D1) is in parallel with AsA (Fig. 4a) whereas the amount of DHA reduced (Fig. 4 b). However, the latter decreased very slightly in D1 as compared with G2 while AsA amount increased almost twofold. This suggests some usage of uronic acids for AsA synthesis. The ratio of AsA/DHA, as the index indicating the general direction of redox reactions, was more in D1 in comparison with G2, as showed above (see Fig. 5). If to take into account the time of sampling, early in June, when water store in soil is rather high, the data suggest that AsA synthesis was more than its reduction.

Late in June the content of uronic acids was higher in mPh than in cPh. As calculated per dry weight it increased from 1.21% in cPh to 1.41% in Cam, whereas the calculation per cell the amount of uronic acids in Cam was lower than in cPh (Fig. 10 a). Both in forming early (in late June) and late (in early August) xylem the functioning conducting phloem cells contained almost equal amounts uronic acids (Fig. 10 a, b). This suggests essential level of uronic acids for AsA synthesis and the maintenance of its concentration both in late June and early August, although AsA oxidation into DHA in cPh was much more active during the formation of early tracheids than during latewood development. The consumption of uronic acids for the maintenance of AsA level in Cam is clearly more active in late June than in early August, the more so that AsA oxidation into DHA differs insignificantly as AsA/DHA ratio shows. During early and late tracheid maturation the content of uronic acids decreased in line with that of AsA.

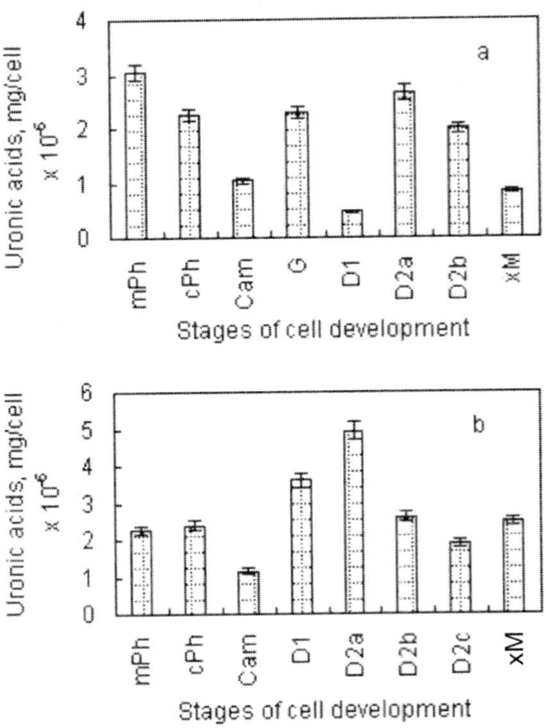

Fig. 10. The content of uronic acids in the cells of mPh (mature phloem), cPh, Cam, G, D1, D2a, D2b, D2c and xM during annual ring formation in Scots pine trunk in late June (a) and in early August. Designations as in Figs. 4 and 7.

Interestingly, according to our previous data (Stasova 1991, Antonova 1999) no galactose, as possible source of uronic acids (Smirnoff 1996), was observed in the carbohydrate composition during xylem cell formation up to the end of their differentiation. As far as growth stage the glucose amount per cell increased from Cam (1,7 mg/10^6) to G2 (30.6 mg/10^6). At the same time, galactose-contained carbohydrates – raffinose and stachyose – in G1 increased almost in tenfold in comparison with Cam and further decreased in G2, especially stachyose (in twofold). This permits to suggest an elimination of galactose residues from the derivatives of sucrose for its using in AsA synthesis at G2 growth step as the amount of sucrose enhanced is threefold.

4. CONCLUSION

In general, during formation of early and latewood in the larch (*Larix sibirica* Ldb.) and Scots pine trunks, each stage of phloem and xylem cell development is characterized by definite levels of AsA, DHA and uronic acids which influences the intensity of cell differentiation. The level of AsA depends on the activity processes in ascorbate-glutathione cycle and in particularly on its accessibility to oxidants agents. The latter depends on localization of AsA either apoplast or symplast and the capacity of the cell to regulate its transport via membranes. As remarked by (Horemans et al., 2000) "regulation of ascorbate transport systems may be central in the regulation of different processes through the cell cycle, expansion of cell wall" and, as can be added, through the lignification.

The seasonal decline in cambial activity correlated with the decreased AsA content in the Cam cells. While the cambium produced the early xylem cells, the AsA content was higher than during the late xylem development. As a result, the number of earlywood cells in the annual ring in pine exceeded that in the latewood. In larch the number of early and late tracheids can be equal in annual ring of wood (Antonova and Stasova 1997). In July, when cambium produces xylem cells further developing in late tracheids (Antonova and Stasova 2002), a water store in the larch trunk appears to be greater as compared with pine. Because of that cambium produces late xylem cells in larch more than in pine trunk.

The beginning of cell enlargement was associated with AsA accumulation, as compared to cambium zone. In the course of the tracheid growth both the AsA content in cells and growth rate of cells were declined. The AsA content in growth zone during early xylem development was higher than in the course of late xylem, and the tracheid diameter in the earlywood was greater than in the latewood.

The lignification in early and late xylem depended on the available AsA. Through maturation stages the ratio AsA/DHA was lowered in early and increased in late xylem, what means the redox state and, as a result, the conditions of monolignol polymerization during lignification in the early and late xylem, are different. We demonstrated the opposite trends in the lignification of the early and late xylem both in larch and pine. While in the earlywood the rate of lignification gradually increased reaching the maximum towards mature tracheids, in the latewood the most active lignin deposition was observed at the initial steps of lignin deposition and then it dropped towards mature cells. Our data suggests that the different patterns of lignification in two wood types are due to the opposite direction of the AsA-dependent reactions and probably to the different activity of laccase and peroxidase as well as to the nature of oxidants along cell developmental stages. It may suggest that laccase controls first steps of lignification and peroxidase acts at next steps. Furthermore, laccase seems to predominantly control the early xylem lignification, peroxidase is apparently most active in the late xylem.

Thus, the amount of AsA at cell developmental stages and its accessibility to be oxidized influences the activity of cambium, of cell growth and lignification what results in the number of early and late tracheids in annual wood layer as well as their radial diameters and cell wall thickness in larch and pine trunks.

ACKNOWLEDGEMENTS

This work was supported by the Russian Foundation for Basic Research, project no. 01-04-48077 and project no. 06-04-49501

REFERENCES

Antonova, G.F. and Shebeko, V.V. (1981). Usage of Cresyl Violet for Studying of Wood Formation, *Chemistry of Wood* (*Rus.*) **N4:** 102-105.

Antonova, G.F. and Shebeko, V.V. (1985). Xylem Formation in Conifers: 3. Dynamics of Tracheid Development in the Zones of Differentiation. *Lesovedenie* (*Rus.*). **N5:** 71-74.

Antonova, G.F. and Stasova, V.V. (1992). Annual Rings Formation in Trunks of the Scots Pine and the Siberian Larch Trees. *Lesovedenie* (*Rus.*). **N5:** 19-27.

Antonova, G.F. and Stasova, V.V. (1993). Effect of Environmental Factors on Wood Formation in Scots Pine Stems. *Trees.* **7:** 214-219.

Antonova, G.F. and Stasova, V.V. (1997). Effects of Environmental Factors on Wood Formation in Larch (*Larix deciduas* Ldb.) Stems. *Trees.* **11:** 462-468.

Antonova, G.F. (1999). Cell Growth in Conifers. Novosibirsk: Nauka.

Antonova, G.F. and Stasova, V.V. (2002). Seasonal distribution of processes responsible for radial diameter and wall thickness of larch (*Larix sibirica* Ldb.) tracheids. Proc. Intern. Symp "Improvement of larch (Larix Sp.) for better growth, stem form and wood quality", 16-21 September 2002, Gap, France. pp. 369-377.

Antonova, G.F., Chaplygina, I.A., Varaksina, T.N. and Stasova, V.V. (2005). Ascorbic Acid and Xylem Development in Trunks of the Siberian Larch Trees. *Russ. J. Plant Physiol.* **52(1):** 83-92.

Antonova, G.F., Stasova, V.V. and Varaksina, T.N. (2009). Ascorbic Acid and Development of Xylem and Phloem Cells in the Pine Trunk, *Russ. J. Plant Physiol.* **56(2):** 190-199.

Antonova, G.F., Jeleznichenko, T.V. and Stasova, V.V. The changes in the content and composition of phenolic acids during xylem cell growth in Scots pine. *Russ. J. Develop. Bio.* (in press).

Arrigoni, O. (1994). Ascorbate System and Plant Development. *J. Bioenerg. Biomembr.* **26:** 407-419.

Asada, K. (2006). Production and scavenging of reactive oxygen species in chloroplasts and their functions. *Plant Physiol.* **141:** 391-396

Baig, M.M., Kelly, S. and Loewus, F.A. (1970). L-Ascorbic acid biosynthesis in higher plants from L-gulono-1,4-lactone and L-galactono-1,4-lactone. *Plant Physiol.* **46:** 277-280.

Balmaeva, L.I. (1986). Influence of unairobiose on some oxydoreductase activity in pine seedlings. In: Physiology of growth and nutrition of coniferous. Krasnoyarsk. pp. 48-58.

Bao, W., O'Malley, D.M., Whetten, R. and Sederoff, R.R. (1993). A Laccase Associated with Lignification in Loblolly Pine Xylem, *Science.* **260:** 672-674.

Carrol, E.W., Schwarz, O.J. and Hickoko, L.G. (1988). Biochemical Studies of Paraquat-Tolerant Mutants of the Fern *Ceratopteris richardii. Plant Physiol.* **87:** 651-654.

Chalapathi Rao, A.S.V. and Reddy, A.R. (2006). Glutathione Reductase: A Putative Redox Regulatory System in Plant Cells In: Madhava Rao, K.V., Raghavendra, A.S. and Reddy, K.J. (Eds.). Physiology and molecular biology of stress tolerance in plants, Springer, The Netherlands, pp. 111-147.

Córdoba-Pedregosa, M.D.C., Córdoba, F., Villalba, J.M. and González-Reyes, J.A. (2003). Zonal Changes in Ascorbate and Hydrogen Peroxide Contents, Peroxidase, and Ascorbate-Related Enzyme Activities in Onion Roots. *Plant Physiol.* **131:** 697-706.

Cosgrove, D. (1986). Biophysical control of plant cell growth. *Annu. Rev.* **37:** 377-405.

Cosgrove, D. (1997). Assembly and Enlargement of the Primary Cell Wall in Plants, *Annu. Rev. Cell Dev. Bio.* **9:** 1031-1041.

De Gara, L., Paciolla, C., Tomassi, F., Liso, R. and Arrigoni, O. (1994). *In vivo* inhibition of galactono-γ-lactone conversion to ascorbate by licorine. *J. Plant Physiol.* **144:** 649-653.

De Gara, L., Paciolla, C., De Tullio, M.C., Motto, M. and Arrigoni, O. (2000). Ascorbate-dependent hydrogen peroxide detoxification and ascorbate regeneration during germination of a highly productive maize hybrid: evidence of an improved detoxification mechanism against reactive oxygen species. *Physiologia Plantarum.* **109:** 7-13.

Debolt, S., Melino, V. and Ford, C.M. (2007). Ascorbate as a Biosynthetic Precursor in Plants. *Ann. Bot.* **99(1):** 3-8.

Forti, G. and Elli, G. (1995). The Function of Ascorbic Acid in Photosynthetic Phosphorylation. *Plant Physiol.* **109:** 1207-1211.

Foyer, C.H. and Halliwell, B. (1976). The presence of glutathione and glutathione reductase in chloroplasts: a proposed role in ascorbic acid metabolism. *Planta.* **133:** 21-25.

Franceschi, V.R. and Tarlyn, N.M. (2002). L-Ascorbic acid is accumulated in source leaf phloem and transported to sink tissues in plants. *Plant Physiol.* **130(2):** 649-656.

Freudenberg, K. (1959). Biosynthesis and constitution of lignin. *Nature.* **183:** 1152-1155.

Fry, S.C. (1986). Cross-linking of matrix polymers in the growing cell wall of Angiosperms. *Ann. Rev. Plant Physiol.* **37:** 165-186.

Fry, S.C. (2004). Oxidative coupling of tyrosine and ferulic acid residues, intra- and extra-protoplasmic occurrence, predominance of trimers and larger products, and possible role in inter-polymeric cross-linking. *Phytochem. Rev.* **3:** 97-111.

Fry, S.C. (1982). Phenolic Components of Primary Cell Wall. *Biochem. J.* **293:** 493-504.

Glass, A.D.M. and Dunlop, J. (1974). Influence of phenolic acids on ion uptake. 4. Depolarization of membrane potential. *Plant Physiol.* **54:** 855-858.

Green, M.A. and Fry, S.C. (2005). Apoplastic degradation of ascorbate: novel enzymes and metabolites permeating the plant cell wall. *Plant Biosystems.* **139:** 2-7.

Gregg, B.M., Dougherty, P.M. and Henessey, T.C. (1988). Growth and wood quality of young loblolly pine trees in relation to stand and climatic factors. *Can. J. For. Res.* **18:** 851-858.

Hancock, R.D., McRae, D., Haupt, S. and Viola, R. (2003). Synthesis of L-ascorbic acid in the phloem. *Plant Biol.* **24:** 3-7.

Horemans, N., Foyer, C.H., Potters, G. and Asard, H. (2000). Ascorbate function and associated transport systems in plants. *Plant Physiol. Biochem.* **38:** 531-540.

Hossain, M.A., Nakano, Y. and Asada, K. (1984). Monodehydroascorbate Reductase in Spinach Chloroplasts and Its Participation in Regeneration of Ascorbate for Scavenging Hydrogen Peroxide. *Plant Cell Physiol.* **25:** 385-395.

Isherwood, F.A. and Mapson, L.W. (1962). Ascorbic acid metabolism in plants: Part II. Biosynthesis. *Ann. Rev. Plant Physiol.* **13:** 329-350.

Kärkönen, A. and Fry, S.C. (2006). Effect of ascorbate and its oxidation products on H_2O_2 production in cell-suspension cultures of *Picea abies* and in the absence of cells. *J. Exp. Bot.* **57:** 1633-1644.

Lee, B-R., Kim, K-Y., Jung, W-J., Avice, J-C., Ourry, A. and Kim, T-H. (2007). Peroxidases and lignification in relation to the intensity of water-deficit stress in white clover (*Trifolium repens* L.) *J. Exp. Bot.* **58(6):** 1271-1279.

Liso, R., Innocenti, A.M., Bitonti, B.M. and Arrigoni, O. (1988). Ascorbic Acid-Induced Progression of Quiescent Centre Cells from G1 to S Phase. *New Phytol.* **110:** 469-471.

Loewus, F.A. (1999). Biosynthesis and Metabolism of Ascorbic Acid in Plants and Analogs of Ascorbic Acid in Fungi. *Phytochem.* **52:** 193-210.

Lyr, H., Polster, H. and Fiedler, H-J. (1967). Gehölzphysiologie. VEB Gustav Fisher Verlag, Jena.

Marjamaa, K., Kukkola, E., Lindell, T., Karhunen, P., Saranpää, P. and Fagerstedt, K.V. (2006). Monolignol Oxidation by Xylem Peroxidase Isoforms of Norway Spruce (*Picea abies*) and Silver Birch (*Betula pendula*), *Tree Physiol.* **26:** 605-611.

McDougall, G.J., Stewart, D. and Morrison, L.M. (1994). Cell Wall-Bound Oxidases from Tobacco *(Nicotiana tabacum)* Xylem Participate in Lignin Formation. *Planta.* **194:** 9-14.

Mukheljee, S.P. and Choudhuri, M.A. (1983). Implication of Water Stress-Induced Changes in the Levels of Endogenous Ascorbic Acid and Hydrogen Peroxide in *Vigna* Seedlings. *Physiol. Plant.* **58:** 166-170.

Noctor, G. and Foyer, C.H. (1998). Ascorbate and glutathione, keeping active oxygen under control. *Annu. Rev. Plant Physio. Plant Mole. Bio.* **49:** 249-279.

Oba, K., Ishikawa, S., Nishikawa, M., Mizuno, H. and Yamamoto, T. (1995). Purification and properties of L-galactono-γ-lactone dehydrogenase, a key enzyme for ascorbic acid biosynthesis, from sweet potato roots. *J. Biochem.* **117:** 120-124.

Otter, T. and Polle, A. (1994). The influence of apoplast ascorbate on the activity of cell wall-associated peroxydases and NADH oxydase on needles of Norway spruce (*Picea abies* L). *Plant Cell Physiol.* **35:** 1231-1238.

Paciolla, C., de Tullio, M., Chiappetta, A., Innocenti, A.M., Bitonti, V.B., Liso, R. and Arrigoni, O. (2001). Short- and Long-Term Effects of Dehydroascorbate in *Lupinus albus* and *Allium cepa* Roots, *Plant Cell Physiol.* **42:** 857-863.

Palma, J.M. (2006). Glutathione reductase from pea leaves: response to abiotic stress and characterization of the peroxisomal isozyme. *New Phytol.* **170:** 43-52.

Pedreira, J., Sanz, N., Pena, M.J., Sánchez, M., Queijeiro, E., Revilla, G. and Zarra, I. (2004). Role of apoplastic ascorbate and hydrogen peroxide in the control of cell growth in pine hypocotyls. *Plant and Cell Physiol.* **45:** 530-534.

Pignocchi, C. and Foyer, C.H. (2003). Apoplastic ascorbate metabolism and its role in the regulation of cell signalling. *Curr. Opin. Plant Bio.* **6:** 379-389.

Pinto, M. and De Gara, L. (2004). Changes in the ascorbate metabolism of apoplastic and symplastic spaces are associated with cell differentiation *J. Exp. Bot.* **55:** 2559-2569.

Pinto, M.C., Mata, A.M. and Lopez-Barea, (1985). The redox interconversion mechanism of *Saccharomyces cerevisiae* glutathione reductase. *Eur. J. Biochem.* **15:** 275-281.

Potters, G., Horemans, N., Bellone, S., Caubergs, R.J., Trost, P., Guisez, Y. and Asard, H. (2004). Dehydroascorbate influences the plant cell cycle through a glutathione-independent reduction mechanism. *Plant Physiol.* **134:** 1479-1487.

Rautenkranz, A.A.F., Li, L., Machler, F., Martinoia, A. and Oertli, J. (1994). Transport of Ascorbic and Dehydroascorbic Acids across Protoplast and Vacuole Membranes Isolated from Barley (*Hordeum vulgare* L. cv. Gerbel) Leaves. *Plant Physiol.* **106:** 87-193.

Romero-Puertas, M.C., Corpas, F.J., Sandalio, L.M., Leterrier, M., Rodriguez-Serrano, M., del Rio, L.A., Ros-Barceló, A., Gómez-Ros, L.V., Ferrer, M.A. and Hernández, J.A. (2006). The Apoplastic Antioxidant Enzymic System in the Wood-Forming Tissues of Trees. *Trees.* **20:** 145-156.

Sánchez, M., Pena, M.J., Revilla, G. and Zarra, L. (1996). Changes in Dehydrodiferulic Acid and Peroxidase Activity against Ferulic Acid Associated with Cell Walls during Growth of *Pinus pinaster* Hypocotyl. *Plant Physiol.* **111:** 941-946.

Sánchez, M., Quejieiro, E., Revilla, G. and Zarra, L. (1997). Changes in Ascorbic Acid Levels in Apoplastic Fluid during Growth of Pine Hypocotyls: Effect on Peroxidase Activity Associated with Cell Walls. *Physiol. Plant.* **101:** 815-820.

Savidge, R. and Udagama-Randeniya, P. (1992). Cell Wall Bound Coniferyl Alcohol Oxidase Associated with Lignification in Conifers. *Phytochem.* **31:** 2959-2966.

Schulze, E-D., Cermak, J., Matyssek, R., Penka, M., Zimmermann, R., Vasicek, F., Gries, W. and Kuèera, J. (1985). Canopy transpiration and water fluxes in the xylem of the trunk of Larix and Picea trees—a comparison of xylem flow, porometer and cuvette measurements. *Oecologia.* **66(4):** 475-483.

Siendones, E., González-Reyes, J.A., Santos-Ocaña, C., Navas, P. and Córdoba, F. (1999). Biosynthesis of Ascorbic Acid in Kidney Bean. L-Galactono-ã-Lactone Dehydrogenase is an Intrinsic Protein Located at the Mitochondrial Inner Membrane. *Plant Physiol.* **120:** 907-912.

Smirnoff, N. and Colombe, S.V. (1988). Drought Influence on the Activity of Enzymes of Chloroplast Hydrogen Peroxide Scavenging System, *J. Exp. Bot.* **39:** 1097-1108.

Smirnoff, N., Conklin, P.L. and Loewus, F.A. (2001). Biosythesis Ascorbic Acid in Plants: A Renaissance. *Annu. Rev. Plant Physiol. Plant Mol. Biol.* **52:** 437-467.

Smirnoff, N. (2001). L-ascorbic acid biosynthesis. *Vitam Horm.* **61:** 241-266.

Smirnoff, N. (1996). The function and metabolism of ascorbic acid in plants. *Ann. Bot.* **78:** 661-669.

Stassolla, C. and Yeung, E.C. (2006). Endogenous Ascorbic Acid Modulates Reactivation in White Spruce Somatic Embryos and Affects Thymidine and Urudine Metabolism. *Tree Physiol.* **26:** 1197-1206.

Stasova, V.V. (1991). The peculiarities of wall tracheid development during wood formation in Scots pine. Dissertation, Ph.D. Krasnoyrsk, pp. 174.

Sterjiades, R., Dean, L.E.D., Gamble, G., Himmelsbach, D.S. and Eriksson, K-EL. (1993). Extracellular Laccases and Peroxidases from Sycamore Maple *(Acer pseudoplatanus)* Cell-Suspension Cultures. *Planta.* **190:** 75-87.

Sudachkova, N.E., Kozhevnikova, N.N. and Lyubarskaya, T.G. (1980). Oxidoreductase of Near Cambial Zone in *Pinus sylvestris.* In: Problems of Forest Biology, Geology, and Cenology. Ed. Isaev, A.S. Novosibirsk: Nauka, pp. 182-190.

Takahama, U. (1993). Regulation of Peroxidase-Dependent Oxidation of Phenolics by Ascorbic Acid: Different Effects of Ascorbic Acid on the Oxidation of Coniferyl Alcohol by the Apoplastic Soluble and Cell Wall-Bound Peroxidases from Epicotyls of *Vigna angularis. Plant Cell Physiol.* **34:** 809-817.

Takahama, U. (1994). Changes Induced by Abscisic Acid and Light in the Redox State of Ascorbate in the Apoplast of Epicotyls of *Vigna angularis. Plant Cell Physiol.* **35:** 975-978.

Takahama, U. and Oniki, T. (1997). A peroxidase/phenolics/ascorbate system can scavenge hydrogen peroxide in plant cells. *Physiol. Plant.* **101:** 845-852.

Tommasi, F., Paciolla, C., De Pinto, M.C. and De Gara, L. (2001). A comparative study of glutathione and ascorbate metabolism during germination of *Pinus pinea* L. seeds. *J. Exp. Bot.* **52(361):** 1647-1654.

Van Rensburg, L. and Krüger, G.H.J. (1995). Evolution of Components of Oxidative Stress Metabolism for Use in Selection of Drought Tolerant Cultivars of *Nicotiana tabacum* L. *J. Plant Physiol.* **143:** 730-737.

Wheeler, G.L., Jones, M.A. and Smirnoff, N. (1998). The biosynthesis pathway of vitamin C in hither plants. *Nature.* **393:** 365-369.

Zahner, R., Lotan, J.E. and Baughman, W.D. (1964). Earlywood-latewood features of red pine grown under simulated drought and irrigation. *For. Sci.* **10(3):** 361-370.

Zahner, R. (1963). Internal Moisture Stress and Wood Formation in Conifers. *Forest Prod.* **13:** 240-247.

Zarra, I. (2004). Role of apoplastic ascorbate and hydrogen peroxide in the control of cell growth in pine hypocotyls. *Plant and Cell Physiol.* **45:** 530-534.

Zarra, I., Sánchez, M., Quejieiro, E., Péna, M.J. and Revilla, G. (1999). The Cell Wall Stiffening Mechanism in *Pinus pinaster* Aiton: Regulation by Apoplastic Levels of Ascorbate and Hydrogen Peroxide. *J. Sci. Food Agric.* **79:** 416-420.

16

Modulation of Glutathione, its Redox Couple and Related Enzymes in Plants Under Abiotic Stresses

Naser A. Anjum[1,*], Iqbal Ahmad[1,2], Mário Pacheco[2], Armando C. Duarte[1], Eduarda Pereira[1], Shahid Umar[3*], Altaf Ahmad[3], and Muhammad Iqbal[3]

[1]*Center for Environmental and Marine Studies (CESAM) & Department of Chemistry, University of Aveiro, 3810-193 Aveiro, Portugal*

[2]*Center for Environmental and Marine Studies (CESAM) & Department of Biology, University of Aveiro, 3810-193 Aveiro, Portugal*

[3]*Department of Botany, Faculty of Science, Hamdard University, New Delhi-110 062, India*

[*]***Corresponding authors**: anjum@ua.pt; s_umar9@hotmail.com*

ABSTRACT

Abiotic stresses are the primary cause of crop failure worldwide. Among major abiotic stresses, drought and salinity stresses are becoming more prevalent with increase in the intensity of agriculture. Whereas, the metal pollutants, derived from a growing number of diverse anthropogenic sources, have had enormous impact on different ecosystems including plant-soil interactions. The majority of these metals are consequently accumulating in agricultural soils causing global concerns because of their detrimental effects on plant-soil ecosystem, food safety issues and potential health risks. Moreover, these stress factors have been shown to largely accelerate the generation of reactive oxygen species (ROS) manifolds and lead to oxidative stress. Plants possess an efficient antioxidant defence system to combat ROS-induced oxidative stress. The plant antioxidant system comprises various antioxidant enzymes and metabolites. As a significant component of ascorbate-glutathione (AsA-GSH) pathway, reduced glutathione (GSH), a peptide containing three amino acids (Glu-Cys-Gly) with several GSH-related enzymes such as GSH reductase (GR, EC 1.6.4.2), GSH peroxidase (GPX, EC 1.11.1.19) and GSH sulphotransferases (GSTs, EC 2.5.1.18), is involved in both the direct and the indirect control of ROS concentrations in cells and thus, protects plants against drought-soil salinity and heavy metals-accrued oxidative stress. The current chapter critically reviews the recent studies on the modulation of total reduced GSH, GSH:GSSG redox couple, the major GSH-related enzymes and their cumulative significance in plant tolerance to drought-soil salinity and heavy metals stress.

Keywords: Glutathione; glutathione metabolism; drought stress; heavy metals; redox couple; salinity; oxidative stress; tolerance.

1. INTRODUCTION

Abiotic stress is the primary cause of crop failure worldwide, reducing average yields for most major crop plants by more than 50% (Boyer 1982; Bray et al., 2000). Drought and salinity are becoming particularly widespread in many regions, and may cause serious salinization of more than 50% of all arable lands by the year 2050. A large percentage of the world's crops are exposed to chronic or sporadic periods of drought and that affect virtually every aspect of growth, physiology and metabolism in plants (Shinozaki et al. 2002; Zhu 2002). Such stresses are becoming even more prevalent as the intensity of agriculture increases. Both the stress factors affect various aspects of plant life and thus prevent them from reaching their full genetic potential, thereby limiting the crop productivity worldwide (Mahajan and Tuteja 2005; Anjum et al. 2008a; Koyro et al. 2008; Khan et al. 2009). Whereas, the environmental pollution by metals has become extensive with rapid increase in the mining and industrial activities during the last two centuries. The metal pollutants, derived from a growing number of diverse anthropogenic sources, have had enormous impact on different ecosystems (Macfarlane and Burchett 2001). The majority of these metals are consequently accumulating in agricultural soils causing global concerns because of their detrimental effects on plant-soil ecosystem, food safety issues and potential health risks (McLaughlin et al. 1999; Macfarlane and Burchett 2001; Pinto et al. 2004; Yadav 2010).

In addition to different responses of plants to drought, salinity and heavy metal stresses, enhancement in the generation of reactive oxygen species (ROS) such as singlet oxygen ($^1O_2$), superoxide radicals ($O^{\bullet -}{}_2$), hydroxyl radicals ($OH^{\bullet}$), hydrogen peroxide (H_2O_2) and peroxide radicals in plants is the major response to these stress factors (Foyer and Noctor 2000; Mittler 2002; Mithofer et al. 2004; Anjum 2006). These species react with lipids, proteins, pigments and nucleic acids and cause lipid peroxidation and membrane damage (Foyer and Noctor 2000; Mittler 2002; Sairam and Tyagi 2004; Sairam et al. 2004). Therefore, these ROS must be eliminated if metabolic disorder and/or oxidative stress are to be avoided in plants. The ascorbate-glutathione (AsA-GSH) pathway is central to the efficient removal of ROS and for the maintenance of cellular homeostasis in plants under drought, salinity and heavy metals stress conditions (Noctor and Foyer 1998; Aravind and Prasad 2005; Anjum 2006; Anjum et al. 2010; Latowski et al. 2010; Mahmood et al. 2010; Sofo et al. 2010) (Fig. 1).

Glutathione (GSH, γ-glutamate-cysteine-glycine) is an important, abundant, ubiquitous and low molecular weight sulphur-containing tri-peptide non-protein thiol, and antioxidant metabolite in most plants (Bergman and Rennenberg 1993; Noctor and Foyer 1998; Anjum et al. 2010). GSH is widespread in bacteria, plants, and animals (Meister and Anderson 1983) and participates in some significant cellular activities including the protection of cells against toxic metals. GSH is present at millimolar concentrations in plant cells and participates in stress resistance and adaptation, such as scavenging of ROS generated during normal cell metabolism and induced by biotic and abiotic factors including drought and salinity (Noctor and Foyer 1998; Cobbett 2000). Plants tend to maintain high ratio of GSH to GSSG in cells but as a component of the AsA-GSH pathway, GSH reacts non-enzymatically with singlet oxygen, superoxide and hydroxyl radical and gets itself converted into its oxidized form (GSSG) (Noctor and Foyer 1998; Kuz´niak and Sklodowska 2001; Noctor et al. 2002; Szalai et al. 2009; Anjum et al. 2010, 2011). Nevertheless, a number of metabolic processes in plants such as higher glutathione reductase (GR) activity, increased GSH synthesis, decreased GSH degradation, or the transport of GSH and GSSG can largely help plants to restore the high GSH to GSSH ratio in cells (Noctor and Foyer 1998; Aravind and Prasad 2005). Further, there are several enzymes that directly and/or indirectly play critical roles in the removal of excess ROS in living organism. Among them, GR, glutathione peroxidase (GPX) and glutathione S-transferase (GST) are the most crucial enzymes in the cellular antioxidant system (Mittler 2002; Yan et al. 2008). Where, GPX is a seleno-enzyme that catalyzes the decomposition of H_2O_2 and other peroxides (e.g., lipid peroxides in cell membranes) using reduced glutathione (GSH) as substrate; GST catalyzes the reaction of GSH conjugation to a wide range of electrophilic metabolites of xenobiotics and (Coleman et al. 1997; Noctor et al. 1998). Therefore, GR, GPX and GST, in conjugation with superoxide dismutase (SOD) form a system to protect the plants against ROS-accrued effects during various stress conditions in plants (Halušková et al. 2009). There are plethora of reports that the GSH pool, its redox couple (GSH: GSSG) and related enzymes are modulated by both drought and salinity stress (Mittler 2002; Anjum et al. 2010, 2011; Latowski et al. 2010; Mahmood et al. 2010; Sofo et al. 2010).

2. GSH POOL AND GSH: GSSG REDOX COUPLE

An optimum level of GSH pool, required for normal metabolic activities is maintained in plants growing under normal conditions. There are reports that the redox state of the GSH pool in plants is remarkably constant but is very sensitive to extreme oxidative stress caused due to a number of stress factors including toxic metals, drought and salinity stress and during the process of non-enzymatic reaction of GSH with singlet oxygen, superoxide and hydroxyl radical GSH, the reduced form is converted to its oxidized form (GSSG) (Vanacker et al. 2000; Noctor et al. 2002; Szalai et al. 2009; Anjum et al. 2010). In fact, the change in the ratio of its reduced (GSH) to oxidized (GSSG) form during the degradation of ROS including H_2O_2 is important in certain redox signalling pathways (Millar et al., 2003). Many researchers including Millar et al. (2003) and Foyer and Noctor (2005) are of the opinion that the GSH/GSSG ratio is indicative of cellular redox balance and that it may be involved in cellular perception of ROS.

2.1 GSH Pool and GSH: GSSG Redox Couple in Response to Drought and Salinity Stress

Bai et al. (2009) examined the response of antioxidative system in the different phases of drought stress and during recovery in desert plant *Reaumuria soongorica*. The reduced GSH content in *R. soongorica* leaves had a decrease under drought stress before 16 days and then an increase. Reduced GSH content increased to 162% of its initial value in plants subjected to drought stress for 91 days. When plants that had been desiccated for 16 and 72 days were rehydrated, leaf reduced glutathione content recovered rapidly. Khan and Panda (2008) reported the changes in GSH level in roots of two *Oryza sativa*

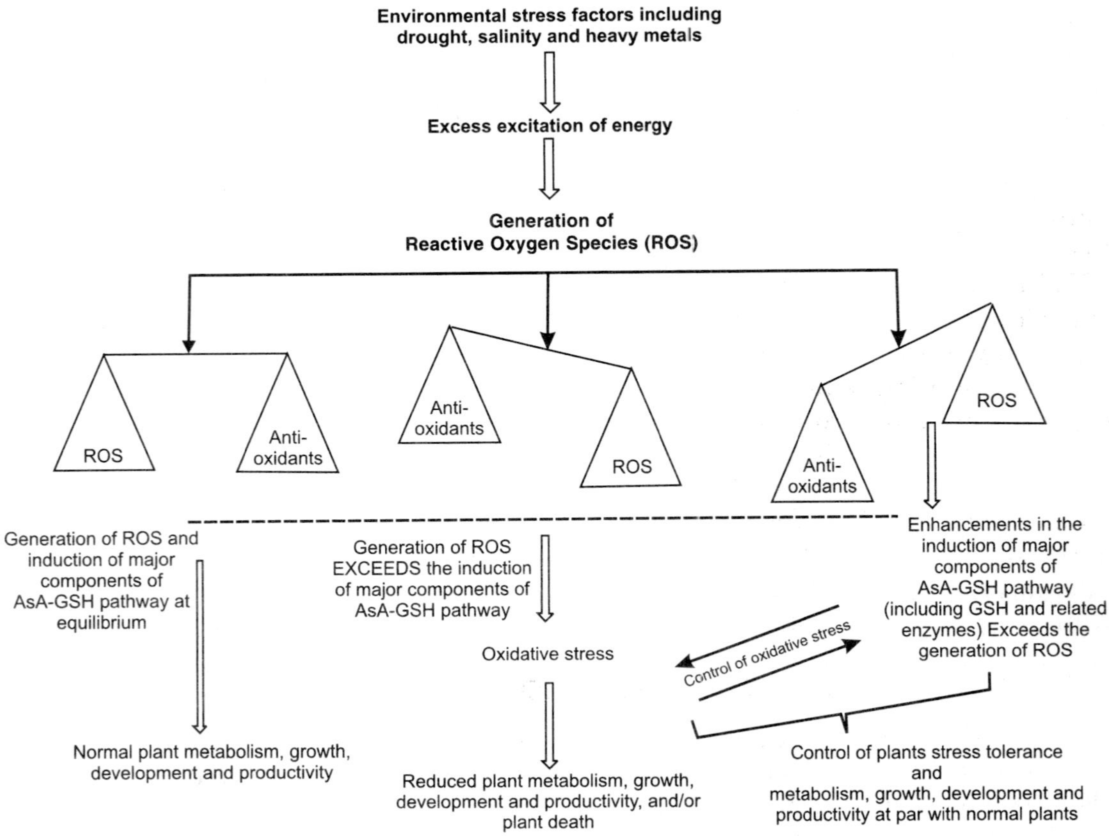

Fig. 1. Schematic representation of the cause of reactive oxygen species generation, major events of the balance between reactive oxygen species and antioxidants, and control of plant metabolism, growth, development and productivity. [AsA – ascorbate; GSH – reduced glutathione; ROS – reactive oxygen species]

cultivars Lunishree and Begunbitchi (differing in salt tolerance) exposed for 24 h to increasing (50, 100 and 150 mmol l^{-1}) concentrations of NaCl. Glutathione content significantly increased in Lunishree and decreased significantly in Begunbitchi. Recovered Lunishree roots showed significant increase in glutathione content with no significant change in Begunbitchi roots. Shehab et al. (2010) reported drought-induced increase in GSH content in *Oryza sativa* treated with drought stress. The increase of GSH content was gradually increased by increasing the time of PEG treatments. Wang et al. (2010) reported response of leaf GSH level in non-tolerant and tolerant *Oryza sativa* cultivars during grain-filling stage. Under water stress, leaf GSH content for each cultivar was found lower than their respective controls. The more tolerant cultivars exhibited higher GSH content compared to less tolerant cultivars. As the stress became more severe, the more drought cultivars had less reduction in GSH than the less tolerant cultivars did. The largest reduction in GSH for all cultivars occurred 21 days after heading with 3.0, 3.4 and 3.5% for Gangyou 527, Yixiangyou 9 and Gangyou 188, respectively. Sharma and Dubey (2005) reported changes in total GSH and GSH: GSSG ratio in *Oryza sativa* seedlings grown for 10 and 20 days and subjected to *in vitro* drought stress of -0.5 and -2.0 MPa for 24 h. Concentration of total glutathione (GSH + GSSG), reduced glutathione (GSH), oxidized glutathione (GSSG) and GSH/GSSG ratio increased in mild drought stressed seedlings compared to controls whereas a higher level of drought stress led to a significant decline in GSH concentration as well as total glutathione pool (GSH + GSSG) in 20 days grown seedlings. Twenty days old seedlings subjected to a higher level of drought stress showed between a 20 and 38% decline in GSH concentration as well as total glutathione pool (GSH + GSSG). Moradi and Ismail (2007) investigated GR activity in three *Oryza sativa* cultivars contrasting in tolerance (tolerant: IR651 and IR632; and sensitive: IR29) of salt stress (EC = 6 and 12 dSm^{-1}) during the vegetative and reproductive stages. GR activity increased in leaves of the tolerant lines with increasing salt stress, particularly in IR651, which showed a >3.5-fold increase in activity under 12 dS m^{-1}. However, GR activity in leaves of IR29 did not change under salt stress. GR activity also increased in both tolerant lines under salinity, with the greatest increase of about 120% in IR651, but showed no change in the sensitive cultivar.

Selote and Khanna-Chopra (2010) studied the changes in GSH and GSSG levels in *Triticum aestivum* seedlings of a drought-resistant cultivar (cv.) C306 subjected to severe water deficit directly or through stress cycles of increasing intensity with intermittent recovery periods. Water-stressed roots of A1 (mild intensity) plants showed no significant change in total cellular glutathione (GSH plus GSSG) and reduced glutathione (GSH) contents. Acclimated (A2) and nonacclimated (D) plants respectively exhibited 27% and 40% decline in total glutathione and 28% and 48% reduction in GSH content as compared to that of control plants during severe water stress conditions. However, acclimated plants (A2) maintained significantly higher GSH/GSSH ratios during severe stress condition when compared with non-acclimated plants (D). Rewatering led to an enhancement in the GSH pool in both the plants, but remained lower than control plants. Ma et al. (2010) studied the response of GSH level and GSH: GSSG ratio in the leaves of 2-year-old potted apple (*Malus domestica*) subjected to drought stress. The content of total glutathione at 6 days of drought stress was, however, not declined compared with that at 4 days, and re-watering did not lead to an increase in GSH content. Under salt stress, the content of GSH in roots and shoots of *Triticum aestivum* has been shown to increase (Meneguzzo et al., 1999). In salt tolerant potato, reduced glutathione significantly elevated, while in sensitive cultivar reduced glutathione content decreased by 3500 (Benavides et al. 2000). Comba et al. (1998) showed that in *Glycine max* root nodules, salt stress resulted in an increase in reduced glutathione content and it remained high after recovery from salt stress. In cucumber, salt stress increased the content of reduced glutathione (Lechno et al., 1997). The salt tolerant genotype of *Triticum aestivum* showed a higher ascorbic acid content when compared with the sensitive one under water deficit (Sairam et al. 1998). In pea, NaCl stress was accompanied by loss in glutathione pool (Hernandez et al. 1999). After 12 h of NaCl treatment of *Nicotiana plumbaginifolia* L., glutathione was found to be largely in oxidized form, suggesting an inability of the recycling process to cope with the stress (Savoure et al., 1999). Sofo et al. (2005) reported the influence of drought stress and rewatering on GSH and GSSG levels in four interspecific *Prunus* hybrids, namely, 'P3605' (*Prunus amygdalus* L. 'Garfi' x *P. persica* L. 'Nemared'), '8–9' (*P. cerasifera* L. 'P2980' x 'P3605'), '7–7' (*P. cerasifera* L. 'P2175' x *Prunus davidiana* L.) and '6–5' (*P. cerasifera* L. 'P2175' x *P. amygdalus* L. 'Garfi'). Drought stress caused marked increases in GSH content in P3605 and 8–9 and, in a lesser extent, 7–7 and 6–5. GSSG levels in P3605, 8–9 and 6–5 showed a slight increase during the progressive water shortage followed by a decrease in the last dates of the drought period, while in 7–7 the rise was more intense. In 7–7, the ratios of GSH to GSSG remained relatively stable during the drought period, whereas in P3605, 8–9 and, in a lesser extent, in 6–5, higher values were found during high levels of water deficit if compared with control plants. Hafsi et al. (2010) reported no significant change in the total glutathione (GSH + GSSG), GSSG, and GSH contents and the GSH/GSSG ratio in *Hordeum maritimum* due to all the treatments applied [(without (+K/-NaCl) or with 100 mM NaCl (+K/ +NaCl), or inK+-free medium containing 100 mMNaCl (-K/+NaCl)].

2.2 GSH Pool and GSH: GSSG Redox Couple in Response to Heavy Metal Stress

Cregnato et al. (2008) studied the changes in GSH pool in seedlings of *Avicennia marina* exposed to concentrations of 0, 50, 100, 200, 400 and 800 μg Zn ml^{-1}, for 96 h, 2 and 8 weeks. Authors reported that the levels of total GSH in leaf tissue did not vary significantly across the treatments, with the values ranging from 3.1 ± 0.3 to 15.6 ± 2.6 μmol g^{-1} wet weight. In addition, no significant dose-dependent relationships were observed between total (GSH + GSSG) glutathione and Zn concentrations in leaf tissue at any time. Cuypers et al. (2001) studied the changes in GSH pool, GSH: GSSG redox couple in roots and primary leaves of ten-day-old bean seedlings exposed to 50 μM Zn concentration on hydroponics. After application of 50 μM Zn to the roots, the oxidized form of the glutathione pool (GSSG: glutathione disulphide) was affected in these organs. Compared to the control plants, the GSSG content of roots was already enhanced 5 h after the start of the treatment. The total (GSH + GSSG) and the reduced glutathione (GSH) showed only limited fluctuations in the treated plants as compared to the controls. Elevation of the GSSG/GSH ratio was observed from 48 h on towards the end of the experiment; at 72 h, this effect was not significant. While, in the primary leaves a significant rise of the total glutathione pool was only observed 96 and 120 h after the application of 50 μM Zn to the nutrient solution. At 168 h, no significant effect could be observed. Similarly, the GSH level was enhanced at 96 and 120 h after Zn application. The amount of GSSG in the zinc treated plants was higher at 24 and beyond 72 h. From this latter moment on, the GSSG/GSH ratio was significantly enhanced until the end of the experiment.

Bruns et al. (2001) studied the influence of Cd (100 μmol l^{-1} Cd(II) stress on the glutathione pool in the water moss *Fontinalis antipyretica.* In Cd(II)-treated plants of *F. antipyretica,* the GSH level was significantly higher than in the controls. During a culture period of 10 days in the presence of 100 μmol l^{-1} Cd(II), the total GSH content of *F. antipyretica* rose from the second day after heavy metal supply. The GSH pool of *F. antipyretica* rise was not linear with the Cd(II) concentrations employed. Thus, a doubling of the Cd(II) concentration in the medium did not induce a GSH content twice as high.

If the moss was transferred into Cd(II)-free medium after 3 days of cultivation, the GSH pool increased further. In different experiments, the GSH content of *F. antipyretica* reached a maximum of between 250 and 500 μg g^{-1} fw, depending on the different samples of plant material used for the experiments and the native GSH level, which could depend on seasonal variations at the sampling site. In *F. antipyretica*, differences in the GSH content of young apical tips and older basal segments were obvious. In the young tips, the GSH content reached 4.6-fold of the GSH pool of older plant segments. In response to 100 μmol/L Cd(II), the GSH of young tips raised from 72 to 357 μg g^{-1} fw within 6 days, where the GSH synthesis the basal segments was significantly slower, with an increase from 46 to 78 μg g^{-1} fw during the same time. An enhanced GSH level during heavy metal exposure is only known from a few plants, but is always accompanied by a significant PC synthesis. In roots of maize seedlings, for example, the GSH content increased in the presence of Cu(II) during Cd(II) supply to about 45% and 270%, respectively (Tukendorf and Rauser 1990). The GSH pool of Cd(II)-tolerant tomato cells also rose 4.5-fold (Chen and Goldbrough 1994) or 6-fold, respectively (Gupta and Goldsbrough 1991). In Poaceae, hydroxymethyl- GSH (γ-glu-cys-ser) accumulated during heavy metal treatment up to 125% (Klapheck et al., 1994), but GSH was also synthesized and declined, whereas PC synthesis began. Similar results were described for the homo-GSH (γ-Glu-Cys-Ala) content of *Pisum sativum*, which increased up to 3-fold (Klapheck et al. 1995).

Dràzÿkiewicz et al. (2003) investigated the response of GSH level in *Arabidopsis thaliana* exposed to Cu concentrations (0, 5, 25, 50 and 100 μM) for 1, 3 and 7 days. GSH level in leaves of *A. thaliana* displayed a time dependent response to the presence of excess Cu in the growth medium. It was smaller than control (49-57% and 52-81% of control) in plants after 1 and 7 days of treatment, but exceeded control after 3-day treatment (173-249% of control). Changes in GSSG content were less visible. Considerable increase of this metabolite content was observed in plants exposed to 50 and 100 μM Cu for 3 days (138% of control). In plants exposed to Cu for 1 day, GSSG content was not quite significantly correlated with Cu concentration in the growth medium. Guo et al. (2007) reported Cd-induced increase in GSH level in roots of *Oryza sativa* exposed to 50 μM Cd for 2, 4 and 6 days. Cd exposure significantly increased the concentration of GSH, by 151% on day 4 and 120% on day 6 compared with control. Gupta et al. (2010) compared the significance of the link of GSH and phytocheltins to the detoxification of Pb stress in *Sedum alfredii* and concluded that the detoxification of Pb in *Sedum alfredii* is not related to the later but to the former. After exposure to increasing Pb concentration, GSH levels increased in both Pb-accumulating ecotypes (AE) and Pb-non-accumulating ecotypes (NAE). Maximum GSH level was recorded in the roots of AE of *S. alfredii*, i.e. 1.561 nmol kg^{-1} fresh weight, as compared to control. In case of shoots, GSH concentration was much higher upon addition of Pb concentration in NAE as compared to AE. Huang and Wang (2010) studied the influence of four different concentrations of Cd^{2+}, Pb^{2+} and Hg^{2+} on GSH level in the leaves of *Kandelia candel* and *Bruguiera gymnorrhiza* seedlings grown for 30 d in the nutrient solution. In *K. candel*, GSH content significantly increased under heavy metal stress as compared to the control except when irrigated with medium containing 0.1 mg l^{-1} Cd, and 1 mg l^{-1} Pb and 0.1 mg l^{-1} Hg. Heavy metals significantly enhanced the content of GSH in the leaves of *B. gymnorrhiza* at all heavy metal stress levels versus the control. In another study, Huang et al. (2010) investigated the effect of multiple heavy metal stress on GSH content in the leaves, stems and roots of *Kandelia candel* and *Bruguiera gymnorrhiza* treated with five different concentrations of a heavy metal mixture (Cd^{2+}, Pb^{2+} and Hg^{2+}). Authors reported a marked increase in GSH content of the two species was observed under heavy metal stress which was attributed to heavy metals-induced synthesis of GSH as (Mishra et al. 2006). In addition, the changes in the GSH/GSSG ratio in leaves of both the test plants were found more obvious than in the stems.

Rehman and Anjum (2010) investigated the significance of GSH in detoxification, multiple metal tolerance and biosorption of Cd in by *Candida tropicalis* isolated from industrial effluents. Cd treatment altered the levels of GSH and GSSG in *C. tropicalis*. Cd at a concentration of 100 mg l^{-1} significantly increased GSH level (135%) and GSH/GSSG ratio (3.5) with respect to the control. The level of non-protein thiols (NPT) also showed a significant increase (134%) in Cd-treated yeast cells. Cd produced an increase in GSH and NPT levels by 135% and 134% at 100 mg l^{-1} concentration, respectively. Authors concluded that Cd-accrued increase in GSH and NPT levels in *C. tropicalis* protected plant against Cd toxicity and helped plant to accumulate high Cd. Zhou et al. (2009) investigated the influence of Hg toxicity (10 μM Hg^{2+} for 24 h) on GSH level in roots of *Medicago sativa*. Authors reported a significant decrease in GSH content in *Medicago sativa* roots due to 10 μM Hg^{2+}. Lomonte et al. (2010) investigated the changes in GSH and GSSG levels in seedlings of *Atriplex codonocarpa* grown for 4 weeks in hydroponic culture containing 0, 0.05, 0.1 and 1 mg l^{-1} Hg. In general, values for total glutathione (GSH + GSSG), GSH and GSH/GSSG ratio were higher in shoots than in roots. Total glutathione decreased with increasing Hg concentration. The decrease was more pronounced in roots, e.g. at 1 mg l^{-1} Hg it was 93%. A similar trend was observed for GSH, as well, in both shoots and roots. The greatest decrease of GSH in roots was 98% at 1 mg l^{-1} Hg. The GSH/GSSG ratio in shoots decreased by about 30% in comparison with the control when *A. codonocarpa* seedlings were treated with 0.05 or 0.1 mg l^{-1} Hg, and it was unaffected at 1 mg l^{-1} Hg. Diwan et al. (2010) investigated the influence of Cr (0, 50, 100, 200 μM) on GSH level in *Brassica juncea* and *Vigna radiata*. Authors reported a higher content of GSH in *V. radiata* than *B. juncea*. Chromium treatments increased the GSH content in the *B. juncea*, compared to control. In *V. radiata*, on the contrary, GSH content decreased with increase in the Cr treatments. Zhou et al. (2008) investigated the changes in GSH level in *Medicago sativa* treated with different concentrations of $HgCl_2$ (0–40 μM) for 7 d. Authors reported a sharp increase in GSH accumulation due to $HgCl_2$ stress. Compared to control, authors noticed a significant difference of GSH accumulation between the treatments at the time point of day 4, indicating the sensitivity of GSH to Hg^{2+} toxicity. Iqbal et al. (2010) studied the changes in GSH and GSSG levels in *Brassica juncea* cultivars, namely, Pusa Jai Kisan and SS2 differing in Cd-tolerance treated with 0, 25 and 50 μmol l^{-1} Cd. With the increase in Cd concentration, both GSH and GSSG contents increased, but the increase in GSH content was higher in Pusa Jai Kisan. In contrast, GSSG content was higher in SS2 and was the maximum at 50 μmol l^{-1} Cd. Pusa Jai Kisan exhibited twofold higher increase in GSH content than SS2 at 50 μmol l^{-1} Cd compared to control. Total GSH content increased in both the cultivars but the increase was greater in Pusa Jai Kisan than in SS2.

Zhang and Ge (2008) investigated the response of GSH in rice seedlings exposed to four Cd concentrations (0, 10, 20 and 40 mg l^{-1}). Exposure to Cd resulted in a dose-dependent increase in cellular GSH content in both shoots and roots of rice, and the two tissues showed the very similar pattern in the response to Cd exposure. GSH content increased with Cd level and there was a significantly positive correlation between GSH content and Cd level. Szollosi et al. (2009) investigated the influence of different Cd concentrations (0, 50, 100 and 200 mg l^{-1}) on GSH level in seeds of *Brassica juncea* germinated in dark for 12, 24, 48 and 96 h. GSH accumulation occurred by increase in concentrations of Cd, which referred to the activity of the non-enzymatic antioxidant system. In control seeds, GSH level decreased with time, but this decrease was not significant. On the contrary, it increased in the treated seeds, particularly at 100 and 200 mg l^{-1} Cd concentrations. Analysis of correlation confirmed that if a stronger Cd stress takes longer time (24, 48 or 96 h), GSH level rises ($r = 0.62$, $r = 0.77$ and $r = 0.92$) and GSH becomes greater within creasing Cd content in seeds ($r = 0.59$). Moreover, after 24 h the value of GSH was 5-fold higher than at the beginning (12 h), later on (after 48 and 96 h) it decreased and the level was half as that at 24 h. Sun et al. (2005) studied the Zn- and Pb-induced changes in GSH level in *Sedum alfredii* collected from an old Pb/Zn mine site in Quzhou, Zhejiang Province (ZJ mine plant), and a clean site in Guangzhou, Guangdong Province (GD control plant), China, respectively. After exposure to increasing Zn, the concentrations of GSH significantly increased in these plants, whereas a marked increase was observed in each of the three tissue parts of the ZJ mine plants. The increase of GSH was almost linearly consistent with Zn accumulation in leaf, stem and root of the of the ZJ mine plant. However, this relationship was weaker in different tissue parts of the GD control plants, except the stem. Lead increased the concentrations of GSH in each of the three tissues of the ZJ mine plants in the descending order of stem > root > leaf. Conversely, the concentrations of GSH showed a slight increase in stem and leaf tissues in the control plants, though significantly increased in the roots. A mild chlorosis of leaf tissue was observed in the ZJ plants at 900 and 1500 μM Pb.

Schroder et al. (2009) investigated the changes in GSH level in three different species: *Phragmites australis*, *Typha latifolia*, and a heterotrophic cell culture of *Picea abies* exposed to 10 - 500 μM $CdSO_4$. Application of 150 μM $CdSO_4$ resulted in a reduction in GSH and GSSG of 75% and 96%, respectively. When a heavy metal mixture containing 80 μM Na_2HAsO_4, 150 μM $CdSO_4$ and 150 μM $PbCl_2$ was applied to the cultures, both GSH and GSSG levels declined by 60% and 99%, respectively. Incubation with 80 μM arsenic alone doubled GSSG values. Incubation with elicitor resulted in the doubling of GSH values within 16 h to 933 nmol g fresh weight^{-1}, but GSSG concentrations remained unchanged. On the contrary, incubation with 117 mM H_2O_2 decreased GSH levels by 30% and doubled GSSG contents. Russo et al. (2008) studied the response of GSH (GSSG and GSH) levels to Cu (100 μM $CuSO_4$) stress in *Brassica nupus*. Control roots had higher levels of both total glutathione (GSSG + GSH) and GSH in comparison with leaves, whereas GSSG had the same percentage. In roots, GSH + GSSG and GSH contents decreased till 10 min of the treatment reaching a transient peak at 15 min and decreasing again thereafter. The percentage of GSSG increased continuously reaching 42% at the end of the experiment. In leaves, total glutathione and GSH showed at 5 min of the treatment increases by 24 and 31%, respectively, in comparison with the control. In comparison with the control, both GSH + GSSG and GSH increased in the stroma after 10 min and reached a double value at 30 min. After 30 min both GSH + GSSG and GSH decreased maintaining anyway values higher by 47 and 55%, respectively in comparison with the control. GR specific activity was in the stroma 10-fold and 100-fold lower than in the leaves and roots, respectively, and the percentage of GSSG in the control stroma resulted twofold higher than in control leaves and roots. Notwithstanding the low GR activity in the stroma, its increase after 10 min of treatment was sufficient to reduce the percentage of GSSG to a value lower than in the roots at the end of the experiment.

Israr et al. (2006a) reported changes in GSH level in seedlings of *Sesbania drummondii* treated with different concentrations (10, 20, 30, 40, 50 and 100 mg l^{-1}) of $HgCl_2$. GSH content as well as GSH/GSSG ratio were increased with increasing Hg concentration in the growth solution. Mercury at a concentration of 50 mg l^{-1} significantly increased GSH level (128.8%) and GSH/GSSG ratio (37.2%), with respect to the control. A slight reduction in GSH as well as in GSH/GSSG ratio was observed at a concentration of 100 mg l^{-1} Hg. In another study, Israr et al. (2006b) investigated the influence of Cd (0-250 μM) on GSH and GSSG in the *Sesbania drummondii* callus. Authors reported that Cd treatments altered the levels of GSH and GSSG level in *Sesbania* callus. Cd at a concentration of 50 μM significantly increased GSH level (94.7%) and GSH/GSSG ratio (36.7%), with respect to the control. A slight reduction in GSH as well as in GSH/GSSG ratio was observed at a concentration of 100 μM Cd. Israr and Sahi (2006) studied changes in GSH and GSSH level in the cell cultures of *Sesbania drummondii* treated with different concentrations of Hg (0-250 μM). Authors reported that the Hg treatments altered the levels of reduced (GSH) and oxidized (GSSG) glutathione in *Sesbania* cells and noticed a direct correlation between the level of GSH as well as GSH/GSSG ratio and Hg concentration (up to 40 μM) in the growth medium. At 40 μM Hg concentration GSH level and GSH/GSSG ratio significantly increased to 93.5% and 41.0%, respectively, with respect to the control. Though, authors reported a dramatic decline in GSH level as well as in GSH/GSSG ratio at a concentration of 50 μM Hg.

Malecka et al. (2009) studied the influence of different concentrations of $Pb(NO_3)_2$ (0.1 and 0.5 mM added to Hogland solution) on changes in GSSG and GSH levels in roots of *Pisum sativum* harvested after 0, 2, 8, 24, 48, 72 and 96 h of cultivation. It was observed that the level of total glutathione (GSH and GSSH) in *Pisum sativum* roots decreased during the first 8 h of $Pb(NO_3)_2$ influence (0.1 and 0.5 mM). At the same time, authors reported an increase in the oxidative stress conditions estimated in terms of increased level of GSSG in the *P. sativum* root cells cultivated in the presence of Pb. However, after 24 h of *P. sativum* exposition to Pb, gradual increase in level of GSH in plants and simultaneous decrease of GSSG level were observed for both concentrations of the metal. In addition, they found that the level of total glutathione (GSH and GSSH) for the *P. sativum* root exposed to 0.5 mM $Pb(NO_3)_2$ was by several times higher than for the control plants. They noted a decrease of total glutathione concentration (GSH and

GSSH) by 22.1% after 96 h and speculated due to inhibitory influence of large amount of reduced form of glutathione (GSSG) on the activity of γ-EC synthetase. During the 96 h exposition of *P. sativum* roots to 0.1 and 0.5 mM Pb, authors noticed a 6-fold and 23-fold increase of GSSG in comparison to the control plants. The most significant decrease of GSH/GSSG ratio was observed during the first 24 h for both Pb concentrations. In the next 3 days of cultivation, the GSH/GSSG ratio raised gradually. Lin et al. (2007) studied the effects of soil-Cd (0, 0.3, 1.0, 3.3, 10, 33 mg Cd^{2+} kg^{-1} dry soil,) on growth, oxidative stress and antioxidant system in *Triticum aestivum* seedlings. The GSH content in leaves continuously decreased with increasing Cd treatments up to 3.3 mg kg^{-1}, which caused a 43% decrease in GSH compared to the control. However, the highest Cd treatment (33 mg kg^{-1}) resulted in a 44% GSH increase. Cd treatment increased the GSSG content of leaves, but significant variance was not observed until the concentration of 10 mg kg^{-1} was reached. The total glutathione content (GSH + GSSG) did not change much at low Cd concentrations, but significantly increased at higher concentrations. The ratio of GSH to GSSG (GSH/GSSG) decreased first, followed by a sharp increase with increasing Cd concentrations.

Kolb et al. (2010) investigated the influence of Cd (50 μM $CdCl_2$) on changes in subcellular GSH contents within glandular trichomes of *Cucurbita pepo*. Cd treatment significantly reduced immunogold labelling of GSH in all cell compartments of mesophyll cells and glandular trichomes. In mesophyll cells, GSH levels were decreased after Cd treatment of about 43% in plastids, 38% in the cytosol, 34% in mitochondria, and 30% in nuclei. In long- and short-stalked glandular trichome cells a much stronger decrease of GSH was observed in nuclei (76% and 70%) and in the cytosol (74% and 76%) in comparison to mesophyll cells whereas the decrease in GSH contents was similar to mesophyll cells in plastids (33% and 45%) and mitochondria (25% and 27%). Interestingly, the ratio of GSH between organelles was strongly changed in glandular trichome cells but not in mesophyll cells. In long- and short-stalked glandular trichomes the GSH pool decreased in the cytosol from 7% to 2.8% and in nuclei from about 11–12% to about 5% of the total GSH pool found in glandular trichomes. Authors also observed that the ratio of GSH increased in mitochondria which contained between 72% and 74% of the total GSH before Cd treatment and about 83.5% to 84.9% of the total GSH after Cd treatment within long- and short-stalked glandular trichome cell. No significant change in GSH pool was observed in plastids. Taking together the above results, authors concluded that Cd exposure led to the withdrawal of a massive amount of GSH from the nuclei in order to assuage the higher demand of GSH in the cytosol for the detoxification of Cd. In addition, it was also suggested that low contents of GSH may limit the operation of the AsA-GSH cycle in the detoxification of Cd in the long term, which could lead to oxidative stress, lipid peroxidation, and, eventually, to plant death. Qadir et al. (2004) reported an increase in GSH content in *B. juncea* cultivars, namely, Vardhan (V_1), Pusa Bahar (V_2), Pusa Bold (V_3), BTO (V_4), Pusa Jai Kisan (V_5), Agrini (V_6), Varuna (V_7), Kranti (V_8), Vaibhav (V_9) and Pusa Basant (V_{10}) exposed to different concentrations of Cd (T_1, 0.5 mM; T_2, 1.0 mM; T_3, 1.5 mM and T_4, 2.0 mM) harvested 24, 48 and 72 h after the treatment (HAT). Authors noticed maximum increase in GSH in V_5 (26–105%) among all the Cd-treated plants during the entire experimental period, while minimum in V_8 (7–65%), over their respective controls. Wu et al. (2004) reported changes in GSH level in plant parts of two genotypes of *Hordeum vulgare*, namely, ZAU 3, a relatively tolerant genotype and Wumaoliuling, a relatively sensitive one exposed 0.5 and 5 μM Cd. Authors observed a Cd-dose-dependent changes in GSH concentration. During the first 5 d exposure, GSH concentration in roots increased in 0.5 μM Cd, but with significant genotypic difference, compared with control. GSH concentration decreased significantly in 5 μM Cd without significant genotypic difference. Authors noticed no significant differences between Cd-treated plants and the control. After 10 d Cd exposure, GSH concentrations in leaves, stems and roots decreased with increasing Cd level in the culture medium. The inhibition effect of Cd on GSH concentration was found as follows: root > stem > leaf. However, authors reported a significant difference between genotypes in Cd-induced reduction of GSH concentration. Wumaoliuling was more severely affected than ZAU 3; at 5 μM Cd treatment, GSH concentration was reduced by 24.3% in stems of ZAU 3 relative to control, while corresponding reduction for Wumaoliuling was 38.2%.

Anjum et al. (2008b) studied the influence of Cd levels (0, 25, 50 and 100 mg Cd kg^{-1} soil) on GSH pool in *Brassica campestris*. With the increasing soil-Cd level the GSH content was found decreased significantly at 100 mg Cd kg^{-1} soil (40.57%) followed by 50 mg Cd kg^{-1} soil (32.17%) and 25 mg Cd kg^{-1} soil (22.60%). In another, study, Anjum et al. (2008c) reported Cd-induced decrease in GSH content maximally in PS-16 and Pusa Vishal (18–37% and 14–30%) and minimum in Pusa 9531 and Pusa 9072 (7–20% and 9–26%) cultivars of *Vigna radiata* exposed to 25, 50, and 100 mg Cd kg^{-1} soil. Recently, Anjum et al. (2011) investigated the influence of Cd (0, 25, 50 and 100 mg kg^{-1} soil) on GSH and GSSG:GSH pools in Pusa and PS 16 cultivars of *Vigna radiata*. Cd at 100 mg kg^{-1} soil resulted in 41.1%, 20.8% and 65.6% decreases in GSH (reduced), total GSH (GSH and GSSH) and GSH: GSSG ratio, respectively in cv. Pusa 9531 as compared to the control, but in cv. PS 16 decreases in these parameters were 57.3%, 11.7% and 76.9%, respectively. Singh et al. (2006) investigated the metabolic adaptations to As-induced oxidative stress in *Pteris vittata* and *Pteris ensiformis*. Authors reported a substantial increase in the reduced glutathione (GSH) upon As exposure at 133 mM in *P. vittata*. Whereas, in *P. ensiformis*, it decreased significantly except for 1 d. Unlike GSH, upon As exposure, GSSG increased significantly in both plants though the magnitude of increase was greater in *P. vittata* than in *P. ensiformis*. Similar to the ratios of AsA/DAsA, the ratios of GSH/GSSG were greater in *P. vittata* in comparison to *P. ensiformis* in all treatments. In the absence of As, the concentrations GSH and GSSG in *P. vittata* were 119–291 and 67–105% greater than those in *P. ensiformis*. Shi et al. (2005) studied the Mn (10 and 600 μM $MnSO_4$)-induced changes in GSH level in *Cucumis sativus*. Excess Mn considerably decreased the concentrations of GSH and total glutathione. Excess Mn increased GSSG concentration. The ratio of GSH/GSSG was significantly reduced by excess Mn. Nagalakshmi and Prasad (2001) investigated the responses of GSH cycle enzymes and GSH metabolism to Cu stress (0, 50, 100 and 200 μM concentrations of $CuSO_4$ for 3 days) in *Scenedesmus bijugatus*. GSH content in the cells after 24, 48 and 72 h of copper treatment decreased progressively in all the concentrations tested. The decrease in GSH content was drastic in the first 24 h of copper treatment and gradual after 24 h.

Authors speculated that Cu-accrued decrease in GSH content may be due to enhanced utilization of GSH and decreased rate of synthesis.

3. GLUTATHIONE REDUCTASE

Glutathione reductase (GR, EC 1.6.4.2, NADPH: oxidized-glutathione oxidoreductase) belongs to a family of flavin-containing pyridine nucleotide-disulphide oxidoreductases. The major function of these homologous dimeric proteins is to catalyze the conversion of GSSG to GSH using NADPH as a coenzyme (Scruton et al. 1990; Rice-Evans et al. 1996). GR is one of the key components of the cell's scavenging system for ROS in plants. It mediates the reduction of GSSG to GSH by using NADPH as an electron donor, and thus maintains the cellular redox levels through maintaining a highly reduced state of GSH/GSSG and ASA/DHA ratios at the intracellular level during oxidative stress. A number of biotic and abiotic stress factors including toxic metals, drought and salinity stresses have been shown to affect the activity of GR (Anjum et al., 2010, 2011).

3.1 Glutathione Reductase in Response to Drought and Salinity Stress

Ahmad et al. (2010) reported the influence of a range of NaCl concentrations: 0 (control), 50, 100, and 150 mM NaCl; the electrical conductivity (EC) of soil saturation extract was 1.9, 2.3, 4.5, and 7.4 mS/cm, respectively, on GR activity in three-month-old *Morus alba* plants. Salt stress caused a significant increase in the levels of GR activity in both mulberry cultivars at all stress levels during 10th and 20th days of stress. The degree of accumulation was more in cv. Local (140%) than cv. Sujanpuri (115%) after 10 days, and after 20 days of stress the increase was 180 and 150% in cv. Local and Sujanpuri, respectively, at 150 mM NaCl. Kim et al. (2005) studied the responses of antioxidant enzymes including GR in shoot and root of *Hordeum vulgare* plants exposed to 200 mM NaCl for 1, 2 and 5 days. In general, the activities of antioxidant enzymes were increased in the root and shoot under saline stress. But the increase was more significant and consistent in the root. The activity GR was increased significantly in the root within 1 day after treatment, and various elevated levels were maintained by 5 days after treatment. The significant increase in the activity GR in the NaCl-stressed *H. vulgare* root was highly correlated with the increased expression of the constitutive isoforms as well as the induced ones. Desingh and Kangaraj (2007) studied GR activity in two *Gossypium hirsutum* varieties (Arya-Anubam and LRA-5166) treated with 0 mM (control), 50 mM, 100 mM, and 150 mM NaCl. They observed a progressive increase in both *G. hirsutum* varieties at all salt concentrations applied. However, GR activity was markedly higher in var. Arya-Anubam than in var. LRA-5166 at all salinity levels. Variety Arya-Anubam showed a significantly increased activity of GR (99.47 μmol mg^{-1} protein min^{-1}) at 150 mM salt concentration when compared to control plants (52.19 μmol mg^{-1} protein min^{-1}). Agarwal and Shaheen (2007) investigated the antioxidant response system and oxidative stress in leaves from bitter gourd (*Momordica charantia*) subjected to NaCl and water stress at three different stages of plant growth: pre-flowering, flowering and post-flowering. Activity of GR increased under NaCl stress and the maximum activity was evident at the flowering stage. Under water stress the GR activity was reduced below the control values. Sharma and Dubey (2005) reported changes in GR activity in *Oryza sativa* seedlings grown for 10 and 20 days and subjected to *in vitro* drought stress of -0.5 and -2.0 MPa for 24 h. GR showed significant increase in activity with the imposition of drought stress. Around a 32 to 102% increase in GR activities were noted in 20 days old mild drought stressed seedlings compared to control grown seedlings. Bai et al. (2009) examined the response of antioxidative system in the different phases of drought stress and during recovery in desert plant *Reaumuria soongorica*. GR activity declined during 16 days drought stress and elevated with further progressing drought stress. When plants subjected to drought stress for 16 days were re-watered, GR activity recovered rapidly, and even under the initial values. Sairam et al. (2005) reported differences in antioxidant activity in response to salinity stress (100 and 200 mM NaCl; ECe = 6.85 and 12.3 dS m^{-1}) in tolerant (Kharchia 65, KRL 19) and susceptible (HD 2009, HD 2687) *Triticum aestivum* genotypes. GR activity was highest in Kharchia 65 and increased up to 200 mM NaCl. KRL 19, HD 2687 and HD 2009 showed increase in GR activity only at 100 mM NaCl and decline at 200 mM NaCl. In HD 2687, the GR activity at 200 mM NaCl decreased even below that in controls. Jung (2004) reported changes in GR activity in young and mature leaves of 4-week-old *Arabidopsis* plants exposed to drought stress up to -2.54 MPa of water potential (Ψw). The two GSSG-specific GR bands were present in Arabidopsis and they increased in drought-stressed mature leaves. After 24 h of rewatering, the staining activity of GR in mature leaves decreased slightly, which was still higher than that of well-watered controls. Terzi and Kadioglu (2006) studied the response of GR activity in the leaf, petiole and root of *Ctenanthe setosa* during drought stress-induced leaf rolling. Most changes in glutathione reductase activity in leaf and petiole versus the control were not statistically significant. It was 0.093 U per mg protein in control leaves and 0.117 U per mg protein in 32-day-drought leaves; GR activity decreased on day 48 of drought, but not significantly. In root, GR activity increased on day 32 of drought, again not significantly. Later increases of GR activity in root were statistically significant.

Demiral and Türkan (2005) studied the changes in GR activity in roots of two *Oryza sativa* cultivars subjected to three salt treatments, 0, 60, and 120 mol l^{-1} NaCl for 7 days. GR activity significantly decreased in roots of both cultivars with increasing salinity. At both 60 and 120 mol l^{-1} NaCl, salt-tolerant Pokkali showed a lesser decline in GR activity than in salt-sensitive IR-28. Yazici et al. (2007) investigated the response of GR activity in 2-month-old purslane seedlings subjected to 0, 70, 140 mM NaCl concentrations. Constitutive GR activity in control groups decreased during the experimental period. GR activity in purslane plants under 140 mM NaCl was 3.5 times higher than in control group whereas 70 mM NaCl did not cause significant change in its activity on day 18. After 30 days, 70 and 140 mM NaCl treatments caused 49% and 97% increments in GR activity. Selote and Khanna-Chopra (2010) studied the changes in GR activity in *Triticum aestivum* seedlings of a drought-resistant cv. C306 subjected to severe water deficit directly or through stress cycles of increasing intensity with intermittent recovery periods. Water stress of mild intensity (A1) and acclimated (A2) plants exhibited 2.5-fold and threefold higher total GR activity, whereas non-acclimated (D) plants observed a reduction when compared to that of control plants during water stress conditions. Both A1 and A2 plants maintained higher GR activity during rewatering, while D plants showed a 1.2-fold increase in the GR activity upon rewatering. Kim et al. (2005) studied the response of GR activity in barley seedlings treated with 200 mM NaCl for 0, 1, 2, 5 days, respectively. GR activity was only increased significantly in the root in response to the NaCl treatment. The activity was increased by about 47% at 1 day after treatment (DAT), and was maintained at the increased levels until 5 DAT. The activity in the root was about three-times higher than that in the shoot. Pérez-Lo´pez et al. (2009) reported the changes in antioxidant metabolism because of the effect of salinity stress (0, 80, 160 or 240 mMNaCl) on protective enzyme activities under ambient (350 μmol mol^{-1}) and elevated (700 μmol mol^{-1}) CO_2 concentrations in two barley cultivars (*Hordeum vulgare* L., cvs Alpha and Iranis). The activity of GR in non-salinized plants was higher in Alpha than in Iranis. High CO_2 alone caused a decrease of GR activity only in Alpha. After subjecting ambient CO_2-grown Alpha plants to salinity treatment, total enzyme activity increased by 8%, 9% and 16% at 80, 160 and 240 mM NaCl, respectively, whereas in Iranis, it increased by 11%, 15% and 32%, respectively. So the relative rise in GR activity was more pronounced in Iranis than in Alpha. Exposure of barley leaves to salinity and elevated CO_2 reduced the rise of GR activity in Alpha but not in Iranis, which exhibited a similar increase. In Alpha, GR activity was always higher under ambient CO_2, whereas no significant differences between CO_2 treatments were observed in Iranis. Hafsi et al. (2010) reported an increase in GR activity in *Hordeum maritimum* due to all the treatments applied [(without (+K/-NaCl) or with 100 mM NaCl (+K/+NaCl), or inK+-free medium containing 100 mM NaCl (-K/+NaCl)], the highest increase being observed -K/+NaCl-treated plants (+271% in comparison with the control). Turhan et al. (2008) investigated the activity of antioxidative enzymes in three strawberry cultivars, namely 'Camarosa', 'Tioga,' and 'Chandler' irrigated with nutrient solution containing 0, 8.5, 17.0, and 34.0 mM sodium chloride (NaCl) for 30 days. In three of the cultivars, the activity of GR remained almost unchanged in both 15 and 30 days of NaCl treatments while only that of GR sharply increased in 17 mM NaCl treatment and remained almost the same level in 34.0 mM NaCl treatment for 30 days in Camarosa. Results of variance analysis indicated that the effects of all the independent variables and their interactions on the GR activity were significant.

Yasar et al. (2008) reported salt stress-induced changes in antioxidant enzymes including GR activity in leaves of a green bean genotype Gevas sirsk 57 (GS57) and cv. Fransiz 4F-89 differing in salt tolerance subjected to three salt treatments (0, 50, and 100 mM NaCl) under controlled climatic conditions for 7 days. High NaCl concentration (100 mM) increased GR activity significantly, but low NaCl concentration (50 mM) caused reduced GR activity in the leaves of GS57 genotype as compared to control plants. On the other hand, cv. 4F-89 exhibited a significant decrease in GR activity with increasing NaCl treatment. Khan and Panda (2008) reported the changes in GR activity in roots of two *Oryza sativa* cultivars Lunishree and Begunbitchi (differing in salt tolerance) exposed for 24 h to increasing (50, 100 and 150 m mol l^{-1}) concentrations of NaCl. GR activity in response to increasing (50, 100 and 150 m mol l^{-1}) NaCl concentrations decreased significantly in Lunishree as well as in Begunbitchi. Recovered roots showed similar results with significant decrease in GR activity against the control values and with the increasing NaCl concentrations. Bian and Jiang (2009) studied the response of GR activity in *Poa pratensis* subjected to soil drying for 5 d and then rewatered for 1 d in growth chambers. Leaf GR activities were not affected by all treatments but root GR activities increased 38% under drought stress, compared to the control. The GR activities of the roots returned back to the control level after recovery. Esfandiari et al. (2007) reported the response of GR activity in *Triticum aestivum* seedlings subjected to 50, 100, 150 and 200 mM NaCl for 10 days. GR activity started to decrease with the rise of NaCl content in comparison with the control in Alvand. However, at 150 and 200 mM, NaCl showed significant difference with the control. Moreover, GR activity fallen at 50 and 100 mM NaCl was not significant relative to control. GR activity had the highest activity at 50 mM NaCl in Sardari. But at 100, 150 and 200 mM, GR activity decreased in comparison with the control and 50 mM NaCl. Furthermore, there is a significant difference between 50 mM with 150 and 200 mM NaCl. Ma et al. (2010) studied the response of GR activity in the leaves of 2-year-old potted apple (*Malus domestica*) subjected to drought stress. GR activity exhibited an obvious increase under the drought treatment compared with control, reached the highest values at 4 days of drought treatment, and decreased at 6 days. Re-watering was accompanied by a further decrease of GR activity.

Meloni et al. (2003) investigated the response of GR activity in two cotton cultivars, Guazuncho and Pora (hybrids between *Gossypium hirsutum* x/*G. arboretum* x/*G. raimondii*) treated with three salt concentrations (50, 100 and 200 mol l^{-1} NaCl) for 21 days. GR activity in NaCl stressed plants was higher than in control plants but only in Pora. The average increase in GR activity was about 53% and it did not increase with NaCl increase in concentration. In Guazuncho, no significant increase in GR activity was observed. Seckin et al. (2010) reported the influence of salinity stress (0, 150 and 300 mM NaCl stress for 7 days) on GR activity in *Hordeum marinum* and *H. vulgare*. In *H. marinum*, GR activity increased significantly under both NaCl concentrations. However, GR activity in *H. vulgare* increased only at 150 mM NaCl. However, the induced level of GR activity was higher in *H. marinum* at 150 mM NaCl as compared to *H. vulgare*. Five and six GR isoenzymes were detected in the leaves of *H. vulgare* and *H. marinum*, respectively, with GR6 being the major one. Vaidyanathan et al. (2003) reported a continous increase in GR activity with increasing levels of NaCl stress (100–300 mM, for 42 h) in two *Oryza sativa* cultivars, salt sensitive Pusa Basmati 1 (PB) and salt-tolerant Pokkali (PK). Anjum et al. (2008a) investigated the response of GR activity in two genotypes (Pusa 9531 and PS 16) of *Vigna radiata* (differing in photosynthetic capacity) grown for 30 days in earthen pots at three field capacities (FC, 100, 75 and 50%). The field capacity-dependent-response of GR activity was noted in both low and high photosynthetic capacity genotypes. GR activity increased by 80.25% and 68% at 50% FC in Pusa 9531 and PS 16, respectively. When grown at 75% FC, GR activity increased by 31.48% and 20% in Pusa 9531 and PS 16, respectively.

3.2 Glutathione Reductase in Response to Heavy Metal Stress

Zhou et al. (2009) investigated the influence of Hg toxicity (10 μM Hg^{2+} for 24 h) on GR activity in roots of *Medicago sativa*. Authors reported a strong inhibition of GR activity in *M. sativa* roots due to 10 μM Hg^{2+} and suggested that the activity of GR is sensitive to

Hg exposure. Huang et al. (2010) investigated the effect of multiple heavy metal stress on GR in the leaves, stems and roots of *Kandelia candel* and *Bruguiera gymnorrhiza* treated with treated with five different concentrations of a heavy metal mixture (Cd^{2+}, Pb^{2+} and Hg^{2+}). The activity of GR initially increased but subsequently decreased in response to heavy metal stress in the both the test plants. At the highest metal concentration, a significant decrease in GR activity was observed in *K. candel* and *B. gymnorrhiza*. Cuypers et al. (2001) studied the changes in GR activity in roots and primary leaves of 10-day-old bean seedlings exposed to 50 μM Zn concentration on hydroponics. In roots, by 24 h after zinc application, the GR capacity was significantly decreased. Compared to the controls, a GR capacity of less than 40% of the control value was maintained from 96 h after the start of the treatment until the end of the experiment. While in leaves, the GR capacity was unaffected until 120 h after transfer of the plants to the Zn-containing nutrient solution. Only at 168 h was a significant increase observed in the GR capacity of the treated plants compared to the controls. Dràzÿkiewicz et al. (2003) investigated the response of GR activity in *Arabidopsis thaliana* exposed to Cu concentrations (0, 5, 25, 50 and 100 μM) for 1, 3 and 7 days. After 1-day-exposition of plants to Cu, GR activity decreased to 60% of control at 25 μM Cu. After 3-day-exposition GR activity enhanced to 150-165% of control. In *A. thaliana* exposed to Cu for 7 days, GR activity did not differ from control. Ekmekci et al. (2009) investigated the changes in GR activity in seedlings of *Zea mays* cultivars, viz., Vero and 3223 when exposed to Pb stress (2, 5 and 8 mM $Pb(NO_3)_2 \cdot 4H_2O$). The concentration of 2 mM Pb induced a slight increase in GR activity in Vero. At 5 mM Pb concentration, Vero showed a significant increase in the GR activity relative to the control. No significant deviation from the control level was observed in 3223 cultivar at 2 and 5 mM Pb concentrations. At highly toxic Pb level, Pb induced a significant increase in GR activity for both cultivars compared with their controls. The increase of GR activity was higher in Vero than in 3223. In addition, Pb has also been reported to cause an increase in GR activity in *Zea mays* cali and *Oryza sativa* by Verma and Dubey (2003). They reported that the activity of GR appears to play a pivotal role in combating Pb-induced oxidative injury in rice plants. Increased GR activity under Pb toxicity helps in recycling oxidized glutathione (GSSG) to reduced glutathione (GSH) to maintain the ratio of oxidized glutathione (GSH/GSSG) and the total glutathione pool (Foyer et al. 1997). Verma and Dubey (2003) studied the influence of concentrations of $Pb(NO_3)_2$ (500 and 1000 μM) on GR activity in seedlings of two *O. sativa* cultivars were raised in sand cultures. In control grown rice seedlings of both the cultivars, the activity behaviour of GR was found different in roots than in shoots during a 5–20 day growth period. During early days of growth of seedlings, up to 10 days, the activity of GR increased in both roots as well as shoots, whereas during 10–20 day period a gradual decline in enzyme activity was noticed in roots and not in shoots where enzyme showed gradual increase in activity up to day 20. But, in both the rice cultivars increasing levels of Pb treatment *in situ* led to a marked increase in enzyme activity. At day 15 of growth, seedlings growing in presence of 1000 μM Pb showed about 52–127% increase in GR activity in roots and 66–114% increase in shoots. Anjum et al. (2008c) reported Cd-induced significant increase in GR activity with increasing Cd concentrations in cultivars of *Vigna radiata* namely Pusa 9531, Pusa 9072, Pusa Vishal and PS-16 exposed to 25, 50, and 100 mg Cd kg^{-1} soil. Authors noticed the maximum increase in GR activity in Pusa 9531 (54–103%) followed by Pusa 9072 (25–92%), and the minimum in Pusa Vishal (20–89%) and PS-16 (16–68%), compared with the control treatment. Yannarelli et al. (2007) studied the response of GR activity and isoforms in leaves and roots of *Triticum aestivum* plants subjected to 100 μM $CdCl_2$ for 7, 14 and 21 days. Cd exposure provoked no changes on GR activity in wheat leaves over the study period. However, in the roots, GR activity increased by 111% and 200% after 7 and 14 days, respectively. Interestingly, control values were regained after 3 weeks. Authors also investigated the pattern of GR isoforms in leaves and roots by activity staining in non-denaturating polyacrylamide gels. Control leaves showed two GR isoforms, which were not modified by Cd treatment at any of the tested times. On the other hand, three GR isoforms were found in roots of control and treated plants. While the lower mobility isoform (GR3) remained unaffected, the other two isoforms activities (GR1 and GR2) were enhanced by Cd exposure for 7 and 14 days. This effect was totally reversed at 21 days, when the pattern of isoforms resembled that obtained in control plants. To confirm the specificity of GR activity staining, leaves and roots of 7 days Cd-treated plants were also analyzed by immunobloting. Both detection methods showed similar results. Reddy et al. (2005) investigated the influence of 0, 200, 500 and 800 ppm concentration of Pb-$Pb(NO_3)_2$ on antioxidant metabolism of *Macrotyloma uniflorum* and *Cicer arietinum*. The activity of GR increased in Pb-stressed plants of *Macrotyloma uniflorum* and *Cicer arietinum*. The degree of elevation in the activity of GR was 3–5 folds in both the plants studied, compared to their controls. However, the per cent increase was relatively more in *Macrotyloma uniflorum* than *Cicer arietinum* under Pb stress conditions. And, the degree of elevation in the activity of GR was more in roots compared to leaves of Pb stressed plants of both the species. Khan et al. (2009) reported significant increase in GR activity in two cultivars (RH30 and Varuna) of *Brassica juncea* exposed to 0 and 200 mg Cd kg^{-1} soil. The activity of GR increased by 4.3% in RH30, while increase in the activity of GR in Varuna was 23.0%. In another study, Anjum et al. (2011) investigated the influence of Cd (0, 25, 50 and 100 mg kg^{-1} soil) on GR activity in Pusa and PS 16 cultivars of *Vigna radiata*. Cd-tolerant cv. Pusa 9531 exhibited significantly higher activity of GR than Cd-susceptible cv. PS 16 due to Cd treatments. Pusa 9531 exhibited increases of 247.4% in GR activity due to maximum level of soil-Cd (100 mg Cd kg^{-1} soil) compared to control plants; enhancements of 140.5% was observed in the activity of GR in cv. PS 16. Russo et al. (2008) studied the response of GR activity to Cu (100 μM $CuSO_4$) stress in *Brassica nupus*. GR activity reached the maximum at 10 min showing a value 2.4-fold higher than in the control; thereafter, it decreased maintaining anyway a value 67% higher than that of the control at the end of the experiment. Zhou et al. (2008) investigated the changes in the activity of GR in *Medicago sativa* treated with different concentrations of $HgCl_2$ (0–40 μM) for 7 d. Authors reported that the GR activity in leaves of alfalfa was not affected by Hg^{2+} exposure at 1–10 μM, as compared to the control, but raising the concentrations of Hg^{2+} up to 20 or 40 μM in the medium induced remarkably increased activities. Also, the time-course study shows that during the initial 6 d of treatment, the GR activity was not significantly affected by 20 μM Hg^{2+}. However, the prolonged exposure of Hg^{2+} resulted in the rapid enhancement of activities. Lomonte et al. (2010) investigated the changes in GR activity in seedlings of *Atriplex codonocarpa* grown for 4 weeks in hydroponic culture containing 0, 0.05, 0.1 and 1 mg l^{-1} Hg. Shi et al. (2005) studied the Mn (10 and 600 μM $MnSO_4$)-induced changes in GR activity in *Cucumis sativus*. Excess Mn significantly increased GR activity in the leaves. GR activity in shoots of seedlings treated with Hg maintained a constant lower value than the control (20–30% lower). In roots, however, it remained unchanged at 0, 0.05 and 0.1 mg l^{-1} Hg and then declined by 65% in comparison with the control at 1 mg l^{-1} Hg. In addition, GR activity was about 3-fold higher in shoots than in roots for all treatments except for 1 mg l^{-1} Hg, where it was about 6-fold higher.

Diwan et al. (2010) investigated the influence of Cr (0, 50, 100, 200 μM) on GR activity in *Brassica juncea* and *Vigna radiata*. Authors reported that GR activity in the *B. juncea* increased in the range of 111–178% with all the Cr treatments, when compared with control. Maximum increase, however, was observed with 100 μM. In *V. radiata*, the GR activity was increased significantly at 1 day after treatment only, which was 56–77% over control. Israr et al. (2006) reported the changes in GR activity in seedlings of *Sesbania drummondii* treated with different concentrations (10, 20, 30, 40, 50 and 100 mg l^{-1}) of $HgCl_2$. GR activity increased significantly up to 100 mg l^{-1} Hg. The activity in the presence 100 mg l^{-1} Hg increased by 70%, with respect to the control. However, no significant change in GR activity was observed among the different treatments of Hg. In another study, Israr et al. (2006) investigated the influence of Cd (0-250 μM) on GR activity in the *Sesbania drummondii* callus. Authors reported that the GR activity increased significantly up to 50 μM Cd and then decreased. The activity in the presence of 50 and 100 μM Cd increased by 62.5 and 56.2%, respectively, with respect to the control. However, no significant change in GR activity was observed among the different treatments of Cd. The enzyme activity was significantly higher at each treatment when compared to control. Israr and Sahi (2006) studied the changes in GR activity in the cell cultures of *Sesbania drummondii* treated with different concentrations of Hg (0-250 μM). Authors reported that the Hg treatments altered the GR activity. GR activity was found increased significantly upto 40 μM Hg and then decreased. Authors noted a 62.8% increase in GR activity at 40 μM Hg concentration, when compared with control. However, they observed a drastic decline in GR activity at 50 μM Hg concentration. Iqbal et al. (2010) studied the changes in GR activity in *Brassica juncea* cultivars, namely, Pusa Jai Kisan and SS2 differing in Cd-tolerance treated with 0, 25 and 50 μmol l^{-1} Cd. GR activity increased linearly in both the cultivars and was the highest at 50 μmol l^{-1} Cd treatment. The increase in GR activity was higher in Pusa Jai Kisan in comparison to SS2. Qadir et al. (2004) reported an increase in GR activity in *B. juncea* cultivars, namely, Vardhan (V_1), Pusa Bahar (V_2), Pusa Bold (V_3), BTO (V_4), Pusa Jai Kisan (V_5), Agrini (V_6), Varuna (V_7), Kranti (V_8), Vaibhav (V_9) and Pusa Basant (V_{10}) exposed to different concentrations of Cd (T_1, 0.5 mM; T_2, 1.0 mM; T_3, 1.5 mM and T_4, 2.0 mM) harvested 24, 48 and 72 h after the treatment (HAT). The increase in GR activity varied from 12 to 35%, 14–36% and 16–37% with the treatment T_4 at 24, 48 and 72 HAT, respectively with maximum in V_5 and minimum in V_1. In V_5, the GR activity increased by 34.4, 35.8 and 37% with the treatment T_4 at 24, 48 and 72 HAT, respectively over their respective controls. In V_1, GR activity increased only by 12, 14 and 16% with the same treatment and at the same time.

4. GLUTATHIONE PEROXIDASE

Glutathione peroxidases (GPXs, EC 1.11.1.9) are ubiquitously occurring antioxidant enzymes, mainly present in cytosol, chloroplast, mitochondria, peroxisome and apoplast, and catalyze the reduction of hydroperoxides, organic and lipid hydroperoxides employing the glutathione GSH pool directly as a reducing agent, thereby protecting the cell against oxidative damage (Halliwell and Gutteridge 1999, Milla et al. 2003, Navrot et al. 2006).

4.1 Glutathione Peroxidase in Response to Drought and Salinity Stress

There are few reports regarding the drought-salinity-induced changes in GPX activity in plants. Galle et al. (2009) investigated the response of total GPX activity in *Triticum aestivum* cv. MV Emese and cv. Plainsman (drought tolerant) and cv. GKE´let and Cappelle Desprez (drought-sensitive) flag leaves under control and drought stress conditions during the grain-filling period. The highest level of extractable GPX activity was found in the Plainsman cultivar. In control plants, the activity of peroxidase was enhanced after anthesis. In control conditions, the extractable GPX activity increased significantly in every investigated cultivar. In MVEmese, Plainsman and Cappelle Desprez, drought stress further increased GPX activity at the first sampling day at anthesis. The GPX activity of the sensitive GKE´let cultivar showed significant decline due to stress. Gapin´ska et al. (2008) studied the changes in GPX activity in five-week-old plants, treated once with NaCl solution and with final concentration 50 mM (mild stress) or 150 mM (severe stress). In the plants treated with 50 and 150 mM of NaCl, GPX activity increased already in the first hour of the experiment to 221 and 219%, respectively. The higher activity of this enzyme was also observed 5 and 14 days after the treatment and it reached 123 and 208% at the lower dose applied and 163 and 189% of the control in the severely stressed roots. Salt stress induced significant changes in GSH activity also on the second day of severe and on seventh day of mild stress, when an increase by 60 and by 34% above the control was found, respectively.

4.2 Glutathione Peroxidase in Response to Heavy Metal Stress

There are contradictory reports regarding the response of GPX to number heavy metals. There are reports of significant increase in GPX activity in leaf tissues different in plant species exposed to Se and Cu, under both laboratory and field conditions (Hartikainen et al. 2000, Boojar and Goodarzi 2007). Dixit et al. (2001) reported that GPX activity in *Pisum sativum* plants exposed to Cd (4 and 40 μM for 7 days) decreased in root tissue but remained almost unaltered in leaves. Similarly, Baccouch et al. (2001) showed that although *Zea mays* plants exposed to Ni (250 μM $NiCl_2$ for 5 days) accumulated the metal in roots (~ 4000 μg Ni g^{-1}), GPX activity in this tissue was unaltered compared to controls. Hartikainen et al. (2000) observed that GPX activity was significantly enhanced in *Lolium perenne* shoots exposed to Se (1–10 μg g^{-1}) after 40 days, but chronic metal exposure decreased enzyme activity. Chen et al. (2010) investigated the modulation of exogenous glutathione in antioxidant defence system against Cd stress in the two

Hordeum vulgare genotypes differing in Cd tolerance. Cd stress induced a consistent increase in root GPX activity while external GSH counteracted Cd-induced alterations of certain antioxidant enzymes and brought root GPX activity of the both genotypes down towards the control level. Halušková et al. (2009) investigated the effect of abiotic stresses including Cd, Co, Pb, Ni, and Zn on GPX activity in barley root tips. Only a slight increase in GPX activity was observed after the application of Pb, Ni, and Zn, while Co did not activate its activity. Similarly to Hg and Cd, Cu treatment caused a strong increase in GPX activity. A strong effect of Cu has also been reported in *Glycine max* roots, where Cu markedly increased GPX activity. Authors speculated that the induction of GST activity by a large variety of compounds and stresses may be related to common signal probably originated through changes in the redox state of cells which is bound to change during different abiotic stresses. Cregnato et al. (2008) GPX activity in leaf tissue was found to respond to Zn exposure. Ninety-six hours after experimental commencement, patterns of GPX activity reflected accumulated leaf Zn, with a significant positive relationship between Zn concentration in leaf tissue and GPX activity. Seedlings exposed to 100 μg Zn ml^{-1} exhibited a 230% increase in enzyme activity (43.7 ± 6.5 nmol min^{-1} g^{-1} wet weight), while seedlings in the presence of 800 μg Zn ml^{-1} showed a 300% increase in GPX activity (48.1 ± 1.2 nmol min^{-1} g^{-1} wet weight) compared to control seedlings. After 2 weeks exposure, there was no significant linear relationship between GPX activity and leaf Zn concentration, but after 8 weeks of treatment, GPX activity was again significantly associated with leaf tissue Zn concentration, with bivariate regression analysis revealing a positive linear relationship between GPX activity and accumulated leaf Zn. Although GPX activity remained responsive in a dose-dependent fashion after 8 weeks of exposure, enzyme activity was lower than in 96 h treated plants. Seedlings exposed to 100 μg Zn ml^{-1} at 8 weeks showed a 140% increase in enzyme activity (21.9 ± 2.8 nmol min^{-1} g^{-1} wet weight), while seedlings in the presence of 400 μg Zn ml^{-1} exhibited a 210% increase in GPX activity (33.3 ± 4.5 nmol min^{-1} g^{-1} wet weight) compared to control plants. Fatima and Ahmad (2005) observed significant increase in GPX activity in *Allium cepa* exposed to similar concentrations of metals (Hg, Pb, Cr, Cu, Zn or Cd at 200–800 μg ml^{-1}), suggesting that the elevated activity of both enzymes was a result of metal-induced free radical generation. Huang et al. (2010) investigated the effect of multiple heavy metal stress on GPX in the leaves, stems and roots of *Kandelia candel* and *Bruguiera gymnorrhiza* treated with five different concentrations of a heavy metal mixture (Cd^{2+}, Pb^{2+} and Hg^{2+}). GPX in the leaves, stems and roots of *K. candel* reached their respective maximal values at the highest metal concentration, whereas GPX activity in roots and stems of *B. gymnorrhiza* was similar to the controls at higher metal concentrations.

5. GLUTATHIONE SULPHOTRANSFERASE

Glutathione S-transferase (GST, EC 2.5.1.18) is one of the GSH-dependent ROS-detoxifying enzymes in plants, which has collectively been shown to constitute 0.1% of the soluble protein in *Zea mays* leaves (Marrs 1996, Dixon et al. 1998, 2000). GSTs, in fact, are ubiquitous in aerobic organisms in several isoforms catalysing the conjugation of GSH to a wide variety of hydrophobic and electrophilic compounds to form less- or non-toxic peptide derivates (Marrs 1996, Frova 2003).

5.1 Glutathione Sulphotransferase in Response to Drought and Salinity Stress

Gapińska et al. (2008) studied the changes in GST activity in five-week-old plants, treated once with NaCl solution and with final concentration 50 mM (mild stress) or 150 mM (severe stress). The severe stress induced in tomato roots enhancement of GST activity to 135% as compared to the control 1 h after NaCl application. That increasing tendency persisted: third hour −150%, first day −175% and the maximum activity was observed on the second day when it reached 301% of the control. Enhanced GST activity, i.e. 189% of the control value lasted up to the end of the experiment. In the plants treated with 50 mM NaCl no significant changes in GST activity were observed up to the second day when the increase by 70% above the control was found and it persisted up to the 14th day of the experiment when it reached 208% of the control.

Rios-Gonzalez et al. (2002) reported the response of the activity of antioxidant enzymes including GST in *Zea mays* and *Helianthus annus* seedlings as affected by salinity (100 mM NaCl) and different nitrogen sources ($NaNO_3/(NH_4)_2SO_4$). Author noticed no activity of GST in *H. annus* but GST activity was found higher in the leaves and roots of *Zea mays* plants grown under ammonium nutrition. The response of this enzyme to salinity differed with the plant organ. Control plants had higher activity in the leaves, while salinity-treated plants expressed the highest activity in their roots. Galle et al. (2009) investigated the response of total glutathione S-transferase activity in *Triticum aestivum* cv. MV Emese and cv. Plainsman (drought tolerant) and cv. GKE´let and Cappelle Desprez (drought-sensitive) flag leaves under control and drought stress conditions during the grain-filling period. The highest level of extractable GST activity was found in the Plainsman cultivar. In control plants, the activity of transferase was enhanced after anthesis; transferase exhibited as lower change in cv. MVEmese and GKE´let. The induction of GST activity following drought stress was detected earlier in resistant cultivars than in sensitive ones. In Cappelle Desprez, no significant changes were detectable, and in GKE´let, GST activity increased only on 12 days post anthesis in the drought-stressed plants.

5.2 Glutathione Sulphotransferase in Response to Heavy Metal Stress

Dixon et al. (1998) and Edwards et al. (2000) have earlier reported the significance of the conjugation of GSH to a wide variety of hydrophobic and electrophilic compounds and confirmed that this conjugation reaction is involved in the detoxification and processing of various xenobiotics but also endogenous toxic and non-toxic metabolites. Although, GSTs are mostly cytoplasmic proteins, but Flury et al. (1996) and Frova (2003) have reported the occurrence of microsomal, plastidic, nuclear and apoplastic isoforms of GSTs and in some cases GSTs in its cytoplasmic protein forms represent more than 1% of soluble proteins in plant cells (Edwards and Dixon 2000). There are also reports on the isolation of 59, 47, 42 and 25 putative GST genes respectively, from *Oryza sativa, Arabidopsis thaliana, Zea mays* and *Glycine max* (McGonigle et al. 2000; Wagner et al. 2002; Soranzo et al. 2004).

Various isoforms of GST isolated from several plant species displayed also significant GPX activity (GST/GPX) toward lipid hydroperoxides catalysing their reduction to the less toxic alcohols (Bartling et al. 1993; Cummins et al. 1999).

Hana et al. (2008) investigated the influence of various concentrations of Cd (0, 20, 40, 80, 100 and 200 μM) on GST activity in *Lycopersicon esculentum.* Authors noticed increase in the activity of GST only at low concentrations of Cd (20 to 80 μM). Zhang and Ge (2008) investigated the response of GST in rice seedlings exposed to four Cd concentrations (0, 10, 20 and 40 mg l^{-1}). The activity of GST differed greatly between shoots and roots exposed to Cd concentrations. In shoots, GST activity increased by around twofold at the 40 mg l^{-1} Cd treatment relative to the control. Significantly positive correlation was observed between the enzyme activity and Cd level. In contrast, GST activity in roots was significantly reduced by Cd treatment. Moreover, GST activity in roots was found dramatically higher than that in shoots. Szollosi et al. (2009) investigated the influence of different Cd concentrations (*vitro* at 0, 50, 100 and 200 mg l^{-1}) on GST activity in seeds of *Brassica juncea* germinated in dark for 12, 24, 48 and 96 h. GST activity seemed to be remarkably time- and dose-dependent. It increased in the control, Cd 50 and Cd 100 seeds with time, but a positive correlation between the duration of treatment and the concentration was found only at 50 and 100 mg l^{-1} Cd levels ($r = 0.86$ and $r = 0.59$). Positive but not significant correlation occurred between Cd concentration measured in seeds and GST activity ($r = 0.04$). At the beginning of the germination (12–24 h), the GST activity was only 50% of that of the control. The highest activity was measured after 24 h in Cd 200 seed sand reached 150% of the control, but it declined later on. At the same time, 48–96 h after Cd exposure there were no significant differences between the control and the treated seeds. Halušková et al. (2009) investigated the effect of abiotic stresses including Cd, Co, Pb, Ni, and Zn on GST activity in *Hordeum vulgare* root tips. Authors reported a strong induction of GST activity in Hg-treated root tips accompanied by a highest root growth inhibition. An association between increased GST activity and root growth inhibition was also observed during other heavy metal treatments. Whereas, Cu exhibited no effect on GST activity. Authors speculated that the induction of GST activity by a large variety of compounds and stresses may be related to common signal probably originated through changes in the redox state of cells which is bound to changes during different abiotic stresses. Reddy et al. (2005) investigated the influence of 0, 200, 500 and 800 ppm concentration of Pb-$Pb(NO_3)_2$ on antioxidant metabolism of *Macrotyloma uniflorum* and *Cicer arietinum.* The activity of GST increased in Pb-stressed plants of *Macrotyloma uniflorum* and *Cicer arietinum.* The degree of elevation in the activity of GST was 3–4 folds in both the plants studied, compared to their controls. However, the per cent increase was relatively more in *Macrotyloma uniflorum* than *Cicer arietinum* under Pb stress conditions. And, the degree of elevation in the activity of GST was more in roots compared to leaves of Pb stressed plants of both the species. Nagalakshmi and Prasad (2001) reported the responses of GSH cycle enzymes and GSH metabolism to Cu stress (0, 50, 100 and 200 μM concentrations of $CuSO_4$ for 3 days) in *Scenedesmus bijugatus.* The activity of GST showed an increase of 17, 35 and 78%, respectively in 50, 100 and 200 μM $CuSO_4$ treatments. During Cu stress protective mechanisms of the cell are employed, leading to activation of protective enzymes like GST.

6. CRITICAL APPRAISAL, SUMMARY AND PERSPECTIVES

The literature reviewed present a clear picture of physiological and biochemical aspects of GSH and its related enzymes such as GR, GPX, and GST in different plants under drought, salinity and heavy metal stress conditions. However, we are still far from understanding the complete circuits of major abiotic stress reactions including the major discretions of GSH and its related enzymes in major crop plants. Only a few components of many pathways related to plant responses to drought, salinity and heavy metal stresses have been the subject of investigations. But it is now clear that as a major and significant component of AsA-GSH pathway, the reduced glutathione (GSH) and its related enzymes are of great significance for the detoxification, chelation and compartmentalization of major heavy metals in plants, and directly and/or indirectly removal of ROS generated due to drought-salinity and heavy metals and protects plants against drought-salinity and heavy metals-accrued oxidative stress to a great extent. GSH:GSSG is one of the important soluble redox couples central to bioenergetics, and to both ROS-producing and ROS-processing metabolism influence to a great extent many of the mechanisms by which plants respond to environmental constraints and fluctuations including potential toxic metals. There seems a lack of reports on exhaustive molecular aspects of drought-salinity and heavy metals-accrued influence on the modulation of GSH, its redox couple (GSH:GSSG) and related enzymes (GR, GPX, and GST) in plants. Therefore, the future priorities should be aimed at the identification of molecules connecting pathways and of key components in each pathway. It is proposed that our understanding of plant stress tolerance can be greatly refined by thorough characterization of individual genes and assessing their contribution to abiotic stress tolerance. In addition, a number of questions are yet be answered such as the regulation of protein SH oxidation, S-glutathionylation and S-nitrosylation, and the molecular characterization of many GSH-dependent enzymes in major economically important plant species growing under the influence of various abiotic stress factors.

ACKNOWLEDGEMENTS

NAA is grateful to the Portuguese Foundation for Science and Technology (FCT) for financial assistance (SFRH/BPD/64690/2009).

REFERENCES

Agarwal, S. and Shaheen, R. (2007). Stimulation of antioxidant system and lipid peroxidation by abiotic stress in leaves of *Momordica charantia*. *Braz. J. Plant Physiol.* **19**: 149-161.

Ahmad, P., Jaleel, C.A. and Sharma, S. (2010). Antioxidant defense system, lipid peroxidation, proline-metabolizing enzymes, and biochemical activities in two *Morus alba* genotypes subjected to NaCl stress. *Russ. J. Plant Physiol.* **57**: 509-517.

Allakhverdiev, S. I., Sakamoto, A., Nishiyama, Y., Inaba, M. and Murata, N. (2000) Ionic and osmotic effects of NaCl-induced inactivation of photosystem I and II in Synechococcus sp. *Plant Physiol.* **123**: 1047-1056.

Anjum, N.A. (2006). Effect of abiotic stresses on growth and yield of *Brassica campestris* L. and *Vigna radiata* (L.) Wilczek under different sulphur regimes' Ph.D. Thesis, Jamia Hamdard, New Delhi.

Anjum, N.A., Umar, S. and Chan, M.-T. (2010). Eds. *Ascorbate-Glutathione Pathway and Stress Tolerance in Plants*, The Netherlands: Springer, ISBN 978-90-481-9403-2.

Anjum, N.A., Umar, S., Ahmad, A. and Iqbal, M. (2008c). Responses of components of antioxidant system in moongbean [*Vigna radiata* (L.) Wilczek] genotypes to cadmium stress. *Commun. Soil Sci. Plant Anal.* **39**: 2469-2483.

Anjum, N.A., Umar, S., Ahmad, A., Iqbal, M. and Khan, N.A. (2008b). Sulphur protects mustard (*Brassica campestris* L.) from cadmium toxicity by improving leaf ascorbate and glutathione. *Plant Growth Regul.* **54**: 271-279.

Anjum, N.A., Umar, S., Iqbal, M. and Khan, N.A. (2008a). Growth characteristics and antioxidant metabolism of moongbean genotypes differing in photosynthetic capacity subjected to water deficit stress. *J. Plant Interact.* **3**: 127-136.

Anjum, N.A., Umar, S., Iqbal, M. and Khan, N.A. (2011). Cadmium causes oxidative stress in moongbean [*Vigna radiata* (L.) Wilczek] by affecting antioxidant enzyme systems and ascorbate-glutathione cycle metabolism. *Russ. J. Plant Physiol.* **58**: 00-00, In Press

Aravind, P. and Prasad, M.N.V. (2005). Cadmium and zinc interactions in a hydroponic system using *Ceratophylum demersum* L.: adaptive ecophysiology, biochemistry and molecular toxicology. *Braz. J. Plant Physiol.* **17**: 3-20.

Ashraf, M. (1994). Breeding for salinity tolerance in plants. *Crit. Rev. Plant Sci.* **13**: 17-42.

Baccouch, S., Chaoui, A. and Ferjani, E.E. (2001). Nickel toxicity induces oxidative damage in *Zea mays* roots. *J. Plant Nutr.* **24**: 1085-1097.

Bai, J., Gong, C.-M., Chen, K., Kang, H.-M. and Wang, G. (2009). Examination of antioxidative system's responses in the different phases of drought stress and during recovery in desert plant *Reaumuria soongorica* (Pall.) Maxim. *J. Plant Biol.* **52**: 417-425.

Bartling, D., Radzio, R., Steiner, U. and Weiler, E.W. (1993). A glutathione S-transferase with glutathione-peroxidase activity from *Arabidopsis thaliana*. *Eur. J. Biochem.* **216**: 579-586.

Benavides, M.P., Marconi, P.L., Gallego, S.M., Comba, M.E. and Tomaro, M.L. (2000). Relationship between antioxidant defence systems and salt tolerance in *Solanum tuberosum*, *Aust. J. Plant Physiol.* **27**: 273-278.

Bergmann, L. and Rennenberg, H. (1993). Glutathione metabolism in plants. In *Sulfur nutrition and assimilation in higher plants*, pp. 109-123, Eds. LJ De Kok (SPB Academic Publishing, The Hague, The Netherlands).

Bian, S. and Jiang, Y. (2009). Reactive oxygen species, antioxidant enzyme activities and gene expression patterns in leaves and roots of Kentucky bluegrass in response to drought stress and recovery. *Sci. Hortic.* **120**: 264-270.

Bohnert, H.J., Nelson, D.E. and Jensen, R.G. (1995). Adaptations to environmental stresses. *Plant Cell.* **7**: 1099-1111.

Boojar, M.M.A. and Goodarzi, F. (2007). The copper tolerance strategies and the role of antioxidative enzymes in three species grown on copper mine. *Chemosphere.* **67**: 2138-2147.

Borsani, O., Valpuesta, V. and Botella, M.A. (2001). Evidence for a role of salicylic acid in the oxidative damage generated by NaCl and osmotic stress in *Arabidopsis* seedlings. *Plant Physiol.* **126**:1024-1030.

Boyer, J.S. (1982). Plant productivity and environment. *Sci.* **218**: 443-448.

Bray, E.A., Bailey-Serres, J. and Weretilnyk, E. (2000). Responses to abiotic stresses. In *Biochemistry and molecular biology of plants*, pp. 1158-1249, Eds. W Gruissem, B Buchannan and R Jones (American Society of Plant Physiologists, Rockville, MD).

Bricker, T.J., Pichtel, J., Brown, H.J. and Simmons, M. (2001). Phytoextraction of Pb and Cd from superficial soil: effects of amendments and croppings. *J. Environ. Sci. Health. Part A Toxic/ Hazard. Subst. Environ. Eng.* **36**: 1597-1610.

Bruns, I., Sutter, K., Menge, S., Neumann, D. and Krauss, G.-J. (2001). Cadmium lets increase the glutathione pool in bryophytes. *J. Plant Physiol.* **158**: 79-89.

Caregnato, F.F., Koller, C.E., MacFarlane, G.R. and Moreira, J.C.F. (2008). The glutathione antioxidant system as a biomarker suite for the assessment of heavy metal exposure and effect in the grey mangrove, *Avicennia marina* (Forsk.) Vierh . *Marine. Poll. Bull.* **56**: 1119-1127.

Chen, F., Wang, F., Wu, F., Mao, W., Zhang, G. and Zhou, M. (2010). Modulation of exogenous glutathione in antioxidant defense system against Cd stress in the two barley genotypes differing in Cd tolerance. *Plant Physiol. Biochem.* **48**: 663-672.

Chen, J. and Goldsbrough, P.B. (1994). Increased activity of γ-glutamylcysteine synthetase in tomato cells selected for cadmium tolerance. *Plant Physiol.* **106**: 233-239.

Choukr-Allah, R. and Harrouni, M.C. (1996). The potential use of halophytes under saline irrigation in Morocco (abstract). In: Symposium on the conservation of mangal ecosystems. Al Ain: United Arab Emirates, 15-17.

Coleman, J.O.D., Blake-Kalff, M. and Davies, E. (1997). Detoxification of xenobiotics by plants: chemical modification and vacuolar compartmentation. *Trends Plant Sci.* **2**: 144-151.

Comba, M.E., Benavides, M.P. and Tomaro, M.L. (1998). Effect of salt stress on antioxidant defence system in soybean root nodules. *Aust. J. Plant Physiol.* **25**: 665-671.

Cummins, I., Cole, D.J. and Edwards, R. (1999). A role for glutathione transferases functioning as glutathione peroxidases in resistance to multiple herbicides in black-grass. *Plant J.* **18**: 285-292.

Cuypers, A., Vangronsveld, J. and Clijsters, H. (2001). The redox status of plant cells (AsA and GSH) is sensitive to zinc imposed oxidative stress in roots and primary leaves of *Phaseolus vulgaris. Plant Physiol. Biochem.* **39**: 657-664.

Demiral, T. and Türkan, Ý. (2005). Comparative lipid peroxidation, antioxidant defense systems and proline content in roots of two rice cultivars differing in salt tolerance. *Environ. Exp. Bot.* **53**: 247-257.

Desingh, R. and Kanagaraj, G. (2007). Influence of salinity stress on photosynthesis and antioxidative systems in two cotton varieties. *Gen. Appl. Plant Physiol.* **33**: 221-234.

Diwan, H., Khan, I., Ahmad, A. and Iqbal, M. (2010). Induction of phytochelatins and antioxidant defence system in *Brassica juncea* and *Vigna radiata* in response to chromium treatments. *Plant Growth Regul.* **61**: 97-107.

Dixit, V., Pandey, V. and Shyam, R. (2001). Differential antioxidative responses to cadmium in roots and leaves of pea (*Pisum sativum* L. cv. Azad). *J. Exp. Bot.* **52**: 1101-1109.

Dixon, D.P., Cummins, I., Cole, D.J. and Edwards, R. (1998). Glutathione-mediated detoxification systems in plants. *Curr. Opin. Plant Biol.* **1**: 258-266.

Dixon, D.P., Lapthorn, A. and Edwards, R. (2000). Plant glutathione transferases. *Genome Biol.* **3**: 1-10.

Dràzÿkiewicz, M., Skórzyńska-Polit, E. and Krupa, Z. (2003). Response of the ascorbate-glutathione cycle to excess copper in *Arabidopsis thaliana* (L.). *Plant Sci.* **164**: 195-202.

Edwards, R. and Dixon, D.P. (2000). The role of glutathione transferases in herbicide metabolism. In *Herbicides and their mechanisms of action*. pp. 33-71, Eds. AH Cobb and AC Kirkwood (Sheffield: Sheffield Academic Press).

Edwards, R., Dixon, D.P. and Walbot, V. (2000). Plant glutathione S-transferases: enzymes with multiple functions in sickness and in health. *Trends Plant Sci.* **5**: 193-198.

Ekmekci, Y., Tanyolac, D. and Ayhan, B. (2009). A crop tolerating oxidative stress induced by excess lead: maize. *Acta Physiol. Plant.* **31**: 319-330.

Esfandiari, E., Shekari, F., Shekari, F. and Esfandiari, M. (2007). The effect of salt stress on antioxidant enzymes' activity and lipid peroxidation on the wheat seedling. *Not. Bot. Hort. Agrobot. Cluj.* **35**: 48-56.

Fatima, R.A. and Ahmad, M. (2005). Certain antioxidant enzymes of *Allium cepa* as biomarkers for the detection of toxic heavy metals in wastewater. *Sci. Total Environ.* **346**: 256-273.

Flury, T., Wagner, E. and Kreuz, K. (1996). An inducible glutathione S-transferase in soybean hypocotyl is localized in the apoplast. *Plant Physiol.* **112**: 1185-1190.

Foyer, C.H. and Noctor, G. (2000). Oxygen processing in photosynthesis: regulation and signaling. *New Phytol.* **146**: 359-388.

Foyer, C.H. and Noctor, G. (2005). Oxidant and antioxidant signaling in plants: a re-evaluation of the concept of oxidative stress in a physiological context. *Plant Cell Environ.* **28**: 1056-1071.

Foyer, C.H., Lopez-Delgado, H., Dat, J.F. and Scott, I.M. (1997). Hydrogen peroxide and glutathione-associated mechanisms of acclamatory stress tolerance and signaling. *Physiol. Plant.* **100**: 241-254.

Frova, C. (2003). Glutathione transferases in the genomics era: New insights and perspectives. *Biomol. Eng.* **23**: 149-169.

Galle, A., Csisza´r, J., Secenji, M., Guo´th, A., Cseuz, L., Tari, I., Gyo¨rgyey, J. and Erdei, L. (2009). Glutathione transferase activity and expression patterns during grain filling in flag leaves of wheat genotypes differing in drought tolerance: Response to water deficit. *J. Plant Physiol.* **166**: 1878-1891.

Gapińska, M., Skłodowska, M. and Gabara, B. (2008) Effect of short- and long-term salinity on the activities of antioxidative enzymes and lipid peroxidation in tomato roots. *Acta Physiol. Plant.* **30**: 11-18.

Geissler, S., Hussin, S. and Koyro, H.-W. (2009). Interactive effects of NaCl salinity and elevated atmospheric CO_2 concentration on growth, photosynthesis, water relations and chemical composition of the potential cash crop halophyte *Aster tripolium* L. *Environ. Exp. Bot.* **65**: 220-231.

Gorham, J., McDonnel, E., Budrewicz, E. and Wyn Jones, R.G. (1985). Salt tolerance in the Triticeae: Growth and solute accumulation in leaves of *Thinopyrum bessarabicum*. *J. Exp. Bot.* **36**: 1021-1031.

Greenway, H. and Munns, R. (1980). Mechanism of salt tolerance in non halophytes. *Annu. Rev. Plant Physiol.* **31**: 149-190.

Guo, B., Liang, Y.C., Zhu, Y.G. and Zhao, F.J. (2007). Role of salicylic acid in alleviating oxidative damage in rice roots (*Oryza sativa*) subjected to cadmium stress. *Environ. Pollut.* **147**: 743-749.

Gupta, D.K., Huang, H.G., Yang, X.E., Razafindrabe, B.H.N. and Inouhe, M. (2010). The detoxification of lead in *Sedum alfredii* H. is not related to phytochelatins but the glutathione. *J. Hazard. Mat.* **177**: 437-444.

Gupta, S.C. and Goldsbrough, P.B. (1991). Phytochelatin accumulation and cadmium tolerance in selected tomato cell lines. *Plant Physiol.* **97**: 306-312.

Hafsi, C., Romero-Puertas, M.C., Gupta, D.K., del Río, L.A., Sandalio, L.M. and Abdelly, C. (2010). Moderate salinity enhances the antioxidative response in the halophyte *Hordeum maritimum* L. under potassium deficiency. *Env. Exp. Bot.* **69**: 129-136.

Halliwell, B. and Gutteridge, J.M.C. (1999). In: *Free radicals in biology and medicine,* p. 936, Eds. B Halliwell and JMC Gutteridge (University Press, Oxford).

Halus˘kova,´L., Valentovic¡ová, K., Huttová, J., Mistrÿk, I. and Tamás, L. (2009). Effect of abiotic stresses on glutathione peroxidase and glutathione S-transferase activity in barley root tips. *Plant Physiol. Biochem.* **47**: 1069-1074.

Hana. S., Rachid, R., Ibtissem, S., Houria, B. and Mohammed-Réda, D. (2008). Induction of anti-oxidative enzymes by cadmium stress in tomato (*Lycopersicon esculentum*). *Afric. J. Plant Sci.* **2**: 72-76.

Hartikainen, H., Kue, T.L. and Piironem, V. (2000). Selenium as an antioxidant and prooxidant in rye grass. *Plant Soil.* **225**: 193-200.

Hernandez, J.A., Campillo, A., Jimenez, A., Alarcon, J.J. and Sevilla, F. (1999). Response of antioxidant systems and leaf water relations to NaCl stress in pea. *New Phytol.* **141:** 241-251.

Hu, Y. and Schmidhalter, U. (2005). Drought and salinity: A comparison of their effects on the mineral nutrition of plants. *J. Plant Nutr. Soil Sci.* **168**: 541-549.

Huang, G.-Y. and Wang, Y.-S. (2010). Physiological and biochemical responses in the leaves of two mangrove plant seedlings (*Kandelia candel* and *Bruguiera gymnorrhiza*) exposed to multiple heavy metals. *J. Hazard. Mat.* **182:** 848-54.

Huang, G.-Y., Wang, Y.-S., Sun, C.-C. and Dong, J.-D. (2010) The effect of multiple heavy metals on ascorbate, glutathione and related enzymes in two mangrove plant seedlings (*Kandelia candel* and *Bruguiera gymnorrhiza*). *Oceanol. Hydrobiol. Stud.* **39**: 11-25.

Iqbal, N., Masood, A., Nazar, R., Syeed, S. and Khan, N.A. (2010). Photosynthesis, growth and antioxidant metabolism in mustard (*Brassica juncea* L.) cultivars differing in cadmium tolerance. *Agric. Sci. China.* **9**: 519-527.

Israr, M. and Sahi, S. (2006). Antioxidative responses to mercury in the cell cultures of *Sesbania drummondii. Plant Physiol. Biochem.* **44**: 590-595.

Israr, M., Sahi, S., Datta, R. and Sarkar, D. (2006a). Bioaccumulation and physiological effects of mercury in *Sesbania drummondii. Chemosphere.* **65**: 591-598.

Israr, M., Sahi, S.V. and Jain, J. (2006b). Cadmium accumulation and antioxidative responses in the *Sesbania drumondii* callus. *Arch. Environ. Contam. Toxicol.* **50**: 121-127.

Jung, S. (2004). Variation in antioxidant metabolism of young and mature leaves of *Arabidopsis thaliana* subjected to drought. *Plant Sci.* **166**: 459-466.

Khan, M.H. and Panda, S.K. (2008). Alterations in root lipid peroxidation and antioxidative responses in two rice cultivars under NaCl-salinity stress. *Acta Physiol. Plant.* **30**: 81-89.

Khan, N.A., Anjum, N.A., Nazar, R. and Iqbal, N. (2009). Increased activity of ATP-sulfurylase and increased contents of cysteine and glutathione reduce high cadmium-induced oxidative stress in mustard cultivar with high photosynthetic potential. *Russ. J. Plant Physiol.* **56**: 670-677.

Kim, S.Y., Lim, J.-H., Park, M.R., Kim, Y.J., Park, T., Seo, Y.W., Choi, K.G. and Yun, S.J. (2005). Enhanced antioxidant enzymes are associated with reduced hydrogen peroxide in barley roots under saline stress. *J. Biochem. Mol. Biol.* **38**: 218-224.

Klapheck, S., Fliegner, W. and Zimmer, I. (1994). Hydroxymethyl-phytochelatins [(γ-glutamylcysteine)$_n$ – serine] are metal-induced peptides of the Poaceae. *Plant Physiol.* **104**: 1325-1332.

Klapheck, S., Schlunz, S. and Bergmann, L. (1995). Synthesis of phytochelatins and homo-phytochelatins in *Pisum sativum* L. *Plant Physiol.* **107**: 515-521.

Kolb, D., Müller, M., Zellnig, G. and Zechmann, B. (2010). Cadmium induced changes in subcellular glutathione contents within glandular trichomes of *Cucurbita pepo* L. *Protoplasma.* **243**: 87-94.

Koyro, H.-W., Geissler, N., Hussin, S., Debez, A. and Huchzermeyer, B. (2008). Strategies of halophytes to survive in a salty environment. In *Abiotic stresses and plant responses.* pp. 83-104. Eds. N.A. Khan, S. Singh and S. Umar, (New Delhi: I.K. International PublishingHouse Pvt. Ltd.).

Kuz´niak, E., Sklodowska, M. (2001). Ascorbate, glutathione and related enzymes in chloroplasts of tomato leaves infected by *Botrytis cinerea*. *Plant Sci.* **160**: 723-731.

Latowski, D., Surówka, E. and Strzałka, K. (2010). Regulatory role of components of ascorbate-glutathione pathway in plant stress tolerance. In *Ascorbate-Glutathione Pathway and Stress Tolerance in Plants*, pp. 1-54, Eds. NA Anjum, S Umar and M-T Chan (The Netherlands: Springer).

Lechno, S., Zamki, E. and Telor, E. (1997). Salt stress-induced responses in cucumber plants. *J. Plant Physiol.* **150**: 206-211.

Lin, R.Z., Wang, X.R., Luo, Y., Du, W.C., Guo, H.Y. and Yin, D.Q. (2007). Effects of soil cadmium on growth, oxidative stress and antioxidant system in wheat seedlings (*Triticum aestivum* L.). *Chemosphere*. **69**: 89-98.

Lomonte, C., Sgherri, C., Baker, A.J.M., Kolev, S.D. and Navari-Izzoc, F. (2010). Antioxidative response of *Atriplex codonocarpa* to mercury. *Env. Exp. Bot.* **69**: 9-16.

Ma, Y.-H., Ma, F.-W., Wang, Y.-H. and Zhang, J.-K. (2010). The responses of the enzymes related with ascorbate-glutathione cycle during drought stress in apple leaves. *Acta Physiol. Plant* DOI 10.1007/s11738-010-0535-5.

MacFarlane, G.R. and Burchett, M.D. (2001). Photosynthetic pigments and peroxidase activity as indicators of heavy metal stress in the grey mangrove, *Avicennia marina* (Forsk) Vierh. *Marine Poll. Bull.* **42**: 233-240.

Mahajan, S. and Tuteja, N. (2005). Cold, salinity and drought stresses: An overview. *Arch. Biochem. Biophy.* **444**: 139-158.

Mahmood, Q., Ahmad, R., Kwak, S.-S., Rashid, A. and Anjum N.A. (2010). Ascorbate and glutathione: protectors of plants in oxidative stress. In *Ascorbate-Glutathione Pathway and Stress Tolerance in Plants*. pp. 209-230, Eds. NA Anjum, S Umar and M-T Chan (The Netherlands: Springer).

Malecka, A., Piechalak, A. and Tomaszewska, B. (2009). Reactive oxygen species production and antioxidative defense system in pea root tissues treated with lead ions: the whole roots level. *Acta Physiol. Plant.* **31**: 1053-1063.

Marrs, K.A. (1996). The functions and regulation of glutathione S-transferases in plants. *Annu. Rev. Plant Physiol. Plant Mol. Biol.* **47**: 127-158.

McGonigle, B., Keeler, S.J., Lau, S.-M.C., Koeppe, M.K. and O'Keefe, D.P. (2000). A genomics approach to the comprehensive analysis of the glutathione S-transferase gene family in soybean and maize. *Plant Physiol.* **124**: 1105-1120.

McLaughlin, M.J., Parker, D.R. and Clarke, J.M. (1999). Metals and micronutrients -food safety issues. *Field Crops Res.* **60**: 143-163.

Meister, A. and Anderson, M.E. (1983). Glutathione. *Annu Rev. Biochem.* **52**: 711-760.

Meloni, D.A., Oliva, M.A., Martinez, C.A. and Cambrai, J. (2003). Photosynthesis and activity of superoxide dismutase, peroxidase and glutathione reductase in cotton under salt stress. *Env. Exp. Bot.* **49**: 69-76.

Meneguzzo, S., Navari-Izzo, F. and Izzo, R. (1999). Antioxidative responses of shoots and roots of wheat to increasing NaCl concentrations. *J. Plant Physiol.* **155**: 274-280.

Milla, M.A.R., Maurer, A., Huete, A.R. and Gustafson, J.P. (2003). Glutathione peroxidase genes in *Arabidopsis* are ubiquitous and regulated by abiotic stresses through diverse signalling pathways. *Plant J.* **36**: 602-615.

Millar, A.H., Mittova, V. and Kiddle, G. (2003). Control of ascorbate synthesis by respiration and its implications for stress responses. *Plant Physiol.* **133**: 443-447.

Mishra, S., Srivastava, S., Tripathi, R.D., Govindarajan, R., Kuriakose, S.V. and Prasad, M.N.V. (2006). Phytochelatin sysnthesis and response of antioxidants during cadmium stress in *Bacopa monnieri* L. *Plant Physiol. Biochem.* **44**: 25-37.

Mithofer, A., Schulze, B. and Boland, W. (2004). Biotic and heavy metal stress response in plants: evidence for common signals. *FEBS Letters.* **566**: 1-5.

Mittler, R. (2002). Oxidative stress, antioxidants and stress tolerance. *Trends Plant Sci.* **7**: 405-410.

Moradi, F. and Ismail, A.M. (2007). Responses of photosynthesis, chlorophyll fluorescence and ros-scavenging systems to salt stress during seedling and reproductive stages in rice. *Ann. Bot.* **99:** 1161-1173.

Nagalakshmi, N. and Prasad, M.N.V. (2001). Responses of glutathione cycle enzymes and glutathione metabolism to copper stress in *Scenedesmus bijugatus*. *Plant Sci.* **160**: 291-299.

Navrot, N., Collin, V., Gualberto, J., Gelhaye, E., Hirasawa, M., Rey, P., Knaff, D.B., Issakidis, E., Jacquot, J.P. and Rouhier, N. (2006). Plant glutathione peroxidases are functional peroxiredoxins distributed in several subcellular compartments and regulated during biotic and abiotic stresses. *Plant Physiol.* **142**: 1364-1379.

Noctor, G. and Foyer, C.H. (1998). Ascorbate and glutathione: keeping active oxygen under control. *Annu. Rev. Plant Physiol. Plant Mol. Biol.* **49**: 249-279.

Noctor, G., Arisi, A.-C.M., Jouanin, L., Kunert, K.J., Rennenberg, H. and Foyer, C.H. (1998). Glutathione: biosynthesis, metabolism and relationship to stress tolerance explored in transformed plants. *J. Exp. Bot.* **49**: 623-647.

Noctor, G., Veljovic-Jovanovic, S., Driscoll, S., Novitskaya, L. and Foyer, C.H. (2002). Drought and oxidative load in the leaves of C3 plants: a predominant role for photorespiration? *Ann. Bot.* **89**: 841-850.

Pe´rez-Lo´pez, U., Robredo, A., Lacuesta, M., Sgherri, C., Mun˜oz-Rueda, A., Navari-Izzo, F. and Mena-Petite, A. (2009). The oxidative stress caused by salinity in two barley cultivars is mitigated by elevated CO_2. *Physiol. Plant.* **135**: 29-42.

Pinto, A.P., Mota, A.M., de Varennes, A. and Pinto, F.C. (2004). Influence of organic matter on the uptake of cadmium, zinc, copper and iron by sorghum plants. *Sci. Total Environ.* **326**: 239-247.

Qadir, S., Qureshi, M.I., Javed, S. and Abdin, M.Z. (2004). Genotypic variation in phytoremediation potential of *Brassica juncea* cultivars exposed to Cd stress. *Plant Sci.* **167**: 1171-1181.

Reddy, A.M., Kumar, S.G., Jyothsnakumari, G., Thimmanaik, S. and Sudhakar, C. (2005). Lead induced changes in antioxidant metabolism of horse gram (*Macrotyloma uniflorum* (Lam.) Verdc.) and bengal gram (*Cicer arietinum* L.). *Chemosphere.* **60**: 97-104.

Rehman, A. and Anjum, M.S. (2010). Multiple metal tolerance and biosorption of cadmium by *Candida tropicalis* isolated from industrial effluents: glutathione as detoxifying agent. *Environ. Monit. Assess.* doi: 10.1007/s10661-010-1480-x.

Rice-Evans, C.A., Miller, N.J. and Paganga, G. (1996). Structure-antioxidant activity relationships of flavonoids and phenolic acids. *Free Rad. Biol. Med.* **20**: 933-956.

Rios-Gonzalez, K., Erdei, L. and Lips, S.H. (2002). The activity of antioxidant enzymes in maize and sunflower seedlings as affected by salinity and different nitrogen sources. *Plant Sci.* **162**: 923-930.

Russo, M., Sgherri, C., Izzo, R. and Navari-Izzo, F. (2008). *Brassica napus* subjected to copper excess: Phospholipases C and D and glutathione system in signaling. *Env. Exp. Bot.* **62**: 238-246.

Sairam, R.K., Srivastava, G.C., Agarwal, S. and Meena, R.C. (2005). Differences in antioxidant activity in response to salinity stress in tolerant and susceptible wheat genotypes. *Biol. Plant.* **49**: 85-91.

Sairam, R.K. and Tyagi, A. (2004). Physiology and molecular biology of salinity stress tolerance in plants. *Curr. Sci.* **86**: 407-421.

Sairam, R.K., Deshmukh, P.S. and Saxena, D.C. (1998). Role of antioxidant system in wheat genotypes tolerance to water stress. *Biol. Plant.* **41**: 387-394.

Savoure, A., Thorin, D., Davey, M., Hua, X.-J., Mauro, S., Van Montagu, M., Inze, D. and Verbruggen, N. (1999). NaCl and $CuSO_4$ treatments trigger distinct oxidative defence mechanisms in *Nicotiana plumbaginifolia* L. *Plant Cell Environ.* **22**: 387-396.

Schroder, P., Lyubenova, L. and Huber, C. (2009). Do heavy metals and metalloids influence the detoxification of organic xenobiotics in plants? *Env. Sci. Poll. Res.* **16**: 795-804.

Scruton, N.S., Berry, A. and Perham, R.N. (1990). Redesign of the coenzyme specificity of a dehydrogenase by protein engineering. *Nature.* **343**: 38-43.

Seckin, B., Turkan, I., Sekmen, A.H. and Ozfidan, C. (2010). The role of antioxidant defense systems at differential salt tolerance of *Hordeum marinum* Huds. (sea barleygrass) and *Hordeum vulgare* L. (cultivated barley). *Env. Exp. Bot.* **69**: 76-85.

Selote, D.S. and Khanna-Chopra, R. (2010). Antioxidant response of wheat roots to drought acclimation. *Protoplasma* DOI 10.1007/s00709-010-0169-x.

Sharma, P. and Dubey, R.S. (2005). Drought induces oxidative stress and enhances the activities of antioxidant enzymes in growing rice seedlings. *Plant Growth Regul.* **46**: 209-221.

Shehab, G.G., Ahmed, O.K. and El-beltagi, H.S. (2010). Effects of various chemical agents for alleviation of drought stress in rice plants (*Oryza sativa* L.). *Not. Bot. Hort. Agrobot. Cluj.* **38**: 139-148.

Shi, Q., Bao, Z., Zhu, Z., He, Y., Qian, Q. and Yu, J. (2005). Silicon-mediated alleviation of Mn toxicity in *Cucumis sativus* in relation to activities of superoxide dismutase and ascorbate peroxidase. *Phytochem.* **66**: 1551-1559.

Shinozaki, K.Y., Kasuga, M., Liu, Q., Nakashima, K., Sakuma, Y., Abe, H., Shinwari, Z.K., Seki, M. and Shinozaki, K. (2002). Biological mechanism of drought stress response. *JIRCAS Working Report*: 1-8.

Singh, S., Eapen, S. and D'Souza, S.F. (2006). Cadmium accumulation and its influence on lipid peroxidation and antioxidative system in an aquatic plant, *Bacopa monnieri* L. *Chemosphere.* **62**: 233-246.

Sofo, A., Cicco, N., Paraggio, M. and Scopa, A. (2010). Regulation of the ascorbate–glutathione cycle in plants under drought stress. In *Ascorbate-Glutathione Pathway and Stress Tolerance in Plants,* pp. 137-190, Eds. NA Anjum, S Umar and M-T Chan (The Netherlands: Springer).

Sofo, A., Tuzio, A.C., Dichio, B. and Xiloyannis, C. (2005). Influence of water deficit and rewatering on the components of the ascorbate-glutathione cycle in four interspecific *Prunus* hybrids. *Plant Sci.* **169**: 403-412.

Soranzo, N., Sari, Gorla, M., Mizzi, L., De Toma, G. and Frova, C. (2004). Organisation and structural evolution of the rice glutathione S-transferase gene family. *Mol. Gen. Genomics.* **271**: 511-521.

Sun, Q., Wang, X.R., Ding, S.M. and Yuan, X.F. (2005). Effects of exogenous organic chelators on phytochelatins production and its relationship with cadmium toxicity in wheat (*Triticum aestivum* L.) under cadmium stress. *Chemosphere.* **60**: 22-31.

Szalai, G., Kellos, T., Galiba, G. and Kocsy, G. (2009). Glutathione as an antioxidant and regulatory molecule in plants under abiotic stress conditions. *J. Plant Growth Regul.* **28**: 66-80.

Szollosi, R., Varga, I., Erdei, L. and Mihalik, E. (2009). Cadmium-induced oxidative stress and antioxidative mechanisms in germinating Indian mustard (*Brassica juncea* L.) seeds. *Ecotoxicol. Environ. Saf.* **72**: 1337-1342.

Terzi, R. and Kadioglu, A. (2006). Drought stress tolerance and the antioxidant enzyme system in *Ctenanthe setosa. Acta Biol. Cracoviensia Series Bot.* **48/2**: 89-96.

Tukendorf, A. and Rauser, W.E. (1990). Changes in glutathione and phytochelatins in roots of maize seedlings exposed to cadmium. *Plant Sci.* **70**: 155-166.

Turhan, E., Gulen, H. and Eris, A. (2008). The activity of antioxidative enzymes in three strawberry cultivars related to salt-stress tolerance. *Acta Physiol. Plant.* **30**: 201-208.

Tuteja, N. (2007). Mechanisms of high salinity tolerance in plants. *Methods Enzymol: Osmosensing and Osmosignaling.* **428**: 419-438.

Vaidyanathan, H., Sivakumar, P., Chakrabarty, R. and Thomas, G. (2003). Scavenging of reactive oxygen species in NaCl stresses rice (*Oryza sativa* L.) - differential response in salt-tolerant and sensitive varieties. *Plant Sci.* **165**: 1411-1418.

Vanacker, H., Carver, T.L.W. and Foyer, C.H. (2000). Early H_2O_2 accumulation in mesophyll cells leads to induction of glutathione during the hyper-sensitive response in the barley powdery mildew interaction. *Plant Physiol.* **123**: 1289-1300.

Verma, S. and Dubey, R.S. (2003). Lead toxicity induces lipid peroxidation and alters the activities of antioxidant enzymes in growing rice plants. *Plant Sci.* **164**: 645-655.

Wagner, U., Edwards, R., Dixon, D.P. and Mauch, F. (2002). Probing the diversity of the Arabidopsis glutathione S-transferase gene family. *Plant Mol. Biol.* **49**: 515-532.

Wang, H.-Z., Zhang, L.-H, Ma, J., Li, X.-Y., Li, Y., Zhang, R.-P. and Wang, R.-Q. (2010). Effects of water stress on reactive oxygen species generation and protection system in rice during grain-filling stage. *Agric. Sci. China.* **9**: 633-641.

Wu, F.-B., Chen, F., Wei, K. and Zhang, G.-P. (2004). Effect of cadmium on free amino acid, glutathione and ascorbic acid concentrations in two barley genotypes (*Hordeum vulgare* L.) differing in cadmium tolerance. *Chemosphere.* **57**: 447-454.

Yadav, S.K. (2010). Heavy metals toxicity in plants: An overview on the role of glutathione and phytochelatins in heavy metal stress tolerance of plants. *South Afr. J. Bot.* **76**: 167-179.

Yan, F., Yang, W.-K., Li, X.-Y., Lin, T.-T., Lun, Y.-N., Lin, F., Lv, S.-W., Yan, G.-L., Liu, J.-Q., Shen, J.-C., Mu, Y. and Luo, G.-M. (2008). A trifunctional enzyme with glutathione S-transferase, glutathione peroxidase and superoxide dismutase activity. *Biochim. Biophys. Acta.* **1780**: 869-872.

Yannarelli, G.G., Ferna´ndez-Alvarez, A.J., Santa-Cruz, D.M. and Tomaro, M.L. (2007). Glutathione reductase activity and isoforms in leaves and roots of wheat plants subjected to cadmium stress. *Phytochem.* **68**: 505-512.

Yasar, F., Ellialtioglu, S. and Yildiz, K. (2008). Effect of salt stress on antioxidant defense systems, lipid peroxidation, and chlorophyll content in green bean. *Russ. J. Plant Physiol.* **55**: 782-786.

Yazici, I., T¨urkan, I., Sekmen, A.H. and Demiral, T. (2007). Salinity tolerance of purslane (*Portulaca oleracea* L.) is achieved by enhanced antioxidative system, lower level of lipid peroxidation and proline accumulation. *Env. Exp. Bot.* **61**: 49-57.

Zhang, C.-H. and Ge, Y. (2008). Response of glutathione and glutathione S-transferase in rice seedlings exposed to cadmium stress. *Rice Sci.* **15**: 73-76.

Zhou, Z.S., Guo, K., Elbaz, A.A. and Yang, Z.M. (2009). Salicylic acid alleviates mercury toxicity by preventing oxidative stress in roots of *Medicago sativa*. *Env. Exp. Bot.* **65**:27-34.

Zhou, Z.S., Wang, S.J. and Yang, Z.M. (2008). Biological detection and analysis of mercury toxicity to alfalfa (*Medicago sativa*) plants. *Chemosphere.* **70**: 1500-1509.

Zhu, J.K. (2002). Salt and drought stress signal transduction in plants. *Annu. Rev. Plant Biol.* **53**: 247-73.

17

Heavy Metals in Soil and Plants: An Overview of Arsenic, Cadmium, Chromium and Mercury

Mohd. Kafeel A. Ansari[1*,$], **Naser A. Anjum**[2], **Altaf Ahmad**[1], **Shahid Umar**[1] and **Muhammad Iqbal**[1]

[1]*Department of Botany, Faculty of Science, Hamdard University, Hamdard Nagar, New Delhi-110 062, India*

[2]*Center for Environmental and Marine Studies (CESAM) & Department of Chemistry, University of Aveiro, 3810-193 Aveiro, Portugal*

[$]*Present address: Centre for Environmental Risk Assessment and Remediation, University of South Australia, Mawson Lakes, SA 5095, Australia*

***Corresponding author:** kafeelansari123@gmail.com*

ABSTRACT

Various anthropogenic activities have dramatically changed the composition and organization of the majority of agricultural soils. Industrial and urban wastes, agricultural application and also mining activities resulted in an increased concentration of different heavy metals (HMs) in soils. In fact, accumulation of HMs in agricultural soils is of increasing concern because of food safety issues, potential health risks, and its detrimental effects on soil ecosystems. Among major heavy metals arsenic (As), cadmium (Cd), chromium (Cr), and mercury (Hg) are of great concern because of their wide occurrence and toxicity to biotic community. Crop plants take up these metals and accumulate them in their edible and non-edible parts in various concentrations. The intake of the edible parts is an important path for HMs in the soil to harm human health. It is, therefore, important to control and limit the HMs accumulation in agricultural products. The current chapter mainly gives an overview of As, Cd, Cr and Hg in the environment, the specific forms in soils, the mode of uptake, translocation and accumulation of these heavy metals by plants, information on how these heavy metals modulate antioxidant enzymes system in different plants, and ends with critical appraisal, conclusion and perspectives.

Keywords: Heavy metals; oxidative stress; tolerance.

1. INTRODUCTION

The term 'heavy metal' (HM) is somewhat imprecise; it includes most metals with an atomic number greater than 20, but excludes alkali metals, alkaline earths, lanthanides and actinides. Passow et al. (1961) defined HMs as those metals that have a density greater than five; there are about thirty-eight such elements. However, in general terms, metals may be found in the lower-left side of the diagonal of semi-metals or metalloids (from Boron to Polonium) in the periodic table, which may act as both metals and non-metals. Metals are distinguished from non-metals by their capacity to lose electrons, forming positively charged ions, or Lewis acids, and by their speciation-dependent affinity for abiotic and biological interaction (Duffus 2002, D'Amore et al. 2005). Some metals are essential for biological life in trace amount, but their excessive amounts are poisonous and can even cause death to most of the living organisms. Accumulation of HM in agricultural soils is of increasing concern because of food safety issues, potential health risks, and its detrimental effects on soil ecosystems (McLaughlin et al. 1999; Cui et al. 2004; Hung and Jin 2008). Heavy metals are frequently used as components of fungicides, pesticides or disinfectants. In agricultural areas, application of phosphate fertilizers, pesticides, and sewage sludge is the most important anthropogenic source of potentially toxic metals (Mermut et al. 1996; Yanai et al. 1998; Madyima et al. 2002).

As, Cd, Cr, and Hg are of great concern because of their toxicity to human health and other organisms, whereas Cu and Ni are essential elements for plants and human. Crop plants take up these metals and accumulate them in their edible and non-edible parts in various concentrations. The intake of the edible parts is an important path for heavy metals in the soil to harm human health. It is, therefore, important to control and limit the HM accumulation in agricultural products (Finster et al. 2004; Glorennec, 2006).

2. HEAVY METALS IN THE ENVIRONMENT

Arsenic (As) is a metalloid with atomic number 33 and atomic weight 74.92 and its position in periodic table is group fifteenth in 4th period. Arsenic occurs naturally in the environment, being approximately 0.8 x 10^{16} kg^{-1} in the atmosphere (Walsh et al. 1979) and 4.01 x 10^{6} in the earth's crust (Matschullat 2000). On an average, shales, granites and sandstones have 13.0, 3.0 and 1.0 mg As kg^{-1} soil, respectively (Onishi 1969). Anthropogenic sources, such as mining and smelting operations, fuel combustion, biosolids, tanning, wood preservatives and pesticides have resulted in a significant increase in As concentration in the environment (O'Neill 1990). Another examples of anthropogenic sources are ore beneficiation, fertilizers and chemical industries, thermal power plants using coal, wood preservation industries using CCA, and incinerations of preserved wood (Jacks et al. 1998; Tamaki et al. 1992). Arsenic concentration presently ranges from 0.2 to 41 mg kg^{-1} in different soils of the world (Kabata-Pendias and Pendias 2001).

Cadmium is a heavy metal with the symbol Cd, having atomic number 48, with an atomic weight 112.411. The soft, bluish-white metal is in group twelve and in 5th period in the periodic table. Cadmium is an element with an average distribution of 0.1 mg kg^{-1} in the earth's crust. It occurs naturally in the geosystem. Cd is released in the environment as a result of rock mineralization process. Igneous and metamorphic rocks are the commonest natural sources in the soil. Page and Bingham (1973) reported that soils derived from igneous rocks have Cd contents of 0.1-0.3 mg kg^{-1}, those on metamorphic rocks contain 0.1-1.0 mg Cd kg^{-1} soil and those derived from sedimentary rocks 0.3-11 mg Cd kg^{-1} soil. Volcanic action, ocean sprays and forest fires are the other natural sources of Cd emission. Anthropogenically, cadmium is released into the environment by power stations, heating systems, metal-work industries, waste incinerators, urban traffic, cement factories and as a by-product of phosphate fertilizers (Sanita' di Toppi and Gabbrielli 1999). Campbell et al. (1983) compared natural and anthropogenic quantities of trace metals emitted to the atmosphere and showed that around 15 times more Cd is emitted by human activities, compared with the natural processes. Wagner (1993) estimated that non-polluted soil solutions contained 0.04 to 0.32 μM Cd concentrations. Soil solutions with Cd concentration from 0.32 to 1.0 μM are regarded to be polluted to a moderate level.

Chromium is a chemical element which has the symbol Cr and atomic number 24 and atomic weight 51.99, first element in Group 6 and period 4th. Naturally occurring Cr in the soil, ranges from 10 to 50 mg kg^{-1}, depending on the parental material. However, in ultramafic soils (serpentine), it can reach up to 125 mg kg^{-1} (Adriano 1986). Water coming from Cr-related industries has severely contaminated soil and water bodies. Worldwide, anthropogenic discharge of Cr in freshwater bodies has been estimated to be 3550 metric tons (Nriagu 1990). In freshwater, Cr concentrations generally range from 0.1 to 117 μg L^{-1}, whereas values for seawater range from 0.2 to 50 μg L^{-1}. Cr concentration varies widely in the atmosphere, from background concentrations of 5.0×10^{-6} – 1.2×10^{-3} μg m^{-3} in air samples from remote areas to 0.015-0.03 μg m^{-3} in air samples collected from the urban areas (Nriagu 1988). Metal industry uses chromium in the formation of alloys (Downing et al., 1986), preferably in stainless steel. During the production of such alloys small particles of chromium escape into the atmosphere in the form of dust (Downing et al., 1986).

Copper is an element with the symbol Cu and atomic number 29 and atomic weight 63.54. Its position in the periodic table is group eleventh in 4th period. Copper occurs naturally in most soils and in fruits and vegetables. Both humans and animals need some copper in their diet. In humans, it helps in the production of blood haemoglobin. High level of Cu in the environment becomes toxic to living beings. Up to a level of 2000 ppm Cu in soil was recorded in mining areas and in the vicinity of Cu smelters (Freedman and Hutchinson 1980; Humphreys and Nicholls 1984). Anthropogenic Cu sources are copper compounds, which are used as insecticides (copper fluoroarsenate), fungicides (copper sulphate), algaecides (copper dimethylgloxime) and molluscicides. The anthropogenic sources of Cu are paint, analytical reagents, and solvents used in rayon manufacture and electroplating processes. Compounds of copper, such as acetoarsenite and cupric arsenite are used as pigments (Patty 1963).

Mercury, also known as quicksilver or hydrargyrum is a chemical element with the symbol Hg. It contains atomic number 80 and atomic weight 200.59. Among the metal contaminants, mercury (Hg) is a major pollutant in the soil and water. Considerable amounts of Hg may be added to agricultural land with the application of sludge, fertilizers, pesticides, lime and manures (Kabata-Pendias and Pendias 1989). In nature, mercury occurs in a variety of physical and chemical forms. Industrial wastes and domestic discharges contribute Hg to aquatic environments (Sinha and Gupta et al., 1996; Wang and Greger et al., 2004). Normal soils typically contain 20-150 ppb Hg, but its level can reach as high as 80% (WHO 1976). The most common mineral form of Hg is the non-toxic, insoluble mercuric sulphide or cinnabar (HgS), a by-product obtained by the processing of complex ores that contain mixed sulphides, oxides, and chloride minerals. The naturally occurring Hg is released by degassing of the earth's crust, volcanoes and the evaporation from oceans (Boening 1999). Anthropogenically, Hg is added to the environment because of its use in industry, medicine, dentistry, batteries, science, and military applications. The burning of fossil fuels and medical-waste incineration accounts for more than 80% of all anthropogenic sources (Raskin et al. 2000).

Nickel (Ni) is also treated as a heavy metal even in higher concentration. Nickel has atomic number 28 and atomic weight 58.69. On an average, Ni constitutes 0.0086 per cent (80 parts per million) of the earth's crust. In plants, nickel forms complexes with several organic compounds such as amino acids (Kramer et al. 1996) and organic acids (Pelosi et al. 1976; Lee et al. 1978). Seawater contains 0.1 to 0.5 μg Ni L^{-1}, mainly in the form of Ni^{2+} cations and the chloro- and carbonato-complexs (Bruland 1983). The natural sources of Ni are volcanoes, wind-blown dust and wildfires. The natural sources release approximately 8.5 million kg of Ni into the atmosphere per year (Public Health Service, Washington 1997). Anthropogenic sources of Ni are nickel metal refining, municipal waste incineration, burning of residual and fuel oil, coal combustion, Ni alloy production and steel production (Public Health Service, Washington 1993). Anthropogenic activities release approximately 56 million kg of Ni into the atmosphere per year (Nriagu et al., 1988).

3. HEAVY METAL SPECIES

Species can be defined as an ensemble of chemically identical molecular entities that can explore the same set of molecular energy levels on the time scale of the experiment. The term is applied equally to a set of chemically identical atomic or molecular structural units in a solid array. Arsenic (As) in the soil normally occurs in the III and V oxidation states (Cullen et al., 1989), in the form of $H_3As^{III}O_3^0$ and as $H_2AsO_4^-$ (pH < 7.0) (Bohn et al. 1979; Sadiq et al. 1983; Montgomery 1985; Cotton and Wilkinson 1988). As-III is more toxic and more mobile in soils than As-V (Ferguson et al. 1972; Deuel et al. 1972; Hess and Blanchar 1976; Hermann and Neumann-Mahlkau 1985). The adsorption optimum for As-III is around pH 7.0, while As-V adsorbs optimally at pH 4.0 (Pierce and Moore 1982). In groundwater, inorganic As commonly exists as As-V (arsenate) and As-III (arsenite). As-III is considered more toxic to plants than As-V (Korte, Fernando 1991).

Baker et al. (1990) reported that cadmium (Cd) never occurs in isolation in natural environments, but mostly as a 'guest' metal in Pb/Zn mineralization. Duxbury (1985) classified cadmium as an element of 'intermediate toxicity'. In the environment, cadmium exists in many forms, such as elemental cadmium, cadmium carbonate, cadmium chloride, cadmium oxide, cadmium sulphate and cadmium sulphide (ATSDR 1993).

Chromium (Cr) exists predominantly in III and VI oxidation states. The hexavalent chromium (Cr-VI) compounds are comparatively more toxic than trivalent chromium (Cr-III) due to high solubility of the former in water (Panda and Patra 1997). Cr-III is less mobile, less toxic and is mainly found bound to organic matter in the soil and aquatic environment (Becquer et al. 2003). Cr-VI compounds are widely used as industrial chemicals and are major contaminants of both soil and water. Both Cr-III and Cr-VI are biologically active oxidation states. Sethunathan et al. (2005) found that soil microorganisms contribute to the reduction of Cr-III to Cr-VI.

Copper (Cu) is present naturally in most soils and in the crop plants. Cu levels in soils higher than 2000 ppm have been recorded in mining areas and in the vicinity of the Cu smelters (Freedman and Hutchinson 1980; Humphreys and Nicholls 1984). Copper becomes a toxic element when in high concentration. It occurs in the environment Cu either metallic or compound forms as (Cu-I) and (C-II). According to Flemming and Travors, 1989, Cu is transition metal with three oxidation states; zero (Cu^0, solid metal), plus one [Cu (I), cuprous ion]; and plus two [Cu (II), cupric ion].

Mercury (Hg) is present in three different oxidation states, Hg° (metallic), Hg_2^{2+} (mercurous), and Hg^{2+} (mercuric) (Hem 1970). Mercury compounds differ greatly in their solubility in water. The solubility in water increases in the following order: elemental mercury (2μ g L^{-1} at 30°C) < mercurous chloride (2 mg L= $^{-1}$ at 25°C). The solubility of methyl mercury chloride is lesser than mercuric chloride (69 g L^{-1} at 20°C), but higher than that of mercurous chloride by about three times, this being related to the high solubility of methyl mercury ion in water (WHO 1990). The mercurous and mercuric states differ considerably from numerous inorganic and organic chemical compounds in their toxic effects. Organic mercury compounds concentrate in organic tissues in the food chain because of inability of organisms to decompose them (Greeson 1970; WHO 1991).

Nickel (Ni) is obtained primarily from sulphide ores (e.g., pentlandite and nickeliferous pyrrhotite) and to a lesser extent from oxide ores (e.g., laterite) (Saager 1984; Mastromatteo, 1986; Schaufelberger 1987). In aqueous solutions, nickel is most often divalent and exists primarily as the hexaquonickel $[Ni (H_2O)_6]^{2+}$ ions. Other valences include -1, 0, +1, +3 and +4 in solution (Nieboer et al., 1984). Ni^{2+} is 4- or 6- coordinated and most commonly occurs in square planar configuration and less often in tetrahedral or octahedral configurations (Coogan et al., 1989). Ni^{2+} exhibits both hard and soft acid properties (Nievoer et al., 1996). Ni also combines with CO at atmospheric pressure to form highly toxic nickel carbonyl $[Ni (CO)_4)]$ (Von Burg 1997).

4. UPTAKE, TRANSLOCATION AND ACCUMULATION OF MAJOR HEAVY METALS

In order to understand the mode of action of heavy metals it is necessary to examine their uptake and translocation. The entry of HM from the external solution into the cell wall is a non-metabolic, passive process, driven by diffusion or mass flow (Marschner 1995). The metal first enters into the apoplast of the root. Then some of the total amount of the HM is transported further into the cells, some in the apoplast and some becomes bound to cell wall substances. The total amount of the HM taken up is dependent on the metal and the plant genotype. Part of HM that is taken up into the apoplast is further transported through the plasma membrane into the cytoplasm. In the cytoplasm, the HM binds to negative charges in various macromolecules that are either soluble or part of the cell structures. Soluble molecules in the cytoplasm like phytochelatins (PCs) form complex with metals and may also function as shuttles to HM transport over the tonoplast into vacuole of the cell.

4.1 Uptake of Arsenic

Arsenic uptake is influenced by its source and solubility (Marcus-Wyner and Raines 1982). Arsenate is a chemical analogue of phosphate, and is taken up via the phosphate transport system (Asher and Reay 1979). Arsenate uptake is active and was taken up at approximately the same rate as arsenite (Abedin et al., 2002). Wang et al. (2002) examined the uptake kinetics of arsenate and arsenite, and the effects of phosphate on arsenic uptake in *Pteris vittata.* Phosphate can decrease metal solubility and mobility, which will reduce HM transport from contaminated soils to surface and groundwater and induce the formation of HM phosphate precipitates (Cotter-Howells and Capron 1996). Typical of hyper accumulators, arsenic concentrations in *P. vittata* are mostly concentrated in the fronds (Ma et al., 2001; Tu et al. 2002; *Zhang* et al., 2002), suggesting an efficient transport of arsenic from roots to fronds in *P. vittata.* The other plant species are also able to reduce arsenic (V), however, these species have a lower As accumulation capacity compared with *P. vittata* (Singh and Ma 2006). In general, plants uptake and mobilize As (V) via the phosphate transport channels (Tripathi et al. 2007; Jose R. Peralta-Videa et al., 2009). Because of their chemical similarity, As (V) competes with phosphate for root uptake and interferes with metabolic processes like ATP synthesis and oxidative phosphorylation (Tripathi et al., 2007). Pickering et al. (2000) has been reported that As (III) can be coordinated to sulphur ligands and transported as As (III)-*tris*-glutathione complex.

4.2 Uptake of Cadmium

Movement of cadmium ions from the soil solution to plant roots takes place by mass flow of the soil solution. The degree to which the higher plants are able to take up cadmium depends on cadmium concentration in the soil and its bioavailability modulated by the presence of organic matter, pH level and the concentration of other metals (Cataldo et al., 1981; Salt et al., 1995). Decrease in soil pH increases the Cd bioavailability and the uptake of cadmium. The optimum Cd mobility is achieved at pH 4.5-5.5 (Bingham 1980; Joner and Leyval 1997). The adsorption of Cd is almost doubled for each increase of 0.5 units in the pH from 4-7 (Anderson 1998). Accumulation of Cd is directly proportional to the concentration of Cd supplied to the growing plant as well as the phenological state at which Cd is supplied. However, the actual accumulations are influenced by the plant species and soil properties (Jemal et al. 1998, Fediue and Erdei 2002). A tenfold accumulation of Cd in roots of *Chicorium intybus* and a fivefold in the root of *Catharanthus roseus* was found as compared to their shoots, suggesting that these plants are the cadmium shoot excluders (Lozano-Rodriguez et al., 1997; Florijin and van Bensicgem 1993). In soybean, about 98% of the accumulated Cd is strongly retained by roots, and only 2% is transported to the shoot system (Cataldo et al., 1993; Prasad 1997).

There are some plants (e.g., *Alyssum* spp., *Thlaspi earrulescens)* that accumulate more cadmium in shoots than in roots. These are called hyperaccumulators (Kramer et al., 1996; Li et al., 2003). In soybean, about 98% of the accumulated Cd is strongly retained by roots, and only 2% is transported to the shoot system (Cataldo et al., 1993; Prasad 1997). In rice, the concentration of Cd in grains is governed somewhat by its uptake and transport from roots to shoots, and to a greater extent, by the transport of Cd from shoots to grain. Liu et al. (2007) in their earlier study, did in China, concluded that about 0.73% of the total Cd taken up by six rice cultivars was transferred to the grain. This represents an average of 1.02 mg kg^{-1}, which is 100 times higher than the concentration allowed by the European Union for Cd concentration in rice grain (Olsson et al., 2005). The above studies were validated later by Jose R. Peralta-Videa et al., 2009.

4.3 Uptake of Chromium

Chromium uptake by plants is mainly non-specific, probably as a result of plant uptake of essential nutrients and water. Plants can absorb both Cr-VI and Cr-III. The former is more easily transported inside the plant than the latter (Mortvedt and Giordano 1975). Cr-VI uptake has been reported to occur by an active mechanism, whereas Cr-III uptake is passive, indicating that the two forms do not share a common uptake mechanism (Skeffington et al. 1976). In contrast, an active uptake of both Cr species, slightly higher for Cr-III than for Cr-VI, has been found in the same crop (Ramachandran et al., 1980). Once in the xylem, CrO_4^{2-} is more easily transported than Cr-III, presumably because the latter is bound by ion exchange on the vessel walls or precipitated as $Cr\ (OH)_3$. Cr-EDTA moves faster than Cr-III ions to the shoots due to the higher solubility of the former in aqueous solution. Cr-III is predominantly transported in the form of organic complexes. For example, the only species of Cr-III, identified in xylem sap of *Leptospernum scoperium*, following the addition of $CrCl_3$, was the Cr-III trioxalate complex. Golovatyj et al. (1999) have shown that Cr distribution in crops has a stable character, which does not depend on soil properties and concentration of this element; the maximum quantity of element contaminant is always contained in roots and a minimum in the vegetative and reproductive organs. In bean, only 0.1% of the Cr accumulated in the seeds as against 98% in the roots (Huffman and Allaway 1973). The reason of this high accumulation in the roots could be that Cr is immobilized in the vacuoles of the root cells, thus rendering it less toxic, which may be a natural toxicity response of the plant (Shanker et al., 2004). The scarce translocation of Cr to harvestable plant parts does not increase by the addition of chelators (Chen and Cutright 2001). According, Cohen et al. (2006), at neutral pH, Cr (VI) compounds are tetrahedral and are transported across cell membranes via similar tetrahedral ion channels, while Cr (III) is octahedral and transported through diffusion across membranes. Montes-Holguin et al. (2006) have been reported that *Convolvulus arvensis* plants treated with potassium dichromate or chromium nitrate had in leaves different per cents of chromium compounds similar to chromium acetate, chromium phosphate, and chromium nitrate. The highest per cent of chromium nitrate type compounds suggests, due to the high solubility of chromium nitrate, that plants with high capability of chromium accumulation can contribute Cr to the food chain as trivalent Cr (Jose R. Peralta-Videa et al., 2009).

4.4 Uptake of Copper

Due to the strong binding of Cu by organic matter and other soil colloids, its mobility is severely restricted and the fraction of total Cu available for plant uptake is usually low. Thus, total Cu concentrations in the soil are not usually related to plant uptake and Cu additions to the soil are frequently not accompanied by a concomitant increase in plant Cu levels, especially in soils with high organic matter content (Berry 1985; Bjerre and Schierup 1985). Cu levels below the deficiency limit were found in the leaves of *Oryza sativa* growing in paddy soils containing Cu levels up to 94 ppm (Fernandes and Henriques 1990). Accumulation of Cu in agricultural soils resulting from continuous use of Cu fungicides and sludge application has been reported in many countries (Merry et al., 1983; Alva et al., 1995; Adriano 2001). One of the major consequences of excessive accumulation of Cu in soils is its toxicity to plants and microbial communities (Hunter et al., 1989). For example, high concentration of Cu in soils has been attributed as one of the reasons for the formation of bare sterile patches in orchards where excessive amounts of Cu fungicides have been used. In many countries, a large number of horticultural soils which have elevated concentration of Cu have been converted to other uses. The hazards of excessive Cu on grazing animals in relation to Cu-induced molybdenum deficiency and the possible reduction of yields of sensitive crops in these soils have been reported (Merry et al., 1983). Copper is strongly bound to iron and aluminium hydrous oxides, silicate clay minerals and organic matter (McBride 1994). Plant availability of Cu in soils is controlled largely by the nature and the amount of these soils components, and soil pH (McLaren et al., 1990; Baker and Sneft 1995; Adriano 2001).

4.5 Uptake of Mercury

The absorption of organic and inorganic mercury from soils by plants is low (Lodenius 1999) and there is probably a barrier to Hg translocation from plant roots to tops. However, Hg-containing pesticide/fungicide spray residues are taken up by plants and translocated to edible portions. Hg salts in the soil may be reduced by biological and chemical reactions to metallic or methylated compounds, which may volatilize and be taken up through the leaves in plants grown in enclosed spaces. Factor affecting the Hg uptake include external Hg concentration and exposure time, the soil or sediment organic content, carbon-exchange capacity, the oxide and carbonate contents, and the redox potential (Crowder 1991; Cho and Park 1999). Accumulation, toxicity response, and distribution of Hg differ between the exposure to elemental Hg^0 through shoots or ionic Hg^{2+} through roots. In rice plants exposed to Hg^0, Hg accumulated in the shoots with no movement to roots. Hg-exposed plants showed accumulation of Hg with movement to the shoots by the 10th day. Inhibition of root and shoot growth occurred at 1.0 mM and above, with very limited tissue damage at higher treatment levels (Suszcynsky and Shann 1995). Moreover, the detailed studies of Hg uptake in model food grain sources have been conducted to establish the impact of Hg and mercurials on human health. For examples, the transfer characteristics of Hg, Pb, Cd, Zn and Cu from soil to vegetables near a zinc smelting plant in Huludao City (China) were investigated, and the concentrations of heavy metals in the soil and vegetable were analyzed (Zheng et al., 2007b; Jose R. Peralta-Videa et al., 2009).

4.6 Uptake of Nickel

Ni uptake depends mainly on plant species and soil conditions, particularly pH and texture (Adriano 1986; Tinker 1986; Welch and Lund 1987; Uren 1993). Bioavailability of Ni increases as soil pH decreases (Adriano 1986; Uren 1993). The source of nitrogen may influence the bioavailability and amounts of Ni taken up by plants. Wiltshire (1974) found that fertilization with nitrate increased Ni levels in the roots of some plant species grown on metalliferous soils. Ni accumulation has been reported in a variety of medicinal plants, grown in the Negev desert (Sathiyamoorthy et al., 1997). The uptake of nickel decreases after the addition of EDTA as chelating agent to the soil or media (Eskew et al., 1984; Albasel and Cottenie 1985; Molas 1997). On the contrary, complexation of Ni by histidine increases its phytoassimilation, but decreases its phytotoxicity (Kramer et al., 1996).

The major way of Ni uptake from soil is its absorption by plant roots. Some of the soil-accumulated metal becomes bound to its organic components and becomes inaccessible to plants, whereas metal ions can enter the roots (Seregin ar.d Kozhevnikova 2006). The pattern of Ni accumulation in cell walls and protoplasts of all root tissues presumes that Ni is translocated via both the apoplast and symplast. When reaching the xylem by radial transport, Ni will reach the shoots with the xylem sap (Cataldo et al., 1988; Temp 1991; Yang 1997; Kramer 1997; Sagner et al., 1998). Toxic metals are typically sequestered in roots with limited translocation of such metals to shoots (Rao and Sresty 2000; Shah et al., 2001; Verma and Dubey 2003; Parida et al., 2003, Sharma and Dubey 2007).

5. MODULATION OF ANTIOXIDANT ENZYMES AS INFLUENCED BY HEAVY METALS

Plants have divergent mechanisms for modulating metal levels to adapt to changes in concentrations of metals in the environment. The metals are toxic to plants if their accumulation levels exceed the detoxification capacity of the plant tissues (Sinha and Gupta et al., 1996). Plant cells possess antioxidants like α-tocopherol, glutathione and ascorbate and antioxidative enzymes such as superoxide dismutase (SOD, EC1.15.1.1), ascorbate peroxidase (APX, EC 1.11.1.11), catalase (CAT, EC1.11.1.6) and glutathione reductase (GR, EC1.6.4.2) that participate in scavenging the reactive oxygen species (ROS). An important feature of HM toxicity is the generation of ROS such as the superoxide anion (O_2^-), singlet oxygen, hydrogen peroxide (H_2O_2) and the hydroxyl radical ($OH^\bullet$), which causes tissue injury after the exposure of plants to HMs (Halliwell

1982; Cho and Park 2000; Qureshi et al., 2005; Anjum et al., 2008a-c, 2011; Ansari et al., 2009; Ishrat et al., 2009) (Fig. 1). Complex antioxidant systems which neutralize and scavenge the ROS are very important for plants in order to protect cellular membranes and organelles from the damaging effects of ROS, generated by the environmental stress. Exposure of plants to heavy metals may result in the production of ROS, such as superoxide anions, hydrogen peroxide (H_2O_2) and hydroxyl radicals, resulting in damage to cell components (Weckx and Clijsters 1996; Conklin 2001; Hartley-Whitaker et al., 2001a). In response to the creation of ROS, plants synthesize enzymatic and non-enzymatic antioxidants (Meharg and Hartley-Whitaker 2002). Through the use of antioxidant enzymes and molecules, plants can manage the detrimental effects of ROS. Specifically, ascorbic acid, which makes up approximately 10% (w/w) of the plant soluble carbohydrate, is an important and major plant antioxidant due to its high abundance (Noctor and Foyer 1998). As an antioxidant, ascorbic acid can manage ROS through direct elimination of superoxide anions, hydrogen peroxide and hydroxyl radicals (Padh 1990). It can also act as a secondary antioxidant through the maintenance of reduced α-tocopherol, another plant antioxidant (Liebler et al. 1986; Noctor and Foyer 1998; Conklin 2001), or obliquely through the action of ascorbate peroxidase (Foyer and Halliwell 1976; Asada 1992). Superoxide dismutase eliminates superoxide anions (O_2^-), yielding oxygen and hydrogen peroxide (H_2O_2), superoxide dismutase is associated with various metal cofactors: CuSOD and ZnSOD are located in cytosol, peroxisome, plastids and root nodules; MnSOD in mitochondria; and FeSOD in plastids. Because SOD can degrade superoxide anions, it can play important role in the defence of cells upon stress (Fridovich 1978; Fridovich 1986; Fridovich 1995). A study on enzymatic antioxidants found that SOD and CAT are induced in the fronds of *P. vittata* upon arsenic exposure. However, they are not induced in the fronds of *P. ensiformis* under the same conditions (Srivastava et al., 2005). This suggests a role for antioxidant enzymes in arsenic tolerance and/or hyperaccumulation by *P. vittata*. Similarly, SOD and CAT activities were found to increase in *Zea mays* embryos upon arsenic exposure (Mylona et al., 1998). In several plants, Cd-induced changes in activities of ROS-scavenging enzymes, including SOD, CAT and APX, have been demonstrated (Shaw 1995; Gallego et al., 1996; Lozano-Rodriguez et al., 1997; Chaoui et al., 1997; Wu et al., 2003). Depending on concentration, cadmium can either inhibit or stimulate the activity of several antioxidative enzymes before the appearance of any visible symptoms of toxicity. In sunflower leaves, Cd decreased the activity of SOD, CAT, APX, GR, and dehydroascorbate reductase (DHAR, EC1.8.5.1), while LOX (lipoxygenase), which mediates lipid peroxidation, increased (Gallego et al., 1996). Pea plants exposed to environmentally relevant (20 μM) and acute (200 μM) concentrations of Cr (VI) for 7 days affected total SOD activity of root mitochondria differently. At 20 μM Cr (VI), SOD activity increased by 29%, whereas 200 μM Cr (VI) produced a significant inhibition (Dixit et al., 2002). A decline in the specific activity of catalase with increase in Cr concentration from 20 to 80 ppm was observed by Jain et al. (2000). Excess of Cr (0.5 mM) restricted the activity of catalase in the leaves of cauliflower (Chatterjee and Chatterjee 2000). In addition to SOD, CAT and glutathione transferase (GST), the enzymes of ascorbate glutathione cycle, APX, GR, Monodihydro-ascorbate reductase (MDHAR) and dihydro-ascorbate reductase (DHAR)

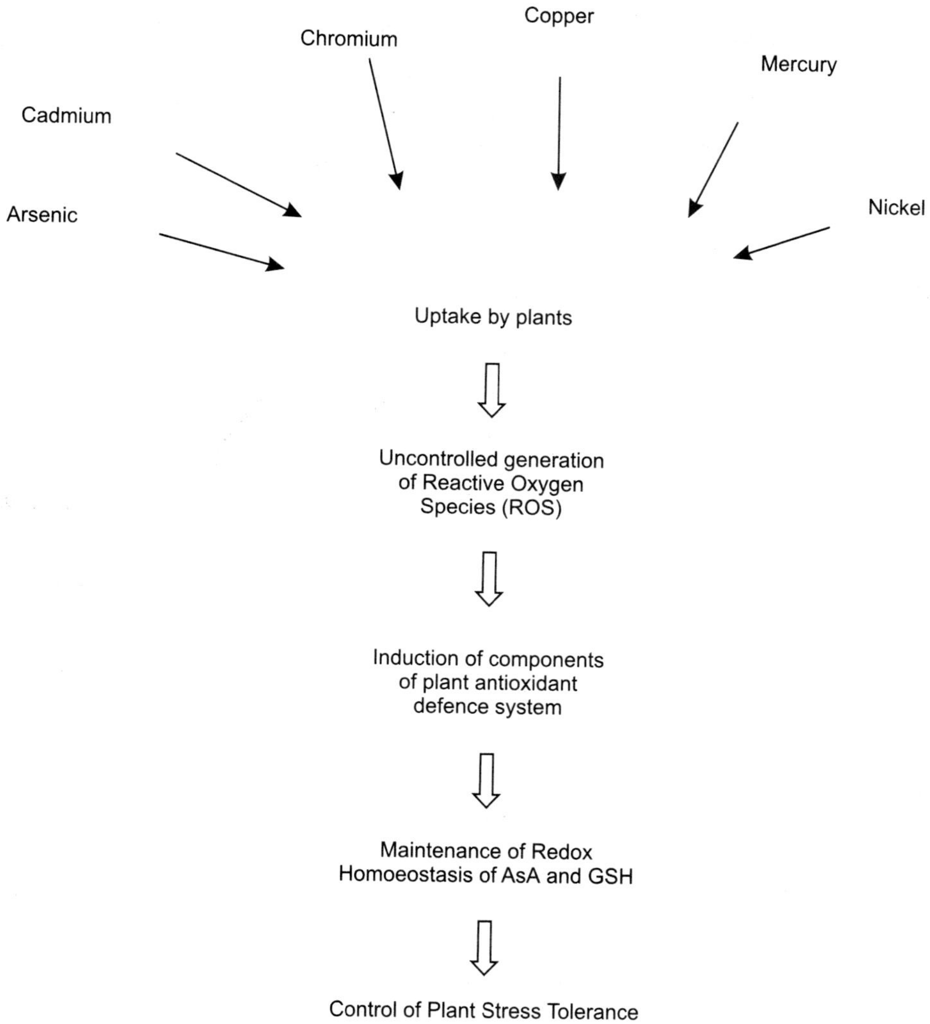

Fig. 1. Schematic summary of uptake of major heavy metals by plants, generation of reactive oxygen species, induction of major components of plant antioxidant defence system for the maintenance of ascorbate-glutathione-homoeostasis and the control of plant stress tolerance. [AsA- ascorbate, GSH- reduced glutathione, ROS- reactive oxygen species]

play important role in overcoming the metal-induced oxidative stress at the cellular and sub-cellular levels (Foyer et al., 1997). Ascorbate and glutathione, other components of the antioxidative system, increase under metal stress (Prasad et al., 1999). Since Cr^{6+} is a strong oxidant, it may cause severe oxidative stress in plant tissue. The antioxidative responses of plants to excess Cr have not been studied critically.

Increasing Cu concentration up to 1.5 mM increased SOD activities in the leaves of maize cultivars of 3223 and 31G98 (Deniz Tanyolac et al., 2007). Activity of APX increased by 25% (at 1 mM Cu), 42 % (at 1.5 mM Cu), 16.7% (at 1 mM Cu) and 558 % (at 1.5 mM Cu) for maize cultivars 3223 and 31 G98, respectively, compared to the control (Deniz

Tanyolac et al., 2007; According to Ansari et al. (2009) Hg-induced generation of ROS triggers activation of components of antioxidative defense system in Indian mustard, which comprises both the enzymatic (e.g. SOD, CAT, APX and GR) and the non-enzymatic low-molecular-weight antioxidants (e.g. ascorbic acid glutathione).

Several studies have indicated that excess Ni concentrations may cause oxidative stress in plants and affect activities of antioxidative enzymes (Baccouch et al., 1998; Madhava Rao and Sresty 2000; Gonnelli et al., 2001; Boominathan and Doran 2002). However, data concerning the effects of Ni on these enzymes are contradictory, since both increases (Baccouch et al., 1998) and decreases (Boominathan and Doran 2002; Pandey and Sharma 2002; Ruchi Maheshwari et al., 2009) in their activities have been noticed. Treatment of wheat plants with high Ni concentration resulted in a decrease in activities of SOD and CAT (Gajewska et al., 2005) in hairy root of *Alyssum beraolonii* and *Nicotiana tobaccum* (Boominathan and Doran 2002) but an opposite effect has been observed in *Zea mays* shoots (Baccouch et al., 1998). One possible mechanism via which the elevated concentrations of heavy metals may damage plant tissues is the stimulation of free radical production, thus imposing oxidative stress (Foyer et al., 1997).

6. SUMMARY

The literature reviewed above gives a clear picture of studies on various aspects of major heavy metals in environments, their uptake, translocation, accumulation and role of antioxidant defence system in various crop plants under various levels of HMs. It also appears that there are only a few reports concerning the effect of HMs on various metabolic activities in plants of economic importance. In fact, the work on various crop plants is invariably aimed only to understand mechanisms of antioxidant machinery going in plants exposed to various abiotic stresses only inside the laboratory conditions. No comparative studies on laboratory versus field conditions have been studied. Availability of information about plant species that are best adapted against the abiotic stress is meagre. Moreover, a deeper understanding of the mechanisms involved against abiotic stress especially due to HMs, is needed to enable the selection, breeding or genetic engineering of plants for enhanced phytoremediation and crop productivity.

ACKNOWLEDGEMENTS

The first author is indebted to American Federation of Muslims of Indian Origin (AFMI), America and Hamdard National Foundation (HNF), New Delhi, India for providing financial support to accomplish this research work.

REFERENCES

Abedin, M.J. and Meharg, A.A. (2002). Relative toxicity of arsenite and arsenate on germination and early seedling growth of rice (*Oryza sativa* L.). *Plant Soil.* **243:** 57-66.

Adriano, D.C. (1986). Arsenic In: Adriano DC (Ed.), Trace Elements in the Terrestrial Environment. Springer-Verlag, New York, pp. 308-321.

Adriano, D.C. (2001). Trace Elements in Terrestrial Environments: Biochemistry, Bioavailability and Risks of Metals, 2nd ed. Springer-Verlag, New York.

Albasel, N. and Cottenie, A. (1985). Heavy metals uptake from contaminated soils as affected by peat, lime, and chelates. *J. Soil Soc. Am.* **492:** 386-390.

Alva, A.K, Graham, J.H. and Anderson, C.A. (1995). Soil pH and copper effects on young 'Hamlin' orange trees. *Soil Sci. Soc. Amer. J.* **59:** 481-487.

Anjum, N.A., Umar, S., Ahmad, A. and Iqbal, M. (2008c). Responses of components of antioxidant system in moongbean [*Vigna radiata* (L.) Wilczek] genotypes to cadmium stress. *Commun. Soil Sci. Plant Anal.* **39**: 2469-2483.

Anjum, N.A., Umar, S., Ahmad, A., Iqbal, M. and Khan, N.A. (2008b). Sulphur protects mustard (*Brassica campestris* L.) from cadmium toxicity by improving leaf ascorbate and glutathione. *Plant Growth Regul.* **54**: 271-279.

Anjum, N.A., Umar, S., Iqbal, M. and Khan, N.A. (2008a). Growth characteristics and antioxidant metabolism of moongbean genotypes differing in photosynthetic capacity subjected to water deficit stress. *J. Plant Interact.* **3**: 127-136.

Anjum, N.A., Umar, S., Iqbal, M. and Khan, N.A. (2011). Cadmium causes oxidative stress in moongbean [*Vigna radiata* (L.) Wilczek] by affecting antioxidant enzyme systems and ascorbate-glutathione cycle metabolism. *Russ. J. Plant Physiol.* **58**: 00-00, In Press

Ansari, M.K.A., Ahmad, A., Umar, S. and Iqbal, M. (2009). Mercury-induced changes in growth variables and antioxidative enzymes in Indian mustard (*Brassica juncea L.*) *J. Plant Interac.* **4:** 131-136.

Asada, K. (1992). Ascorbate peroxidase-a hydrogen peroxide-scavenging enzyme in plants, *J. Plant Physiol.* **85:** 235-241.

Asher, C.J. and Reay, P.F. (1979). Arsenic uptake by barley seedlings. *Aust. J. Plant Physiol.* **6:** 459-466.

ATSDR. (1993). Toxicological Profile for Cadmium, Agency for Toxic Substances and Disease Registry, US Public Health Service, *Atlanta,* GA.

Baccouch, S., Chaoui, A. and El-Ferjani, E. (1998). Nickel toxicity: Effects on growth and metabolism of maize. *J. Plant Nutr.* **21:** 577-588.

Baker, A.J.M. and Walker, P.L. (1990). Ecophysiology of metal uptake by tolerant plants. In: Shaw, A J (Ed.), Heavy Metal Tolerance in Plants Evolutionary Aspects. CRC Press, Boca Raton, FL, 155-177.

Baker, D.E. and Senft, J.P. (1995). Copper. In: Alloway, B.J. (Ed.), Heavy Metals in Soils, second ed. Blackie Academic and Professional, London, pp. 179-205.

Becquer, T., Quantin, C., Sicot, M. and Boudot, J.P. (2003). Chromium availability in ultrafamic soils from New Caledonia. *Sci. Total Environ.* **301:** 251-261.

Berry, C.R. (1985). Growth and heavy metal accumulation in pine seedlings grown with sewage sludge. *J. Environ. Qual.* **14:** 415-419.

Bingham, F.T, Page, A.L, and Strong, J.E. (1980). Yield and cadmium content of rice grain in relation to addition rate of cadmium, copper, nickel and zinc, with sewage sludge and liming. *Soil Sci.* **130:** 32-38.

Bjerre, G.K. and Schierup, H.H. (1985). Uptake of six heavy metals by oat as influenced by soil type and additions of cadmium, lead, zinc and copper. *Plant Soil.* **88:** 57-69.

Boening, D.W. (1999). An evaluation of bivalves as biomonitors of heavy metal pollution in marine waters. *Environ. Monit. Assess.* **55:** 459-470.

Bohn, H.L., Mc Neal, B.L. and O'Connor, G.A. (1979). Soil Chemistry, John Wiley, New York.

Boominathan, R. and Doran, P.M. (2002). Ni-induced oxidative stress in roots of the Ni hyperaccumulator, *Alyssum bertolonii. New Phytol.* **156:** 205-215.

Bruland, K.W. (1983). Trace elements in sea water. In: Chemical Oceanography, Vol. 8, Eds. Riley JP and Chester R, Acad. Press, 157-220, Chapter 45.

Campbell, W.F., Miller, R.W., Reynolds, J.H. and Schreeg, T.M. (1983). Alfalfa, sweetcorn and wheat responses to long-term application of municipal waste water to cropland. *J. Environ. Qual.* **12:** 243-249.

Cataldo, D.A., Garland, T.R. and Wildung, R.E. (1981). Cadmium distribution and chemical fate in soybean plants. *J. Plant Physiol.* **68:** 835-839.

Cataldo, D.A, Mc Fadden, K.M., Garland, T.R. and Wildung, R.E. (1988). Organic constituents and complexation of nickel (1I), iron (111), cadmium (IIL and plutonium (IV) in soybean xylem exudates. *Plant Physiol.* **86:** 734-739.

Chaoui, A., Mazhoudi, S., Ghorbal, M.H. and Ferjani, E.E. (1997). Cadmium and zinc induction of lipid peroxidase and effects on antioxidant enzyme activities in bean (*Phaseolus vulgaris* L.). *Plant Sci.* **127:** 139-147.

Chatterjee, J. and Chatterjee, C. (2000). Phytotoxicity of cobalt, chromium and copper in cauliflower, *Environ. Pollut.* **109:** 69-74.

Chen, H. and Cutright, T. (2001). EDTA and HEDTA Effects on Cd, Cr, and Ni Uptake by *Helianthus annuus, Chemosphere.* **45:** 21-28.

Cho, U.H. and Park, J.O. (1999). Changes in hydrogen peroxide content and activities of antioxidant enzymes in tomato seedlings exposed to mercury. *J. Plant Biol.* **42:** 41-48.

Cho, U.H. and Park, J.O. (2000). Mercury-induced oxidative stress in tomato seedlings. *Plant Sci.* **156:** 1-9.

Cohen, S.M, Arnold, L.L, Eldan, M., Lewis, A.S. and Beck, B.D. (2006). Methylated arsenicals: The implications of metabolism and carcinogenicity studies in rodents to human risk assessment. *Crit. Rev. Toxicol.* **36:** 99-133.

Conklin, P.L. (2001). Recent advances in the role and biosynthesis of ascorbic acid in plants. *Plant Cell Environ.* **24:** 383-394.

Coogan, T.P., Latta, D.M., Snow, E.T. and Costa, M. (1989). Toxicity and carcinogenicity of nickel compounds. *CRC. Crit. Rev. Toxicol.* **19:** 341-384.

Cotter-Howells, J. and Capron, S. (1996). Remediation of contaminated land by formation of heavy metal phosphate. *Appl. Geochem.* **11**: 335-342.

Cotton, F.A. and Wilkinson, G. (1988). Advanced Inorganic Chemistry, 5th ed., John Wiley and Sons.

Crowder, A. (1991). Acidification, metals and macrophytes. *Environ. Pollut.* **71:** 171-203.

Cui, Y.L., Zhu, Y.G., Zhai, R.H., Chen, D.Y., Huang, Y.Z., Qiu, Y. and Liang, J.Z. (2004). Transfer of metals from soil to vegetables in an area near a smelter in Nanning, China, *Environ. Internat.* **30:** 785-791.

Cullen, W.R. and Dodd, M. (1989). Arsenic speciation in clams of British Columbia. *Appl. Organometallic Chem.* **3:** 79-87.

Duffus, J.H. (2002). "Heavy metals" a meaningless term? (IUPAC technical report). *Pure Applied Chem.* **74:** 793-807.

D'Amore, J.J., Al-Abed, S. R., Scheckel, K.G. and Ryan, J.A. (2005). Methods for speciation of metals in soils: a review. *J. Environ. Qual.* **34:** 1707-1745.

Deniz, T. and Yasemin, E.S. (2007). Ünalan. Changes in photochemical and antioxidant enzyme activities in maize (*Zea mays* L.) leaves exposed to excess copper. *Chemosphere.* **67:** 89-98.

Deuel, L.E. and Swoboda, A.R. (1972). Arsenic solubility in a reduced environment. *Soil Sci. Soc. Am. Proc.* **36:** 276–278.

Dixit, V., Pandey, V. and Shyam, R. (2002). Chromium ions inactivate electron transport and enhance superoxide generation *in vivo* in pea (*Pisum sativum* L. cv: Azad) root mitochondria. *Plant Cell Env.* **25:** 687-693.

Downing, J., Deeley, P. and Fichte, R. (1986). Chromium and Chromium Alloys in Ullmanns, fifth ed. Encyclopedia of Industrial Chemistry, vol. A7. VCH. Verlagsgesellschaft, Weinheim, pp. 43-65.

Encyclopedia of Industrial Chemistry, vol. A7. VCH. Verlagsgesellschaft, Weinheim, pp. 43-65.

Duxbury, Y. (1985). Ecological aspects of heavy metal responses in microorganisms, *Ad. Microb. Ecol.* **8:** 185-135.

Eskew, D.L., Welch, R.M. and Norvell, W.A. (1984). Nickel in higher plants: further evidence for an essential role. *J. Plant Physiol.* **76:** 691-693

Fediuc, E. and Erdei, L. (2002). Physiological and biochemical aspects of cadmium toxicity and protective mechanisms induced in *Phragmites australis* and *Typha latifolia*. *J. Plant Physiol.* **159:** 265-271.

Ferguson, J.F. and Gavis, J. (1972). A review of the arsenic cycle in natural water. *Water Res.* **6:** 1259-1274.

Fernandes, J.C. and Henriques, F.S. (1990). Metal levels in soils and cattail (*Typha latifolia* L.) plants in a pyrites mine area at Lousal, Portugal. *Int. J Environ. Stud.* **36:** 205-210.

Finster, E.M., Gray, A.K. and Binns, J.H. (2004). Lead levels of edibles grown in contaminated residential soils: a field survey. *Sci. Total Environ.* **320:** 245-257.

Flemming, C.A., and Trevors, J.T. (1989). Copper toxicity and chemistry in the environment: A review. *Water, Air and Soil Pollution.* **44:** 143-158.

Florijin, P.J. and Van Beusichem, M.L. (1993). Uptake and distribution of cadmium in maize inbred line. *Plant Soil.* **150:** 25-32.

Foyer, C.H. and Halliwell, B. (1976). The presence of glutathione and glutathione reductase in chloroplast: a proposed role in ascorbic acid metabolism. *Planta.* **133:** 21-25.

Foyer, C.H., Lopez-Delgado, H., Dat, J.F. and Scott, I.M. (1997). Hydrogen peroxide- and glutathione-associated mechanisms of acclimatory stress tolerance and signalling. *J. Plant Physiol.* **100:** 241-254.

Freedman, B. and Hutchinson, T. (1980a). Effects of smelter pollutants on forest leaf litter decomposition near a nickel–copper smelter at Sudbury, Ontario. *Can. J. Bot.* **58:** 1722-1736.

Fridovich, I. (1978). The biology of oxygen radicals. *Sci.* **201:** 875-880.

Fridovich, I. (1986). Superoxide dismutase. *Adv. Enzymol.* **58:** 61-67.

Fridovich, I. (1995). Superoxide radical and superoxide dismutases. *Annu. Rev. Biochem.* **64:** 97-112.

Gajewska, E., Andrzejewska, R., Slaba, M., Marciniak, M. and Skłodowska, M. (2005). Nickel-induced stress in wheat roots: growth and biochemical responses. *J. Plant Physiol.* **27:** 44-45.

Gallego, S.M., Benavides, M.P. and Tomaro, M.L. (1996). Effect of heavy metal ion excess on sunflower leaves: evidence for involvement of oxidative stress. *Plant Sci.* **121:** 151-159.

Glorennec, P. (2006). Analysis and reduction of the uncertainty of the assessment of children's lead exposure around an old mine. *Environ. Res.* **100:** 150-158.

Golovatyj, S.E., Bogatyreva, E.N. and Golovatyj, S.E. (1999). Effect of levels of chromium content in a soil to its distribution in organs of corn plants. *Soil Res. Fert.* **99:** 97-204.

Gonnelli, C., Galardi, F. and Gabbrielli, R. (2001). Nickel and copper tolerance and toxicity in three Tuscan populations of *Silene paradoxa*. *Physiol. Plant.* **113:** 507-514.

Greeson, P.E. (1970). Biological factors in the chemistry of mercury. In: Mercury in the Environment, United States Geological Survey, Professional Paper 713, United States Geological Survey, Washington, DC, pp. 32-34.

Halliwell, B. (1982). The toxic effects of oxygen on plant tissues. In: Oberley, L.W. (Ed.), Superoxide Dismutase, vol. I. CRC Press, Boca Raton, pp. 89-123.

Hartley-Whitaker, J., Ainsworth, G., Vooijs, R., Ten Bookum, W., Schat, H. and Meharg, A.A. (2001). Phytochelatins are involved in differential arsenate tolerance in *Holcus lanatus. Plant Physiol.* **126:** 299-306.

Hem, J.D. (1970). Mercury in the Environment, Professional Paper 713, Washington, DC Mercury in the Environment, Professional Paper 713, Washington, DC, U.S. Geological Survey, pp. 9-24.

Hermann, R. and Neumann-Mahlkau, P. (1985). The mobility of zinc, cadmium, copper, lead, iron and arsenic in ground water as a function of redox potential and pH. *Sci. Total Environ.* **43:** 1-2.

Hess, R. E. and Blanchar, R.W. (1976). Arsenic stability in contaminated soils. *Soil Sci. Soc. Am. J.* **40:** 847-852.

Huang, S. W., and Jin, J. Y. (2008). Status of heavy metals in agricultural soils as affected by different patterns of land use. *Environmental Monitoring and Assessment.* **139** (1–3): 317–327.

Huffman, Jr. E.W.D. and Allaway, W.H. (1973). Growth of plant in solution culture containing low levels of chromium. *J. Plant Physiol.* **52:** 72-75.

Humphreys, M.O. and Nicholls, M.K. (1984). Relationships between tolerance to heavy metals in *Agrostis capillaris* L. (*A. Tenuis* Sibth.). *New Phytol.* **98:** 177-190.

Hunter, B.A., Johnson, M.S. and Thompson, D.J. (1989). Ecotoxicology of Cu and Cd in a contaminated grassland ecosystem. IV. Tissue distribution and age accumulation in small mammals. *J. Appl. Ecol.* **26:** 89-99.

Jacks, G. and Bhattacharya, P. (1998). Arsenic contamination in the environment due to the use of CCA-wood preservatives. In: Arsenic in Wood Preservatives. Part 1. Kemi report (3) **98:** 7-75.

Jain, C.K. and Ali, I. (2000). Arsenic: occurrence, toxicity and speciation. *Water Res.* **34:** 4304-4312.

Jemal, F., Didierjean, L., Ghrir, R., Ghorbal, M.H., Burkard, G. (1998). Characterization of cadmium binding peptides from pepper (*Caspicum annuum*). *Plant Sci.* **137:**143-54.

Joner, E.J. and Leyval, C. (1997). Uptake of 109Cd by roots and hyphae of a *Glomus mosseae/ Trifolium subterraneum mycorrhiza* from soil amended with high and low concentrations of cadmium. *New Phytol.* **135:** 353-360.

Jose, R., Peralta, V., Martha, L.L., Mahesh, N., Geoffrey, S. and Jorge, G.T. (2009). The biochemistry of environmental heavy metal uptake by plants: Implications for the food chain. *The Internat. J. Biochem. Cell Biol.* **41:** 1665–1677

Khan, I., Ahmad, A. and Iqbal, M. (2009). Modulation of antioxidant defence system for arsenic detoxification by Indian mustard. *Ecotoxic. Environ. Saf.* **72:** 626-634.

Kabata-Pendias, A. and Pendias, H. (1989). Trace Elements in the Soil and Plants (1st ed). CRC Press, Florida.

Kabata-Pendias, A. and Pendias, H. (2001). Trace Elements in Soils and Plants (3rd ed), CRC Press, Boca Raton: 413

Korte, N.E. and Fernando, Q. (1991). A review of arsenic (III) in groundwater. *Crit Rev. Environ. Control.* **21:**1-39.

Kramer, U., Cotter-Howells, J.D., Charnock, J.M., Baker, A.J.M. and Smith, A.C. (1996). Free histidine as a metal chelator in plants that accumulate Nickel. *Nature.* **379:** 635-38.

Lee, J., Reeves, R.D., Brooks, R.R. and Jaffré, T. (1978). The relation between nickel and citric acid in some nickel-accumulating plants. *Phytochem.* **17:** 1033-1035.

Liu, J., Goyer, R. and Waalkes, M.P. (2007). Toxic effects of metals. In: C.D. Klaassen, Editor, *Casarett and Doull's Toxicology – the Basic Science of Poisons* (7th edition), McGraw-Hill, pp. 931-979.

Liebler, D.C., Kling, D.S. and Reed, D.J. (1986). Antioxidant protection of phospholipid bilayers by tocopherol. Control of α-tocopherol status and lipid peroxidation by ascorbic acid and glutathione. *J. Biol. Chem.* **261:** 12114-12119.

Li, M., Xia, T., Jiang, C.S., Li, L.J., Fu, J.L. and Zhou, Z.C. (2003). Cadmium directly induced the opening of membrane permeability pore of mitochondria which possibly involved in cadmium-triggered apoptosis. *Toxicol.* **194:** 19-33.

Lodenius, M., Seppanen, A. and Autio, S. (1999). Leaching of mercury from peat soil. *Chemosphere.* **16:** 1215-1220.

Lozano-Rodriguez, E., Hernandez, L., Bonay, P. and Carpena-Ruiz, R. (1997). Distribution of cadmium in shoot and root tissues of maize and pea plants: physiological disturbances. *J. Exp. Bot.* **48:** 123-128.

Ma, L.Q., Komar, K.M., Tu, C., Zhang, W., Cai, Y. and Kennelley, E.D. (2001). A fern that hyperaccumulates arsenic. *Nature.* **409:** 579.

Madhava Rao, K.V. and Sresty, T.V.S. (2000). Antioxidative parameters in the seedlings of pigeonpea (*Cajanus cajan* (L.) Millspaugh) in response to Zn and Ni stresses. *Plant Sci.* **157:** 113-128.

Madyima, S., Chimbari, M., Nyamangara, J. and Bangira, C. (2002). Cumulative effects of sewage sludge and effluent mixture application on soil properties of a sandy soil under a mixture of star and kikuyu grasses in Zimbabwe. *Phys. Chem. Earth.* **27:** 747-753.

Marschner, H. and Guo, Y. (1995). Uptake, distribution, and binding of cadmium and nickel in different plant species. *J. Plant Nutr.* **8:** 2691-2706.

Merry, R.H., Tiller, K.G. and Alston, A.M. (1983). Accumulation of copper, lead and arsenic in some Australian orchard soil. *Aust. J. Soil Res.* **21:** 159-161.

McBride, M.B. (1994). Environmental Chemistry of Soils, Oxford University Press, New York, NY.

McLaren, R.G., Hogg, D.S. and Swift, R. S. (1990). Some factors, affecting the availability of native and applied soil copper in New Zealand soils. *For. Ecol. Managmt.* **37:** 131-132.

Montes-Holguin, M.O., Peralta-Videa, J.R., Meitzner, G., Martinez-Martinez, A., de la Rosa, G., Castillo-Michel, H.A. and Gardea-Torresdey, J.L. (2006). Biochemical and spectroscopic studies of the response of *Convolvulus arvensis* L. to chromium(III) and chromium(VI) stress. *Environ. Toxicol. Chem.* **25:** 220-226.

Montgomery, J.M. (1985). Water Treatment Principles and Design, John Wiley & Sons, New York.

Mastromatteo, E. (1986). Nickel. *Am. Ind. Hyg. Assoc. J.* **47:** 589-601.

Matschullat, J. (2000). Arsenic in the geosphere—A review. *Sci. Total Environ.* **249:** 297-312.

McLaughlin, M.J., Parker, D.R. and Clarke, J.M. (1999). Metal and micronutrients-food safety issues. *Field Crops Res.* **60:** 143-163.

Meharg, A.A. and Hartley-Whitaker, J. (2002). Arsenic uptake and metabolism in arsenic resistant and nonresistant plant species. *New Phytol.* **154:** 29-43.

Mermut, A.R., Jain, J.C., Song, L., Kerrich, R., Kozak, L. and Jana, S. (1996). Trace element concentrations of selected soils and fertilizers in Saskatchewan, Canada. *J. Environ. Qual.* **25:** 845-853.

Molas, J. (1997b). Range of tolerance, limits and toxicity symptoms of ionic and chelatic nickel forms in cabbage plants. *Zesz Problem Post Nauk Roln.* **448b:** 203-209 (in Polish).

Mortvedt, J.J. and Giordano, P.M. (1975). Response of corn to zinc and chromium in municipal wastes applied to soil. *J. Environ. Qual.* **4:** 170-174.

Matschullat, J. (2000). Arsenic in the geosphere — a review. *Sci. Total Environ.* **249:** 297–312.

Mylona, P.V., Polidoros, A.N. and Scandalios, J.G. (1998). Modulation of antioxidant responses by arsenic in maize. *Free Radical Biol. Med.* **25:** 576-585.

Marcus-Wyner, L. and Rains, D.W. (1982). Uptake, accumulation, and translocation of arsenical compounds by cotton. *J. Environ. Qual.* **11:** 715-719.

Mustafa, G., Singh, B. and Rai, S. K. (2004). Cadmium adsorption and desorption behaviour on goethite at low equilibrium concentrations: effects of pH and index cations. *Chemosphere.* **57**: 1325-1333.

Nieboer, E., Yassi, A.A. and Muir, D.C.F. (1984). The Technical Feasibility and Usefulness of Biological Monitoring in the Nickel Producing Industry. Special Report, McMaster University, Hamilton, Ontario, Canada. Available through NiPERA, Durham, NC.

Nieboer, E. and Fletcher, G.G. (1996). Determinants of reactivity in metal toxicology. In: Chang LW Ed. Toxicology of metals. Boca Raton, Florida, CRC Lewis Publishers, pp. 113-132.

Noctor, G. and Foyer, C.H. (1998). Ascorbate and glutathione: keeping active oxygen under control. *Annu. Rev. Plant Physiol. Plant Mol. Biol.* **49:** 249-279.

Nriagu, J.O. and Pacyna, J.M. (1988). Quantitative assessment of worldwide contamination of air, water and soil by trace metals. *Nature.* **333:** 134-139.

Nriagu, J.O. (1988). A silent epidemic of environmental metal poisoning. *Environ. Pollut.* **50:** 139-161.

Nriagu, J.O. (1990). Global metal pollution. *Environment.* **32:** 7-33.

Olsson, I.M, Eriksson J., Oborn, I., Skerfving, S. and Oskarsson, A. (2005). Cadmium in food production systems–a health risk for sensitive population groups. *Ambio.* **34:** 344-351.

O'Neill, P. (1990). Arsenic. In: B.J. Alloway, Editor, *Heavy Metals in Soils*, Blackie, London, pp. 83-99.

Onishi, H. (1969). Arsenic. In: Wedepohl, Ed. Handbook of geochemistry, Chapt. 33, Springer, Berlin.

Patty, F. (1963). Industrial Hygiene and Toxicology. Volume II: Toxicology, 2nd ed., New York: Interscience Publishers. p. 1035.

Parida, B.K., Chhibba, I.M. and Nayyar, V.K. (2003). Influence of nickel contaminated soils on fenugreek (*Trigonella corniculata* L.) growth and mineral composition. *Sci. Hortic. (Amsterdam).* **98:** 113-119.

Padh, H. (1990). Cellular function of ascorbic acid. *Biochem. Cell Biol.* **68:** 1166-73.

Pelosi, P., Fiorentini, R. and Galoppini, C. (1976). *Agric. Biol. Chem.* **40:** 1641.

Pierce, M. L. and Moore, C.B. (1982). Adsorption of As(III) and As(V) on amorphous iron hydroxide. *J. Water Research.* **16:** 1247.

Page, A.L. and Bingham, F.T. (1973). Cadmium residues in the environment. *Residue Rev.* **48:**1-44.

Panda, S.K. and Patra, H.K. (1997). Physiology of chromium toxicity in plants–a review. *J. Plant Plant Physiol. Biochem.* **24:**10-17.

Passow, H, Rothstein, A and Clarkson, TW. (1961). The general pharmacology of the heavy metals. *Pharmacol. Rev.* **13:** 185-224.

Pandey, N. and Sharma, C.P. (2002). Effect of heavy metals Co^{2+}, Ni^{2+} and Cd^{2+} on growth and metabolism of cabbage. *Plant Sci.* **163:** 753-758.

Pickering, I.J., Prince, R.C., George, M.J., Smith, R.D., George, G.N. and Salt, D.E. (2000). Reduction and coordination of arsenic in Indian mustard. *Plant Physiol.* **122:** 1171-1177.

Prasad, K.V.S.K., Pardha Saradhi, P. and Sharmila, P. (1999). Concerted action of antioxidant enzymes and curtailed growth under zinc toxicity in *Brassica juncea. Environ. Exp. Bot.* **42:** 1-10.

Prasad, M.N.V. (1997). Trace metals. In: Prasad MNV (Ed) Plant ecophysiology. John Wiley & Sons, Inc., New York, pp. 207-249

Qureshi, M.I., Israr, M., Abdin, M.Z. and Iqbal, M. (2005). Responses of *Artemisia annua* L. to lead and salt-induced oxidative stress. *Environ. Exp. Bot.* **53:** 185-193.

Ramachandran, V., D'Souza, T.J. and Mistry, K.B. (1980). Uptake and transport of chromium in plants. *J. Nucl. Agric. Biol.* **9:** 126-128.

Raskin, I. and Ensley, B.D. (2000). Phytoremediation of toxic metals: using plants to clean up the environment, John Wiley and Sons, N. York, p. 303.

Ruchi, M. and Dubey, R.S. (2009). Nickel-induced oxidative stress and the role of antioxidant defence in rice seedlings. *Plant Growth Regul.* **59:** 37-49.

Sadiq, M., Zadi, T.H. and Mian, A.A. (1983). Environmental behaviour of arsenic in soils: theoretical. *Water Air Soil Pollut.* **20:** 369-377.

Saager, R. (1984). From Antimony to Zirconium. Bank Vontobel.

Skeffington, R.A., Shewry, P.R. and Peterson, P.J. (1976). Chromium uptake and transport in barley seedlings (*Hordeum vulgare* L.). *Planta.* **132:** 209-214.

Salt, D.E., Prince, R.C., Pickering, I.J. and Raskin, I. (1995). Mechanisms of cadmium mobility and accumulation in Indian mustard. *J. Plant Physiol.* **109:** 1427-1433

Singh, N. and Ma, L.Q. (2006). Arsenic speciation, and arsenic and phosphate distribution in arsenic hyperaccumulator *Pteris vittata* and nonhyperaccumulator *Pteris ensiformis. Environ. Poll.* **141:** 238-246.

Sanita di Toppi, L. and Gabbrielli, R. (1999). Response to cadmium in higher plants. *Environ. Exp. Bot.* **41:** 105-130.

Sathiyamoorthy, P., Van Damme, P., Oven, M. and Golan-Goldhirsh, A. (1997). Heavy metals in medicinal and fodder plants of the Negev desert. *J. Environ. Sci. Health.* **32:** 2111-21123.

Schaufelberger, D., Gupta M.P. and Hostettmann, K. (1987). Flavonol and secoiridoid glycosides from *Coutoubea spicata. Phytochem.* **26:** 2377-2379.

Sethunathan, N.M., Megharaj, M., Smith, L., Kamaludeen, S.P.B., Avudainayagam, S. and Naidu R. (2005). Microbial role in the failure of natural attenuation of chromium (VI) in long-term tannery waste contaminated soil. *Agric. Ecosys. Environ.* **105:** 657-661.

Rao, K.V.M. and Sresty, T.V.S. (2000). Antioxidative parameters in the seedlings of pigeon pea (*Cajnus cajan* (L.) Millspaugh) in response to Zn and Ni stresses. *Plant Sci.* **157:** 113-128.

Seregin, I.V. and Kozhevnikova, A.D. (2006). Physiological role of nickel and its toxic effects on higher plants. *Russ. J. Plant Physl.* **53:** 257-277.

Shah, K., Kumar, R.G., Verma, S. and Dubey, R.S. (2001). Effect of cadmium on lipid peroxidation, superoxide anion generation and activities of antioxidant enzymes in growing rice seedlings. *Plant Sci.* **161:** 1135-1144.

Sharma, P. and Dubey, R.S. (2007). Involvement of oxidative stress and role of antioxidative defense system in growing rice seedlings exposed to toxic concentrations of aluminum. *Plant Cell Rep.* **26:** 2027-2038.

Shanker, A.K., Djanaguiraman, M., Sudhagar, R., Chandrashekar, C.N. and Pathmanabhan, G. (2004). Differential antioxidative response of ascorbate glutathione pathway enzymes and metabolites to chromium speciation stress in green gram (*Vigna radiata* (L.) R. Wilczek. cv CO 4) roots. *Plant Sci.* **166:** 1035-1043.

Sinha, S., Gupta, M. and Chandra, P. (1996). Bioaccumulation and biochemical effects of mercury in the plant *Bacopa monnieri* L. *Environ. Toxicol. Water Qual.* **11:** 105-112.

Skiffington, R.A., Shewry, P.R. and Peterson, P.J. (1976). Chromium uptake and transport in barley seedlings (*Hordeum vulgare* L.). *Planta.* **132:** 209-214.

Srivastava, M., Ma, L.Q. and Cotruvo, J.A. (2005). Uptake and distribution of selenium in different fern species. *Internat. J. Phytoremed.* **7:** 33-42.

Suszeynsky, E.M. and Shann, J.R. (1995). Phytotoxicity and accumulation of mercury in tobacco subjected to different exposure routes. *Environ. Toxicol. Chem.* **14:** 61-67.

Tamaki, S. and Frankenberger, W.T. (1992). Environmental biochemistry of arsenic. *Rev. Environ. Contam. Toxicol.* **124:** 79-110.

Tripathi, R.D., Srivastava, S., Mishra, S., Singh, N., Tuli, R., Gupta, D.K. and Maathuis, F.J.M. (2007). Arsenic hazards: Strategies for tolerance and remediation by plants. *Trends Biotechnol.* **25:** 158-165.

Tinker, P.B. (1986). Trace elements in arable agriculture. *J. Soil Sc.* **37:** 587-601.

Tu, C., Ma, L.Q. and Bondada, B. (2002). Arsenic accumulation in the hyperaccumulator Chinese brake and its utilization potential for phytoremediation. *J. Environ. Qual.* **31:** 1671-1675.

Uren, N.C. (1993). Mucilage secretion and its interaction with soil, and contact reduction. *Plant Soil.* **156:** 79-82.

Verma, S. and Dubey, R.S. (2003). Lead toxicity induces lipid peroxidation and alters the activities of antioxidant enzymes in growing rice plants. *Plant Sci.* **164:** 645-655.

Von Burg, R. (1997). Toxicology update. Nickel and some nickel compounds. *J. Appl. Toxicol.* **17:** 425-431.

Wagner, G.J. (1993). Accumulation of cadmium in crop plants and its consequences to human health. *Adv. Agronom.* **51:** 173

Wang, Y.D. and Greger, M. (2004). Clonal differences in mercury tolerance, accumulation, and distribution in willow. *J. Environ. Qual.* **33:** 1779-1785.

Wang, J., Zhao, F..J, Meharg, A.A., Raab, A., Feldman, J. and McGrath, S.P. (2002). Mechanisms of arsenic hyperaccumulation in *Pteris vittata.* Uptake kinetics, interactions with phosphate, and arsenic speciation. *J. Plant Physiol.* **130:** 1552-1561.

Welch, J.E. and Lund, L.J. (1987). Soil properties, irrigation water quality, and soil moisture level influences on the movement of nickel in sewage sludge-treated soils. *J. Environ. Qual.* **16:** 403-410.

WHO. (1990). Beryllium, Environmental Health Criteria, 106, Geneva.

Wu, F.B., Zhang, G.P. and Dominy, P. (2003). Four barley genotypes respond differently to cadmium: lipid peroxidation and activities of antioxidant capacity. *Environ. Exp. Bot.* **50:** 67-78.

WHO. (1976). Mercury. pp. 1-31 *in* Environmental Health Criteria 1.World Health Organization, Geneva.

WHO. (1991). Evaluation of Certain Veterinary Drug Residues in Food (38th Meeting of JECFA), WHO Technical Report Series, World Health Organization, Geneva.

Weckx, J.E.J. and Clijsters, H.M.M. (1996). Oxidative damage and defense mechanisms in primary leaves of *Phaseolus vulgaris* as a result of root assimilation of toxic amounts of copper. *Plant Physiol.* **96:** 506-512.

Walsh, P.R., Duce, R.A. and Fasching, J.L. (1979). Considerations of the enrichment, sources, and flux of arsenic in the troposphere. *J. Geophya. Res.* **84:** 1719-1726.

Wiltshire, G.H. (1974). Growth of plants on soils from two metalliferous sites in Rhodesia. *The J. Ecol.* **62:** 501-525.

Yang, Z. (1997). PAML: a program package for phylogenetic analysis by maximum likelihood. Comput. *Appl. Biosci.* **13:** 555-556.

Yanai, J., Yabutani, M., Yumei, K., Biao, H., Guobao, L. and Kosaki, T. (1998). Heavy metal pollution of agricultural soils and sediments in Liaoning Province, China. *Soil Sci. Plant Nutr.* **44:** 367-375.

Zhang, W., Cai, Y., Tu, C. and Ma, L.Q. (2002). Arsenic speciation and distribution in an arsenic hyperaccumulating plant. *Sci. Total Environ.* **300:** 167-177.

Zheng, L.G., Liu, G.J. and Wong, M.H. (2007b). The distribution, occurrence and environmental effect of mercury in Chinese coals. *Sci. the Total Env.* **384:** 374-383.

Index